GEOTHERAPY

GEOTHERAPY

Innovative Methods of Soil Fertility Restoration, Carbon Sequestration, and Reversing CO_2 Increase

EDITED BY
Thomas J. Goreau • Ronal W. Larson • Joanna Campe

CRC Press
Taylor & Francis Group
Boca Raton London New York

CRC Press is an imprint of the
Taylor & Francis Group, an **informa** business

CRC Press
Taylor & Francis Group
6000 Broken Sound Parkway NW, Suite 300
Boca Raton, FL 33487-2742

Printed on acid-free paper
Version Date: 20141030

International Standard Book Number-13: 978-1-4665-9539-2 (Hardback)

Visit the Taylor & Francis Web site at
http://www.taylorandfrancis.com

and the CRC Press Web site at
http://www.crcpress.com

Table of Contents

Richard Grantham: Father of Geotherapy, 1922–2009

It must have been in the late 1970s when I first met Richard Grantham. He came to give a lecture at the research laboratory where I worked and described how he had used a computer analysis of the genetic sequences present in a virus capable of causing cancer to identify a gene that had its origins in the virus's host, not in the ancestors of the virus. I was fascinated. Some years before, I had attempted to develop computer programs for biological problems, although I am a dreadful mathematician, and I asked him if, with the consent of my director, I could come and collaborate with his group one day a week. He agreed, and this arrangement lasted throughout our working careers. His group developed many of the basic computer techniques employed today for the accession and analysis of genetic sequences—it was an exciting time and I learned much. A lot of the time was spent discussing projects and future research directions together in English, because although Richard, like me, was very fluent in French (we both had French wives), we sometimes found it useful to revert to our native language for the more abstruse concepts. As time passed, these discussions took a more philosophical turn and Richard became progressively more concerned with the degradation of the biosphere and the impact of humanity on the natural world.

The contributions that Richard and his colleagues made to the dawning science of the molecular biology of genetic sequences, and their relevance to evolutionary processes, were many and varied. He is most often remembered in the context of his *genome hypothesis*, which allowed that the DNA sequence in the gene itself carried to some degree a *signature* characteristic of the species (or species group) from which that gene derived. Classically at that time, it was considered that Darwinian selection could only operate at the level of proteins because the genes themselves were only *visible* to selection pressures through their expressed products. However, most of the amino acids that compose proteins can be specified by more than one sequence of three nucleotides (codon) in the genetic DNA. Richard realized that the different alternative codons were not used randomly in the genes from different species and that there was a pattern of use that appeared to be common to different genes from the same species. It must be realized that, for a paper published in 1980 (1), the total number of sequences available in his pioneering database was 160. As of April 2011, GenBank contained 135,440,924 sequences! It is not surprising that his initial insight and analysis have now been greatly developed by later work, but it remains amazingly relevant today. For a deeper view of his contributions, I recommend a visit to http://post.queensu.ca/~forsdyke/granth01.htm where Donald Forsdyke presents some of his works. Professor Forsdyke's own contributions to this important field may be found on the same website.

During our discussions, we found that we had both read and been affected by Rachel Carson's *The Silent Spring* and we quickly discovered Aldo Leopold's *A Sand County Almanac: And Sketches Here and There*, then the influential books by Van Rensselaer Potter (see References). These convinced

Richard that he should concern himself actively with the problems of human impact on the natural world, particularly the concept to global bioethics. Incidentally, we both deplored the hijacking of the simple term *bioethics* by the medical community—despite clear priority for the term for ecological concerns. Among other important inputs, I would mention especially John Rawls' *A Theory of Justice* and Julia Annas' *The Morality of Happiness*, which helped us consider the knotty problems of the impact of bio-remediation and the instauration of a *sustainable* lifestyle on a the real human population.

Despite the hundreds of hours of discussion and exchange with Richard, I realized that I would never get more than a glimpse of the complex and profound person I was privileged to encounter. We became friends, but I never knew much of his background or personal life. He had served as a bomber pilot during World War II in Italy and returned to Europe to study evolutionary biology at the University of Montpelier in the 1960s. I learned little about his youth in central California or of his family. Any diversion in our conversations in those directions was quickly brought back to the central theme of his projects and philosophical analyses. On occasion we would find time to walk through the woods together—sometimes hunting mushrooms, and always alert for whatever wildlife we should chance to meet. His knowledge was wide and eclectic and it was a joy to accompany him on these diversions.

Richard corresponded quite extensively with Van Rensselaer Potter on themes of mutual interest, and I had the pleasure of participating in their exchanges. I think that both of them appreciated finding someone of similar stature to converse with, and Richard met Van on a rare visit to the United States. He brought back some very helpful insights and several pleasant reminiscences. His mind turned progressively to a very practical problem: The damage is done—how can we mitigate its effects? To this end, he imagined grandiose schemes like the irrigation of North African dessert areas with desalinated water to reconstitute the forests that existed there not so long ago (at least in geological timescales). Geotherapy became his watchword. The earth is already very sick, and heroic remedies are necessary to give even a chance of the survival of our *patient*. As an evolutionary biologist, he was only too aware of the perils of an overreaching species. It was clear to both of us that the carrying capacity of the planet for *Homo sapiens* had been exceeded for a long while. Richard did not despair of a downturn to more supportable human numbers and balanced survival, given that we could buy time for the message to be heard and understood by our species as a whole. I confess to being less optimistic… We both gave much time and thought to possible remedial approaches—which are, indeed, both marginally possible and vitally urgent—but the scale of the problem and the intransigence of human politics remained discouraging. We both feared that the natural solution would be the elimination, probably through disease, of much of the human population—and that our experience of studies of such events in other species suggested that the most frequent outcome is extinction and not simple population reduction! Humanity is perhaps the first species to be able to see the consequences of its proliferation, even if dimly, but will it be able to act on those warnings? We both agonized on the unfolding of the AIDS epidemic and contributed to studies of the virus and its consequences. We have perhaps postponed that threat. Then came other novel pathogens, and more will certainly come. Each may be the last. The question is simply one of survival, not of preservation of comforts, but life or death. Will we measure up? He and Professor Potter jointly published a call to action that can be seen at http://sunsite.utk.edu/FINS/Sustainable_Development/Fins-SD-03.htm.

The term *geotherapy* was formally introduced at a conference entitled "The Colloquium on Modelling and Geotherapy for Global Changes" held at the Université Claude Bernard in Lyon in May 1991, and sponsored by the CNRS. There Richard and other colleagues sought to define the needs and responsibilities encompassing human responses to the environmental degradation that was so obvious to those who would look around them. The final declaration may be seen at http://www.angelfire.com/mac/egmatthews/geotherapy/foundation.html and other references at http://www.angelfire.com/mac/egmatthews/geotherapy/geotherapy.html or http://jebin08.blogspot.com/2007/06/thinkable-why-geotherapy-should-not-be.html

A very interesting artistic reaction to the notion of geotherapy can be found at http://www.thewip.net/contributors/2010/05/geotherapy_artist_mara_haselti.html—I am not sure what Richard's reaction would have been, but knowing his deep interest in, and appreciation for, matters artistic, I am sure he would have been intrigued!

Over the last years of his life, Richard often received me at his apartment to continue our discussions. His interest in the problems of the planet remained present to the last, and he retained his knowledge of evolution and of humanity, although much else faded. I have been privileged to know and to spend time with an exceptional man. I hope that his message will be considered by others and that *H. sapiens* may manage to stop before the crumbling edge of the precipice. It is very close to our feet right now.

Tim Greenland
Universite Claude Bernard Lyon I
Villeurbanne, France

REFERENCES

Annas, J., *The Morality of Happiness*, Oxford University Press, Oxford, U.K., 1993.

Carson, R., *The Silent Spring*, Houghton Mifflin, Boston, MA, 1962.

Leopold, A., *A Sand County Almanac*, Oxford University Press, Oxford, U.K., 1949.

Potter, V.R., *Bioethics: Bridge to the Future*, Prentice-Hall, Englewood Cliffs, NJ, 1971.

Potter, V.R., *Global Bioethics: Building on the Leopold Legacy*, Michigan State University Press, East Lancing, MI, 1988.

Rawls, J., *A Theory of Justice*, Harvard University Press, Cambridge, MA, 1971.

Dick Grantham: A Reminiscence

Dick Grantham was a protean thinker who exemplified the saying that great minds focus on ideas, average minds on events, and mediocre minds on personalities. A pioneer from the early days of understanding the genetic code, he switched at the end of his career to solving the global environmental crisis. He immediately grasped that new restorative approaches were needed to return to health a patient battered by abuse; that ethics and education, as well as technology, were central; and that politics and shortsighted economic greed were consciously preventing a rational, sustainable future for our planet and all the organisms living on it.

Dick first contacted me by letter in the mid-1980s right after I had published several papers on greenhouse gas emissions from Amazonian forests, pastures, and degraded lands, and on stabilization of atmospheric CO_2, but he was also fascinated by my work developing the method for predicting coral bleaching from satellite data and on restoring coral reefs through sea water electrolysis. We quickly developed an intense correspondence on a huge range of scientific issues, focused on geotherapy, conducted by what is now called *snail mail*, and by periodic phone calls. When we first got access to fax machines in the late 1980s, we would then fax letters back and forth, marveling that we did not have to wait a week or more for the letters to arrive, as that was before the days of e-mail, which we would have greatly appreciated. We were then in contact on more or less a weekly basis, often daily, discussing recent scientific papers and arguing out their implications.

It took about a year to get the support from the French national scientific organization to hold the first world geotherapy conference and track down leading scientists, historians, and philosophers to brainstorm how we could act effectively to reverse course. Fundamental and stimulating papers were presented, but unfortunately there was no funding to publish them, so only the consensus "Declaration of Geotherapy Principles" was circulated (see Chapter 1). That meeting, held in 1991, aimed to present a scientific consensus on sustainable global management that could be presented to governments at the United Nations Conference on Sustainable Development (UNCSD, the "Earth Summit") in Rio de Janeiro in 1992, where the United Nations Framework Convention on Climate Change (UNFCCC) was approved by the world's governments.

Soon after the workshop, the daily flood of faxes and phone calls abruptly stopped from one day to the next. I knew Dick had serious medical problems, but he did not talk about them, so I was not sure how serious they were, so I sadly assumed he had suffered a stroke. Letters, faxes, and phone calls were not answered, and I thought he had died, although I could find no obituaries. I knew Dick had counted on me and our colleagues to continue the geotherapy mission, and while I continued on my own in my own small way, the heart of the movement was gone. Dick had been an indefatigable correspondent to global environmental policy and opinion makers, while I was working in the field, often out of reach in remote places for months at a time.

It took me nearly 15 years to find out what had happened to him. He had suddenly been struck by a Parkinson's disease–like syndrome that rendered writing, reading, and speaking difficult, although his mind was still razor sharp. When I finally managed to track down Tim Greenland through the Internet years later, I was astonished to learn that Dick was still alive, but incapacitated—able to think but unable to communicate because he could not control his hands well enough to write, though he still read even if rather slowly. With only poor control of his lips, he tended to mumble with some difficulty so only a few could understand him. When Tim told Dick that I sent my best wishes and wanted him to know that we were still pursuing his goal, even if delayed, Dick smiled broadly and nodded. A year or so later, he died.

This book is the sadly delayed result of Dick Grantham's vision. Had he not been incapacitated at the critical moment, a version of this book would have come out more than 20 years earlier, and his relentless energy would have resulted in it being sent to all major environmental decision makers. We owe it to him to continue his planet-saving work. We have lost 20 years, but the fact is that the world was not ready at the time, and only now is the litany of climate-change–related disasters getting worse and worse finally starting to motivate the educated and responsible public, although politicians and industry are still in denial. We hope this book will help start a global geotherapy movement. That will be Dick Grantham's best memorial.

Thomas J. Goreau

Preface

Rather than a conglomeration of mineral and organic substances along with water and air, soil is a living entity that affects and is affected by its environment through numerous biogeochemical processes. Thus, it is hardly a new observation that soil is the foundation of all life in the terrestrial biosphere and supports numerous ecosystem functions and services. A strong soil-civilization nexus also exists, because most prehistoric civilizations developed and flourished on fertile soils (e.g., Mesopotamia, Indus). Evidently, some of these once thriving civilizations vanished when soils that supported them were ignored, and degraded because the precious resource was taken for granted. It is argued, for example, that the Sumerian civilization collapsed because of soil salinization, siltation of canals and waterlogging. Whereas societies with fertile soil resources can adapt to some environmental stresses (e.g., drought stress, heat wave, climate change), those dependent on marginal and degraded soils neither have the resilience nor the capacity to adapt.

Basic ecosystem services provisioned by soils include food, feed, fiber, fuel, water resources, industrial raw materials, construction materials, etc. Thus, sustainable management of soils is critical to achieving food and nutritional security, adapting to and mitigating climate change, enhancing rural income, eliminating poverty, empowering minorities, and promoting the overall sustainable development. Indeed, soils are essential to advancing the Millennium Development Goals and must be integral to any agenda of addressing food security, climate security, water security, energy security, etc. A healthy soil is essential to a healthy population. In addition, soil is also important for numerous spiritual, cultural, and aesthetical uses and to the overall societal well-being. However, competing uses of soil and over exploitation can exacerbate degradation and jeopardize sustainability. Thus, soil use must be prioritized critically and objectively.

In view of the important role that soil plays in human well-being and nature conservancy, the 68th UN General Assembly, held on December 20, 2013, recognized December 5 as World Soil Day and the year 2015 as the International Year of Soils. These official recognitions by the UN acknowledge the importance of soils at all levels of the modern civilization.

World Soil Day, aimed at enhancing the awareness regarding the importance of soil to societal well-being and nature conservancy, coincides with the birth date of His Majesty King Bhumibol Adulyadej, the king of Thailand, who is a strong supporter of conservation and protection of world soil resources.

The fact that the UN General Assembly has designated 2015 as the Year of Soils is indicative of the recognition by the world body of the importance of soils in addressing some critical global issues of the twenty-first century.

It is in this context that this book, *Geotherapy*, is timely and highly pertinent to *carbon civilization*, or *anthropocene*. The present era is called Carbon Civilization (similar to the Stone Age or Neolithic civilization, water or hydraulic civilization, etc.) because all basic necessities are based on carbon as the principal source of energy. The major source of carbon-based energy is the fossil fuel, and its combustion in 2013 emitted ~10 Gt of C (~37 Gt of CO_2) into the atmosphere. The era is also called Anthropocene because humans are as big a force, as any geological perturbation, in changing planet Earth.

Civilization refers to the history of human relations with its own kind, other species, nature, and actual or imaginary forces perceived to be responsible for the historic, current, and future order. Science as a tool (servant) is beneficial to civilization and nature. However, science as a master of civilization can disturb the natural order, impact civilization and planetary processes. Yet, it is not the science per se, but the use of *technology without wisdom* that is the cause of any disorder or drastic perturbations. Humans, as the term *anthropocene* implies, have the capacity to perturb the delicate balance between soil and nature and also restore the balance. Humans prosper when they cooperate among themselves and work with nature, and suffer when they disturb it, are at war with themselves and work against nature. One of the consequences of the drastic anthropogenic perturbation of the biosphere is the depletion of the ecosystem and soil carbon pools. Rather than commencing with the onset of industrial revolution around 1750, anthropocene began with the beginning of settled agriculture 10 to 12 millennia ago. Over this period, more carbon may have been emitted into the atmosphere from deforestation and land use conversion than from fossil fuel combustion until the end of the twentieth century. Thus, recarbonization of the biosphere in general and that of the soil carbon pool in particular is important to maintenance and enhancements of ecosystem functions and services. It is precisely in this context that *Geotherapy* is timely and very pertinent.

Global issues (e.g., food security, climate change, water and air purification) are complex issues and require a wide range of options. Each of these options has trade-offs (pros and cons) that must be critically and objectively addressed. Sequestration of carbon in soils and other components of the biosphere (e.g., trees, wetlands), besides being cost-effective, has numerous ancillary or cobenefits. Carbon sequestration in soil is a "win-win option" and is the common thread that links all the chapters into a coherent and informative book.

Geotherapy is comprised of 34 chapters prepared by highly recognized professionals. Chapter 1 outlines the basic concepts of geotherapy and the inherent capacity of the biosphere to store soil carbon and strengthen numerous ecosystem services such as moderating climate change while enhancing soil, food, water, and air. Several chapters address processes and techniques of soil carbon restoration through biogeochemical cycling, biochar, slow-release fertilizers, weathering of minerals (olivine) and rock (basalt) powders, amendments and biofertilizers, establishment of vetiver and other perennials, etc. Chapter 33 addresses an important theme of carbon farming and the associated policies and the accounting system. In addition to creating another income stream, building up soil carbon also enhances resilience to drought by increasing soil-water storage and improving soil quality. Chapter 34 reinstates the importance of soils in reducing the net rate of increase of atmospheric CO_2 and restoring soil resilience.

Furthermore, *Geotherapy* supports the truism that "If soils are not restored, crops will fail even if rains do not; hunger will perpetuate even with emphasis on biotechnology and genetically modified crops; civil strife and political instability will plague the developing world even with sermons on human rights and democratic ideals; and humanity will suffer even with great scientific strides. Political stability and global peace are threatened because of soil degradation, food insecurity, and desperateness. The time to act is now" (Lal, 2004).

Rattan Lal

REFERENCE

Lal, R., Soil carbon sequestration impacts on global climate change and food security, *Science*, 304:1623–1627, 2004.

Foreword

Farmers and Their Soils (Figure F.1)

This book, *Geotherapy*, with its multiplicity of perspectives focuses on three fundamental issues: healing the planet, stabilizing climate, and feeding the human community. All are linked and interwoven in a story that begins with the soil and its stewardship and extends outward to encompass the whole planet. The authors are from all of the continents except for Antarctica, and their perspectives are as diverse as the communities in which they live. The chapters that follow represent the unfolding of a new paradigm that is deeply rooted in the teachings of nature.

Perhaps this is no better exemplified than through the ecological wisdom unfolding at a small independent research institute in the Midwest of the United States. At The Land Institute in Kansas, Wes Jackson and his colleagues are undertaking one of the most important agricultural research projects in the world. They have gone back to first principles and are breeding new perennial rather than annual grain crops. They are crossing high-yielding annual crops like wheat and corn with their wild perennial relatives. They expect that it will take over half a century for the research to begin to make a significant impact on agriculture. After decades of plant breeding, they are now producing a grain they call Kernza™. It is a highly nutritious perennial intermediate wheatgrass that has an excellent taste. It is also a tough plant that is new to the human food inventory. The breeders at The Land Institute feel they still need another ten years to produce a commercial version of the crop, but each year the Kernza seeds get larger.

The goal of The Land Institute is to restore the prairie and produce foods while creating deep, rich organic soils that not only capture carbon dioxide from the atmosphere but also sequester carbon. In such prairie land, much of the carbon is stored as a stable and long-lived organic material. The institute envisions a new agriculture that will heal the earth and help stabilize the climate, while providing the seeds and grains to help supply the food needs of the human family.

Wes Jackson is an ecologist, geneticist, and a plant breeder. His original insight came from an understanding of the uniqueness of the prairie where he grew up. The prairie is one of the best soil-building ecosystems on the planet. Its diversity is not just in plant species but also in the architecture of their root systems. Some perennials send their roots down ten meters or more into the soil. At The Land Institute, the surrounding prairie provides the operating instructions—or roadmap—for the new generation of plant breeders. Staff at the institute discovered that their systems had to mimic the prairie as closely as possible in order to tap into the uniqueness of that environment. Their ingenuity lies in the idea that high-yielding perennial grains can be developed through cross-breeding annual seed crops with their perennial relatives.

At The Land Institute, I saw a photographic exhibit that had a powerful influence on me. The photographer, James Richardson, had taken pictures from around the world of farmers on their land. They were from Africa, the Middle East, Asia, and the Americas. The top half of each photograph was composed of a close-up of the farmers standing with their crops. For the lower half, the photographer had excavated the soil where they stood down to several meters. It appeared that they were standing on the edge of a cliff, but it was really a cross-section of the soil beneath them. The contrast between the soil cross-sections from around the world was startling and revealing. A farmer from the prairie stood over deep, rich organic soil. Others, including African and Middle Eastern farms, had soils that looked like old bones: dry, hard, and bleached in color; these soils were tired and parched—the legacy of centuries of working the land without giving new life back to the soil. The tragic end-of-the-road look to them did not speak to a viable and durable future for those people.

(a) (b) (c) (d)

FIGURE F.1 Farmers and their soils: (a) James Duggan, Kansas; (b) Hassoun Hariri, Syria; (c) Mariama Abdouleye, Niger; and (d) Farhad Khalwai, Syria. (From www.JimRichardsonPhotography.com.)

It was hard for me to take my eyes off those photos. They helped me realize that the great work ahead for humanity has to be the rebuilding of soils on a global scale. Fortunately there are soil-building techniques, some ancient and others modern, that have been developed in many parts of the world and can be applied to the task. Many of them have been derived from the uniqueness of local ecosystems. Others are more universal in nature such as remineralization and the addition of a charcoal material called biochar to the soil. They have the potential to be incorporated on agricultural, forested, and even wild lands. The benefits would be huge.

If we could double, triple, or quadruple the organic carbon content of soils in the decades ahead, the productivity of the land would rise and the capacity of the soils to hold and attract moisture would increase dramatically. This book provides a road map toward this future. If the ideas laid out in *Geotherapy* are implemented, the capacity of the land to support economic activity would also diversify and expand. If the scale of the projects is large enough, soil building as much as any other program currently being considered could help lower atmospheric carbon dioxide and create a more stable climate. It could also engage a large percentage of the people around the world. Our future as a civilization may depend on returning to our roots and the mineral as well as the organic materials that sustain them.

How to instigate such massive change is a major issue. Most people do not understand that our future depends on new soils. This idea, no matter how viable, would have a hard time gaining traction in industrial and urban societies.

Humanity's survival has always been carbon based. The carbon we needed throughout most of our history was the slow carbon that was embodied in trees and other plants as well as animals. With the Industrial Revolution we shifted to fast carbon in the form of oil and natural gas. Fast carbon is mainly finite and nonrenewable.

Perhaps we could use carbon as a universal currency. People around the world could be paid to capture and sequester carbon, particularly in soils. One possible option would be to create a worldwide carbon currency. This would define, create, and make known the value of sequestering carbon. Carbon sequestration is a global public good. My colleague Dan Kirk thinks that publically funded carbon sequestration is the way to make it happen. He envisions a global trust organized on a country-by-country basis that would be a regulatory system that primarily rewards positive changes rather than impose punishments. It would thus encourage people to make positive changes. It also equalizes the playing field across the globe. A ton of carbon in Iowa would have the same value as a ton of carbon in Africa. Participating countries would benefit from direct but also indirect payments from carbon sequestration. These include ecosystem services that would improve food security and ameliorate the destructive effects of weather extremes. The system would allow polluters to be charged by the various national entities administering the carbon currency initiative.

If successful, such a measure would produce a tectonic shift in the way the world conducts its business, from farming to aerospace. Since the stakes are too important for these ideas not to be developed further, a carbon- and soil-based economy should become central to modern economic systems.

Geotherapy is an encyclopedia of ideas and tools for transforming the planet. The book is written by genuine earth stewards, most of who have field experience and a broad theoretical knowledge of the larger workings of the planet. Their experience spans many different climates, soils, and cultures. This practical, get-your-hands-in-the-soil manual is a good place to start to transform the world.

John Todd
Todd Ecological Design
Woods Hole, Massachusetts

Executive Summary of Chapters

Chapter 1 introduces the goals of the book and the concept, methods, and urgency of geotherapy; the process of healing our biosphere's ability to store soil carbon and prevent climate change impacts, while greatly enhancing our food, water, soil, air, and waters. The organization and origin of the book are presented, including sections on processes, field results, and policy related to soil fertility restoration and carbon storage.

In Chapter 2, the carbon cycle is reviewed with regard to the potential of stabilizing CO_2 at safe levels. Topics included are as follows: why and how policymakers have severely underestimated the long-term impacts of climate change, how the carbon management mechanisms they have developed are so flawed as to be counterproductive, how they could be corrected, where the carbon can be most effectively stored, how quickly that could be done, and the practical and policy actions needed to get there.

In Chapter 3, global biochar carbon sink estimates to date have been conservative, considering only agricultural crop residues, but there are many more opportunities. These are examined, including bioenergy, biofuels, forestry, etc., and found to greatly raise the potential to remove significant amounts of CO_2 into the range of 10 GTC/year.

Chapter 4 reviews more than a century of work showing that rock powders release nutrients that increase growth rates of forest trees and agricultural crops, improve the nutritional quality of food, and improve resistance of crops to pests and diseases. Policy steps needed for wider use as a natural slow-release fertilizer are proposed.

The fundamental chemistry of soil fertility is reviewed in Chapter 5 with a focus on soil organic carbon and biochar, and how they act to increase soil carbon, fertility, productivity, and nutritional quality of produce.

Weathering of rock minerals releases nutrients to soils, controlling soil fertility, and consuming carbon dioxide. Both the reactions and potential ways to increase the efficiency of reactions taking place in soils are reviewed in Chapter 6. The weathering of concrete debris is also shown to be a significant potential carbon dioxide sink in urban soils.

Green olivine is the most abundant mineral on our planet, mostly found deep in the earth. It is the fastest mineral to react with CO_2, making it a great potential carbon sink. There is enough available to effectively reduce global atmospheric CO_2, if the mineral can be ground into a fine enough powder. In chapter 7, many ingenious applications are proposed, using olivine powder weathering for carbon dioxide removal while serving a myriad of other purposes now provided by other materials.

In Chapter 8, the economics of enhanced olivine weathering as a carbon dioxide sink is evaluated and found to be about an order of magnitude lower than carbon capture and sequestration or geoengineering proposals.

While overgrazing destroys soil carbon, carefully managed grazing does the opposite: restores soil carbon, productivity, water storage, and biodiversity. Chapter 9 discusses the large areas where these methods can be applied and could provide a major global carbon sink, while restoring the land and creating economic opportunities.

A thorough review of how carbon is made and stored in soil, what it does there, and practical methods to enhance soil fertility through carbon-smart farming methods using biochar are discussed in Chapter 10. These absorb carbon dioxide from the atmosphere by enhancing interactions between biochar, fungi, bacteria, and plants. The appendices provide a detailed tutorial in soil biology, chemistry, and sustainable management.

Charcoal has been used to enhance soil fertility for hundreds of years. In Chapter 11, the history of research on biochar is reviewed and USDA trials summarized. When properly used, biochar provides agricultural production and greenhouse gas emission benefits, but improper use can be counterproductive. Therefore training is needed for effective use.

Field trials testing biochar effects on soil fertility are surveyed and links to detailed results are provided in Chapter 12. These provide a useful review of current research activity in the field for those wanting to explore deeper.

Practical uses of biochar for soil fertility enhancement and optimization of its use are presented in Chapter 13. The bulk of biochar benefits for plant nutrition and water uptake, and for soil water percolation and retention, can be achieved by applying small amounts of biochar directly to seedlings and plants compared to wide-scale broadcasting. The most cost-effective dosage of biochar is determined through experiments.

Japan has been the historic leader in charcoal use in agriculture, research on applications, and development of carbon-negative food certification accounting. The long and diverse history of charcoal use, and challenges for the future of soil carbon trading are outlined in Chapter 14.

Chapter 15 discusses the role of biochar in farming systems. Vietnam integrates biochar made from rice straw and sugar cane residues with biodigester effluents to greatly increase crop productivity. Properly made Biochar works better than ordinary charcoal. Biochar fed to cattle increases growth and reduces methane emissions. Bioenergy from rice straw biochar could, along with hydroelectricity, make Vietnam's energy completely sustainable.

Chapter 16 focuses on the Vetiver grass, the world's best erosion control plant, which is an ideal biochar feedstock. When grown along contours, the deep roots completely stop soil erosion. This not only prevents loss of soil carbon, but intensive agricultural terraces are created on previously barren slopes, and Vetiver, one of the most productive plants known, is ideal for bioenergy and biochar. Vetiver should play a central role in tropical land rehabilitation strategies.

Chapter 17 discusses the benefits of using basalt powder. Basalt dust application increased tree height on poor soils in Panama more than twice, and biomass growth more than six times, over five years, elevating levels of all plant nutrients in the soil. Additional biochar and compost should have produced even better results. The method is highly cost effective for restoring fertility of poor tropical soils.

Chapter 18 deals with basalt dust and biochar interactions. Different combinations of biochar and basalt dust were used for the growth of beets and radishes. Basalt dust more than doubled growth of beets and radishes, but beet growth was inhibited in the first crop by immature biochar due to high C/N ratios since no compost was added. Earthworm activity doubled with biochar but increased more than 20 times with basalt dust. Long-term interaction studies are needed to optimize applications because both materials have long-lasting effects and take years to mature.

Chapter 19 presents soil remineralization trials. Local high-school students tested andesitic volcanic ash from Montserrat on the dry limestone island of Barbuda. They found dramatic increases in plant growth and color, so the method could be used to greatly increase food production in soils poor in limestone.

Limestone soils are notoriously infertile, but experimental tests in the Bahamas show that additions of charcoal, ground glass, of basalt powder greatly increase vegetable growth and survival. Methodology and results of those tests and their potential for allowing crops to be grown in places where they cannot grow now are reviewed in Chapter 20.

Rock powders have been developed and used for decades in Scotland to greatly stimulate vegetable and flower production on soils regarded as barren, and for education and training in rock powder use. The applications and dramatic results are reviewed in Chapter 21.

Brazilian research on rock powders (stonemeal) shows them to be highly effective, long lasting, and cheaper than chemical fertilizers. By increasing productivity and profits by farmers, they are expected to play a key role in Brazil's agricultural development. Brazilian tests of rock powders and their policy implications are reviewed in Chapter 22.

Chapter 23 deals with the chemical composition of litter fall and inputs of C, N, and mineral elements in a secondary forest of South Cameroon. In tropical ecosystems most nutrients are in the vegetation rather than the soil, which makes leaf litter recycling the major soil nutrient input. Forest soils showed declining nutrient inputs following deforestation. Nutrient inputs (including trace elements not in most fertilizers) are needed to compensate loss of this source after forests are converted to agriculture.

Chapter 24 is a case study of pyroclastic rocks as natural fertilizers. Volcanic ashes in Cameroon were analyzed for elemental composition, found to improve soil quality, and tested as a slow release fertilizer for local crops with beneficial results.

Chapter 25 deals with the effects of powders of basalt, tuff, granites, and pyroclastic materials on the yield and quality of carrots and cabbages grown on tropical soils in the Northwest region of Cameroon. Various local rock powders and volcanic ash increased short-term production of carrots and cabbages up to three times in local soils, with the results varying between crops. It was observed that rock powder and volcanic ash use had long-lasting and cost-effective benefits.

Chapter 26 discusses rock fertilizers as an alternative to conventional fertilizers. Basalt rocks in Cameroon were found to have good properties for rock powder agricultural supplements. Tests show dramatic increases in corn productivity with use of basalt powder. Rock powder application was found more economically beneficial than conventional fertilizers due to lower cost and longer lasting effects.

Chapter 27 deals with seawater concentrate for abundant agriculture. Many soils are deficient in trace elements. Seawater contains all elements plants need except for nitrogen and phosphorus. Dilute seawater additions greatly stimulate plant growth, but too much sodium and chloride can be a serious problem for plants. Seawater extracts with those elements removed are powerful fertilizers, especially with compost and biochar, and can be sprayed on leaves for rapid uptake.

Chapter 28 looks at superior food production using sea salt and plant extracts. Sea salts, combined with organic extracts from plants and algae containing growth hormones, are shown to be effective liquid spray fertilizers for a wide range of tropical crops in Panama and for temperate crops in Canada and Romania. Treated plants grew faster, had larger crops, were more drought and disease resistant, and their produce was nutritionally richer than controls.

Chapter 29 presents biochar-based amendment enhancing tomato transplant growth and early fruiting. Probiotic microorganisms growing in biochar are symbiotic with plant roots and greatly stimulate plant growth. Biochar inoculated with beneficial fungi and bacteria significantly increased tomato yields and made the bearing season earlier and longer compared to biochar without inoculum.

Chapter 30 presents organic restoration methods to upgrade soil. Many important elements are not taken up effectively unless they are bound in organic molecules. Experiments show organically bound elements have strong effects on plant growth and yields and on soil biology. A variety of experiments are described from the point of view of a soil-building horticulturalist.

Chapter 31 presents SEM-EDX observations of diatomaceous earth at radioactive paddy soils in Fukushima, Japan. Rock powders can provide healthy elements but they can also remove high levels of harmful elements depending on their chemistry. Rock powders experimentally used to treat radioactive rice paddy soils near Fukushima have been shown to absorb many harmful radionuclides.

Chapter 32 deals with recycling waste nutrients using biochar and limestone. Loss of nutrients from soils impoverishes agriculture and harms aquifers, rivers, lakes, and oceans. Biochar and limestone absorb dissolved nitrogen and phosphorus from wastewaters and convert them into effective fertilizers. Closing the nutrient recycling loop as sewage and farm wastewaters are cleaned benefits both soil and water.

Chapter 33 presents commercially viable carbon farming from an Australian perspective. Australia is the leader in promoting policies and accounting systems to increase soil carbon, as it suffers the results of global warming and droughts. It may provide a model for global efforts to reduce CO_2. Many farmers and ranchers have already developed methods to build up soil carbon at very high rates, and are reaping benefits in the form of increased production, resilience to drought, increased water storage, and restoration of dried-up springs and streams.

Chapter 34 concludes the book with general thoughts on regreening the earth and growing our way out of global crises. Soils are the only place CO_2 can be stored fast enough to make a considerable difference in decreasing global atmospheric CO_2 to safe levels. The techniques to achieve this are currently available but are not being used. Soil carbon increases must lie at the core of UN climate change negotiations, and urgent support is needed for large-scale efforts to increase soil carbon, in all forms, in all habitats, and by all methods.

All images contained in this book are available in full color on the enclosed CD.

Acknowledgments

This book pays homage to the late Dick Grantham for his invention of geotherapy and his encouragement to pursue it when he no longer could. Special thanks go to the brilliant scholars who taught me biogeochemistry, G. E. Hutchinson, T. F. Goreau, H. A. Lowenstam, R. A. Berner, and H. D. Holland. Key parts of this book have benefited from the critical review and comments by my daughter, Marina Goreau.

Thomas J. Goreau

Thanks to Tom Goreau for the invitation to join this book project.

Ronal W. Larson

I would like to thank John D. Hamaker for his seminal multidisciplinary thesis, and Don Weaver, who, alongside Hamaker, co-authored *The Survival of Civilization*. Without this book the grassroots movement for remineralizing the earth would not exist, with its specific focus on agriculture and forests, and stabilizing the climate. My gratitude goes to Christian Campe for decades of support that made my dedication to Remineralize the Earth possible, as well as to my daughter, Gaea Campe. Thanks to Ken Whitman of Natural Vitality for ongoing support through their Natural Revitalization program. Thanks to Dr. Tom Goreau for making this book possible, and for his vision, guidance, and expertise with editing; to Idalia Buddington and Maya Berenholz of Hampshire College; Kendyl Eskeli of UMass and Maddie Stone of the University of Pennsylvania for their assistance in editing, as well as coworker Dasha Gaian.

Producing a volume with so many authors, with such different backgrounds, from all over the world, writing in very different styles has been an interesting exercise. Although the figures in the book are in black and white, the accompanying CD contains all figures in full color.

The editors specially thank CRC Press and the editorial staff for the amazing job they have done in helping to ensure that all this material is printed to high-quality specifications and for their patience in supporting the book when some authors dragged out their contributions long past the original deadlines, and the eagle-eyed editorial proofreaders catching small errors that had escaped us. In particular, four individuals who played a key role in the approval and production of this book, John Sulzycki, Kathryn Everett, and Todd Perry from Taylor & Francis, and Paul Abraham of SPi Global, deserve a special thanks.

Joanna Campe

How This Book Came About

This book is the result of the Special Session on Geotherapy: Global Restoration Needs for Stabilizing CO_2 and Climate, held at the World Conference on Ecological Restoration in Merida, Yucatan, Mexico, in August 2011. That session focused on the need for global-scale ecological restoration as the sine qua non for climate change stabilization.

The original plan was to publish papers presented at both of the conference's parallel marine and soil restoration sessions together as a single volume on innovative methods of ecosystem restoration, but it was quickly realized that the amount of material would be too large for one volume and that it made more sense to publish two separate volumes, one focused on the land and one on the oceans.

The first book, *Innovative Methods of Marine Ecosystem Restoration*, was published in December 2012 by CRC Press. The terrestrial companion volume took longer because we realized there was a need to recruit many new chapters from leading restorationists to cover the range of ground needed. The editors have spent the last three years working with authors from around the globe.

Note on Products

Several chapters test the effects of various commercial products. These results are provided for informational purposes only and should not be taken to imply that the publishers and the editors of this book endorse these specific products. There are many people developing innovative products with some differences in composition and preparation, not all of them are discussed here, but all will have a role to play. Readers interested in more information about these products should contact the authors of the relevant chapters or the companies that manufacture them.

Contributors

Monique Abossolo Angue
Department of Earth Sciences
University of Yaoundé I
Yaounde, Cameroon

Bonaventure Alemanji
Department of Geology
Higher Teachers' Training College Bambili
University of Bamenda
Bamenda, Cameroon

Carlos A. Arango
CYASA
Panama City, Panama

Jean Jacques Braun
Institute of Research for Development
Yaounde, Cameroon

Jennifer A. Brodie
REMIN
Aberdeen, Scotland

Joanna Campe
Remineralize the Earth
Northampton, Massachusetts

Barry Carter
Agricultural Consultant
Baker City, Oregon

Gabriel Despaigne-Ceballos
Technosub Marine Services
Panama City, Panama

Gabriel Despaigne-Matchett
Green Association of Panama
Panama City, Panama

Cindy Eiritz
Healthy Soils Australia
Canberra, Australia

Paul F. Gonang Achoumele
Department of Geology
Higher Teachers' Training College Bambili
University of Bamenda
Bamenda, Cameroon

Marina Goreau
Global Coral Reef Alliance
Cambridge, Massachusetts

Thomas J. Goreau
Global Coral Reef Alliance
Cambridge, Massachusetts

Ioan Hossu
Industrias y Manufacturas Bionaturales S.A.
Panama City, Panama

Yasuhito Ishigaki
Medical Research Institute
Kanazawa Medical University
Uchinada, Japan

Seth Itzkan
Planet-Tech Associates
Somerville, Massachusetts

Kirk D.S. Jones
Department of Civil and Environmental Engineering
University of Vermont
Burlington, Vermont

Pierre Kamga
Agricultural Research Institute for Development
Bamenda, Cameroon

Erich J. Knight
Shenandoah Gardens
McGaheysville, Virginia

Ronal W. Larson
Yahoo Biochar-Policy Internet Discussion List
Boulder, Colorado

Jim Laurie
Biodiversity for a Livable Climate
Lexington, Massachusetts

Othon Henry Leonardos
Center for Sustainable Development
University of Brasília
Brasilia, Brazil

Felix Lufkin
Remineralize the Earth
Northampton, Massachusetts

Dieudonne Z. Manjo
Department of Geology
Higher Teachers' Training College Bambili
University of Bamenda
Bamenda, Cameroon

David A.C. Manning
Department of Civil Engineering and Geosciences
Newcastle University
Newcastle upon Tyne, United Kingdom

Linda Mazoh
Department of Geology
Higher Teachers' Training College Bambili
University of Bamenda
Bamenda, Cameroon

Steven R. McGreevy
Center for Research and Development
Research Institute for Humanity and Nature
Kyoto, Japan

Ronald Morse
Department of Horticulture
Virginia Tech
Blacksburg, Virginia

D. Nathaniel Mulcahy
World Stove
Gloucester, Massachusetts

John Mussington
Sir McChesney George Secondary School
Barbuda Research Complex
Barbuda, Antigua and Barbuda

Hideaki Nakagawa
Medical Research Institute
Kanazawa Medical University
Uchinada, Japan

Christophe Ngansop
Ministry of Agriculture and Rural Development
Yaounde, Cameroon

Jon Nilsson
CharGrow
Asheville, North Carolina

David Guimolaire Nkouathio
Department of Earth Sciences
University of Dschang
Dschang, Cameroon

Jeff Novak
Coastal Plain Soil, Water and Plant Conservation Research
Agricultural Research Service
United States Department of Agriculture
Florence, South Carolina

Damien Henri Odigui Ahanda
Department of Earth Sciences
University of Yaoundé I
Yaounde, Cameroon

Masayuki Okuno
Department of Earth Sciences
Faculty of Natural System
Kanazawa University
Kanazawa, Japan

Elise Pinners
The Vetiver Network International
Nairobi, Kenya

T.R. Preston
Center for Research and Technology Transfer
Nong Lam University
Ho Chi Minh City, Vietnam

Fred Provenza
Utah State University
Logan, Utah

Fred Riger
Grand Bahama Scuba
Grand Bahama, Bahamas

Rebecca Ryals
The Ecosystems Center
Marine Biological Laboratory
Woods Hole, Massachusetts

Adam D. Sacks
Biodiversity for a Livable Climate
Lexington, Massachusetts

Olaf Schuiling
Department of Geochemistry
University of Utrecht
Utrecht, the Netherlands

Elaine Senechal
New Harmony Farm
Newbury, Massachusetts

Akira Shibata
Regional Information Research Center
Ritsumeikan University
Kyoto, Japan

Roque Solis
Congresso General Kuna
Coedup, Panama

Kurt Spokas
Soil and Water Management Unit
Agricultural Research Service
United States Department of Agriculture
St. Paul, Minnesota

Erin Stack
New Harmony Farm
Newbury, Massachusetts

Richard S. Stein
Department of Chemistry
University of Massachusetts
Amherst, Massachusetts

Teruaki Takehara
Medical Research Institute
Kanazawa Medical University
Uchinada, Japan

Jianwu Tang
The Ecosystems Center
Marine Biological Laboratory
Woods Hole, Massachusetts

Kazue Tazaki
Department of Earth Sciences
Faculty of Natural System
Kanazawa University
Kanazawa, Japan

Jean Pierre Tchouankoue
Department of Earth Sciences
University of Yaoundé I
Yaounde, Cameroon

Richard Teague
Texas A&M University
Vernon, Texas

Arliane Nicole Tetchou Tchekambou
Department of Earth Sciences
University of Yaoundé I
Yaounde, Cameroon

Samuel Tetsopgang
Department of Geology
Higher Teachers' Training College Bambili
University of Bamenda
Bamenda, Cameroon

Suzi Huff Theodoro
Postgraduate Program in Environment and Rural Development
University of Brasília
Brasilia, Brazil

Oliver Tickell
The Ecologist
Resurgence Trust
Devon, United Kingdom

Tom Vanacore
Rock Dust Local
Bridport, Vermont

Tadeusz S. Wysocki
South Meadow Farm and Research Center
Wales, Massachusetts

David Yarrow
The Earth Restoration & Regeneration Alliance (TERRA): Turtle Eyeland
Columbia, Missouri

Arthur Zeigler
Ambrosia Technology
Raymond, Washington

1

Introduction: Geotherapy, the Down-to-Earth Solution to Global Warming

Thomas J. Goreau, Ronal W. Larson, and Joanna Campe

CONTENTS

Goals of This Book

This book presents innovative soil carbon restoration solutions that address humanity's major ecological, environmental, energy, and economic problems. It is a call to action to restore our planet's inherent but badly damaged life support systems (which regulate temperature, sea level, storms, water, and soil) in an integrated way to solve the problems we have created through greed, ignorance, and mismanagement.

Soils are the *only* place where we can store our waste carbon, at a cost we can afford, on a time scale that will not only solve the climate change problem but also our major food, energy, and water problems. The *down-to-earth solution* takes carbon from the atmosphere where it does the most harm and puts it in the soil where it does the most good.

Information on the innovative methods presented in this book for soil fertility restoration, carbon sequestration, and reversing atmospheric CO_2 increase are urgently needed: the public is not aware that these solutions exist and governments are not using them where they are most critical—to reverse global warming and sea level rise.

Although the techniques to resolve these global problems are well known, they are not being used due to failure of the United Nations Framework Convention on Climate Change to recognize that *soil is a major carbon sink and the only one that we can cost-effectively turn from part of the problem to part of the solution.*

This book shows that it is not possible to stabilize CO_2 and our planet's renewable natural resources at safe levels without systematic, global-scale restoration of soil fertility and carbon storage.

Our destruction to the natural biogeochemical regulation mechanisms that stabilize climate, atmosphere chemistry, water, and soil provided immediate gratification (food, energy, profits) but ignored their long-term costs. Our crisis took centuries to be created and cannot be undone as fast as needed. But if we do not reverse our misuse of carbon very quickly on a sufficient scale, it will be impossible to prevent extreme damage from runaway global climate change that will start sooner, last longer, and be far more extreme than policymakers realize.

Policymakers are complacent in part because, as shown in this book, they have been misled by incorrect interpretation of inadequate models that miss most of the real responses to global warming. They either do not grasp the long-term consequences of their local actions for the entire planet on a geological time scale or are in denial.

The simple and effective solutions presented here mimic and extend natural processes but are being ignored by global policy makers. Yet pointless, unproven, and expensive dead ends are being investigated and implemented due to an irresponsible focus on isolated components of the CO_2 cycle instead of a holistic understanding of the entire cycle. These red herrings will be ruinously expensive, are almost certainly ineffective, waste valuable and irreplaceable time, and probably create worse problems than they solve.

Reversing soil carbon loss and converting the land from decline into increasing growth are simple to learn and easily implemented! *The solution lies right beneath our feet, and almost everyone can contribute, using the approaches described in this book.*

The focus of this book is on practical methods of soil fertility and carbon storage that contribute direct solutions to global climate change. We avoid high-tech *geoengineering* solutions that put bandages over symptoms without addressing causes. We also avoid methods such as genetic bioengineering to increase plant productivity and produce plants better adapted to stress, but which can lead to monoculture, genetic pollution, sterile soils, and massive inputs of fossil fuel energy, pesticides, and fertilizers that impede or prevent the natural soil nutrient recycling mechanisms whose reinvigoration is the theme of this book.

For generations, humans have reaped all available benefits at the expense of the environment, and now the best land, raw materials, and other ecological resources are gone. Our goal must be to restore much greater biodiversity and productivity in all habitats, restoring barren and underproductive areas to bear again, so that we do not need to keep constantly destroying more pristine habitat to maintain food supplies for growing populations. We need more productive, diverse, synergistic, value-added ecosystems, not monocultures that displace native forests or compete with food resources. Change in global carbon practices will create a rising tide that lifts all boats.

Organization of This Book

This book has 34 chapters containing real results by pioneering restoration ecologists around the world who have greatly improved the productivity and carbon storage of sites with very different geology, climate, soil type, vegetation, and land management practices. The first section focuses on global principles related to climate change and planetary sustainability in the context of global soils, the biosphere, and the atmosphere. Papers focusing on land management, the use of biochar, rock powders, sea salts, probiotic microorganisms, and organic amendments follow; local case studies showing practical work to reverse degradation and greatly increase productivity and carbon sequestration follow; and finally, papers on the implications for the global climate change negotiations follow.

The field results in this book come from Africa, South America, Asia, Europe, North America, and Australia and show these practical, cost-effective methods to reverse global warming can be applied anywhere, in any soil, any climate, any crop, and any geological setting, although naturally the details will vary accordingly. Moreover, if done in an integrated and timely way, they can avert, at minimum cost, the worst consequences of runaway climate change we are now rushing headlong into and condemning generation after generation to suffer.

The authors of this book are doers, not talkers. They span a huge range of people developing practical solutions, including scientific researchers, farmers, activists, and nonprofit groups that are out there with their hands in the soil. They work with a wide range of materials, but almost all their results have been produced by small, dedicated teams with little or no funding.

This book focuses on using biochar, rock dusts, composts, seawater minerals, plant extracts, and probiotic microbes as natural, slowly released, long-lasting, and balanced nutrient sources to stimulate balanced soil biology. This is an alternative to the chemical/industrial fertilizer model that uses expensive soluble chemicals with very high fossil fuel energy content. Because chemical fertilizers are not balanced to meet the full spectrum of plant nutritional needs, many critical elements are missing.

As a result, plants fertilized by traditional methods cannot use most of the expensive fertilizer provided. The excess is wasted, contaminating groundwater, rivers, lakes, and the sea. Industrial chemical agriculture also heavily relies on chemical pesticides, sterilizing soil and harming beneficial bacteria, fungi, and invertebrates that normally break down dead organic matter and recycle nutrients back to plants. The resulting monoculture is dependent on expensive imported chemicals in which the soil is merely the physical support for the plant rather than a stable biodiverse ecosystem where each part feeds the others. Worse, it is bleeding carbon dioxide to the atmosphere as soil carbon is degraded, part of the problem that should be part of the solution. In contrast, the materials described in this book are natural, not energy intensive, balanced, long-lasting, and promote greatly increased biological cycling in soils, while reducing greenhouse gas emissions, both from industrial processes and from the soil itself.

Geotherapy versus Geoengineering and Bioengineering

The technology to exploit nature got us into the runaway climate change situation we currently find ourselves in, and there is no going back now to a pretechnological world because there are already too many people for our planet's natural life support systems to maintain everybody at an adequate and equitable quality of life. *We need to develop technologies of ecological mass restoration to replace technologies of ecological mass destruction we have practiced for so long.*

As with every technology, there are questions of which technologies are appropriate and how and by whom they will be used. Will new techniques benefit humanity as a whole or only those who control them? Throughout history, scientific inventions that could be used for great good, like metals to make plows to grow more food, have almost inevitably been used first for evil, making swords to kill each other. Some people feel science and technology itself is to be blamed for these misuses, but that is a serious error: these abuses are not inherent in the technology itself, but they result from the goals of those who are in control of it, who decide how, when, where, and by whom they are used.

A very vocal group is now promoting complex, expensive, high-technology *geoengineering* solutions to climate change, such as putting mirrors in space to reflect sunlight, shooting sulfuric acid aerosols by rockets into the atmosphere to reduce light and heat reaching the surface, dumping fertilizers in the ocean, pumping deep ocean water to the surface to cool it, or extracting CO_2 from smokestacks to pump into the deep sea or underground (and hope it never comes back up). What distinguishes all of these proposals, smoke and mirrors, dumping, and pumping, are expensive unproven experimental technologies that may be incapable of doing what is claimed or simple-minded theoretical suggestions that could be fundamentally flawed, which have vast but seriously underestimated cost and unknown side effects that may be worse than the problem they are trying to solve. These proposals are gambles that probably will not work and probably will be ruinously expensive to implement. Many of them are irresponsible or impossible schemes when examined closely, but they attract great publicity and funding because governments, and the energy industry that writes their policies, are grasping at straws to ensure no change to business as usual. They leap to back any proposal that promises to allow unabated release of CO_2 from fossil fuel burning and the profits that follow.

Another set of approaches involve bioengineering (*genetic engineering*), selecting or manipulating genes to change the plant in the hope that new genes may prove more efficient or that the architecture of the plant itself can be rearranged. In general, the green revolution led to the selection, breeding, cloning, or genetic engineering of plants with greater productivity, but in most cases, these have proven to reduce genetic and species diversity and led to plants that require ever greater amounts of fertilizer and pesticides, at increasing cost, to maintain production, which often sterilize the soil beneath them. Bioengineering is expensive, is slow to produce results and acceptance, and may eventually lead to real advances, but is not covered in this book since many of the costs and benefits remain to be proven, and there are many other publications focusing on them since they are well funded by governments and the large chemical and agricultural companies.

This book focuses on making the soil *itself* richer, not manipulating the plant, the atmosphere, or the ocean as bioengineering and geoengineering seek to do. We do not cover tilling methods or efforts to affect plant growth by altered physical conditions, such as water, sound, light, electricity, and magnetism.

There are a century and a half of claims in these areas, but we have found little that is well verified, and much more careful work is needed to know if there is a real basis for claims made by their proponents. These methods, if proven to work, will work best if the soil has the minerals the plants need, and so would add to the methods shown in this book.

We call the methods discussed in this book *geotherapy*, the philosophical antithesis of geoengineering, because it is based on re-invigorating the natural biogeochemical mechanisms by which our planet regulates climate. The concept of geotherapy employs a conscious medical metaphor: like a good doctor dealing with a sick patient, we must first correctly diagnose the cause of imbalance in the patient's health before we can correct it and then prescribe a conscious therapy to restore the patient to health (Grantham, 1992a,b). Geotherapy was invented by the late Richard Grantham, a pioneering molecular biologist who turned to bioethics when he retired, to whom this book is dedicated.

Geotherapy Principles

Richard Grantham, Tim Greenland, and Tom Goreau were principal organizers of the First World Conference on Geotherapy, held at Universite Claude Bernard Lyon 1 in Lyons, France, in May 1991.

The Colloquium on Modelling and Geotherapy for Global Changes, sponsored by France's Centre National de la Recherche Scientifique (CNRS), included distinguished scientists from many disciplines and continents. They discussed global environmental problems and the urgent need to resolve them and concluded their panel discussion on May 17, 1991, with the following declaration of geotherapy principles:

- Accelerating environmental degradation threatens the habitability of the biosphere. We believe that corrective action is possible and urgent.
- Our goal is long-term survival in an acceptably maintained ecosystem.
- We, as human beings, take full responsibility for our actions by not sacrificing natural resources for short-term gains and by working to make the world a better living place.
- This choice will influence our biological and cultural evolution; we cannot avoid it without grave consequences.
- A global bioethic should further develop a guide and motivate geotherapy and our cultural evolution.
- A root problem is excessive demographic growth; the Earth's carrying capacity is being exceeded. With the present style patterns of development, pollution of all kinds will increase as long as the population increases.
- We declare that scientists of all walks of life should adopt the aforementioned goals and participate in meeting at all levels to apply these principles.

These goals have only become more urgent with time. They must be well thought out and presented to global decision makers, in particular to the United Nations Framework Convention on Climate Change and the United Nations Conference on Sustainable Development, to be implemented.

REFERENCES

Grantham, R., 1992a, Geotherapy as evolutionary choice, *Global Environmental Change*, 2:2–4.

Grantham, R., 1992b, Geotherapy and global bioethics: Towards acceptable survival, *Global Environmental Change*, 2:258–261.

2

Global Biogeochemical Restoration to Stabilize CO_2 at Safe Levels in Time to Avoid Severe Climate Change Impacts to Earth's Life Support Systems: Implications for the United Nations Framework Convention on Climate Change

Thomas J. Goreau

CONTENTS

Introduction

This chapter focuses on (1) why and how policymakers have severely underestimated the long-term impacts of climate change, (2) how the carbon management mechanisms they have developed are flawed and counterproductive, (3) how they could be corrected, (4) where the carbon can be most cost-effectively stored, (5) how quickly that could be done, and (6) the practical and policy actions needed to get there.

Atmospheric carbon dioxide (CO_2) concentrations are currently at levels the Earth has not experienced for around 25 million years (Pagani et al., 2005; Zhang et al., 2013), a time when there were no polar ice caps and temperatures and sea levels were much higher than today, and CO_2 is headed much higher. Humanity faces unpalatable short- versus long-term trade-offs between reducing emissions of CO_2 versus short-lived climate pollutants to minimize future global warming impacts (Shoemaker et al., 2013). Supply-side approaches amount to reshuffling Titanic deck chairs if they overlook demand-side solutions. CO_2 cannot be reduced to safe levels in time to avoid serious long-term impacts unless the other side of atmospheric CO_2 balance is included by increasing sinks while simultaneously decreasing sources (Goreau, 1987, 1990, 1995).

Down to Earth underground solutions to global warming (Figure 2.1) have been ignored by policy makers' exclusive focus on source reductions and geoengineering. They need to look at the other side of the carbon coin, implementing solutions that can work in time to make a difference by removing carbon from the atmosphere, where it does the most harm, and putting it in the soil, where it does the most good. The answer lies at our feet.

Reducing emissions is of course essential, but without greatly increasing carbon sinks at the same time, it alone is not adequate to do the job (Goreau, 1995). Increasing sinks is not an excuse to avoid emissions reductions but is the *sine qua non* for making emissions reductions work effectively and fast enough to prevent severe climate change impacts.

The excess CO_2 in the atmosphere largely came from fossil fuel combustion and from soil and vegetation carbon loss. Soil holds around four times more carbon than the atmosphere or vegetation and could hold more again (Lal, 2004; Lehmann and Joseph, 2009; Woolf et al., 2010). The dynamic time response spectrum of CO_2 sources and sinks shows the fastest way to decrease CO_2 is to increase photosynthesis and biomass storage in tropical soil carbon, which could resolve the problem in decades, but CO_2 source reductions alone will take centuries to millennia to have an effect, so impacts will be far worse (Goreau, 1995).

Transferring carbon from atmosphere to soil would also greatly increase soil productivity, biomass, and groundwater resources, while increasing rainfall and reducing temperature through increased evapotranspiration. Effective methods to greatly increase soil carbon by intensifying natural biogeochemical recycling work in all ecosystems, agricultural land, degraded land, pastures, and forests. Not only is there 2–4 times more carbon in soil than in the atmosphere or vegetation, this article shows that the

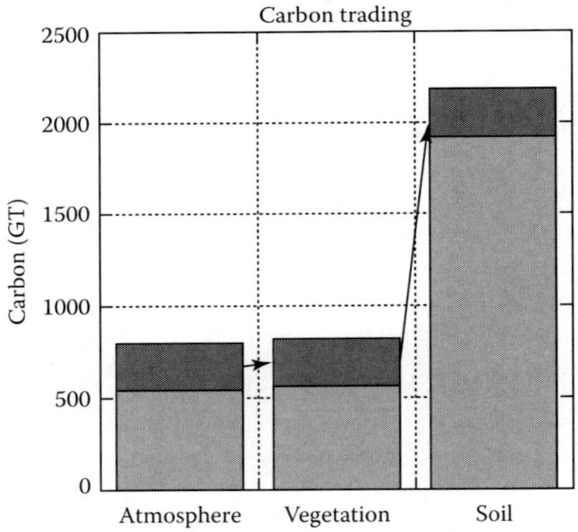

FIGURE 2.1 Carbon contents of atmosphere, vegetation, and soil, showing transfer of the anthropogenic excess, shown in light gray. In the left column, dark gray represents the preindustrial CO_2 level, and the current value is the dark gray plus light gray column. To get transfer of all the excess to the current biosphere (dark gray, middle column) would be an equally large increase, but if that is transferred to the current soil carbon (dark gray, right column), the proportional increase is small. Soil carbon estimates to a depth of 2.3 m are based on Shangguan et al. (2014). Note that this is an underestimate because it does not include carbon in deeper soil horizons. Also note that wetland carbon and marine mangrove and sea grass carbon may also be equivalent in magnitude to soil, but current estimates are too poor to show.

soil is the only place we can store the excess CO_2 quickly enough to prevent severe future impacts from global warming. The *Down to Earth* solution to global warming, putting it in the ground, will save money in the long run and provide many additional benefits to humanity.

Without restoring global soil fertility, and the biogeochemical planetary life-support systems only healthy growing plants can provide, it will be not possible to stabilize CO_2 at safe levels and avert the worst impacts of runaway global warming and sea level rise; that is the bad news (Pimentel and Pimentel, 1996; Flannery, 2005; Lovelock, 2006; Romm, 2007; Ehrlich and Ehrlich, 2013). The good news is that this can *still* be done in time to avoid the worst, but not all, of the impacts, in a way that minimizes costs and maximizes benefits, if we act right away. To understand how, we must begin with the entire global carbon cycle and descend to the atomic-scale details of how carbon works in soils, oceans, and plants, before returning again to the global scale.

CO_2 as Planetary Currency

CO_2 is central to the physics, chemistry, biology, geology, and climate of our planet, quite apart from being both food for plants and the largest waste product of our energy and food (for just a small part of the important literature, see Woodwell et al., 1978; Holland, 1984; Solomon et al., 1985; Post et al., 1990; Schlesinger, 1990; Gerlach, 1991; Jenkinson et al., 1991; Sundquist, 1993; Greenland, 1995; Berner, 2003, 2004, 2006; Houghton, 2003; Retallack, 2003; Sundquist and Visser, 2003; Royer et al., 2004, 2007; Beerling and Berner, 2005; Pagani et al., 2009). CO_2 is the major greenhouse gas (GHG) causing global warming, confirmed again by the latest Intergovernmental Panel on Climate Change (IPCC) report (Hegerl and Stott, 2014). CO_2 therefore serves as the most fundamental natural *currency* unit of the major planetary biological, environmental, energy, and ecological life-support services (Daily, 1997). We have not learned how to assign a monetary value to most of these benefits or the costs from their degradation. Until we do, we cannot assign a realistic price for the carbon these processes produce or consume as $XCO_2 = Y\$$. The obvious idea of CO_2 as a universal environmental currency has come independently to almost everyone who has thought about the issue, so many it is impossible to know who was first or who should be credited. It is made yet again very clearly in the Foreword of this book by John Todd. It is simply an idea whose time has come, whether or not we call payment for carbon services a *tax*, a *fee*, a *surcharge*, or some other euphemism.

We pay directly for the costs of fossil fuel extraction, processing, distribution, and profits in order to produce that CO_2, but we completely ignore the many real indirect costs such as global climate change, simply foisting them off on future generations. Like a profligate heir squandering an inherited family fortune created through no work of his/her own on wasteful personal gratification, we are condemning our descendants to beggary and servitude if we do not clean up our act and stop fouling our own nest. But instead of paying to undo the environmental damages of fossil fuels, we effectively subsidize it, making it artificially cheap. If we paid those costs in the fossil fuel energy price, renewable energy would not need subsidies to compete. A recent estimate of the real costs of fossil fuel CO_2 emissions puts the minimum cost at \$125/ton of CO_2 and could be a lot higher (van den Bergh and Botzen, 2014).

Failure to set a carbon price in a way that accurately reflects *all* costs of these complex and interlinked services, whenever and wherever they take place, guarantees a failure of market mechanisms to address the complex and intertwined issues. These are the economists' *externalities*, which society is left to pay (Stern, 2006; Dasgupta and Ehrlich, 2013). This failure is evident in the collapse of the European carbon market price, which was doomed at its very conception because the EU began by giving away CO_2 pollution permits free to the biggest polluters, devaluing its price, and because, for political reasons, they gave away permits to emit more CO_2 than polluters could actually produce, thereby ensuring carbon price collapse. With pollution permits so cheap as to be practically free, there was little incentive to clean up. While the Europeans tried to set up a carbon market, and blew it, other rich countries simply refused to require any carbon accounting and trading. The poorest countries are most vulnerable to climate hazards (Ehrlich et al., 2012), none more so than the low-lying developing states, so they pay the disproportionate cost of market failures from which they derive no benefit. This is clearly no way to effectively regulate GHG emissions.

Global Carbon Cycle

A complete and holistic understanding of the complete global biogeochemical CO_2 cycle is essential for any solution that solves the CO_2 waste problem by taking it from the atmosphere, where it does the most harm, and putting it in the soil, where does the most good (Figure 2.2).

Global warming over the last century and a half is only the very start of much larger future change to come unless we quickly restore global ecosystems to recycle CO_2 into permanent storage and stabilize CO_2 and our planet's critical life-support systems. Large-scale restoration of degraded habitats is the most important option for CO_2 sequestration: so little intact wild habitats now remain that strict conservation of them is incapable of storing the carbon needed. Without large-scale restoration of habitats that are already damaged or degraded, the requisite amount of carbon cannot be stored. Every degraded habitat, forests, pastures, marshes, swamps, mangroves, sea grasses, coral reefs, and even deserts, needs to be involved. The areas are vast and the home to the bulk of species biodiversity (Figure 2.3).

Tragically, restoration of ecosystems in general, and of their soil carbon in particular, is completely missing from the United Nations Framework Convention on Climate Change (UNFCCC) negotiations. UNFCCC's continued failure to grapple with the real issues and solutions guarantees continued runaway climate change. Soil carbon simply is not on the list of accepted carbon sinks in the UNFCCC negotiations, even though it has around four times more carbon than the atmosphere (Figure 2.1). This fundamental error makes the rest of their carbon accounting basically irrelevant.

Unfortunately, the entire CO_2 discussion is bedeviled by selective and misleading analysis since very few scientists, much less decision makers, have the full grasp of the *entire* carbon cycle system essential for scientifically sound policy. The polemics of the CO_2 issue are full of selective or false arguments that usually consider only one process at a time, not the entire interlinked cycle. Confusion is caused by

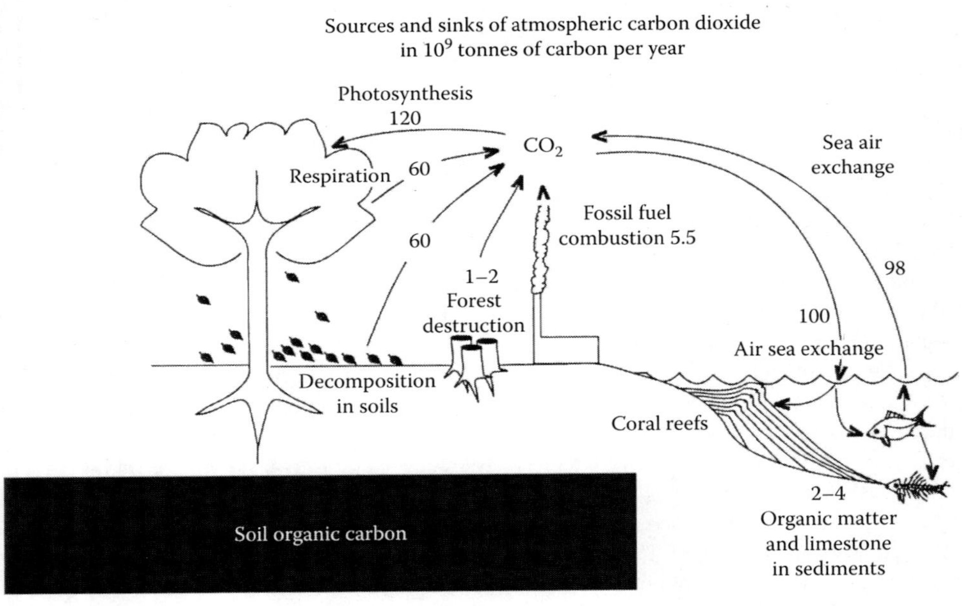

FIGURE 2.2 Major fluxes in the global carbon dioxide cycle. (After Goreau, T.J., *Ambio*, 19, 230, 1990, drawing by Dr. Peter Goreau, adapted by T. Goreau.) Of these fluxes, only fossil fuel inputs are accurately quantified because coal, oil, and natural gas commerce is well measured. The others have uncertainties that will slightly affect the quantitative details of the discussion in the following, but not the fundamental features. This figure, while capturing the essence of the entire cycle, does so by simplifying smaller or slower subcycles that affect atmospheric CO_2 more slowly. This chapter shows that only increasing CO_2 transfer from atmosphere to biomass and soil carbon can reduce CO_2 levels fast enough to solve the problem in time.

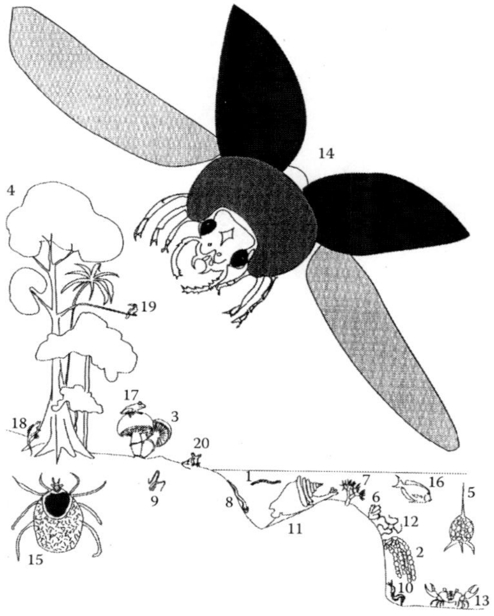

FIGURE 2.3 Global biodiversity with each major class of organism shown in a size proportional to the number of species. (Figure by Dr. Peter. D. Goreau.).

mistaking gross carbon sources and sinks with net ones, and even confusing sources with sinks. Global policy is being made with fundamental errors of incomplete and partial accounting of GHG sources and sinks that ignores or confuses the major global carbon flows.

Unless we start large-scale global restoration efforts to increase atmospheric carbon sinks *now*, they will come too late to prevent catastrophic overshoot, of CO_2, temperature, and sea level (Hansen et al., 2007, 2008, 2010; Rogelj et al., 2011, 2013; Matthews and Solomon, 2013; Stocker, 2013). The global restoration profession must lead in urging world leaders to start immediate massive restoration efforts in all degraded ecosystems. These must simultaneously maximize biodiversity, biogeochemical function, nutrient recycling, and natural carbon sequestration, and so cannot be based on monocultures (single species plantations). The sooner they are started, the less they will cost, the shorter they will be needed, and the greater the ecological, environmental, and economic benefits to humanity and the other species we share the planet with.

CO_2 Futures

If we burn all the oil, natural gas, and coal we can extract from the ground, we will increase atmospheric CO_2 from 5 to 10 times, but in about 100 years, we will have used up most of the available oil and gas, and in about 300 years, the coal too. No matter which fossil fuel *business as usual* scenario we adopt, a century, perhaps three, of vastly increasing atmospheric CO_2 lies ahead (Figure 2.2) unless we switch to non-fossil-fuel energy resources. We would have to eliminate almost all fossil fuel use to bring CO_2 back down to safe levels, and that would take centuries.

We have not yet felt the vast bulk of the temperature and sea level response that will hit us, because most of the effects that today's level of CO_2 will cause lie far in the future. Technologies needed to stabilize CO_2 at safe levels, including sustainable biomass energy and biochar to sequester carbon (increasing soil carbon, fertility, water-holding capacity, and forest and agricultural productivity) discussed in this book, are mature and well known (Green Disc, 2014) but, due to the lack of support at policy and funding levels, are not being used on the scale needed as cost-effective problem-solving tools.

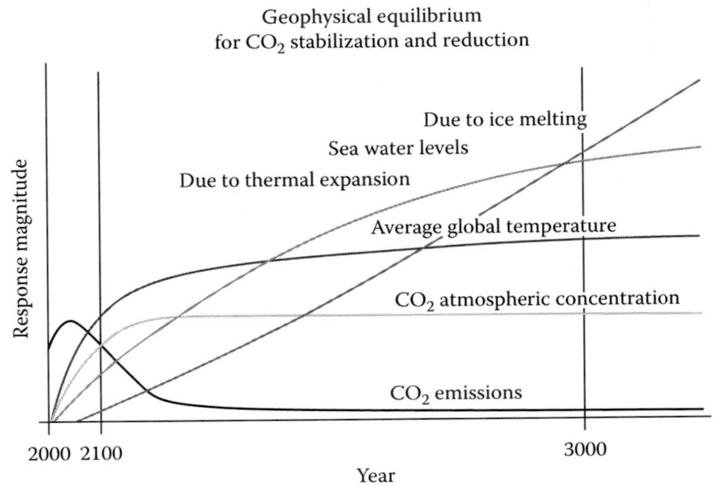

FIGURE 2.4 Time scale of CO_2 temperature and sea level responses to fossil fuel emissions. (Adapted from Climate Change, Synthesis Report. A contribution of working groups I, II and III to the third assessment report of the Intergovernmental Panel on Climate Change, Figure SPM-5, 2001, Cambridge University Press.)

It takes about 1000 years for excess atmospheric CO_2 to dissolve and mix into the deep ocean, so the CO_2 concentration in the atmosphere will stabilize on the time scale of ocean mixing, in a few thousand years. The same is true of temperature (Figure 2.2). Right now we feel only a small part of the heat increase of the planet as sensible temperature, because the excess planetary heat is flowing into the largest heat sink, the deep sea (Levitus et al., 2012), and being used to melt ice (which does not change the temperature, even as it adds heat to the system). These last two important effects are not included in the IPCC (Figure 2.4). Until the deep sea warms up, and all the ice is melted, we would not feel the full effect of global warming of atmospheric and land surface temperature (SST), and that will take thousands of years.

After a thousand years, CO_2, temperature, and sea level rise from thermal expansion of seawater will start to level off, but sea level rise from ice melting will continue to increase for thousands of more years. Sea level rise will not level off until all the ice melts, which will raise sea level by up to 50 or more meters after several thousand years. We do not really know how long because that depends on just how quickly the glaciers flow, and as is well known, they move at a glacial pace, you can outwalk them. There is clear evidence that at the end of the last Ice Age (10,000–18,000 years ago), there were times of very rapid sea level rise, possibly with surges (Hanebuth et al., 2000; Overpeck et al., 2006; Rohling et al., 2008). During that time, changes in sea level were likely due to catastrophic surges of melting ice suddenly sliding into the sea, because instead of glacial ice being frozen to the rocks beneath them, the bottom of the ice was lubricated by a layer of melt water, as they are increasingly now in Greenland and Antarctica. No one can predict when such surges in sea level will take place due to glacial melt (especially due to climate change), but large and sudden sea level rises are not just possibilities, but certain based on our current trajectory. This unpredictable surge and melt pattern will make stable life along coastal shorelines impossible when melting accelerates.

Increased atmospheric CO_2 and temperatures will remain in the atmosphere until the carbon dioxide is removed through a chemical reaction with rocks, called weathering. High CO_2, temperature, and sea level is expected to persist on a time scale in the range of tens to hundreds of thousands of years unless rock weathering is greatly accelerated above natural levels (Norris et al., 2013). That can be induced with the methods described using rock powders in many chapters in this book. Increased biological soil carbon sinks also greatly increase rock weathering carbon sinks.

The weathering time scale is determined by the time it takes atmospheric CO_2 dissolved in rainwater to react with minerals in rocks and soils, being converted into dissolved bicarbonate and acid (which dissolves more minerals):

$$H_2O + CO_2 = HCO_3^- + H^+$$

The bicarbonate then is washed by rivers into the ocean and builds up in the deep sea until it is ultimately precipitated as limestone and finally buried deep in marine sedimentary rocks, which takes hundreds of thousands of years:

$$2HCO_3^- + Ca^{++} = CaCO_3 + H_2O + CO_2$$

About half the marine carbon involved in limestone production is buried and the other half is released back to the atmosphere (Stumm and Morgan, 1996).

Using fossil fuels now will therefore cause environmental impacts to everyone, everywhere, for perhaps around a million years or so. These long-term impacts are largely *not* included in the IPCC climate change models because IPCC was mandated for political reasons to use models to predict impacts for 20, 50, or 100 year horizons, and so they miss almost all of climate responses to increasing CO_2 that take thousands of years to develop. The IPCC was ordered by UNFCCC to make projections only over short time periods that coincidentally ignore almost all of the anticipated damage. In effect, they were instructed to come up with a totally irrelevant *answer* to the wrong question, which is of little use for serious scientifically sound long-term planning.

One leading thinker on these issues states pithily: "I don't believe that humans can destroy the earth's capacity for renewal. I do believe that full recovery from our abuse will come in geologic time (millions of years), not human time (200,000 years), certainly not agricultural time (10,000 years). We want to stay here, but we want all the other species embedded in the same ecosphere within their ecosystem to stay with us too" (Jackson, 2010).

Climate Change Sensitivity: IPCC Models Underestimate Reality

All components of the earth atmosphere/ocean/biosphere/soil/ice system are affected by complicated positive and negative feedbacks (interactions that increase and reduce their impacts) linked to exchange of CO_2 and by the responses to the temperature changes that CO_2 causes. The climate change prediction models are far too simple to describe the many known feedback mechanisms accurately in quantitative terms, even if they give qualitative insight into them. We can gain the best insight into the long-term responses of the climate system from actual real-world data: the paleoclimatological records preserved in ice, sediments, corals, tree rings, and permafrost.

Changes from Ice Age to interglacial conditions have been clearly controlled for millions of years by small changes of the earth orbit, as explained by Milankovitch theory (eccentricity, obliquity, and inclination), that affect the amount and timing of sunlight and heat reaching different parts of the earth. These cycles are completely and accurately predictable from physics and explain the timing of long-term climate change with extreme precision (except blips caused by unpredictable massive volcanic eruptions or asteroid impacts). The amplitude of these changes in solar warming input is only enough to explain long-term temperature changes of around 0.1°C from Ice Age to interglacial orbital configurations, while the actual changes at the earth's surface between Ice Age and interglacial conditions are around 5°C–10°C. This means that the earth climate system contains very large positive feedbacks that amplify these small input heat changes about 50–100 times.

Unfortunately, the bulk of these feedback mechanisms that we know exist in the system are not well enough known to accurately include in climate change models, not even the most mathematically sophisticated models running on the largest and most expensive computers in the world (Soden and Held, 2006; Torn and Harte, 2006; Hansen et al., 2007). Those feedbacks that are well known, like the fact that dark open seawater absorbs much more heat than water covered with shiny reflective white ice (ice–albedo feedback) or the fact that increased temperature causes more evaporation of water into the atmosphere that absorbs more heat (water vapor feedback), are readily described and modeled, but only around a quarter or a third of the positive feedbacks that we know the system actually has is included in the models; the rest cannot be described well enough to model accurately yet (Stocker et al., 2013).

For example, respiration rates are strongly affected by the facts that the land is warming faster than the ocean (Seneviratne et al., 2014), nighttime temperatures are rising faster than daytime temperatures (Karl et al., 1991; Peng et al., 2013), and tropical sea surface winter temperatures are rising faster than summer temperatures (Goreau et al., 1993). As a result night and winter respiration are increasing faster than day and summer photosynthesis (Wan et al., 2009; Xia et al., 2009, 2014), acting as another positive feedback not in the models. These changes act to decrease net primary production and increase the amplitude of seasonal atmospheric CO_2 variations, especially in boreal climates (Graven et al., 2013). Furthermore, these feedbacks are highly nonlinear, so they do not simply add up (Zickfeld et al., 2011) and are very hard to model accurately. The crucial result of this neglect of feedbacks is that even the best climate models underestimate long-term impacts by perhaps around a factor of 3 or 4. IPCC was mandated by the UN to use climate change models, not real-world paleoclimate data, to predict future changes, so IPCC predictions, being based on incomplete models, are intrinsically guaranteed to underestimate real long-term impacts. How much is estimated from the paleoclimate data further in this chapter?

Some people have proposed that negative feedbacks will stabilize CO_2 and prevent further temperature rise, although there is no evidence for this in the fossil record. Fossil fuel industry lobbyists like to call CO_2 *plant food* and suggest that the more we add to the atmosphere, the faster the plants will grow and that there is no need for concern. For greenhouse plants that are saturated with fertilizers, adding CO_2 to greenhouse air can cause faster growth. But this does not happen in nature because plant growth under real conditions is always limited by lack of nutrients, water, or light, not by insufficient CO_2. That can only happen under artificially manipulated conditions. When CO_2 is added to forest air, there is no increase in plant growth or litter production (Bader et al., 2013). Plants take up all the CO_2 they need, as determined by nutrient levels in the soil, and once they have taken up what they can use, they shut their stomata, the holes in the leaves through which CO_2 enters. This reduces the amount of water lost by transpiration, so it makes plants use water more efficiently and increases the amount of water, CO_2, and dissolved inorganic carbon (bicarbonate) stored in the soil, increasing chemical weathering (Bader et al., 2013). However, there is a possible negative consequence of increased CO_2, because increased soil microbial nitrification can inhibit mycorrhizal fungi that feed plant roots, increase the amount of nitrogen loss from soil as nitrate that pollutes streams, and increase soil nitrogen limitation and soil acidification (Bader et al., 2013).

Underestimation of positive feedbacks is far from the most serious flaw in IPCC model projections. What really reduces their projections to serious underestimates is the fact that IPCC was ordered to make predictions of climate change over 10, 20, 50, at most 100 year periods. This time frame was set by political convenience and has no relationship to the real time scale of the phenomena being modeled. All future impacts longer than a century are effectively discounted to zero. Since the responses take thousands of years, 90% or so of the responses are not included in the models as a result. Consequently, IPCC's mandate was to come up with the wrong answer to an irrelevant question, and not to the questions we really need to answer: what are the long-term impacts of climate change and how long will they last? Policymakers who use IPCC projections are therefore dealing with quasi-irrelevant and arbitrary estimates of impacts, not those based on long-term effects. Policy based on bad science cannot work.

The real-world long-term sensitivity of temperature and sea level to CO_2 can be accurately determined from the long-term data in Antarctic ice cores, deep-sea sediments, and corals, and they tell a very different story than the models. Immediately after the first IPCC projections (Houghton et al., 1990) were made, the first paper contrasting the climate sensitivity of IPCC projections versus the climate record revealed fundamental flaws in IPCC's climate sensitivity estimates, which were much lower than the climate data indicated (Goreau, 1990). Table 2.1 from a briefing to delegations from the Association of Small Island States to the United Nations Conference on Sustainable Development in Rio de Janeiro in 1992, reveals the differences clearly. These are not small errors in the IPCC estimated climate change sensitivities, they are up to ten, a hundred, and a thousand times too low for the sensitivity of temperature to CO_2, of sea level to temperature, and of sea level to CO_2, respectively (Figure 2.5a through c).

TABLE 2.1

Table of Sensitivities Based on Changes in Global CO_2, Temperature, and Sea Level, Comparing IPCC Model Projections with Actual Data

Factor	IPCC Projections	Ice Age–Interglacial
CO_2 (ppm)	350	100
Temperature (°C)	1–4	5–10
Sea level (m)	0.1–0.5	100–150

Source: After Goreau, T.J., *Nat. Resour. Forum*, 2, 5, 1992.

Notes: Briefing to association of small island states delegations at the United Nations Conference on Sustainable Development, Rio de Janeiro, 1992.

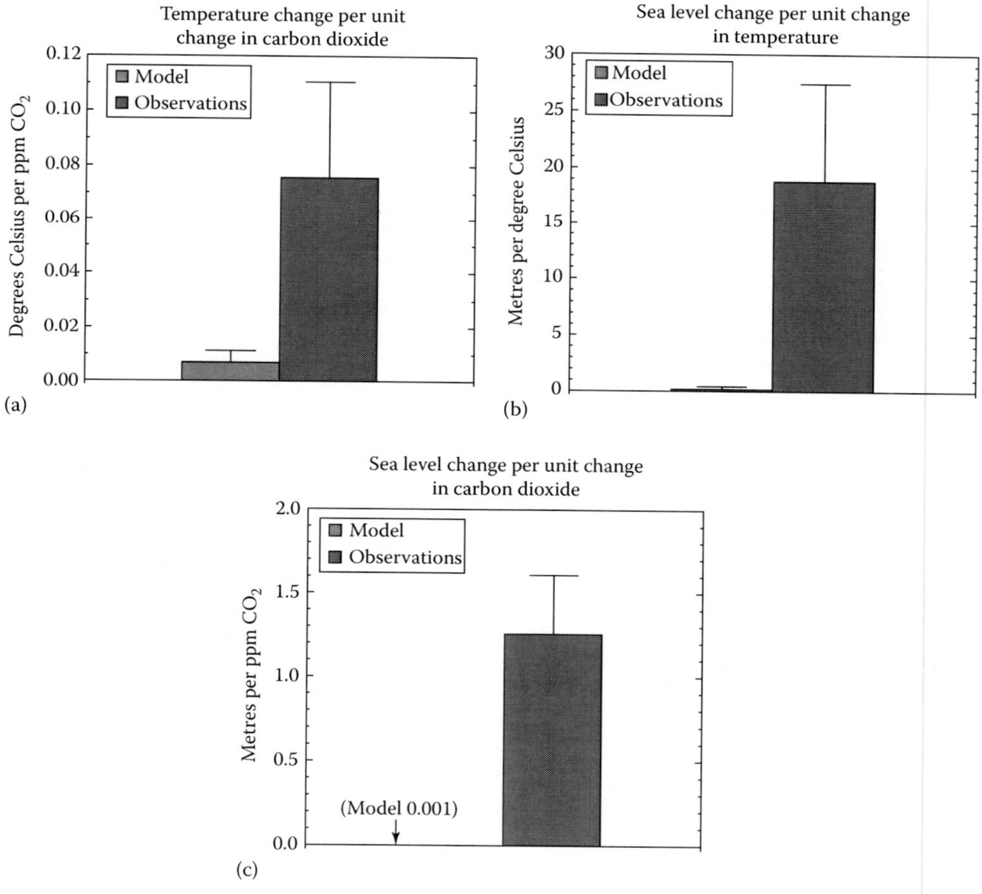

FIGURE 2.5 (a–c) IPCC model projections for sensitivity of temperature, sea level, and CO_2 to each other compared to real climate data throughout the last 800,000 years, based on data from Table 2.1.

By using the Antarctic ice core records of temperature and CO_2, and the deep-sea stable isotope records of sea level, supplemented by fossil coral samples, we can calculate the real sensitivities of global temperature and sea level over the last 800,000 years. The data show that temperature is many times more sensitive to CO_2 than the values assumed in the IPCC models. Nor do the IPCC models give the right value of temperature for the current level of CO_2, indicating that the first IPCC models seriously underestimate global temperature response to CO_2 (Goreau, 1990; Figure 2.6).

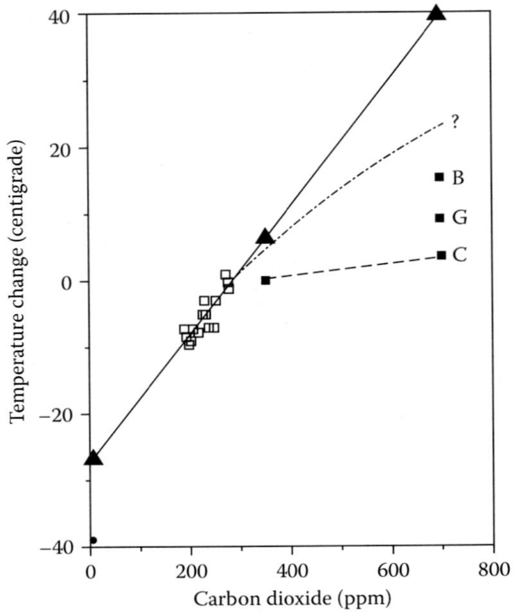

FIGURE 2.6 Global temperature versus CO_2. (After Goreau, T.J., *Ambio*, 19, 230, 1990.)

Straight black line is the line of best fit of the data based on values 10,000 years apart, the central triangle is for the atmospheric CO_2 in 1990, the top right triangle is the linear regression value for 700 ppm, the bottom left triangle is the regression temperature for no CO_2 in the atmosphere, and the straight dashed line (C) is the IPCC projection for increase of CO_2 from 350 to 700 ppm. The IPCC projection (dashed line) is seen to greatly underestimate the changes. B and G are increased model estimates of geophysical and biogeochemical feedbacks (Lashof, 1989), but it is clear that even these are insufficient to reflect the real feedbacks in the system. The real value is likely much higher due to unknown feedbacks, and most likely curved as in the dashed dot line to the (question mark), but it should also curve down to the circle at the zero axis, which is the temperature of the earth with no CO_2. This curvature results from saturation of the CO_2 thermal infrared absorption lines with increasing CO_2 concentration (Goody and Yung, 1989). At the time this data was plotted, only the ice core data from the first Ice Age cycle (160,000) years had been analyzed. Much more data over the last 800,000 years that has accumulated since then; the results are basically identical to the first estimates in this figure.

Since 1990, IPCC models have greatly increased in complexity, but with only minor increases in model-predicted sensitivity of CO_2 to temperature. There has also been a huge increase in ice core temperature and CO_2 data, providing a much larger database for calculating climate sensitivity. Recent determinations of real-world climate sensitivity using 800,000 years of data are in (Rohling et al., 2009; Grant et al., 2012; Skinner, 2012; Hansen et al., 2013). The long-term equilibrium sea level for today's CO_2 levels is 23 m (75 ft) above today's sea level, and equilibrium temperature is around 16°C (27°F) above today's value (Figure 2.7) (Rohling et al., 2009). Both will be higher if CO_2 rises above the present level. Eventually, the sea level rise will stop because there is no more ice to melt and runaway warming is now racing toward that point when the Arctic Ocean was 19°C warmer than today (Ballantyne et al., 2010) and when CO_2 is thought to be around 390 ppm, below the current value of 400 ppm, and rising.

Rohling was not aware of the earlier work shown in Figure 2.5, and he plotted his curve the other way (CO_2 vs. temperature rather than temperature vs. CO_2 in Goreau (1990)), but when the graphs are flipped and scaled to overlay the axes, the results are essentially identical, confirming that IPCC models seriously underestimate how global temperature and sea level responds to CO_2.

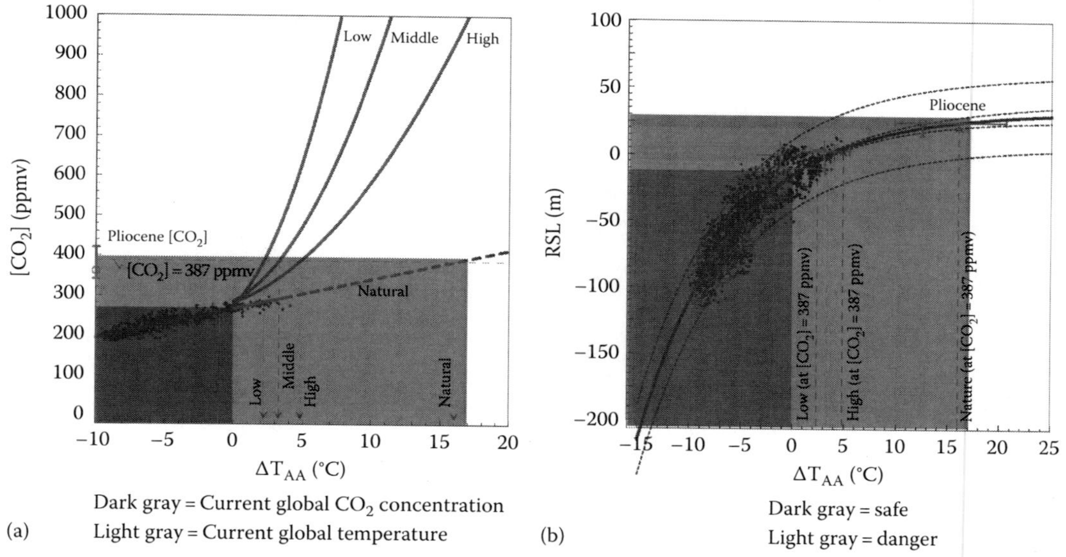

FIGURE 2.7 Identifying safe CO_2 limits from temperature and sea level responses. (a) Global CO_2 versus global mean temperature over the last 800,000 years, (b) global sea level changes versus mean temperature over the last 800,000 years. (Data from Rohling, E.J. et al., *Nat. Geosci.*, 1, 38, 2008.)

What Is the Safe Level of CO_2?

When CO_2 was around 280–290 ppm 125,000 years ago (Luthi et al., 2008), it was 1°C–2°C warmer than today and the sea level was 7 m higher than today. This suggests the safe level of CO_2 to avoid dangerous changes in sea level that will imperil billions of people living in coastal areas should be distinctly lower than 290 ppm. But there is an important caveat, because too low CO_2 values can trigger an ice age. The safe level of CO_2 today is higher than it was 125,000 years ago because at that time the Earth's orbital configuration was in a full interglacial (warm) mode, whereas now, if humans were not interfering with the system, the current orbital configuration would be moving toward an Ice Age. Slightly higher CO_2 levels are needed to compensate for the difference in the orbital characteristics at different times. The new ice age that would be expected from the current orbital configurations without human interference is now likely impossible because the huge excess CO_2 that we have now put in the atmosphere prevents it. Ruddiman (2005) convincingly claims that the earth would have already started an Ice Age about 5000 years ago if it were not for the massive CO_2 inputs from deforestation for agriculture that turned the natural decline in CO_2 levels (and temperature) into a dramatic increase starting about 8000 years ago (and accelerating further due to the industrial revolution). As a result, the safe level under today's conditions is likely around 270 ppm or so. Unless we can get to those levels very quickly, the long-term climate change forecast is far more alarming than any policymaker grasps.

The widely discussed *goals* for CO_2 discussed in the UNFCCC negotiations are dangerously higher than the safe levels. Some countries call for levels of 450 or more ppm, and the most conservative goals are 350 ppm (Hansen et al., 2008). Only Micronesia has called for a preindustrial CO_2 goal (i.e., around 280 ppm) at UNFCCC, but their call was ignored by the other small island developing states, which accepted 350 ppm as being a better target than 450 ppm. A level of 350 ppm amounts to a death sentence for all coral reefs and low-lying coastal areas, so low-lying island nations that support such goals are effectively agreeing to a suicide pact. The safe level should be based on the science of how the climate system responds to CO_2, and not politically convenient, arbitrary, and fictional goals.

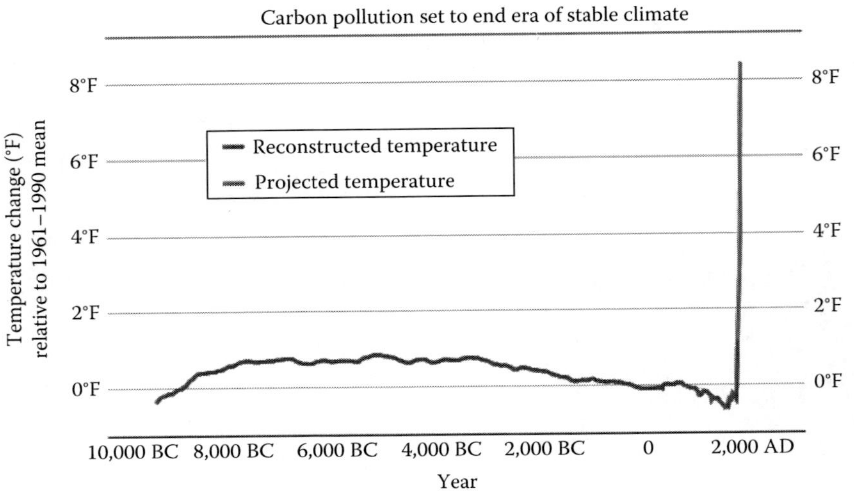

FIGURE 2.8 11,300 years of temperature.

Figure 2.7 shows how to determine the safe level of CO_2 that avoids serious change to current conditions using the data by Rohling and colleagues 2008.

We can use Rohling's graphs to determine the *safe* levels of CO_2 that do not cause dangerous increases in temperature and sea level—dark gray is safe and light gray is dangerous—with regard to change from today's conditions (Figure 2.7). These graphs suggest that the warming *in the pipeline* due to the current levels of CO_2, even if all production of fossil fuel CO_2 were completely stopped, is far higher than IPCC projections. We are already completely beyond the CO_2 range of the last 3 million years. We are headed directly for temperatures similar to the Pliocene, before the ice ages began, when the Arctic was up to 19°C warmer (Ballantyne et al., 2010; Brigham-Grette et al., 2013) and sea levels were round 25 m higher (Fagan, 2013). The graphs show that current CO_2 level corresponds to a long-term increase of temperature of around 17°C and of sea level by more than 23 m, based on climate changes over the last several million years. This will result in increases that are beyond our experience and the capability of most to adapt (Figure 2.8), and averting it requires a safe level of CO_2 around 270 ppm.

The reconstructed line shows global average temperature from global long-term climate records over the last 11,300 years since climate and sea level stabilized at the start of the current interglacial period (Marcott et al., 2013). The projected line is the mean IPCC estimate based on CO_2 increases, which is argued here to be a serious underestimate, so it should be even steeper than shown.

Limitations of Climate Prediction Models

All climate models agree that it will get warmer as CO_2 increases because they all contain the same fundamental physical laws, especially Lambert's law, which says that if we increase the concentration of a gas that absorbs heat, the gas will get hotter. This fundamental physical law is inviolable and allows no exceptions, just as the law of gravity means we can drop a billion apples, and every single one will fall down. All climate models come to precisely the same qualitative conclusion, but since they differ slightly in mathematical assumptions in the models redistributing that heat, they give slightly different predictions for how hot it will get and how long it will take to reach that level.

A fundamental weakness is the model's treatment of heat flow to the deep sea, which is by far the major heat sink in the earth climate system. Until the deep sea warms up fully, the surface ocean, atmosphere, and land surface will not reach their maximum temperatures, so the response contains an intrinsic thousand-year time lag. Most climate change models do not include realistic ocean models due to the fundamentally different space scales over which ocean and atmosphere processes operate.

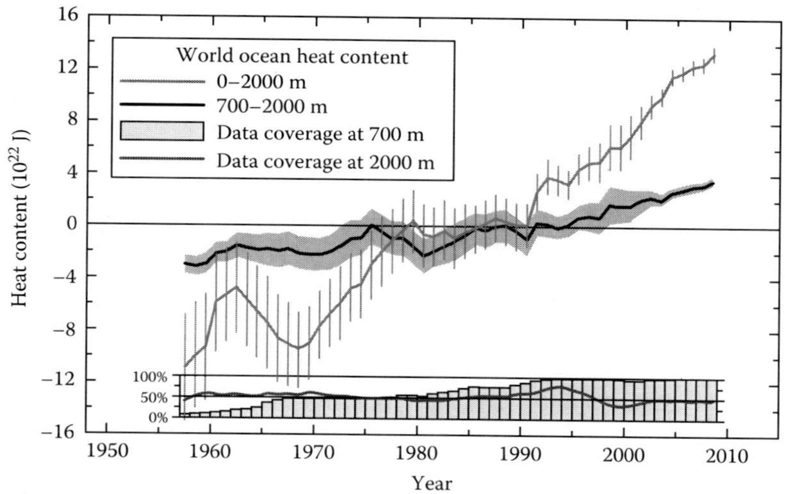

FIGURE 2.9 Heat content of the oceans. (From Levitus, S. et al., *Geophys. Res. Lett.*, 39, L10603, 2012.)

The increased global heat caused by the greenhouse effect is very real and fully measureable, but most is flowing rapidly into the deep sea, not accumulating at the surface. The increase in total heat in the earth system is increasing linearly and inexorably (Figure 2.9), even though we are not yet feeling it as SST. Despite many incorrect claims that global warming has stopped, in fact, there has been no slowdown in global warming (Levitus et al., 2000, 2001, 2005, 2012; Antonov et al., 2005; Hansen et al., 2010; Meehl et al., 2011; Balmaseda et al., 2013; Guemas et al., 2013; Otto et al., 2013; Rosenthal et al., 2013; World Meteorological Organization, 2013), precisely as expected from the physical effects of CO_2 increases.

A large part of the apparent slowdown of SST rise in this period was due to increased upwelling of cold water in the Southern and Eastern Tropical Pacific (Kosaka and Xie, 2013) caused by global-warming-induced increases in wind speed in the Southeast Pacific (England et al., 2014). The East Pacific has been in an unusually prolonged La Niña phase since 1998. The last La Niña was so strong that the global sea level rise was temporarily cancelled and ocean levels actually fell (Boening et al., 2012). This was due to record rainfall in Australia and South America, and much in Australia originated from evaporation of seawater that fell in inland drainage basins where it could not run back to the sea (Fasullo et al., 2013). After this ponded water evaporated, global sea level rise again accelerated to record levels. When the next El Niño hits, an upward jump in global SST and sea level will follow.

Any slowdown in the rate at which heat is flowing to the deep ocean will result in a dramatic jump in sea SSTs. There is clear evidence that during the warmest part of the last interglacial, when temperatures were about a degree warmer than now, a period that is the best model for the conditions we are soon headed for, the rate at which deep ocean water formed underwent sudden and prolonged shutdowns for century long periods (Galaasen et al., 2014). If that were to happen now, a sharp rise in SST would be expected.

Another process slowing the temperature increase is that melting of ice adds heat to the system without raising the temperature. Once the ice is gone, the temperature increase will jump dramatically upward. This effect has been neglected in climate models to date.

A further factor is all the pollution we put in the atmosphere causing smog that reflects sunlight. Dirty fossil fuels with high sulfur content add sulfuric acid to the atmosphere, which forms aerosols that cool the earth by reflecting sunlight. When we burn fossil fuels this cooling effect lasts days to weeks until the sulfuric acid rains out (Smil, 2001), but the warming effect of the CO_2 that accompanies it lasts for centuries. One would have to add exponentially increasing sulfuric acid pollution just to stay in place and keep balancing the temperature effect of the original CO_2 input with sulfuric acid aerosols, as proponents of geoengineering propose. Even if we were to shoot sulfuric acid by rockets into the upper atmosphere, it might last only a year or so before raining out, causing acid rain damage to forests and agriculture, and one needs to keep adding more and more to mask the warming of the CO_2, while adding more CO_2 to do so. This is a race that can never be won. Incredibly, the fossil fuel industry pays lavishly for research on such schemes that cannot possibly work even if they were technically feasible and affordable, but some academics and politicians treat this as a serious option, most likely to get funding from oil and coal companies.

Even if the models all agree on the fact that it will get hotter, they do not agree on what will happen to rainfall. Some models predict increases, others decreases, and others no change. They do not even agree with each other qualitatively due to the models being far too primitive to predict the real-world distribution and effects of clouds. One can look up at the complex pattern of clouds in the sky and realize no model is complicated enough to realistically describe what we see every day above our heads.

Qualitatively, the warmer it gets, the more evaporation will take place, so atmospheric moisture and rain will dramatically increase on a global average. Real-world measurements show strong global increases of atmospheric moisture and average rainfall with a water cycle increase of around 8%/°C (Durack et al., 2012) or 16%–24% if temperature rises 2°C–3°C. Tropical extreme rainfall events get up to 6%–14% more frequent per °C of temperature increase (O'Gorman, 2012). Water vapor in the air over the tropical ocean increases about 7%/°C (Held and Soden, 2006). Changes in surface ocean salinity show wet areas are getting wetter while dry areas are getting drier (Boyer et al., 2005). Periods in the geological past with much higher CO_2 and temperature had much higher rainfall and erosion, especially before plants evolved and began to stabilize and build soil (Beerling and Berner, 2005; Beerling, 2007; Gibling and Davies, 2012). What is clear is that global warming will make climate much wetter along coasts near sources of moisture, while inland areas in the centers of large continents will get drier because soil moisture evaporation increases strongly with temperature, but the rain falls out in the mountains before it reaches the continental interiors. The result is that wet areas will get wetter and dry areas drier, as data shows it is actually happening (Jung et al., 2010).

Analysis of global rainfall data from 1979 to 2010 (Chou et al., 2013) shows that wet seasons are getting wetter (0.94 mm/day/century), dry seasons are getting drier (−0.53 mm/day/century), and the annual range is increasing even more (1.47 mm/day/century), just as expected from fundamental physical principles. The largest variations of rainfall were over the oceans, rather than land, while for temperature, the opposite was true. This results from the fact that the ocean is all water, while only small parts of the land are covered with water that can evaporate, and since water has much higher thermal capacity, it changes temperature more slowly than land does.

On land, most areas got wetter in the rainy season, in particular the Sahel, Southern Africa, Northern Australia, Northwest India, Pakistan, the Amazon, and the Southern Caribbean. However, some areas got *less* rain in the wet season, in particular, the Near East, the Mediterranean, North China, East Africa, Central Brazil, Southern Australia, the northern Caribbean, the Coral Triangle (Indonesia, Philippines, New Guinea, the global center of marine species diversity, the Amazon and Congo of the ocean), the central Andes, and the Western and Southern United States. This has enormous implications for agriculture in those places and alarming ones for some of the world's major food producing areas that are getting less rain in the growing season. Another interesting trend is the large increase in wet season (summer) rain in the Arctic regions, which will further melt ice.

While dry season rainfall decreased most over the oceans, South America, India, Pakistan, the Caribbean, and the Western and Southern United States were especially affected, leading to increased droughts. Polar regions also got drier in the dry season, meaning less snowfall, leading to reduction in snow and ice thickness. An unusual exception is the Coral Triangle, which became slightly wetter in the dry season, probably from increased ocean evaporation.

The seasonal range shows increased seasonal ranges in rainfall over most of the oceans, which will cause water availability for aquifers to decrease in the dry season and increased droughts, with more runoff and erosion of soil onto coral reefs in the rainy season. The Coral Triangle is an unusual exception where the range decreases slightly.

So even if the models are essentially useless to predict future trends, the measured current rainfall trends, which are our best guide to future trends, are cause for serious concern regarding disruption to worldwide agriculture and ecosystems from rainfall changes, as the latest IPCC states.

How High Can the Sea Level Go?

There is no limit to how high temperature can go if CO_2 reaches above the levels of the distant geological past before the ice ages began around 3 million years ago, when periods of high CO_2 resulted in temperatures up to 15°C or 20°C higher than today (Ballantyne et al., 2010). One recent study from such a period of high CO_2 around 55 million years ago found SSTs as high as 74°C, or 165°F! On the other hand, there is a real limit as to how high sea level can rise, which is limited by the total amount of ice in the continental ice caps (mainly Antarctica, plus a much smaller contribution from Greenland) and mountain glaciers (mainly in the Himalayas, the Andes, and Alaska). Once that is all melted, sea level cannot rise further. The data suggest that at essentially all the ice will eventually melt at a global atmospheric CO_2 concentration of about 400 ppm, what we *already* have at time of writing this chapter.

Rohling et al. (2009) calculated the sensitivity of sea level to CO_2 (Figure 2.7). This indicates that the equilibrium level of today's CO_2 level is 23 m above today's level (75 ft), and if CO_2 increases further, it will be significantly higher. Projected increases in CO_2 expected in the coming decades of this century could lead to equilibrium sea levels 50 m or more above today's levels and even higher (80–100 m) if runaway ice cap melting takes place. These are vastly beyond the levels that IPCC model projections present, of 5–10 cm, with extremes of 1 m sea level rise from more recent IPCC models. The geological record represents the entire long-term system response, not the short-term initial transient response in IPCC models. As was pointed out earlier, it takes thousands of years before the responses are fully developed. But they do represent the real long-term changes to which we have committed future generations, a situation that would flood billions of people from their homes in coastal areas, perhaps half the human population. These changes will be vastly greater than the feeble sea level changes IPCC projects, which people falsely think can be adapted to by building 1 m high sea walls, not realizing that is but a temporary fix, and they will need to build higher walls every generation, for hundreds of generations.

The logical fallacy of relying on IPCC projections as a guide to future changes is like someone kicking a football, measuring how far the ball has moved in one tenth of a second, and then thinking that is how far the ball will go! This is precisely the thinking that currently guides global climate change policy, based on IPCC projections for time scales that miss almost all of the inevitable response, giving the wrong answer to the wrong question.

Seeing the Past, Present, and Future

The sensitivity of sea level to CO_2 and temperature is clearly visible to those who know how to read the landscapes left on coastlines all around the world from the last time in earth history when sea levels, and temperatures, were higher than today. That happened at the last interglacial epoch 120,000–130,000 years ago, when the configuration of the earth's orbit, amplified by natural internal climate feedback factors, caused some of the highest temperatures in the last 3 million years (van de Berg et al., 2011). At that time, global temperatures were around 1°C–2°C warmer than today, and sea levels were about 7 m higher than they are now due to the melting of polar ice caps (Rohling et al., 2008; Kopp et al., 2009; Dutton and Lambeck, 2012). Yet at that time, CO_2 levels were around 280 parts per million (ppm). The current CO_2 value at time of writing is more than 40% higher than those values. And, unlike now, the feedbacks that affected CO_2 then were purely natural, since the human race was only recently evolved, very small in number, confined to southern and eastern Africa, and had not developed the technology to cut down and burn the forests on a large scale and significantly change CO_2.

We can see the precise sea level from those days perfectly preserved at a handful of places, probably no place better than on the North Coast of Jamaica. Stable sea levels create a sea level notch in limestone rocks, and Jamaica has ideal conditions for its formation and preservation, although one can find examples in other places, such as Bonaire and Barbuda. Jamaica has been tectonically stable in the last 130,000 years, and the same level is found at sites all around the island and correspond closely to those in Bonaire and Barbuda, so it is the result of global sea level changes, not of local vertical tectonic movements caused by earthquakes or uplift.

FIGURE 2.10 Fossil sea level notch at Discovery Bay, Jamaica. (Photographed by T.J. Goreau.)

Figure 2.10 was photographed inside the sea level notch at Discovery Bay, Jamaica, looking outward with today's ocean, some 7 m lower, clearly visible in the background. Inside the sea level notch, one sees large masses of rock (visible at right) that fell from the limestone cliff above into the mouth of the notch. Fossil corals and encrusting marine organisms are found below the notch itself, but not above, and also on the fallen rocks below this level, showing that the rock fell from the cliff into seawater at the time that the notch formed and was colonized by marine organisms below the high water mark (Goreau, 1990).

Discovery Bay, Jamaica, is a place where one can see the past and the present, and if you know what to look for, you can see our future as well. At the same time that sea level formed this notch in Jamaican limestone cliffs, hippopotamuses and crocodiles lived in subtropical swamps where London, England, now stands, their fossils are in the British Museum of Natural History. Since conditions at that time represented CO_2 levels of around 280 ppm, and we are already at 400 ppm, it is clear that what happened at that time is a considerable underestimate of what is going to happen from today's CO_2 levels, even if all fossil fuel use stops overnight.

A geological cross section of the North Coast of Jamaica shows the fossil and recent sea level notches, and the coral reefs that grew in front of them. Figure 2.11 shows these formations, mapped by the author when he was 14. Much higher up the cliff, around 50 m or so high, are the traces of several fossil sea level notches that probably date from the Pliocene period, before the ice ages started, when CO_2 levels were a bit less than the 400 ppm they now are and the Arctic was up to 19°C warmer (Ballantyne et al., 2010; Brigham-Grette et al., 2013)! That is where our current course of action will inevitably lead us if we do not quickly reduce CO_2 levels.

The Jamaican fossil reef from 125,000 years ago has two distinct layers. The lower layer has the corals in position of growth, then there is a thin sharp layer of mud about a centimeter thick, and above that, all the corals are broken and lying on their sides. These coral reefs appear to have died from excessively high temperatures (Kiessling et al., 2012) and then been devastated by huge waves, from hurricanes stronger than any we have experienced (another well-founded prediction for global warming according to Emanuel, 1987), or tsunamis. This sharp boundary is found all around Jamaica and can also be seen at other sites all around the Caribbean that the author has looked at, suggesting the latter. Given that we are already at the maximum temperature that corals can tolerate (Goreau and Hayes, 1994, 2005) and

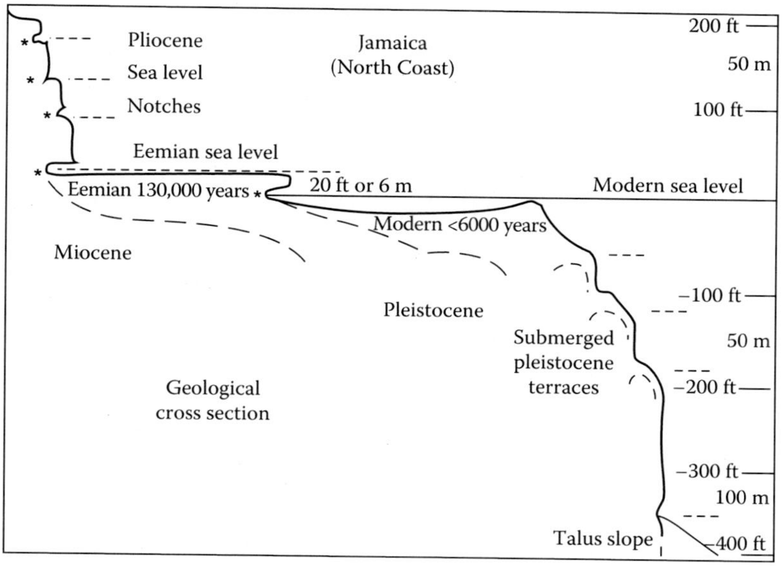

FIGURE 2.11 Geological cross section of North Jamaica. (After Goreau, T.J., 1965.)

that the temperatures at that time were above that lethal threshold, then it is likely that these reefs had already been killed by high temperatures (Kiessling et al., 2012; O'Leary et al., 2013) before they were physically damaged, and it was the dead coral skeletons that were flattened.

At present, we are at CO_2 levels much higher than in those days but we have not yet felt the response of the latest increase (Figure 2.12).

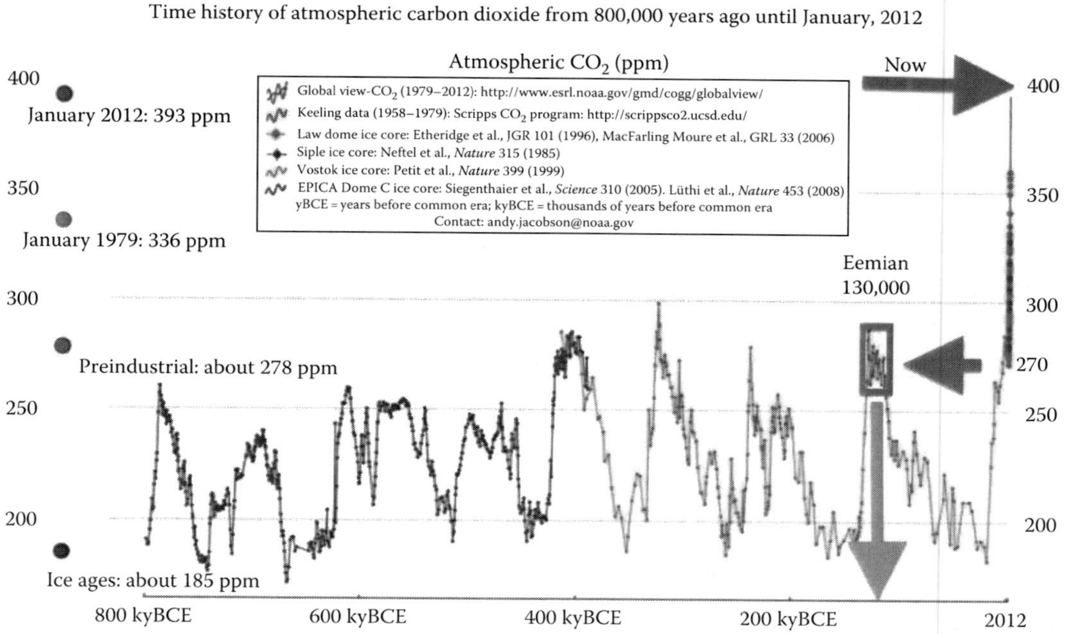

FIGURE 2.12 CO_2 changes measured in Antarctic ice cores over the last 800,000 years. CO_2 levels at the last interglacial, when the elevated sea level notches formed, were around 270 ppm, much less than the current value of 400 ppm. (After NASA, http://www.youtube.com/watch?v=vA7tfz3k_9A&feature=player_embedded#at=203.)

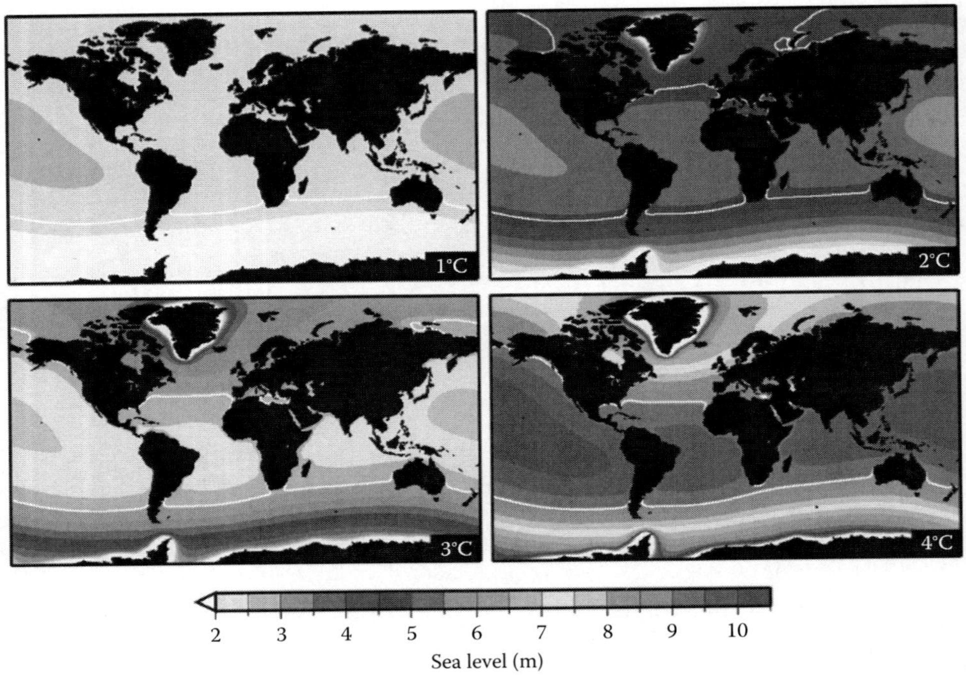

FIGURE 2.13 Long-term sea level changes (2000 years) for various global warming scenarios. (After Levermann, A. et al., *Proc. Natl. Acad. Sci.*, 2013.)

Long-term sea level rise estimates based on paleoclimate data (Levermann et al., 2013) are in accord with much higher climate change sensitivity than IPCC model projections (Figure 2.13).

The Other Half of the Global Carbon Dioxide Problem

Almost all discussions of CO_2 stabilization focus only on emissions reductions. We show in the following that such supply-side *solutions* cannot possibly work in time and miss the most important opportunities to remove excess CO_2:

The rate of change of atmospheric carbon dioxide is associated with changes in its major sources and sinks, which are biotic, not industrial. The literature focuses on comparing quantities of carbon in different pools rather than fluxes between them, which are poorly known, except for combustion. This emphasizes large pools, which slowly exchange with the atmosphere (fossil fuels, soil humus, the deep ocean, limestones, volcanic gases) over smaller ones, which exchange much more rapidly (the tropical biota). Hence it artificially obscures the other key to the atmospheric CO_2 problem, which is to protect the metabolism of the tropical biota, the primary determinant of atmospheric composition.

Current land 'development' practices that ignore changes in community ecophysiology have a large hidden price tag. They diminish the capacity of the biosphere to modulate atmospheric CO_2, which increases the magnitude and duration of climatic alteration (Goreau and de Mello, 1987, 1988; Goreau, 1992) beyond those forecast by conventional climate models.

There are two real limiting options. First, destruction of the biota and conversion of productive ecosystems to degraded secondary habitats increases combustion-derived atmospheric CO_2, which dissolves in the ocean on a 103 (thousand) year time scale, and becomes incorporated into marine sedimentary carbonates and organic matter on a 105 (hundred thousand) year time scale. Avoidance of the worst risks of the Greenhouse Effect will require global limits and national quotas on combustion.

The second, missing option is to conserve the remaining undisturbed habitats that cycle carbon rapidly (like tropical rain forests and coral reefs) and to undertake pan-tropical replanting and fertilization to restore productivity to degraded areas. Atmospheric CO_2 is held down by rapidly recycling through the biota, and the ultimate fate of most added carbon is soil and sediment organic carbon, further enhancing soil productivity.

To escape the Greenhouse Problem by renewable resource-based land management is much the cheapest option in the long run, and has many advantages. Solutions equitable to all countries demand cooperation. The price of fossil fuel should include its long-term environmental costs as an added "energy-growth" tax, which transfers income from fuel burners to environmentally sound tropical development. The sooner such a course is adopted, the greater the benefit and the lower the cost (Goreau, 1987).

The key question then is what time scales are involved in the trajectories of CO_2, temperature, and sea level in both of these scenarios, and if the impacts can still be meaningfully reduced in the rapidly shrinking window of time remaining. Had the world acted in 1987 when that paper was published, much of what will happen could have been averted, and the longer we wait, the more difficult and expensive it will be to accomplish. To understand the time responses, we need to look more closely at the dynamic time scales with which each major carbon pool interacts with atmospheric CO_2, including all their sources and sinks.

Carbon Cycle Response Times: Dynamic Time Scales of CO_2 Sources and Sinks

Every process that produces or consumes CO_2 results in a flux of CO_2 molecules between the atmosphere and each other carbon pool that interchanges with it (Goreau, 1992). We can define a matrix of time scales that define how long each process takes to significantly change atmospheric CO_2 and combine these time constants into the complete spectrum of all of the source and sink time constants, which then allows us to determine how quickly the atmosphere will respond to changes in fluxes (Goreau, 1995). It is crucial to use gross fluxes, not net fluxes, which are the difference of two separate processes with different time constants, but one can treat any combination of source and sink fluxes. These fluxes are shown in Table 2.2 (Goreau, 1995) and their time scales in Figure 2.14. We can use these figures to estimate the

TABLE 2.2

Major Atmosphere CO_2 Exchange Fluxes and Time Constants

Process	Flux (%/Year)	Time (Years)
Sources		
Ocean–atmosphere	12.99	7.7
Plant respiration	7.81	12.8
Decomposition	7.81	12.8
Fossil fuel combustion	0.714	140
Deforestation	0.338	295.8
Volcanism	0.012	8,500
Sinks		
Photosynthesis (land)	−15.625	−6.4
Atmosphere–ocean	−12.99	−7.7
Sedimentary burial	−0.39	−256.4
Soil formation	−0.052	−1,925
Weathering	−0.01	−10,000
Net balance		
All sources	29.674	3.37
All sinks	−29.067	−3.44
Net change	0.607	165

Source: After Goreau, T.J., Tropical ecophysiology, climate change, and the global carbon cycle, in J. Pernetta, R. Leemans, D. Elder, and S. Humphrey (Eds.), *Impacts of Climate Change on Ecosystems and Species: Environmental Context*, International Union for the Conservation of Nature, Gland, Switzerland, pp. 65–79, 1995.

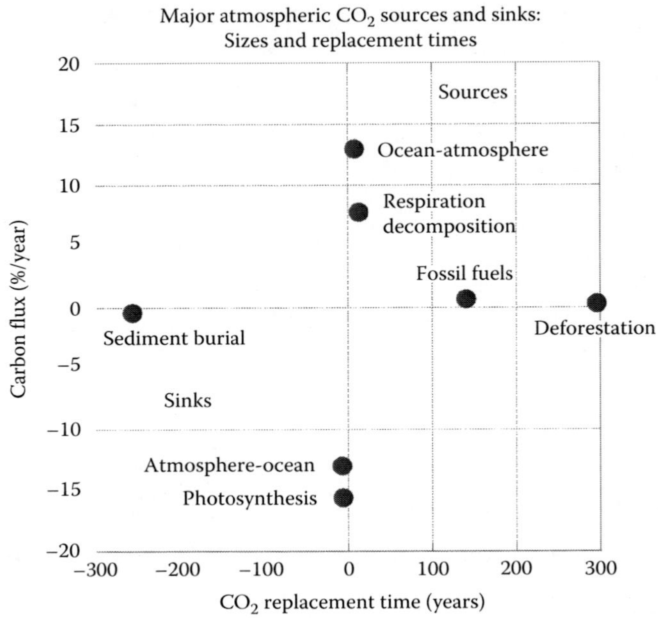

FIGURE 2.14 Time response spectra of all major sources and sinks of atmospheric CO_2, showing the percentage of CO_2 replaced each year and the length of time each source or sink takes to cycle CO_2 to or from the atmosphere. (After Goreau, T.J., Tropical ecophysiology, climate change, and the global carbon cycle, in J. Pernetta, R. Leemans, D. Elder, and S. Humphrey, Eds., *Impacts of Climate Change on Ecosystems and Species: Environmental Context*, International Union for the Conservation of Nature, Gland, Switzerland, pp. 65–79, 1995.) Note that only a few processes can act on a time scale of less than a century (other smaller and slower acting sources and sinks are not shown on this scale). Also note that respiration and decomposition are shown as equal. In reality, only fossil fuel sources are well known; the others have a fair range of estimates and will be refined in the future when better numbers are available. However, this will not result in important changes in the conclusions. In particular, as discussed further on, a large part of decomposition is actually underground plant respiration, so respiration is in reality larger and decomposition smaller. Photosynthesis is the only major source or sink that can be readily modified by human activity. Fossil fuels are too small a global source to be able to change CO_2 much on a time scale of a century even if all fossil fuel sources were eliminated immediately.

response time scales of various CO_2 source and sink options. For example, all sources acting together or all sinks acting together will process atmospheric CO_2 in about 3 years, but since sources outweigh sinks by only a small amount, the current rate that the net of all sources and sinks would change CO_2 by the same amount is around 165 years. Here, we focus on the *fast carbon cycle* and ignore the *slow carbon cycle* exchanges that take much longer to affect CO_2, such as with ocean bicarbonate, which takes 200,000 years to turn over, or sedimentary organic matter and limestone, which affect CO_2 over many millions of years (Holland, 1984; Walker, 1986; Berner, 2004).

These atmospheric CO_2 response spectra allow one to see at a glance which processes are the most effective. We now use these data to systematically examine the rate at which various supply and demand-side measures act to affect atmospheric CO_2:

1. Source side reductions only: We cannot practically reduce the rate at which CO_2 is released to the atmosphere by the oceans and by respiration and decomposition, so our only leverage on the supply side is on fossil fuel combustion and reducing emissions from deforestation and land degradation. Even if we *completely stop* these sources right now (or had 100% carbon capture and sequestration [CCS] of all fossil fuel CO_2, which is impossible technically as well as unaffordable if it were), it would take about 140 years for the atmosphere excess to respond by decreasing by half if no other sources and sinks changed. If we reduce emissions by only 50%,

it will take twice as long. Stopping deforestation will take several times longer to have an impact. It is clear that supply-side measures alone will be very slow at best, centuries to thousands of years, and cannot possibly act quickly enough to avert the worst damage.

2. Sink side increases only: We cannot practically change the rates at which CO_2 is transferred from atmosphere to the ocean any more than we can change the converse flux, because we cannot control the wind and waves. Photosynthesis recycles all of atmospheric CO_2 in only about 7 years. Almost all of it is returned in short order via respiration and decomposition, so the strategy that is needed is to increase photosynthesis (which is limited not by CO_2 but by nutrients, which is a key point of this book). Even more important is to find ways to store that photosynthetic carbon so it does not go right back to the atmosphere. What we can do though is to convert photosynthetic carbon into forms of carbon, like biochar, that are very slow to be oxidized or decomposed. If we could turn half of photosynthesis carbon into biochar, we could remove most of the excess CO_2 in the atmosphere in around 15 years. Figure 2.14 shows that this is the fastest (and very likely the cheapest) method for getting rid of the excess CO_2. The more biochar we make, the more root biomass, fungi, and soil organic matter we grow, storing even more carbon, and the faster we can avert the worst consequences of runaway warming and the more cost effective the investment will be. By accelerating the rate of rock mineral weathering by plant roots pumping CO_2 into the soil, we also add an additional CO_2 sink. While that is likely smaller than what we can do with biochar, the chapters in this book show that combining these two carbon sinks in the right ratio maximizes the increase in soil fertility and photosynthesis, further speeding up the removal of excess CO_2, and has the capacity to take up the excess CO_2.
3. Combined strategies: By combining source reductions with sink increases, we can accelerate the transformation and avert more damage to the global environment and economy. We can stabilize CO_2 at safe levels in the minimum possible time only by combined strategies. It is therefore important to reduce fossil fuel sources as quickly as we can, since relying on increased sinks alone will be slower and more damaging in the long run. Combining processes is much faster (Goreau, 1995), but it must be noted that time constants are not additive, they add the same way as parallel electrical resistances, the inverse of the net time constant of several processes acting together is the sum of the inverse time constants of all the process (Goreau, 1995).

Polluter Pays: Carbon Taxes and Fees

The only practical way to avert the consequences of the existing CO_2 excess is to reduce CO_2 quickly enough to reduce the impacts, and this must now be done very quickly, within a few decades, if we are to avert the most dire and drastic consequences. Had the world community acted seriously when UNFCCC was signed in 1992, the problem could already be well on the way to solution, but instead the problem was ignored and allowed to get worse, and these systematic failures to act have made solutions far more difficult and expensive. The window of opportunity is constantly shrinking and will soon close.

All economic experts agree that prices of products should include all their real costs and that *the polluter pays principle* is the fundamental basis of waste management. Any other strategy saddles other people with the costs of benefits they do not obtain, what economists call *externalities* (DasGupta, 2002; Stern, 2006). A realistic price must include all the real costs, and we have failed to do so: recent estimates place the real costs of fossil fuel CO_2 emissions at a minimum of \$125/ton, possibly higher (van den Bergh and Botzen, 2014). If fossil fuels and nuclear power paid for the real costs of the energy they produce, renewable energy would be far cheaper and no subsidy would be needed to compete, but the fossil fuel and nuclear industry are massively subsidized directly and indirectly through giveaways and tax breaks totaling trillions of dollars (Trucost, 2013). On top of this, most of the value of the oil companies is now in underground carbon reserves that cannot be burned, creating an enormous financial bubble of overstated assets (Carbon Tracker, 2013).

If polluters do not clean up their own mess, they leave it indirectly to public taxation to pay the cleanup as a public service. The more directly this carbon transfer is done, the cheaper and the faster the result. Transaction costs are minimized and efficiency increased by eliminating fund skimming by intermediaries who do not actually solve the problem, passing the problem on to the next level while taking a cut of the funds for themselves. There is great pressure to control such funds by politicians and government bureaucracies in order to pay for the deficits they create through unproductive expenditure. The desire to intercept tax revenues for other purposes prevents interactions in which the *polluter pays* up front as directly as possible to the person who cleans up their mess (if they are incapable of doing so themselves).

The polluter should pay up front at the time they purchase fossil fuels, and not saddle future generations with the costs of a previous generation's gratification. Apart from being inequitable and unjust, any other approach results in more expensive and complicated *ad hoc* arrangements, which are usually inadequate to resolve the problems created. The simplest way is a direct carbon tax paid by the purchaser in proportion to the carbon content, which is easily verifiable at the source. There is no guarantee that any *post hoc* market mechanism will settle on the correct price for pollution and plenty of empirical evidence that it does not. Figure 2.15 shows the European carbon trading price. Since then, the price has collapsed further as the result of the failure of the EU to reform the carbon trading system, even though it is the centerpiece of their global climate change strategy. It is clear this strategy has failed politically, economically, and environmentally, and a much more serious strategy is needed.

At present, the world subsidizes fossil fuels to the rate of 500 billion dollars/year (Whitely, 2013), or about \$15/ton of CO_2 emitted. This subsidy to pollute far exceeds the \$3/ton EU carbon tax, so that the current structure of carbon trading rewards pollution. If this perverse subsidy rewarding polluters were instead spent on actually absorbing carbon, most of the costs of CO_2 stabilization might be met.

There are two major options by which a carbon tax can be administered. One is as direct as possible a transfer from carbon fuel purchasers to those who verifiably absorb their carbon. That way, the polluter directly pays for the removal of his/her pollution, and this is precisely the sense in which a carbon tax was originally proposed (Goreau, 1987, 1990; Myers and Goreau, 1991), not as a tax that would disappear into general tax revenue and be spent on unrelated problems.

Unfortunately, the term carbon tax was later used on a much larger scale by economists to convey a diametrically opposite concept, trivializing the term as originally intended and making it ineffective at

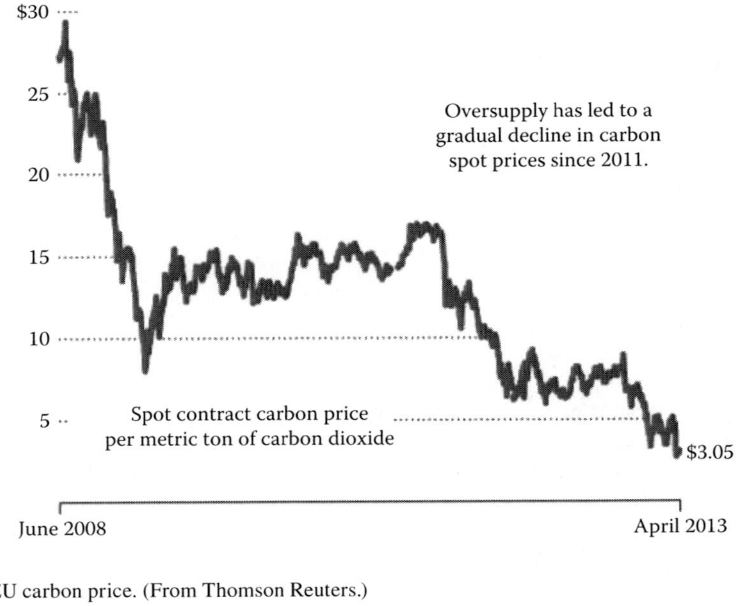

FIGURE 2.15 EU carbon price. (From Thomson Reuters.)

solving the problem. Economists seeking market mechanisms ask: how much do we need to tax *goods* and *bads* (products with negative societal consequences) so the increase in price reduces their consumption and the damage they cause? Obviously real *goods* that only provide positive benefits should be encouraged as much as possible, and any tax on them reduces their benefit value, which is why many economists believe that true *goods* should not be taxed, only *bads* that reduce the capacity of society to use its resources to best effect, saying "tax bads, not goods" (Stiglitz, 2006).

However, energy use is not what economists call *elastic*, that is to say that an increase in price results in a proportional decrease in consumption. You can raise fossil fuel prices in Los Angeles 10 fold, and people will still need to drive, so large increases in prices are needed for small decreases in consumption. The result would be small contribution to real pollution reduction, while increasing energy prices so much that all users would suffer brutal declines in living standards as most of their money goes to buy petroleum. This is the most expensive and ineffective solution.

But the problem is worse than that, because carbon taxes as proposed by economists and politicians are designed to go into general government revenue, they would be spent without solving the problem by directly absorbing the CO_2. The economists' and politicians' current concept of a carbon tax is an expensive boondoggle that impoverishes people without solving the problem and prevents direct solutions transferring the CO_2 to where it will provide the greatest benefit to society.

Equally destructive is the economists' habit of *discounting the future*, making severe impacts count less and less the further in the future they occur. Either the discount rate for CO_2 pollution should be zero or the discount rate should be small and last long enough to extend its duration the entire time length of the impacts in order to be meaningfully calculated (Stern, 2006; van den Bergh and Botzen, 2014).

Where Can We Store Excess CO_2: Geoengineering

There are two options: (1) to enhance natural sinks and storage mechanisms (geotherapy) or (2) to create artificial ones (geoengineering). We discuss geoengineering first, because it gets all the attention and press. It is important that these diametrically opposed concepts not be confused, because they are different in all respects. Geoengineering sinks usually involve (1) chemicals to absorb CO_2, (2) pumping CO_2 someplace where it would not come back (CCS), or creating shields to reflect sunlight from reaching the earth's surface, whether clouds, aerosols, smoke, or mirrors (The Royal Society, 2009). Chemical absorbents are costly and relatively inefficient, although some good new ones may be developed. But it is unlikely that these can make much difference to the atmosphere as a whole at an affordable cost, so they are largely confined to trying to strip CO_2 out of the exhausts of major power plants (Rau et al., 2007; Rau, 2008, 2009, 2011). In this case, they reduce some of the new CO_2 inputs, but cannot remove the excess CO_2 already in the atmosphere like plants do.

CCS, in order to be effective, must extract a very small amount of CO_2 from power plant exhaust gases (typically no more than a few percent of the total gas) from a vastly greater amount of nitrogen and oxygen, purify it, compress it, and pump it large distances to a hole in the ground or the deep sea, where it is hoped that it will never leak back. If concentrated CO_2 is pumped into the deep ocean, it will immediately bubble to the surface, and high-pressure pumping into underground layers, such as done in fracking, fractures overlying confining layers. These conditions make this option very unlikely to be efficient (proper testing has not yet occurred due to prohibitive testing costs) or cost effective, a source of earthquakes, and a threat to overlying aquifers. Concentrating CO_2 and shooting it out into space would be prohibitively expensive, even if all rocket launches actually succeeded, more so when they do not.

CCS has been widely touted by fossil fuel producers to be the answer that will allow us to keep burning fossil fuels as if there were no tomorrow without paying an environmental price, and CCS research is lavishly funded by the fossil fuel industry. All claims that CCS could be feasible and affordable have been clearly rejected by serious investigators, who suggest that the real costs would be something like around $1000/ton (House et al., 2011; Michel, 2013; Scott et al., 2012). The CCS cost could be more expensive than the benefits from burning the fossil fuel, while also consuming additional fossil fuels for CO_2 extraction, concentration, and transportation. Almost all CCS projects that have been started have been abandoned as uneconomic (House et al., 2011; Scott et al., 2012; Michel, 2013). Even if they could affordably capture *all* of the CO_2 emissions from large stationary sources like big power plants, the inability to trap smaller household cooking, heating, and transportation fossil fuel emissions would cause the net impact on global atmosphere CO_2 to be at best a small and slow source reduction (Michel, 2013).

Another popular high-tech *solution*, nuclear power, is also less effective than it seems. Nuclear power does in fact have a large carbon footprint from power plant manufacture and fuel processing. Nuclear power is intrinsically inefficient because it can only be used to make electricity and bombs, and electrical transmission line losses require thousands of nuclear power plants spaced every few hundred kilometers apart. It makes no thermodynamic sense to use power sources with potential energy of millions of degrees only to heat up water by a few tens of degrees to boil water for electrical power plants, since this wastes almost all of the energy potential. Even if there were no safety and terrorism issues, and no problems of storing nuclear wastes safely for a quarter of a million years afterward, the costs of nuclear power do not include the costs of decommissioning power plants, which are passed on to the taxpayers later. It can cost hundreds of billions of dollars to shut down nuclear power plants. If these real costs were included in the electricity price, nuclear power would not be cost competitive, but instead, taxpayers heavily subsidize them. Furthermore, the total amount of the uranium resource is limited and would not last much longer than petroleum, for about a human generation or so. Somewhat more power could be obtained using thorium, and even more making plutonium from uranium and thorium, but this would require breeder reactors every few hundred kilometers making plutonium and shipping it in large quantities to plutonium power plants. Plutonium is perhaps the most toxic substance known; a teaspoonful in a reservoir could kill a city of millions and an open invitation to mad terrorists and mad generals to build bombs. The precautionary principle tells us that this is a future worth avoiding.

Several geoengineering approaches have been proposed to use the oceans to get rid of CO_2 and heat. One proposal involves building huge pumps to pump cold deep ocean bottom water to the surface, in order to mix atmospheric CO_2 down into deep water (Lovelock and Rapley, 2007). The proponents unfortunately do not realize that deep water has much higher CO_2 content than surface water. Ocean bottom water forms in polar regions since CO_2 is far more soluble in cold water than warm water, and in addition, bottom water has much more added CO_2 from decomposition of organic matter in deep waters and sediments. The net effect of this proposal would be to greatly increase the rate at which CO_2 is added to the atmosphere, the opposite result of what they claim.

Another geoengineering fix is to fertilize the oceans in the hope that plankton will grow, take up CO_2 from the surface, and sink to the bottom without decomposing. In fact, almost all the organic matter plankton biomass produced is rapidly returned as CO_2 within days to weeks as the plankton respire, are eaten, or die and rot, and most of this recycling takes place in surface waters. Most of the rest is decomposed on its fall down to the deep sea, on the bottom mud, or inside the sediments. Only a very tiny fraction of the carbon, typically one tenth of one percent, are preserved, so in fact, the net sink would be very small indeed (Betts and Holland, 1991). The only way to bury carbon without it rapidly rotting is to turn the entire ocean into an anoxic *dead zone* devoid of oxygen. This is indeed the condition that was present when much of our oil supply formed around a hundred million years ago, when the water was very hot due to high atmospheric CO_2, with the result that very little oxygen could dissolve in hot water, and was rapidly consumed by bacteria. To make the ocean an effective carbon sink by turning it into a dead zone would kill the fish, almost all marine life, and toxic hydrogen sulfide produced by bacteria would give the ocean a rotten egg stink. Adding nutrients to warm surface waters would cause massive algae blooms that would kill coral reefs, the most nutrient-sensitive ecosystem (De Georges et al., 2010). Ocean fertilization hardly seems a sensible or well-thought-out proposal because the negative effects would greatly outweigh the benefits. All geoengineering proposals seem to involve unproven technology, whose effectiveness is questionable, whose costs have been greatly understated, and whose side effects may be serious and unpredictable.

Where Can We Store Excess CO_2: Geotherapy

For enhancing natural carbon sinks, there are two major options, sea and land.

Despite the many suggestions that the ocean absorbs CO_2, there are at least three independent lines of evidence that suggest that the ocean is a net source, not a sink of atmospheric CO_2, because more organic carbon is decomposed in the ocean than is produced or imported:

1. The amount of organic carbon entering the ocean from marine primary productivity and from rivers and coastal wetlands exceeds the amount of organic carbon being buried in marine sediments (Hedges, 1992; Smith and McKenzie, 1993; Berner, 2004).
2. The concentration of CO_2 in the surface ocean exceeds that in the atmosphere, except in polar waters (Takahashi et al., 2002), so the net flow of CO_2 must be from ocean to atmosphere. Tropical waters in particular are large sources of CO_2 to the atmosphere because CO_2 solubility decreases as temperature increases, so global warming acts to increase atmospheric CO_2. CO_2 is much more soluble in cold water, and so the effects of ocean acidification will be felt first in cold and deep waters and only last in hot tropical waters like coral reefs. Coral reefs are already at their upper temperature limit (Goreau and Hayes, 1994; Goreau et al., 2005; 2012), and the most vulnerable of all ecosystems to global warming (Mora et al., 2013a), and they will the first ecosystem to be severely impacted in coming decades when mean temperatures exceed the extreme hottest years of the past record (Mora et al., 2013b). Corals will die from further temperature increases decades to centuries before ocean acidification begins to dissolve their dead skeletons, making the long-term acidification effects from increased CO_2 a straw man threat to coral reefs compared to the much more immediate effects of the rising temperature CO_2 causes. Controlling CO_2 in time to prevent coral reef extinction from global warming will automatically take care of the global acidification problem, but controlling CO_2 in time to avert ocean acidification will ensure that coral reefs will die from high temperatures. For coral reefs, the focus on acidification effects instead of the high-temperature effects is a dangerously irresponsible red herring that acts as a deliberate delaying tactic and will produce disastrous policy based on bad science. As global warming proceeds, CO_2 solubility decreases so warming waters release CO_2 to the atmosphere, a positive feedback. As surface water warms and the surface layer becomes too thick for nutrient upwelling to reach the surface, the phytoplankton concentration and productivity decreases (Boyce et al., 2010), so less CO_2 is removed to marine organic matter, yet another positive feedback.
3. The concentration of oxygen in the surface coastal waters is almost everywhere less than the atmosphere, although in low-productivity open ocean areas, it is near to equilibrium with the atmosphere (Levitus, 1982), so the net flow of oxygen is from the atmosphere to the ocean. I have made hundreds of measurements of oxygen in tropical coral reefs, sea grasses, and coastal zones, and the concentrations were almost everywhere about 90% of atmospheric values (being higher only over dense shallow sea grass beds in full sunlight). As the ocean has more respiration than photosynthesis, it is inefficient at storing carbon, unless a major part is converted into anoxic *dead zones*. Tropical waters are especially low in oxygen because high temperature reduces its solubility, so global warming amplifies oxygen depletion. The long-term consequence of increased CO_2 will be to increase anoxia that could cause severe ocean dead zones lasting around 100,000 years (Shaffer et al., 2009), similar to the worldwide ocean anoxic events during high temperatures in the Cretaceous, 80–120 million years ago (Wilson and Norris, 2001), and other times in the geological past. Eventually, the increased carbon storage in these zones acts to reduce the amount of CO_2 in the atmosphere, but that is on a geological time scale, not a human one, so it would be foolish to wait for it.

This leaves the land as the only feasible prospect, to which we now turn. There are at least three land sinks, biomass, soil, and rock weathering, which by themselves could be easily managed to absorb excess CO_2 and can do so even faster in combination. The intensity of the carbon cycle per unit area on land is more than three times as great as the ocean, because roughly equal gross fluxes of CO_2 from the atmosphere take place on land and sea (Figure 2.1 and Table 2.2) but the ocean has more area than land. Land photosynthesis is far more efficient at storing carbon because plant biomass lasts much longer than marine biomass, being stored as biomass carbon for years to centuries before it is returned to the atmosphere from land, versus only days to weeks from the ocean. The amount of organic carbon stored in soils is around 2–4 times higher than the atmosphere or biomass and that also lasts for centuries, unless we cause it to erode and decompose due to bad management. There is good reason to believe that soil carbon estimates are too low for several reasons: poor estimates for wetlands, especially in the tropics and Arctic, and the fact that soil carbon assays usually include only shallow soil due the difficulty of digging and so miss large amounts of carbon in deeper layers (Jobbagy and Jackson, 2000; Harper and Tibbett, 2013; Scharlemann et al., 2014; Shangguan et al., 2014). Thus, soil storage is the most sensible option, not the ocean, or outer space, or deep underground.

How Long Can We Store Carbon in Soil?

We have treated our soil like dirt, and let it degrade and be eroded (Carter and Dale, 1974; Hyams, 1976; Nye and Greenland, 1965; Brady, 1990; Hillel, 1991; Lal et al., 1995; Richards, 2003; Diamond, 2005; Montgomery, 2007), even though restoring soil fertility is the key to our sustainable future. Cultivation of soils is estimated to have released 150 × 10^9 tons of soil organic carbon (SOC) to the atmosphere as CO_2, and it is only in the 1960s that fossil fuel emissions exceeded net soil CO_2 sources (Bohn, 1978). Reversing this loss and managing soils so there is only a modest increase (roughly 10%) of global soil carbon over current levels could reestablish atmospheric CO_2 at preindustrial levels (Aniwaer and Goreau, 2009) in as little as a few decades (Goreau, 1995). As some of the examples given further on in this chapter show, many people have achieved much higher rates of carbon sequestration per unit area than is needed on a global scale, done so in a large variety of habitats, and it is likely that this could be improved with sound scientific carbon management.

SOC is composed of a huge range of compounds, some of which are very rapidly consumed by bacteria, fungi, and soil invertebrates, and those that are so resistant to decomposition that they last centuries (Pauli, 1967; Alexander, 1977; Brady, 1990; Schmidt et al., 2011; Huang et al., 2012a,b). Managing soil to increase SOC can create enormous sinks if managed properly (Lal, 2004; Post et al., 2012). SOC can be made even more effective by producing biochar. Biochar consists of a wide range of compounds, depending on the composition of the starting material and the physical conditions (temperature and pressure) under which it is made. The key feature is that a major part of it is largely pure elemental carbon rather than organic carbon (carbon combined with oxygen, hydrogen, nitrogen, sulfur, phosphorus, and other trace elements), and the composition ranges between these two extremes depending on age and conditions of formation. *Terra preta* (black earth) soils in the Amazon are among the most fertile soils known and contain up to 35% black carbon (Glaser et al., 2000). Char materials are extremely abundant in grassland soils, where they make up to 40%–50% of SOC (Mao et al., 2012).

Elemental carbon, and organic carbon that approaches it in composition, is extremely resistant to chemical and biological destruction, practically indestructible (Lehmann and Joseph, 2009; Ladygina and Rineau, 2013). However, fine particles can be windblown as black carbon in the atmosphere, where it can cause complicated effects acting to increase global warming (Bond et al., 2013). Research on elemental *black carbon* found that it was globally distributed, and an excellent tracer of the distribution of fires, both natural and man-made, faithfully documenting their history worldwide, carbon transport by soil erosion, water, or wind (Goldberg, 1985; Suman et al., 1997). These studies allowed the frequency of natural and human-caused fires to be determined and directions of ancient winds to be mapped from their distribution. The geological record suggests that black carbon, once formed, is virtually indestructible, or it would not serve as good a tracer for sources and transport mechanisms as long as it does.

Recent research shows that large portions, in some cases or most, of the long-lived *recalcitrant* carbon in soil are in fact biochar from forest and grass fires. In some cases, this may have been deliberate management to clear land or increase productivity, repeated use of the same sites for hunter campfires for thousands of years, or inadvertent results of use of fire to manage grassland for wild game, or for livestock, which is widespread. Large parts of Australia, South America, and Africa are deliberately burned each year. Such fires are the most inefficient way to produce biochar, because almost all the carbon is converted to CO_2 and only a percent or so winds up as biochar. The use of biochar made in kilns instead of traditional burning could therefore increase soil carbon inputs by around 50-fold over traditional open air burning and rapidly build up carbon-rich soil.

Because biochar is such an efficient absorber of water and nutrients and because biochar is densely intergrown by fungi symbiotic with plant roots, mature biochar greatly stimulates nutrient and water uptake by plants, causing much more rapid growth (Glaser et al., 2002; Steiner, 2006; Warnock et al., 2007). At the same time, it increases the release of CO_2 from root metabolism, while decreasing the release of the powerful GHG nitrous oxide (Stewart et al., 2013). But these intricate symbiotic soil interconnections involving fungi are damaged or destroyed by chemical fertilizers and by fungicides (Stamets, 2005).

There is much controversy over how long biochar lasts in soils, largely due to use of materials of varying properties and formation conditions. While there is some evidence for loss of char from soils with time, especially of lower-molecular-weight volatile compounds, in most cases, these are very low (Ladygina and Rineau, 2013). Where they are appreciable, it is likely that these studies refer to very low-grade char that is much lower in total carbon and contains large amounts of organic carbon molecules that can be easily broken down by bacteria and fungi. Low-temperature chars have a much lower elemental carbon content, and hence degrade much faster, than high-temperature chars (Lehmann and Joseph, 2009; Bates, 2010; Taylor, 2010). Char contains a mixture of compounds from small to large. The smaller-size materials can be readily transported as colloidal suspensions in water and be transported via rivers to the sea or blown by the wind.

Ocean sediment carbon contains about 10%–30% black elemental carbon (Masiello and Druffel, 1998). Much of this is derived from erosion of black carbon in soils (Dittmar et al., 2012). Recent measurements have mischaracterized this fine charcoal fraction transported via rivers into the ocean as *soluble black carbon* that makes up 10.6% of river organic carbon (Jaffe et al., 2013). But rather than being soluble, these are largely insoluble microscopic or nanocolloidal particles that are so small they pass through the small holes in filters. These materials are a major source of marine *dissolved* organic carbon (DOC), which is highly resistant to decomposition, is evenly mixed throughout the ocean, and lasts thousands of years. This DOC contains a *black carbon* component that is much older, about 18,000 years old and therefore very resistant to decomposition (Ziolkowski and Druffel, 2010), subject only to photooxidation in surface waters (Stubbins et al., 2012). Unfortunately, the data of Jaffe et al. have been widely misinterpreted to mean that soil biochar is *soluble* and washes away. In fact, this transportable component hardly decomposes, and from the point of view of atmospheric carbon sinks, it hardly matters whether it is stored in soil or in the ocean as long as it does not rapidly get oxidized back to CO_2.

Several lines of geological evidence suggest that biochar has a very long lifetime, practically infinite for all practical purposes, and provides accurate historical records of past fires (Goldberg, 1985; Suman et al., 1997). The art of making *terra preta do indio* (black earth of the Indians, in Portuguese) by Amazonian Indians in pre-Columbian times was lost following genocide by European diseases some 500 years ago, but *terra preta* deposits remain for thousands of kilometers along the banks of the Amazon River and its major tributaries (Sombroek, 1966, 1984; Sombroek et al., 2003). The demise of mammoths and mastodons in North America 12,800 years ago is marked by a layer of charcoal across North America and Europe that are associated with forest fires triggered by a small asteroid impact (Wittke et al., 2013). The settlement of Australia by the Aboriginals some 40–50,000 years ago is marked by a dramatic increase in soil charcoal in lake sediments well above the natural background from wild fires (Kershaw et al., 2002), indicating that systematic burning of the land for wildlife began shortly after the large megafauna were hunted to extinction, triggering major vegetation changes (Lopes dos Santos et al., 2013). All these lines of evidence related to human-caused fire suggest charcoal can last in soils for up to 50,000 years, but even older evidence can be found.

The charcoal layer from worldwide forest fires that followed the asteroid impact that killed the dinosaurs 65 million years ago is perfectly preserved (Wohlbach et al., 1990; Kruge et al., 1994). One of my students once brought this material to my laboratory to look at in the microscope and we were amazed that we could see clearly every plant cell in the charcoal ash. Similar charcoal layers mark the

Permian extinction event 250 million years ago, although those fires are thought to have been caused by massive lava eruptions (Grasby et al., 2011). The oldest known perfectly preserved charcoal known goes back to nearly the evolution of the first forests, some 420 million years ago, and becoming more common after 350 million years ago (Scott and Glasspool, 2006; Scott, 2009). This suggests a lifetime of elemental carbon charcoal in soils of at least 400 million years and indicates that elemental carbon has a lifetime that is infinite for all our practical carbon storage needs and far longer than any other form of carbon.

How Much CO_2 Can We Store: Carbon Sequestration Capacity

Many different soil and biomass sinks are capable of storing enough carbon by themselves, and in combination, to absorb all the excess CO_2 above safe levels, and stabilize climate change.

The first and most obvious is biomass. Massive and accelerating increase of CO_2 from fossil fuel combustion over the last 150 years, since the industrial revolution, caused the famous so-called *hockey stick* curve of temperature increase (Mann et al., 2008), which has been fully confirmed in all of its details as it has been greatly extended in length (Marcott et al., 2013). However, a major part of the excess atmospheric CO_2 to date comes from the much slower destruction of the great forests of Asia, Europe, and North America since the start of agriculture 10,000 years ago, with South America, Africa, and Southeast Asia coming to the fore only after the forest vanished in northern temperate zones and from erosion and decomposition of soil carbon following deforestation (Bohn, 1978; Houghton, 2003; Lal, 2004; Ruddiman, 2005; Perlin, 2005; Lal, 2011). The amount of forest left is constantly decreasing, while the fossil fuel output is rising exponentially, so land degradation is now a minor and diminishing global source of atmospheric CO_2. It has been obvious to anyone who looks at forests and deforested areas that if the latter could be reforested, vast amounts of carbon could be stored, and if the forests were to be fully restored, most of the existing excess could be absorbed. If the fertility of the soil were to be restored or further enhanced by using rock powders and biochar, then we propose that *all* of the excess could be taken up, the core point of this entire book.

The obvious concept of large-scale ecosystem reforestation to store carbon has occurred independently to many. As an example, we look at one of the first such estimates. Myers and Goreau (1991) looked only at the areas of deforested, badly degraded land in the tropics that had been abandoned and useless after deforestation because of the lack of productivity and that had been overrun by invasive weeds. They asked:

1. Is there enough land available to absorb global fossil fuel CO_2 production using the productivity of typical tropical replanted forests?
2. If the polluter were to pay a carbon tax used directly for the replanting, what would it cost them using typical tropical reforestation costs?

They found that

1. There was more than enough tropical land that had been abandoned and become economically useless after deforestation. In fact, there was enough in Brazil alone.
2. The real costs for tropical tree planting came to about \$2–3/ton of carbon, so average cost per capita for global citizens to remove their own excess carbon through tropical reforestation would be about \$2–3/year or for highly fossil-fuel-intensive countries like the United States, about \$10–15/person/year (Myers and Goreau, 1991).

A fossil fuel carbon tax earmarked for tree planting would be very affordable, even in very poor countries, and could solve the CO_2 pollution problem at real cost without being inflationary. This is in sharp contrast to the economists' concept of a punitive carbon tax to drive prices up and decrease consumption of fossil fuels, and hence emissions, to the needed level, which would be ruinously expensive for the consumer. This distinction highlights the fundamental differences between CO_2 demand-side measures like reforestation versus supply-side measures like punitive carbon taxes. The key is for the tax to reflect only the real costs of planting, administration, and verification and that it is not diverted by politicians as discretionary revenue for projects that do not directly solve the CO_2 problem.

Many other independent estimates have been made of reforestation carbon sink costs, but most have been based on reforestation costs in North America and Europe, where there is much less degraded land available, and planting costs are much higher, so they result in cost estimates several times higher than Myers and Goreau. But even so, they are quite affordable, being around 10 times less than CCS estimated costs. On a complete global scale, the real cost will vary from place to place depending on land availability and labor and tree nursery costs, so the Myers and Goreau estimates, which are for poor tropical countries, are a lower end estimate, while those from North America and Europe are an upper end estimate, and the true global average is somewhere in between.

There are vast opportunities to increase soil carbon everywhere (Figures 2.16 to 2.18), as living roots, as fungi, as soil organic matter, and as biochar, and we must do so as fast as we can, everywhere to preserve a sustainable future and grow our way out of the climate change crisis.

It is clear that the greatest benefit at the least cost will be obtained by focusing on the tropics because the growth rates of the trees are much higher (Purseglove, 1968; 1972), and the labor costs lower, so tropical restoration should be the first priority. On the other hand, despite their slower growth, cold boreal zone forests like those of Siberia and Canada grow more slowly but are more efficient carbon sinks, since the cold greatly reduces respiration, so they hold onto their carbon much longer. Boreal zone forest replanting is a key part of the solution along with equatorial zone forests because cold forests are more efficient sinks, but hot forests are more effective at recycling CO_2 and reducing its lifetime in the atmosphere (Figure 2.19). Vast areas of our planet are in need of reforestation to conserve soil and water, quite apart from their role as carbon sinks. Global warming, because it reduces carbon storage efficiency, makes it harder to store carbon as living biomass.

Cold forests store about 70% of photosynthesis carbon as living biomass, but hot forests store only 20%. Much less efficient carbon storage in tropical soils is due to the very high rates of CO_2 released

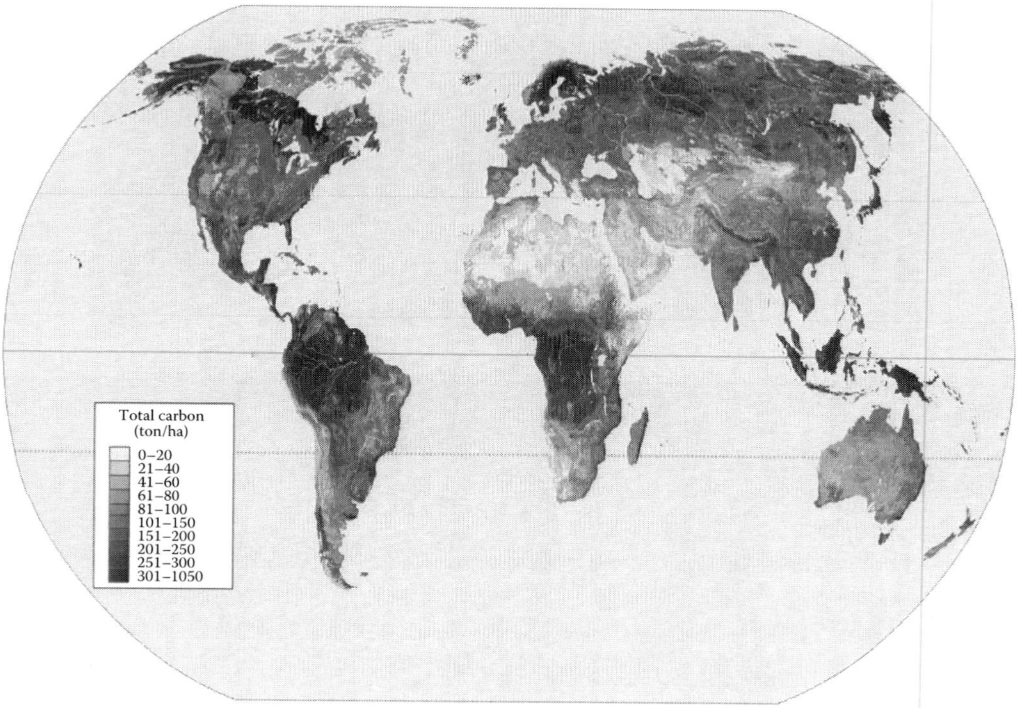

FIGURE 2.16 Soil organic carbon. (UNEP/WCMC 2008.)

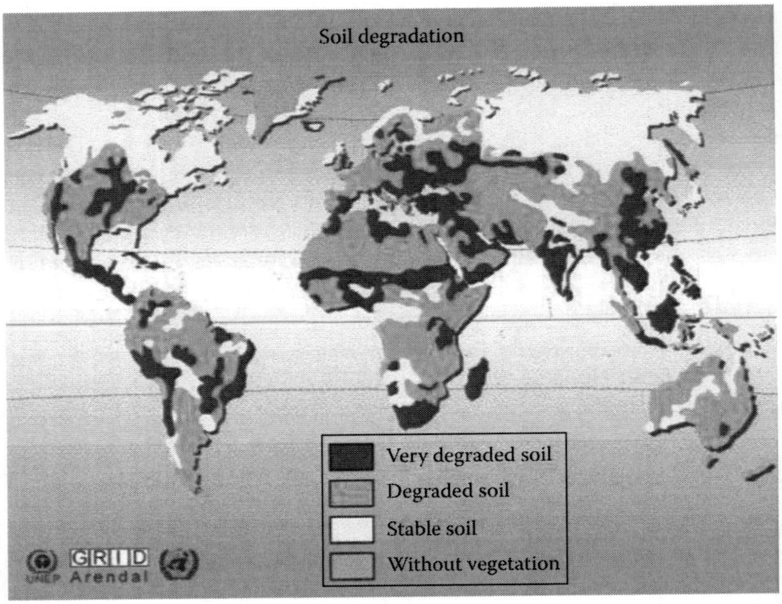

FIGURE 2.17 World map of degraded soils.

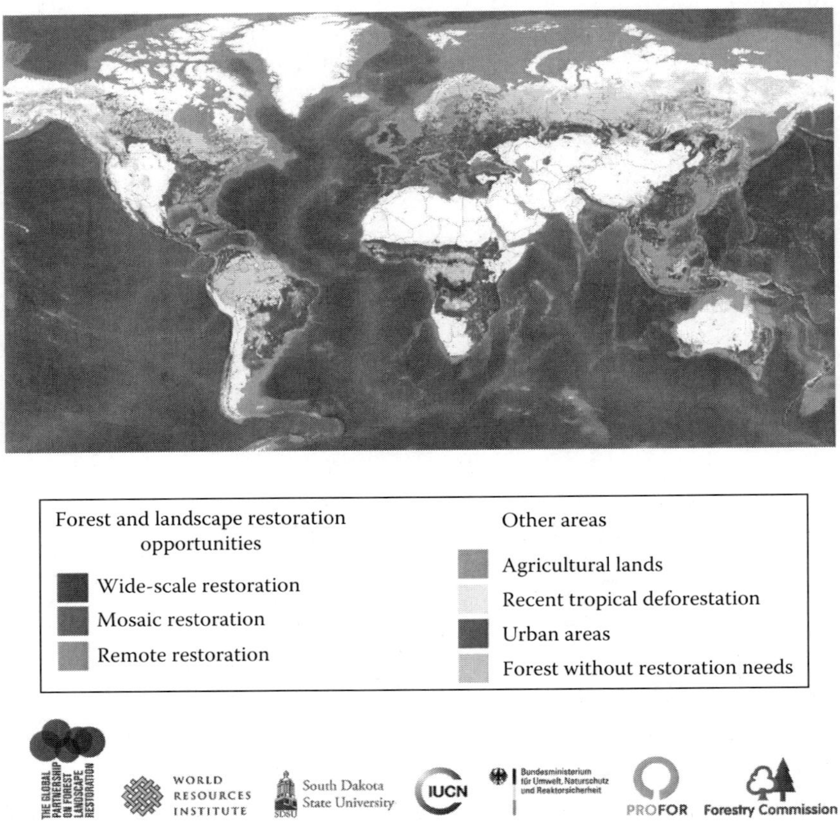

FIGURE 2.18 Forest opportunities cover much of the planet.

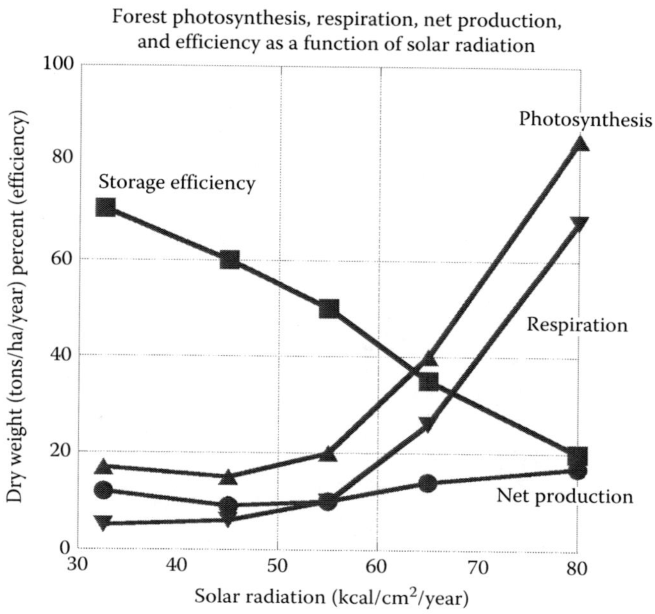

FIGURE 2.19 Gross photosynthesis (upward triangles), respiration (downward triangles), net production (filled circles), and net ecosystem efficiency (P-R)/P as a function of solar radiation. Equatorial forests are at the right, boreal forests at the left. Note that photosynthesis (gross production) and respiration increase sharply with sunlight due to the increased temperature, but net production is relatively constant. (Data from Jordan, C.F., Productivity of tropical rain forest ecosystems and the implications for their use as future wood and energy sources, in F. Golley, Ed., *Tropical Rain Forest Ecosystems: Structure and Function*, Elsevier, Amsterdam, the Netherlands, pp. 117–135, 1983; Larcher, W., *Physiological Plant Ecology*, Springer Verlag, Berlin, Germany, 1980; graphed in Goreau, T.J., Tropical ecophysiology, climate change, and the global carbon cycle, in J. Pernetta, R. Leemans, D. Elder, and S. Humphrey, Eds., *Impacts of Climate Change on Ecosystems and Species: Environmental Context*, International Union for the Conservation of Nature, Gland, Switzerland, pp. 65–79, 1995.)

from soils due the increased respiration and decomposition caused by higher temperatures, which will increase further with global warming (Lloyd and Taylor, 1994; Kirschbaum, 1995, 2000, 2006; Meir et al., 1996; Boone et al., 1998; Schlesinger and Andrews, 2000; Fang and Moncrieff, 2001). Cold forests are more efficient carbon sinks, but hot forest carbon cycling regulates atmospheric CO_2 lifetimes by rapid recycling. Both ecosystems are crucial for global climate management, but for different reasons. There are huge areas of both in need of restoration for soil and water conservation reasons alone (Figure 2.18). The possibility of restoration should never be an excuse for not conserving all primary forests, since they are more species rich and big old trees store carbon at the greatest rate (Stephenson et al., 2014).

The transformation from barren soil to forests provides amazing benefits. In the early 1800s, all of the forests in the mountains above Rio de Janeiro were cut down to plant coffee, the soil washed away, and bare rock mountains remained (those famous bare peaks overlooking the city had forests to their very top when Pedro Cabral *discovered* them for Portugal in 1504). As a result of deforestation, the year-round forest springs and streams dried up and the city was without water in the dry season. There was no doubt that the cutting of the forest had destroyed their water supply, and so the city systematically replanted the hill slopes where soil remained at the base of the mountains. A handful of slaves, employed by the city, spent their careers climbing up into the few forests that remained because they had been too steep to cut down and plant coffee, and these green heroes carefully transplanted tree seedlings, trying to maximize biological diversity (IBDF, 1991). When the trees came back, so did the water supply (Bandeira, 1994). This forest, the Floresta da Tijuca Park, is the beautiful green forest that overlooks Rio de Janeiro (Figure 2.20), and it is entirely artificial, although few people are now aware of that. In the 2012 Rio+20 UN Conference on Sustainable Development, the official brochure to delegates incorrectly described it

FIGURE 2.20 Floresta da Tijuca was planted by hand, and restored the water supply to Rio de Janeiro. (Photo by Alberto Peterson.)

as *natural forest*, implying it had been successfully conserved, when in fact it had been totally destroyed and successfully restored. Around a hundred and fifty years after replanting, William de Mello of the Universidade Federal Fluminense and I measured the CO_2 emissions from the soils of the Tijuca Forest, and we found that it had completely recovered its biogeochemical function in terms of GHG cycling by comparing our data with that from virgin Amazonian jungles (de Mello and Goreau, 1998).

Millions of individuals and thousands of groups have similarly learned on their own how to restore barren areas back to lush and complex ecosystems, storing carbon, nutrients, and water. Moreover, these unsung heroes have done so in every single ecosystem, habitat, climate zone, and soil type. Their work proves that mature technology exists to restore all our ecosystems if we chose to apply it on the scale needed. It is impossible to begin to list all these groups, but here, we give one example, in the same Brazilian Coastal Rainforest habitat as Floresta Tijuca, the Minas Gerais family farm of the famous photographer Sebastião Salgado. The family that herded their cattle said that 40 years ago, there had not been a single stick of shade to hide from the sun, but after the barren site was systematically replanted with trees, a forest now stands, and springs and rivers have returned to the formerly barren landscape. For before and after photographs, see http://www.institutoterra.us/about/history_land.html.

Panama is an unusual example of how deforestation can be turned into massive reforestation through suitable policy. Twenty or thirty years ago, the banks regarded forests as economically worthless and would only give bank loans to people who had first cut down all their trees to *make worthless land valuable*. Now, the banks pursue the opposite strategy, they would not give money to people who deforest, but they will give loans to replant degraded land. A systematic national policy of large-scale tree nurseries make tree seedlings cheap and widely available, and all over Panama, one sees massive forests arising where the land had been previously covered with useless weeds. The major problem is that almost all of the seedlings being planted are Asian teak, *Tectona grandis*, the most valuable tropical hardwood, so biodiversity does not increase as it should. But the Panamanian example does show how quickly effective policy can reverse the situation, even if they could be improved by variety.

Even more carbon can be stored in soils if we manage them to increase carbon content, and this can be done in pasture and agricultural habitat as well as forests. Soil holds around four times more carbon than the atmosphere or the biosphere, even though we have already lost around half of the soil carbon from areas deforested for agriculture or pastureland (Lal, 2004). It is easy to manage the soil to build up organic matter instead of destroying it (Kimble et al., 2007; Magdof and van Es, 2009), and there are enormous benefits in increased agricultural and forest and livestock production and soil water retention.

There are many paths to doing so. One major key is to build up soil root biomass and the biomass of root-associated symbiotic fungi, a major source of soil carbon (Clemmensen et al., 2013), especially by the use of perennial plants instead of annual crops that die and have to be replanted every year, which have much less root biomass, fungi, and SOC (Jackson, 2010). By managing succession of species in ecosystems, large increases in soil carbon storage and biomass can result (Wolkovich et al., 2010). It takes clever thinking to manage livestock so that they increase plant growth and soil carbon, fertilizing them with manure and then leaving them alone until they recover and preventing overgrazing that has the opposite effects, but such methods have been developed and work very well (Savory, 2009; see Chapter 9) although there is not yet enough data on the large increases of soil root and carbon that result. *Carbon farmers* do the same with soil, managing their fields so the SOC content is increasing instead of decreasing, which has the added benefit of increasing their agricultural productivity at lower cost (Lal et al., 1995).

These methods, applicable in all human-managed environments, are reported to be capable of sequestering from a few to several tens of tons per hectare per year and a maximum of over 30 tons/ha/year measured. If managed responsibly, pastures alone, agriculture alone, or reforestation alone could absorb much of the CO_2 increase. But there are even more opportunities, such as biochar, which greatly increases carbon storage and its storage lifetime, as well as soil water-holding capacity, with many additional benefits. Of course biochar should not be produced from plantations that compete with food production land, and the biochar should be made from agricultural wastes returned to the soil on the land from which they came to increase future production or be made from weedy biomass on land that is being restored to more economically productive and biodiverse habitats. Those possibilities are immense in every single ecological habitat type. The goal should be to be able to produce enough biochar by restoring weedy and degraded areas to greatly increase the quantity and quality of their productivity, while having a surplus of biochar to apply to agricultural and pasture soils where the produce carbon is exported. With a strategy that combines all of these approaches in all habitats, there should be no difficulty to meet the global goals to absorb all the excess.

How much carbon would we need to add to soil to do the job needed? If we focus not merely on removing the year by year CO_2 increase but on the much larger task of removing all the excess CO_2 above safe levels already in the atmosphere and get back to preindustrial levels, then global soil carbon content would need to rise from the current average value by around 10% (Aniwaer and Goreau, 2009). This is doable (Woolf et al., 2010; Chapter 3), although the increases would be greater in some places than others. Figure 2.1 shows the atmospheric, biomass, and soil carbon pools and the amount of changes needed to stabilize CO_2 at safe levels.

Additional and very important opportunities also occur in wetlands, which are major carbon stores because waterlogged conditions promote bacterial decomposition that consumes the oxygen from the water, greatly reducing decomposition rates. Wetland peats are vast carbon stores, around half or more of all soils, and instead of increasing these sinks, we are destroying them, draining them, and allowing the soil to oxidize to CO_2 (Moore et al., 2013), so the soil vanishes and atmospheric CO_2 increases, as well as methane. The Everglades have lost up to 5–10 m of soil in some locations since the soil carbon has been drained, dried, and decomposed for agriculture (Stephens and Speor, 1970). Global-warming-caused melting of frozen Arctic peats in Siberian and Canadian wetlands and submerged peats will cause huge new sources of CO_2 and methane to go into the atmosphere from buried peats in the Arctic Ocean where the overlying ice has melted away (Shakhova et al., 2010). This is another positive feedback increasing global warming, especially when added to the huge melting permafrost peat sources. There is an urgent need to not only stop wetland destruction but to restore them. This can be done by managing water flow to retain water rather than drain it and needs to be part of a global carbon strategy. This will of course at the same time provide additional water reservoirs. Since wetland peat soils can be nearly half carbon, wetland restoration is the most effective way to store soil carbon.

Besides freshwater wetlands, enormous carbon possibilities exist in marine wetlands, mangroves, and salt marshes (so-called blue carbon). Here too, these ecosystems are being cut down and their huge peat carbon deposits being destroyed for *development*, further accelerating global warming as the peat carbon oxidizes. The potential carbon stores of marine wetlands alone, if sensibly managed, could probably store a significant amount of the global excess (Mitra et al., 2005) if we were growing wetlands instead of destroying them. The same applies to marine coastal forests, mangroves, salt marshes, etc., which store huge amounts of peat that are being oxidized following drainage. Restoration of these ecosystems could provide huge carbon storage benefits (Nelleman et al., 2008; Fourqurean et al., 2012). Wetlands can be intense sources of the major GHGs CO_2, CH_4, and N_2O, but each gas shows a different environmental pattern in their changes (Goreau and de Mello, 2002, 2007) so maximizing carbon storage in wetlands may also involve complex trade-offs with GHG emissions. Wetlands show elevated GHG emissions compared to dry soil, but it is likely that the increased burial of soil peat carbon in these habitats is greater than the increased flux, since a much smaller portion of the carbon is decomposed; however, detailed studies of soil carbon storage with regard to GHG fluxes still remain to be done.

An additional important advantage of increasing soil carbon storage is preventing loss of nitrogen and phosphorus. Nitrogen loss from soil is a major source of both air and water pollution. Soil-derived atmospheric N_2O is a major GHG and regulator of the stratosphere ozone layer (Goreau, 1982), NH_3 causes aerosol smog pollution, NO and HONO also produce ozone pollution in the lower atmosphere, and acid rain, while ammonium, nitrate, nitrite, and phosphate leached from soils, contaminate ground waters, rivers, lakes, and the ocean. By overfertilizing the water, they cause severe harmful algae blooms that kill coral reefs and fisheries, cause red tides, and create dead zones (Sutton et al., 2013). Storing the nutrients in the soil will not only make the soil more fertile, it will prevent air and water pollution now causing severe consequences worldwide. Biochar is an ideal medium to adsorb these nutrients, retain them in the soil, and prevent water pollution (see Chapter 32).

In summary, there are more than enough places where the carbon could be stored cost effectively, with enormous direct economic benefits, while preventing the costs of runaway global climate change. To solve the CO_2 problem, we need to systematically restore plant productivity, soil carbon storage, and the integrated global biogeochemical ecosystem functioning of all major habitats that have been damaged. It is not simply enough to pay a corporation to cut down other people's forests, convert it to biochar, dump it all on their land, and then demand carbon credits. Without global-scale restoration, it is simply impossible to store enough carbon to meet the global needs, so large-scale restoration of destroyed, degraded, and damaged ecosystems is the *sine qua non* for stabilizing climate at safe levels. The methods to do so already exist. All that is lacking is the will on the part of policy makers and funding agencies to solve global problems on a global scale.

How Fast Can We Store CO_2: Carbon Sequestration Rates

There is clear evidence that with good management, soil carbon can readily be increased by tons, of carbon per hectare per year, and to tens of tons in best cases (Lal et al., 1995; Lenton, 2010; Lal, 2011; Stockman, 2011; Powell and Lenton, 2012; Kang and Banga, 2013; Schwartz, 2013; Soil Carbon Sequestration Conference, 2013; Ohlson, 2014; see Chapter 33). It should be noted that most of these estimates consider only opportunities from conventional crop agriculture or pasture and ignore the vast opportunities for forestry and restoration of unused degraded lands.

Restoration of the vast areas of land that has been deforested, degraded, abandoned, and overrun by weeds offers enormous opportunities to increase both biomass and soil carbon (Figures 2.17 and 2.18). This would turn useless land into economically and biologically productive carbon sinks.

Biochar can be applied at rates of up to hundreds of tons of carbon per hectare per year, but in fact, this is much more than is needed for benefits to plants, cannot be supplied from in situ carbon biomass carbon resources, and would require import of carbon from elsewhere sustainably (see Chapter 13). Such carbon transportation could defeat the purpose of increasing carbon stocks everywhere.

Scientifically sound soil carbon management should be based on accelerating natural processes that build soil carbon. The semiarid Loess Plateau grassland soils of northern China are some of the most fertile in the world but are rapidly depleted and eroded by conventional farming methods. When these soils are abandoned due to loss of fertility and are allowed to recover, the SOC doubles from around 3% after abandonment to about 6% in 20–30 years, with an increase soil nitrogen, water-holding capacity, and total microbial populations (Xiao et al., 2013). A natural succession in soil microbial communities and the biochemistry of the soil takes place during which soil CO_2 metabolism remains fairly constant, so there is a decline in soil respiration per unit of microbial biomass, making them less efficient as they approach steady state (Xiao et al., 2013). If this process were increased by adding soil carbon in both metabolizable forms (such as crop litter) and recalcitrant forms (such as biochar), the potential to significantly increase soil carbon in a few decades could very likely be further enhanced.

There is therefore no technical or economic barrier to solve the excess CO_2 problem in timely fashion, and there are incalculable economic and environmental benefits that will reward current and future generations. The problem is only lack of understanding and lack of will by policymakers and funding agencies to solve the problem as fast as possible so as to minimize the costs and maximize the benefits. They have frittered away decades asleep at the wheel since the problem and solutions were understood, and there is now very little time left for serious action before the inevitable crash.

Managing Soil Carbon Sinks

Reversing soil loss and soil carbon loss is the art of maximizing plant growth and biomass, underground root biomass, symbiotic soil fungi, and bacteria that live with roots and provide them with nutrients (Malavolta, et al., 1962), as well as the soil organisms that recycle them (Darwin, 1881; Pauli, 1967; Brady 1990) while storing a major part of SOC and biochar carbon. The tools that will greatly amplify these *carbon farming* methods will include selection of perennial crops so annual roots do not die and need to regrow each year. The use of biochar combined with rock dust can store soil carbon in virtual perpetuity, while accelerating soil CO_2 reactions to form bicarbonate (which eventually will neutralize ocean acidification) by weathering rocks to provide plant root nutrients that are inadequate in most soils (Brady, 1990).

To understand this process better, we need to understand the dynamics of CO_2 in soil. If the major source of CO_2 in soils was from the atmosphere, and its major sink was weathering, then CO_2 concentrations in soil would be less than in the atmosphere. In fact, it is typically from 10 to up to a hundred times higher. Figure 2.21 shows the rate at which CO_2 flows out of soils in just minutes into chambers placed at the soil surface in a temperate mountain hardwood forest (Goreau, 1982), much higher rates are found in tropical forests. Plants greatly increase soil CO_2 through root respiration, which serves to increase the rate of chemical weathering and nutrient release from soil minerals.

Amazonian Soil Metabolism and the Lifetime of CO_2

It is generally assumed that soil CO_2 is largely derived from the microbial decomposition of dead organic matter provided by leaf litter falling from plants to the ground. But measurements of CO_2 emissions from soils to the atmosphere in virgin Amazonian jungles that had just been clear-cut, and left to dry in preparation for burning and conversion to cattle pasture, showed otherwise. If decomposition of dead organic matter were the source of soil CO_2, the flux out of the soil surface would greatly increase after the living forest was chopped down and turned into dead organic matter. Instead, the release of CO_2 rapidly fell immediately after felling, and bare soils had only about 25% as high CO_2 release (Figure 2.22 after data in Goreau and de Mello, 1987, 1988). It is important to note that this site was not burned; the measurements were made in sites where the biomass was left to dry prior to burning.

This indicates that around 75% of CO_2 in soil comes from the respiration of living plant roots and the respiration of fungi, bacteria, and soil invertebrates directly feeding from living root organic carbon or from metabolizable carbon compounds that are released from roots. The data shows that forest metabolism above- and belowground is more symbiotically integrated, and larger amounts of carbon are flowing through tropical forest soils than had been realized. In Figure 2.2, most of what is labeled belowground *decomposition* is therefore actually metabolism of living plants and plant-dependent organisms. If we restored tropical rain forests instead of destroying them, forests would reduce the atmospheric lifetime and warming impacts of CO_2. The master planters of Floresta da Tijuca showed it can be done.

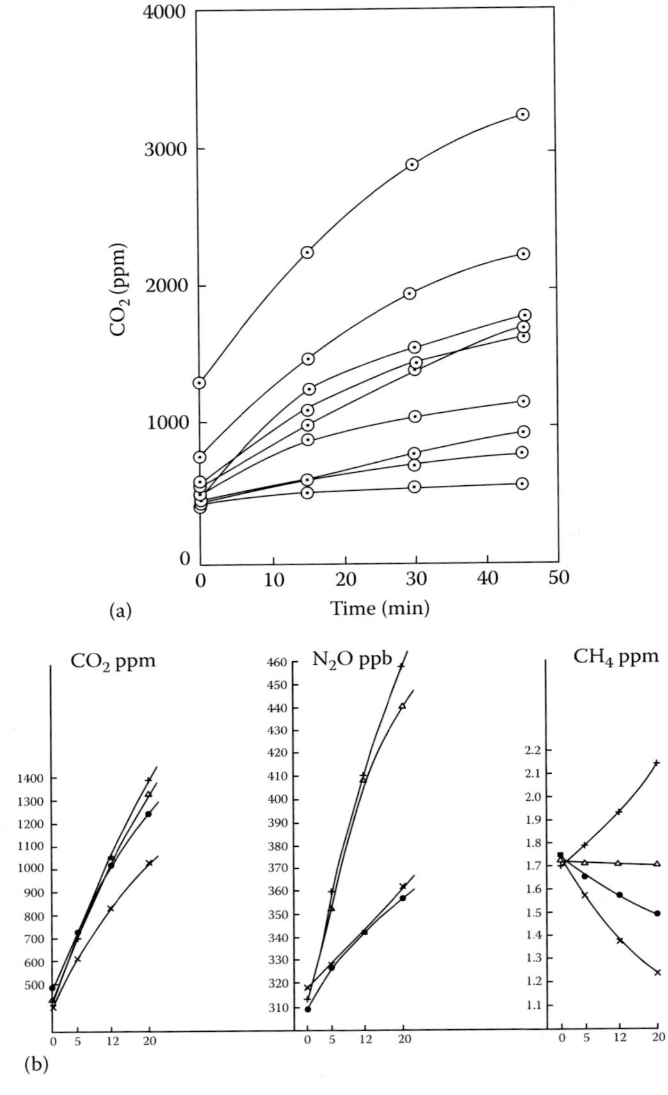

FIGURE 2.21 (a) CO_2 concentrations in chambers above the soil, showing soil levels are up to 10 times higher than atmospheric. These are measurements made at different seasons and locations at the Hubbard Brook Experimental Forest in the White Mountains National Park, New Hampshire, United States. All chambers started out with higher than atmospheric CO_2 concentrations (at that time around 350 ppm). (From Goreau, T.J., Biogeochemistry of nitrous oxide, PhD thesis, Harvard University, Cambridge, MA, 1982.) (b) Changes in greenhouse gas concentrations over Amazonian forest soils (Goreau & De Mello, 1987). The site that released methane was next to a termites nest.

We also looked at pastures that had been burned and seeded to grass, both before and after cattle started to graze it. It was also found that after the cut-down forest was burned, only the leaves and fine twigs were turned into CO_2 and a black surface layer of charcoal, but the fires burned quickly and did not burn the vast majority of tree trunks and branches. These lay superficially charred on the ground after the fire had passed and the ground was seeded to grass. But within a few years, the charred wood biomass vanished entirely because it was completely consumed by termites that built their nests over it. Measurements of the gases emitted from Amazonian termite nests showed very high methane-to-carbon dioxide ratios (Goreau and de Mello, 1987). This methane is produced by symbiotic bacteria, living in the guts of the termites, and the amount produced after deforestation is so large that it acts to amplify global

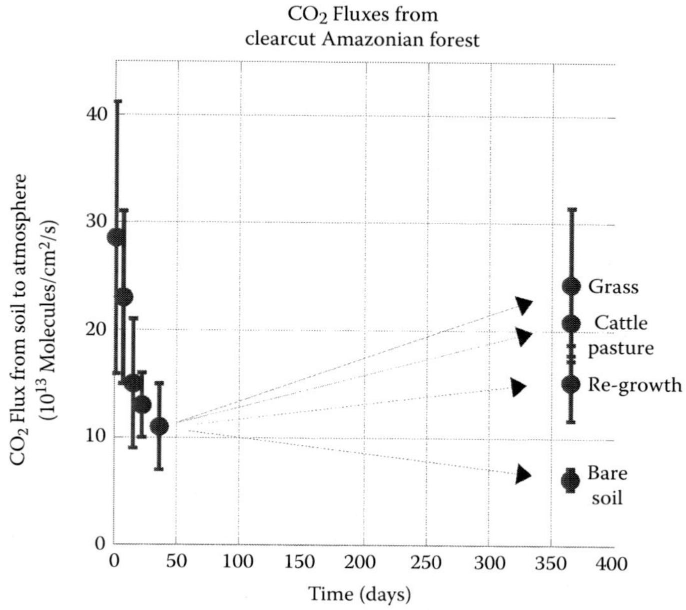

FIGURE 2.22 CO_2 fluxes (means plus or minus standard deviations) from clear-cut Amazonian forest soils versus time. (After Goreau, T.J. and de Mello, W.Z., Effects of deforestation on sources and sinks of atmospheric carbon dioxide, nitrous oxide, and methane from Central Amazonian soils and biota during the dry season: A preliminary study, in D. Athie, T.E. Lovejoy, and P. de M. Oyens, Eds., *Proceedings of the Workshop on Biogeochemistry of Tropical Rain Forests: Problems for Research*, Centro de Energia Nuclear na Agricultura and World Wildlife Fund, Piricicaba, Sao Paulo, Brazil, pp. 51–66, 1987.) Fluxes were measured in multiple replicate plots in undisturbed rainforest and at various times after the forest had been clear-cut. The cut trees and vegetation simply lay where they fell and were left to dry as was so it could be burned at the end of the dry season with seeding of grass at the start of the next rainy season. The data at the left follow the decrease in CO_2 emissions with time. The data at the right are comparisons with plots cut and cleared the previous year; one had grass with no cattle, one had grass being actively grazed, one was regenerating in place with no burning (due to early rains), and the fourth was weeded once and left bare (nothing more grew there after the first weeding). The arrows show the trajectory of CO_2 fluxes for various land management for various forms of ecosystem recovery. This should be a curve, not a straight line, but there are no more data available to refine it. More data are needed to follow the changes in such plots, but it is clear that the deforested ecosystems all cycle CO_2 at lower rates than intact forest.

warming because methane is a much stronger GHG than CO_2. We also compared sites that were regenerating after clear-cutting without burning, due to the rainy season coming before they could light it. Methane release was also found from cow manure (Goreau and de Mello, 1987; Steinfeld et al., 2006).

Since most CO_2 emitted from soils comes from the respiration of the living biota, and not decomposition of dead organic matter, CO_2 is being recycled back to the atmosphere much faster than had been realized. Tropical forests with very rapid CO_2 cycling rates act to control the lifetime of CO_2 in the atmosphere. If rapid CO_2 recycling ecosystems are replaced with land uses that recycle CO_2 much more slowly, the lifetime of CO_2 in the atmosphere will increase, and each molecule of CO_2 will absorb heat longer and produce more global warming per molecule (Goreau and de Mello, 1988; Goreau, 1990), another major positive feedback mechanism not included in global climate change models. Since respiration of plant roots controls CO_2 concentrations in soils, it controls the rate at which minerals are released by weathering, and hence the magnitude of weathering carbon sinks.

Much of this mineral weathering and nutrient uptake is performed by mycorrhizae, fungi that are symbiotic with the plant roots. Mycorrhizae are fed organic carbon by the roots, while providing them nutrients in exchange (Taylor et al., 2009) and can make up a major part of soil carbon (Rillig et al., 2001; van der Heijden et al., 2008; Maser et al., 2010). Healthy growing plants are the key to increase weathering sinks, but deforestation and conversion to annual crop greatly reduces the rate at which such nutrients are released to soil. Restoring plant growth restores the carbon sink, but deforestation reduces the weathering carbon sink and therefore acts as yet another positive feedback mechanism for global warming not in the models.

Mycorrhizae types greatly influence soil bacterial activity (Averill and Finzi, 2011; Waring et al., 2013) and influence soil carbon storage by affecting the efficiency with which nitrogen is recycled in soils (Averill et al., 2014). The type of mycorrhizae most common in tree roots is unable to take up nitrogen, which is processed in the soil, but the type more common in swamp plants and nitrogen-poor environments is able to take up nitrogen directly and therefore short-circuits the bacterial nitrogen cycle, reducing loss of nitrogen to groundwater and to the atmosphere. Habitats dominated by the latter group of mycorrhizae store much more soil carbon for the same amount of nitrogen (Averill et al., 2014).

The effects of these differences are very clear in the two major Amazonian soil types, yellow clay soils and white sand soils. Trees on the yellow soils that cover most of the Amazon Basin have mycorrhizae that are efficient in taking up phosphorus but not nitrogen, and such tropical forest soil biomes are the largest single source of atmospheric nitrous oxide, a potent GHG and ozone layer regulator (Goreau and de Mello, 1987, 1988). No litter builds up on the soil surface in these forests because it is immediately removed by termites and broken down by symbiotic fungi underground. On the other hand, the few percent of the Amazon Basin with white sand soils has a stunted tree vegetation of different species adapted to extreme nutrient scarcity. Surprisingly, there is thick surface leaf litter layer, and the roots of the plants do not grow down into the sand, but they grow *up* into the leaf litter, where their symbiotic mycorrhizae are extremely efficient at taking up nitrogen. The bacterial nitrogen cycle is so efficiently short-circuited that there is no release of nitrous oxide from the soil (Goreau and de Mello, 1988). These areas, despite extremely low productivity, are so efficient at storing carbon that the rivers draining them have black waters, stained by DOC compounds leached out of the soil, and these small areas are the major source of organic matter exported in the Amazon River.

Soil Carbon, Water, and Temperature Cycles under Extreme Conditions

The Amazon soil GHG study also revealed important interactions between the soil carbon and the soil water cycle. Around 85% of global water that is evaporated to the atmosphere from soils goes via transpiration through plants rather than direct evaporation (Jasechko et al., 2013). Plants dominate water and heat flow to the atmosphere in all except plant-free deserts, even though evaporation greatly accelerates with increased temperature. We compared GHG emissions from virgin Amazonian jungle with a nearby site where the forest had been completely cut down 1 year before, but due to early rains, they could not burn the biomass, so it was overgrown by weeds; the ranchers could not convert it to pasture and lost their deforestation investment. We made measurements of temperature and GHG emissions at both sites around the clock.

In the forest site, the temperature range between the afternoon maximum and early morning minimum was about 1.5°C, and it was always delightful in the equatorial forest even at the warmest time of day. The nearby clear-cut plot was much hotter, and the daily temperature range was nearly 10 times higher, and nearly 15°C (Figure 2.23a, Goreau and de Mello, 1987). During the short Amazon dry season, the forest soil was moist and the ground vegetation was green and succulent. But in the nearby clear-cut, the soil was bare, dry, and baked to a concrete-like hardness, and the vegetation was brown. In the city of Manaus, center of a 50 km deforested area, the grass was brown and dying, as if in severe drought.

Where the forest is intact, the ground and air are cooled by shading and by evapotranspiration of moisture through the tree leaves, but clear-cut areas lack these cooling mechanisms, making them much hotter. Degradation of the carbon cycle causes perturbation of the water and heat cycles that act to greatly increase temperatures and reduce water soil water availability. Every increase in soil roots, fungi, soil organic matter, and biochar increases soil's water-holding capacity and stabilizes temperatures. The increased evapotranspiration of the vegetation cools the ground while it increases rainfall. It was noted in Brazil as far back as the 1600s that where the forest was cut to grow sugar cane, the land became much hotter and the small streams dried up. After Honduras was rapidly deforested, air temperatures rose very rapidly, around 10 times faster than global warming. Much local heating has

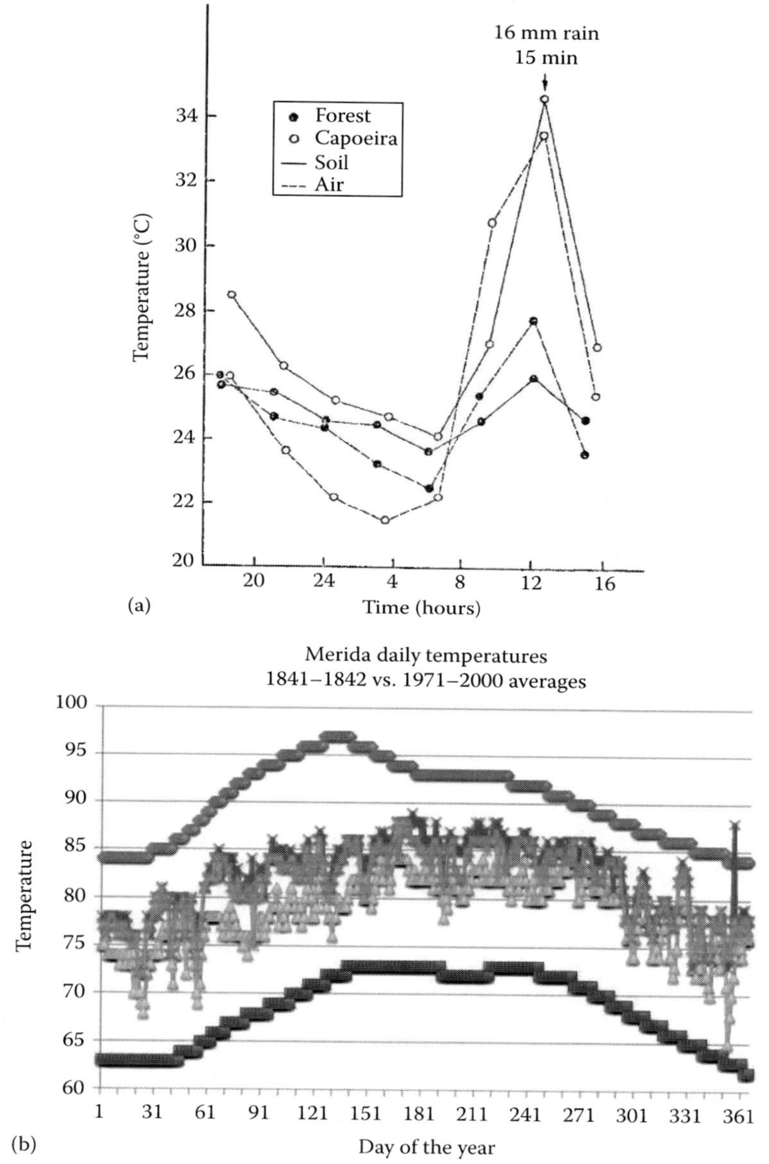

FIGURE 2.23 (a) Daily temperature cycles in virgin Amazonian rainforest versus in a nearby clear cut patch. (After Goreau, T.J. and de Mello, W.Z., Effects of deforestation on sources and sinks of atmospheric carbon dioxide, nitrous oxide, and methane from Central Amazonian soils and biota during the dry season: A preliminary study, in D. Athie, T.E. Lovejoy, and P. de M. Oyens, Eds., *Proceedings of the Workshop on Biogeochemistry of Tropical Rain Forests: Problems for Research*, Centro de Energia Nuclear na Agricultura and World Wildlife Fund, Piricicaba, Sao Paulo, Brazil, pp. 51–66, 1985.) (b) Merida, Yucatan, temperature changes over 170 years. Daily measurements from 1841 to 1842 at 6 AM, 12 noon, and 6 PM are graphed from data tables in Stephens (1843). The average 1970–2000 maxima (top light gray) and minima (bottom dark gray) are from URL.

been caused by loss of the evapotranspiration cooling mechanism that living vegetation provides, and restoration of the forests would in turn help cool the climate.

Increased soil carbon storage brings huge further benefits in terms of cooling off the temperature while increasing our water resources, quite apart from biological production benefits. As the result of greater soil moisture, the growing season lasts longer, and groundwater is recharged, restoring springs and rivers. Wetland habitats urgently need to be restored because the large amounts of organic peat they build up are the most important carbon storage ecosystems in the world per unit area. Recreating them will not only be the most effective carbon storage possible; it will also stabilize our water resources and reduce erosion at the source, turning brown land green and turning brown water blue.

Temperature extremes are becoming worse with time. This is dramatically shown in daily temperatures measured at 6 AM, noon, and 6 PM in Merida, Yucatan, in 1841–1842 (Stephens, 1843), when compared to the average maxima and minima from 1971 to 2000 (Figure 2.23b). The surrounding area was forested in 1841, but deforested by 1970. Not only does the daily temperature range seem to have increased greatly as the result of losing the cooling and regulation of temperature by forest transpiration, but the peak of the hot season now comes suddenly 3 months earlier.

Turning the UNFCCC into an Effective Tool

To solve the problem, the UNFCC must take the lead. In 1989, I was Senior Scientific Affairs Officer for climate change and biodiversity at the United Nations Centre for Science and Technology for Development and helped review the first draft version of the UNFCCC. We tried to make sure that the draft treaty was both scientifically sound and complete, so it could serve as a strategy to effective action to halt climate change and degradation. As an internal UN technical expert, I helped ensure inclusion of the essential elements in the first draft: (1) complete global accounting of *all* GHG sources and also of *all* of their sinks; (2) a formal stated goal to protect Earth's most high-temperature-sensitive ecosystems, especially the most fragile and threatened of all, coral reefs; (3) monitoring of the most temperature-sensitive ecosystems to determine if they were being pushed beyond their limits; and (4) trigger mechanisms to reduce GHG emissions if such harm was shown.

Every one of these crucial elements of an effective and sound treaty in the initial drafts was eliminated or watered down in the political horse trading of governments interested in minimizing their moral responsibilities and financial liabilities to the planet as a whole and to future generations in particular. These failures, selective and incomplete accounting, confusion of gross and net fluxes, failure to include the 70% of the planet covered by oceans, lack of a clear goal to protect climatically sensitive ecosystems, failure to monitor ecosystem performance and health, and lack of corrective mechanisms, turned the treaty into being incapable of solving the problem. Had their accountant or finance ministers done such an incompetent job with their money, they would be immediately fired or shot, depending on their country.

A couple of typical examples of the tricky and scientifically unsound way the treaty was gutted are important, because few people understand how governments of the world acted in these crucial negotiations:

1. In the very last moments of the last negotiating session before approval of the draft treaty at the UN General Assembly in 1989 (actually, after those moments, because they were mandated to finish Friday night at midnight, and when it was clear that they were not going to make it, having wasted so much time with meaningless posturing and ritual phrases, they stopped the clock at few minutes before midnight, and it was actually around 2 AM the next day when they finally passed the resolution), the US delegation suddenly called for all countries to get credit for photosynthesis on their land, knowing that photosynthesis absorbs much more CO_2 than fossil fuel emissions, but completely ignoring respiration and decomposition that returns essentially all of it back to the atmosphere! The result of this dishonest accounting would be that almost all countries except totally barren deserts could falsely claim to be CO_2 sinks entitled to credits.
2. A movement by Indonesia and other coral reef countries tried to claim that coral reefs are a sink of atmospheric CO_2, when in fact they are a small source, only about 1% as much as fossil fuels. The use of electrolysis to grow coral reefs (Goreau, 2012a,b; 2014) or precipitate limestone from the ocean (Rau, 2011) has been coupled to mistaken claims that limestone is CO_2 sink.

> Although limestone deposition is a carbon sink from the ocean, in fact limestone deposition releases an equal amount of CO_2 to the atmosphere to maintain charge and pH balance, so it is actually a source of atmospheric CO_2 and not an atmosphere CO_2 sink at all (Stumm and Morgan, 1996). To place this into perspective, about half of the limestone burial in the sea took place in coral reefs (Milliman, 1974, 1993), before global warming killed most corals, and at that time, the global source of CO_2 from coral reefs was only about 1% as much as we release every year from fossil fuel combustion (Ware et al., 1991), showing how human activity has overwhelmed the natural carbon balance.

Instead of effective proposals to solve the problem, governments have refused to think globally and are simply protecting the interests of their politicians and energy industry. Even in the poorest countries, the energy industry pays politicians' election campaigns, pocket money, and slush funds and writes national energy policy. The result is a blizzard of smoke screen proposals that do little or nothing to solve the problem but can be claimed to be a symbolic step in the right direction (sometimes falsely) for PR purposes. These slick proposals are developed by high-paid technical consultants and cunningly misnamed to make it appear that they do the opposite of their likely effects. They amount to stratagems to continue burning fossil fuels as if there were no tomorrow, while tossing a bit of money to people in poor countries so that they do not also destroy their forests the way the rich countries did, in effect forgoing the same *development* strategy of destroying and capitalizing their natural resources. This is widely viewed by developing countries as a trick to keep them poor.

Examples are the current *carbon trading markets* and the so-called clean development mechanism (CDM) by which rich countries pay poor countries to absorb their carbon pollution. CDM is not clean because it does not decrease gross pollution and is most unlikely to result in real development of poor countries, but it can be claimed to be a mechanism. Another is reducing emissions from deforestation and forest degradation (REDD), in which rich countries pay poor countries not to cut their trees down. Since much of the forests are in the lands of indigenous peoples, this has directly motivated governments to act to seize control of forests from their native inhabitants, so governments can pocket the carbon credits that would result. The indigenous peoples of Panama, where most of the forest is on indigenous lands, have rejected REDD (http://www.redd-monitor.org/2012/08/30/coonapip-panamas-indigenous-peoples-coordinating-body-denounces-un-redd/) as a trick that will not prevent pollution, and that will be used by their own national government to steal their lands and REDD carbon credits. Many other indigenous forest people, in South America, Africa, and Asia, for example, the native forest people of Malaysian Borneo and Indonesian Kalimantan, feel the same way.

The worst of these proposals, CCS, envisions pumping CO_2 from power plants into the deep sea or the ground, in the hope that it never comes back. This is the favorite of the energy industry and of all countries that produce, or consume, fossil fuels, because it means no change in business as usual. At every UN climate change meeting, the oil-producing and coal-producing countries demand that more money be spent on CCS, even though CCS has no hope of being affordable or adequate to solve the problem even if it were technically feasible.

At present, UNFCCC is incapable as serving as an effective global problem-solving strategy because it relies on scientifically unsound accounting, goals, and models. It needs to be reformed with honest and complete accounting of all GHGs; meaningful goals to stabilize climate at safe levels that prevent extinction of our most productive, diverse, and temperature-sensitive ecosystems; and a mechanism that allows CO_2 emitters to directly pay for their CO_2 absorption, focused on verifiable and real removal and storage, without the funds being diverted by unproductive middlemen who do not solve the problem. UNFCCC's flaws can easily be remedied and turned into an effective problem-solving tool if (1) soil is recognized as a carbon sink; (2) there is complete accounting of *all* GHG sources and sinks; (3) the goal of the convention is stated to protect earth's most climatically sensitive ecosystems; (4) international and funding agencies fund large-scale increases in soil carbon, in all forms, by all methods, in all habitats; and (5) restoration of degraded habitats to regulate climate becomes a key strategy in reversing global warming.

Conclusions

Atmospheric CO_2, sea level, temperature, water, soil, agriculture, forestry, and fisheries cannot be stabilized at safe levels without restoring the integrated biogeochemical functioning of global ecosystems that we have degraded and destroyed in many places. Soil is the only place where carbon can be quickly and affordably stored, with many additional benefits. The fundamental solution is simple: manage the carbon cycle to store carbon in soil (Figure 2.24) instead of using it up (Figure 2.25a through c).

In Figure 2.25a, the carbon is in balance in that photosynthesis carbon is returned directly to the atmosphere by aboveground respiration and by underground respiration and soil decomposition of litter fall. There is no net storage of carbon in soil once the ecosystem is mature. In Figure 2.25b, photosynthesis is low because of soil exhaustion (unless chemical fertilizers are added, which acts to increase decomposition of soil carbon even faster), and respiration is also lower (although likely with a higher respiration/photosynthesis ratio and lower carbon storage efficiency than (Figure 2.25a) caused by higher heat stress). Decomposition is greatly reduced because litter fall of crops, biomass, wood, fuels, and meat of grazing animals are removed from the ecosystem, resulting in increased emissions of CO_2 where they are consumed, reducing soil carbon cycling at the site and greatly increasing decomposition of SOC, which declines rapidly at first (Nye and Greenland, 1965). In Figure 2.25c, the system is managed to maximize soil fertility using biochar, rock dust, and compost increasing photosynthesis, respiration, and litter fall carbon to the soil. SOC increases, and the addition of biochar made from local weeds stabilize it in the soil.

All biomes, ecosystems, and countries can and should, indeed must, be involved, marine as well as terrestrial (Goreau and Trench, 2012; Goreau et al., 2009; Goreau 2010). Millions of individual soil carbon restorationists, working on their own in every ecosystem and every part of the world have taken barren severely degraded sites and restored them to stunning productivity, enriching biomass, soils, and water supplies by paying careful attention to retaining and recycling essential nutrients. Soil can store excess atmospheric CO_2 in a time scale of decades and avert damage that would otherwise continue for thousands of years. The ecosystem service benefits of increased biodiversity, productivity, food, water, and soil resources additional to climate stabilization far outweigh the costs of reversing runaway climate change and are dwarfed by the costs we will have to pay if we do not. If we seriously break our critical biogeochemical life-support systems, we may not be able to repair them (Figure 2.26).

FIGURE 2.24 Brazilian archaeologists, led by Dr. Eduardo Goes Neves (at left), look down on a freshly excavated pit in Amazonian Terra Preta black earth soils. The soils contain high levels of biochar, fish bones, and broken pottery, which sticks out of the sides. (Photograph by James Richardson.)

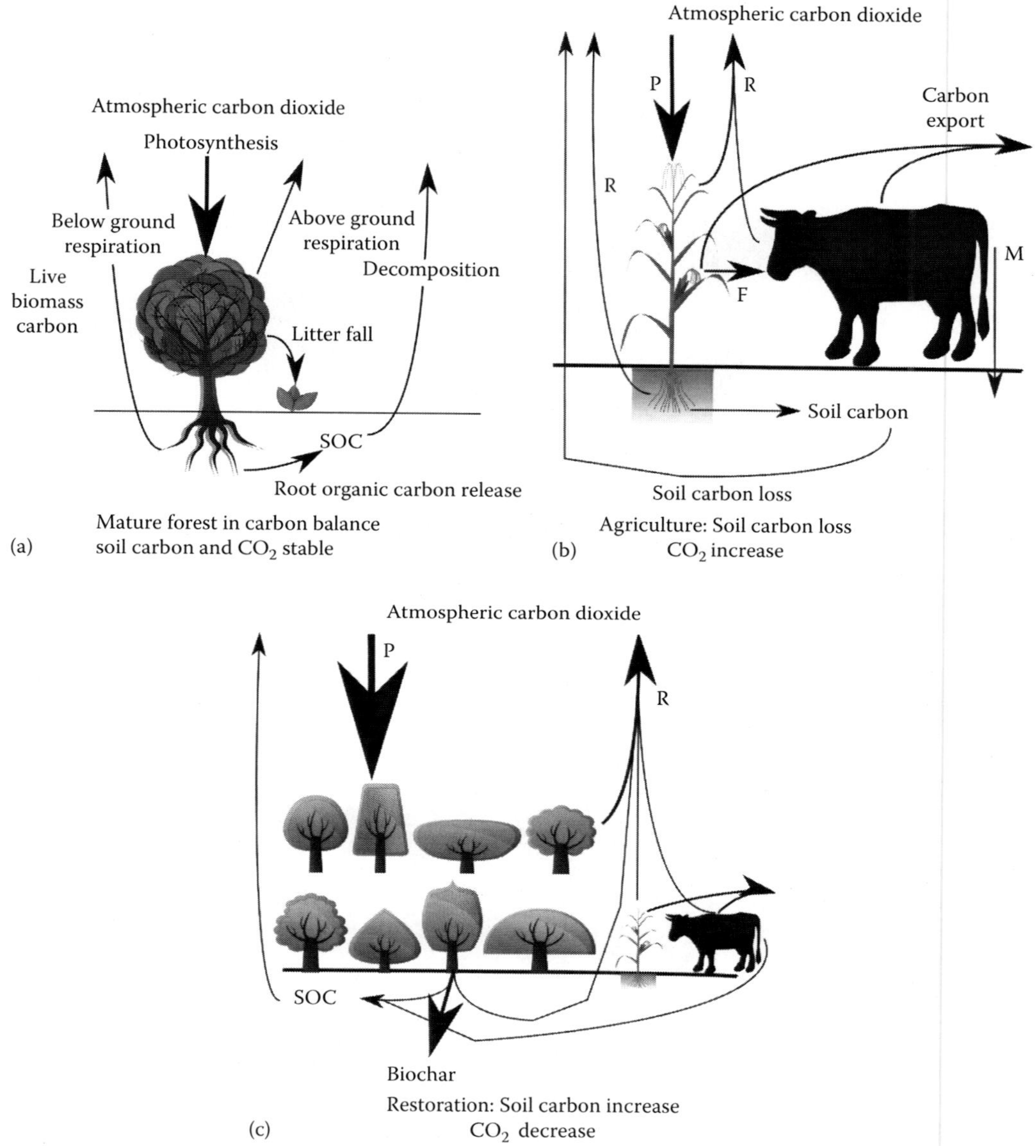

FIGURE 2.25 The carbon cycle in (a) undisturbed forest, (b) conventional agriculture and pasture, and (c) restored ecosystems managed to store carbon.

The tools to solve the problem are already at our disposal, as described in this book, if we choose to use them. All countries should use them including the developed countries that have historically produced most of the CO_2 excess. At the same time, it is crucial that the world's largest CO_2 polluter changes toward a carbon neutral or negative path. This is indeed possible if China steps aside from following the old dirty *brown* Western path of development and switches to leading a *green* low-carbon energy revolution. There are hopeful signs in this regard (Liu et al., 2013), and it is urgent that they be rapidly implemented and applied to the other developing countries so that sustainable development can be achieved without runaway climate change.

FIGURE 2.26 Prevention is cheaper than cure. "All the king's horses and all the king's men couldn't put Humpty Dumpty together again." (From Harvard Lampoon, 1968.)

REFERENCES

Alexander, M., 1977, *Introduction to Soil Microbiology*, 2nd edn., Krieger, Malabar, FL.

Aniwaer, A. and T. Goreau, 2009, Charcoal to reverse global warming and increase soil fertility: Costs and benefits, 5pp., in T. J. Goreau and S. T. Nielsen (Eds.), *The Green Disc: New Technologies for a New Future*, Gibby Media Group, Spokane, WA.

Antonov, J. L., S. Levitus, and T. P. Boyer, 2005, Thermosteric sea level rise, 1955–2003, *Geophysical Research Letters*, 32:L12602, doi:10.1029/2005GL023112.

Averill, C. and A. Finzi, 2011, Plant regulation of microbial enzyme production *in situ*, *Soil Biology and Biochemistry*, 43:2457–2460.

Averill, C., B. L. Turner, and A. Finzi, 2014, Mycorrhiza-mediated competition between plants and decomposers drive soil carbon storage, *Nature*, 505:543–545.

Bader, M. K. F., S. Leuzinger, S. G. Keel, R. T. W. Siegwolf, F. Hagedorn, P. Schleppi, and C. Korner, 2013, Central European hardwood trees in a high CO_2 future: Synthesis of an 8-year forest canopy CO_2 enrichment project, *Journal of Ecology*, 101:1509–1519, doi:10.1111/1365–2745.12149.

Ballantyne, A. P., D. R. Greenwood, J. S. Sinninghe-Damste, A. Z. Csank, J. J. Eberle, and N. Rybczynski, 2010, Significantly warmer Arctic temperatures during the Pliocene indicated by multiple independent proxies, *Geology*, 38:603–606.

Balmaseda, M. A., K. E. Trenberth, and E. Kallen, 2013, Distinctive climate signals in reanalysis of global ocean heat content, *Geophysical Research Letters*, 40:1754–1759, doi:10.1002/grl.50382.

Bandeira, C. M., 1994, *Parque Nacional da Tijuca*, Makron Books do Brasil, Rio de Janeiro, Brazil.

Bates, A., 2010, *The Biochar Solution: Carbon Farming and Climate Change*, New Society Publishers, Gabriola Island, British Columbia, Canada.

Beerling, D. J., 2007, *The Emerald Planet: How Plants Changed Earth's History*, Oxford University Press, Oxford, U.K.

Beerling, D. J. and R. A. Berner, 2005, Feedbacks and the coevolution of plants and atmospheric CO_2, *Proceedings of the National Academic of Sciences*, 102:1302–1305.

Berner, R. A., 2003, The long-term carbon cycle, fossil fuels, and atmospheric composition, *Nature*, 426:323–326.

Berner, R. A., 2004, *The Phanerozoic Carbon Cycle: CO_2 and O_2*, 150pp., Oxford University Press, Oxford, U.K.

Berner, R. A., 2006, Geocarbsulf: A combined model for Phanerozoic atmospheric O_2 and CO_2, *Geochimica et Cosmochimica Acta*, 70:5653–5664.

Betts, J. N. and H. D. Holland, 1991, The oxygen content of ocean bottom waters, the burial efficiency of organic carbon, and the regulation of atmospheric oxygen, *Paleogeography, Paleoclimatology, and Paleoecology*, 97:5–18.

Boening, C., J. K. Willis, F. W. Landerer, R. S. Nerem, and J. Fasullo, 2012, The 2011 La Niña: So strong, the oceans fell, *Geophysical Research Letters*, 39:L19602, doi:10.1029/2012GL053055.

Bohn, H. L., 1978, On organic soil carbon and CO_2, *Tellus*, 30:472–475.

Bond, T. C., S. J. Doherty, D. W. Fahey, P. M. Forster, T. Berntsen, B. J. DeAngelo, M. G. Flanner et al., 2013, Bounding the role of black carbon in the climate system: A scientific assessment, American Geophysical Union, *Journal of Geophysical Research: Atmospheres*, 118:5380–5552, doi:10.1002/jgrd.50171.

Boone, R. D., K. J. Nadelhoffer, J. D. Canary, and J. P. Kaye, 1998, Roots exert a strong influence on the temperature sensitivity of soil respiration, *Nature*, 396:570–572.

Boyce, D. G., M. R. Lewis, and B. Worm, 2010, Global phytoplankton decline over the past century, *Nature*, 466:591–596.

Boyer, T. P., S. Levitus, J. I. Antonov, R. A. Locarnini, and H. E. Garcia, 2005, Linear trends in salinity for the world ocean, 1955–1998, *Geophysical Research Letters*, 32:L01604, doi:10.1029/2004GL021791.

Brady, N. C., 1990, *The Nature and Property of Soils*, 10th edn., Macmillan, New York.

Brigham-Grette, J., M. Melles, P. Minyuk, A. Andreev, P. Tarasov, R. DeConto, S. Koenig et al., 2013, Pliocene warmth, polar amplification, and stepped Pleistocene cooling recorded in NE Arctic Russia, *Science*, 340:1421–1427.

Carbon Tracker, 2013, *Unburnable Carbon 2013: Wasted Capital and Stranded Assets*, 39pp., www.carbontracker.org. Accessed August 23, 2014.

Carter, V. G. and T. Dale, 1974, *Topsoil and Civilization*, University of Oklahoma Press, Norman, OK.

Chou, C., C. W. Lan, C. H. Chung, Y. C. Liao, and C. J. Lee, 2013, Increase in the range between wet and dry season precipitation, *Nature Geoscience*, 6:263–267, doi:10.1038/NGEO1744.

Clemmensen, K. E., A. Bahr, O. Ovaskainen, A. Dahlberg, A. Ekblad, H. Wallander, J. Stenlid, R. D. Finlay, D. A. Wardle, and B. D. Lindahl, 2013, Roots and associated fungi drive long-term carbon sequestration in boreal forest, *Science*, 339:1615–1618.

Daily, G. C. (Ed.), 1997, *Nature's Services: Societal Dependence on Natural Ecosystems*, Island Press, Washington, DC.

Darwin, C., 1881, *The Formation of Vegetable Mould through the Action of Worms with Observations on their Habits*, John Murray, London, U.K.

Dasgupta, P., 2002, *Human Well-Being and the Natural Environment*, Oxford University Press, Oxford, U.K.

Dasgupta, P. and P. R. Ehrlich, 2013, Pervasive externalities at the population, consumption, and environment nexus, *Science*, 340:324–328.

de Georges, A., T. J. Goreau, and B. Reilly, 2010, Land-sourced pollution with an emphasis on domestic sewage: Lessons from the Caribbean and implications for coastal development on Indian Ocean and Pacific Coral Reefs, *Sustainability*, 2:2919–2949.

de Mello, W. Z. and T. J. Goreau, 1998, Fluxos de oxido nitroso (N_2O) e dioxido de carbono (CO_2) em solos de remanescentes da floresta atlantica no Sudeste do Brasil, *Geochimica Brasiliensis*, 12:17–27.

Diamond, J., 2005, *Collapse: How Societies Choose to Fail or Survive*, Penguin, London, U.K.

Dittmar, T., C. E. de Rezende, M. Manecki, J. Niggemann, A. R. C. Ovalle, A. Stubbins, and M. C. Bernardes, 2012, Continuous flux of black carbon from a vanished tropical biome, *Nature Geoscience*, 5:618–622.

Durack, P. J., S. E. Wijffels, and R. J. Matear, 2012, Ocean salinities reveal strong global water cycle intensification during 1950–2000, *Science*, 336:455–458.

Dutton, A. and K. Lambeck, 2012, Ice volume and sea level during the last interglacial, *Science*, 337:216–219.

Ehrlich, P. E. and A. H. Ehrlich, 2013, Can a collapse of global civilization be avoided? *Proceedings of the Royal Society London B*, 280:20122845.

Ehrlich, P. E., P. M. Kareiva, and G. Daily, 2012, Securing natural capital and expanding equity to rescale civilization, *Nature*, 486:68–73.

Emanuel, K. A., 1987, The dependence of hurricane intensity on climate, *Nature*, 326:483–485.

England, M. H., S. McGregor, P. Spence, G. S. A. Meehl, A. Timmermann, W. Cai, A. Sen Gupta, M. J. McPhaden, A. Purich, and A. Santoso, 2014, Recent intensification of wind-driven circulation in the Pacific and the ongoing warming hiatus, *Nature Climate Change*, 4:222–227, doi:10.1038/NCLIMATE2106.

Fagan, B., 2013, *The Attacking Ocean: The Past, Present, and Future of Rising Sea Levels*, Bloomsbury, London, U.K.

Fang, C. and J. B. Moncrieff, 2001, The dependence of soil CO_2 efflux on temperature, *Soil Biology and Biochemistry*, 33:155–165.

Fasullo, J. T., C. Boening, F. Landerer, and R. S. Nerem, 2013, Australia's unique influence on global sea level in 2010–2011, *Geophysical Research Letters*, 40:4368–4373, doi:10:1002/grl.50834.

Flannery, T., 2005, *The Weather Makers: The History and Future Impact of Climate Change*, Text Publishing Company, Melbourne, Victoria, Australia.

Fourqurean, J. W., C. M. Duarte, H. Kennedy, N. Marba, M. Holmer, M. A. Mateo, E. T. Apostolaki et al., 2012, Seagrass ecosystems as a global significant carbon stock, *Nature Geoscience*, 5:505–509.

Galaasen, E. V., U. S. Ninnemann, N. Irvali, H. F. Kleiven, Y. Rosenthal, G. Kissel, and D. A. Hodell, 2014, Rapid reductions in North Atlantic Deep Water during the peak of the last Interglacial period, *Science*, 343:1129–1132.

Gerlach, T. M., 1991, Present day co_2 emissions from volcanoes, *EOS*, 72:249–255.

Gibling, M. R. and N. S. Davies, 2012, Paleozoic landscapes shaped by plant evolution, *Nature Geoscience*, 5:99–105, doi:10.1038/NGEO1376.

Glaser, B., E. Balashov, L. Haumaier, G. Guggenberger, and W. Zech, 2000, Black carbon in density fractions of anthropogenic soils of the Brazilian Amazonian region, *Organic Geochemistry*, 31:669–678.

Glaser, B., J. Lehmann, and B. Zech, 2002, Ameliorating physical and chemical properties of highly weathered soils in the tropics with charcoal: A review, *Biology and Fertility of Soils*, 35:219–230.

Goldberg, E. D., 1985, *Black Carbon in the Environment*, Wiley, New York.

Goreau, T. J., 1965, Stratigraphy of Jamaican fossil coral reefs, unpublished field study.

Goody, R. A. and Y. L. Yung, 1989, *Atmospheric Radiation: Theoretical Basis*, 520pp., Oxford University Press, Oxford, U.K.

Goreau, T. J., 1982, *Biogeochemistry of Nitrous Oxide*, PhD thesis, Harvard University, Cambridge, MA.

Goreau, T. J., 1987, The other half of the global carbon dioxide problem, *Nature*, 328:581–582.

Goreau, T. J., 1990, Balancing atmospheric carbon dioxide, *Ambio*, 19:230–236.

Goreau, T. J., 1992, Control of atmospheric carbon dioxide, *Natural Resources Forum*, 2:5–11.

Goreau, T. J., 1995, Tropical ecophysiology, climate change, and the global carbon cycle, pp. 65–79, in J. Pernetta, R. Leemans, D. Elder, and S. Humphrey (Eds.), *Impacts of Climate Change on Ecosystems and Species: Environmental Context*, International Union for the Conservation of Nature, Gland, Switzerland.

Goreau, T. J., 2010, Coral reef and fisheries habitat restoration in the Coral Triangle: The key to sustainable reef management, Invited Keynote Talk, World Ocean Congress, Manado, Sulawesi, Indonesia, in J. Jompa, R. Basuki, S. M. Tesoro, and E. T. Lestari (Eds.), *Proceedings of the COREMAP Symposium on Coral Reef Management in the Coral Triangle*, pp. 244–253.

Goreau, T. J., 2012a, Marine electrolysis for building materials and environmental restoration, pp. 273–290, in J. Kleperis and V. Linkov (Eds.), *Electrolysis*, InTech Publishing, Rijeka, Croatia.

Goreau, T. J., 2012b, Marine ecosystem electrotherapy: Practice and theory, in T. J. Goreau and R. K. Trench (Eds.), *Innovative Technologies for Marine Ecosystem Restoration*, CRC Press, Boca Raton, FL.

Goreau, T. J., S. Alexander, and A. Kornevall, 2009, Global restoration of planetary support systems and increasing CO_2 sinks, 4pp., in T. J. Goreau and S. T. Nielsen (Eds.), *The Green Disc: New Technologies for a New Future*, Gibby Media Group, Spokane, WA.

Goreau, T. J. and W. Z. de Mello, 1987, Effects of deforestation on sources and sinks of atmospheric carbon dioxide, nitrous oxide, and methane from Central Amazonian soils and biota during the dry season: A preliminary study, in D. Athie, T. E. Lovejoy, and P. de M. Oyens (Eds.), *Proceedings of the Workshop on Biogeochemistry of Tropical Rain Forests: Problems for Research*, Centro de Energia Nuclear na Agricultura and World Wildlife Fund, Piricicaba, Sao Paulo, Brazil, pp. 51–66.

Goreau, T. J. and W. Z. de Mello, 1988, Tropical deforestation: Some effects on atmospheric chemistry, *Ambio*, 17:275–281.

Goreau, T. J. and W. Z. de Mello, 2002, Nitrous oxide, methane, and carbon dioxide fluxes from South Florida habitats during the transition from dry to wet seasons: Potential impacts of Everglades drainage and reflooding on the atmosphere, pp. 281–306, in K. G. Porter and J. W. Porter (Eds.), *The Everglades, Florida Bay, and the Coral Reefs of the Florida Keys: An Ecosystem Sourcebook*, CRC Press, Boca Raton, FL.

Goreau, T. J. and W. M. de Mello, 2007, Minimizing net greenhouse gas emissions from mangrove and wetland soils, pp. 239–248, in Y. Tateda et al. (Eds.), *Greenhouse Gas and Carbon Balances in Mangrove Coastal Ecosystems*, Gendai Tosho, Tokyo, Japan.

Goreau, T. J. and R. L. Hayes, 1994, Coral bleaching and ocean "hot spots", *Ambio*, 23:176–180.

Goreau, T. J. and R. L. Hayes, 2005, Global coral reef bleaching and sea surface temperature trends from satellite-derived Hotspot analysis, *World Resource Review*, 17:254–293.

Goreau, T. J., R. L. Hayes, J. W. Clark, D. J. Basta, and C. N. Robertson, 1993, Elevated sea surface temperatures correlate with Caribbean coral reef bleaching, pp. 225–255, in R. A. Geyer (Ed.), *A Global Warming Forum: Scientific, Economic, and Legal Overview*, CRC Press, Boca Raton, FL.

Goreau, T. J., R. L. Hayes, and D. McAllister, 2005, Regional patterns of sea surface temperature rise: Implications for global ocean circulation change and the future of coral reefs and fisheries, *World Resource Review*, 17:350–374.

Goreau, T. J., W. Hilbertz, A. A. A. Hakeem, T. Sarkisian, F. Gutzeit, and A. Spenhoff, 2012, Restoring reefs to grow back beaches and protect coasts from erosion and global sea level rise, in T. J. Goreau and R. K. Trench (Eds.), *Innovative Technologies for Marine Ecosystem Restoration*, CRC Press, Boca Raton, FL.

Goreau, T. J. and R. K. Trench (Eds.), 2012, *Innovative Technologies for Marine Ecosystem Restoration*, CRC Press, Boca Raton, FL.

Grant, K. M., E. J. Rohling, M. Bar-Matthews, A. Ayalon, M. Medina-Elizalde, C. B. Ramsey, C. Satow, and A. P. Roberts, 2012, Rapid coupling between ice volume and polar temperature over the past 150,000 years, *Nature*, 491:744–747.

Grasby, S. E., H. Sanei, and B. Beauchamp, 2011, Catastrophic dispersion of coal fly ash into oceans during the latest Permian extinction, *Nature Geoscience*, 4:104–107, doi:10.1038/NGEO1069.

Graven, H. D., R. F. Keeling, S. C. Piper, P. K. Patra, B. B. Stephens, S. C. Wofsy, L. R. Welp et al., 2013, Enhanced seasonal exchange of CO_2 by northern ecosystems since 1960, *Science*, 341:1085–1089.

Greenland, D. J., 1995, Land use and soil carbon in different agroecological zones, pp. 9–24, in R. Lal, J. Kimble, E. Levine, and B. A. Stewart (Eds.), *Soil Management and Greenhouse Effect*, CRC Press, Boca Raton, FL.

Green Disc, 2014, *The Green Disc: New Technologies for a New Future*, 2nd ed., Gibby Media, Spokane WA.

Guemas, V., F. J. Doblas-Reyes, I. Andreu-Burillo, and M. Asif, 2013, Retrospective prediction of the global warming slowdown in the past decade, *Nature Climate Change*, 3:649–653, doi:10.1038/nclimate1863.

Hanebuth, T., K. Stattegger, and P. M. Grootes, 2000, Rapid flooding of the Sunda Shelf: A Late-glacial sea-level record, *Science*, 288:1033–1035.

Hansen, J., R. Ruedy, M. Sato, and K. Lo, 2010, Global surface temperature change, *Reviews of Geophysics*, 48(RG4004):1–29.

Hansen, J., M. Sato, P. Karecha, D. Beerling, R. Berner, V. Masson-Delmotte, M. Pagani, M. Raymo, D. L. Royer, and J. C. Zachos, 2008, Target atmospheric CO_2: Where should humanity aim? *Journal of Open Atmospheric Science*, 2:217–231, http://arxiv.org/pdf/0804.1126.pdf.

Hansen, J, M. Sato, P. Karecha, G. Russell, D. W. Lea, and M. Siddall, 2007, Climate change and trace gases, *Philosophical Transactions of the Royal Society A*, 365:1925–1954.

Hansen, J., M. Sato, G. Russell, and P. Kharecha, 2013, Climate sensitivity, sea level, and atmospheric CO_2, http://arxiv.org/ftp/arxiv/papers/1211/1211.4846.pdf.

Harper, R. J. and M. Tibbett, 2013, The hidden organic carbon in deep mineral soils, *Plant Soil*, 368:641–648.

Hedges, J. I., 1992, Global biogeochemical cycles: Progress and problems, *Marine Chemistry*, 39:67–93.

Hegerl, G. and P. Stott, 2014, From past to future warming, *Science*, 343:844–845.

Held, I. M. and B. J. Soden, 2006, Robust responses of the hydrological cycle to global warming, *Journal of Climate*, 19:5686–5699.

Hillel, D., 1991, *Out of the Earth: Civilization and the Life of the Soil*, University of California Press, Berkeley, CA.

Holland, H. D., 1984, *The Chemical Evolution of the Atmosphere and Oceans*, Princeton University Press, Princeton, NJ.

Houghton, R., G. Jenkins, and J. Ephraums, 1990, *Climate Change: The IPCC Scientific Assessment*, Cambridge University Press, Cambridge, U.K.

Houghton, R. A., 2003, The contemporary carbon cycle, Vol. 8, pp. 473–513, in H. D. Holland and K. K. Turekian (Eds.), *Treatise on Geochemistry.*

House, K. Z., A. C. Baclig, M. Ranjan, E. A. van Nierop, J. Wilcox, and H. J. Herzog, 2011, Economic and energetic analysis of capturing CO_2 from ambient air, *Proceedings of the National Academy of Sciences*, 108:20428–20433, doi:10.073/pnas.1012253108.

Huang, P. M., Y. Li, and M. E. Sumner (Eds.), 2012a, *Handbook of Soil Sciences, Properties and Processes*, 2nd edn., CRC Press, Boca Raton, FL.

Huang, P. M., Y. Li, and M. E. Sumner (Eds.), 2012b, *Handbook of Soil Sciences, Resource Management and Environmental Impacts*, 2nd edn., CRC Press, Boca Raton, FL.

Hyams, E., 1976, *Soil and Civilization*, Harper & Row, New York.

IBDF (Instituto Brasileiro de Desenvolvimento Florestal), 1991, *Plano de Manejo: Parque Nacional da Tijuca*, IBDF/FBPCN, Brasilia, Brazil, http://www.parquedatijuca.com.br/manejo.php. Accessed August 23, 2014.

Jackson, W., 2010, *Consulting the Genius of the Place: An Ecological Approach to a New Agriculture*, Counterpoint, Berkeley, CA.

Jaffe, R., Y. Ding, J. Niggemann, A. V. Vahatalo, A. Stubbins, R. G. M. Spencer, J. Campbell, and T. Dittmar, 2013, Global charcoal mobilization from soils via dissolution and riverine transport to the oceans, *Science*, 340:345–347.

Jasechko, S., Z. D. Sharp, J. J. Gibson, S. J. Birks, Y. Yi, and P. J. Fawcett, 2013, Terrestrial water fluxes dominated by transpiration, *Nature*, 496:347–350.

Jenkinson, D., D. Adams, and A. Wild, 1991, Model estimates of CO_2 emissions from soil in response to global warming, *Nature*, 351:304–306.

Jobbagy, E. G. and R. B. Jackson, 2000, The vertical distribution of soil organic carbon and its relation to climate and vegetation, *Ecological Applications*, 10:423–436.

Jordan, C. F., 1983, Productivity of tropical rain forest ecosystems and the implications for their use as future wood and energy sources, pp. 117–135, in F. Golley (Ed.), *Tropical Rain Forest Ecosystems: Structure and Function*, Elsevier, Amsterdam, the Netherlands.

Jung, M., M. Reichstein, P. Ciais, S. I. Seneviratne, J. Sheffield, M. L. Goulden, G. Bonnan et al., 2010, Recent decline in the global land evapotranspiration trend due to limited moisture supply, *Nature*, 467:951–954.

Kang, M. S. and S. S. Banga (Eds.), 2013, *Combating Climate Change: An Agricultural Perspective*, CRC Press, Boca Raton, FL.

Karl, T. R., G. Kukla, V. N. Razuvayev, M. J. Changery, R. G. Quayle, R. R. Heim, D. R. Easterling, and C. B. Fu, 1991, Global warming: Evidence for asymmetric diurnal temperature change, *Geophysical Research Letters*, 18:2253–2256.

Kershaw, A. P., J. S. Clark, A. M. Gill, and D. M. DaCosta, 2002, A history of fire in Australia, pp. 3–25, in R. A. Bradstock, J. E. Williams, and A. M. Gill (Eds.), *Flammable Australia in the Fire Regimes and Biodiversity of a Continent*, Cambridge University Press, Cambridge, MA.

Kiessling, W., C. Simpson, B. Beck, H. Mewis, and J. M. Pandolfi, 2012, Equatorial decline of coral reefs during the last Pleistocene interglacial, *Proceedings of the National Academy of Sciences*, 109:21378–21383, doi:101073/pnas.1214037110.

Kimble, J. M., C. W. Rice, D. Reed, S. Mooney, R. F. Follett, and R. Lal (Eds.), 2007, *Soil Carbon Management: Economic, Environmental, and Social Benefits*, CRC Press, Boca Raton, FL.

Kirschbaum, M. U. F., 1995, The temperature dependence of soil organic matter decomposition, and the effect of global warming on soil organic C storage, *Soil Biology and Biochemistry*, 27:753–760.

Kirschbaum, M. U. F., 2000, Will changes in soil organic carbon act as a positive or negative feedback on global warming? *Biogeochemistry*, 48:21–51.

Kirschbaum, M. U. F., 2006, The temperature dependence of organic-matter decomposition—Still a topic of debate, *Soil Biology and Biochemistry*, 38:2510–2518.

Kopp, R. E., F. J. Simons, J. X. Mitrovica, A. C. Maloof, and M. Oppenheimer, 2009, Probabilistic assessment of sea level during the last interglacial stage, *Nature*, 462:863–867.

Kosaka, Y. and S. P. Xie, 2013, Recent global-warming hiatus tied to Equatorial Pacific surface cooling, *Nature*, 501:403–407, doi:10.1038/nature12534.

Kruge, M. A., B. A. Stankiewicz, J. C. Crelling, A. Montanari, and D. F. Bensley, 1994, Fossil charcoal in Cretaceous-Tertiary boundary strata: Evidence for catastrophic firestorm and megawave, *Geochimica et Cosmochimica Acta*, 58:1393–1397.

Ladygina, N. and F. Rineau (Eds.), 2013, *Biochar and Soil Biota*, CRC Press, Boca Raton, FL.

Lal, R., 2004, Soil carbon sequestration impacts on global climate change and food security, *Science*, 304:1623–1627.

Lal, R., 2011, Sequestering carbon in soils of agro-ecosystems, *Food Policy*, 36:S33–S39.

Lal, R., J. M. Kimble, E. Levine, and B. A. Stewart (Eds.), 1995, *Soil Management and Greenhouse Effect*, CRC Press, Boca Raton, FL.

Lampoon, H., 1968, *LIFE Magazine Parody Issue*, Harvard Lampoon, Cambridge MA.

Larcher, W., 1980, *Physiological Plant Ecology*, Springer Verlag, Berlin, Germany.

Lashof, D. A., 1989, The dynamic greenhouse: Feedback processes that may influence future concentrations of atmospheric trace gases and climate change, *Climatic Change*, 14:213–242.

Lehmann, J. and S. Joseph, 2009, *Biochar for Environmental Management*, Routledge, London, U.K.

Lenton, T. M., 2010, The potential for land-based biological CO_2 removal to lower future atmospheric CO_2 concentration, *Carbon Management*, 1:145–160.

Levitus, S. 1982, *Climatological Atlas of the World Ocean*, 173pp., NOAA Professional Paper 13, NOAA, Washington, DC.

Levitus, S., J. Antonov, T. P. Boyer, and C. Stephens, 2000, Warming of the world ocean, *Science*, 287:2225–2229.

Levitus, S., J. I. Antonov, T. P. Boyer, O. K. Baranova, H. E. Garcia, R. A. Loacarnini, A. V. Mishonov et al., 2012, World ocean heat content and thermosteric sea level change (0–2000 m), 1955–2010, *Geophysical Research Letters*, 39:L10603, doi:10.1029/2102GL051106.

Levitus, S., J. L. Antonov, and T. P. Boyer, 2005, Warming of the world ocean, 1955–2003, *Geophysical Research Letters*, 32:L02604, doi:10.1029/2004GL021592.

Levitus, S., J. L. Anotonov, J. Wang, T. L. Delworth, K. W. Dixon, and A. J. Broccoli, 2001, Anthropogenic warming of Earth's climate system, *Science*, 292:267–270.

Levermann, A., P. U. Clark, B. Marzeion, G. A. Milne, D. Pollard, V. Radic, and A. Robinson, 2013, The multimillenial sea-level commitment of global warming, *Proceedings of the National Academy of Sciences*, 110:13745–13750, www.pnas.org/cgi/doi/10.1073/pnas.1219414110.

Liu, Z., D. Guan, D. Crawford-Brown, Q. Zhang, K. He, and J. Liu, 2013, A low-carbon roadmap for China, *Nature*, 500:143–145.

Lloyd, J. and J. A. Taylor, 1994, On the temperature dependence of soil respiration, *Functional Ecology*, 8:315–323.

Lopes dos Santos, R. A., P. De Deckker, E. C. Hopmans, J. W. Magee, A. Mets, J. S. Sinninghe Damste, and S. Schouten, 2013, Abrupt vegetation change after the late Quaternary megafaunal extinction in south-eastern Australia, *Nature Geoscience*, 6:627–631, doi:10.1038/NGEO1856.

Lovelock, J., 2006, *The Revenge of Gaia: Why the Earth is Fighting Back and How We Can Still Save Humanity*, Penguin, London, U.K.

Lovelock, J. E. and C. G. Rapley, 2007, Ocean pipes could help the earth to cure itself, *Nature*, 449:403.

Luthi, D., M. Le Floch, B. Bereiter, T. Blunier, J.-M. Barnola, U. Seigenthaler, D. Raynaud et al., 2008, High resolution carbon dioxide concentration record 650,000–800,000 years before present, *Nature*, 453:379–382.

Magdof, F. and H. van Es, 2009, *Building Soils for Better Crops: Sustainable Soil Management*, 310pp., Sustainable Agriculture Research and Education Program, U.S. Department of Agriculture, Washington, DC.

Malavolta, E., H. P. Haag, F. A. F. Mello, and M. O. C. Brasil, 1962, *On the Mineral Nutrition of Some Tropical Crops*, International Potash Institute, Bern, Switzerland.

Mann, M. E., Z. Zhang, M. K. Hughes, R. S. Bradley, S. K. Miller, S. Rutherford, and F. Ni, 2008, Proxy-based reconstructions of hemispheric and global surface temperature variations over the past two millennia, *Proceedings of the National Academy of Sciences*, 105:13252–13257.

Mao, J. D., R. L. Johnson, J. Lehmann, D. C. Olk, E. G. Neves, M. L. Thompson, and K. Schmidt-Rohr, 2012, Abundant and stable char residues in soil: Implications for soil fertility and carbon sequestration, *Environmental Science and Technology*, 46:9571–9576.

Marcott, S. A., J. D. Shakun, P. U. Clark, and A. C. Mix, 2013, A reconstruction of regional and global temperature for the past 11,300 years, *Science*, 339:1198–1201.

Maser, C., A. W. Claridge, and J. C. Trappe, 2010, *Trees, Truffles, and Beasts: How Forests Function*, Rutgers University Press, New Brunswick, NJ.

Masiello, C. A. and E. R. M. Druffel, 1998, Black carbon in deep sea sediments, *Science*, 280:1911–1913.

Matthews, H. D. and S. Solomon, 2013, Irreversible does not mean unavoidable, *Science*, 340:438–439.

Meehl, G. A., J. M. Arblaster, J. T. Fasullo, A. Hu, and K. E. Trenberth, 2011, Model based evidence of deep-ocean heat uptake during surface-temperature hiatus periods, *Nature Climate Change*, 1:360–364.

Meir, P., J. Grace, A. Miranda, and J. Lloyd, 1996, Soil respiration in a rainforest in Amazonia and in cerrado in central Brazil, pp. 319–329, in J. H. C. Gash, C. A. Nobre, J. M. Roberts, and R. L. Victoria (Eds.), *Amazonian Deforestation and Climate*, Wiley, Chichester, U.K.

Michel, J. H., 2013, Lost hopes for CCS—Added urgency for renewable energy, Air Pollution and Climate Series 28, 72pp., Air Pollution and Climate Secretariat, Göteburg, Sweden, http://www.airclim.org/sites/default/files/documents/apc-28.pdf.

Montgomery, D. R., 2007, *Dirt: The Erosion of Civilizations*, University of California Press, Berkeley, CA.

Moore, S., C. D. Evans, S. E. Page, M. H. Garnett, T. G. Jones, C. Freeman, A. Hooijer, A. J. Wiltshire, S. H. Limin, and V. Gauci, 2013, Deep instability of deforested tropical peatlands revealed by fluvial organic carbon fluxes, *Nature*, 493:660–663.

Mora, C., A. G. Frazier, R. J. Longman, R. S. Dacks, M. M. Walton, E. J. Tong, J. J. Sanchez et al., 2013b, The projected timing of climate departure from recent variability, *Nature*, 502:183–187.

Mora, C., C. L. Wei, A. Rollo, T. Amaro, A. R. Baco, D. Billett, L. Bopp et al., 2013a, Biotic and human vulnerability to projected changes in ocean biogeochemistry over the 21st century, *PLOS Biology*, 11(10):e1001682.

Milliman, J. D., 1974, *Marine Carbonates*, Springer Verlag, Berlin, Germany.

Milliman, J. D., 1993, Production and accumulation of calcium carbonate in the ocean: Budget of a nonsteady state, *Global Biogeochemical Cycles*, 7:927–957.

Mitra, S., R. Wassmann, and P. L. G. Vlek, 2005, An appraisal of global wetland area and its organic carbon stock, *Current Science*, 88:25–35.

Myers, N. and T. J. Goreau, 1991, Tropical forests and the greenhouse effect: A management response, *Climatic Change*, 19:215–225; reprinted 1992 in N. Myers (Ed.), *Tropical Forests and Climate*, Kluwer Academic Publishers, Dodrecht, the Netherlands.

Nellemann, C., E. Corcoran, C. M. Duarte, L. Valdes, C. DeYoung, L. Fonseca, and G. Grimsditch (Eds.), 2008, *Blue Carbon: The Role of Healthy Oceans in Binding Carbon*, UNEP, GRID-Arendal, Arendal, Norway.

Norris, R. D., S. K. Turner, P. M. Hull, and A. Ridgwell, 2013, Marine ecosystem responses to Cenozoic global change, *Science*, 341:492–498.

Nye, P. H. and D. J. Greenland, 1965, *The Soil under Shifting Cultivation*, Commonwealth Agricultural Bureau, London, U.K.

O'Gorman, P. A., 2012, Sensitivity of tropical precipitation extremes to climate change, *Nature Geoscience*, 5:697–700.

Ohlson, K., 2014, *The Soil Will Save Us: How Scientists, Farmers, and Foodies are Healing the Soil to Save the Planet*, Rodale Press, New York.

O'Leary, M. J., P. J. Hearty, W. G. Thompson, M. E. Raymo, J. X. Mitrovica, and J. M. Webster, 2013, Ice sheet collapse following a prolonged period of stable sea level during the last interglacial, *Nature Geoscience*, 6:796–800.

Otto, A., F. E. L. Otto, O. Boucher, J. Church, G. Hegerl, P. M. Forster, N. P. Gillett et al., 2013, Energy budget constraints on climate response, *Nature Geoscience*, 6:415–416, doi:10.1038/ngeo1836.

Overpeck, J. T., B. L. Otto-Bliesner, G. H. Miller, D. R. Muhs, R. B. Alley, and J. T. Kiehl, 2006, Paleoclimatic evidence for future ice-sheet instability and rapid sea level rise, *Science*, 311:1747–1750.

Pagani, M., K. Caldeira, R. A. Berner, and D. J. Beerling, 2009, The role of terrestrial plants in limiting atmospheric CO_2 decline over the past 24 million years, *Nature*, 460:85–88.

Pagani, M., J. C. Zachos, K. H. Freeman, B. Tipple, and S. Bohaty, 2005, Marked decline in atmospheric carbon dioxide concentrations during the Paleogene, *Science*, 309:600–603.

Pauli, F. W., 1967, *Soil Fertility: A Biodynamical Approach*, Adam Hilger, London, U.K.

Peng, S., S. Piao, P. Ciais, R. B. Myneni, A. Chen, F. Chevallier, A. J. Dolman et al., 2013, Asymmetric effects of daytime and night-time warming on Northern Hemisphere vegetation, *Nature*, 501:88–92.

Perlin, J., 2005, *A Forest Journey: The Story of Wood and Civilization*, The Countryman Press, Woodstock, VT.

Pimentel, D. and M. Pimentel (Eds.), 1996, *Food, Energy, and Society*, 2nd edn., University of Colorado Press, Denver, CO.

Post, W. M., I. C. Izauralde, T. O. West, M. A. Liebig, and A. W. King, 2012, Management opportunities for enhancing terrestrial carbon sinks, *Frontiers in Ecology and Environment*, 10:554–561.

Post, W. M., T.-H. Peng, W. R. Emmanuel, A. W. King, V. H. Dale, and D. L. De Angelis, 1990, The global carbon cycle, *American Scientist*, 78:312–326.

Powell, T. W. R. and T. M. Lenton, 2012, Future carbon dioxide removal *via* biomass energy constrained by agricultural efficiency and dietary trends, *Energy and Environmental Science*, 5:8116–8133, doi:10.1039/c2ee21592f.

Purseglove, J. W., 1968, *Tropical Crops: Dicotyledons*, Vol. 2, Longmans, London, U.K.

Purseglove, J. W., 1972, *Tropical Crops: Monocotyledons*, Longmans, London, U.K.

Rau, G. H., 2008, Electrochemical splitting of calcium carbonate to increase solution alkalinity: Implications for mitigation of carbon dioxide and ocean acidity, *Environmental Science & Technology*, 42:8935–8940.

Rau, G. H., 2009, Electrochemical CO_2 capture and storage with hydrogen generation, *Energy Procedia*, 1:823–828.

Rau, G. H., 2011, CO_2 mitigation via capture and chemical conversion in seawater, *Environmental Science & Technology*, 45:1088–1092.

Rau, G. H., K. G. Knauss, W. H. Langer, and K. Caldeira, 2007, Reducing energy-related CO_2 emissions using accelerated weathering of limestone, *Energy*, 32:1471–1477.

Retallack, G., 2003, Soils and global change in the carbon cycle over geological time, Vol. 5, pp. 581–605, in H. D. Holland and K. K. Turekian (Eds.), *Treatise on Geochemistry*. Elsevier, New York.

Richards, J. F., 2003, *The Unending Frontier: An Environmental History of the Early Modern World*, University of California Press, Berkeley, CA.

Rillig, M. C., S. F. Wright, K. A. Nichols, W. F. Schmidt, and M. S. Torn, 2001, Large contribution of arbuscular mycorrhizal fungi to soil carbon pools in tropical forest soils, *Plant and Soil*, 233:167–177.

Rogelj, J., E. W. Hare, J. Lowe, D. P. van Vuuren, K. Riahi, B. Matthews, T. Hanaoka, K. Jiang, and M. Meinshausen, 2011, Emissions pathways consistent with a 2°C global temperature limit, *Nature Climate Change*, 1:413–418.

Rogelj, J., D. McCollum, A. Reisinger, M. Meinshausen, and K. Riahi, 2013, Probabilistic cost estimates for climate change mitigation, *Nature*, 493:79–83.

Rohling, E. J., K. Grant, M. Bolshaw, A. P. Roberts, M. Siddall, C. Hemleben, and M. Kucera, 2009, Antarctic temperature and global sea level closely coupled over the past five glacial cycles, *Nature Geoscience*, 2:500–504, doi:10.1038/ngeo557.

Rohling, E. J., K. Grant, C. Hemleben, M. Siddall, B. A. A. Hoogakker, M. Bolshaw, and M. Kucera, 2008, High rates of sea level rise during the last Interglacial period, *Nature Geoscience*, 1:38–42, doi:10.1038/ngeo.2007.28.

Romm, J., 2007, *Hell and High Water: The Global Warming Solution*, Harper, New York.

Rosenthal, Y., B. L. Linsley, and D. W. Oppo, 2013, Pacific Ocean heat content during the past 10,000 years, *Science*, 342:617–621.

Royer, D. L., R. A. Berner, I. P. Montañez, N. J. Tabor, and D. J. Beerling, 2004, CO_2 as a primary driver of Phanerozoic climate, *Geological Society of America Today*, 14:4–10.

Royer, D. L., R. A. Berner, and J. Park, 2007, Climate sensitivity constrained by CO_2 over the past 420 million years, *Nature*, 446:530–532.

Ruddiman, W. F., 2005, *Plows, Plagues, and Petroleum: How Humans Took Control of Climate*, Princeton University Press, Princeton, NJ.

Savory, A., 1999, *Holistic Management: A New Framework for Decision Making*, Island Press, Washington, DC.

Scharlemann, J. P. W., E. V. J. Tanner, R. Hiederer, and V. Kapos, 2014, Global soil carbon: Understanding and managing the largest terrestrial carbon pool, *Carbon Management*, 5:81–91.

Schlesinger, W. H., 1990, Evidence from chronosequence studies for a low carbon storage potential of soils, *Nature*, 348:232–234.

Schlesinger, W. H. and J. A. Andrews, 2000, Soil respiration and the global carbon cycle, *Biogeochemistry*, 48:7–20.

Schmidt, M. W. I., M. S. Tom, S. Abiven, T. Dittmar, G. Guggenberger, I. A. Janssens, M. Kleber et al., 2011, Persistence of soil organic matter as an emergent property, *Nature*, 478:49–56.

Scott, A. C., 2009, Forest fire in the fossil record, pp. 1–37, in A. Cerda and P. R. Robichaud (Eds.), *Fire Effects on Soils and Restoration Strategies*, Science Publishers, Enfield, NH.

Scott, A. C. and I. J. Glasspool, 2006, The diversification of Paleozoic fire systems and fluctuations in atmospheric oxygen concentration, *Proceedings of the National Academy of Sciences*, 103:10861–10865.

Scott, V., S. Gilfillan, N. Markusson, H. Chalmers, and R. S. Haszeldine, 2012, Last chance for carbon capture and storage, *Nature Climate Change*, 3:105–111, doi:10.1038/NCLIMATE1695.

Schwartz, J., 2013, *Cows Save the Planet: And Other Improbable Ways of Restoring Soil to Heal the Earth*, Chelsea Green, White River Junction, VT.

Seneviratne, S. I., M. G. Donat, B. Mueller, and L. V. Alexander, 2014, No pause in the increase of hot temperature extremes, *Nature Climate Change*, 4:161–163.

Shaffer, G., S. M. Olsen, and J. O. P. Pedersen, 2009, Long-term ocean oxygen depletion in response to carbon dioxide emissions from fossil fuels, *Nature Geoscience*, 2:105, doi:10.1038/NGEO420.

Shahkova, N., I. Semiletov, A. Salyuk, V. Yusupov, D. Kosmach, and O. Gustafsson, 2010, Extensive methane venting to the atmosphere from sediments of the East Siberian Arctic shelf, *Science*, 327:1246–1250.

Shangguan, W., Y. Dai, Q. Duan, B. Liu, and H. Yuan, 2014, A global soil data set for earth system modeling, *Journal of Advances in Modeling Earth Systems*, 6:249–263, doi:10.1002/2013MS000293.

Shoemaker, J. K., D. P. Schrag, M. J. Molina, and V. Ramanathan, 2013, What role for short-lived climate pollutants in mitigation policy? *Science*, 342:1323–1324.

Skinner, L., 2012, A long view on climate sensitivity, *Science*, 337:917–919.

Smil, V., 2001, *Cycles of Life: Civilization and the Biosphere*, Scientific American Library, New York.

Smith, S. V. and J. T. Hollibaugh, 1993, Coastal metabolism and the ocean organic carbon balance, *Reviews of Geophysics*, 31:75–89.

Soden, B. J. and I. M. Held, 2006, An assessment of climate feedbacks in coupled ocean-atmosphere models, *Journal of Climate*, 19:3354–3360.

Soil Carbon Sequestration for Climate, Food, and Ecosystem Services Conference, 2013, Book of Abstracts, 186pp., Reykjavik, Iceland.

Solomon, A., D. Trabalka, D. Reichle, and L. Voorhees, 1985, The global cycle of carbon, pp. 3–24, in D. Trabalka (Ed.), *Atmospheric Carbon Dioxide and the Global Carbon Cycle*, U.S. Department of Energy, Washington, DC.

Sombroek, W., 1966, *Amazonian Soils*, Centre for Agricultural Publications and Documentation, Wagenigen, the Netherlands.

Sombroek, W., 1984, Soils of the Amazonian Region, pp. 521–535, in H. Sioli (Ed.), *The Amazon*, Springer Verlag, Amsterdam, the Netherlands.

Sombroek, W., M. de L. Ruivo, P. M. Fearnside, B. Glaser, and J. Lehmann, 2003, Amazonian dark earths as carbon stores and sinks, pp. 125–139, in J. Lehmann et al. (Eds.), *Amazonian Dark Earths: Origin, Properties, Management*, Kluwer, Dordrecht, the Netherlands.

Stamets, P., 2005, *Mycelium Running: How Mushrooms Can Help Save the World*, Ten Speed Press, Berkeley, CA.

Steiner, C., 2006, *Slash and Char: An Alternative to Slash and Burn*, Cuvillier Verlag, Gottingen, Germany.

Steinfeld, H., P. Gerber, T. Wassenaar, V. Castel, M. Rosales, and C. de Haan, 2006, *Livestock's Long Shadow: Environmental Issues and Options*, FAO, Rome, Italy.

Stephens, J. C. and W. H. Speir, 1970, Subsidence of organic soils in the USA, *International Association of Scientific Hydrological Symposium*, Tokyo, Japan, Publication No. 89, pp. 523–534.

Stephens, J. L., 1843, *Incidents of Travel in Yucatan*, John Murray, London, U.K.

Stephenson, N. L., A. J. Das, R. Condit, S. E. Russo, P. J. Baker, N. G. Beckman, D. A. Coomes et al., 2014, Rate of tree carbon accumulation increases continuously with tree size, *Nature*, 507:90–93.

Stern, N., 2006, *The Economics of Climate Change*, Cambridge University Press, Cambridge, U.K.

Stewart, C. E., J. Zheng, J. Botte, and M. F. Cotrufo, 2013, Cogenerated fast pyrolysis biochar mitigates greenhouse gas emissions and increases carbon sequestration in temperate soils, *Global Change Biology Bioenergy*, 5:153–164.

Stiglitz, J. E., 2006, A new agenda for global warming, *Economists Voice*, Berkeley Electronic Press, Berkeley, CA, http://web.uvic.ca/~egugl/indexecon325_files/pigou-global_warming.pdf.

Stocker, B. D., R. Roth, F. Joos, R. Spahni, M. Steinacher, S. Zaehle, L. Bouwman, Xu-Ri, and I. C. Prentice, 2013, Multiple greenhouse gas feedbacks from the land biosphere under future climate change scenarios, *Nature Climate Change*, 3:666–672.

Stocker, T. W., 2013, The closing door of climate targets, *Science*, 339:280–282.

Stockmann, U., 2011, *Managing the Soil-Plant System to Mitigate Atmospheric CO_2: Discussion Paper for the Soil Carbon Sequestration Summit*, 55pp., University of Sydney, Sydney, New South Wales, Australia.

Stubbins, A., J. Niggemann, and T. Dittmar, 2012, Photo-lability of deep sea organic black carbon, *Biogeosciences*, 9:1661–1670.

Stumm, W. and J. J. Morgan, 1996, *Aquatic Chemistry: Chemical Equilibria and Rates in Natural Waters*, 3rd edn., Wiley Interscience, New York.

Suman, D., T. Kuhlbusch, and B. Lim, 1997, Marine sediments: A reservoir for black carbon and their use as spatial and temporal records of combustion, Vol. 51, pp. 271–293, in J. Clark et al. (Eds.), *Sediment Records of Biomass Burning and Global Change*, NATO ASI Series, Series I, Springer, Berlin, Germany.

Sundquist, E. and K. Visser, 2003, The geologic history of the carbon cycle, Vol. 8, pp. 425–472, in H. D. Holland and K. K. Turekian (Eds.), *Treatise on Geochemistry*.

Sundquist, E. T., 1993, The global carbon dioxide budget, *Science*, 259:934–941.

Sutton, M. A., A. Bleeker, C. M. Howard, M. Bekunda, B. Grizzetti, W. de Vries, H. J. M. van Grinsven et al., 2013, *Our Nutrient World: The Challenge to Produce More Food and Energy with Less Pollution*, Global Overview of Nutrient Management, Centre for Ecology and Hydrology, Edinburgh on behalf of the Global Partnership on Nutrient Management and the International Nitrogen Initiative, Edinburgh, U.K.

Takahashi, T., S. C. Sutherland, C. Sweeney, A. Poisson, N. Metzl, B. Tilbrook, N. Bates et al., 2002, Global sea-air CO_2 flux based on climatological surface ocean pCO_2, and seasonal biological and temperature effects, *Deep Sea Research II*, 49:1601–1622.

Taylor, L. L., J. R. Leake, J. Quirk, K. Hardy, S. A. Banwart, and D. J. Beerling, 2009, Biological weathering and the long-term carbon cycle: Integrating mycorrhizal evolution and function into the current paradigm, *Geobiology*, 7:171–191.

Taylor, P. (Ed.), 2010, *The Biochar Revolution: Transforming Agriculture and the Environment*, Global Publishing Group, Mt. Evelyn, Victoria, Australia.

The Royal Society, 2009, *Geoengineering the Climate: Science, Governance, and Uncertainty*, 83pp., Policy Document 10/09, The Royal Society, London, U.K.

Torn, M. S. and J. Harte, 2006, Missing feedbacks, asymmetric uncertainties, and the underestimation of future warming, *Geophysical Research Letters*, 33:L10703.

Trucost, 2013, Natural Capital at Risk: The Top 100 Externalities of Business, 43pp., www.trucost.com. Accessed August 23, 2014

van de Berg, W. J., M. van den Broeke, J. Ettema, E. van Meijgaard, and F. Kaspar, 2011, Significant contribution of insolation to Eemian melting of the Greenland ice sheet, *Nature Geoscience*, 4:679, doi:10.1038/NGEO1245.

van den Bergh, J. C. J. M. and W. J. W. Botzen, 2014, A lower bound to the social cost of CO_2 emissions, *Nature Climate Change*, 4, doi:10.1038/NCLIMATE2135.

van der Heijden, M. G. A., R. D. Bardgett, and N. M. Van Straalen, 2008, The unseen majority: Soil microbes as drivers of plant diversity and productivity in terrestrial ecosystems, *Ecology Letters*, 11:296–310.

Walker, J. C. G., 1986, Global geochemical cycles of carbon, sulfur, and oxygen, *Marine Geology*, 70:159–174.

Wan, S., J. Xia, W. Liu, and S. Niu, 2009, Photosynthetic overcompensation under nocturnal warming enhances grassland carbon sequestration, *Ecology*, 90:2700–2710.

Ware, J. R., S. V. Smith, and M. Reaka-Kudla, 1991, Coral reefs: Sources or sinks of atmospheric CO_2? *Coral Reefs*, 11:127–130.

Waring, B. G., C. Averill, and C. V. Hawkes, 2013, Differences in fungal and bacterial physiology alter soil carbon and nitrogen cycling: Insights from meta-analysis and theoretical models, *Ecology Letters*, 16:887, doi:10.1111/ele.12125.

Warnock, D. D., J. Lehmann, T. W. Kuyper, and M. C. Rillig, 2007, Microbial responses to biochar in soil—Concepts and mechanisms, *Plant and Soil*, 300:9–20.

Whitely, S., 2013, *Time to Change the Game: Fossil Fuel Subsidies and Climate*, 38pp., Overseas Development Institute, London, U.K.

Wilson, P. A. and R. D. Norris, 2001, Warm tropical ocean surface and global anoxia during the mid Cretaceous period, *Nature*, 412:425–429.

Wittke, J. H., J. C. Weaver, T. E. Bunch, J. P. Kennett, D. J. Kennett, A. M. T. Moore, G. C. Hillman et al., 2013, Evidence for deposition of 10 million tonnes of impact spherules across four continents 12,800 years ago, *Proceedings of the National Academy of Science*, 110:E2088–E2097, www.pnas.org/cgi/doi/10.1073/pnas.1301760110.

Wohlbach, W. S., I. Gilmour, and E. Anders, 1990, Major wildfires at the Cretaceous-Tertiary boundary, *Geological Society of America Special Papers*, 247:391–400.

Wolkovich, E. M., D. A. Lipson, R. A. Virginia, K. L. Cottingham, and D. T. Bolger, 2010, Grass invasion causes rapid increases in ecosystem carbon and nitrogen storage in a semiarid shrubland, *Global Change Biology*, 16:1351–1365.

Woodwell, G. M., R. H. Whittaker, W. A. Reiners, G. E. Likens, C. C. Delwiche, and D. B. Botkin, 1978, Biota and the world carbon budget, *Science*, 199:141–146.

Woolf, D., J. E. Amonette, F. A. Street-Perrot, J. Lehmann, and S. Joseph, 2010, Sustainable biochar to mitigate global climate change, *Nature Communications*, 1:56, doi:10.1038/ncomms1053.

World Meteorological Organization, 2013, *The Global Climate: 2001–2010, A Decade of Climate Extremes*, WMO-No. 1103, Geneva, Switzerland.

Xia, J., J. Chen, S. Piao, P. Ciais, Y. Luo, and S. Wan, 2014, Terrestrial carbon cycle affected by non-uniform climate warming, *Nature Geoscience*, 7:173–180.

Xia, J., Y. Han, Z. Zhang, and S. Wan, 2009, Effects of diurnal warming on soil respiration are not equal to the summed effects of day and night warming in a temperate steppe, *Biogeosciences*, 6:1361–1370.

Xiao, L., G. B. Liu, S. Xue, and C. Zhang, 2013, Soil microbial community composition during natural recovery in the Loess Plateau, China, *Journal of Integrative Agriculture*, 12:1872–1883.

Zhang, Y. G., M. Pagani, Z. Liu, S. M. Bohaty, and R. DeConto, 2013, A 40-million-year history of atmospheric CO_2, *Philosophical Transactions of the Royal Society A*, 371, doi:10.1098/rsta.2013.0096.

Zickfeld, K., M. Eby, H. D. Matthews, A. Schmittner, and A. J. Weaver, 2011, Nonlinearity of carbon cycle feedbacks, *Journal of Climate*, 24:4255–4275.

Ziolkowski, L. A. and E. R. M. Druffel, 2010, Aged black carbon identified in marine dissolved organic carbon, *Geophysical Research Letters*, 37:L16601, doi:10.1029/2010GL043963.

3

Potential Annual and Cumulative Carbon Dioxide Removal via Biochar

Ronal W. Larson

CONTENTS

Introduction to Biochar's Role in Carbon Dioxide Removal

Biochar is charcoal placed in soil for both soil and climate change improvement through carbon dioxide removal (CDR). Most of this chapter is on the latter biochar attribute—an atmospheric benefit. Biochar is the only CDR option where the *carbon dioxide* (transformed to char) is placed in soil. For biochar alone, the atmospheric CO_2 has had two prior stages: a photosynthesis stage producing biomass and a pyrolysis stage producing charcoal (WSU, 2012). This chapter cannot discuss either of these in depth, although there is information in the following on both.

By far, the most complete reference for biochar is the website for IBI—the International Biochar Initiative (IBI, 2014). This chapter will introduce some of the complexities covered there, but here the discussion is mainly on the question of whether biochar is potentially a large or a small future industry. IBI has only limited coverage of this topic. The evidence is that it can be large—very large—as large as the entire present fossil fuel industry. IBI has produced several useful reports on the present biochar industry, the most recent being IBI (2014).

Biochar's potential is an important, controversial, and understudied topic. The most respected set of documents on climate, the five International Panel on Climate Change (IPCC) documents that have come out roughly every 7 years, have mostly ignored biochar, along with its CDR alternatives. For biochar, this is not so surprising as the term *biochar* was only officially adopted in 2007, while the other main CDR approaches have much longer histories. Unfortunately, CDR is also going by the names *negative-emissions technologies* (NETs) and *greenhouse gas removal* (GGR) (Oxford, 2013; Tyndall, 2013).

The three main competitors to biochar seem to be (1) afforestation/reforestation (UNFCCC, 2013), (2) biomass energy with carbon capture and sequestration (BECCS), and (3) direct air capture (DAC—artificial trees). Readers will have to research those CDR alternatives separately. The first has the CO_2 tied up in biomass, so can be thought of as a precursor of biochar. The latter two place liquified CO_2 deep underground. Wiki can provide plentiful information on each—as it can for biochar.

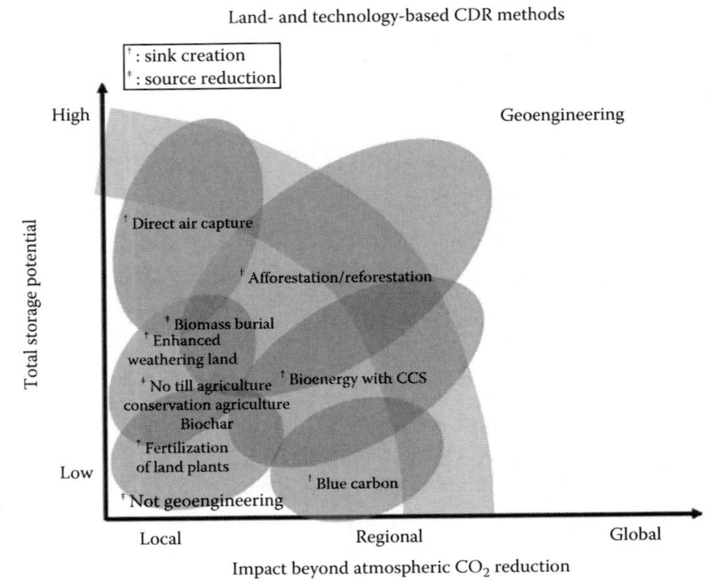

FIGURE 3.1 One comparison of different CDR alternatives. (From IPCC, 2011, http://www.ipcc-wg3.de/publications/supporting-material-1/EM-GeoE-Meeting-Report-final.pdf.)

The major climate change literature resource globally is the many publications of the IPCC. The most recent, AR5 Volume 1 (IPCC, 2013), was the first to have a full chapter on the word CDR—Chapter 6—with a small amount on biochar. There is a bit more in both Chapter 7, solar radiation management (SRM), and Chapter 12, which includes irreversibility. In total, unfortunately, there is little in Volume I to interest readers of this chapter. But we start with the IPCC to best indicate biochar's present status; biochar is neither very well described nor appreciated in the world's premier climate publication.

In preparation for the release of AR5, a special 2011 meeting on geoengineering was held in Lima, Peru. The report on its findings (IPCC, 2011) was certainly not a ringing endorsement for biochar, but neither was it for any CDR alternative. Biochar was represented by its principal spokesperson, Professor Johannes Lehmann, who also had a poster there (Lehmann et al., 2011). Figure 3.1, taken from that 2011 meeting report, shows the low ranking of biochar. The report emphasizes the continuing concern for governance, stating through this figure that biochar does not need such. This chapter attempts to disprove that Figure's indication that biochar has low storage (CDR) potential.

Readers are urged to look at Volume III of AR5, which was being released after this chapter was completed. It is rumored to call for greater attention to the potential magnitude topics of this chapter. It is highly unlikely that its treatment of biochar will look anything like this chapter, but it could be more positive than what Figure 3.1 presents. In particular, future IPCC documents may begin to give credit to the soil improvement aspects of biochar, which are the main topic of this book. Biochar has come a long way since the 2011 date of this figure—we believe much further than any CDR alternative.

The following five sections (1) extend this introduction to biochar potentials, (2) provide reasons for their underestimation, (3) identify opportunities for larger estimates, (4) discuss the major gaps in knowledge, and (5) give the summary.

What Else the CDR Literature Says about Biochar Potential

The range of annual CDR estimates for biochar has been large—from zero to about 10 Gt of carbon per year (hereafter abbreviated as Gt C/year). Zero only because biochar is sufficiently new that it was not included in all the early technical articles. Values close to 10 have been suggested (Lehmann

TABLE 3.1

Biochar Summary Evaluation Table

	Biochar	
Effectiveness	Limited by plant productivity and conflicts over land use with agriculture and biofuels	Low
	Burning biochar (in place of fossil fuels) may be preferable to burying it	
Affordability	Similar to biofuels (NB costs of fertilizers and transportation)	Low
Timeliness	Slow to reduce global temperatures (CDR method)	Low
	Substantial prior research required to investigate efficacy and impacts	
Safety	Potential land-use conflicts (food vs. growth of biomass for fuel)	Medium
	Long-term effects on soils not yet known	

Source: Data from Shepherd, J.G., 2009, *Geoengineering the Climate: Science, Governance and Uncertainty*, Royal Society, London, U.K.; http://royalsociety.org/uploadedFiles/Royal_Society_Content/policy/publications/2009/8693.pdf.

et al., 2006; Lenton and Vaughan, 2009; Lenton, 2010; Vaughan and Lenton, 2011), but these did not contain significant detail. Valuable biochar potential discussion, especially by list owner Tom Miles and moderator Erich Knight, occurs at Yahoo lists (Biochar, 2014) and (Biochar-policy, 2014). CDR's (including biochar's) potential is best followed at the Google list (Geoengineering, 2014).

The most influential statement to date about biochar's potential has been a 2009 study on both the CDR and SRM forms of geoengineering by the Royal Academy (Shepherd, 2009). Again, there are no specific annual quantities, but as seen in Table 3.1, the rating was *low* for effectiveness, probably meaning a potential less than 1 Gt C/year. These four ratings came from a panel of about a dozen geoengineering experts, none working in the biochar field. The contention of this chapter is that the panelists would likely now score biochars four ratings higher—as biochar has become better known. So we will not further discuss these values, which were surpassed by several CDR alternatives.

The main global biochar assessment with any quantitative detail is found in Woolf et al. (2010), from which the data for Figure 3.2 is drawn. The left side of Figure 3.2 shows the first 30 years of the 100 year modeling of the largest of three biochar scenarios they discussed. At 30 years, their data and this plot shows about 12.1 Gt C and 1.21 Gt C/yr for the cumulative (circles) and annual (triangles) carbon sequestration values. The annual value of 1.21 is obtained by dividing the graph value of 12.1 (a coincidence) by 10. For the same quantities, after 100 years, they projected for these same two quantities about 130 Gt C (a bit more than 10 times larger) and 1.7 Gt C/yr (only about 40% larger). These two year-100 values are the main numbers that are quoted from their article. Their annual increment (the first derivative, "+") drops off (and eventually even goes negative at about year 66) because of the assumption of late negative

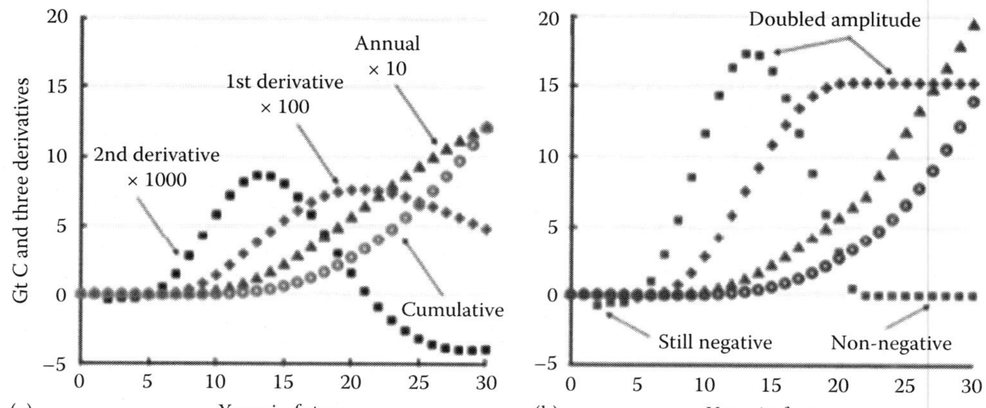

FIGURE 3.2 Biochar growth curves. (a) direct- and (b) modified method. (From Woolf, D., et al., *Nature Communications*, 1, 56, 2010.).

growth after year 21 that they made about the second derivative (squares). Not calculated or discussed by Woolf et al., these extra time derivatives are easily derived from the Excel spread sheets in their Supplementary Information. This chapter supports their estimate as realistic for the future condition that they specified: a conservative one. However, these first and second time derivatives that drive the two main curves of interest seem overly conservative in light of later parts of this chapter. Note their article shows contributions from different crops, while Figure 3.2 contains none of that useful detail.

One way to discuss annual percentage growth is to focus on the broad peak in the first derivative, which shows about (7.7/100 = 0.077 Gt C/yr) new annual biochar capability (the "+" symbol) at year 21 (again, note the need to divide this graph value by 100). This is when one is already adding about (7/10= 0.7 Gt C/yr) (the "triangle"), giving an annual production growth rate of about 11%. Alternatively, with cumulative removal ("o") at about 3 Gt C by year 21, we see this cumulative quantity is growing at about an 0.7/3= 23% rate. One reason for this larger rate is the first half-dozen years where Woolf et al. found emissions rather than sequestration, as shown for the negative portion of the second derivative curve ("squares"). That left-most curve describes the growth curve of the industries that are manufacturing char-making equipment. To emphasize, in the left scenario, no new factories for making char-making equipment (not for making char) are needed after 21 years, and some even are disappearing. By their assumption, those in production become less efficient with time, no land is available, and/or char producers simply have no new customers.

Figure 3.2b representing my modifications differs by assuming that the second derivative is twice as large and never goes negative (older factories to make char-making equipment are now replaced and land is always available). This leads to the first derivative also being twice as high and staying constant, and to the annual sequestration being about 65 % larger at year 30 (and now having exactly linear growth—growing to an unrealistically large year-100 total of 12.6 Gt C/yr). The cumulative total is seen to change very little in this first 30 years (14 vs. 13 Gt C), but by the 100th year is 519 Gt C— about 3 times larger (and also growing much too rapidly) than in the Figure 3.2a. This tripling shows the power of exponential behavior when the forcing itself (a constant annual demand, shown as the first derivative, for additional conversion equipment and land) continues unabated. Figure 3.2 is only intended to show the impact of one relatively small change in the first 30 years. Figure3.2b is not a proposed model scenario.

One way to better understand the influence of the four curves making up these two scenarios is to consider the new annual land needed after year 21 for the right-hand scenario. That graph shows the first derivative plateauing at about 15.3/100 = 0.153 Gt C/yr per year. If we assume 4 tons additional carbon in biochar per hectare per year, then for the right scenario we need to annually add about 153 Mt C/yr / 4 t C/ha-yr = about 38.25 Mha or 0.382 Million sqkm (a circle of radius about 350 km). To put this area into perspective, let us compare to the most productive 80% (10 Gha = 0.1 G sqkm = 100 Million sqkm) of the earth's land surface. Then, in percentage terms, each annual land area addition is (.382 million sqkm / 100 million sqkm = 0.0038 (or .38% per year—a tiny annual growth number even for this right hand scenario but of course considerable after 100 years). However, as some this new small assumed annual land acquisition can be reduced each year as the photosynthesis and conversion processes become more efficient through experience and learning. The assumption of 4 t C/ha-yr of annual production is a key one, further discussed below.

In year 100, the maximum amount of land being utilized for production in the conservative left scenario in the 66th year with the same 4t C/ha-yr assumption is 1.8 Gt C/yr / 4t C/ha-yr = 0.45 Gha. Since the late years of the right-hand scenario are unrealistic, we can instead say that having 4 Gha and 4t C/ha available would allow a peak annual sequestration of 16 Gt C-yr—much more than is proposed later in this chapter. This chapter is not the place to present details of more realistic scenarios. However, not much increase in the first and second derivatives of the Figure 3.2b will be all that is needed—if then also cut back. Such a scenario has apparently not received attention in any biochar literature.

There are now many dozens of videos available on the different aspects of biochar. An important one was released in early 2014, based primarily on presentations made at the 2012 USBI conference held in Sonoma, California (USBI, 2014). A few important parts of the video were updated to 2014. The 30-min video answers several questions raised in this chapter. The video does not have an excessive commercial flavor—all presented to a mostly technical audience at this conference. All seven of the presenters have a commercial background.

The middle part of the video contains considerable information about the company Cool Planet (Cool Planet, 2013; 2014). The most surprising statement was a claimed potential 100 ppm drop in CO_2 using only 3% of global land area within 40 years (mostly using miscanthus grass as a biomass feedstock). Another surprise was a claimed 4–8-year doubling time for out-year effects, providing exponential, not linear, growth in conversion facilities and lands. A final surprise was a statement that this same company, Cool Planet, could provide 100 quads of carbon-negative energy when they have grown from 400 US plants in 2018 to 2000 plants worldwide (in an unspecified future year). This company has influential financial backers presumably checking such statements (Cool Planet, 2013; 2014).

The purpose of this chapter is to explore the validity of replacing Woolf et al.'s conservative assumption with an aggressive CDR scenario. We are, in essence, supporting Cool Planet and the roughly 10 Gt C/year values suggested by Professors Lehmann and Lenton. This aggressive scenario does not seem to have appeared elsewhere, although many portions have citations given in the following.

Reasons for Underestimations

1. Biochar is a very new topic. Only in 2013 did Cool Planet appear with plans consistent with the future roughly 10 times than that in Woolf et al. The number of researchers and technical papers in the biochar field has grown by at least a factor of 10 since the 2009 time frames cited earlier. The 2014 IBI annual industry report confirms the growth one expects with new high-potential activities. Long ago, biochar failed the not-ready test for one early supporter (Carbon War Room, 2014; Virgin Earth, 2014). Perhaps now, biochar might look more ready.
2. Biochar is a very complicated topic. It can be associated with every possible form of energy and is applicable in every country, every soil type, and every form of vegetation. At least three sources of funding (atmosphere, soils, energy [A–S–E]) will influence every biochar company (Downie et al., 2009; Downie, 2011).
3. Biochar is still expensive. Carbon credits seem likely, with only voluntary credits now available. But this is the case for all the CDR options. Biochar is probably moving faster than others in reducing costs and has multiple ways of sharing the costs (Kleiner, 2009).
4. Those interested in CDR are generally unaware of biochar's soil augmentation benefits—as no other CDR approach includes soil benefits (McLaren, 2011). A recent example is Keller et al. (2013), who analyzed five CDR and SRM approaches, but not biochar.
5. Those interested in soils are generally interested in local applicability, not global and climate-connected issues. Few biochar proponents have had a reason to project global totals, as their focus has been regional, if not local.
6. Too many climate policy analysts leave out CDR as an alternative different from, and additive to, mitigation and adaptation—for which there are numerous projections.
7. Sufficient soil benefits of biochar from a buyer's/user's perspective are not yet assured. Funding for R&D has been too low to have yet obtained all the needed data (Thies and Rillig, 2009).
8. A majority of the US population is not concerned about climate issues in general. This has prevented needed political action (Blunden et al., 2013; Yale, 2014).
9. The differences within CDR itself have not been well understood historically. Recent papers that are helpful in understanding the pros and cons of biochar vs. other CDR approaches are as follows:
   a. Bellamy et al. (2013): A comparison between biochar and DAC. Biochar won handily using a panel voting approach on eight attributes.
   b. Meadowcroft (2013): A single person's view over a wide range of CDR options, ending up favoring BECCS, which had been independently selected as part of a larger study. Biochar didn't fare too badly.
   c. Heyward (2013): This social scientist did a good job in separating CDR from SRM, to fall between mitigation and adaptation. She also introduced a fifth category called *rectification*—making amends to those most harmed by climate change. Hers was an ethical response that can help biochar.

   d. Moss et al. (2013): This article went deeply into the adaptation side of climate responses, from a social science perspective, without ever mentioning CDR. But biochar fit well into the recommendations given there for future action. The societal failings of many biofuel projects should cause caution as biochar is introduced on ever-larger scales. The world of *adaptability* seems better prepared to address these societal topics than can mitigation.
   e. Bracmort and Lattanzio (2010): SRM and CDR were compared and partially separated in a policy sense in a report from the U.S. Congressional Research Service (CRS). Biochar has been covered separately by the first author (Bracmort, 2010).
   f. Hogeboom (2013): A popular press article mentioning seven CDR approaches, with the majority of the discussion on SRM. Numerical values are given for three criteria: cooling factor, readiness, and cost. Biochar community rankings would be very different from the early ones assigned by Oxford (2009), which were used by Hogeboom.
10. There has been overt opposition to biochar:
    a. BiofuelWatch (BFW), various dates: BFW's main activity is opposition to all biofuels. This London-based group sees biochar as another biofuel. Their being against large corporations resonates with many, despite the likelihood that biochar seems to be the CDR approach most likely to have many small independent businesses. Most of BFW's seemingly authoritative negative statements are in fact made by well-known strong biochar supporters. See also Paul et al. (2009). Rebuttals have been offered by Walker (2011) and WorldStove (various years).
    b. H.O.M.E. (2013) provides BFW views on two nonbiochar CDR alternatives, but with sprinkled references to biochar, as they respond to views of a pair of social scientists (Smith and Torn, 2013).
    c. Morton (2009) is a website where a knowledgeable climate reporter for the *Guardian* newspaper engages in a dialog on geoengineering with Mr. James Thomas, a spokesperson for ETC and H.O.M.E., both strong allies of BFW. None of the half-dozen views analyzing biochar seem valid to this author.
11. Counteracting these biochar opponent's views, outside of biochar circles, is rare. However, positive views are occurring, as seen in the following:
    a. Buck (2010, 2011a,b, 2014) has followed technologies like biochar for some time, with a more positive social science perspective. This follows from having great concern for those to be harmed by climate change.
    b. The work by Neville (2012) is a doctoral thesis about biofuels, analyzing the changing views by a range of interest groups. Biochar would seem to not likely have the same problems.
12. Land grabbing is an emotional issue to which biochar has been tied. Similar social science views on purported land grabbing and job loss related to biofuels (no articles specifically available on biochar) are found and sometimes rebutted (Magdoff, 2008; Lawrence, 2010; Robbins, 2011; Garrett-Peltier, 2012; Hurni et al., 2013). Jokimaki (2014) provides a list of papers dealing with both benefits and harms from GHG emissions from bioenergy-related land use.
13. CDR is too often lumped with SRM. SRM's many detractors have also discouraged studies that could have counteracted the underestimates claimed by this chapter. These detractors are mostly motivated by ethical concerns over SRM (Morrow et al., 2009; Buck, 2010; Preston, 2011; Boucher and Gough, 2012; Hale, 2012; Wong, 2013; Montana, 2014).
14. SRM proponents have tended to view CDR as competition rather than allies. Biochar would be moving faster if the SRM community tied itself to CDR. The SRM approaches are not the subject matter of this chapter, but some SRM proponents have endorsed biochar (AMEG, 2013).
15. Energy issues are often ignored by CDR analysts. And even when considered, biomass is perceived as being less desirable than wind and solar energy. Biomass is seen, with justification based on faulty past practice, as being less renewable. For instance, the American Solar Energy Society, which does support small wind along with half a dozen solar energy approaches, has no program in biomass, much less biochar (Larson, 2007a,b, 2006, 2009, 2010, 2011, 2013).

The same is true in some renewable energy umbrella organizations (Ren21, 2013). This energy topic is considered further in the Opportunities section.

16. There are not enough analysts providing estimates. Funders, for valid reasons, have been more interested in biochar's soil technical issues. Questions of maximum potential have not yet been asked by funders.
17. Scientists tend to be conservative in their estimates and CDR analysts are scientists.
18. Advocates for competitors to biochar (such as BECCS and DAC) are as vocal as biochar advocates, but they have been active longer, influencing the choice of CDR approaches to analyze. Biochar's potential albedo impacts are raised in this category (Verheijen et al., 2013).
19. The criteria used in comparing the CDR options have been widely different, but most disfavor biochar. Specifically, the benefits that biochar can supply to soils are rarely considered since, among the major competitors, these apply only to biochar.
20. Credit is not given for nontechnical advantages that biochar has—such as rapid current growth rate of the biochar industry, plenty of skilled personnel, and numerous country and regional support organizations.
21. Several new specific market areas, listed next under opportunities, have only become active in the last several years. Cool Planet is the major example.
22. Emphasis has been on the use only of residues. Many CDR advocates would say that climate urgency requires the more aggressive vision of the next section on opportunities.
23. There has been insufficient recognition of how much soil carbon has been lost historically (Houghton, 2003).
24. The word *irreversibility* implies hopelessness about being able to successfully carry out CDR that is inappropriate (Solomon, 2009; Matthews and Solomon, 2013).
25. Those providing assessments are often too little aware of the views of biochar experts. Favorable public opinion for biochar has been expressed in cases where biochar expertise was available for in-depth answers (Corner et al., 2012; Bellamy et al., 2013).
26. As with most renewable energy technologies, biochar requires an upfront large capital investment expense. In assessments, this has tended to favor SRM over CDR and a continuation of fossil fuels.
27. The biochar industry is so small that there have been few funds available to make a strong case for biochar. Marketing messages have been rudimentary (Nerlich and Jaspal, 2012).
28. Major authorities on long-term climate change global economic costs have been unfamiliar with biochar (Sugiyama et al., 2008; Nordhaus, 2013; Stern, 2013).
29. Most climate models have a simplified interaction with the biosphere (Smith, 2013). Biochar is not known to appear as a variable in any climate model.

Group A: Opportunities

It would be premature to identify in detail specific submarket opportunities that would comprise a total future biochar industry. However, this section will be more specific than the previous on how biochar can supply a large removal—perhaps 10 Gt C/year. It is useful to break this analysis into four parts, with the first being on the land-availability side shown in Table 3.2.

The base case (#1) is that by Woolf et al. (2010). This has been widely cited by CDR analysts as providing a maximum annual amount of biochar. But that paper also made clear that theirs was a conservative estimate. The authors studied three cases, which are simplified to be between 1 and 2 wedges in this table. To emphasize that this is a broad-brush treatment, entries are being kept to single-figure accuracy. Readers should look at the Woolf article in detail to see how the word *conservative* has influenced their numerical results. Roberts et al. (2010) also gives excellent and favorable cost details on agricultural residue use for biochar.

The second row of the table (an aggressive version of Woolf et al.) suggests that doubling is possible—by looking at both more land and by many other means discussed in the following. There are many

TABLE 3.2
Supply-Side Availability for CDR Needs

#	Category	Min Gt C	Max Gt C	Area (Gha)
1	Base: Woolf (conservative)	1	2	0.5
2	Woolf, aggressive	1	1	0.5
3	Agroforestry	2	4	1
4	Pasture conversion	2	4	1
5	Dry-land conversion	1	3	1
6	Fire-based resources	1	2	1
7	Oceans (not in land total)	2	4	*1–2'*
8	Species improvement	1	2	
	Total	11	22	5

reasons for saying that doubling of the first row numbers is possible—which is what the rest of this section is about. Stavi and Lal (2011) and Slade et al. (2011) support this more aggressive vision. The results of Tables 3.2 through 3.6 must not be simply summed. A good resource overall for biochar technical soil articles that support an aggressive stance is at Cornell (2014).

Row 3 for agroforestry is quite different. This suggests biochar production can be as active in forestry as in agriculture. The postulated new gigahectare of land can in part be current forest, but it can also be a formerly forested and now heavily degraded land. A minimum biochar output of 2 Gt C from the 1 Gha is roughly equivalent to saying that the annual dry biomass productivity is 4 t C/ha, assuming that half the input carbon can be captured as char (Lehmann, various; IBI, 2014). The global average is more than this today (60 Gt C/year of NPP from about 10 Gha)—which further justifies the upper limit of 4 Gt C/year. The assumption of 1 Gha means that approximately 1/4 to 1/3 of present forest land would be employed. This can mean some diversion from the pulp and paper and similar industries but also different species and different management practices (Haberl et al., 2007; Jandl et al., 2007a,b; Jindal et al., 2008; Kalbitz and Kaiser, 2008; World Bank, 2011).

Row 4 similarly assumes a similar supply of land that is now pasture or rangelands. This does not imply that there is a great reduction in the production of meat since this land can serve a dual purpose. As for agroforestry, this category was ruled out by Woolf et al., as one of their conservative, minimum deviation from existing land use, assumptions. Stavi (2012) presents a positive view of this application of char on 0.55 Gha of degraded lands. Stavi and Lal (2012) extend that positive view to a wider range of lands, including pastures.

Row 5 completes the look at land types different from agriculture land—with a third category of land that today only supports species that can survive long dry spells. Fortunately, there is a type of photosynthesis called *crassulacean acid metabolism* (CAM) that can provide growth even under hot dry conditions for species like agave. Much reduced evaporation is possible since CAM plants only take in CO_2 during the much cooler nighttime hours. Here the average minimum production is only half of that assumed earlier, but not quite so large a drop in the *max* column—based on the encouraging literature on CAM (Borland et al., 2009).

The sixth row, on fire-related resources, is similarly conservative on the yield for a new assumed 1 Gha of land. This is for land available as we attempt to restrict fires. Part of this land is in Africa where fires are set annually to make land more productive. With the alternative of harvesting the grasses for energy and biochar, even greater soil productivity seems assured. Another large area subjected to large annual forest fire loss is in the northern forests of Canada and Russia. Here, there can be harvesting of already fire-killed trees, but also there can be preventive removal of strips of forest to keep lightning-set fires from spreading. In this category, we can also include early removal of trees stricken by insects, which pose a serious fire hazard. Lamers et al. (2014) discuss use of damaged forests for bioenergy, but without including a biochar option. That favorability should increase with biochar.

The seventh row is for oceans. Up to this point, we have postulated an order of magnitude more land—from half a gigahectare to 5 Gha. Mostly, this is possible without changing the Woolf analysis; to be conservative, they had ruled out land-use changes. And perhaps for good reason, certainly there will be objections for the use of each of categories 2–6. Should that be the case, there is even more reason to consider ocean-based resources. Roughly three times more surface area is available. Many ocean plant species grow rapidly, and use of the oceans for biomass growth can clean up dead zones, which are growing in size and number. The alternative ocean CDR approaches discussed do not provide the added benefits of biochar (Kim et al., 2012; N'Yeurt et al., 2012; Capron, 2012; OCB, 2014; Tulip, 2014; WHOI, 2014).

The eighth and last supply side category relates to the species being used to produce char. Woolf et al. assumed no improvement over time in either input biomass per unit land or the later growth of biomass after the char was placed in the soil. They were working primarily with agricultural crops such as corn, wheat, and rice. The productivity of energy species has not been similarly improved, but should be. Biomass crops such as miscanthus have had major yield improvements recently and could thereby favorably impact the very rough guesses shown in Table 3.2. With species improvement, the required areas for each land category could be well less than the species improvement justifying the additional 1–2 Gt C/year shown.

Group B: Energy Consideration Opportunities

Effective energy markets are needed to realize the possible biochar sinks given earlier. It may be impossible to achieve those targets unless there are uses for the pyrolysis gases. In most energy future studies, bio-energy markets have been presumed to develop for mitigation, not CDR, reasons. In Table 3.3, we are adding new topics to the general bioenergy markets already assumed by analysts. These are mostly for electrical generation and thermal applications—markets already occupied by biomass. More land will be required with biochar.

The ninth category of biofuels (bioliquids) has been perhaps the most important addition for biochar since Woolf et al.'s estimate. One reason is the rapid expansion of Cool Planet (2014), which has quickly become the largest biochar company. See also Laird et al., 2009; Tilman et al., 2006, 2009; EU, 2012; and Thornley, 2012 for other nonchar biofuel projections.

Category 10 (combined heat power and biochar [CHPB]) ensures the application of using waste heat from electrical generation to also accomplish space heating, as practiced widely in Scandinavia. With biochar added, this is *killing three birds with one stone*. Note also that biomass is stored energy—and can be a perfect low-cost backup for intermittent, non-dispatchable wind and solar energy (Campbell et al., 2013; Larson, 2007a,b, 2006, 2009, 2010, 2011, 2013). Biochar can be a key part of these markets if (and probably only if) carbon removal subsidies are in place.

Category 11 (cooking) is included because biochar discussion is still occurring on the *stoves* list (Stoves, 2014). Stove discussion, ongoing since 1996, saw some reduction in 2006 when the *biochar* list grew out of the stoves list (Biochar, 2014). There is no agreement on the ability of char-making stoves (most are called TLUDS—which stands for top-lit updraft) to outperform other stove types, but the trend is in that direction (ETHOS, 2014). There is a curious juxtaposition of trends here, where charcoal is today blamed for forest degradation; but the same product (as biochar) can aid forest recovery (Larson, 2007a,b, 2006, 2009, 2010, 2011, 2013; Bailis, 2012).

Category 12 on hydrothermal carbonization (HTC) was also barely on the horizon in 2009. It is still not a certain part of biochar, since its in-soil lifetime is appreciably shorter than regular biochar. It is

TABLE 3.3

Demand-Side Needs: Energy

#	Category	Min Gt C	Max Gt C
9	Bioliquid focus	2	5
10	CHPB focus	2	3
11	Developing country cooking	1	2
12	HTC	1	2
13	Coupling with BECCS	1	3
	Total	7	15

mentioned here because it can fit so well with wet biomass—perhaps from oceans—but certainly from municipal and animal waste (Titirici et al., 2007).

Category 13, the coupling of biochar with BECCS, is mentioned because the two are often seen only as competing approaches. Every BECCS plant could produce char, albeit with a needed doubling of input material. Conversely, some biochar electrical generating facilities could capture the exiting carbon dioxide and feed that into the underground CCS systems needed for BECCS (Hale 2013a,b; Hale and Grundy, 2009). We say *some* since most biochar facilities are apt to be small enough and in remote locations that would not warrant a CCS collection system (which is already questionable at larger scale). Often the BECCS projections for annual possible CDR are larger than those for biochar. This seems highly unlikely when every possible BECCS facility is also a possible biochar facility. Biochar also can be produced along with biofuels and/or thermal facilities much too small for economic CO_2 capture with BECCS. The supposed larger percentage capture of CO_2 by BECCS also ignores the out-year soil and biomass increases discussed in Category D.

To repeat, these five energy categories totaling 7–13 annual gigatons are not the total CDR contributions. These are listed because the energy side of biochar is often not considered in CDR analyses, where the norm is energy consumption, not energy production.

Group C: Other Demand-Side Opportunities

Four nonenergy demand items are shown in Table 3.4, in an *other* demand grouping. As in the group B listing, these are not in addition to the eight supply categories in group A. Rather, they are justifications that are often mentioned as important advantages, but not necessarily as the demand-side drivers they may turn out to be.

In category 14, food is found in many biochar documents as the main reason for interest in biochar. Biochar advocates often state that reliance on chemical fertilizers has health impacts, as many micronutrients are now missing that could be returned with significant use of biochar (Chan and Xu, 2009). Biochar has many advocates in permaculture with this outlook (Bates, 2013). These advocates understand the CDR and carbon-neutral energy attributes of biochar, but food quantity and quality benefits that follow from healthier soil are the main drivers. There is also a food aspect when we talk of reclaiming *lost* land—seen in the previous A group (Bryan et al., 2008). There is also the aspect of finding species that can supply food as well as feedstock; more biochar may be produced simply because there need not be competition between food and energy (Forster et al., 2012). Aquaponics may be an additional biochar opportunity.

Category 15 is included to emphasize that biochar has important water-conserving attributes. Irrigation and drought avoidance costs will be decision variables for many biochar users. This may help most in differentiating biochar from BECCS (Laird, 2008; Ulyett et al., 2014).

Category 16 (fertilizers) is much the same as categories 14 and 15. Fertilizer expenditures are, for many farmers, their major annual expense. All the same arguments hold with the right combination of soil and crop, so this category can also mean extra biochar beyond the usual CDR estimates (Wood and Cowie, 2004; British Biochar, 2014).

Category 17 (miscellaneous) covers everything else that has been stated as positive for biochar. Many of the items in Table 3.5 have already been mentioned earlier, especially the first six and all the co-benefit items relating to energy.

TABLE 3.4

Demand-Side Drivers: Other

#	Category	Min Gt C	Max Gt C
14	Food	1	2
15	Water	1	2
16	Fertilizer	1	2
17	Miscellaneous (jobs, waste, etc.)	1	2
	Total	4	8

TABLE 3.5

Comparison of 27 Benefits for Six Biomass CDRs

	Biochar	Afforestation	BECCS	Compost	Holistic Management	Burial/Oceans
1. *Continuing out-year benefits*	Yes	Yes	No	Limited	Yes	No
2. *Produces energy*	Yes	Delayed	Yes	No	No	No
3. Sequesters CO_2	Yes	Decades	Yes	Short term	Yes	Yes
4. Improves soils	Yes	No	No	Yes	Yes	No
5. Conserves water	Yes	No	No	Maybe	Yes	No
6. Conserves fertilizers	Yes	No	No	Some	No (?)	No
7. Rural economic Development	Yes	Yes	Some	Some	Yes	No
8. Livelihoods	Yes	Yes	Low	Yes	Some	Low
9. Water quality	Yes	Yes	No	Possibly	Yes	No
10. Air quality	Yes	Some	No	No	Some	No
11. Biodiversity	Yes	Yes	No	Some	Yes	No
12. Displace fossil petroleum	Yes	No	No	No	No	No
13. Displace natural gas	Yes	No	Rarely	No	No	No
14. Displace coal	Yes	No	Yes	No	No	No
15. Encourage efficient CHP	Yes	Some	Rarely	No	No	No
16. More efficient biomass use	Yes	No	Not rural	No	No	No
17. Improve food security	Yes	No	No	Yes	Some	No
18. Less N_2O release	Yes	No	No	No	No	No
19. Reduced ocean dead zones	Yes	No	No	No	No	No
20. Applies everywhere	Yes	Many	Limited	Many	No	No
21. Public interest/ involvement	Yes	Yes	Little	Yes	Growing	Little
22. Corporate and NGO interest	Yes	Yes	Yes	Unkn.	Unkn.	Unkn.
23. Protect indigenous peoples	Yes	Decades	No	Little	Maybe	No
24. Can utilize ocean resources	Yes	No	Limited	No	No	No
25. Work with rock dust	Yes	Some	No	Some	No	No
26. Backup to wind/solar	Yes	No	Some	No	No	No
27. Reduced ocean acidification	Yes	Yes	Yes	Yes	Yes	Yes

Remaining are these in the above co-benefit list:

#4, which includes erosion control as well as improvement of soil. Perhaps a majority of all lost soil carbon has occurred because of soil erosion (Foereid et al., 2011).

#7 and 8 for rural economic development and jobs. The Cool Planet website emphasizes both (Wilson, 2013; Cool Planet, 2013; 2014).

#9 and 10 for water and air quality, the former being important for capture of nutrients now causing ocean dead zones, as noted earlier. Air quality includes concerns over black carbon (Ippolito et al., 2012; Bond et al., 2013).

#11 on biodiversity, where biochar can restore some lands and preserve existing forests and other sensitive lands by demanding that the resource be sustainably produced. This topic has been especially prominent in UN discussions (Chum et al., 2011; CBD, 2012; Cowie et al., 2012; IBI Certification, 2013).

#20–23, which are all people oriented. A few groups see a large-business danger in biochar and harm to indigenous people, but the majority seem to feel biochar can move faster and have an overall larger impact because of its positive ratings in these four areas (Hertsgaard, 2014).

#24 (ocean resources) may offer a large potential (Bird et al., 2011; Bird and Ascough, 2012; N'Yeurt et al., 2012; Capron, 2012; Tulip, 2014).

#25 on rock dust is important in this book with several chapters promoting rock dust. Again listed here because this *other* item could soon drive a few hundred million tons of char.

#27 (ocean acidification) has been saved for last because it is the dominant characteristic that separates CDR from SRM (and a possible new subcategory of geoengineering called earth radiation management [ERM] that does not impact biochar at all). This could be the main reason why CDR proceeds while SRM does not (Robock, 2008).

This list of 27 biochar opportunities has been extended to a total of 50 (Schmidt, 2012), but he includes "55" in the title so as to accommodate his expectations of growth. Examples that are not in Table 3.5 include feeding char to animals and using char to improve biogas production in digesters. Table 3.5 should not be perceived as being exhaustive.

The entries for the other five columns of Table 3.5 are personal opinions, restricted to only the biological types of CDR. These are presented to show in an overall sense why biochar is so often ranked highly for the comparison category of *co-benefits*. These 27 rows are at least half in this *miscellaneous* category. Whether they will be important in decisions on biochar is critically dependent on the criteria to be used by CDR decision makers. Too often, these totally exclude the concept of co-benefits, focusing only on the third row: sequestering CO_2. The first two rows are first and in italics because they were the *only* co-benefit entries shown in one study (Meadowcroft, 2013). His interest was in BECCS—which did not have so many co-benefits.

Group D: Improved Accounting

Table 3.6 identifies the final four categories where biochar differs from all the other CDR approaches by having sequestration value (1) in advance of, (2) because of, and (3) following placement in soil. Most existing analyses, including that by Woolf et al., provide sequestration credit only for #2. Unlike opportunity groups B and C, some of these can be added to group A.

Category 18 shows that some biochar producers will be using a species that needs several years of growth prior to its harvesting, drying, and pyrolysis. CO_2 will have been withdrawn from the atmosphere exactly as in a reforestation/afforestation project, where carbon credits are already being provided. That process will presumably become much larger as REDD+ takes on increasing importance (Redd-Monitor, 2013;

TABLE 3.6

Improved Accounting

#	Category	Min Gt C	Max Gt C
18	Pre-pyrolysis CDR	1	2
19	Out-year aboveground	2	3
20	Out-year belowground	2	3
21	N_2O and CH_4	2	5
	Total	7	13

Stavi and Lal, 2013; UNFCCC, 2013). Although the withdrawal would not have been long term, there was as much benefit for the short time as any other form of CDR. To accelerate biochar as a CDR approach taking account of this 18th category, means will have to be found to think in years rather than centuries and millennia. The harvesting of the aboveground biomass will also leave behind new biochar-generated underground carbon from the atmosphere that has similarly benefited the climate. This belowground carbon removal benefit could continue for centuries. This is not to suggest exactly how to provide fair treatment for this unique form of CDR, but to note that such an accounting procedure will strongly benefit biochar introduction.

Categories 19 and 20 (out-year above- and belowground new biomass) are different in that most analysts do discuss some continuing sequestration benefit beyond the char placed in the soil (Roberts et al., 2010). Life cycle accounting is used by (Hammond, 2009; Shackley, 2010, 2011a,b; Hammond et al., 2011). Amazonian Terra Preta soils today have greater aboveground annual growth and belowground soil carbon resulting in much higher sales price than nearby nonaltered soil (Glaser et al., 2001; Novotny et al., 2009). But this won't show up as an accepted carbon credit if the analysis only assumes crop residues as the starting material and no change anywhere to annual growth thereafter. That is, some biochar carbon-credit analyses are the same as if the carbon was simply placed in a cave. Soil carbon being about double the aboveground carbon makes the accounting problem even more serious, as new fungi growth is being ignored in much present carbon accounting (Clemmensen et al., 2013). There needs to be a new vision where the loss of soil carbon (Guo and Gifford, 2002) and biochar's soil benefits are given greater credence (Hammes and Schmidt, 2009; Gunther, 2012).

Category 21 (for non-CO_2 warming gases) is quite different (CDIAC, 2014). Less nitrous oxide is released in many biochar field studies (van Zwieten et al., 2010; Kookana et al., 2011). N_2O is about 300 times more influential in warming than CO_2, so reduced N_2O emissions are a further advantage for biochar that needs a stronger consideration (Singh et al., 2010). Methane emissions are also reduced with biochar (Feng et al., 2012).

To summarize for the four categories in group D, unlike the nine categories in groups B and C that were not intended to increase the group A totals, these in group D could even double the group A totals. Stated alternatively, an annual impact of 10 Gt C removal might only require a group A total of 5 Gt C actually going in the ground. Much more fieldwork is needed to support this suspicion.

Summarizing the comments about Groups A to D, there are at least 20 reasons for believing that biochar can supply a sizable CDR sink. These reasons include the use of land other than agricultural land, the need for energy from biomass, and the need for ancillary benefits that biochar can supply. The initial application CDR numbers in group A must be derated due to limited recalcitrance (Marschner et al., 2008; ACR, 2013). That derating could easily be overly cautious. Such derating needs to be counterbalanced in most cases by what happens in the centuries after char is placed in soil. Among the numerous CDR approaches, biochar is the only one with this attribute—for future biomass both above- and belowground.

Gaps in Knowledge

A main gap is being able to predict biochar's recalcitrance. Some critics/opponents claim that biochar may have only a short lifetime in certain soils. The lack of better predictive capability is presently hindering the availability of the carbon credits that can foster the use of biochar (ACR, 2013). Several valuable articles on char lifetime can be found at EISS (2014). Work in Australia has especially concentrated on recalcitrance (Singh and Cowie, 2010; Sohi et al., 2010).

There is a serious gap in knowledge on the importance of early action. Not strictly a biochar issue, clearly a majority of decision makers have not yet agreed to support CDR, much less biochar (Krugman, 2013). Some of the strongest technical support for a more rapid return to lower atmospheric CO_2 is in Hansen et al. (2008, 2013). IPCC may be removing this gap in its forthcoming Vol. III report of AR5. Legal action could speed action (Grossman, 2003).

Although there are numerous technical reports on successes and failures, there has been no full report yet on how to optimize the growth of above- and belowground biomass (especially fungus) following biochar application. With the exception of the most degraded lands, Terra Preta's double and triple soil productivity have not yet been duplicated anywhere for modern biochar (Criscuoli et al., 2014; IBI, 2014).

Numerous items in the earlier sections still are only partially understood in hard economic terms. Examples are how to optimize application rates to minimize costs for irrigation and fertilizers, two high-cost items for every farmer (IBI, 2014). Biochar is no longer in an infant stage, but it is far from mature.

Credits could be supplied for out-year sequestration if policy studies demonstrated how to include their value in a fair manner. Large initial payments could be replaced by annual payments—payments possibly associated with tax credits (Reed, 2011).

The gaps in knowledge are being filled in technical reports at a rate better than one a day, with many examples available through the IBI (2014) and Eurochar (2014) websites. A gap-filling example is Matovic (2011). Some are being filled in publications available to the general public (Levitan, 2010).

Summary and Conclusions

1. *Projections.* Given the short time biochar has been under study and the complexities in finding optimum combinations of char temperature, soil, and plant species, the recent growth and future prospects for a biochar industry should be viewed as excellent. There are enough high projections (such as 10 Gt C/year) to be positive about the growth of a biochar industry. In the opinion of this chapter, no alternative should be projecting CDR higher than biochar. However, biochar has a definite upper land-availability limit; all forms of CDR are needed and some (such as rock dust) can be mutually supportive.

   Should the claims of this chapter seem reasonable, the position of biochar in Figure 3.1 should cover the full abscissa range and should have an ordinate carbon reduction potential as high as any CDR approach. Similarly, the four entries in Table 3.1 are arguably at least *high*. Figure 3.2 would of course have much higher totals.

   Hopefully this chapter has shown that a peak decline rate approximately equal to the present atmospheric carbon growth rate (10 Gt C/year) can be achieved. A decline of 100 ppm CO_2 becomes (conservatively) approximately 400 Gt C when we include the amount that will ultimately be returned from the oceans. At an average decline rate of 8 Gt C/year, 50 years would be required. Char application not only will need time to reach peak, but also the application needs to be phased down as well—to avoid removing too much carbon.

   Spreading 400 Gt C over an assumed 4 Gha means 100 t C/ha—a number that is roughly 10 times than that used in many biochar experiments. But this factor of 10 is an overly pessimistic measure of the difficulty we face. Using the multiplier ideas of this chapter, only half as much char is likely needed to remove the 100 ppm CO_2; we need only an average incremental char amount of 50 t C/ha on these 4 Gha. There are many places on earth with much more than 50 t C/ha combined above and below the surface. In fact, the 200 t C/ha average globally from about 2000 Gt C spread over the most productive 10 Gha is today four times the needed additional 50 Gt C/ha of biochar. And, as with Terra Preta, we are talking about reclaiming lost land—land that often had this much carbon in the past.

   Summing, we can do all these with biochar: removing 100 ppm CO_2, adding 400 Gt C in 50 years, and an added 100 Gt C/ha over only 4 Gha (combined above- and below ground level). This sequestration is an increase of only about 400 Gt C/2000 Gt C = 20%, with only a 10% increase for the biochar itself. Dividing the full needed 20% by 50 years gives 0.4% per year average change in total soil and plant biomass. The productivity gains of the world's major agricultural crops have been appreciably larger over the past 50 years, and biochar promises to continue that improvement as well.

   Thus the biochar process promises to be positive, low cost (maybe even cost free, when savings are considered) and not a burden on society. But upper limits for biochar become of greater concern if we inordinately delay a start, which would then require increasingly more than the above assumed 100 ppm CO_2.

2. *Underestimation reasons.* The stated 29 underestimation reasons should be viewed as a positive—none are permanent. The main one could be the global governance failure to place a high emphasis on the early need for CDR. Another is a failure to give appropriate credit to out-year

sequestration. Existing low estimates for biochar are linked to a common biochar industry assumption that it is best to be conservative; that is not the assumption of this chapter.

3. *Opportunities.* Biochar is not lacking for opportunities—as shown by the score of opportunity entries. Substantial private and corporate R&D is ongoing and increasing, despite limited governmental funding. Commercialization progress may appear to be too slow, but much is happening that is very encouraging.
4. *Gaps in knowledge.* It is still not possible for soil scientists to recommend a specific best-economics combination of char production temperature and amount for a particular soil and plant species. It is also not yet possible to do more than hint at the out-year added carbon negativity following placement of a specific biochar at a specific time in a specific location. There is much to learn, but also much is being learned.
5. *Overall.* Biochar as an industry could become as large as the fossil fuel industry. Biochar could approach that of the world's largest industry—the food industry, with which it can be in more than full harmony. Biochar offers hope to reverse global warming—more hope than any other CDR approach. The hope biochar offers is needed in society today.

REFERENCES

ACR, 2013; Methodology for biochar projects, American Carbon Registry, http://americancarbonregistry.org/carbon-accounting/standards-methodologies/methodology-for-emissions-reductions-from-biochar-projects/acr_tct_biochar-meth_public-comment-draft.pdf. Accessed date: 1 October, 2014.

AMEG (2014); Arctic Methane Emergency Group, http://ameg.me. Accessed date: 21 August, 2014.

Bailis, R. et al., 2012, Innovation in charcoal production: A comparative life-cycle assessment of two kiln technologies in Brazil, *Energy for Sustainable Development*; 17(2):189–200.

Bates, A., 2013, *The Biochar Solution: Carbon Farming and Climate Change*, New Society Publishers, Gabriola Island, British Columbia, Canada.

Bellamy, R., J. Chilvers, N.E. Vaughan, and T.M. Lenton, 2013, "Opening up" geoengineering appraisal: Multi-criteria mapping of options for tackling climate change. *Global Environmental Change*, 23(5):926–937.

Biochar, 2014; A major biochar Internet list. https://groups.yahoo.com/neo/groups/biochar/info. Accessed date: 21 August, 2014.

Biochar-policy, 2014; A list specializing in biochar policy. https://groups.yahoo.com/neo/groups/biochar-policy/info. Accessed date: 21 August, 2014.

Bird, M. and P. Ascough, 2012, Isotopes in pyrogenic carbon: A review, *Organic Geochemistry* 42(12):1529–1539; http://eprints.gla.ac.uk/40429/1/40429.pdf. Accessed date: 21 August, 2014.

Bird, M. I., C. M. Wurster, P. H. de Paula Silva, A. M. Bass, & R. De Nys, 2011, Algal biochar–production and properties. *Bioresource technology*, 102(2):1886–1891.

Blunden, J., D.S. Arndt, K.M. Willett, A.J. Dolman, B.D. Hall, P.W. Thorne, A. Slnchez-Lugo et al., 2013, State of the climate in 2012, *Bulletin of the American Meteorological Society*; 94(8), S1–S238. See especially pp. 41–46; http://www.kcet.org/news/rewire/2012stateoftheclimate-web.pdf.

Bond, T.C., S.J. Doherty, D.W. Fahey, P.M. Forster, T. Berntsen, B.J. DeAngelo, M.G. Flanner et al., 2013, Bounding the role of black carbon in the climate system: A scientific assessment, *Journal of Geophysical Research: Atmospheres* 118(11):5380–5552; http://escholarship.org/uc/item/1c68530v.

Borland, A.M., H. Griffiths, J. Hartwell, and J.A.C. Smith, 2009, Exploiting the potential of plants with crassulacean acid metabolism for bioenergy production on marginal lands, *Journal of Experimental Botany* 60(10):2879–2896; http://jxb.oxfordjournals.org/content/60/10/2879.full.

Boucher, P. and C. Gough, 2012, Mapping the ethical landscape of carbon capture and storage, *Poiesis & Praxis* 9(3–4):249–270; http://link.springer.com/article/10.1007/s10202-012-0117-2/fulltext.html.

Bracmort, K., 2010, *Biochar: Examination of an Emerging Concept to Mitigate Climate Change*, Congressional Research Service, Washington, DC. Open source.

Bracmort, K. and R.K. Lattanzio, 2010, *Geoengineering: Governance and Technology Policy,* Congressional Research Service, Library of Congress, Washington, DC; http://agriculturedefensecoalition.org/sites/default/files/file/geo_current_116/116D_2010_Geoengineering_Governance_and_Technology_Policy_Report_by_Congressional_Research_Service_August_16_2010.pdf. Accessed date: 28 August, 2014.

British Biochar, 2014; Biochar—a sustainable waste management solution and effective soil amendment for cacao growers in Central America. http://www.britishbiocharfoundation.org/wp-content/uploads/Belize-biochar-project.pdf. Accessed date: 21 August, 2014.

Bryan, E., A. Wisdom, Y. Mahmud, and C. Ringler, 2008, *Global Carbon Markets: Are There Opportunities for Sub-Saharan Africa*? International Food Policy Research Institute, Washington, DC; http://books.google.com/books?hl=en&lr=&id=_tYVnOCxd3gC&oi=fnd&pg=PR5&dq=jindal+carbon+africa&ots=3AInwJ8RXa&sig=_h7NaeWdDb00nI7r1Vd5Ig8f6dk#v=onepage&q=jindal%20carbon%20africa&f=false.

Buck, H.J., October 2010, What can geoengineering do for us? Public participation and the new media landscape, in: *A Workshop on the Ethics of Geoengineering: Investigating the Moral Challenges of Solar Radiation Management,* University of Montana, Missoula, MT, pp. 18–20; http://umt.edu/ethics/ethicsgeoengineering/Workshop/articles1/Holly%20Buck.pdf.

Buck, H.J., 2011a; A framing analysis for biochar. http://groups.yahoo.com/neo/groups/biochar-policy/conversations/topics/3222. Accessed date: 21 August, 2014.

Buck, H.J., 2011b, *Climate Engineering in the New Media Landscape: Culture, Power, and Climate Control*; http://lup.lub.lu.se/luur/download?func=downloadFile&recordOId=1940495&fileOId=1940496.

Buck, H.J., 2014; Village science meets global discourse. http://geoengineeringourclimate.com/2014/01/14/village-science-meets-global-discourse-case-study/. Accessed date: 21 August, 2014.

Campbell, J.E., D.B. Lobell, R.C. Genova, A. Zumkehr, and C.B. Field, 2013, Seasonal energy storage using bioenergy production from abandoned croplands, *Environmental Research Letters* 8:035012, 7pp., doi:10.1088/1748-9326/8/3/035012; http://iopscience.iop.org/1748-9326/8/3/035012/pdf/1748-9326_8_3_035012.pdf.

Capron, M., 2012; http://www.oceanforesters.org/uploads NegativeCarbonViaOceanAfforestation2012Authors.pdf. Accessed date: 21 August, 2014.

Carbon War Room, 2014; Biochar. http://www.carbonwarroom.com/sectors/agriculture/biochar. Accessed date: 21 August, 2014.

CBD, 2012; Geoengineering in relation to the convention on biological diversity, Convention on Biological Diversity. http://www.cbd.int/doc/publications/cbd-ts-66-en.pdf; also http://jgshepherd.com/wp-content/uploads/2011/01/RS-Geoengineering-overview.pdf. Accessed date: 21 August, 2014.

CDIAC, 2014; Frequently asked global change questions, Carbon Dioxide Information Analysis Center, 2014; http://cdiac.ornl.gov/pns/faq.html. Accessed date: 22 August, 2014.

Chan, K.Y. and Z. Xu, 2009, Biochar: Nutrient properties and their enhancement, in: Lehmann, J. and S. Joseph, Eds. *Biochar for Environmental Management: Science and Technology*, pp. 67–84, Earthscan, London, U.K.

Chum, H., A.P.C. Faaij, J.R. Moreira, H.M. Junginger, and M.K. Patel, 2011, Bioenergy, IPCC special report on renewable energy sources and climate change mitigation, pp. 1–188; http://srren.ipcc-wg3.de/report/IPCC_SRREN_Ch02.pdf.

Clemmensen, K.E., A. Bahr, O. Ovaskainen, A. Dahlberg, A. Ekblad, H. Wallander, J. Stenlid, R.D. Finlay, D.A. Wardle, and B.D. Lindahl, 2013, Roots and associated fungi drive long-term carbon sequestration in boreal forest, *Science* 339(6127):1615–1618; https://www.sciencemag.org/content/339/6127/1528.full.pdf.

Cool Planet, 2013; Rick Wilson Presentation at Biochar 2013 Conference. http:// scholarworks.umass.edu/biochar/2013/Policy/1/. Accessed date: 22 August, 2014.

Cool Planet, 2014; Corporate website. http://www.coolplanet.com/; http://scholarworks.umass.edu/biochar/2013/Policy/1/. Accessed date: 22 August, 2014.

Cornell, 2014; Website for numerous published papers by Cornell University staff working on biochar topics. http://www.css.cornell.edu/faculty/lehmann/. Accessed date: 22 August, 2014.

Corner, A., K., Parkhill and N. Pidgeon, 2012; Experiment Earth? Reflections on a public dialogue on geoengineering, School of Psychology, Cardiff University. http://www.iagp.ac.uk/sites/default/files/IAGPWP_Cardiff_Reflections%20on%20ExpEarth_0.pdf. Accessed date: 27 August, 2014.

Cowie, A.L., A.E. Downie, B.H. George, B-P. Singh, L. Van Zwieten, and D.O Connell, 2012, Is sustainability certification for biochar the answer to environmental risks? *Pesquisa Agropecuuria Brasileira* 47(5):637–648.

Criscuoli, I., G. Alberti, S. Baronti, F. Favilli, C. Martinez et al., 2014, Carbon sequestration and fertility after centennial time scale incorporation of charcoal into soil, *PLoS One* 93:e91114, doi:10.1371/journal.pone.0091114; http://www.plosone.org/article/info%3Adoi%2F10.1371%2Fjournal.pone.0091114.

Downie, A., Biochar production and use: Environmental risks and rewards, Doctoral thesis, University of New South Wales, Paddington, New South Wales, Australia, (downloadable).

Downie, A., A. Crosky, and P. Munroe, 2009, Physical properties of biochar, in: *Biochar for Environmental Management: Science and Technology*, pp. 13–32.

EISS, 2014, Biochar in soils: New insights, *European Journal of Soil Science* 65(1):1–195.

Ernsting, A. and R. Smolker, 2011; A critical review of biochar science and policy. http://www.biofuelwatch.org.uk/2011/a-critical-review-of-biochar-science-andpolicy. Accessed date: 21 August, 2014.

ETHOS, 2014; Website on stoves conferences maintained by Prof. Mark Bryden. http://www.vrac.iastate.edu/ethos/conference.php. Accessed date: 23 August, 2014.

EU, 2012; Proposal for a Directive of the European Parliament and of the Council, European Commission. http://ec.europa.eu/energy/renewables/biofuels/doc/biofuels/com_2012_0595_en.pdf. Accessed date: 23 August, 2014.

Eurochar, 2014; Development of technologies for long-term carbon sequestration, Eurochar. http://www.eurochar.org/. Accessed date: 23 August, 2014.

Feng, Y., X. Yanping, Y. Yongchang, X. Zubin Xie, and X. Lin, 2012, Mechanisms of biochar decreasing methane emission from Chinese paddy soils, *Soil Biology and Biochemistry* 46:80–88.

Foereid, B., J. Lehmann, and J. Major, 2011, Modeling black carbon degradation and movement in soil, *Plant Soil* 345:223–236, doi:10.1007/s11104-011-0773-3.

Forster, P., L. Jackson, S. Lorenz, E. Simelton, E. Fraser, and K. Bahadur, 2012, Food Security: Near future projections of the impact of drought in Asia, Report no. 011, Centre for Low Carbon Futures.

Garrett-Peltier, H., 2012, *The Employment Impacts of a Low-Carbon Fuel Standard for Minnesota,* Political Economy Research Institute at the University of Massachusetts, Amherst, MA; http://www.peri.umass.edu/fileadmin/pdf/published_study/MNcleanfuels_PERI_Sept14_2012.pdf. Accessed date: 23 August, 2014.

Geoengineering, 2014; https://groups.google.com/forum/#!forum/geoengineering. Accessed date: 23 August, 2014.

Glaser, B., L. Haumaier, G. Guggenberger, and W. Zech, 2001, The "Terra Preta" phenomenon: A model for sustainable agriculture in the humid tropics, *Naturwissenschaften* 88:37–41.

Goreau, 2014; Website for Dr. Thomas Goreau, Principal Editor and the author of several chapters in this book. www.globalcoral.org/author/tgoreau/. Accessed date: 23 August, 2014.

Grossman, 2003, Warming up to a not-so-radical idea: Tort-based climate change litigation, *Columbia Journal of Environmental Law* 28:1; http://www.mdcampbell.com/pdfs/grossmanclimatechange.pdf. Accessed date: 23 August, 2014.

Gunther, 2012; Rethinking carbon dioxide: From a pollutant to an asset. Environment 360, Yale University. http://e360.yale.edu/feature/geoengineering_carbon_dioxide_removal_technology_from_pollutant_to_asset/2498/. Accessed date: 27 August, 2014.

Guo, L.B. and R.M. Gifford, 2002, Soil carbon stocks and land use change: A meta analysis, *Global Change Biology* 8(4):345–360.

H.O.M.E., 2013; Tree plantations and bioenergy with carbon capture: Far from "safe" geoengineering. http://www.handsoffmotherearth.org/2013/06/beccs-and-tropicalafforestation-review. Accessed date: 21 August, 2014.

Haberl, H., K.H. Erb, F. Krausmann, V. Gaube, A. Bondeau, C. Plutzar, S. Gingrich, W. Lucht, and M. Fischer-Kowalski, 2007, Quantifying and mapping the human appropriation of net primary production in earth's terrestrial ecosystems, *Proceedings of the National Academy of Sciences* 104(31):12942–12947; http://www.pnas.org/content/104/31/12942.full.

Hale, B., 2012, Getting the bad out: Remediation technologies and respect for others, *The Environment: Philosophy, Science, and Ethics* 2012:223.

Hale, B., 2013a; Remediation vs. Steering, Chapter 10 in Basr & Sandler *Designer biology*. http://sciencepolicy.colorado.edu/admin/publication_files/2013.28.pdf. Accessed date: 23 August, 2014.

Hale, B., 2013b; Can we remediate wrongs?, Chapter 9 in Kahn & Leonard. *Consequentialism and Environmental Ethics*. http://sciencepolicy.colorado.edu/admin/publication_files/2013.39.pdf. Accessed date: 23 August, 2014.

Hale, B. and W.P. Grundy, 2009, Remediation and respect: Do remediation technologies alter our responsibility? *Environmental Values* 18(4):397–415; http://128.138.136.233/admin/publication_files/resource-2781–2009.59.pdf.

Hammes, K. and M.W.I. Schmidt, 2009, Changes of biochar in soil, in: Lehmann, J. and S. Joseph, Eds. *Biochar for Environmental Management: Science and Technology*, pp. 169–181, Earthscan, London, U.K.

Hammond, J.R., 2009, *The Best Use of Biomass? Greenhouse Gas Lifecycle Analysis of Predicted Pyrolysis Biochar Systems*; https://www.era.lib.ed.ac.uk/bitstream/1842/3114/1/Hammond%20JAR%20Msc%2009.pdf.

Hammond, J., S. Shackley, S. Sohi, and P. Brownsort, 2011, Prospective life cycle carbon abatement for pyrolysis biochar systems in the UK, *Energy Policy* 39(5):2646–2655.

Hansen, J., P. Kharecha, M. Sato, V. Masson-Delmotte, F. Ackerman, D. Beerling, P. Hearty et al., 2013, Assessing dangerous climate change: Required reduction of carbon emissions to protect young people, future generations and nature, *PloS One* 8(12).

Hansen, J., M. Sato, P. Kharecha, D. Beerling, R. Berner, V. Masson-Delmotte, M. Pagani, M. Raymo, D.L. Royer, and J.C. Zachos, 2008, Target atmospheric CO_2: Where should humanity aim?, http://arxiv.org/pdf/0804.1126v3.pdf; Supplementary material http://arxiv.org/pdf/0804.1135v3.pdf. Accessed date: 23 August, 2014.

Hertsgaard, M., 2014; As uses of biochar expand, climate benefits still uncertain, Environment 360, Yale University. http://e360.yale.edu/feature/as_uses_of_biochar_expand_climate_benefits_still_uncertain/2730/. Accessed date: 27 August, 2014.

Heyward, S., 2013; Situating and abandoning geoengineering: A typology of five responses to dangerous climate change, *Political Science & Politics* 46(1):23–27.

Hogeboom, 2013; Are ideas to cool the planet realistic? *News-Science and the Environment*, BBC, 20 September, 2013. http://www.bbc.co.uk/news/science-environment-24033772. Accessed date: 27 August, 2014.

Houghton, R.A., 2003, Revised estimates of the annual net flux of carbon to the atmosphere from changes in land use and land management 1850–ough, *Tellus B* 55(2):378–390; http://onlinelibrary.wiley.com/doi/10.1034/j.1600-0889.2003.01450.x/full.

Hurni, H., T. Breu, P. Messerli, and B. Portner, 2013, Lead article in Chapter 4: Key Implications of Land Conversions in *Agriculture, Trade and Environment Review 2013*, UNCTAD http://unctad.org/en/PublicationsLibrary/ditcted2012d3_en.pdf. Accessed date: 23 August, 2014.

IBI, 2014; Biochar bibliography, international biochar initiative. http://www.biochar-international.org/biblio. Accessed date: 23 August, 2014.

IBI Certification, 2013; IBI biochar certification program manual, international biochar initiative. http://www.biochar-international.org/sites/default/files/IBI_Biochar_Certification_Program_Manual_August_28.pdf. Accessed date: 23 August, 2014.

IPCC, 2011; IPCC Expert Meeting on Geoengineering Lima, Peru, 20-22 June 2011 Meeting Report, http://www.ipcc-wg3.de/publications/supporting-material-1/EM-GeoE-Meeting-Report-final.pdf. Accessed date: 23 August, 2014.

IPCC, 2013, SPM summary for policymakers; IPCC-AR5. http://www.climatechange2013.org/images/report/WG1AR5_SPM_FINAL.pdf. Accessed date: 23 August, 2014.

Ippolito J.A., D.A. Laird, and W.J. Busscher, 2012, Environmental benefits of biochar, *Journal of Environmental Quality* 41:967–972, doi:10.2134/jeq2012.0151; https://www.soils.org/publications/jeq/pdfs/41/4/967.

Jandl, R., M. Lindner, L. Vesterdal, B. Bauwens, R. Baritz, F. Hagedorn, D.W. Johnson, K. Minkkinen, and K.A. Byrne, 2007a, How strongly can forest management influence soil carbon sequestration? *Geoderma* 137(3):253–268.

Jandl, R., L. Vesterdal, M. Olsson, O. Bens, F. Badeck, and J. Rock, 2007b, Carbon sequestration and forest management, CAB Reviews: Perspectives in Agriculture, Veterinary Science, Nutrition and Natural Resources 2(017); http://cabiblog.typepad.com/hand_picked/files/pav2017.pdf.

Jindal, R., B. Swallow, and J. Kerr, 2008, Forestry-based carbon sequestration projects in Africa: Potential benefits and challenges, *Natural Resources Forum* 32(2):116–130. Blackwell Publishing Ltd; http://www.katoombagroup.org/documents/cds/uganda_2011/PES%20in%20ESA/Forestry-based%20carbon%20sequestration%20projects%20in%20Africa.PDF.

Jokimaki, A., 2014; Papers on GHG emissions from bioenergy related land-use, AGW Observer, March 2014. http://agwobserver.wordpress.com/2014/03/10/papers-on-ghg-emissions-from-bioenergy-related-land-use/. Accessed date: 27 August, 2014.

Kalbitz, K. and K. Kaiser, 2008, Contribution of dissolved organic matter to carbon storage in forest mineral soils, *Journal of Plant Nutrition and Soil Science* 171(1):52–60; http://www.soil-science.com/fileadmin/downloads/som/session3_3.pdf.

Keller, D., Y. Feng, and A. Oschlies, 2013, Potential climate engineering effectiveness and side effects during a high CO_2-emissions scenario: A comparative assessment, *Nature Communications* 5: article 4304; http://www.nature.com/ncomms/2014/140225/ncomms4304/full/ncomms4304.html.

Kim, S.-S., V.L. Hoang, G.-H. Choi, J. Kim, and H.C. Woo, 2012, Pyrolysis characteristics and kinetics of the alga *Saccharina japonica*, *Bioresource Technology* 123(11):445–451.

Kleiner, K., 2009, The bright prospect of biochar, *Nature Reports Climate Change* 2009:72–74; http://www.nature.com/climate/2009/0906/full/climate.2009.48.html.

Kookana, R.S., A.K. Sarmah, L. van Zwieten, E. Krull, and B. Singh, 2011, Biochar application to soil: Agronomic and environmental benefits and unintended consequences, *Advances in Agronomy* 112:103; http://s3.amazonaws.com/academia.edu.documents/30651117/1-s2.0-B9780123855381000032-main.pdf?.

Krugman, P., 2013; http://www.nybooks.com/articles/archives/2013/nov/07/climate-change-gambling-civilization/.

Laird, D.A., 2008, The charcoal vision: A win-win-win scenario for simultaneously producing bioenergy, permanently sequestering carbon, while improving soil and water quality, *Agronomy Journal* 100(1):178–181; http://carbon-negative.us/docs/CharcoalVision.pdf.

Laird, D.A., R.C. Brown, J.E. Amonette, and J. Lehmann, 2009, Review of the pyrolysis platform for coproducing bio-oil and biochar, *Biofuels, Bioproducts and Biorefining* 3(5):547–562.

Lamers, P., M. Junginger, C.C. Dymond, and A. Faaij, 2014, Damaged forests provide an opportunity to mitigate climate change, *GCB Bioenergy* 6(1):44–60; http://onlinelibrary.wiley.com/store/10.1111/gcbb.12055/asset/gcbb12055.pdf?v=1&t=htbgv2uo&s=0420c1c4043fffbe9d89c81bb85848f5f2d68a84.

Larson, R.W., 2006, Editorial in Nov-Dec. Issue of Solar Today Magazine; the text was being electronically transcribed in early 2014. See Larson, 2007a for equivalent.

Larson, R.W., 2007a, Using biochar for cost-effective CO_2 sequestration in soils, *Proceedings of ISES World Congress 2007*, Vols. I and V, pp. 2462–2467, Springer, Berlin, Heidelberg, 2009; http://www.denverclimatestudygroup.com/08-7THMeeting200709-reviewANDTerraPreta/20070911b%20-%20LARSON%2002%20Terra%20Preta%20ISES-3.pdf & http://www.denverclimatestudygroup.com/08-7THMeeting200709-reviewANDTerraPreta/20070911b%20-%20LARSON%2002%20ISES_terrapreta.pdf.

Larson, 2007b; RD&D issues for agrichar, emphasis on energy and developing countries, Abstract for 2007 Conference. http://www.biochar-international.org/images/IAI_2007_Conference_Booklet.pdf. Accessed date: 26 August, 2014.

Larson, R.W., 2009; A powerpoint at group meeting. http://www.denverclimatestudygroup.com/19-18TH_Meeting20091013BIOCHAR_RON-LARSON/RonLarsonBiochar_B.pdf. Accessed date: 26 August, 2014.

Larson, R.W., 2010, Biochar GHG reduction accounting, in: *Potential Biochar Greenhouse Gas Reductions*, pp. 51. http://biochar-us.org/pdf%20files/biochar_report_lowres.pdf. Accessed date: 26 August, 2014.

Larson, R.W., 2011, A better backup for solar/wind? Think biochar, *Solar Today* 25(7):10.

Larson, R.W., 2013; Several presentations at the USBI2013 Conference, Amherst MA. http://scholarworks.umass.edu/biochar/2013/Policy/8/. Accessed date: 26 August, 2014.

Larson, R.W. and J. Levine, 2008; Present value of future biochar economic benefits, Poster for 2008 conference. http://www.biochar-international.org/images/levine_biochar_newcastle_glc_rl_jl_v2_2_.pdf. Accessed date: 26 August, 2014.

Lawrence, R., 2010, *How Good Politics Results in Bad Policy: The Case of Biofuel Mandates*; http://www.hks.harvard.edu/var/ezp_site/storage/fckeditor/file/pdfs/centers-programs/centers/mrcbg/publications/fwp/mrcbg_fwp_2010-12_Lawrence_biofuel.pdf.

Lehmann, J., J. Gaunt, and M. Rondon, 2006, Bio-char sequestration in terrestrial ecosystems—A review, *Mitigation and Adaptation Strategies for Global Change* 11:403–427.

Lehmann, J. and Joseph, S., 2009, Biochar systems, in: Lehmann, J. and S. Joseph, Eds. *Biochar for Environmental Management: Science and Technology*, pp. 147–168, Earthscan, London, U.K.

Lehmann, J. et al., 2011, Biochar for climate change mitigation: Prospects and limitations; p52-53 IPCC Expert Meeting on Geoengineering, Lima, Peru, 20–22 June 2011 meeting report. http://www.ipcc-wg3.de/publications/supporting-material-1/EM-GeoE-Meeting-Report-final.pdf. Accessed date: 23 August, 2014.

Lenton, T.M., 2010, The potential for land-based biological CO_2 removal to lower future atmospheric CO_2 concentration, *Carbon Management* 1(1):145–160 (downloadable).

Lenton, T.M. and N.E. Vaughan, 2009, The radiative forcing potential of different climate geoengineering options, *Atmospheric Chemistry and Physics* 9:5539–5561.

Levitan, T.M., 2010; http://e360.yale.edu/feature/refilling_the_carbon_sink_biochars_potential_and_pitfalls/2349/.

Magdoff, F., 2008, The political economy and ecology of biofuels, *Monthly Review-New York* 60(3):34; http://monthlyreview.org/2008/07/01/the-political-economy-and-ecology-of-biofuels.

Marschner, B., S. Brodowski, A. Dreves, G. Gleixner, A. Gude, P. Grootes, U. Hamer et al., 2008, How relevant is recalcitrance for the stabilization of organic matter in soils? *Journal of Plant Nutrition and Soil Science* 171(1):91–110.

Matovic, D., 2011, Biochar as a viable carbon sequestration option: Global and Canadian perspective, *Energy* 36(4):2011–2016.

Matthews, H.D. and S. Solomon, 2013, Irreversible does not mean unavoidable, *Science* 340(6131):438–439.

McLaren, D., 2011, *Negatonnes. An Initial Assessment of the Potential for Negative Emission Techniques to Contribute Safely and Fairly to Meeting Carbon Budgets in the 21st Century*, Friends of the Earth, London, U.K.; http://www.foe.co.uk/resource reports/negatonnes.pdf 2011 or http://www.foe.co.uk/sites/default/files/downloads/negatonnes.pdf; or https://www.see.ed.ac.uk/~shs/Climate%20change/Carbon%20sequestration/negatonnes.pdf.

Meadowcroft, J., 2013, Exploring negative territory carbon dioxide removal and climate policy initiatives, *Climatic Change* 118(1):137–149; http://www.cmc-nce.ca/wp-content/uploads/2013/09/Exploring-negative-territory.pdf.

Montana, 2014; Biochar, afforestation, resources, *Ethics of Geoengineering Online Resource Center*, University of Montana. http://www.umt.edu/ethics/resourcecenter/bibliography/alternatives.php. Accessed date: 23 August, 2014.

Morrow, D.R., R.E. Kopp, and M. Oppenheimer, 2009, Toward ethical norms and institutions for climate engineering research, *Environmental Research Letters* 4(4):045106; http://www.princeton.edu/step/people/faculty/michael-oppenheimer/research/Morrow-et-al-2009-Toward-ethical-norms-and-institutions-for-climate-engineering-research.pdf.

Morton, O., 2009; Why people disagree about geoengineering, Morton blog entry, October 26, 2009. http://heliophage.wordpress.com/2009/10/26/why-people-disagree-about-geoengineering/#comments. Accessed date: 23 August, 2014.

Moss, R.H. Meehl, G.A., Lemos, M. C., et al., 2013, Hell and high water: Practice-relevant adaptation science, *Science* 342(6159):696–698.

Nerlich, B. and R. Jaspal, 2012, Metaphors we die by? Geoengineering, metaphors, and the argument from catastrophe, *Metaphor and Symbol* 27(2):131–147; https://www.academia.edu/1123104/Metaphors_we_die_by_Geoengineering_metaphors_and_the_argument_for_catastrophe.

Neville, K., 2012, The contentious political economy of biofuels: Transnational struggles over food, fuel, and the environment, Doctoral Thesis, Political Science, University of British Columbia, Kelowna, British Columbia, Canada; http://www.ligi.ubc.ca/?p2=/modules/liu/scholars/profile.jsp&id=10.

Nordhaus, W.D., 2013, *The Climate Casino*, Kindle Location 78. Yale University Press, New Haeven, CT.

Novotny, E.H., M.H.B. Hayes, B.E. Madari, T.J. Bonagamba, E.R. de Azevedo, A.A. de Souza, G. Song, C.M. Nogueira, and S. Mangrich, 2009, Lessons from the Terra Preta de ari, of the Amazon region for the utilisation of charcoal for soil amendment, *Journal of the Brazilian Chemical Society* 20(6):1003–1010.

N'Yeurt, A.R. et al., D.P. Chynoweth, M.E. Capron, J.R. Stewart, and M.A. Hasan, 2012, Negative carbon via Ocean Afforestation, *Process Safety and Environmental Protection* 90(6):467–474; http://www.oceanforesters.org/uploads/NegativeCarbonViaOceanAfforestation2012Authors.pdf.

OCB, 2014; Websites for the US Office of Carbon and Biogeochemistry. http://www.us-ocb.org/ and http://www.whoi.edu/OCB-OA/FAQs. Accessed date: 23 August, 2014.

Oxford, 2009; What is geoengineering, Oxford Geoengineering Programme, Oxford Martin School, University of Oxford. http://www.geoengineering.ox.ac.uk/what-is-geoengineering/what-is-geoengineering/. Accessed date: 23 August, 2014.

Oxford, 2013; What is geoengineering, Oxford Geoengineering Programme, Oxford Martin School, University of Oxford. http://www.climate.ox.ac.uk/ai1ec_event/oxford-conference-on-negative-emissions-technologies/?instance_id=. Accessed date: 23 August, 2014.

Paul, H., A. Ernsting, S. Semino, S. Gura, and A. Lorch, 2009, Agriculture and climate change: Real problems, false solutions, preliminary report by Econexus, Biofuelwatch, Grupo de Reflexion Rural and NOAH, Bangkok, Thailand; http://www.powerbase.info/images/d/d5/Agriculture_climate_change_September_2009.pdf.

Preston, C.J., 2011, Re-thinking the unthinkable: Environmental ethics and the presumptive argument against geoengineering, *Environmental Values* 20(4):457–479; https://www.see.ed.ac.uk/~shs/Climate%20change/Geo-politics/Preston%20ethics.pdf.

Reed, D., 2011; IBI update on UNFCCC negotiations and the potential role for biochar. http://www.biochar-international.org/sites/default/files/IBI_UNFCCC_Policy_Update_January_2011.pdf. Accessed date: 23 August, 2014.

Redd-Monitor, 2013; Guest post: LULUCF, loopholes and REDD, website for redd-monitor. http://www.redd-monitor.org/2010/08/06/lulucf-loopholes-and-redd/#more-5353. Accessed date: 23 August, 2014.

Ren21, 2013; Renewables 2013 Global status report, REN21, http://www.ren21.net/Portals/0/documents/Resources/GSR/2013/GSR2013_lowres.pdf. Accessed date: 23 August, 2014.

Robbins, P., 2011, *Political Ecology: A Critical Introduction*, Vol. 16, Wiley, West Sussex, U.K.; http://geopuc.geo.puc-rio.br/media/6resenha9.pdf.

Roberts, K.G., B.A. Gloy, S. Joseph, N.R. Scott, and J. Lehmann, 2010, Life cycle assessment of biochar systems: Estimating the energetic, economic, and climate change potential, *Environmental Science & Technology* 44(2):827–833; ftp://ftp.dpvta.uniud.it/peressotti/Biochar/2010/Cornell-Roberts2010-Life%20cycle%20assessment%20of%20biochar%20systems-Estimating%20the%20energetic,%20economic%20and%20climate%20change%20potential.pdf.

Robock, A., 2008, 20 reasons why geoengineering may be a bad idea, *Bulletin of the Atomic Scientists* 64(2):14–18, 59.

Schmidt, H.-P., 2012, 55 uses of biochar, *Ithaka* 2012(1):286–289.

Shackley, S., S. Carter, K. Sims, and S. Sohi, 2011b, Expert perceptions of the role of biochar as a carbon abatement option with ancillary agronomic and soil-related benefits. *Energy & Environment* 22(3):167–188.

Shackley, S., J. Hammond, J. Gaunt, and R. Ibarrola, 2011a, The feasibility and costs of biochar deployment in the UK, *Carbon Management* 2(3):335–356; http://www.geos.ed.ac.uk/homes/sshackle/CostsBiochar.pdf.

Shackley, S., S. Sohi, P. Brownsort, S. Carter, J. Cook, C. Cunningham, J. Gaunt et al., 2010, *An Assessment of the Benefits and Issues Associated with the Application of Biochar to Soil,* Department for Environment, Food and Rural Affairs, UK Government, London, U.K.; http://www.geos.ed.ac.uk/homes/sshackle/SP0576_final_report.pdf.

Shepherd, J.G., 2009, *Geoengineering the Climate: Science, Governance and Uncertainty,* Royal Society, London, U.K.; http://royalsociety.org/uploadedFiles/Royal_Society_Content/policy/publications/2009/8693.pdf.

Singh, B.P. and A. Cowie, August 2010, The mean turnover time of biochar in soil varies depending on biomass source and pyrolysis temperature, *19th World Congress of Soil Science*, Soil Solutions for a Changing World, Brisbane, Queensland, Australia; www.iuss.org/19th%20WCSS/symposium/pdf/0326.pd.

Singh, B.P., B.J. Hatton, B. Singh, A. Cowie, and A. Kathuria, 2010, Influence of biochars on nitrous oxide emission and nitrogen leaching from two contrasting soils, *Journal of Environmental Quality* 39:1224–1235.

Slade, R., R. Saunders, R. Gross, and A. Bauen, 2011, Energy from biomass: The size of the global resource—An assessment of the evidence that biomass can make a major contribution to future global energy supply; www.ukerc.ac.uk/support/tiki-download_file.php?fileId=2095.

Smith, B., 2013; The biosphere in the climate system, Spring Working Meeting 2013, MERGE. http://www4.lu.se/upload/cec/MERGE/SpringWorkingMeeting2013/1_smith_130305.pdf. Accessed date: 25 August, 2014.

Smith, L.J. and M.S. Torn, 2013, Ecological limits to terrestrial biological carbon dioxide removal, *Climatic Change* 118:892013, doi:10.1007/s10584-012-0682-3; http://link.springer.com/content/pdf/10.1007%2Fs10584-012-0682-3.pdf.

Sohi, S.P., E. Krull, E. Lopez-Capel, and R. Bol, 2010, A review of biochar and its use and function in soil, *Advances in Agronomy* 105:47–82; http://www.soil-science.com/fileadmin/downloads/som/session5_4.pdf.

Solomon, S., G.-K. Plattner, R. Knutti, and P. Friedlingstein, 2009, Irreversible climate change due to carbon dioxide emissions, *Proceedings of the National Academy of Sciences of the United States of America* 106(6):1704–1709; http://www.pnas.org/content/106/6/1704.full.

Stavi, I., 2012, The potential use of biochar in reclaiming degraded rangelands, *Journal of Environmental Planning and Management* 55(5):657–665.

Stavi, I. and R. Lal, 2011, Challenges and opportunities of soil organic carbon sequestration in croplands, in: *Biodiversity, Biofuels, Agroforestry and Conservation Agriculture,* pp. 149–174, Springer, Dordrecht, the Netherlands.

Stavi, I and R. Lal, 2013, Agroforestry and biochar to offset climate change: A review, *Agronomy for Sustainable Development* 33(1):81–96.

Stern, N., 2013, The structure of economic modeling of the potential impacts of climate change: Grafting gross underestimation of risk onto already narrow science models, *Journal of Economic Literature* 51(3):838–859; http://personal.lse.ac.uk/sternn/128NHS.pdf.

Stoves (2014); http://stoves.bioenergylists.org. Accessed date: 21 August, 2014.

Sugiyama, M., R.J. Nicholls, and A. Vafeidis, 2008, *Estimating the Economic Cost of Sea-Level Rise*, MIT Joint Program on the Science and Policy of Global Change, Boston, MA; http://web.mit.edu/globalchange/www/MITJPSPGC_Rpt156.pdf.

Thies, J. and M. Rillig, 2009, Characteristics of biochar: Biological properties, in: Lehmann, J. and S. Joseph, Eds. *Biochar for Environmental Management: Science and Technology*, pp. 85–106, Earthscan, London, U.K.

Thornley, P., June 2012, Biofuels review, Report for Government Office for Science, d Technologyhnologye; climate change: Grafting gross undele Consumption Institute, University of Manchester, Manchester, U.K., Report number BA07067/2012/rep001r04.

Tilman, D., J. Hill, and C. Lehman, 2006, Carbon-negative biofuels from low-input high-diversity grassland biomass, *Science* 3145805:1598v1600, doi:10.1126/Science.1133306.

Tilman, D., R. Socolow, J.A. Foley, J. Hill, E. Larson, L. Lynd, S. Pacala et al., 2009, Beneficial biofuels—The food, energy, and environment trilemma, *Science* 325(5938):270.

Titirici, M.-M., A. Thomas, and M. Antonietti, 2007, Back in the black: Hydrothermal carbonization of plant material as an efficient chemical process to treat the CO_2 problem? *New Journal of Chemistry* 31(6):787–789.

Tulip, 2014; Ocean based algae production system, Mr. Tulip's Web site. http://rtulip.net/ocean_based_algae_production_system_provisional_patent. Accessed date: 25 August, 2014.

Tyndall, 2013; The Radical Emission Reduction Conference: 10–11 December 2013. Tyndall Center for Climate Change Research, University of East Anglia (UEA). http://www.tyndall.ac.uk/radical-emission-reduction-conference-tyndall-centre-event-confronting-challenge-climate-change. Accessed date: 25 August, 2014.

Ulyett, J., R. Sakrabani, M. Kibblewhite, and M. Hann, 2014, Impact of biochar addition on water retention, nitrification and carbon dioxide evolution from two sandy loam soils, *European Journal of Soil Science* 65(1):96–104.

UNFCCC, 2013, *Afforestation and Reforestation Projects under the Clean Development Mechanism: A Reference Manual*; http://cdm.unfccc.int/filestorage/e/x/t/extfile-20140117095534160-196615_UNFCCC_CDM_AFFORESTATION_BRO_web.pdf/196615_UNFCCC_CDM_AFFORESTATION_BRO_web.pdf?t=T2h8bXpzZjh5fDDZEzX9iXRjlKJOOm7RNFnz.

USBI, 2014; http://climatestate.com/2014/02/17/biochar-the-next-stage-in-climate-action/; http://www.youtube.com/watch?v=JPJsYZLU_sM.

van Zwieten, L., S. Kimber, S. Morris, K. Chan, A. Downie, J. Rust, S. Joseph, and A. Cowie, 2010, Effects of biochar from slow pyrolysis of papermill waste on agronomic performance and soil fertility, *Plant Soil* 327:235–246.

Vaughan, N.E. and T.M. Lenton, 2011, A review of climate geoengineering proposals, *Climate Change*, 109(3–4):745.

Verheijen, F.G.A., S. Jeffery, M.van der Velde, V. Penener, M. Beland, A.C. Bastos, and J.J. Keizer, 2013, Reductions in soil surface albedo as a function of biochar application rate: Implications for global radiative forcing, *Environmental Research Letters* 8(4):044008; http://iopscience.iop.org/1748-9326/8/4/044008/article.

Virgin Earth, 2014; Virgin earth challenge, Wikipedia. http://en.wikipedia.org/wiki/Virgin_Earth_Challenge. Accessed date: 25 August, 2014.

Walker, 2011; Biochar remains promising, website of Mr. Gareth Renowden. http://hot-topic.co.nz/biochar-remains-promising/. Accessed date: 25 August, 2014.

WHOI, 2014; http://www.whoi.edu/OCB-OA/page.do?pid=112161#0.

Wilson, R., 2013; Intro to Cool Planet; http://scholarworks.umass.edu/biochar/2013/Policy/1/. Accessed date: 25 August, 2014.

Wong, P.H., 2013; http://blog.practicalethics.ox.ac.uk/2013/07/pak-hang-wong-on-virtuous-climate-making-towards-a-virtue-theoretic-approach-to-geoengineering/. Accessed date: 25 August, 2014.

Wood, S. and A. Cowie, 2004, A review of greenhouse gas emission factors for fertiliser production, IEA Bioenergy Task 38, Greenhouse Gas Balances of Biomass and Bioenergy Systems. www.ieabioenergy-task38.org/publications/. Accessed date: 25 August, 2014.

Woolf, D., J.E. Amonette, F.A. Street-Perrott, J. Lehmann, and S. Joseph, 2010, Sustainable biochar to mitigate global climate change, *Nature Communications* 1:56 (downloadable).

World Bank, 2011, Wood-based biomass energy development for sub-Saharan Africa; http://siteresources.worldbank.org/EXTAFRREGTOPENERGY/Resources/717305-1266613906108/BiomassEnergyPaper_WEB_Zoomed75.pdf. Accessed date: 25 August, 2014.

WorldStove, various years, this site on BFW was inaccessible through its recently modified home page; readers should search on the terms World Stove Biofuelwatch.

WSU, 2012; Report on pyrolysis, Washington State University. https://fortress.wa.gov/ecy/publications/publications/1107017.pdf. Accessed date: 25 August, 2014.

Yale, 2014; Yale project on climate change communication, Yale University. http://environment.yale.edu/climate-communication/. Accessed date: 25 August, 2014.

4

Potential of Remineralization as a Global Movement

Joanna Campe

CONTENTS

Overview

Remineralization, amending soils with highly mineralized rock dust, sea-based products, and other natural materials, is urgent to restore ecological balance and stabilize climate. Soil remineralization (SR) has the potential to play a critical role in sustainable agriculture, ecological restoration, carbon sequestration, and climate stabilization while addressing major social issues of overcoming poverty and hunger.

Though simple and inexpensive, remineralization could revolutionize agriculture by breaking the cycles of soil destruction, chemical dependency, and nutritional degradation. In addition to rock dust, soils can be fertilized with diluted seawater. Seawater can also be concentrated into a commercial product, diluted, and applied as sprays, irrigation systems, or directly to soils. Land-based sea solids from geological deposits, like those found in Utah or Baja California in Mexico, are also excellent sources. Some dam sediments may also contain nutrients similar to the Nile River silt that fertilized soils along the Nile River. Companies currently formulate rock dust and sea-based mineral products for agriculture, landscaping, and forestry. Some form of remineralization is possible almost everywhere using an available natural resource or commercial product.

This chapter looks at the remineralization movement through the lens of its history, current developments, and potential future directions in order to address global ecological and social challenges.

Historic Perspective of Remineralization

> Our most optimistic expectations are no less than the realization of an old dream: "What will fertilizing with stone dust accomplish? It will turn stones into bread… make barren regions (fruitful) (and) feed the hungry."
>
> **Chesworth et al., 1983**

Remineralization has been researched and explored by three distinct groups:

First, German nutritional biochemist Julius Hensel pioneered SR in the 1880s with his book *Bread from Stones*. From this book, a modest agricultural movement came into being (Hensel, 1982). Following his contribution, many scientists did research since the late 1930s in Germany and Central Europe on SR for agriculture and forests, including one 60-year study of a forest described later in this chapter.

The technology was not available at the turn of the century to produce finely ground rock dust, so SR, as promoted by Hensel, was not then feasible on a large scale. However, SR was revived about 40 years ago in Europe. Many rock dust products for agriculture, forestry, and sewage sludge treatment have been created and successfully marketed in Germany, Austria, and Switzerland in the last several decades. Companies including Lava-Union (Germany), Sanvita (Austria), and Bernasconi (formerly known as Zimmerli, Switzerland), as well as the trade association, the National Stone Industry (Die Naturstein Industrie) based in Bonn, Germany, have also done a great deal of research.

Dr. Peter von Fragstein at the University of Kessel, Germany, has researched remineralization as a slow-release fertilizer with many different rock types. Rock dust sprayed on plants deters insects in the short term and silica in rock dust strengthens plant tissues (which contain silica granules called phytoliths) in the long term and makes them less susceptible to drought, insects, and diseases (Fragstein, 1995).

The Carl Duisberg Gesellschaft e.V., in partnership with the Technische Universität Berlin, the University of Abidjan, the Société pour le Développement Minier de la Côte d'Ivoire (SODEMI), and the German Federal Ministry of Cooperation, organized regional workshops in the Ivory Coast in 1991 entitled *Rock Fertilizers: A Chance for West Africa's Food Production*.

The second is the more recently developed field of agrogeology. This research has been carried out mainly in Canada, Brazil, Tanzania, the Canary Islands, and West Africa, especially on laterite soils. Under intense tropical rainfall, NPK fertilizers can be washed out in only a few weeks, so they cannot be stored by many tropical soils and are especially harmful to groundwater. Rock fertilizers not only supply nutrients throughout longer periods to cultivated plants but also improve the ion exchange capacity of soils by forming new clay minerals during weathering of rock powder fertilizer. Researchers documenting these effects include William Fyfe, Ward Chesworth, Peter van Straaten, Othon Leonardos, and Suzi Theodoro. Leonardos initiated his own remineralization work and only later came together with agrogeologists at UC Berkeley. The first International Conference on Rocks for Crops, made up of agrogeology researchers, was held in Brasilia in 2004.

Third, the grassroots movement concerned with the premise of John Hamaker in the book The Survival of Civilization, coauthored with Don Weaver, asserts that SR is not only the key to restoring soils and forests but, in the larger context, absolutely necessary and urgent to reduce levels of carbon dioxide in the atmosphere and stabilize climate. Especially recommended are rock gravels and glacial moraine deposits, which provide natural mixtures of rocks with the broadest possible spectrum of minerals and trace elements, as well as many hard silicate rock types such as basalt and slower-weathering granite.

The movement that began with Hamaker's writing in the early 1970s expanded in the 1980s into a global grassroots community of ecologically concerned individuals and organizations, farmers and gardeners, scientists, and policymakers. Two examples of organizations developing from this movement are Remineralize the Earth (RTE), in Northampton, MA, and the Sustainable Ecological Earth Regeneration (SEER) Centre, a permaculture-based center with ongoing demonstration plots in Scotland.

To facilitate networking and the flow of information and promote remineralization as advocated by Hamaker and Weaver in The Survival of Civilization, in 1986 Joanna Campe started *Soil Remineralization, A Network Newsletter*, renamed *Remineralize the Earth Magazine* in 1994. The magazine networked people all over the world, collecting a wealth of anecdotal results from farmers and gardeners, as well as research from silviculture and forests. Hamaker's coauthor Don Weaver continues to network through his publications, lectures, videos, and articles and also contributes to the RTE website.

The mission of RTE, a US Internal Revenue Service 501(c)(3) nonprofit organization formed in 1995, is to educate and facilitate a worldwide movement of dedicated local, national, and international policies for application of land (rock dust), sea-based minerals, and other natural means to promote sustainable farming and forestry and ensure healthier soils, food, and natural environments. Through research, education, information databases, advocacy, partnerships, and implementation, RTE supports efforts regenerating soils and forests around the world, increasing the nutritional quality and yield of food production and stabilizing climate.

Some of the more important initiatives of RTE include the following:

> On May 24, 1994, the United States Department of Agriculture (USDA) (Beltsville, MD), the United States Bureau of Mines (USBM) (Washington, DC), the National Stone Association (NSA) (Washington, DC), the National Aggregate Association (NAA) (Silver Spring, MD), and RTE cosponsored A Forum On Soil Remineralization and Sustainable Agriculture at the USDA Agricultural Research Station in Beltsville, MD.
>
> **Campe et al., 2012**

The forum brought together the by-product rock mineral fines generating industry and proponents of SR to explore environmentally sound uses of rock dust (referred to by industry as mineral fines) to identify the state of the science supporting their use and the gaps in knowledge that need to be filled.

The USDA began a series of demonstration trials with rock mineral fines (from Georgia, Maryland, and New York) and other industrial by-products. Dr. Ronald Korcak, research leader of the fruit lab, directed the trials over a 3-year period. Compost with rock dust was researched under the direction of Dr. Larry Sikora. The USBM designed a prototype geographic information system (GIS) database to target soils most in need of SR and their distance from regional sources of rock fines to calculate transport costs and marketability of specific mineral fines. The NAA created a Task Force on Remineralization exploring the possibilities for creating sustainable products for agriculture, forestry, and other uses. Research projects were undertaken at universities, as part of research and development programs of some of the largest aggregate companies in the United States, Europe, and Australia and through organizations such as Men of the Trees in Australia (Campe et al., 2012).

RTE was invited by the State Department and the American Council on Renewable Energy (ACORE) to sponsor an official side event at the Washington International Renewable Energy Conference (WIREC) (2008).

In September 2009, a conference on remineralization entitled *Primeiro Congresso Brasileiro de Rochagem* was held in Brasilia. Hosted by researchers from the University of Brasilia, the conference was attended by 170 participants. The second congress was held in April 2013.

Recent developments are updated in the "Current Developments (2008–2013)" section of this chapter.

Soil Formation and Fertility

SR creates fertile soils by returning mineral elements to the soil, much the same way the Earth does: by weathering minerals from rocks. Normally, this is a slow physical and chemical process, limited by the surface area of the rock exposed to water and CO_2. During ice ages, glaciers crushed rocks in their path, producing fine rock debris that was carried by water and ice to form sediment deposits at the end of the glacier called moraines. After the ice ages, winds blew the dust, called *loess*, all over the globe. Loess soils, most abundant in China, Eastern Europe, North America, and Argentina, are the foundation of the highly productive agriculture of those regions.

Volcanoes erupt, spewing forth minerals from deep within the Earth, and minerals are contained in alluvial deposits. As rocks weather, their minerals are released into new soils. These are extremely productive where fresh volcanic materials are weathered but decrease in fertility with age as soluble minerals are leached from the soils, especially in warm areas with high rainfall. This is why young volcanic areas, such as Java, Costa Rica, and Hawaii, are so fertile, and why older tropical areas, such as Brazil, Africa and Australia, have such poor, mineral-deficient soils. Such infertile soils would benefit enormously by having volcanic rock powders added to them to increase their fertility.

Reversing Desertification and Climate Change

Remineralization of denuded soils is one of the most effective techniques for dealing with global climate change. Denuded soils are often deficient in specific minerals that are limiting factors in biological systems. By remineralizing these soils with often local and inexpensive mineral amendments, their biological vitality can be greatly increased.

This principle of limiting factors is well understood, and the tests and materials are well developed and inexpensive. Forests can grow much more rapidly when limiting minerals are replaced, and eroded and desertified lands can be brought back to life with simple proper remineralization protocols. Most importantly for humanity, agricultural lands can have yields, disease resistance, pest resistance, and nutritive levels increased dramatically through proper remineralization.

Proper SR, with attention paid to amendments of deficient minerals and biological digestion of these minerals to make them available for plant life, is key in increasing soil carbon. Numerous studies have shown that increased primary productivity can result in increased soil carbon sequestration. Allan Yeomans in *Priority One* proposes to bring atmospheric carbon to preindustrial revolution levels in 5 years through targeted use of SR of the world's agricultural lands (Yeomans, 2005).

Rock Powders

> … a one-pound stone might have a surface area of 12 square inches. Ground to about 200 mesh, it would have a surface area of about 8 acres. One ton would therefore have a surface area of 16,000 acres. The significant thing about that 16,000 acres is that it is all freshly-broken stone with the useful elements exposed right on the surface. These elements are readily available for extraction by the microorganisms.
>
> **John Hamaker**
> *The Survival of Civilization*

FIGURE 4.1 Brazilian basalt rock dust of three different finenesses.

Igneous rocks contain a broad spectrum of up to 100 elements and trace elements necessary for the well-being of all life and the creation of fertile soils. Glacial moraine sediments or mixtures of rock types applied to soils create a sustainable and superior alternative to the use of ultimately harmful chemical fertilizers, pesticides, and herbicides. These minerals are slowly released and long lasting, as they must be weathered out of the mineral crystals that contain them. The finer grained the material, the more rapidly it becomes available to soil and plant roots (Figure 4.1).

Rock powders act as fertilizers because their small grain size allows the mineral elements they contain to be readily released by the action of water, carbon dioxide, and organic acids released by plant roots and beneficial microorganisms. Finer materials have higher surface area and release their minerals more quickly. The greatest benefits are obtained on the most nutrient-deficient soils. Rock dust amendments therefore are a strategy for improving soil fertility that mimics natural processes that are slow acting in many places because elements essential to plant growth are deficient in the soil. Biochar can serve an important role to enhance the effect of rock powders because though biochar is not a fertilizer *per se*, it acts to retain soil water and nutrients, making them available to plant roots and symbiotic fungi that provide plants with much of their nutrition.

Rock powders provide slow, natural release of elements and trace minerals, increasing the nutrient intake of plants, providing higher yields and giving a higher Brix reading, which measures dissolved solids in the sap of plants, correlating with greater nutritive value. They also rebalance soil pH, increase the growth of microorganisms and earthworm activity, build soil organic matter, prevent soil erosion, and increase the water storage capacity of the soil. Finally, they increase resistance to insects, disease, frost, and drought; enhance flavor in crops; and decrease dependence on fertilizers, pesticides, and herbicides.

Current Agricultural Practices Transformed through Remineralization: Chemical Fertilizers versus Igneous Rock

> There's no secret formula, the product is simply ancient volcanic rock containing mineral nutrients, which has been ground into a fine powder.
>
> **Sam Catalano**
>
> *MinPlus™, Queensland, Australia*

Contrary to opinion among some agronomists that rock-forming minerals have a soil lifetime far beyond a crop cycle, many rock minerals solubilize rapidly both in water and in weak organic acids, promptly releasing nutrients and increasing the pH of solution until the system is saturated at nutrient levels equivalent to *glacial milk* and groundwater composition (Leonardos et al., 1987).

In contrast to chemical fertilizers, rock powders are a natural material, not a synthetic one, so production costs are much lower than for chemicals that must be extracted and refined into pure form or synthesized like ammonium or urea.

Millions of tons of rock dust appropriate for soil and forest regeneration are stockpiled by the gravel and stone industries worldwide as a nontoxic and inexpensive by-product that can be used regionally. While rock powder is a bulky item and transportation to remote locations constitutes a large portion of the cost, this is also true for chemical fertilizers or organic mulches.

Industrial agriculture primarily uses soluble chemical fertilizers that are rapidly released. In particular, excess nitrogen lost from soil contaminates ground waters and rivers, requiring frequent applications. Chemical fertilizers are expensive due to large amounts of fossil fuel energy used in manufacture, especially for nitrogen fertilizers. They consist largely of nitrogen, phosphorus, and potassium but generally lack the other essential trace elements needed by plants.

Organic agriculture uses composted organic material, containing nitrogen and phosphorus in organic form, to replace chemical fertilizers. These are released more slowly than inorganic chemical forms and so are longer lasting. However, organic materials are also often deficient in minerals and trace elements. This is where rock powders can make an enormous difference.

Fine materials (<200 mesh, 0.075 mm) can be added periodically. They can also be mixed with coarser rock material that breaks down over time, so one application can last several years, depending on the material added. Crops will improve over the years until a new application is needed. Rock powders can also be applied like chemical fertilizers, with small amounts added more often, but labor costs are lower if rock dust is added in larger amounts less often. In the United States, aggregate mineral fines that are not marketed specifically for agriculture are available from $0 to $8 per ton.

In contrast to chemical fertilization, which leads to soil acidification, nutrient loss, and sometimes reduced microbial activity, rock powder amelioration of soil helps to improve conditions for microorganisms of all kinds, such as mycorrhizae, bacteria, and algae.

Conventional agricultural practices deplete soil fertility by not replacing the nutrients removed via crops and erosion, while rock powders provide a natural way to replenish lost soil fertility.

In Australia, banana growers Kevin and Gary Harding did trials with rock dust from Pin Gin Hill quarry in 1985. Using experience gained from these trials, they radically altered their fertilizer application, and in August 1990 planted out a 4 ha block using the product MinPlus rock dust as the main fertilizer.

The Harding Brothers Farm is west of Innisfail in Queensland, Australia. Bananas were grown in pairs at 1.8 m spacing in loamy river flat soils of basalt origin. The rows were 5 m apart, giving a density of 2200 trees/ha. Each 2 ha block of bananas was cropped for 4 years and then followed by plowing in sorghum and chickpea.

The trial results reduced urea and potash applications by 80%, giving significant savings in fertilizer cost and minimizing environmental damage caused by runoff contamination. Dolomite application was halved, and at 16 months, there was no sign of magnesium deficiency, which was a problem prior to using MinPlus, a locally available product. The banana plants were healthier with vigorous root systems and 25% heavier yields. There was a 20% increase in the growth rate, resulting in a faster turnoff of fruit. The sum of these benefits resulted in an incredible 80% increase in production with a substantial decrease in fertilizer cost.

When a cyclone destroyed the trees of surrounding banana farms and the trees had to be replanted, the Harding Farm did not have to replant because the larger root systems held them in the soil (Edwards, 1991) (Figure 4.2).

FIGURE 4.2 Harding brothers showing deep root system of remineralized bananas.

Agriculture and Nutrient Density

In the Hamaker and Weaver book *The Survival of Civilization*, John Hamaker applied glacial moraine dust to Michigan soils and raised corn yields from 25 to 65 bushels/acre, while also greatly increasing its nutritional value. Protein increased by 28%, calcium by 47%, phosphorus by 57%, magnesium by 60%, and potassium by 90% (Hamaker, 1982) (Figure 4.3).

SEER Centre in Scotland

In Scotland, Cameron and Moira Thomson used local quarry dust to revitalize the initially lifeless soil of the SEER Centre, leading to sustained bumper crops of giant, nutrient-dense vegetables

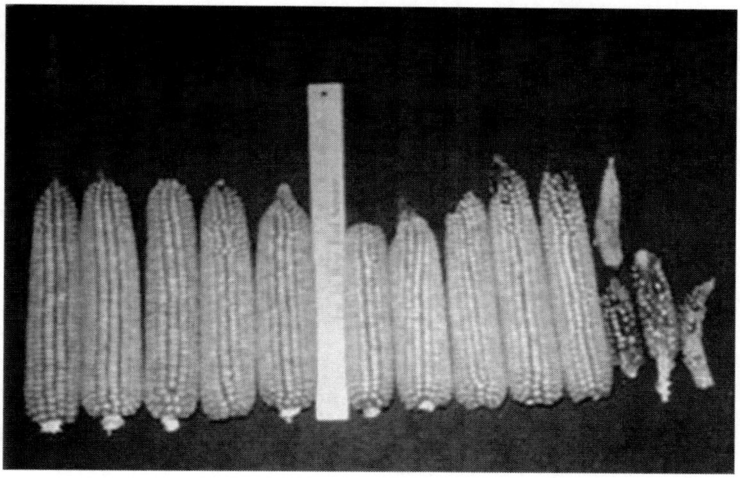

FIGURE 4.3 Control with no inputs on the right, first year after remineralization in the middle, and second year with better results on the left with no additional rock dust applied.

FIGURE 4.4 Large turnip grown at SEER Centre. (Data from Boyer, B., 2010. Transforming barren land into fertile ground. *Remineralize the Earth*. http://remineralize.org/blog/magazine/transforming-barren-land-into-fertile-ground.)

FIGURE 4.5 Large nutrient-dense vegetables grown with Rockdust at SEER Centre.

(Figures 4.4 and 4.5). Beginning with 220 t of rock dust and an award-winning, locally made compost, they created two deep terraces of remineralized soil in which to grow vegetables.

After 13 years of pioneering work, in 1997 the Thomsons established the SEER Centre as a nonprofit organization committed to regeneration of soils to benefit the grower and the environment. The SEER Centre, which focuses on remineralization in the context of permaculture, is a demonstration center open at different stages of the growing season. Cameron and Moira lecture and have received much attention, beginning with local media and later in national and international media. BBC has aired television news items and specials focusing on the SEER Centre. The SEER Centre Trust owns a trading subsidiary, Rockdust Limited, which sells their own product, called Rockdust, all over the United Kingdom and beyond.

SEER Centre now has up to 11 thriving terraces and a large bed where they grow a full range of produce as well as bushes, hedges, and fruit trees (Boyer, 2010).

Finely Crushed Basalt in Queensland, Australia

An extensive research study in Queensland, Australia, carried out by the Rural Industries Research and Development Corporation on a finely crushed basalt rock dust product, MinPlus, has shown MinPlus's capacity to rejuvenate the mineral and chemical properties of highly weathered soils; increase negative charge; increase cation exchange capacity (CEC); enhance ability of soils to retain cations in the soil; enhance ability of soils to supply cations to plants; reduce soil acidity similar to the action of lime, but sustained against soil leaching in high rainfall terrains over a long time period; neutralize soil acidity without liberation of greenhouse gases; reduce soil phosphorus fixation capacity so phosphate fertilizers are kept in soil solution and readily available for plant uptake; provide silicon in a form available for uptake into plant cells, which is thought to result in improved plant vigor and quality; and enhance plant growth and increase crop production for depleted soils.

MinPlus improved plant growth in acidic, highly weathered soils of low fertility, especially those of the humid tropics; reduced fertilizer use in acidic, iron-rich soils with strong phosphorus fixation properties; ensured more sustainable uses of soil resources and fertilizers; replaced high-cost imported fertilizers; reduced environmental hazards from decreased use of agricultural lime; and reduced environmental hazards from decreased erosional losses of nutrients from farmlands and their deposition in sensitive natural wetland and near-shore marine systems (Coventry et al., 2001).

Bob Cannard and Alice Waters, Founders of California Cuisine

> In natural process gardening, you use the study of nature, and you see that up in the high mountains the rocks start big and they tumble against one another and pulverize each other, along with all of the softer elements, and most everything goes into the solution of the floodwater. It carries what I call the "mineral tea" from the mountain to the flood plain and brings a new mineral recharge every time there is an incidence of flooding.
>
> **Bob Cannard**
> *Green String Institute*

Bob Cannard is an agronomist and farmer who as director of the Green String Institute trains growers and as a farmer supplies up to 70 varieties of fruits and vegetables to Alice Waters, world-renowned chef of Chez Panisse and California cuisine. He also manages vineyards, such as Cline Vineyards in Northern California.

Phylloxera, a fungus that gravely affects vineyards and can cause great losses of grapevines, has never been a problem for Cannard's managed vineyards due to addition of rock dust, which strengthens resistance to fungus, insects, and disease. He explains that there are about 70 common elements and rock dust holds thousands of different mineral compounds, and "it's those compounds that biological systems use as the catalyst to control all activities of life." He uses about 100 lb/acre/year of the crushed rock directly applied to the crop. He also utilizes minerals from seawater as well as compost teas.

Dan Barber, one of the most renowned chefs in the United States and a proponent of the Real Food Movement, studied with Cannard. As a result of his emphasis on flavor and nutrient-dense food, he uses a Brix meter in his kitchen to check the quality of the produce for his famed Blue Hill Restaurant at Stone Barns Center in Pocantico Hills, New York. A Brix meter is a refractometer that measures sugar content and can also indicate high mineral content in fruits and vegetables. Barber, in an interview with Organic Connections Magazine, stated that "In almost every way, I think good, thoughtful chefs that care about what they are doing have an imperative for good ingredients and good flavor." In his work as both activist and chef, Barber emphasizes how crucial getting minerals and trace elements back into the soil is to growing nutrient-dense food.

Jamie Oliver is another chef extraordinaire that supports nutrient-dense food and remineralization. Aside from his work as a chef, this British restaurateur has become a media personality known for food-focused television shows, cookbooks, and more recently his campaign against processed foods in national schools. He is part of a growing number of chefs who have a developing concern for enhancing flavor and observe that remineralization greatly improves the taste and nutrition of their food.

Remineralization for more nutrient-dense food is also happening in large-scale organic agriculture. The largest carrot grower in the world, California-based Grimmway Farms, producer of two organic brands, Cal-Organic and Bunny Luv carrots, has remineralized thousands of acres. Purewal Blueberries in British Columbia, the largest blueberry grower in the world, also remineralizes with rock dust. Both companies utilize a source from the American Southwest, although there are ideal glacial deposits available in Canada.

Community Gardens and CSAs

RTE, as part of a grant from its sponsor, the mineral health supplement company Natural Vitality, remineralized 7 acre of the Florence Organic Community Gardens, part of the nonprofit Grow Food Northampton. This is one of many similar projects happening on the local level all across the United States. Community-supported agriculture (CSA) projects, of which there are around 13,000 now throughout the United States, are often committed to growing organic nutrient-dense food, some of which utilize remineralization (USDA, 2007).

Nutrient-dense foods have higher levels of vitamins, carbohydrates, minerals, enzymes, antioxidants and trace minerals, making them ideal for improving health and protecting against disease. The base minerals in our soils are sometimes insufficient for producing rich crops due to years of misuse. The first agricultural communities formed in areas that were regularly flooded by sediment-laden rivers such as the Nile or Amazon. This flooding enriched the earth with the minerals and microorganisms needed for healthy, nutrient-rich plants.

Journalist Graham Harvey, author of We Want Real Food: Why Our Food is Deficient in Minerals and Nutrients—and What We Can Do About It, postulates that the cause of poor health and social disorder is not just fast foods, but the source of our foods. Industrial farming depletes the soil of minerals; even organic carrots growing in poor soil can be tasteless and lacking in essential elements. The book advocates remineralization to increase nutrient density and vitality of the foods we eat. He says that in the last 50 years, vegetables sold in the United Kingdom have lost 24% of their magnesium, 27% of the iron, and 46% of the calcium. Beef and milk are also depleted in mineral nutrients to a similar degree (Harvey, 2006). This book started an advocacy campaign in the United Kingdom that reached the attention of parliament.

In the United States, the RTE Real Food Campaign sought to bring SR into the general public consciousness by making information available to consumers, farmers, and food retailers regarding links between remineralization and nutrient density. Dan Kittredge, director of the Real Food Campaign, ran a series of grower-training workshops in the Northeast to teach farmers how to put nutrients back in their soil. Farmers in the workshops learned to analyze the soil on their land and to identify which factors prevent it from producing high-quality crops.

Kittredge has become the executive director of the Bionutrient Food Association (BFA). He has continued to find aspects of current agricultural practices that might be limiting agricultural productivity and facilitates year-long practicums in biology and organic farming for over 700 northeastern growers. With a solid foundation of educated and actively engaged farmers, the BFA is having a great impact on the northeastern food community and has recently restructured its organization's strategic plan to expand out to engage with a wider community (Kittredge, 2012).

BFA is currently focused on raising funds to research development of a device that would allow consumers to check the nutrient density of their food before they buy it. The basic technology for these scanners is near-infrared (NIR) spectroscopy. With recent developments in miniaturization, these devices can now be manufactured to fit in one's hand. BFA plans to develop a supermarket model of the scanner that could be utilized by farmers to measure nutrient density of their crops in the field while consumers would be able to accurately measure the nutrient density of foods found in supermarkets.

Current supermarket prototype models of the scanner cannot yet provide accurate readings for elements at a concentration of less than 0.1%. The average vegetable is 92% water. After that come macronutrients such as carbon and nitrogen contained in carbohydrates and protein (in large enough quantities to be read by NIR), followed by micronutrients, including vitamins, minerals, and antioxidants (most of which are too low to detect). Kittredge is now working with the Linus Pauling Institute at Oregon State University to run thousands of assays on key foods in the hopes of establishing the algorithms necessary to develop a workable supermarket scanner (Underwood, 2013).

Rock Dust and Pest Control

> Plants have complete immunological systems, but to function properly these immune systems need full support from the plant's environment. Rock dust is the best source of the mineral compounds that are the catalysts of all enzymatic systems that allow the plant to function properly.
>
> **Bob Cannard**
> *Green String Institute*

Chemical pesticides have a fundamental weakness: they encourage proliferation of resistant pests. A farmer spraying insecticides on his or her fields is unintentionally breeding weak plants and strong insects. Any responsible approach to short-term pest control must minimize this effect and should be safe, practical, and cheap (Ebeling, 1971).

Rock dust emerges from these criteria as a promising alternative. Rock powder fits into the category of *inert dusts*, which were identified as effective insecticides by the end of the nineteenth century. Inert dusts operate primarily by absorbing or disrupting the waxy superficial layer of insects that prevents them from drying out. They also physically repel pests, disrupt their reproductive cycles, and abrade them. These small particles disable and discourage insects by causing various forms of mechanical discomfort (Quarles, 1992).

Rock dust is capable of destroying and disabling insects to an extent that can limit their population and reduce crop damage. It does not obliterate insect life in a way that produces ecological imbalances nor does it rapidly breed-resistant pests. This moderate efficacy facilitates a healthy agricultural ecosystem that will gradually require fewer palliative solutions to pest problems. Its value as a fertilizer, along with its low cost and local availability, makes rock dust preferable to diatomaceous earth (Figures 4.6 and 4.7) (Fields and Zlatko, 2002).

Dan Kittredge, executive director of the BFA, describes agricultural pests as biological recyclers that encourage healthy crops by destroying unhealthy plants and returning their nutrients to the soil. Rock dust

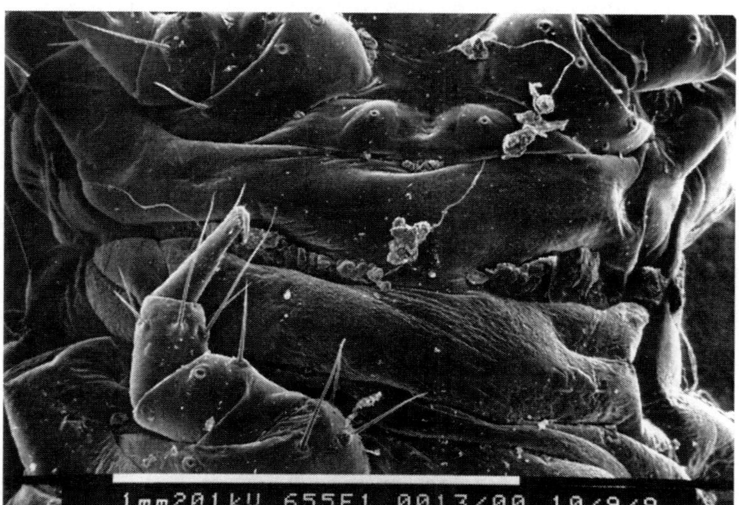

FIGURE 4.6 Insects are deterred by the rock dust, creating a healthy balance rather than wiping them all out. (Data from Keim, R., 2010. Rock dust and pest control. *Remineralize the Earth*. September 18, 2010. http://remineralize.org/blog/magazine/rock-dust-and-pest-control.)

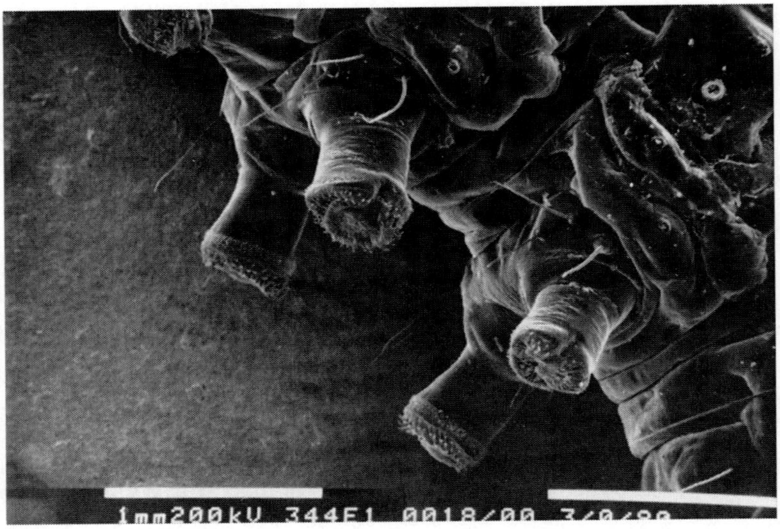

FIGURE 4.7 This microscopic photo shows a rock dusted insect disabled as the dust penetrates segments and organs. (Data from Keim, R., 2010. Rock dust and pest control. *Remineralize the Earth*. September 18, 2010. http://remineralize.org/blog/magazine/rock-dust-and-pest-control.)

is an ideal soil amendment for promoting improved plant immunity to pests and disease. Kittredge explains that silica in rock dust strengthens plant cell walls, making cells less susceptible to the penetrating hyphae of pathogenic fungi. He also emphasizes the role of rock dust's abundant mineral nutrition, which enables the plant to properly synthesize complete proteins claiming that "insects do not have the capacity to break down complete proteins," and that healthy plant tissues cause hungry insects to look elsewhere.

Forestry

> Mineral fertilizers are in common use in agriculture all around the world. They tend to be readily available and relatively inexpensive. Improving forest health and, therefore, carbon storage through additions of these mineral fertilizers is clearly feasible under certain conditions. Studies are now needed to determine the limits of feasibility of soil remineralization techniques in improving tree health and augmenting carbon sequestration in forested areas, especially those forests that are showing decline.
>
> **Lee Klinger, PhD**

SR has been shown in scientific studies to increase yields as much as two to four times for agriculture and four to eight times for forestry wood volume, with immediate results and long-term effects from a single application.

Eliot Coleman did a survey of literature on granites, feldspars, micas, and basalts on the use of ground rock powders in agriculture and writes on basalt, an abundantly available rock dust for forests:

> (Hilf) notes that the beneficial effect of basalt in improving water retention and supplying deficient mineral nutrients brought about a striking increase of growth in most of the hardwoods. [...] Albert found that, "basaltic stone debris" would increase the fertility of poor forest soils. The immediate and lasting effect is ascribed to the chemical nutrients made available, addition of active mineral colloids, neutralizing effect on acid humus, and trace elements such as iodine, flourine [sic], boron, manganese and copper. [...] Gerth determined that a powder of basic igneous rocks can be usefully employed as a soil amendment for forest seedlings on poor soils, providing the size of the rock grains does not exceed 2 mm. [...] Nemec applied crushed basalt to poor podzolic forest soil

in which a spruce plantation, after 25 years of growth, had attained an average height of only 16.5 in. (an average annual height increment of only 0.65 in. per year). After 2 years of basalt application the average annual height increment was increased to 17.5 in. per year (Coleman, 1975).

In Central Europe, Germany and Austria

Long-term experiments in central Europe have shown that in a forest where pine seedlings were remineralized, the wood volume was four times higher than in the corresponding untreated area after 24 years. One application lasted for 60 years before the benefits tapered off (Sauter and Foerst, 1986). This result was traced back to the increasing content of plant-available potassium, calcium, and phosphorus.

Sauter and Foerst (1986) mention that even in antiquity, weatherable rock types were used to improve soil quality and that the first accurate, verifiable research on the fertilization effect of silicate rock dust dates back to the middle of last century.

In the forestry sector, in 1907, an experiment using basalt gravel was done in today's Poland. In 1932, Wiedemann reported that growth in fertilized plots was four times higher than controls. In 1943, Wittich ascribed this to an increase in available potassium, calcium, and phosphorus. The growth-enhancing effect diminished only 60 years later (Sauter and Foerst, 1986).

An approximate calculation shows that the potential exchange capacity of an average forest soil can only be increased long term by 1%–1.5% with 10 t/ha silicate rock dust. The efficacy of the silicate rock dusts depends not only on its fineness but largely on its chemical and mineral composition. Alkaline rocks such as basalt and diabase, which contain high levels of potassium, calcium, magnesium, and phosphorus, are advantageous.

Depending on nutrient content, doses between 5 and 10 t/ha were recommended for extensive basalt dust application and compared to nutrients in conventional forest fertilization. For single tree crop treatments, 0.5–1 kg/plant should suffice. A mixture of silicate rock dust with conventional fertilizers, for example, calcium carbonate or Thomas phosphate, can also be useful.

Depending on the chemical composition of the dust, positive effects should be achieved with 5–10 t/ha. In the trials created by the FVA with a basalt dust, the dosage was between 7 and 14 t/ha. (For agriculture, several authors have recommended quantities of 0.3–1 t/ha/year.)

Gerke (1979) cites several rock dust trials in Czechoslovakia where single tree fertilization in crops produced good results (Sauter and Foerst, 1986). Rock dust had been applied because, at the end of the 1950s, the scarce amounts of commercial fertilizer had to be used entirely for agriculture. The application rates on soils that could not be tilled were up to 3 kg/plant. However, under normal site conditions, growth improvements should be achievable with much lower quantities (0.5–1.0 kg/plant). The silicate dust can be mixed with the excavated soil or be placed on top of the soil (Sauter and Foerst, 1986).

In Austria, Robert Schindele, owner of a gravel quarry, discovered that trees and plants on his property growing alongside gravel roads were much stronger than plants and trees that were not. He deduced that this was because cars driving on the gravel produced a cloud of rock dust that settled on nearby vegetation and positively affected their growth. This led him to promote and publicize the rock dust that he produced from his quarry. It generated such success that he received constant media attention and people lined up to buy his rock dust (Tompkins and Bird, 1989).

Schindele spoke passionately about saving the dying forests with rock dust. Many news headlines in the late 1980s focused on forests declining in Europe and the United States due to acid rain and accumulation of heavy metals in forest soils. Remineralization neutralizes acidity, creating a buffer so that the plants and trees do not take up heavy metals. Dr. Robert Bruck, PhD, a well-known forestry pathologist, observed remarkable recovery of trees in the Black Forest negatively impacted by air pollution from application of mineral slurries and/or dusts.

Unfortunately, Schindele did not make the rock dust available for forestry and instead created a health supplement product that is sold in thousands of pharmacies in Europe and is still marketed today. The popularity of this mineral supplement points to concerns regarding diminished food quality in industrial agricultural systems.

Applying rock lime to trees was a traditional practice in Austria, although overuse can destabilize the acid–alkaline balance. Hard silicate rock dust tends to neutralize or balance the pH. When the pH is

FIGURE 4.8 Spruce branch with rock dust taken for mineral analysis in basalt quarry, Huhnerberg 1986.

FIGURE 4.9 Spruce branch without rock dust taken for mineral analysis in basalt quarry, Huhnerberg 1986.

balanced, uptake of many toxic heavy metals into the trees is reduced. Magnesium plays an important role in the life of plants and soil. The needles of plants near the Huhnerberg basalt quarry were much stronger and sharper because of the element silica, which gives the needles a solid structure and tight surface (Leipold, 1986) (Figures 4.8 and 4.9).

Sanvita, a subsidiary company of the largest producer of diabase in Austria, directed by Engineer Georg Abermann, documented a dramatic recovery in Brixlegg Forest in 1987. Brixlegg Forest, in close proximity to the nearby copper smelting plant, was dying due to a high concentration of heavy metals, including copper, cadmium, and lead, in the soil. Specialists were sure a rock dust application would not help. The forest floor was lifeless and trees were stunted. Six months after the application, there was huge improvement. Wild strawberries started to grow in the soil and plants began to take over the forest floor, indicating development of a healthy forest ecosystem (Campe, 1987).

Since then, diabase application has been repeated on forest tracts in Germany, Austria, and Switzerland with repeated success in recovering from acid rain and *Waldsterben* (dying forests) syndrome. Abermann suggested in 1987 that diabase might also be an effective defense against insect infestations (Campe, 1987) (Figure 4.10).

FIGURE 4.10 Hydraulic spraying of Sanvita rock dust in Brixlegg Forest.

In Australia

The Men of the Trees organization, based in Western Australia, orchestrates planting of ⅓ million trees per year, mostly in arid climates in Australia and Africa. In 1992, Barrie Oldfield directed SR trials with many species of trees in Australia. He found five times greater growth of tree seedlings for remineralized *Eucalyptus gomphocephala* over untreated controls. The time it took for potting out to take place was shortened from 5 months to 6 weeks, with local granite dust from Pioneer Quarry in Western Australia (Oldfield, 1992) (Figure 4.11).

In North Carolina

Between 1989 and 1991, Dr. Robert Bruck, PhD, of North Carolina State University conducted two series of experiments on Grandfather Mountain near Asheville, North Carolina, where 6000 red spruce and Fraser fir seedlings had been planted at 5500 ft elevation. Red spruce and Fraser fir are dominant boreal montane ecosystem species in the high Appalachians of the northeastern and southeastern United States (Bruck, 1992).

FIGURE 4.11 Men of the Trees rock dust trials 1991. Nursery manager Monica Durcan compares the *E. gomphocephala* seedlings grown with the rock dust soil amendment and inoculant with the control seedlings (right). (Data from Oldfield, B., 1992. Soil improvement: The step beyond conservation. *Remineralize the Earth*, 3: 8–10.)

These trees have undergone tremendous stress due to air pollution over the past several decades. By 1991, airborne and satellite surveys indicated that as much as 40% of this ecosystem had already died. There had been severe depletion of micronutrients as well as calcium and magnesium from the soil due to incessant acid rain and cloud deposition. Many trees became yellow, followed by the abscission of needles and death.

The data of Dr. Bruck's forestry trials (first, in the greenhouse, and second, in the field) are summarized below.

Greenhouse Experiments

The study showed increased growth in 500 five-year-old red spruce and Fraser fir trees treated with Planters II rock dust applied at the following rates: 50, 25, 10, and 0 g/pot. Planters II is a 250-million-year sedimentary basin deposit protected by a magma crust during formation of the Rocky Mountains near Salida, CO.

After a 6-month period, observations were made on root color, tree height, diameter, growth, and survival. All rock-dust-treated trees survived, compared with 87% of Fraser fir and 77% of spruce control trees. Growth rate increase varied based on application rates: 37%, 18%, and 5% for red spruce and 39%, 21%, and 14% for Fraser fir for the 50, 25, and 10 g/pot treatments over controls.

Field Experiments

The plots were split in half and 175 lb/acre of Planters II was applied to each split-plot of Fraser fir and red spruce, respectively; the other split-plot received no treatment. After 12 weeks, fertilized red spruce were 27% taller than the nontreated controls; the height of Fraser fir was 19% greater than that of untreated controls. The general appearance of treated trees has very dark-green color and good flushing of tissues, whereas untreated controls appeared to be somewhat yellow and chlorotic (unexpected) (Bruck, 1992).

In California

> The role of forests as either sources or sinks in the carbon cycle depends upon the degree to which plants and soils are managed, protected, or disturbed. Land management methods that improve the health of trees and fertility of soils are likely to create forests that both sequester carbon and provide a sustainable, renewable set of resources for any given region.
>
> **Lee Klinger, PhD**

Lee Klinger, PhD, an atmospheric scientist, has done a significant amount of research on the ways in which forest health and sustainability can benefit from remineralization (Klinger, 2005).

Forests contain a large amount of carbon; hence, they play an integral role in the CO_2 cycle. When forests are unhealthy and dying, which has been the case for many of the world's forests during the last several decades, they release CO_2, but remineralized and healthy forests accumulate carbon, one of the main potential strategies for reversing increase of atmospheric CO_2.

According to Klinger, a tree disease called sudden oak death is devastating oak trees around California, whose long tradition of enforcing rigid fire protection legislature has placed an enormous amount of stress on the oak trees, making them more susceptible to disease. Oaks have a cyclical relationship with fire and benefit from periodic burning. These burnings function both to limit competition and to replenish the soil with fresh nutrients.

There has been debate in California about fire mimicry as a way to manage forest health due to air quality concerns, and the strategy Klinger suggests includes "…thinning stands, clearing brush, pruning trees and amending soils with mineral fertilizers." For foresters, this is much more environmentally sound, as it tends to "mimic the effects of fire, but without all the smoke and flames."

Klinger writes that "Cornell scientists tracking the efforts of a declining sugar maple forest treated with calcium-rich minerals (wollastonite) dropped from helicopters found improved crown density of mature trees, as well as enhanced seedling numbers, survivorship and size in the treated sugar maple stands compared to untreated stands."

From his work restoring trees from sudden oak death to *sudden oak life*, Klinger has shown that applying calcium-rich rock powder such as AZOMITE® Soil Sweetener increases tree health and productivity

FIGURE 4.12 Coast live oak in San Simeon, CA, tended using fire mimicry methods, which include clearing of shrubs around the base, removal of mosses and lichens from the trunk, limewash application to trunk, and amending soils with calcium-rich minerals. Photo on the left was taken on January 4, 2006, at the time of the initial work; photo on the right was taken on January 14, 2014. (Photos by Lee Klinger.)

significantly over time. Calcium is vital for trees and plants that need it for building healthy wood and bark, since "forests in decline are often found to have acidic soils that are depleted in calcium."

Improving forest health and, therefore, carbon storage through additions of rock dust mineral fertilizers is clearly feasible under certain conditions. Forests tend to grow much more rapidly when limiting factor minerals are replaced, and eroded and desertified lands can be brought back to life with simple remineralization protocols. Forests, like agricultural lands, can dramatically increase yields, disease resistance, pest resistance, and nutritive levels through proper remineralization (Figure 4.12).

Studies are now needed to determine the feasibility of SR techniques for improving tree health and augmenting carbon sequestration in forested areas, especially in those forests that are showing decline. Further information can be found on Klinger's website: www.suddenoaklife.org.

Bioremediation of Radiation for Soils and Human Health

Schindele's Minerals (previously called Superbiomin) were used in the wake of the Chernobyl disaster to treat radiation sickness. The minerals provide reactive surface area for some kinds of radioactive ions and so may adsorb them. According to Peter Tompkins and Christopher Bird in Secrets of the Soil (1989), the University of Vienna found that Schindele's rock dust worked as a treatment for radiation sickness, which was confirmed by a Soviet institute for atomic physics in the Ukraine. As a result, the Soviets sent a truck to pick up 2000 kg of this rock dust, which was used to treat victims immediately after the accident (Campe, 2011).

Schindele's Minerals consist of 30% orthoclase (potassium feldspar), 20%–30% plagioclase feldspar, 20%–30% quartz, 15%–20% biotite, 5%–10% kyanite (disthene), garnet, and sillimanite, as well as trace amounts of iron, zircon, and rutile.

Given the recent disaster in northeast Japan, it is timely and urgent to further investigate the bioremediation potential of rock powders on behalf of both disaster victims and others who will be affected as the radiation circulates around the globe and throughout the biosphere. A modestly funded trial study could be created to monitor a group of subjects with high levels of radiation, giving about 1–3 tsp/day of an appropriate source of rock dust (or trial various sources) over a period of 10–14 days or more, compared with a control to see if it can be shown that radiation levels are lowered by rock dust supplementation. For decontamination, rock dust may potentially be a simple, natural, and inexpensive means of alleviating the impact of radiation for soils and people. Because it effectively treated victims of the Chernobyl disaster, remineralization should be explored as a means to decontaminate radioactive soils and treat radiation sickness in Japan. There has been some work done near Fukushima with rock dust, biochar, and microorganisms, as documented elsewhere in this book.

Even before Chernobyl, some dairy farmers in Austria were using remineralization as a part of their operations. They added rock dust to liquid manure along with compost, removing odors and greatly increasing soil biota. Austrians would stand in long lines in order to buy these safe, remineralized products (Campe, 2011).

Current Developments (2008–2013)

Real Food Campaign (BFA)

The Real Food Campaign worked directly with farmers in the field—discovering, uncovering, and sharing information that could improve farm operations and ultimately lead to greater crop quality and improve the health of those who eat the crops. The Real Food Campaign, an RTE project, evolved into the BFA in 2012, a nonprofit research, education, and advocacy organization (Dan Kittredge, Executive Director), whose objective is to improve nutritional quality of the food supply. BFA identifies interdisciplinary parameters that result in optimum soil vitality, educates growers on the principles necessary to achieve this result in the field, and advocates for and facilitates the mainstream availability and consumption of nutrient-dense food (Kittredge, 2012).

Remineralize the Earth's Agroforestry Project in Costa Rica: Food, Fuel, and Income to Sustain Local Communities

A model project designed by John Todd of Ocean Arks International intercropped native commercial hardwood trees, fruit trees, and *Jatropha curcas* (a biofuel-producing oil plant) on abandoned cattle pastures in the Guanacaste region using basalt rock dust from a local quarry. Over a thousand trees of more than 20 species were planted. Half of the trees received 3 kg of basalt rock powder. *Jatropha* receiving rock dust grew more rapidly and produced more and larger seeds. Achieving these results in Cameroon and other parts of Africa with sustainably intercropped *Jatropha* as a biofuel could replace firewood harvesting and expensive imported kerosene in rural areas. It could provide sustainable livelihoods in areas now severely damaged by drought, desertification, and hunger. Marginal soils planted with *Jatropha* sustainably could be restored to agricultural food production in a few years (Todd, 2007) (Figures 4.13 and 4.14).

Panama Study (SER 2011)

RTE participated in the fourth World Conference on Ecological Restoration in 2011 in Merida, Yucatan, Mexico. Over 1200 people, mostly restoration scientists and practitioners, attended the conference, organized by the Society for Ecological Restoration. RTE's study compared tree growth rates over 5 years in highly infertile soils in Panama with those in basalt rock dust (Goreau, Chapter 17). The study documented a significant increase in height and biomass volume of trees remineralized with basalt rock dust in Panama. No chemical fertilizer or compost was added. The trees in basalt dust grew nearly 10 times faster in terms of biomass. It was concluded that even better results would have been achieved if biochar had also been used. Several studies are now being set up in various countries, including Brazil, China, and the United States, to test various mixtures of rock dust, biochar, and compost.

Goreau started this project as an *experiment of opportunity*. He had been present when a basalt quarry in Panama opened and noticed small *Acacia mangium* tree seedlings had been planted in three easily

FIGURE 4.13 Remineralized *Jatropha* produces a higher yield of oil-bearing seeds. (Data from Todd, J., 2008. *Annals of Earth*, 26: 14–15.)

FIGURE 4.14 William Turley with 1-year-old remineralized *Jatropha* plant; RTE Costa Rica Agroforestry Project. (Data from Todd, J., 2008. *Annals of Earth*, 26: 14–15.)

classifiable soil types to create the parameters for the study: trees growing on nonarable local soils, trees grown over the basalt outcrop of the quarry, and trees grown in a basalt rock dust. He observed the trees over 5 years and measured tree height and survivability and did soil chemical analysis of the three areas.

The results were an 8-fold increase in biomass, 2.17-fold increase in tree height, and four times higher survival rate. Rock powders alone were found to provide a wide range of essential minerals depleted in local soils as a slow-release fertilizer and to serve as a CO_2 sink (see Chapter 17 in this book).

The study predicted that the effects could be increased by addition of biochar and compost and recommended large-scale research projects to further optimize use of various rock powders, plants, soil types, climate regimes, and management practices (Gaian, 2011).

Brazil

Brazil has a far-sighted top-down agricultural policy dedicated to creating accessibility to small family farms. Their emphasis on sustainable development, economic empowerment, and social justice, with the end goal of creating a local nutrient-dense food supply for all, suggests commitment to improving heath and generating livelihoods within communities. Brazil has large areas of nutrient-deficient soils, but it also has large resources of rock types available to regenerate soils for agriculture, agroforestry, and production of sustainable biofuels.

A research project in Bahia shows remineralization to be an effective adjunct strategy for remote impoverished communities to produce higher yields of quality crops while remaining independent from chemical fertilizers and government subsidies (Gaian, 2010).

Brazilian Model

The concept of using rock dust and other natural means to restore depleted tropical soils, increase food security, and provide income for small agrarian communities has been tested and proven feasible by a team of Brazilian scientists led by Dr. Suzi Huff Theodoro, in the Sustainable Development Program at the University of Brasilia (Theodoro et al., 1996). This project, first initiated by Professor Othon Leonardos, specialized in remineralization research and project implementation over the past 15 years. The goal of the project was to generate effective strategies for impoverished family farmers to produce higher yields of quality crops without dependence on chemical fertilizers and government subsidies (Theodoro, 2006).

Plots were established in 10 remote impoverished communities (*Quilombos*) founded by escaped slaves of African descent in the state of Bahia. While a few of the farmers traveled to the University of Brasilia for training, most farmers were introduced to project techniques and procedures by Theodoro's team on their own farms. Rock powders donated by a local quarry were mixed with organic matter, manure, and other affordable, natural, locally available materials and applied to the soils where primary food crops for the region were planted. In the first 2 years of testing, yields increased 10%–30%. Soil pH shifted from highly acidic to mildly acidic to mildly alkaline, and nutrient availability in the soil was greatly improved. However, farmers noted that the most significant aspect for them was the improved capacity of remineralized soil to hold water, as shortages and droughts are common during the dry season (Figures 4.15 and 4.16).

Dr. Theodoro is coordinating a project funded by the Brazilian National Research Council (CNPq), which aims to integrate Brazilian and African research. The project includes researchers from Cameroon, South Africa, and Angola. One of the researchers, Jean Pierre Tchouankoue, Professor of Earth Sciences at the University of Yaounde in Cameroon, notes that the soil, geology, and climate of Bahia are similar to those of Cameroon, as well as the cultural and socioeconomic conditions. Projects in Brazil and Cameroon have the same purpose to assist impoverished communities of family farmers (see Chapter 26 in this book).

FIGURE 4.15 Jean Pierre Tchouankoue, PhD, from the University of Yaounde (right) with Quilombo farmer in Bahia, Brazil. (Data from Gaian, D., 2010. Remineralization in rural Brazil. *Remineralize the Earth*. http://remineralize.org/blog/magazine/remineralization-in-rural-brazil.)

FIGURE 4.16 Suzi Huff Theodoro and Jean Pierre Tchouankoue advising farmers of Quilombo communities. (Data from Gaian, D., 2010. Remineralization in rural Brazil. *Remineralize the Earth*. http://remineralize.org/blog/magazine/remineralization-in-rural-brazil.)

At the first stonemeal congress, held in Brasilia in 2009, it became clear that periodic conferences were needed to share and consolidate cutting-edge science, open up debate, and foster education and outreach for the important and developing technology of utilizing rock dust and agrominerals for agriculture.

The second Brazilian stonemeal conference (II Congresso Brasileiro de Rochagem) took place on May 12–17, 2013. This conference focused on deepening scientific and technological debate around the use of rock dust, ores, and powdered minerals for SR and on establishing a regulatory framework for registry and commercialization of rock dust and its products. It introduced the strategy of creating agromineral hubs, organized on a cooperative basis, to raise awareness of remineralization technology, make rock dust available to farmers on a regional basis, and increase production for national agriculture and raising livestock.

Recently, there have been two main pieces of legislation regarding remineralization in the Brazilian Congress. On December 10, 2013, Law No. 12.890/2013, sponsored by Senator Rodrigo Rollemberg, was approved by the congress with the sanction of President Dilma Rousseff, amending an earlier law (Law No. 6.894/1980) (Temer et al., 2013). It alters the original legislation to include rock dust as a category of fertilizer that can be used, commercialized, and inspected in Brazil. Remineralization is now recognized as a new category of agricultural input, which paves the way for the technical committees of the Ministry of Agriculture to establish specific regulatory standards for this type on input, in addition to those that already exist for organic fertilizers. It is in this project that Theodoro and others have concentrated the most efforts and met with a successful outcome, paving the way for the integration of remineralization into Brazil's agriculture sector.

The second law put before the congress, Project 2.727/2011, plans to create a national program for remineralization with specific criteria for the use of rock dust in agriculture. If enacted, this law would offer tax breaks to rock dust companies as an indirect incentive, rather than a subsidy. It is a more complicated project because it has required Theodoro and others to rectify and acknowledge the following:

Economic incentives will be needed to improve the program, proposed in this project. However, as agricultural activity is private, such incentives, if provided, could compromise the budget of the public sector. In addition, the specification about which kinds of rock dust can be used to remineralize soils is not clear.

The Geological Service of Brazil is mapping the various agrogeology zones (regions) and will be testing regional rock dusts through seed germination in the lab (using a mineral solution), greenhouse pot tests, and field tests.

Theodoro expects the use of rock dust will be predicated on certification for rock dust companies, based on these criteria: chemical and mineral composition of the rocks, biological activity, cultivation systems being used, climate conditions, and regional issues. This will include a consideration of geological rock type, chemical analysis for major and trace elements in the rocks, petrographic analysis of the mineral composition, weatherablity and granulometry (measurement of particle size distribution in a collection of grains), rock solubility, as well as physical qualities such as abrasion and pH.

Research undertaken to study the potential uptake of toxic elements by plants will determine limits to be set on potentially hazardous elements. Sodium has already been identified as an element that needs to be limited. Rock dust cannot be utilized if it has a pH under 7. There will also need to be safety measures in place to prevent silicosis in the manufacturing process.

Brazilian research has emphasized phosphorus, a mineral highly deficient in most Brazilian soils, and most recent research suggests a gradual increase in plant available phosphorus through smaller rock dust applications. A study in the Amazon is focusing on phosphates as well as sedimentary and magnetic rocks that could contain phosphates.

In Rio Grande do Sul, the Brazilian National Agricultural Research Agency Embrapa has a team of researchers led by geologist Magda Bergmann PhD in consultation with agrogeologist Peter van Straaten PhD working at three locations: Rio Grande do Sul, Santa Catarina, and Paraná. In Rio Grande do Sul, they are using aerogeophysics to map and study agrominerals. They have collected and created a bank of rock dust and soil samples from different regions and are establishing critical levels of mineral input for soils, plants, waterways, animals, and people. In order to study the agronomic efficiency of rock dust for plants, 20 ha of land have been dedicated to experiments.

The Geological Service of Brazil is doing research in many different regions with specific rock dusts and crops. Currently, Theodoro is conducting research on tropical soils funded by CNPq for corn, beans, carrots, garlic, and tropical fruits. She is using compost with many types of rock dust to determine availability of nutrients over time and the effect on nematodes. She has established that remineralization reduces the need for water and irrigation (see Chapter 22 in this book).

There is now a third generation of research taking place among geologists, agronomists, and members of other specialized fields such as soil microbiology, food safety, and security. This research is being financed on a federal level by many different institutions including government agencies, universities, companies, organizations, and trade associations. There are an increasing number of students at several universities doing their PhD thesis research on various aspects of SR, particularly as it relates to geology and agronomy.

Recognition and interest has come about through the results obtained by several research institutions (such as Embrapa, University of Brasilia, UFV, and CPRM) confirming that remineralization is an alternative and/or complement to standard fertilizers in research and demonstration trials. Stonemeal technology has already become a reality in several sustainable agricultural production systems across the country, particularly among family farmers in the south of Brazil. Since it is in agreement with agroecological principles, it has become a viable alternative for regenerating degraded soils. Agroecology is an interdisciplinary science founded in Brazil that applies ecological concepts and principles to the design, development, and management of sustainable agricultural systems. Agroecology science encompasses mineral cycles, energy transformation, biological processes, and socioeconomic relationships.

There has been considerable funding for remineralization research in Brazil for many years. CNPq, the National Council for Scientific and Technological Development, has funded remineralization research since the mid-1990s, when Embrapa joined the research consortium. Funding has come from the Ministry of Mining and Energy and the Ministry of Science and Technology. Research centers have access to this funding, including the Center of Mineral Technology of Rio de Janeiro (CETEM), the universities, and Embrapa. Funding agencies are becoming aware that remineralization helps provide incentives for poor family farmers to enter the market, to create food security, and to transition toward sustainable agriculture, which is so important when the world is looking to preserve the environment and reverse climate change.

Embrapa, the Brazilian Enterprise for Agricultural Research, is a state-owned company affiliated with the Brazilian Ministry of Agriculture. It is devoted to pure and applied research. Embrapa conducts research in many disciplines, including remineralization. The mission is to provide feasible solutions for the sustainable development of Brazilian agriculture. Eder de Souza Martins, one of Embrapa's leading

FIGURE 4.17 Bougainvillea before rock dust application, December 21, 2009.

researchers, is currently researching remineralization in the following regions: southwest Goiás, south of Minas Gerais, north of Minas Gerais, the central region of Tocantins, federal district, and Brazil's southern region (Paraná and Rio Grande do Sul). Field experiments are being conducted with rock dusts derived primarily from regional mining waste, gravel, and limestone. They are developing an experimental protocol for the validation of SR in commercial areas. They are exploring longer-term experiments, but for now, the minimum time for these experiments is 5 years (2009). Already in their third year, they are studying annuals (soy, corn, sunflower), semi-perennials (cane sugar, bananas), and perennials (eucalyptus), among others crops. They are also interested in carrying out large-scale research for the study of carbon sequestration for remineralization.

Joanna Campe, executive director of RTE, has for the last several years spent a few months each year near Brasilia. There, Walter Schroth has remineralized hundreds of trees on his tree farm and remineralized the garden with local basalt rock dust. The garden was composed of gravel mulch and the rock dust was applied on top of it on December 21, 2009, allowing rain to distribute the rock dust through the gravel and into the soil. After 6 weeks, the *bougainvillea* was flowering so heavily that it was bending over and new supports were needed to hold up its tremendous growth. The house owner had never seen such rapid and abundant growth in the garden before. The large new branches were home to several birds' nests, creating a new habitat within a very short time period (Figures 4.17 and 4.18).

Latin America

Mexican Government Remineralization Program

Zacatecas is the state with the highest production of beans in Mexico, accounting for 35% of national production. In the span of only 5 years, yields of beans dropped from 1200 to 1400 kg/ha in 2002 to 300 kg/ha in 2007—a clear example of soil depletion and overcultivation leading to increasingly low crop production. In 2008, 8000 ha in different states of the Mexican Republic were remineralized. In 2009, Jorge Villaseñor Garibi, an RTE coordinator, and his company Agro Insumos Nova Terra, SA, participated in a governmental program to support the farmers of Zacatecas, where, for the first time in Mexico, urea was replaced by rock dust. As a consequence, 14,200 ha were remineralized in 2009. The results were outstanding, with bean yields three times higher than the previous year's. Production of corn, grape, peach, nopal, and several varieties of chili pepper also increased. These promising results convinced many farmers of the benefits of remineralizing their soils. Many continued using rock dust even after the governmental campaign ended. Garibi presented the governor of Zacatecas, Amalia García Medina, with an honorary certificate on behalf of RTE in recognition of her efforts to promote the use

FIGURE 4.18 Bougainvillea 6 weeks after rock dust application, February 9, 2010.

of rock dust. Five more such certificates were given to state officials that participated in the initiative, including the Secretary of Agricultural Development (Figure 4.19).

In 2011, a government campaign supported farmers in the township of Zapopan, Jalisco, to use rock dust. The government has committed itself to making Zapopan a green city, and in the summer of 2010, it provided farmers with diatomaceous rock dust to fertilize 1250 ha of crops including corn, nopal, agave, and fruit trees. Farmers are being trained to use rock dust in 28 towns throughout Zapopan as an integral part of

FIGURE 4.19 Remineralizing nopal cactus.

FIGURE 4.20 Teacher shows children how to remineralize tree seedling in forest project in Zapopan, Mexico.

the program to support the remineralization of depleted soils. Farms and crops, however, were not the sole beneficiaries: the governmental initiative included the fertilization of various green areas of Zapopan, such as public parks, gardens, and a forest area of about 1500 ha, estimated to contain 700 trees. On June 4th—National Environment Day—around 500 children from several elementary schools in Zapopan attended a workshop in which they were taught to remineralize trees with rock dust. At its end on June 5, 2012, members of Zapopan's government received formal recognition for this program (Ruiz Castro, 2010) (Figure 4.20).

Bernardo Castro Medina founded Eco-Agro in 1997, a company formed as a farmer's collaborative dedicated to research and commercialization of organic alternatives to industrial fertilizers. The company is providing farmers with organic fertilizers supplemented with hard silicate rock dust. Based in Guamúchil, Sinaloa, Eco-Agro has been at the forefront of campaigns promoting the use of rock dust in and out of Mexico, highlighting both its fertilizing properties and its role in increasing resistance of plants to disease and insect infestation. Eco-Agro organic fertilizer is made up of worm leachate containing 5%–6% mycorrhizae, which provide nitrogen to plants, and 5%–6% silicate rock dust, containing 52% silica with more than 70 minerals and trace elements. Rock dust is an essential ingredient in this method of composting; it not only accelerates the process by inducing the growth of microorganisms but also provides depleted soils with a wide range of minerals and trace elements for healthy nutrient-rich plant growth. It has also been shown to be beneficial for worms, allowing them to grow larger. Eco-Agro runs several projects in Sinaloa, particularly corn and sorghum with excellent results. Eco-Agro's website offers testimonials of farmers using rock dust as well as videos in Spanish that explain the production of rock dust and the rationale behind its use (Ruiz Castro, 2012).

Colombia

The Western Andes Range has rich volcanic mineral deposits. Agrempacados makes a diabase rock dust for Colombian farmers, specifically small producers, to improve production and nutritional properties of their crops. Corn, which has received the most study, showed an increase in germination rate of 20%–50% and an increase in production of 10% compared to areas without rock dust. Corn from fields treated with rock dust also exhibited improved resistance to humidity. There have been good results for tomatoes, bananas, avocados, vegetables, and flowers. Coffee plantations that had stopped producing regained fertility in 1 year after they were treated with diabase rock dust (Ruiz Castro, 2011).

Eugenio Gras and Andrés Piñeiro, world-renowned experts on organic farming and members of COAS (Council for Sustainable Agriculture and Permaculture), present courses on remineralization in Central America, South America, and Australia within the context of organic farming and permaculture. Agrempacados has sponsored several seminars and funded two diploma studies, which were completed in 2008 and 2010.

Cameroon

Experimenting with rock fertilizer technology on small-scale farms in Cameroon, contributing to food security and sustainable agriculture, is a current project proposal, which will be implemented by RTE and the Research and Education Centre for Development (CREPD). Other partners include Peace Corps/Cameroon and AAFCOOP. RTE and CREPD propose a pilot project utilizing sustainable agroforestry and agricultural techniques and remediating soils with SR.

The Cameroon volcanic line (CVL) offers a favorable setting for empirical observations on the relationship between geology and soil fertility (see Chapter 26 in this book). The plains and valleys within the CVL make up the breadbasket of Cameroon and Nigeria, supporting a total population of about 200 million inhabitants, due to the high fertility of its volcanic soils. Close to this zone are large grasslands unsuitable for food crop growth but still densely populated by small-scale rural farmers that struggle to survive.

Direct benefits anticipated are improved soil fertility resulting in long-term economic benefits from using local rock powder; a sustainable economic and environmental alternative to imported fertilizers and pesticides; increased knowledge of small-scale rural farmers in sustainable agriculture, agroforestry, remineralization, and business techniques; increased crop nutrient density; and reduced food insecurity in rural areas. Indirect benefits may be seen in an improvement in the local economy and for those that are most vulnerable due to poor nutrition.

Two hundred farmers, both men and women, will be invited to participate. For each farmer, it is anticipated that four family members will indirectly benefit from the project. They will receive education and training in agroforestry, agriculture, and rock dust use. At harvest, these farmers will receive the food produced for their families and sell it if they so choose. Three communities will have 12 plots, 4 plots per community, measuring 25 m^2 each. One plot per community will be a control and the remaining three will investigate how rock dust impacts various plants.

Prior to the project's initiation, a baseline study will measure soil, plant and crop growth, pest resistance, soil water retention, crop nutritional content, plant and crop productivity, and root system health. These baseline results will be compared to the untreated and treated plots. Farmers will be evaluated pre- and posttraining to ascertain changes in their knowledge of agroforestry, rock dust, and agricultural techniques. Farmers will be surveyed to determine how they are using the food produced. Data will be collected from the staple crops and fruit trees, as determined by the participating farmers to determine the following information: crop production, soil fertility, water retention, pest resistance, and root system health. Data will be collected mid-growing season and at harvest. As these staple crops have a growing season of a few months, results should be seen within the first year. The agroforestry system results will be seen in the second and third stages of the project. Based on the merits of this pilot study, training will be extended to more farmers, include marketing and small business development, and more plots of land will be incorporated into the project with a wider variety of crops and plants. A final evaluation at the end of the 2 years will show what impacts have been made, as well as the expected and unexpected outcomes.

Future Directions

Carbon Farming Initiatives and Climate

The soil stores an estimated 1500 Pg of organic carbon, which is twice that of the atmosphere 770 Pg (McCarl et al., 2007). Soil carbon sequestration is often beneficial for soil health as it can increase CEC, structural stability, and water retention capacity. When soil is managed correctly, carbon is stored in the soil naturally. Carbon influences moisture content. Soil carbon increases the soil's water-holding capacity, providing plants with a valuable source of water.

Australia has taken the lead on carbon sequestration by creating a system of carbon credits for agriculture to store carbon or reduce greenhouse gas emissions on the land. This encourages sustainable farming and provides a source of funding for landscape restoration projects.

Agriculture and Forests

Brazil is paving the way for SR on a national scale through its development of research and policy strategies, plans to create a regulatory agency, and distribution hubs that will make rock dust available to farmers throughout the country. The government is beginning similar initiatives through the development of projects in Africa.

Inspired by lectures of visiting professor Peter van Straaten on use of agrominerals in 2006, Vincent Kato started the Agro-Geology Association of Uganda. The Agro-Geology association includes geologists, farmers, agronomists, chemists, mineral processors, and miners carrying out studies with partner organizations. Over 100 Ugandan researchers from different disciplines are studying the inherent value of rocks for crops. Through aerogeophysics, they are mapping potassium-rich zones for growing bananas and studying the use of volcanic rocks rich in potassium, a key element for bananas, the major crop. They are also using airborne data to determine how volcanic regions relate to agricultural productivity (Van Straaten, 2006).

Jean Pierre Tchouankoue, PhD, senior lecturer in Earth Sciences at the University of Yaounde in Cameroon, has been working since the 1990s with a small research group of postgraduate students investigating the potential of pyroclastic basalts for agriculture in Cameroon. They have focused their research on maize (corn).

In Mexico, two rock dust companies have been very successful: one company working on a cooperative basis with farmers, while the other has worked with government partnerships and subsidies on the local, state, and national levels to create projects that have covered large tracts of industrial agricultural land as an alternative to chemicals and pesticides and working with public spaces and forest stands.

RTE is partnering with the University of Massachusetts, Rock Dust Local, and the Pioneer Valley Biochar Initiative, to find the most effective ratios of compost, rock dust, and biochar and to take the results one step further by analyzing not only plant and soil nutrients and biological activity but also the effects on soil carbon storage and greenhouse gas emissions (see Chapter 18 in this book).

Reforestation with rock dust or application to existing trees has the potential to create healthy drought and fire-resistant forests that represent a net carbon sink, especially in places such as Australia, Greece, Los Angeles, and New Mexico where prevention of forest fires is of paramount importance. The rock dust can be applied as part of community projects by volunteers, hydraulically from trucks or sprayed by plane or helicopter.

There are numerous uses for the application of remineralization, from bioremediation of chemicals and radiation in soils, pest control as an alternative to pesticides, and for dairy and livestock operations, all of which have the potential to create a more humane and sustainable relationship between people, flora and fauna, and the biosphere. Remineralization is perfect for landscaping residences and urban landscapes, community gardens, and initiatives of all kinds. From carbon credit programs, such as the one for agriculture in Australia, to production of sustainably remineralized biofuels and the remineralization of forests, carbon can be sequestered on a massive scale to help stabilize the climate.

Conclusion

> The world is in transition from a hydrocarbon-based society (fossil oil, gas and coal) to a carbohydrate society (crops, crop and forestry residues, hybrid trees and grasses, yard and garden trimmings, and the biomass part of garbage heading for the dump). More biomass is needed. The transition requires clean water and air, nutrients, and environmentally enhancing agricultural/forestry practices. These are growers' responsibilities. This includes the need to restore the mineral base of the soils— minerals that have been depleted since the last ice age, as The Survival of Civilization points out. Biorefineries "recycling CO_2" in every county and bioregion, will contribute to national and energy security, new industries, good jobs, democratization of energy supplies, and to global leadership by example in a world of people and other living things, intuitively ennobling that human synergy with nature is an absolute imperative.
>
> **Bill Holmberg**
> *ACORE*

The techniques of SR are simple, easily and intuitively learned, and can be rapidly scaled to the community level. The materials are readily available and are an inexpensive by-product that can be found wherever there is building and road construction using stone aggregates or concrete. No extra energy is needed to grind them up since it is a waste by-product of gravel plants. Silicate rocks and seawater are the most abundant resources on earth.

Remineralization is nature's way of regenerating soils and is needed on a large scale as mismanagement is causing us to lose soils far faster than they can naturally regenerate, and this innovative method will contribute to the increased storage of carbon in our soils and forests.

REFERENCES

Boyer, B., 2010. Transforming barren land into fertile ground. *Remineralize the Earth*. Web. April 7, 2013. http://remineralize.org/blog/magazine/transforming-barren-land-into-fertile-ground.

Bruck, R., 1992. Preliminary results for the soil remineralization forestry trials on Grandfather Mountain in North Carolina. *Remineralize the Earth*, 3: 9–10.

Campe, J., 1987. *Soil Remineralization, A Network Newsletter*, volume II, issue 7.

Campe, J. April 6, 2011. From Chernobyl to Japan: Treating radiation sickness with rock dust. *Remineralize the Earth*. Web. April 3, 2013. http://remineralize.org/blog/magazine/from-chernobyl-to-japan-treating-radiation-sickness-with-rock-dust.

Campe, J., D. Kittredge, and L. Klinger, 2012. The potential of remineralization with rock mineral fines to transform agriculture, forests, sustainable biofuels production, sequester carbon, and stabilize the climate. *Remineralize the Earth*. April 18, 2013. http://remineralize.org/administrator/components/com_jresearch/files/publications/Rio%20Summit-RTE-2012.pdf.

Chesworth, W., F. Macias-Vazquez, D. Acquaye, and E. Thompson, 1983. Agricultural alchemy: Stones into bread, Episodes, 1983 (1), p. 3.

Coleman, E. 1975. The use of ground rock powders in agriculture: A survey of the literature on granites, feldspars, micas, and basalts. Small Farm Research Association, Harborside, ME, 22pp.

Coventry, R.J., G.P. Gillman, M.E. Burton, D. McSkimming, D.C. Burkett, and N.L.R. Horner, 2001. Rejuvenating soils with Minplus™: A rock dust and soil conditioner to improve the productivity of acidic, highly weather soils. A report for the Rural Industries Research and Development Corporation. James Cook University, Townsville, Queensland, Australia.

Ebeling, W., 1971. Sorptive dusts for pest control. *Annual Review of Entomology*, 16: 123.

Edwards, T.D., 1991. Remineralization trials: Minplus and bananas (a cost benefit study), *Pacific Mineral Developments*. April 18, 2013. http://remineralize.org/administrator/components/com_jresearch/files/publications/Remin%20Trials_Edwards.pdf.

Fields, P. and K. Zlatko, 2002. Post-harvest insect control with inert dusts. *Encyclopedia of Pest Management*, 48: 650.

Fragstein, P.V., 1995. Silicate rock dusts as natural fertilizers. *Remineralize the Earth*. 1, 73–76.

Gaian, D., 2011. Outstanding results of Dr. Goreau's rock dust trials in Panama at SER. *Remineralize the Earth*. Web. April 6, 2013. http://remineralize.org/blog/latest-news/outstanding-results-of-dr-goreaus-rock-dust-trials-in-panama-at-ser2011.

Gaian, D., 2010. Remineralization in rural Brazil. *Remineralize the Earth*. Web. April 6, 2013. http://remineralize.org/blog/magazine/remineralization-in-rural-brazil.

Hamaker, J.D. and D. Weaver, 1982. *The Survival of Civilization*. Hamaker-Weaver Publishers, Burlingame, CA.

Harvey, G., 2006. We Want Real Food: Graham Harvey. Constable, London, U.K.

Hensel, J., 1982. *Bread from Stones: A New and Rational System of Land Fertilization and Physical Regeneration*. Republished by Acres USA, Austin, TX.

Keim, R., 2010. Rock dust and pest control. Remineralize the Earth. September 18, 2010. Web. April 7, 2013. http://remineralize.org/blog/magazine/rock-dust-and-pest-control.

Kittredge, D., 2012. Bionurient Food Association. Web. April 18, 2013. http://bionutrient.org/about-us.

Klinger, L.F., 2005. Bryophytes and soil acidification effects on trees: The case of sudden Oak death©. *Remineralize the Earth* (*Combined Proceedings International Plant Propagators' Society*), 55: 493–503.

Leipold, F., 1986. The effects of basalt dust emissions on spruce trees at the Basalt Quarry Albert in the Huhnerberg compared to spruce trees outside the range of emissions. Web. April 18, 2013. http://remineralize.org/administrator/components/com_jresearch/files/publications/Effects%20of%20Basalt_Leipold.pdf.

Leonardos, O.H., W.S. Fyfe, and B.I. Kronberg, 1987. The use of ground rocks in laterite systems: An improvement to use of conventional soluble fertilizers? *Chemical Geology*, 60: 361–370 (Amsterdam, the Netherlands).

McCarl, B.A., F.B. Metting, and C. Rice, 2007. Soil carbon sequestration. *Climatic Change*, 80: 1–3.

Oldfield, B., 1992. Soil improvement: The step beyond conservation. *Remineralize the Earth*, 3: 8–10.

Quarles, W., May/June, 1992. Diatomaceous earth for pest control. *The IPM Practitioner*, 14: 5–6.

Ruiz Castro, P.A. March 15, 2012. Rock dust, mother Earth's milk. *Remineralize the Earth*. Web. April 6, 2013. http://remineralize.org/blog/magazine/rock-dust-mother-earths-milk.

Ruiz Castro, P.A. March 17, 2011. RTE coordinator Daniel Giraldo promotes the use of rock dust in Colombia. Remineralize the Earth. Web. April 6, 2013. http://remineralize.org/blog/magazine/rte-coordinator-daniel-giraldo-promotes-the-use-of-rock-dust-in-colombia.

Ruiz Castro, P.A. September 15, 2010. TE's coordinator Jorge Villasenor Garibi researcher and promotes the use of rock dust throughout Mexico. *Remineralize the Earth*. Web. April 6, 2013. http://remineralize.org/blog/magazine/rtes-coordinator-jorge-villasenor-garibi-researches-and-promotes-the-use-of-rock-dust-throughout-mexico.

Sauter, V.U. and K. Foerst, 1986. Hinweise fur den Einsatz von Silikatgesteinsmehlen zur Waldbodenverbesserung. *Der Forst- und Holzwirt*, 42: 27–30.

Temer, M., A. Andrade, E. Lobao, and I. Monica Vieira Teixeira, December 11, 2013. Legislação Informatizada—LEI N° 12.890, DE 10 DE DEZEMBRO DE 2013—Publicação Original. *Câmara dos Deputados*. 54ª Legislatura—3ª Sessão Legislativa Ordinária. Web. December 12, 2013. http://www2.camara.leg.br/legin/fed/lei/2013/lei-12890-10-dezembro-2013-777603-publicacaooriginal-142208-pl.html.

Theodoro, S.H. and L. Othon, 2006. The use of rocks to improve family agriculture in Brazil. *Annals of the Brazilian Academy of Sciences*, 78: 721–730.

Theodoro, S.H., L. Othon, H. Rego, K.G. Rocha, and L. Eduardo, 1996. Stonemeal technology: Changing the poverty scenario in Brazil.

Todd, J., 2007. Creating world class soils. *Annals of Earth*, 25(1): 22–24.

Todd, J., 2008. Restoring the land: Building durable economies. *Annals of Earth*, 26: 14–15.

Tompkins, P. and C. Bird, 1989. *Secrets of the Soil*. Harper & Row Publishers, Inc., New York.

Underwood, A. March 1, 2013. Organic vs. conventional.... Bionutrient Food Association. Published in *Scientific American*. Web. April 8, 2013. http://www.scientificamerican.com/article.cfm?id=organic-conventional-which-scanner-healthier.

USDA, 2007. Selected Practice of 2007. *Census of Agriculture- State Data*, p. 606.

Van Straaten, P., 2006. Farming with rocks and minerals: Challenge and opportunities. *Anais da Academia Brasileira de Ciencias*, 78: 731–747.

Van Straaten, P., Rocks for crops: Agrominerals of sub-Saharan Africa. ICRAF. Web. April 3, 2013. http://www.uoguelph.ca/~geology/rocks_for_crops/.

World International Renewable Energy Conference, 2008. Proceedings and Green Pledge, www.wirec2008.gov/wps/portal/wirec2008; www.ren21.net/wiap/wirec.asp; www.ren21.net/wiap/detail.asp?id=327.

Yeomans, A.J., 2005. *Priority One: Together We Can Beat Global Warming*. Keyline Publishing, Arundel, Queensland, Australia, 500pp. http://www.yeomansplow.com.au/priority-one.htm.

Organizations and Links

Remineralize the Earth

RTE raises public awareness of the role that SR with finely ground rock dust, sea minerals, and other natural means plays in agriculture by increasing the growth, health, and nutrient value of all plant life; reversing global desertification through agroforestry and reforestation; producing sustainable biofuels; sequestering carbon; and stabilizing climate. RTE promotes partnerships with foundations and educational institutions, government agencies, international development agencies, the private sector, and NGOs.

www.remineralize.org

Bionutrient Food Association

The Real Food Campaign (BFA) organizes networks and promotes healthier nutrient-dense foods for our food supply through remineralization.

www.realfoodcampaign.org

Rocks for Crops

Rocks for Crops works in Africa to improve agriculture through the utilization of natural geological materials.

www.uoguelph.ca/rocks

The first International Rocks for Crops Conference 2004, Brazil, abstracts and papers can be found at

www.uoguelph.ca/rocks

Primeiro *Congresso Brasileiro de Rochagem* 2009 presented research on advances in rock powder fertilization in agriculture.

www.congressorochagem.com.br

Renew the Earth

Renew the Earth encourages international exchange of ideas, information, and actions in the transition from a fossil fuel based economy to one increasingly reliant on sustainable energy and water.

www.renew-the-earth.org

SEER Centre

SEER Centre is committed to regeneration of soils with rock dust to remineralize the soil to benefit the grower and the environment and operates a demonstration garden and farm.

www.seercentre.org.uk

Ocean Arks International

Ocean Arks International specializes in ecological water purification technologies and disseminates ideas and practices of ecological sustainability and design worldwide. They explore linkages between ecology and economics to create new kinds of work and enterprises out of environmental protection and restoration initiatives.

http://www.oceanarks.org

Sudden Oak Life

http://www.treesearch.fs.fed.us/pubs/26610.
http://www.suddenoaklife.org.
http://suddenoaklifeorg.wordpress.com/files/2009/07/tep-paper-final.pdf.
http://www.hubbardbrook.org/research/longterm/calcium/fahey07.htm.
http://www.azomite.com.

5

Curing Sick Soils through Chemistry

Richard S. Stein and Tadeusz S. Wysocki

CONTENTS

Chemistry and the Environment

It is unfortunate that the term *chemistry* has negative implications. Some believe that the use and abuse of chemistry bears much of the responsibility of environmental deterioration. We differ and do not believe that we have to suffer from competition between science and the environment, but we believe that we must learn to use science, chemistry in particular, to solve environmental problems. We shall deal with one of these global problems, that of *sick soils*, describing the problems and ways in which chemistry may helps us solve them.

Problem: *Sick Soils*

Much of our lives depend on rather thin layer of soil, which we need for growing food. This layer is less than a foot thick and is a small fraction when compared to the almost 8000 mile diameter of our planet earth.

It may be thought of like the skin of an apple or the shell of an egg. The soil's nature is sensitive to our actions and to those of the climate.

The increase in the frequency in the past few years of heat waves, droughts, and climate variability has led to a significant decrease in food production and plant yields. Worldwide yield reduction in wheat, corn, soybeans, rice, and canola has caused a noticeable increase in prices for food and commodity items.

Some of the prime agricultural lands were used for growing corn for ethanol gasoline additives, and canola and soybeans are being grown for making biodiesel. The resulting drop in productivity is occurring at a time of when the food needs of the earth are increasing making for an extremely unstable situation. Numerous international and national news and weather reports recently have shown some of the devastation that this *climate chaos* can cause.

In addition to declining food production, there are reports that the soil is being degraded to the point that dust storms similar to those in the 1930s are occurring with increasing frequency. In the *dust bowl* in the prairie states led to abandonment of farm and the migration of farmers to other parts of the country.

While changing farming practices have partially help remedy this, there is a danger of a recurrence. Deforestation and degradation of the ecosystem have resulted in fewer plants capturing carbon dioxide by photosynthesis in a time when there has been an increase in fossil fuel use.

The atmospheric level of carbon dioxide and other greenhouse gases (GHGs) continues to climb.

Nature of the Sickness

For a soil to be fertile, it must be capable of delivering water and nutrients to the root zone of the growing plant for it to produce biomass. This ability of the soil depends upon chemistry, biology, and the physical properties such as texture and permeability. Soil is a mixture of sand, silt, and clay, inorganic substances, and various organic compounds that help deliver and store water and nutrients (both macro and micro) to the growing plant. It also provides a home for soil microbes that significantly increase the organic carbon level in the soil. These microbes help breakdown minerals into a form that can be utilized by the plant. A loss in agricultural productivity has been associated with the decrease in organic carbon that has been seen in degraded, depleted, or sterilized soils.

Plants grow through a process of photosynthesis whereby carbon dioxide and water are combined using energy from sunlight to form carbohydrates consisting of organic molecules like sugars, cellulose, and starches. This carbohydrate building process is retarded by the lack of adequate water availability in places where

1. There is increased evaporation where conditions are becoming hotter and drier
2. There is the disappearance of glacier meltwater, which recharges the aquifers, due to rising temperature, combined with reductions in snowfall and rainfall
3. There is decrease in the soils' *field capacity for water* due in part to declining organic content and increased soil density

A worldwide drop in the water table has aggravated the water shortage problem due to

1. Poor irrigation practices
2. Lack of water conservation
3. Competition with the water needs of growing cities and industries

Along with the lack of water availability, there has been a loss of soil nutrients. Plants need nitrogen, phosphorus, potassium, calcium, magnesium, and sulfur as macronutrients and eight micronutrient elements to grow vigorously to produce nutritious food. Poor farming practices have depleted a majority of these nutrients or made them unavailable to the growing plant.

What Malthus Said

Rev. Thomas Malthus published in 1798 *An Essay on the Principles of Population* (Malthus, 1798) in which he postulated that if the needs of a growing population exceeded the ability of resources to supply for those needs, then growth was not sustainable. He predicted that this could not continue and eventually lead to catastrophe with starvation and disease.

It appeared that he was overly pessimistic. The agricultural revolution, which employed synthetic fertilizers to greatly increased production of food, delayed this predicted catastrophe. However, it seems that this gain was only temporary in that the needs kept growing faster than what technology advances could provide.

More recent discussions of the problem were offered in the book called *Limits to Growth* (Meadows et al., 1974).

We currently face major challenges to advance our technology, including better ways to produce enough food for the growing population. Here, we need technology to help us solve a serious environmental problem.

What to Do

The challenge demands a change in agricultural practices. There must be more farming and more farmers. There is a need to adopt ways that prevent soil degradation and maintain and improve soil quality, to find ways of growing plant species that are more accommodating to environmental changes, and to find more efficient ways to irrigate land with a limited water supply. The contamination of the land by pollutants must cease, and one must find ways to restore unsuitable land into agricultural productivity.

Specific means for achieving these goals will be discussed.

Ways to Farm

There has been a trend to replace the smaller family farms with industrial-scale agriculture. Within a century, the percentage of the US population involved with farming decreased from about 50% to about 1%–2%. This has resulted in increase farming mechanization and the development of larger and more specialized monoculture farms. These are more often profit motivated, leading a greater use of synthetic fertilizers, herbicides, and pesticides, with less concern for the stewardship of the land. There is a decrease in soil organic compounds, combined with the increase in fertilizer runoff that allows nutrients to flow into stream, rivers, and lakes, causing the growth of nuisance algae and weeds that eventually leads to *dead zones*. The cause of our frequent *red tide algae blooms* along our ocean coastlines has been attributed to the outflow from these freshwater *dead zones*.

Another consequence of larger specialized farms is that food is shipped over greater distances, using more fuel to get to market, so the farmer gets a smaller share of the profits. A remedy is to return to smaller, more localized farms. The trend shows that an increasing number of *farmers' markets* and their growing popularity reflect this need. The food grown often caters to local needs, is fresher, and is of better quality and nutrition while having a lower-carbon food print. Added advantages are that it provides for local jobs, while providing greater food security.

Soil Restoration

To increase the organic content of the soil, one may resort to adding compost or to plow in farm wastes such as corn stover or *green manure* cover crop. While this helps, it is only temporary and usually decomposes within 5–10 years back into the atmosphere as carbon dioxide, methane, and water. It is better to enhance soil carbon using a much longer lasting additive like horticultural charcoal (biochar) as will be discussed later.

The detrimental effects of synthetic fertilizers can be partially offset by organic farming techniques where one replaces the synthetics with biomass-derived additives such as compost. Annual additions must be continued to offset their decomposition as discussed earlier. The biochar approach may be better since it has the potential advantage of being a one-time addition due to its longevity and durability. Combined with compost and mineral supplements that even the most depleted soils can be made fertile.

A large fraction of land has become unsuitable for farming due to soil pollution and contamination. These *brownfields* may be remediated by absorbents that prevent the transfer of toxic agents from polluted water and soil into the growing plant. Biochar may have this function.

Water Needs

Increased evaporation of soil moisture caused by global warming (aka climate change) especially in areas of decreased rainfall can reduce the amount of available water. Three measures to deal with it are

1. Employment of more efficient irrigation methods
2. Selection of crops that demand less water
3. Use of additives that retain and release water in a more controlled manner

A common way to irrigate fields is to spray the water into the air resulting in an appreciable fraction of the water being lost to the air by evaporation before it reaches the plant (see Figure 5.1). Another way is to flood the fields from open aqueducts (see Figure 5.2). This also evaporates the water on its way to the fields and eventually to the plant. A better alternative is to use *drip irrigation* where water is directly delivered to the plants' roots via a small plastic tube (see Figure 5.3).

The purity requirements of irrigation water are less stringent than those required for potable drinking water.

Water and sewage treatment facilities could provide the *blackwater* and *gray water* that could be further treated through membranes and absorbents to produce suitable water for irrigation.

Most of the world's water is in the oceans where it contains about 3½% salt. The salt is removed by distillation, desalinization, or reverse osmosis; however, this requires large amounts of energy that often may be too expensive to justify using it. It is employed in Arab states where energy is readily available and for purposes and locations (cruise ships) that justify the cost.

Most water returns to the earth as rain after it evaporates from the ocean.

More efficient ways of collecting and delivering rainwater are needed.

Shallow freshwater wells close to the sea are prone to saltwater contamination as the water table drops or the sea level rises.

The tsunami-flooded rice fields near Fukushima were able to grow rice when biochar and compost were added to the soil. The control field failed to produce any rice.

FIGURE 5.1 Spray irrigation from Google Images of National Geographic's *irrigation sprinklers* in Capetown, South Africa.

FIGURE 5.2 Canal and flooded field irrigation from Google Images.

FIGURE 5.3 Drip irrigation—irrigation equivalent from Dick Stein's NEACT PowerPoint. (Agriculture Research Service.)

Absorbents such as biochar, bentonite clays, and water-holding polymers can help because they absorb large amounts of water and then slowly release it. Their use allows farming in areas where less water and irrigation is available. This has permitted farming in arid sandy regions such as Africa, the Sahara Desert, Arab states, and even Cape Cod.

Biochar

We have repeatedly referred to the use of biochar. This material is made by heating or roasting clean biomass in a limited amount of air so as to *carbonize* it into a form of horticultural or agricultural charcoal. It differs from normal charcoal in the fact that it is not used as an energy source for heating and cooking but as a beneficial soil additive. Its preparation is such as to yield a low-density, highly porous material that has a very high surface area. While mostly carbon, some reactive chemical groups remain, the amount of which can be affected by preparation conditions. Its use was developed by the indigenous peoples of Amazon several thousand years ago to feed large city states with populations of over 100,000 as documented by Lehmann and collegues in 2003 and reported by Charles C. Mann in a book published in 2005 entitled *1491: New Revelations of the Americas before Columbus* (Mann, 2005).

It allowed the Amazonians to have a flourishing society based on high agricultural productivity in an area of very poor *rain forest* soils.

They cultivated many varieties of crops that provide a majority of the world's food supply.

Corn, potatoes, tomatoes, cocoa, peppers, squash, pumpkins, and beans that we enjoy today would not have been possible without this ancient society's agricultural technology.

Charles C. Mann later published in 2011 a sequel to that of *1491* entitled *1493 Uncovering the New World That Columbus Created* (Mann, 2011), which talks about the blight responsible for the Irish great potato famine, which was a microbe found in the natural guano fertilizer found in South America.

It is interesting that crops were successfully cultivated for thousands of years in an area where the blight microbe originated.

Was biochar the agent that allowed cultivation in blighted soils?

Anecdotal reports of good harvests of potato and tomato in blighted soil indicate that this may be true, and a report from Israel (Elad et al., 2010) warrants further controlled studies.

The preparation of biochar can be a *low-tech* procedure not requiring specialized industrial-scale equipment. This can involve recovery from campfires or *ricks* by quenching the glowing embers with water or a more sophisticated means such as the use of a retort (see Figure 5.4) or kilns (see Figure 5.5), or even a continuous screw feed devices such as a forest residue carbonizer (see Figure 5.6), which does not require constant attention. It has the advantages that these devices can be fairly simple and inexpensive enough for individual farmers or collectives to make and use.

In making biochar, energy must be applied for the initial heating until the biomass comes up to temperature to start the pyrolysis reaction. First, water vapor comes off from residual moisture in the feedstock; then, the breakdown of carbohydrates starts releasing more water vapor. At this point, the reaction starts producing its own heat and is known in science as an exothermic reaction. During this phase,

FIGURE 5.4 Adam retort.

FIGURE 5.5 Missouri kiln.

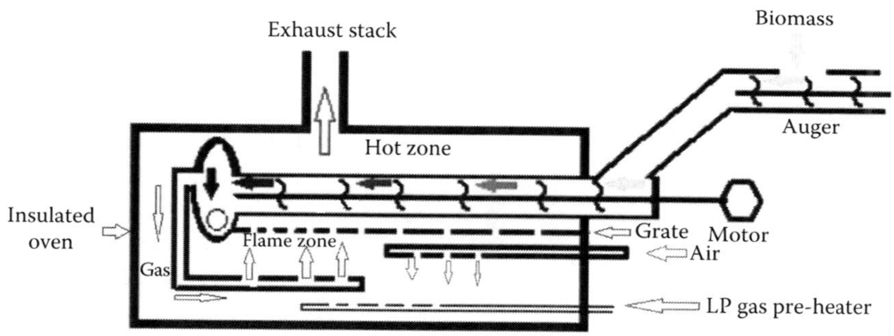

FIGURE 5.6 Continuous wood residue carbonizer. (Based on Dargan, E.E. and Smith, W.R., *Forest Prod. J.*, 9, 395, 1959.)

the flammable gases, such as hydrogen, carbon monoxide, methane, and larger organic molecules, are evolved, which could be used in various ways. The heat can be used directly to dry the fresh biomass feedstock or to heat barns and greenhouses. This dried feedstock can be cleanly *gasified* to supply fuel gas for emergency electrical power generation, while also producing biochar.

There have been considerations of condensing and collecting the volatile products and purifying them into salable products rather than burning them.

However, the economics of this time does not seem attractive. Such procedures are probably too complex for use by small-scale producers of biochar such as farmers or farm groups.

Of course, carbon dioxide (CO_2) is produced when the carbon monoxide and methane are burned off to provide the heat necessary for starting the pyrolysis process. Since 30%–50% of the carbon remains within the biochar, and represents a net removal of CO_2 from the atmosphere it should be considered carbon negative. Because the net effects are removal of the CO_2 from the atmosphere, and safely sequestering it into the soil as carbon, it has been referred to as *coal mining in reverse*. For every ton of carbon that is buried, some three and half tons of carbon dioxide gas are safely sequestered.

The pioneering work in Japan by Dr. Makoto Ogawa in the early 1980s allowed KANSO Coal Company to supply fuel to the KANSAI Electric Company and was able to meet the Kyoto Protocol requirements by utilizing the waste heat to produce biochar to help improve degraded lands in New Zealand back into agricultural productivity.

Dr. Ogawa presented in 2008 a paper summarizing the uses and benefits of biochar in the soil over the past 500 years (Ogawa, 2008).

This improved fertility allowing greater yields of biomass that could be converted into biochar capturing more carbon dioxide through photosynthesis. This extra biomass in the first and subsequent years more than made up the temporary loss of energy that would have been gotten if the biochar was to burned completely to ash.

The use of biochar has the environmentally desirable value of sequestering carbon dioxide. However, current policies awarding *carbon credits* for sequestering carbon dioxide do not appear to justify the cost of producing or using biochar, but its value as a beneficial agricultural soil amendment and as a means of increasing yields in farms and forests would do so.

Biochar is relatively inert, so it is not a *fertilizer* but serves as a substrate or *sponge* to soak up water and nutrients, making them available to the growing plant, and it can also serve as a home for soil microbes (bacteria, fungi, and yeasts) that break down rock minerals that aid in plant growth. It functions best when it is charged with nutrients from composts, barnyard wastes, and fertilizers or is inoculated with microbes such as arbuscular mycorrhizal fungi (AMF) helping to maintain the symbiotic *circle of life*.

It has the advantage, in contrast to other biomass additives, of remaining in the soil for many years (perhaps centuries) without significant breakdown.

After the initial application, only smaller additions may be needed in the following years. Thus, its cost can amortized over its long lifetime and could possibly become a one-time addition to the soil.

Biochar has the following attributes:

1. It rapidly absorbs considerable amounts of water due to its porosity, pore sizes, and capillary structure. As the density of the biochar decreases, there is a corresponding increase in water-holding capacity that is slowly released to the growing root system:
   a. This allows the reduction and frequency of irrigation in arid and semiarid regions.
   b. It was found by this author that a 10% w/w addition of biochar to beach sand increased the water-holding capacity by 300% making the mixture having the same water-holding capacity as fertile soil (Wysocki) when compared with soil that is found in the Connecticut River (Pioneer) Valley at Hatfield. The water holding results fall with the values found in the USDA Soil Conservation Service publications.
   c. It allows plants to survive in drought conditions.
2. It helps moderates the cation and anion exchange capacity of the soil to prevent severe fluctuations in the level of macro- and micronutrients minimizing stress to the plant system.
3. It appreciably increases agricultural productivity, especially in degraded or depleted soils by several hundred percent.
4. It has the capability of binding certain pollutants, preventing them from entering the *food chain* poisoning food and forage crops. Small anecdotal studies confirm these reports and show promise for more detailed controlled studies.

Bottom Line, Is It Worth Doing?

The farmer is mostly interested in the economics and sustainability of their operation, so must consider whether the cost of making or securing biochar and applying it to the soil is justified by the increase in agricultural yields and improved quality of the harvest. There is not a simple answer and the conclusion depends upon how the biochar is obtained and used, the nature of the soil, and the value of the agricultural product that is grown. There are many cases where this is certainly worthwhile.

A farmer on Cape Cod, who makes his own biochar, has used it profitably for growing turnips in rather sandy soil.

Biochar is of further benefit to the public by reducing atmospheric CO_2 and GHGs and can help with contaminated soil remediation. The farmer and forester cannot cover the cost completely.

At present, there appears to be insufficient public support for doing this, but we believe that biochar's value to society, to the farmer, and to the world renders the use of biochar as a very desirable soil amendment.

Thus, biochar, in addition to helping to cure sick soils, serves as a desirable contributor to improving the environment. It is not the *silver bullet* to solve all of our serious problems, but we believe it can make an important contribution.

REFERENCES

Dargan, E.E. and W.R. Smith. Progress in charcoal production: Continuous residue carbonization. *Forest Products Journal*, 9:395–397, 1959.

Elad, Y., D.R. David, Y.M. Harel, M. Borenshtein, H.B. Kalifa, A. Silber, and E.R. Graber. Induction of systemic resistance in plants by biochar, a soil–applied carbon sequestering agent. *Phytopathology*, 100:913–921, 2010.

Lehmann, J., J. Pereira da Silva Jr., C. Steiner, T. Nehls, W. Zech, and B. Glaser. Nutrient availability and leaching in an archaelogical Anthrosol and Ferralsol of the Central Amazon basin: Fertilizer, manure, and charcoal amendments. *Plant and Soil*, 249:343–357, 2003.

Malthus, T.R. *Essays on the Principles of Population*. London, U.K.: J. Johnson, 1798.

Mann, C.C. *1491: New Revelations of the Americas before Columbus*. New York: Knopf, 2005.

Mann, C.C. *1493: Uncovering the New World that Columbus Created.* New York: Knopf, 2011.

Meadows, D.H., D.L. Meadows, J. Randes, and W.W. Behren III. *Limits to Growth*. New York: Universe Books, 1974 (updated in 2004).

Ogawa, M. Introduction to the pioneer works of charcoal uses in agriculture, forestry and others in Japan, presented in 2008.

USDA Soil Conservation Service (in co-operation with Massachusetts Agricultural Experiment Station), Soil Survey of Hampshire County, Massachusetts—Table 15, Physical and Chemical Properties of the Soils.

Wysocki Jr., T.S. South Meadow Farm & Research Center, Wales, MA. USA-Logbook TYM-C, pp. 74–75.

6

Rates and Mechanisms of Functional Mineral Reactions in Soils

David A.C. Manning

CONTENTS

Introduction

The supply of nutrients to plants and crops is a natural process of silicate mineral weathering and is coupled to parallel processes that fix atmospheric CO_2 as soil carbonate minerals. In this chapter, evidence is reviewed that allows estimation of the rates at which these processes take place, showing that they can be rapid on a human time scale. For example, in certain urban soils, the measured rate of removal of CO_2 is 12.5 t/ha/month, through the precipitation of calcium carbonate. Similarly, the oxidation of compost and soil organic matter leads rapidly to the formation of calcium carbonate in artificial soils. The ability of silicate minerals in soils to supply plant nutrients has been demonstrated in a number of studies and relies on the interaction between the kinetics of mineral dissolution and reaction with plant root exudates. Both nutrient supply and carbon capture through carbonate precipitation are coupled reactions; however, their benefits do not necessarily occur in the same location within a soil horizon.

Minerals in Soils

Soils are composite materials that contain a range of components of varying reactivity. We normally think of soils in terms of a matrix of minerals within the pore spaces of which occur water, gas, organic matter, and living organisms (plant roots, microbes, etc.). The pore space is where *all the action is* (Ritz and Young, 2011), with reactions between the soil solution and living organisms that are rapid in the sense that in a healthy soil, they support a dynamic biological community that we observe as vegetated land. The mineral matrix is composed of geological minerals—some of these are residual, derived from the rocks that weathered to form the soil, and others are pedogenic, produced as a consequence of soil processes. Far from acting simply as an inert physical matrix, geological minerals play a dynamic role in the reactions that take place within and sustain a healthy soil.

Examples of the common rock-forming minerals that weather within the soil system, and others that form in soils, are summarized in Table 6.1.

TABLE 6.1

Examples of Mineral Participants in Soil Reactions

Mineral Family	Example	Chemical Composition
Rock-forming minerals		
Olivine	Forsterite	Mg_2SiO_4
Pyroxene	Enstatite	$Mg_2Si_2O_6$
	Diopside	$CaMgSi_2O_6$
Amphibole	Anthophyllite	$Mg_7Si_8O_{22}(OH)_2$
	Tremolite	$Ca_2Mg_5Si_8O_{22}(OH)_2$
Mica	Biotite	$KFe_3AlSi_3O_{10}(OH)_2$
	Muscovite	$KAl_3Si_3O_{10}(OH)_2$
Silica	Quartz	SiO_2
K-feldspar	Orthoclase	$KAlSi_3O_8$
Plagioclase	Albite	$NaAlSi_3O_8$
	Anorthite	$CaAl_2Si_2O_8$
Minerals that form in soils		
Carbonates	Calcite	$CaCO_3$
Clays	Kaolinite	$Al_2Si_2O_5(OH)_4$
Oxyhydroxides	Gibbsite	$Al(OH)_3$
	Boehmite	$AlO \cdot OH$
	Goethite	$FeO \cdot OH$

Plants can be regarded as a living interface between geological materials and the hydrosphere/atmosphere. Through photosynthesis, plants convert solar energy into biomass. Although the product of this process is seen aboveground as plant tissue, it is important to recognize that plant roots pump organic compounds, such as low-molecular-weight organic acids (Kuzyakov and Domanski, 2000; Ryan et al., 2001), into the soil via root exudates (Figure 6.1), with a number of functions. On the one hand, plant root exudates can reduce soil pH, enhancing mineral weathering, and on the other hand, multidentate ligands (e.g., citrate) exuded by plants can complex nutrient elements for uptake as well as neutralize potentially harmful cations (such as Al; Dakora and Phillips, 2002; Mariano and Keltjen, 2003).

The composition of a plant includes inorganic as well as organic compounds. Excluding water, the bulk of plant tissue is dominated by photosynthetically fixed carbon compounds. However, plant metabolism

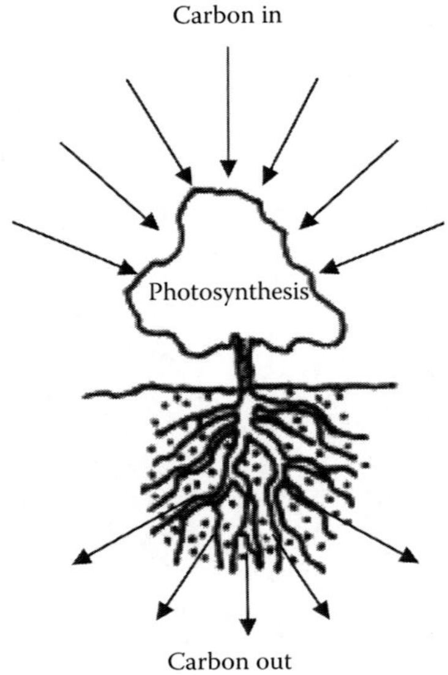

FIGURE 6.1 The action of plants as a carbon pump between the atmosphere and soil.

depends on a range of inorganic (mineral) nutrients. Other than C, H, O, and N, all chemical components within a plant must come ultimately from a geological source, without exception. In some cases, these nutrients are supplied using artificial fertilizers, but these are derived from mined materials. N fertilizer production is utterly dependent on mined energy, which amounts to as much as 90% of the cost of a N fertilizer product (Lægrid et al., 1999).

Thus, plants actively remove mineral-derived materials from the soil. In the reverse sense, plants control the chemistry of the soil in which they grow through the precipitation of newly formed minerals, especially carbonates and also oxalates (Manning, 2000; Braissant et al., 2003; Burford et al., 2003). In part, this is simply an ultimate consequence of plant root exudation; all organic matter that enters a soil finally degrades into carbon dioxide and water, and equilibrium between these involves the formation of dissolved carbonate species that, under appropriate conditions, can precipitate, if sufficient Ca or Mg is in solution to cause saturation with respect to a Ca or Mg carbonate mineral.

The coupled reactions that involve plants and soils have implications for both soil fertility and carbon capture. This chapter addresses the mineralogical reactions that are mediated by plants and considers how they can be managed to improve the sustainability of the plant–soil system.

Mineral Reactions in Soils

The primary rock-forming silicate minerals vary in their stability once exposed to Earth's surface conditions. Those produced by igneous processes, typically occurring in volcanic rocks and plutonic rocks brought to the surface by erosion, may be the least stable.

Table 6.2 gives some examples of weathering reactions and includes the thermodynamic free energy in each case. Weathering reactions for those minerals that are typical of basaltic rocks, such as olivine, pyroxene, and the plagioclase anorthite, have the lowest (most negative) free energies of reaction (ΔG_r), which means that the reactions will proceed to the right. The sequence of stability is summarized graphically in Figure 6.2 and corresponds to the Goldich reaction series, an empirically determined weathering sequence (Goldich, 1938). Importantly, in Figure 6.2, the free energy of reaction is expressed following normalization of the number of product atoms (Curtis, 1976); this allows the weathering of complex minerals (such as tremolite) to be compared on equivalent terms to minerals with a simpler formula (such as olivine). As shown in Figure 6.2, the Goldich series is underpinned by equilibrium thermodynamics.

In Figure 6.2, data are shown for three different reactions:

1. To produce quartz and kaolinite (filled circles)
2. To produce dissolved silica (H_4SiO_4; open circles)
3. To produce the aluminum hydroxide gibbsite ($Al(OH)_3$; crosses)

TABLE 6.2

Weathering Reactions for Selected Silicate Minerals, with Calculated Free Energies of Reaction

Mineral	Weathering Reaction (to Produce Quartz, Q, and Kaolinite)	ΔG_r kJ/mole
Forsterite	$Mg_2SiO_4 + 4H^+ = 2Mg^{2+} + SiO_{2(Q)} + 2H_2O$	−187.7
Enstatite	$Mg_2Si_2O_6 + 4H^+ = 2Mg^{2+} + 2SiO_{2(Q)} + 2H_2O$	−181.0
Diopside	$CaMgSi_2O_6 + 4H^+ = Ca^{2+} + Mg^{2+} + 2SiO_{2(Q)} + 2H_2O$	−169.0
Tremolite	$Ca_2Mg_5Si_8O_{22}(OH)_2 + 14H^+ = 2Ca^{2+} + 5Mg^{2+} + 8SiO_{2(Q)} + 8H_2O$	−556.8
Biotite	$2KFe_3AlSi_3O_{10}(OH)_2 + 14H^+ = 2K^+ + 6Fe^{2+} + Al_2Si_2O_5(OH)_4 + 4SiO_{2(Q)} + 7H_2O$	−390.8
Muscovite	$2KAl_3Si_3O_{10}(OH)_2 + 2H^+ + 3H_2O = 2K^+ + 3Al_2Si_2O_5(OH)_4$	−29.4
Orthoclase	$2KAlSi_3O_8 + 2H^+ + H_2O = 2K^+ + Al_2Si_2O_5(OH)_4 + 4SiO_{2(Q)}$	−52.0
Albite	$2NaAlSi_3O_8 + 2H^+ + H_2O = 2Na^+ + Al_2Si_2O_5(OH)_4 + 4SiO_{2(Q)}$	−85.4
Anorthite	$CaAl_2Si_2O_8 + 2H^+ + H_2O = Ca^{2+} + Al_2Si_2O_5(OH)_4$	−106.1

Source: Using data from Robie, R.A. and Hemingway, B.S., *Thermodynamic Properties of Minerals and Related Substances at 298.15 K and 1 Bar (105 Pascals) Pressure and at Higher Temperatures,* US Geology Survey Bulletin, Washington, DC, Vol. 2131, 470pp., 1995.

Note: These reactions correspond to the acid dissolution mechanism; Palandri and Kharaka (2004).

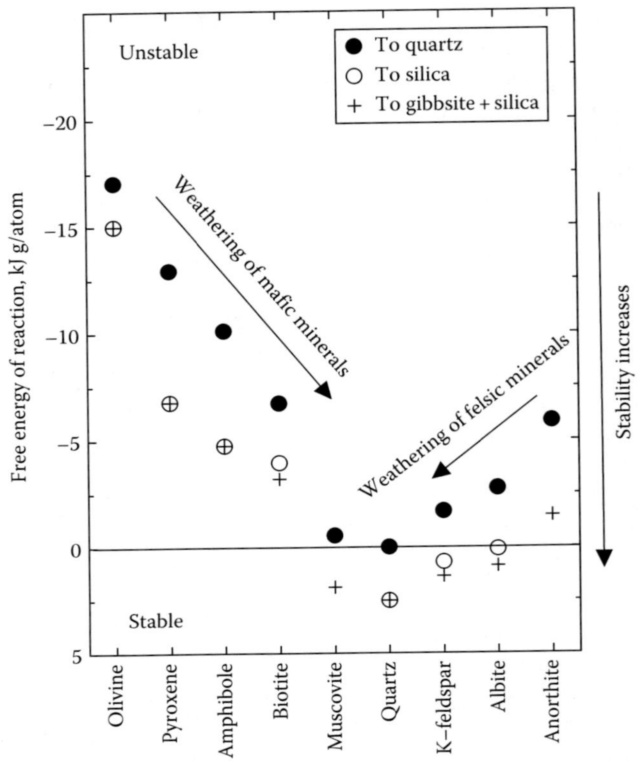

FIGURE 6.2 Comparison of free energies of weathering reactions for common silicate mineral families, with different products (normalized to the number of participating atoms; see text for explanation).

These correspond to reactions that take place in different soil systems.

Reactions that produce quartz and kaolinite are typical of those that approach chemical equilibrium (i.e., ΔG_r is zero; Figure 6.2). Those that produce dissolved silica (H_4SiO_4) and gibbsite ($Al(OH)_3$) have less negative or positive values for ΔG_r. Reactions with positive values for ΔG_r will not occur, under equilibrium conditions, as the reactants are predicted to be more stable than the products. The dissolution of quartz to give H_4SiO_4 has a positive value for ΔG_r, indicating that quartz is predicted to be a stable phase at Earth's surface conditions.

These theoretical observations readily explain the mineral assemblages in soils developed in many northern, temperate zone, soils. In contrast, deeply leached tropical soils developed on silicate rocks are dominated by iron oxides and aluminum oxides—silica and major cations other than Fe and Al have been removed by leaching, before chemical equilibrium could be reached.

The interplay between soil mineralogy and mineral dissolution rates is illustrated in Figure 6.3. This shows that dissolution rates vary from one mineral to another, again resembling the Goldich series. Experimentally determined dissolution rates depend greatly on pH (Figure 6.3), being two or three orders of magnitude greater for the acid dissolution mechanism (as in the reactions listed in Table 6.2: NB the neutral dissolution mechanism involves no protons).

Figure 6.4 shows the relationship between pH and dissolution rate for the olivine forsterite and the two end-member plagioclase feldspars (albite, $NaAlSi_3O_8$, and anorthite $CaAl_2Si_2O_8$). The three minerals show contrasting dissolution behavior. Forsterite shows a steady increase in reaction rate as pH reduces, while albite shows a minimum at pH 6–7, reflecting the amphoteric nature of Al. Anorthite shows considerable scatter in dissolution rate, which is up to one order of magnitude greater than that of albite.

Rates of reaction for silicate minerals are very low. To place these in context, the equivalent times required to completely dissolve a 1 mm sphere are summarized in Table 6.3.

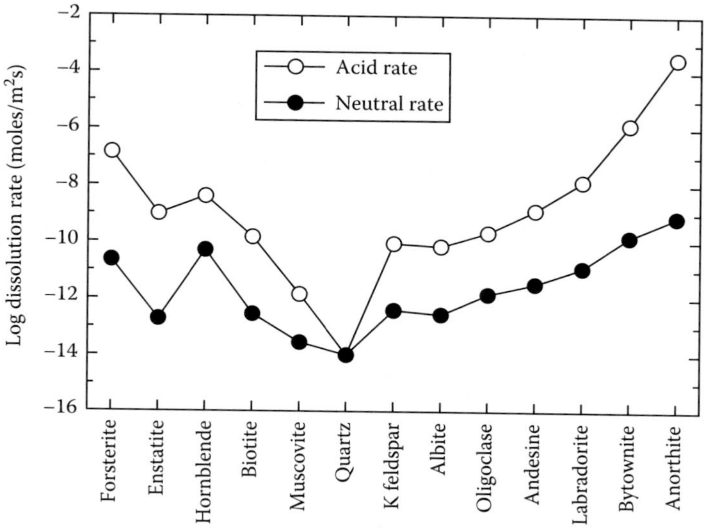

FIGURE 6.3 Dissolution rates under acidic (pH 5) and neutral conditions (pH 7) for minerals in the Goldich series. (Reproduced from Manning, D.A.C., Minerals in soil development, in: *Environmental Mineralogy II,* D.J. Vaughan and R.A. Wogelius, eds., European Mineralogical Union Notes in Mineralogy, London, U.K., Vol. 13, pp. 103–122, 2013. With permission; Data from Palandri, J.L. and Kharaka, Y.K., A compilation of rate parameters of water-mineral interaction kinetics for application to geochemical modeling, US Geological Survey Open File Report 2004-1068, Menlo Park, CA, 64pp., 2004.)

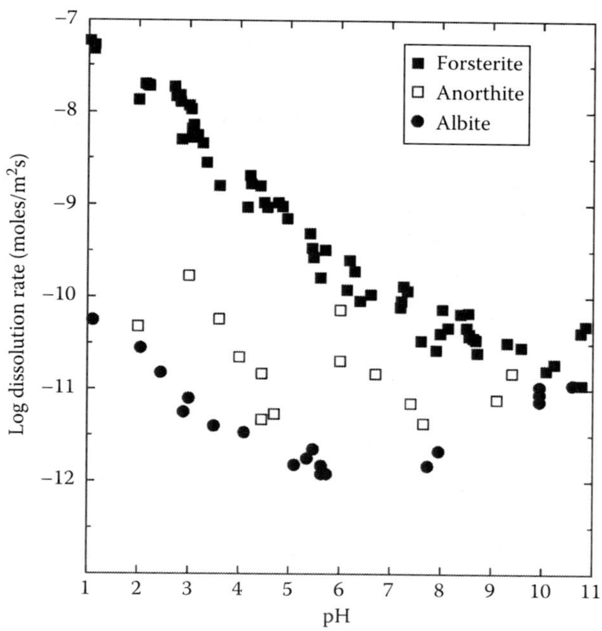

FIGURE 6.4 Dissolution rates for olivine (forsterite) and plagioclase (anorthite and albite). (Data from Palandri, J.L. and Kharaka, Y.K., A compilation of rate parameters of water-mineral interaction kinetics for application to geochemical modeling, US Geological Survey Open File Report 2004-1068, Menlo Park, CA, 64pp., 2004.)

TABLE 6.3
Approximate Lifetime in Soils (Dilute Solution, pH 5) of a 1 mm Sphere of Different Rock-Forming Minerals

Mineral Group	Mineral	Lifetime (Years)
Quartz	Quartz	34,000,000
Clay	Kaolinite	6,000,000
Mica	Muscovite	2,600,000
K-feldspar	Microcline	921,000
Mica	Biotite	900,000
Na-plagioclase	Albite	575,000
Ca-plagioclase	Bytownite	40,000
Mg pyroxene	Enstatite	10,100
Ca-Mg pyroxene	Diopside	6,800
Mg olivine	Forsterite	2,300
Carbonate	Dolomite	1.6
Carbonate	Calcite	0.1

Sources: Based on Drever, J.I., *The Geochemistry of Natural Waters: Surface and Groundwater Environments*, 3rd edn., Prentice-Hall, Englewood Cliffs, NJ, 436pp., 1997; Lasaga, A.C. et al., *Geochim. Cosmochim. Acta*, 58, 2361, 1994; Reproduced from Manning, D.A.C., Minerals in soil development, in: *Environmental Mineralogy II*, D.J. Vaughan and R.A. Wogelius, eds., European Mineralogical Union Notes in Mineralogy, London, U.K., Vol. 13, pp. 103–122, 2013. With permission.

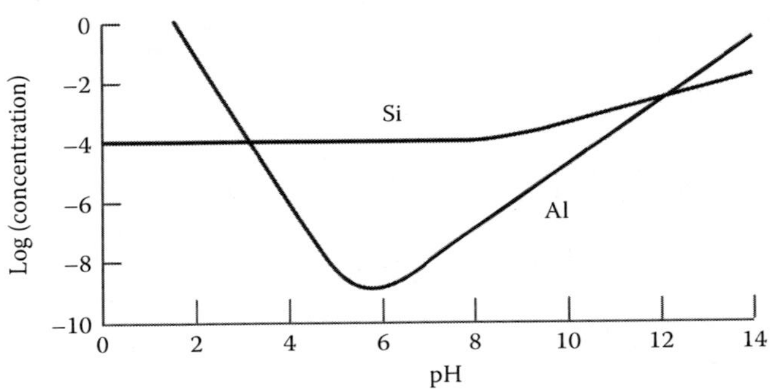

FIGURE 6.5 Comparison of the solubilities of Al and Si as controlled by quartz and gibbsite.

Overall, the key reactions that affect silicate mineral weathering involve the systematic destruction of a silicate mineral lattice. Ionic-bonded cations are released to the soil solution. Si may react to form quartz, if kinetics permit, or it may be removed in solution as the dissolved H_4SiO_4 complex. Al is the least soluble of the products of silicate mineral weathering under near neutral pH conditions (Figure 6.5).

Importantly, Figure 6.5 shows that the solubility of quartz is up to five orders of magnitude greater than that of gibbsite at near neutral to slightly acidic pH. This explains why deeply leached tropical soils lack quartz—its comparatively high solubility removes it in solution, whereas the highly insoluble gibbsite remains. It is also consistent with the observation that natural waters commonly contain 5–10 mg/L Si in solution, whereas Al concentrations are measured in the microgram per liter range.

Carbon Capture Reactions

The formation of pedogenic carbonates is widely observed in nature. The process of their formation involves precipitation of calcium carbonate in a soil; investigation of the carbon and oxygen stable isotope compositions shows that the carbon is derived via photosynthesis (Salomans and Mook, 1976; Salomans et al., 1978; Cerling, 1984). The occurrence of pedogenic carbonates is typically associated with arid climates, but it has been shown (Jenny, 1980; Gunal and Ransom, 2006) that with increasing rainfall the depth of their formation increases so that in some cases they may form below depths normally mapped in soil surveys.

The formation of pedogenic calcium carbonates in urban soils has recently been reported (Renforth et al., 2009; Washbourne et al., 2012). In urban soils, it is quite common to find fine-grained material derived from the demolition of buildings. In cases where a building is being demolished, it is normal for the work to involve crushing of concrete to produce a secondary aggregate product. This process generates fine-grained material that may have no market, in which case it is very often incorporated into the soil as part of the development. The materials involved are rich in artificial calcium silicates and calcium hydroxide minerals. These react with carbonate in the soil solution to produce calcium carbonate as a pedogenic carbonate.

The process of carbonation of urban soils involves two steps. First, carbonate in solution is produced according to the following reactions:

$$CO_2 + H_2O = H_2CO_3$$

$$H_2CO_3 = H^+ + HCO_3^-$$

$$HCO_3^- = H^+ + CO_3^{2-}$$

These equilibria are pH dependent; carbonate dominates above pH 9 and bicarbonate between pH 6 and 9 (Drever, 1997). The origin of the carbon has many possible sources, including formation in the soil by microbial respiration that decomposes both plant root exudates and soil organic matter (Ryan et al., 2001).

Precipitation of calcium (or magnesium) carbonates depends on saturation being achieved and is in accordance with reactions of the following type:

$$Ca^{2+} + HCO_3^- = CaCO_3 + H^+$$

Although this reaction generates protons, the pH of the soil solution may be buffered by their reaction with soil organic matter (that contains carboxyl groups) or by the buffering capacity of mineral reactions (such as calcium hydroxide—which gives a very high pH). The source of the calcium for carbonate precipitation is the calcium hydroxide or silicate minerals that are derived ultimately from Portland cement.

The rate of dissolution of artificial calcium silicates is not well constrained, limiting our ability to model the formation of pedogenic carbonates in urban soils (Renforth and Manning, 2011). Nevertheless, an estimate of the rate of their formation can be derived from careful field investigations. For example, Washbourne et al. (2012) report the amounts of calcium carbonate in a soil developed on a site where buildings were demolished at a known date. In this case, absolute quantities of newly formed $CaCO_3$ in soils reached up to 25% by weight, and the rate of removal of CO_2 from the atmosphere since demolition and formation of the soil was estimated to be 12.5 t CO_2 per hectare per month over a 3-year period. Similar processes occur when steel slags are exposed to the atmosphere, with associated rapid carbonation.

The amount of material that is available for carbonation in this way is substantial. Renforth et al. (2011) have shown that the global production of calcium silicates from different industrial sources is 7–17 billion tons per annum, with a total carbonation potential of 190–332 million tons of C, or 700–1200 million tons CO_2, per year. This contributes to the need to remove 7 billion tons of C from the atmosphere annually, partially mitigating anthropogenic emissions (Pacala and Sokolow, 2004).

Although carbonation of cement-derived materials in urban soils provides a demonstrable carbon capture mechanism, all it does is compensate for the CO_2 lost from limestone when it is calcined in the manufacture of Portland cement or in the production of steel. The carbonation process that takes place in soils helps to close the carbon loop in a life cycle analysis. To be carbon negative, primary rock-forming silicates would need to show evidence of carbonation. In nature, this is widely observed as carbonate precipitation in soils developed on basic igneous rocks (Kalin et al., 1997; Khalaf et al., 2000; Durand et al., 2007). Artificially, the use of olivine for carbon sequestration by reaction in soils has been proposed by a number of authors (e.g., Köhler et al., 2010; Hartmann et al., 2013). Part of this process of enhanced weathering involves modification of ocean chemistry by increasing the flux of basic cations to the ocean, which in turn increases carbonate stability and helps compensate for ocean acidification.

Reactions That Yield Plant Nutrients

The dominant mineral plant nutrients that are derived from silicate minerals are the alkalis and alkaline earths: K, Mg, Na, and Ca, in addition to other trace mineral nutrients including silica. As shown earlier, the weathering reactions shown by silicate minerals involve the release of cations and silica. The stability of silicate minerals in soils is controlled by their thermodynamic properties, and the rates of reaction also vary. Thus, Fe- and Mg-bearing silicate minerals decompose rapidly compared with K-bearing silicates.

The availability of K from silicates has been the subject of research dating back to the early twentieth century (e.g., Goldschmidt, 1922). In a recent review, Manning (2010) demonstrates that it is the rate of reaction that determines the availability of K from a potassium silicate mineral, and this varies considerably. In plant growth trials with K-feldspar, in general, the observed response has not been significant statistically (Sanz Scovino and Rowell, 1988; Harley and Gilkes, 2000). In contrast, a response with increased plant growth has been observed when nepheline-bearing rocks are used as a source of K. This is due to the dissolution rate of nepheline being up to two orders of magnitude greater than that of K-feldspar, although its K content is lower (Figure 6.6; Tole et al., 1986).

In addition to work with the framework silicates (feldspar, nepheline, etc.), recent studies have considered micas. In a study of the hydroponic growth of olives, Karimi et al. (2011) show that the use of a growing medium made from a crushed sandstone with 2% K_2O, containing glauconite mica, enhances plant growth. Similarly, the use of the lithium mica zinnwaldite (10% K_2O) in trials involving barley gives a similar result to equivalent applications of KCl (Madaras et al., 2012). These observations are consistent with a different mechanism for the release of K from mica; cation exchange between the weathering mica and the solution releases K more rapidly than dissolution of framework silicates.

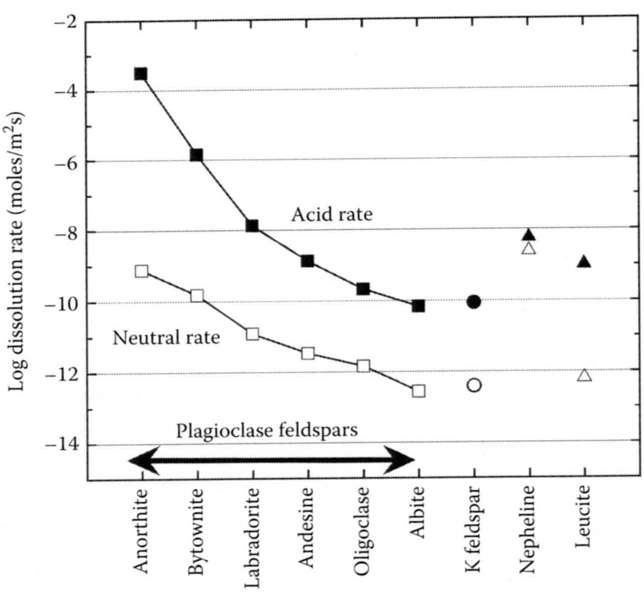

FIGURE 6.6 Comparison of dissolution rates for feldspars and feldspathoids (nepheline, $Na_{0.75}K_{0.25}AlSiO_4$, and leucite, $KAlSi_2O_6$). (Data from Palandri, J.L. and Kharaka, Y.K., A compilation of rate parameters of water-mineral interaction kinetics for application to geochemical modeling, US Geological Survey Open File Report 2004-1068, Menlo Park, CA, 64pp., 2004.)

Implications for Geotherapy

The rates and mechanisms of mineral reactions in soils provide considerable scope for the management of the soil system to achieve specific desired outcomes. Soil is a dynamic system that has been managed for perhaps 10,000 years; Ruddiman (2003) has suggested that the onset of anthropogenic global warming dates back to the start of agriculture. We know that carbon fluxes between the coupled plant–soil system and the atmosphere exceed those between the ocean and the atmosphere (Lal, 2003). The experience of the human race in managing soils has given the landscape we occupy today; our understanding of mineral reactions in soils enables us to change management practice to achieve specific desired goals.

From the point of view of carbon capture, carbonate precipitation has to be considered alongside other approaches to enriching soil carbon. The addition of composts provides improvements in soil fertility and water holding capacity, but only approximately 10% of C within compost is retained within the soil (Diacono and Montemurro, 2010). Microbial respiration and other processes return the majority of compost-derived C to the atmosphere, as CO_2. The use of biochar provides a long-term sink for C, but the process has yet to be widely adopted for a number of regulatory and economic reasons.

The key to understanding how to exploit carbonation of silicate minerals lies in the ability to provide Ca or Mg. There is no shortage of carbon entering the soil via plant root exudates and other pathways (Manning and Renforth, 2013). Thus, any strategic plan to encourage soil carbonate precipitation should focus on sourcing of calcium and magnesium silicate minerals for application to a soil. Kohler et al. (2010) and other authors (Schuiling and Krijgsman, 2006) take the approach of using olivine as a source of Mg, which involves mining of olivine from a rather limited (at present) number of locations that are geographically dispersed. Although attractive as a concept, the feasibility of this approach has been challenged (Hangx and Spiers, 2009). An alternative is to use *local rocks for local people*, in which Ca- and Mg-bearing silicate rocks might be mined for application to soils close to the geological source. This aligns with the mining of aggregates, where local sources predominate given the high cost of transporting a low-price material. As with the crushing of concrete, the process of hard rock aggregate crushing produces substantial quantities of *quarry fines* that may not otherwise have a ready market. In the interests of sustainability, aggregate production and carbon capture might become closely aligned.

The use of calcium and magnesium silicates in soils to enhance carbon capture is directly analogous to the use of reed beds for the passive remediation of contaminated waters. Once soils are prepared for the purpose of carbon capture, they function with no artificial energy inputs. Clearly the ability to capture carbon will decrease to a point where the mineral carbonation reactions have gone to completion, but there is sufficient information available from field studies of concrete carbonation to allow field-based predictions to be made.

The artificial design of land to capture carbon leads to the concept of *Carbon Capture Gardens*. In this, communities are engaged with the design of the landscaping so that an amenity is created that has functions extending beyond carbon capture. Local communities welcome the opportunity to contribute to the mitigation of atmospheric CO_2 emissions and at the same time regenerate land that might otherwise be unattractive.

From the perspective of soil fertility, there is abundant scope for much more *clever* use of natural rock minerals. As Leonardos et al. (1987) state: "Unfortunately, the standard concept and technology of soil fertilizer … is behind that of the superphosphate concept developed by J. B. Lawes in England, 150 years ago… Had this technology been originally developed for the deep leached laterite soils of the tropics instead for (*sic*) the glacial and rock-debris-rich soils of the northern hemisphere our present fertilizers might have been quite different." The key issue that arises from this statement concerns the mineralogy of the soil that requires fertility management. In the soils of temperate zones, rock weathering is less intense, and in those areas with present-day or past glacial activity, mechanical weathering gives soils rich in rock debris. The reactions that dominate here will be those that lead to quartz as a stable weathering product. In deeply leached tropical soils that are dominated by Fe and Al oxides, quartz is not stable and is removed from the soil through dissolution (Figure 6.5). Here, the weathering rates of silicate minerals may be sufficient to provide cations to the soil solution at rates that match the availability required by plants.

Conclusions

The concept of geotherapy as an approach to mitigating the effects of the rising human population helps to address the demands for food production and to compensate for emissions from fossil fuel combustion. Careful use of minerals has a vital role to play. To achieve the full potential of minerals in this way, there needs to be a careful understanding of their reactions within soil systems that is based on fundamental chemistry. Both the thermodynamic stability and dissolution rate of different mineral species differ as a consequence of their composition and crystal structure. The important point to note is that mineral reactions in soils can be rapid on a human time scale. Provided the conditions are right, artificial manipulation of the soil by addition of appropriate minerals has the potential to (1) improve soil fertility by addition of plant nutrients and (2) remove CO_2 from the atmosphere to compensate for fossil fuel combustion, by adding cations such as Ca and Mg that may precipitate as carbonate minerals or contribute to ocean alkalinity. The value of these mineralogical reactions needs to be understood in the specific context in which they are applied as soil processes are controlled by a range of natural climatic and artificial management practices.

REFERENCES

Braissant, O., Cailleau, G., Dupraz, C., and Verrecchia, E. P. (2003) Bacterially induced mineralization of calcium carbonate in terrestrial environments: The role of exopolysaccharides and amino acids. *Journal of Sedimentary Research* 73, 485–490.

Burford, E. P., Fomina, M., and Gadd, G. M. (2003) Fungal involvement in bioweathering and biotransformation of rocks and minerals. *Mineralogical Magazine* 67, 1127–1155.

Cerling, T. E. (1984) The stable isotopic composition of modern soil carbonate and its relationship to climate. *Earth and Planetary Science Letters* 71, 229–240.

Curtis, C. D. (1976) Stability of minerals in surface weathering reactions: A general thermochemical approach. *Earth Surface Processes* 1, 63–70.

Dakora, F. D. and Phillips, D. A. (2002) Root exudates as mediators of mineral acquisition in low-nutrient environments. *Plant and Soil* 245, 35–47.

Diacono, M. and Montemurro, F. (2010) Long-term effects of organic amendments on soil fertility. A review. *Agronomy for Sustainable Development* 30, 401–422.

Drever, J. I. (1997) *The Geochemistry of Natural Waters: Surface and Groundwater Environments*, 3rd edn. Prentice-Hall, Englewood Cliffs, NJ, 436pp.

Durand, N., Gunnell, Y., Curmi, P., and Ahmad, S. M. (2007) Pedogenic carbonates on Precambrian silicate rocks in South India: Origin and paleoclimatic significance. *Quaternary International* 162–163, 35–49.

Goldich, S. S. (1938) A study in rock weathering. *Journal of Geology* 46, 17–58.

Goldschmidt, V. M. (1922) Oversikgtskart over utbredelsen av de forskjellige kalimineraler I norsk fjeldgrund. *Norsk Landmansblad* 41, 268–269.

Gunal, H. and Ransom, M. D. (2006) Clay illuviation and calcium carbonate accumulation along a precipitation gradient in Kansas. *Catena* 68, 59–69.

Hangx, S. J. T. and Spiers, C. J. (2009) Coastal spreading of olivine to control atmospheric CO_2 concentrations: A critical analysis of viability. *International Journal of Greenhouse Gas Control* 3, 757–767.

Harley, A. D. and Gilkes, R. J. (2000) Factors influencing the release of plant nutrients from silicate rock powders: A geochemical overview. *Nutrient Cycling in Agroecosystems* 56, 11–36.

Hartmann, J., West, J., Renforth, P., Köhler, P., De La Rocha, C., Wolf-Gladrow, D., Dürr, H., and Scheffran, J. (2013) Enhanced chemical weathering as a geoengineering strategy to reduce atmospheric carbon dioxide, a nutrient source and to mitigate ocean acidification. *Reviews of Geophysics* AGU, 51, 113–149.

Jenny, H. (1980) *The Soil Resource*. Springer Verlag, New York, 377pp.

Kalin, R. M., Dardis, G., and Lowndes, J. (1997) Secondary carbonates in the antrim basalts: Geochemical weathering at 35KyBP. In: *Geofluids II Conference Extended Abstracts* (J. P. Hendry, P. F. Carey, J. Parnell, A. H. Ruffell, and R. H. Worden, eds.), Queen's University of Belfast, Belfast, U.K., pp. 22–25.

Karimi, E., Abdolzadeh, A., Sadeghipour, H. R., and Aminei, A. (2011) The potential of glauconitic sandstone as a potassium fertilizer for olive plants. *Archives in Agronomy and Soil Science* 58, 983–993.

Khalaf, S., Revel, J. C., Guiresse, M., and Kaemmerer, M. (2000) Some calcareous soils developed on Recent Quaternary basalt in southeast Syria. In: *Global Climate Change and Pedogenic Carbonates* (R. Lal, J. M. Kimble, H. Eswaran, and B. A. Stewart, eds.), Lewis Publishers, Boca Raton, FL, pp. 213–224.

Köhler, P., Hartmann, J., and Wolf-Gladrow, D. (2010) Geoengineering potential of artificially enhanced silicate weathering of olivine. *Proceedings of the National Academy of Sciences of the United States of America* 107(47), 20228–20233, doi:10.1073/pnas.1000545107, hdl:10013/epic.36220.

Kuzyakov, Y. and Domanski, G. (2000) Carbon input by plants into the soil. Review. *Journal of Plant Nutrition and Soil Science* 163, 421–431.

Lægrid, M., Bøckman, O. C., and Kaarstad, O. (1999) *Agriculture, Fertilizers and the Environment*. CABI Publishing, Wallingford, U.K., 294pp.

Lal, R. (2003) Global potential of soil carbon sequestration to mitigate the greenhouse effect. *Critical Reviews in Plant Sciences* 22, 151–184.

Lasaga, A. C., Soler, J. M., Ganor, J., Burch, T. E., and Nagy, K. L. (1994) Chemical weathering rate laws and global geochemical cycles. *Geochimica et Cosmochimica Acta* 58, 2361–2386.

Leonardos, O. H., Fyfe, W. S., and Kronberg, B. I. (1987) The use of ground rocks in laterite systems—An improvement to the use of conventional soluble fertilizers. *Chemical Geology* 60, 361–370.

Madaras, M., Mayerová, M., Kulhánek, M., Koubová, M., and Faltus, M. (2012) Waste silicate minerals as potassium sources: A greenhouse study on spring barley. *Archives in Agronomy and Soil Science* 59, 671–683.

Manning, D. A. C. (2000) Carbonates and oxalates in sediments and landfill: Monitors of death and decay in natural and artificial systems. *Journal of the Geological Society of London* 157, 229–238.

Manning, D. A. C. (2013) Minerals in soil development. In: *Environmental Mineralogy II* (D. J. Vaughan and R. A. Wogelius, eds.), European Mineralogical Union Notes in Mineralogy, London, U.K., Vol. 13, pp. 103–122.

Manning, D. A. C. and Renforth, P. (2013) Passive sequestration of atmospheric CO_2 through coupled plant-mineral reactions in urban soils. *Environmental Science & Technology* (special issue on Carbon Sequestration), 47, 135–141, doi:10.1021/es301250j.

Mariano, E. D. and Keltjen, W. G. (2003) Evaluating the role of root citrate exudation as a mechanism of aluminium resistance in maize genotypes. *Plant and Soil* 256, 469–479.

Pacala, S. and Socolow, R. (2004) Stabilization wedges: Solving the climate problem for the next 50 years with current technologies. *Science* 305, 968–972.

Palandri, J. L. and Kharaka, Y. K. (2004) A compilation of rate parameters of water-mineral interaction kinetics for application to geochemical modeling. US Geological Survey Open File Report 2004-1068, Menlo Park, CA, 64pp.

Renforth, P. and Manning D. A. C. (2011) Laboratory carbonation of artificial silicate gels enhanced by citrate: Implications for engineered pedogenic carbonate formation, *International Journal of Greenhouse Gas Control* 5, 1578–1586.

Renforth, P., Manning, D. A. C., and Lopez-Capel, E. (2009) Carbonate precipitation in artificial soils as a sink for atmospheric carbon dioxide. *Applied Geochemistry* 24, 1757–1764.

Renforth, P., Washbourne, C.-L., Taylder, J., and Manning, D. A. C. (2011) Silicate production and availability for mineral carbonation, *Environmental Science & Technology* 45, 2035–2041.

Ritz K. and Young, I. M., (Ed.). (2011) *Architecture and Biology of Soils: Life in Inner Space*. CABI Publishing, Wallingford, U.K., P. 244.

Robie, R. A. and Hemingway, B. S. (1995) *Thermodynamic Properties of Minerals and Related Substances at 298.15 K and 1 Bar (105 Pascals) Pressure and at Higher Temperatures*, US Geological Survey Bulletin, Washington, DC, Vol. 2131, 470pp.

Ruddiman, W. F. (2003) The anthropogenic greenhouse era began thousands of years ago. *Climatic Change* 61, 261–293.

Ryan, P. R., Delhaize, E., and Jones, D. L. (2001) Function and mechanism of organic anion exudation from plant roots. *Annual Reviews in Plant Physiology and Plant Molecular Biology* 52, 527–560.

Salomons, W., Goudie, A., and Mook, W. G. (1978) Isotope composition of calcrete deposits from Europe, Africa and India. *Earth Surface Processes* 3, 43–57.

Salomons, W. and Mook, W. G. (1976) Isotope geochemistry of carbonate dissolution and reprecipitation in soils. *Soil Science* 122, 15–24.

Sanz Scovino, J. I. and Rowell, D. L. (1988) The use of feldspars as potassium fertilizers in the savannah of Columbia. *Fertilizer Research* 17, 71–83.

Schuiling, R. D. and Krijgsman, P. (2006) Enhanced weathering: An effective and cheap tool to sequester CO_2. *Climatic Change* 74, 349–354.

Tole, M. P., Lasaga, A. C., Pantano, C., and White, W. B. (1986) The kinetics of dissolution of nepheline ($NaAlSiO_4$). *Geochimica et Cosmochimica Acta* 50, 379–392.

Washbourne, C.-L., Renforth, P., and Manning, D. A. C. (2012) Investigating carbonate formation in urban soils as a method for capture and storage of atmospheric carbon. *Science of the Total Environment* 431, 166–175.

7

The Green Cookery Book: Recipes against Climate Change and Ocean Acidification

Olaf Schuiling

CONTENTS

Carbon dioxide has always leaked out of the earth, mainly from volcanoes. If all that CO_2 had remained in the earth's atmosphere, its CO_2 pressure would now be about 100 bars (instead of the current value of 0.0004 bars), the surface temperature of the earth would be close to 500°C due to the greenhouse effect, and the oceans would be converted to steam.

Throughout geological history, the weathering of basic silicates has been the negative feedback mechanism by which CO_2 was removed. Without weathering, there could be no life on the earth. Weathering neutralizes carbonic acid ($H_2CO_3 \rightarrow CO_2 + H_2O$) to bicarbonate ($HCO_3^-$) solutions. These are then carried to the sea by rivers, where they are converted to carbonates by marine organisms (corals and shellfish), forming solid rocks (limestones and dolomites) in which the CO_2 is sustainably stored. Carbonate rocks contain 99.94% of all the CO_2 near the surface of the earth. So rocks are the main reservoir for CO_2, much larger than the oceans, biosphere, and atmosphere together. A generalized description of these processes can be found in Hartmann et al. (2013).

This is how CO_2 is stored:

Or this way:

Humanity is now causing a spike in the CO_2 input by burning fossil fuels, and natural weathering cannot cope with this greatly increased rate. The olivine option provides a logical answer because it enhances the natural feedback mechanism to restore the balance. It can be applied in many sectors of society. Here you can find how and where people can use it. This cookery book contains the recipes for the most important applications of the olivine option. The olivine option provides a multipronged attack to counteract climate change and ocean acidification.

Thanks to this wide range of applications virtually *everybody* can assist in their own way to achieve climate stability. Many of the mentioned applications are devised in such a way that they profit from natural conditions and solve other problems at the same time, thereby reducing costs. Some will even be financially self-supporting, without recourse to subsidies or carbon credits.

The applications have been arranged by sectors. These are agriculture, forestry, roads, buildings, playgrounds, coastal protection, suppression of poisonous dinoflagellate and cyanobacterial blooms, diatom farms for biodiesel production, mining, mineral waters, olivine as a green fuel, olivine in environmental applications, natural emissions of CO_2 for carbon capture, and miscellaneous.

For people who are not familiar with the olivine option or with the volumes of olivine required, nor with the mining and milling costs involved, the following information is provided. Olivine, $(Fe, Mg)_2SiO_4$, is the most abundant mineral in the earth, since the earth's mantle consists mainly of olivine (and at great depth, its high-pressure forms). If all that olivine had stayed in the mantle, it would be difficult to mine because the continental crust has a thickness of 30 km. Fortunately, large slabs of olivine rock have been pushed to the earth's surface by plate tectonic movements, so in many countries in every continent, huge massifs of olivine rocks are found close to the surface, where they can be mined in open pit mines. There are thousands of times more olivine available than we will ever need to counteract climate change and ocean acidification. To reach a new balance in the annual anthropogenic production of CO_2 of somewhat more than 30 Gt, we need about 25 Gt of olivine through:

$$Mg_2SiO_4 + 4H_2O + 4CO_2 \rightarrow 2Mg^{2+} + 4HCO_3^- + H_4SiO_4$$

At a specific mass of 3400 kg/m^3, this is equivalent to the mining every year of an olivine volume of slightly over 7 km^3, large, but within the range of modern mining. The largest mine on the earth, the Bingham mine in Utah, has an excavated volume of 25 km^3. It is evident that we should not start one single huge olivine mine, but a number of large new olivine mines strategically spread in order to reduce transport costs.

The Bingham mine, 25 km^3.

The costs of mining and milling bulk rock in open pit mines are well known in the mining sector. In large mines, mining bulk rock costs approximately €4/t, and milling these rocks to a grain size of fine sand costs €2/t (Steen and Borg, 2002). Four tons of olivine can sequester 5 t of CO_2.

Dunites and peridotites (the major olivine rocks) are widespread over at least 1 million km^2 of the land surface. Several hundred million people live on these rocks for many generations, grow their crops on them, graze their animals on them, and drink the waters from these rocks without any harm. Weathering of olivine has been the main mechanism that has throughout geological time kept the CO_2 levels of the atmosphere within livable bounds, so it is highly unlikely that it would all of a sudden become an environmental danger when we use it to counteract climate change and ocean acidification.

Inventory of Olivine Applications

Agriculture

Wet grasses, like rice, sugar cane, bamboo, and reed, need silica. Three properties of olivine are important for agriculture: it remediates soil acidity, it supplies magnesium (essential for chlorophyll), and it provides silica, which is important for *wet* grasses like rice and sugar cane, which make silica deposits, or phytoliths, in their leaves. The normal weathering reaction (Schuiling and Praagman, 2011) is

$$Mg_2SiO_4 + 4CO_2 + 4H_2O \rightarrow 2Mg^{2+} + 4HCO_3^- + H_4SiO_4$$

Soil acidity is a very severe problem in the Mekong delta, which has many acid sulfate soils, which negatively affect rice production (van Breemen, 1976). The same holds for estuaries along the East coast of Australia that also suffer from acid sulfate soils, sometimes completely destroying the sugar cane plants. Spreading olivine sand can cure these problems and increase the productivity.

The problem of soil acidity, which is usually treated by adding limestone, can be solved better by using olivine sand. Olivine spreading is CO_2 negative, so it not only solves the problem of soil acidity but also helps to counteract climate change. Liming works faster, but needs to be repeated more often, whereas olivine spreading can control soil acidity for a longer period.

On a more modest scale, many people's lawns are on acidic soils and full of moss. Here again a dose of olivine sand can help to improve the quality of their lawn. Pot experiments carried out at the Wageningen Agricultural University with grass grown with different doses of olivine added have shown that olivine has a positive effect on the growth of the grass (ten Berge et al., 2012).

On the smallest scale, even pot plants in apartments can profit from potting soil enriched with olivine, which is available in garden centers in the Netherlands.

Another way to make farmland more productive is the combination of olivine spreading with biochar (pyrolized wood). This should be an ideal combination by which the pH of the farmland is raised, its water retention is improved, and CO_2 is captured, because this biochar is another way to capture CO_2.

The growing season in alpine regions and in high latitudes is short because snow remains on the fields till late spring. Farmers can extend the growing season by covering the snow with a thin layer of olivine or serpentinized olivine grains. Snow reflects sunlight, whereas dark silicates absorb it. In serpentinized olivine rocks, tiny magnetite crystals are formed that give the rock a blackish color by which sunlight is absorbed even better.

The dark grains are heated by absorbing sunlight, causing the underlying snow to melt faster, and the growing season is extended. After the snow has disappeared, the olivine grains will mix with the topsoil and capture CO_2. In Switzerland, this general approach to make the snow melt faster is already long established; instead of olivine or serpentine, only a black shale powder is used, the so-called Bündner Schiefer. That way, they miss the additional advantage of carbon capture.

A completely different agricultural application of olivine is in biodigesters. Organic wastes are digested and produce a biogas that consists of roughly two-thirds CH_4 and one-third CO_2, with traces of H_2S. For commercial applications, it would be an advantage if the CO_2 content could be lowered and the H_2S removed. Test runs at the Wageningen Agricultural University and the Technical University of Delft have shown that both aims can be reached by adding some fine-grained olivine to the digesters. The digester does not smell any more, and part of the CO_2 gas has been converted to bicarbonate in the digestate. There is even some evidence that the absolute amount of methane increases, probably by the following reaction with the iron member of the olivine:

$$6Fe_2SiO_4 + CO_2 + 14H_2O \rightarrow 4Fe_3O_4 + CH_4 + 6H_4SiO_4$$

This reaction is often observed to take place in nature. Notorious examples are the Yanartasi (*the rock that always burns*) in Turkey, and los Fuegos eternos (*the eternal fires*) in the Philippines.

The Yanartasi (the rock that always burns).

Using part of CO_2 to produce methane serves two purposes. The absolute amount of produced methane in the digester increases, and the amount of CO_2 in the biogas decreases, so both the quality and the quantity of the methane are improved. The tiny magnetite crystals formed by the reaction catalyze the formation of methane.

Forestry

The first application of olivine is similar to the spreading of olivine over farmland to reduce soil acidity, because many forest soils also suffer from soil acidity. Spreading, however, is difficult in a standing forest, so it is best to wait until the trees are felled and then to spread olivine when the grounds is prepared for the new tree planting. Governments could use the fact that they decide on concessions for forest exploitation to impose on the concessionary the condition to prepare the soil with olivine as a means both to capture CO_2 and to preserve soil fertility. This last aspect may be strengthened by adding other types of rock meal to the olivine sand before spreading.

The parts of the trees that have no commercial value (bark, branches, sawdust) should be converted to biochar, which can either be used on the spot or be transported to arable land to improve its soil structure and water retention. By mixing the biochar with olivine, an even better result can be obtained.

Roads, Biking, and Walking Paths

The green highway: Although highways are evidently great producers of CO_2 by traffic, they can also serve to capture CO_2. Their top layer is often composed of a mixture of bitumen and sand, which makes braking more effective. This layer is slowly worn down by the traffic, the sand grains are pulverized, and the tiny slivers are blown away by the wind. Quartz and olivine have closely the same hardness, making it easy to replace the quartz sand by olivine sand. Here again, the layer is worn down by the traffic, the tiny slivers are blown away, but contrary to quartz slivers, they weather fast and neutralize CO_2.

Olivine can also be used in sound walls: It sounds strange to use olivine, meant to weather as fast as possible as a construction material, but there are two reasons to do so. When dunite (the rock type that consists of >90% of olivine) is quarried, the operation will automatically produce some fine materials that will weather fast, either on location or blown elsewhere with the wind. After the useful life of the construction has come to an end, it will be demolished and usually reduced to powder that will weather fast (Renforth et al., 2009).

The banks of roads and wadis: These can be covered with a layer of olivine that will react with the rainwater that drains from the road or land. They may also have another unexpected application (see the section "Mining").

Biking and foot paths: These can be covered with olivine sand that will give them an attractive greenish color, fitting better in the landscape than road surfaces of bitumen or slabs of concrete.

Inspection paths along railways: The first tests have been successfully completed by the Dutch railway system. The olivine sand on the inspection path did in fact capture CO_2 during every rainfall.

Driveways: It is a nice view to have one's driveway covered with green olivine grit, and it shows that the owner has an environmental conscience. In the section "Coastal Protection," a way is described to produce nicely rounded olivine grit in the most environmentally friendly way.

Parking lots: Parking lots should be constructed on a layer of olivine sand, through which the rainwater can percolate. Measures should be taken to make sure that the rainwater drains through the top layer.

Salt or olivine: A recurrent problem is hazardous road conditions in winter, due to snow or ice. Spreading salt is an effective way to solve the problem, but has environmental disadvantages. On secondary roads one can avoid these by spreading olivine sand over the snow or ice layer, to provide more grip for cars.

Buildings (Including Roofs)

Many offices and schools suffer from the sick building syndrome. When people come in the morning, the internal atmosphere has a CO_2 concentration of around 400 ppm, but at the end of the day, that has gone up to 1500 or 1600 ppm. People get drowsy, they lose concentration, and their productivity goes down. A simple solution, to open all doors and windows, is usually out of question for reasons of energy loss, noise, and dust. The best solution is to install a system for internal air circulation, by which the internal atmosphere is passed through a CATO reactor (CATO stands for *C*lean *A*ir *T*hrough *O*livine). This is a trough filled with a suspension of fine olivine grains in water. Along the bottom runs a pipe with many tiny holes open to the trough. The air is circulated through that pipe under a slight overpressure. Bubbles pass through the suspension and not only give off their CO_2 but also trap pollen and allergenic dust particles. In order to prevent scaling by silica or magnesium carbonates, the water from the reactor must regularly be pumped in an outside large reservoir (a pond), and the reactor refilled with new water. In the pond, silica will be used by diatoms. Diatoms are a favorite fish food.

The total roof surface of towns is large and constantly increasing. Large roofs need ballast to prevent the roof being blown off in a storm, and pieces of olivine (specific mass of olivine is around 3.4 g/cm^3, considerably higher than most rocks) can serve well as ballast. For carbon capture, one can additionally cover the roofs with olivine sand and make it into a roof garden by planting some sedum on it. Under vegetation, olivine weathers faster. Due to the higher pH imposed by the olivine on the percolating rainwater, metals like copper, lead, and zinc are immobilized, saving expensive remediation measures in urban sewage treatment plants. Many roofs in Amsterdam have already been covered by olivine.

Serpentinite is a very common rock type. One can consider it as a kind of clay mineral, and it can be used as such. It is known that calcined serpentine reacts very fast (faster even than olivine) with CO_2 and water, contrary to other baked clays. It is suggested here to test if serpentine can be used as a raw material to make roofing tiles. These roofing tiles will already contribute to carbon capture during their time on the roof, but when crushed after they are broken or after the house is torn down, they will very quickly react with CO_2 and water, thus making a contribution to CO_2 removal from the atmosphere.

Playgrounds

Gravel on tennis courts can be replaced by olivine gravel: It serves the purpose of capturing CO_2 even better than on other open spaces because in very dry times the tennis courts are kept wet by sprinkling. This increases the time during which reaction between olivine CO_2 and water can take place. Although the total contribution of green tennis courts to carbon capture is limited, their frequent exposure on television during tennis matches will help the introduction of olivine options.

Golf courses: A double possibility, the greens can be covered with a thin layer of olivine sand that will improve the quality of the grass, and the paths for golf carts can be covered with olivine sand.

Astroturf playing fields: During their installation, a mixture of sand and rubber shreds or pieces of cork is added to the mats. After 10–12 years, these are replaced, and the old mats are thrown away. The following variation is in line with the cradle-to-cradle principle. Instead of normal sand, olivine sand is used, so the Astroturf fulfills an environmental role right from the beginning. When it must be replaced, the mats are not thrown away, but used for roof covering. The heavy sand holds the mats down, and the *grass* protects the sand from being blown away. This serves as an insulating layer for the house, keeping heat out in summer, and conserving heat in winter.

Sandboxes: At nursery schools or gardens, sandboxes with olivine sand can be installed (the first ones have been installed in schools in the Netherlands). Olivine sand gives a good feel. If necessary, the grains can first be tumbled, or collected from olivine beaches, where they are rounded by tumbling and abrasion by the surf. Sand sculpturing with olivine sand can produce spectacular results.

Sand sculpture of the Nativity, from 48 t of olivine sand. The Hague (the Netherlands), Christmas 2012.

Coastal Protection

Longshore currents can erode coasts: A curved olivine dam can be constructed on rapidly eroding coastal sections. Such a dam should be built with fairly coarse olivine blocks, making it permeable. When the main current along the coast is from south to north, the general direction of the dam should be northwest, curving in a northerly direction. This way the current is gently deflected away from the coast and loses part of its strength, because the water in the current that passes through the olivine blocks in the dam will lose momentum. When a dam is placed perpendicular to the coast, and at right angles to a longshore current, sand accumulates on its front side and is eroded on the backside. As the flow of water on the backside will be partially pushed back by the water that forced its way through the permeable dam, it is expected that erosion will be less in this setup.

A second way to reduce erosion is the following: The seafloor of the part behind such a dam should be covered with olivine sand. Olivine is significantly heavier than quartz (3400 vs. 2650 kg/m^3), so it offers a larger resistance to erosion. It will react with seawater, capturing CO_2 and readjusting the pH of the local water. Such a sector on the lee side of the olivine dam may even serve as a refuge for marine fauna suffering from ocean acidification.

Covers of dikes should preferably be made of blocks of olivine rocks: Admittedly, the amount of olivine from coarse blocks that will react with seawater will be minimal, but mining of olivine rock itself is carbon-negative, because the dust produced during olivine mining will weather fast, whereas mining granite or basalt carries a CO_2 penalty. Basalts and granites also weather, but their total concentration of calcium and magnesium is considerably lower than that of dunites (olivine rocks), and their rate of weathering is much slower.

Olivine reefs: One can construct olivine reefs at places where these can serve best to keep currents and waves away from the shore in order to reduce erosion. The pH of the pore waters inside these reefs will rise as a consequence of their reaction with olivine, and this will probably lead to calcite precipitation, cementing the reefs and making them better erosion resistant. Very likely, such olivine reefs will become attractive places for marine organisms to settle and for fish to serve for hiding and hatching. If these reefs attract an abundance of marine life forms, they may become an interesting target for tourists, and of course, as holds also for the other proposed measures, they help to protect coasts against erosion, they will capture CO_2, and they help to restore the pH of the sea.

Deposit olivine grit on a rough part of the coast: In the surf zone, angular pieces will become quickly rounded and polished by repeated collisions.

At the start, the olivine grit is angular and has a rough surface.

After 10 days of gentle rotation, the grains are smooth and rounded.

The water is clear at the start, but becomes opaque from all the micron-sized slivers of olivine.

During this abrasion process, they lose small slivers of olivine that rapidly weather and add alkalinity to the sea (Schuiling and de Boer, 2011).

The surf is the largest ball mill on the earth and is free of charge!

Tourists are free to collect some nicely polished olivine marbles as a memory of this new way of coastal protection. People who want roundish olivine grit on their driveways can use the surf to obtain nicely rounded grit. Another application of this tumbled olivine grit might be for chicken farms, to provide grit to the chicken. Sharp angular pieces can damage the chicken stomach, and a short round of cost-free rounding on the beach can avoid this risk.

Suppression of Dinoflagellate and Cyanobacterial Blooms

Toxic dinoflagellate blooms, the feared red tides, occur more and more frequently in coastal waters. A similar problem is faced in freshwaters that are often threatened by cyanobacteria that make them unsuitable for swimming. Dogs die when they swim in such waters. Both marine and freshwater diatoms (siliceous algae) are very common, they grow very fast, and make up 20%–25% of the marine biomass. They are a favorite fish food. Their existence is dependent on the availability of dissolved silica in the water, which they need to make their silica skeletons. As long as silica is available, they grow so fast that they can outcompete dinoflagellates and cyanobacteria. In the coastal sea off China, the following sequence of events unrolls every year. In spring, there is a diatom bloom. As the reservoir of silica that was built up during winter comes to an end, the diatoms die, and a toxic dinoflagellate bloom takes over.

The feared red tide.

Some species of dinoflagellates produce a strong neurotoxin, killing fish and making the waters unfit for touristic uses. All these planktonic species are dependent on the availability of sufficient macronutrients, mainly phosphate and nitrogen, which are brought in from land, derived from fertilizers and untreated wastewaters. As long as there are plenty of diatoms, there will be neither dinoflagellate blooms in the marine realm nor cyanobacterial blooms in freshwater lakes or canals. The most sustainable way by which large volumes of CO_2 can be captured is by weathering of olivine sand on beaches or in drains through which freshwater lakes are fed. This will continue to deliver silica to the water, in summer even more than in winter, because the rate of weathering of the olivine is dependent on the ambient temperature.

Diatom Farms for Biodiesel Production

A very different way of using diatoms is the following. About 50% of their body weight is lipids (fatty acids), which makes them attractive as a raw material for biodiesel production.

Diatoms have beautiful silica skeletons. (Picture made available by Ian Powers.)

In order to make this commercially attractive, ways must be found to farm them and harvest them. The technology to make biodiesel out of algae already exists. One can set up a diatom farm with silica supply as follows. Make an artificial lagoon in front of a beach, by damming off a section of coastal water. Cover the beach with a deposit of olivine sand (~0.5 m thick) between the high tide level and the ebb level. Make a U-tube through the dam, permitting water to flow into the lagoon at high tide, and flow out at ebb tide. This will cause a diurnal wetting and draining of the beach, and deliver the products of olivine weathering to the lagoon. Magnesium, of course, is not an issue because it is already a major component of seawater. Silica is the determining factor, but the bicarbonate will also be used by the diatoms for photosynthesis. Other deficient nutrients must be added to the water, maybe in the form of digestate from anaerobic digesters, or as struvite ($NH_4MgPO_4 \cdot 6H_2O$) produced from such digestates by the struvite process (Schuiling and Andrade, 1999; Andrade and Schuiling, 2001). The lagoon side of the U-tube is covered by a perforated metal disk, which acts as a support for a plankton net that serves to catch diatoms that otherwise might be carried out of the lagoon with the ebb stream. These are collected on the plankton net. Hopefully, a more direct way to catch the diatoms is as follows. Make one section of the lagoon deeper than the rest. Dead diatoms will sink in this pit. Their siliceous skeleton makes them fairly heavy. From time to time, a slurry of dead diatoms is sucked up from this hole and serves as raw material for biofuel production. Biofuel production on land from corn, sugar cane, or oil palms is getting a bad reputation. It uses large areas of land that were meant for food production, it uses large quantities of scarce irrigation water, and it consumes large amounts of fertilizer. Biofuel from marine algae carries none of these negative properties.

Mining

For all these activities, large volumes of olivine must be mined or recovered from olivine-rich mine tailings. Olivine rocks host a number of other commodities, like chromite, magnesite, platinum group elements, sometimes diamonds, and, of course, gemstone-quality olivine crystals (peridot). When reworking the tailings to recover the olivine, it may be worthwhile to go for the second harvest of the materials that were mined.

Olivine rocks may also host economical concentrations of other minerals. In this respect, chromite is the main candidate. If the deposit is mined for olivine, and the rock is crushed to produce olivine sand, it may be worthwhile to recover the chromite as well, as the technology using washing tables is simple and effective to separate chromite from olivine. This will improve the olivine grade of the product and bring some extra cash from the chromite by-product.

Although somewhat speculative at this moment, the recovery of another metal from olivine rocks may become economically far more important. All olivine rocks, and the serpentinites derived from these rocks, contain between 0.25% and 0.3% nickel, which replaces some of the magnesium ions in their crystal lattices. Such concentrations are far below ore grade. Baker and Brooks (1989) found that there are a number of plants (presently there are about 400 species known) that very efficiently extract nickel from such rocks and from the soils on such rocks.

The ash of these so-called nickel hyperaccumulating plants may contain over 10% of nickel, far more than the richest nickel ore. Moreover, it is easier, less polluting, and less energy-intensive to recover the nickel metal from these plant ashes than from regular nickel ores. In simple laboratory tests, small nickel ingots were produced from the plant ashes. Sowing these plants on appropriate soils and harvesting them at the end of the growing season makes for an environmentally friendly way of recovering nickel. Because these plants extract nickel from the olivine lattice, for every ton of nickel in the plants, 330 t of olivine must weather, equivalent to a capture of 400 t of CO_2. Weathering proceeds faster under vegetation. The introduction of this method (Schuiling, 2013) could revolutionize the nickel mining industry. It is also possible that once this method of farming nickel becomes established, governments will require nickel mining companies to realize part of their nickel production in this way, to save the environment and extend the life of their nickel reserves. These countries can claim such measures as their share in global CO_2 sequestration. Soils on olivine rocks are usually poor soils, so switching to this technology will not harm world food production. On the contrary, it may provide income for poor farmers that will permit them to buy some fertilizer and thus increase their food production.

Alyssum has been sowed on the tailings of a former asbestos mine on Cyprus (with flowers) to stabilize the slopes and recover nickel from the serpentinized harzburgite. (Picture made available by the Cypriot Geological Survey.)

Interest is growing in using unproductive soils in the same way even if they have no relation with olivine rocks. One can classify not only the cover of waste deposits and sites of urban sewage treatment plants, but also road banks as unproductive soils. Before the financial crisis, many municipalities had bought agricultural land to convert it into industrial sites or use it for housing projects. Due to the financial crisis, many of these projects were halted, but the municipalities still have to pay the interest on their loans. If covered by a layer of olivine mixed with topsoil and planted with nickel hyperaccumulator plants, they can bring a modest income to the managers of such sites.

A major problem in the mining industry is acid mine drainage (AMD). Sulfide minerals exposed to the atmosphere are altered very fast and produce sulfuric acid. This will damage, and often completely destroy the vegetation, and it makes waters unsuitable for irrigation. If olivine- or serpentine-rich rocks are found in the neighborhood, these can serve very well to neutralize the acid waters, and once neutralized, these waters can serve for irrigation and can sequester some CO_2.

Mineral Waters

As said before, the weathering of olivine (and serpentine) produces magnesium bicarbonate waters, and several well-known mineral waters, like the Greek mineral water Loutraki, issue from springs in olivine rocks. In a FAO handbook from 2002, the effects of different chemical species on human health are described, and for magnesium-rich waters, the verdict is very positive. Such waters are effective against cardiovascular diseases, and a too low intake of magnesium causes premature aging.

In a simple experiment, Spa mineral water bottled under CO_2 pressure was converted to Loutraki water, a magnesium bicarbonate mineral water.

A simple experiment, visualized earlier, can demonstrate the origin of magnesium bicarbonate mineral waters. When Spa Red, a Belgian table water bottled under high CO_2 pressure with a pH of 3.9, is reacted with olivine sand, the pH rises. After 1 month, it had reached 8.3, and the chemistry of this water had become almost identical to that of Loutraki, the best-known Greek mineral water. Not surprisingly, the Loutraki spring issues from an olivine-rich rock-type near Corinth (Schuiling et al., 2012).

This public health aspect is almost an invitation to produce more of these waters to improve public health, also in countries where there are no olivine rocks. This can be done with the help of olivine hills. Such hills can be constructed and operated in the following way. First, install an impermeable bottom in the shape of a very flat slightly inclined gutter. Cover this with several meters of olivine sand.

Artist impression of an olivine hill.

The grain size of the olivine must be selected in such a way that water entering the olivine hill from above should take at least 1 day to reach the bottom. This will provide sufficient time for the water to react with the olivine. The olivine must be covered by a layer of topsoil, which will be planted with shrubs and grass. This serves a double purpose: a vegetation cover will prevent the erosion of the hill during heavy rainfall, and the vegetation causes high CO_2 concentrations in the soil atmosphere. The decay of dead plant material in the soil and respiration of soil fauna both produce CO_2. Rainwater will first pass through this soil and take up CO_2. Then, it will pass through the olivine and is converted to magnesium bicarbonate water. After this water has reached the impermeable bottom, it will slowly migrate to a tap at the low end of the gutter, where it can be collected.

A board placed next to the tap should offer an explanation of the olivine option, how this particular water forms, and what the favorable effects of this mineral water are on public health.

In countries with a poor-quality drinking water, like Bangladesh where much of the drinking water contains high concentrations of arsenic that has led to widespread illness and death, olivine hills constructed near villages will provide healthy water to the population. In times of inundations, such hills can serve as a refuge for the population.

People who want to serve their own brand of mineral water to their guests can build a small replica in their garden. Although the CO_2 capture of such a small olivine hill is modest, it will make more people aware of the possibilities of the olivine option against climate change and ocean acidification, and provides a modest improvement to public health.

Olivine hills can even be planted with nickel hyperaccumulator plants, and the owner can proudly produce his own *nickels*.

Olivine as a Green Fuel

Supergreen energy. The weathering reaction of olivine is exothermal. It releases a lot of heat, known as supergreen energy, because it is heat that is produced at the same time CO_2 is sequestered, whereas green energy is the energy that is produced without CO_2 emission. Unfortunately, the heat is produced at such a slow rate that it is normally impossible to recover it, unless one uses large and very well-isolated volumes of olivine. If the reaction is

$$2Mg_2SiO_4 + CO_2 + H_2O \rightarrow MgCO_3 + Mg_3Si_2O_5(OH)_4$$

a stochiometric mixture of the reactants would heat itself to ~700°C if perfectly isolated. If one relates this to the energy produced by burning coal normalized to the same amount of CO_2, it turns out that the total energy produced by burning coal could be increased by 50%! Moreover, it would offer the possibility to store the CO_2 produced in the coal-fired power plant in a safe and sustainable way.

With such high stakes, it is tempting to look for ways to implement this idea. Rocks are excellent thermal insulators. It would be an advantage to have underground cavities that are no longer in use and even may have a negative value. One can think of cavities produced by solution mining of salt. For safety reasons, to prevent subsidence or even caving in, these cavities must remain filled with saturated brine and indefinitely pressurized. Filling them with fine-grained olivine will push out an equivalent volume of brine, which would otherwise be lost. This additional volume of brine recovered is the first financial advantage. Hot CO_2 will then be injected into the mixture of olivine grains with interstitial brine. Laboratory tests have shown that olivine reacts faster in a brine than in freshwater. Because the volume of serpentine + magnesite is larger than the volume of the olivine, the pore space between the olivine grains begins to be gradually closed, and the solid reaction product can support the cavity, obviating the need for further pressurizing and monitoring. Once the reaction is underway, the produced heat can be tapped by exchanging it from the solid rock to water circulating in pipes installed in the system. The warm water can be used for space heating or for the heating of swimming pools.

It would be even better to recover the heat as high-enthalpy steam to run turbines. This may become practical if a thermal power plant is located close to a dunite massif, like near Orhaneli, Northwest Turkey, where a lignite-powered plant is located near a dunite body that is already used for small-scale olivine mining. The following suggestions are speculative. If one excavates a large room inside this dunite body, constructs a set of perforated pipes in this room, and fills it with olivine rubble obtained from roof caving and connects the piping system to a well-isolated pipe through which the hot flue gases from the power plant are brought in, one can set the reaction in motion. The room must first be closed to the outside. As rocks are good thermal insulators, the reaction will go faster and faster as the temperature goes up. The high-temperature steam may eventually be used for power production, while at the same time CO_2 is captured.

A system that allows better control is the following. The area between Aachen and Köln in Germany is underlain by thick lignite coal seams, at depths around 500 m. These seams are mined in deep opencast mines, and as the seams are excavated at the front side, the mined-out opposite side is filled again with the original overburden, and the area is rehabilitated.

Lignite mine in Germany

This way, the mine moves slowly through the landscape. Instead of filling the entire cavity with the original overburden, one can fill the lower 250 m with olivine sand and cover it up with the remainder of the overburden. Before doing this, a set of perforated pipes should be installed in the olivine layer that permits injection of CO_2/water mixes. After the olivine and the overburden have been emplaced, the system is thermally well isolated. Again, as the system heats up by the ongoing reaction, the steam pressure will also go up, but the overburden will provide a containing pressure of more than 60 bars, equivalent to a steam temperature of 275°C. The tapping of this high-enthalpy steam makes it possible to run a steam turbine.

Olivine in Environmental Applications

Sulfide mines suffer from AMD. Remaining sulfides in the mine tailings or in the mine itself oxidize fairly rapidly in contact with the air, producing sulfuric acid. It is one of the major environmental problems of mining. Some mine locations have the good luck that their mine is situated in a limestone area, which neutralizes the acid (making gypsum), but in many cases, AMD remains a problem. Olivine may act as a useful neutralizing agent. It is abundantly available, and when applied in sufficiently high doses, it may impose a pH at which many toxic heavy metals become immobilized and are no longer bioavailable.

A completely different environmental application, this time of serpentine, which is very commonly formed by a reaction of olivine with (hot) fluids, is its use in fighting forest fires. Serpentine can be considered as a kind of clay mineral, and baking clay for brickmaking requires a lot of heat. Normally, that is an unfavorable property, but there are cases where one tries to remove as much heat as possible from a fire, namely, in forest fires. To check the efficiency of using serpentine powders to extinguish fires, a set of test fires were subjected to spraying with serpentine slurry, and the effect compared to spraying with water. It turned out that spraying with serpentine slurries was a very effective and rapid way to quench the fire. In just a few seconds of spraying with a modest dose of serpentine slurry, the fire was quenched permanently. With the spraying of water, it took much longer before the fire was quenched, and after the spraying was stopped, the fire started again. Heat withdrawal from the fire, however, is probably not the major positive effect. When the fire is sprayed with a serpentine suspension, the baked serpentine forms a thin layer on the burning material, preventing oxygen from reaching the burning material and preventing inflammable gases from escaping from the burning material.

After the fire, the calcined serpentine reacts very rapidly with water and CO_2. So, at the first rain shower, part of the CO_2 that was produced by the fire is already compensated. Forest fires are the world's second biggest emitter of CO_2, after the burning of fossil fuels (van der Werf et al., 2009). Even in the United States, with its many cars and trucks, forest fires are responsible for as much CO_2 emissions as all the traffic.

Using Natural Emissions of CO_2 for Carbon Capture

Although anthropogenic CO_2 emissions far outweigh natural CO_2 emissions, these form an easy prey for carbon capture. It is self-evident that for the climate, it makes no difference whether man-made or naturally emitted CO_2 is sequestered. The atmosphere is a well-mixed reservoir on the scale of a few months, so it makes no difference where the CO_2 is sequestered.

The first example is in the shallow sea near the Greek island of Milos, where each year 2.2 million tons of CO_2 bubble out of the seafloor. The CO_2 is very hot just before it reaches the surface (the bubbles themselves have the same temperature as the seawater). If one would create an olivine island over a location with active bubbling of CO_2, the temperature of the olivine would start to rise. This would set a convection system in motion. Warm water is lighter than cold water, so it rises to the top of the island and will be compensated by a sideways inflow of cold seawater. When a depression is made in the top of the island, this will fill with warm water, making it a touristic attraction. Tourists, even in winter, can take a warm bath while looking over a cold blue Aegean Sea. The exothermal heat released by the olivine reaction with water and CO_2 adds to the heating effect.

This heat of reaction plays a major role in the following application. In northeast Greece, there are a number of geothermal wells that are used for the heating of greenhouses. The geothermal fluid contains high concentrations of CO_2. When a hill of olivine sand is set up over the well, the amount of available energy is increased by the heat of reaction, and CO_2 is captured.

A location where natural CO_2 emissions have caused a major catastrophe is Lake Nyos in Cameroon, where in 1986, a dense cloud of CO_2 was emitted from the lake. It rolled down the mountain and killed 1800 people by asphyxiation. To prevent a future disaster, a kind of geyser system was installed to prevent the accumulation of CO_2 at the bottom of the lake. This is an expensive operation, and it is doubtful whether future funding will remain available. A more permanent solution, requiring no daily supervision and maintenance, would be to spread a layer of fine olivine sand on the bottom of the lake (Schuiling, 2011). This will react with the bubbles of CO_2 and transform them into an innocuous bicarbonate solution, thus avoiding the danger of a buildup of CO_2 pressure. As the lake serves as a source of drinking water for the population, they can benefit from the advantages of healthy magnesium bicarbonate waters.

Mofettes. In many places, mofettes are found, places where CO_2 escapes from the ground. These are harmful, and animals, but occasionally humans as well, can die from asphyxiation when inhaling the concentrated CO_2. The author had a very unpleasant experience, when he sniffed just a little bit of concentrated CO_2 from a mofette in Belgium. Understandably, in several countries, the mofettes are fenced off to prevent accidents. If the mofette area would be dug out somewhat, and the hole filled with olivine sand, the risk of accidents could be reduced, provided that the hole is deep enough to keep the olivine sand in a wet condition. Apart from risk reduction, it is also a small contribution to carbon capture.

Miscellaneous

Olivine grains can be spread on the bottom of sewage pipes: Temperatures in sewage pipes are usually above average, and CO_2 is plentiful, thanks to the decay of labile organic substances. This will speed up the reaction rate of the olivine. It is expected that olivine, as it raises the pH, will slow down the corrosion of the concrete. Replacement of sewage pipes is a matter of millions of euro even for an average-sized city, so extending the lifetime of the pipes even by just a few years can save a lot of money.

Slowing down lava flows with serpentinite blocks: Lava flows may destroy villages or even small towns downstream, so it is understandable that attempts are made to stop them. In an earlier paper, it was calculated that throwing blocks of limestone in the lava flow has a much larger cooling effect than throwing in blocks of concrete (Schuiling, 2008). This is caused by the endothermic reaction of the dissociation of the limestone. A similar argument is presented here on the effect of serpentinite blocks, in which the dehydration of the serpentine withdraws more heat from the lava than unreactive rocks of the same weight.

A lava flow from Etna; can it be stopped?
(Courtesy of Tom Pfeiffer, www.VolcanoDiscovery.com.)

It will depend on the availability of limestones or serpentinites near the volcano and their distance from the flow whether one or the other is more convenient to use. Both are common rock types. Serpentine ($Mg_3Si_2O_5(OH)_4$) can be compared to clay minerals. Making bricks from clay requires a lot of heat, and the same holds for serpentine. The heat of dehydration is in the order of 140 kJ/mol, or 500 J/g. So, with 1 g of serpentine, one can cool 6 g of lava by 100°C. One gram of serpentinite when heated from room temperature to temperatures around 1000°C will additionally cool 10 g of lava by 100° for every gram of serpentinite added. When normalized to equal weights, decarbonation reactions with dolomites or limestones withdraw more heat from lava than dehydration reactions of serpentine, but even with serpentinites, their contribution is still large. The use of serpentine, however, has some additional advantages in terms of CO_2 sequestration. Contrary to carbonates, their use is not producing CO_2 emissions, and after the lava has cooled down and solidified, any calcined serpentine will react fast with CO_2 and water, thereby sequestering CO_2 in a sustainable manner.

Conclusions

Almost all of the CO_2 that has ever leaked out of the earth has been removed by its reaction with basic minerals. This reaction is known as weathering. The weathering products have formed carbonate sediments. These are nature's sustainable storehouses of CO_2. Mankind has enormously increased the input of CO_2 by the burning of fossil fuels, and weathering cannot cope with this increase. In this chapter, it was described how the rate of weathering can be increased by the mining, milling, and spreading of olivine, and how this olivine option can be applied in many sectors of society.

REFERENCES

Andrade, A. and Schuiling, R.D. (2001) The chemistry of struvite crystallization. *Mineralogical Journal (Ukraine)* 23(5/6), 37–46.

Baker, A.J.M. and Brooks, R.R. (1989) Terrestrial higher plants which hyperaccumulate chemical elements—A review of their distribution, ecology and phytochemistry. *Biorecovery* 1, 81–126.

FAO Document Repository. (2002) Magnesium. In: *Human Vitamin and Mineral Requirements*, Chapter 14.

Hartmann, J., West, A.J., Renforth, P., Köhler, P., Wolf-Gladrow, D., De la Rocha, C., Dürr, H.H., and Scheffran, J. (2013) Enhanced chemical weathering as a geoengineering strategy to reduce atmospheric carbon dioxide, a nutrient source and to mitigate ocean acidification. *Reviews of Geophysics* 51, 113. doi: 10.1029/2012RG000404.

Renforth, P., Manning, D.A.C., and Lopez-Capel, E. (2009) Carbonate precipitation in artificial soils as a sink for atmospheric carbon dioxide. *Applied Geochemistry* 24, 1757. doi:10.1016/j.apgeochem.2009.05.005.

Schuiling, R.D. (2008) How to stop or slow down lava flows. *International Journal of Global Environmental Issues* 8(3), 282–285.

Schuiling, R.D. (2011) LANCELOT (Lake Nyos Carbon Emission Lowering by Olivine Treatment). *Natural Hazards* 56(3), 559–563.

Schuiling, R.D. (2013) Farming nickel from non-ore deposits, combined with CO_2 sequestration. *Natural Science* 5(4).

Schuiling, R.D. and Andrade, A. (1999) Recovery of struvite from calf manure. *Environmental Technology* 20, 765–768.

Schuiling, R.D. and de Boer, P.L. (2011) Rolling stones; fast weathering of olivine in shallow seas for cost-effective CO_2 capture and mitigation of global warming and ocean acidification. *Earth System Dynamics Discussions* 2, 551–568. doi:10.5194/esdd-2-551-2011.

Schuiling, R.D., Hoogesteger, A.W., and Praagman, E. (2012) From spa to corinth a road for CO_2 sequestration. *International Journal of Earth Sciences and Engineering* 5(6).

Schuiling, R.D. and Praagman, E. (2011) Olivine Hills: Mineral water against climate change. In: S. Brunn, ed., *Engineering Earth: The Impact of Megaengineering Projects*. Springer, Dordrecht, the Netherlands, pp. 2201–2206, Chapter 122.

Steen, B. and Borg, G. (2002) An estimate of the cost of sustainable production of metal concentrates from the Earth's crust. *Ecological Economics* 42(3), 401–413.

ten Berge, H.F.M., van der Meer, H.G., Steenhuizen, J.W., Goedhart, P.W., Knops, P., and Verhagen, J. (2012) Olivine weathering in soil, and its effects on growth and nutrient uptake in Ryegrass (*Lolium perenne* L.): A pot experiment. *Plos One* 7(8), e42098. doi:10.1371/journal.pone.0042098.

van Breemen, N. (1976) *Genesis and Soil Solution Chemistry of Acid Sulphate Soils in Thailand*. Agricultural Research Report 848, Pudoc, Wageningen, the Netherlands.

Van der Werf, G.R., Morton, D.C., DeFries, R.C., Olivier, J.G.J., Kasibhatla, P.S., Jackson, R.B., Collatz, G.J., and Randerson, J.T. (November 2009) CO_2 emissions from forest loss. *Nature Geoscience* 2, 737–738.

8

Olivine: Time for Action!

Oliver Tickell

The *olivine option* is admirably set out in the previous chapter by Olaf Schuilling, who has devoted much of his professional life to this excellent cause. To summarize, in nature, weathering of olivine is mainly responsible for the long-term withdrawal of carbon dioxide from the atmosphere. The aqueous reaction of carbon dioxide and olivine chemically sequesters carbon dioxide as bicarbonate. Over the long term, the carbon dioxide is fated to settle at the bottom of the oceans as carbonate rock.

So by artificially accelerating the rock-weathering process, carbon dioxide can be safely, and for our purposes permanently, removed from atmosphere. This can be done at low cost, on the order of $10 per ton, by the simple expedient of mining and grinding olivine-bearing rock, and spreading it on land and coasts, where a combination of chemical weathering, physical abrasion, and biological activity can weather away grains of around 0.1 mm in size within a 10-year period.

We also know that the process is intrinsically safe: nature is doing this all the time anyway with no ill effects. Indeed its acceleration is likely to prove beneficial quite aside from the CO_2 sequestration: combating acidification on land and in sea and providing valuable nutrients. The chemical products of olivine weathering are the benign, acid buffering ions of magnesium and bicarbonate; iron (in varying quantities), a frequently limiting nutrient for marine phytoplankton; and the very weak acid, silicic acid, an essential nutrient for grasses on land and diatoms: widespread phytoplankton in both freshwater and marine environments. Any additional silicic acid has the effect of prolonging diatom blooms, providing additional fish food and increasing the drawdown of carbon to the ocean depths.

But there is a mystery: why has this tremendous opportunity to safely and cheaply sequester carbon dioxide received virtually no attention? Papers have been published in scientific journals and in the popular press, including detailed plans and cost/benefit analyses. The cost is roughly a tenth of the far better-known and assiduously promoted *carbon capture and storage* (CCS) approach, which is to chemically extract CO_2 from power station flue gases and inject it into underground reservoirs. CCS would be expensive in fuel, adding 25%–40% to the fuel burn for a given power output. It would also be costly: the World Coal Association reckons that costs begin around $80–120 per ton of CO_2 and will decline to half that level with increased investment (http://www.worldcoal.org/coal-the-environment/carbon-capture-storage/financing-ccs/). Recent studies suggest that those numbers are overly optimistic, and the real costs would be around four times higher.

As for nuclear power, it is astonishingly difficult to discover its true cost. The best estimate we have comes from negotiations now in progress between the UK Government and the French parastatal corporation Electricite Du France (EDF) over the construction of a twin-reactor station at Hinckley Point, Somerset. The fact that EDF is demanding a guaranteed electricity price approaching £100 per MWh for 40 years (15.3 US cents per KWh) is now widely known, and this is our best indicator of nuclear power's real cost, although it is unlikely to finance the entire decommissioning costs, which will be set at a fixed price underwritten by taxpayers. CO_2 emissions from modern gas-burning power stations run at about 330–360 kg/MWh (say 345), and the electricity sells wholesale for about £50/MWh. This compares to life cycle emissions for nuclear power of 90–140 kg/MWh (say, 115 [http://timeforchange.org/co2-emission-nuclear-power-stations-electricity]), so a difference of 230 kg/MWh.

Using nuclear power to save that 230 kg/MWh relative to combined cycle gas turbine gas–fired plant therefore costs £50, implying a carbon price of £217 per ton. Note that this is to disregard other public subsidies to nuclear power like construction cost guarantees, insurance limitation, decommissioning cost guarantees, and waste management cost guarantees. The direct price subsidy alone makes nuclear power a very expensive way to cut CO_2 emissions—around twice as costly as CCS.

So here is the mystery: why are governments (around 25 worldwide in all are currently intent on building new nuclear power stations) so keen to put serious money into very expensive ways of sequestering CO_2, or avoiding its emission? Why do they completely ignore this effective, low cost, low risk alternative? The main reason is surely the very large vested interests that have been created over the years by government largesse with public funds. And here is how it works: the greater the public funds that an industry has captured, the harder it fights to retain the flow of funds—and the more money it has to fight with. Hence, the nuclear industry is top dog in this game.

The CCS lobby is also powerfully supported. It is not just chemical engineers who like it, but the entire fossil fuel industry, and the coal industry in particular. They see in CCS just the right *magic bullet*—it would not only allow them to carry on producing coal and gas for the power sector, but will actually increase the amount of fuel CCS power stations burn by around a third. So far the only industrial scale CCS taking place is designed to squeeze more gas and oil from depleting wells rather than reduce emissions. Now the aim is to establish the claim that coal can be climate friendly, with new coal power stations that should never be built carrying the *CCS-ready* label—a feint at environmental responsibility that carries no obligation ever to follow through.

Now you might think the same would apply to the olivine option. Though not my intention, the large-scale sequestration of CO_2 using olivine spread on land and coasts would surely allow the fossil fuel industry (and fossil fuel users) to stay in business while sequestering their CO_2 emissions at modest cost—tilting the market only slightly in favor of low-carbon energy producers. It would also create an enormous opportunity for mining companies, as the olivine mining industry leapt from nowhere to become the world's single biggest mining activity. Shippers, truckers, farmers, machinery manufacturers, the steel industry, and many other industries besides would all stand to profit.

The scale of the enterprise is readily calculated. The world currently emits some 32 Gt of CO_2 per year. So to sequester the whole lot at $10 per ton implies an expenditure of $320 billion per annum. This is a colossal sum, but nonetheless an affordable one, representing 0.4% of the estimated gross world product of around $75 trillion for 2013. The pie is certainly big enough for the industries that would benefit to get very excited about—and to be well worth their while to campaign for. Yet there is no organized, effective, well-funded lobby arguing for the olivine option. And sadly that means from the point of view of governments and politicians that it might as well not exist.

So why might that be? In my view, it is because the olivine option has its origins in academia and civil society. It has never received industry support, and no commercial lobby *owns* it. Many industries stand to benefit, both directly and indirectly. But perhaps that is the problem—no single industry has a paramount interest in promoting it. There is no equivalent of the coal industry for CCS or the nuclear industry for more atomic power stations.

But there must be a way forward. The answer surely lies in building an effective global organization for the promotion of the olivine option. It needs to include environmental and other campaigning bodies to represent social and environmental interests. It must include scientists: geologists, climatologists, biologists, agronomists, and foresters. But most importantly, it must attract industrial supporters who will see the investment of some millions of dollars as a very good long-term investment, which could see the establishment of a new global industry worth hundreds of billions—and one in which they would take a major share.

The basis of such an organization already exists: the Netherlands-based Olivine Foundation, which operates the website www.smartstones.nl. It is doing a fine job given its resources. But in the absence of serious commercial or industrial support, it will inevitably struggle to influence national and global climate policy. Now is the time for progressive companies to step up to the line and help to establish a transformational new industry—one that could just save the world.

9

Reestablishing the Evolutionary Grassland–Grazer Relationship to Restore Atmospheric Carbon Dioxide to Preindustrial Levels

Adam D. Sacks, Richard Teague, Fred Provenza, Seth Itzkan, and Jim Laurie

CONTENTS

Introduction

Overview

The quantity of carbon in soils is directly related to the diversity and health of soil biota. Since virtually all organic carbon sequestered in soils is extracted from the atmosphere by photosynthetic organisms and converted to complex molecules by bacteria and fungi, in synergy with insects and animals, we propose an effective and sustainable method for increasing soil organic carbon (SOC) by restoring degraded and desertified grasslands worldwide. This approach to grassland management uses livestock according to the principles of holistic planned grazing (HPG), the effectiveness of which has been demonstrated on millions of hectares on four continents, primarily in semiarid and arid areas, since the 1970s. We maintain that it has the potential to remove excess atmospheric carbon resulting from anthropogenic soil loss over the past 10,000 years as well as all industrial-era greenhouse gas (GHGs) emissions. This sequestration potential, when applied to up to five billion hectares of degraded range and agricultural soils (former wild grasslands), could, in theory, return 10 or more gigatons of excess atmospheric carbon to the terrestrial sink annually and lower GHG concentrations to preindustrial levels in decades. It is a low-tech approach that bears minimal risk of unplanned or unintended consequences that are the norm with technological *fixes* to large-scale environmental problems. Other benefits include restoring grassland and water catchment health and providing jobs and high-quality protein for millions of people, especially in rural communities.

Harmful Changes in Earth's Climate and Carbon Fluxes

Harmful changes in the Earth's climate have accelerated faster than the scientific community had anticipated. While there are good reasons that scientific assessments tend to be conservative (Brysse et al. 2013), surprise about the extent and rapidity of change are becoming the norm in global warming circles (Carey 2012; Lyall 2013; National Climate Assessment and Development Advisory Committee 2013; National Wildlife Federation 2013).

Over the past billion years, the natural flux of carbon from geological stores (i.e., Earth's mantle/volcanoes, fossil fuels) was roughly 0.03 Gt/year (DePaolo 2012). Since the human use of fossil fuels, the flux has grown by a factor of 300 to greater than 9.0 Gt/year (DePaolo 2012). This can *only* be the result of human activity. The disruption of global biogeochemical processes is directly linked to excessive use of fossil fuels resulting in large increases in GHGs, including CO_2. This threatens not only climate, ocean, atmosphere, and terrestrial chemistry and reduces water and air quality (Galloway et al. 2008; Kerr 2010).

In theory, a major part of human-induced climate change can be controlled by balancing the sources and sinks of atmospheric carbon dioxide (CO_2). While 20%–35% of carbon dioxide may remain in the atmosphere for up to 20,000 years, that time may be reduced by neutralizing calcium carbonate through slow weathering of rock (Archer et al. 2009). That long time frame does not address current anthropogenic climate change.

Another factor that is insufficiently considered is the *air conditioning* effect of water. A healthy biosphere is essential in water cycles as evapotranspiration by plants absorbs heat and condensation releases it. Temperature extremes are thereby moderated. This effect is widely observed in changes of land use, such as deforestation and the plowing of grasslands for agriculture, which dramatically alter rainfall patterns and transform formerly healthy biodiverse soils into semideserts. Unfortunately, global climate institutions generally address the effect of climate change on the water cycle and not about the reverse process, the profound effects of the water cycle on climate change (Kravcik 2009).

Reestablishing both carbon and water cycles are essential in addressing climate change because the two cycles work in tandem. While in-depth discussion of the water cycle is outside the scope of this chapter, readers are strongly encouraged to explore water-cycle restoration as a collateral path to reversing global warming (Kravcik et al. 2007; Schwartz 2013, pp. 74–94).

Failure of Emission Reduction Strategies

Twenty-five years ago, it seemed straightforward that reducing emissions would lower the amount of carbon in the atmosphere. Substituting noncarbon energy sources would be both technologically feasible and culturally acceptable as resulting in the least possible economic and social disruption. Accordingly, most efforts to combat global warming have been directed only at controlling sources, and primarily only fossil-fuel sources at that, ignoring the extensive quantities of carbon emitted from degraded soils. Efforts to address sinks, such as restoring forests, have been well intended but with marginal effect (e.g., see Xu 2011).

The result is that the emissions-reduction strategy has, to date, been a decisive failure (Table 9.1). It is beyond the scope of this chapter to investigate the causes of failure, which are primarily social, political, and economic. Generally, the obstacles to emissions reduction have been due to political inertia and economic vested interests, as well as difficulties in changing carbon-emitting cultural habits based on unsustainable use of fossil-fuel energy, for example, use of motorized ground and air travel, and intensive synthetics-based agricultural productivity in all developed parts of the globe. Advancement of energy-conserving practices and suitable technologies would have made far more significant progress in the presence of strong political and popular will. Despite almost three decades of attempts on the part of governments, nongovernmental organizations, the scientific community, and citizens worldwide, the total atmospheric carbon burden has increased by approximately 42 parts per million (ppm) to our current level of 393 ppm (NOAA 2013), and the rate of increase is accelerating (International Energy Association 2012).

Alternative Focus on Ecosystem Carbon Sinks

Since it is increasingly apparent that emissions reductions will not take place soon, there have been alternative proposals for reducing GHGs by actively removing them from the atmosphere. The advantages of such undertakings include the opportunity to bypass the impossibly slow international emissions-reduction agreement process and the possibility of reversing global warming *in spite of* persistent ongoing emissions. Unfortunately, most of these proposals rely on high-tech geoengineering schemes, which are largely untested for effectiveness, fraught with unintended consequences, and potentially very expensive in both direct (economic) and indirect (environmental and social) costs (Humphreys 2011; American Meteorological Society 2013; World Economic Forum 2013).

Unfortunately, natural carbon sinks are currently decreasing because agricultural land, grazing lands (Follett et al. 2001; Conant 2010), coral reefs and rain forests are being degraded at an increasing rate.

TABLE 9.1

Accelerating Atmospheric Carbon Dioxide Increase

Decade	Total Increase (ppm)	Annual Increase (ppm/year)
2003–2012	20.74	2.07
1993–2002	16.73	1.67
1983–1992	15.24	1.52
1973–1982	13.68	1.37
1963–1972	9.00	0.90

Note: Atmospheric CO_2 is accelerating upward from decade to decade. For the past 10 years, the average annual rate of increase is 2.07 ppm. This rate of increase is more than double the increase in the 1960s and 100× that of natural glaciation cycles. (http://co2now.org/Current-CO2/CO2-Trend.)

To balance atmospheric carbon flux, we believe that it is important to both restore and protect these ecosystems in addition to making drastic cuts in fossil-fuel use (Goreau 1992). Efforts to limit emissions from fossil-fuel combustion alone are incapable of stabilizing levels of carbon dioxide in the atmosphere, especially with increasing permafrost and seabed methane emissions, along with destructive positive feedback loops such as dramatic melting of Arctic ice. It is imperative that we consider and implement optimal strategies for removing carbon from the atmosphere.

Grassland Potential and Restorative Grazing

We believe that the most feasible and cost-effective approach to carbon sequestration is restoring the massive sink in degraded grassland soils. Numerous instances around the world attest that degraded grasslands can be restored to benefit biodiversity and ecosystem health using nontoxic low-technology methods while sequestering carbon (Savory and Butterfield 1999; Tainton et al. 1999; Teague et al. 2011; Savory 2013; Schwartz 2013). Although many attempts have been made to adopt technical solutions to reverse this degradation, most involve large amounts of capital and expensive technology, require energy inputs from unsustainable sources, have often been culturally inappropriate, and have not been successful in creating large-scale sustained improvements to the landscape.

That terrestrial ecosystems are an important global carbon (C) sink is well known (Prentice et al. 2001; Schimel et al. 2001), and the size of this sink is related to the rangelands and grasslands of the world (Pacala et al. 2001) (see Table 9.2).

TABLE 9.2

Estimated Range of Total Carbon Storage by Ecosystem

		Global Carbon Stocks (GtC)			
Ecosystem Type	**Total Land Area (10^4 km^2)**	**Vegetation (Low-High)**	**Soils (Mean)**	**Total (Low–High)**	**Carbon Stored/Area (tC/ha) (Low–High)**
Forests					
High-latitude	10.3	46–115	266	312–380	303–370
Mid-latitude	5.9	37–77	84	122–161	206–373
Low-latitude	12.8	48–265	131	180–396	140–310
Sub-total	29.0	132–457	481	613–938	211–324
Grasslands					
High-latitude	10.9	14–48	281	295–329	271–303
Mid-latitude	20.1	17–56	140	158–197	79–98
Low-latitude	21.7	40–126	158	197–284	91–131
Sub-total	52.6	71–231	579	650–810	123–154
Agroecosystems					
High-latitude	3.4	8–18	45	52–62	156–187
Mid-latitude	12.7	21–52	134	155–186	122–147
Low-latitude	9.5	20–72	85	105–157	110–164
Sub-total	25.6	49–142	264	313–405	122–159
Other					
High-latitude	18.6	3–31	65	69–96	37–52
Mid-latitude	11.1	9–25	61	70–86	64–78
Low-latitude	8.8	4–16	34	38–50	43–55
Sub-total	38.5	16–72	160	177–232	46–60
Grand total	145.7	268–901	1,484	1,752–2,385	120–164

Source: White, R. et al., *Pilot Analysis of Global Ecosystems: Grassland Ecosystems,* World Resources Institute, Washington, DC, 2000, p. 51.

Notes: Soils are a significantly more stable long-term carbon sink than aboveground biomass. Globally grasslands store more carbon in soils than any other ecosystem, although forests store more tons per hectare.

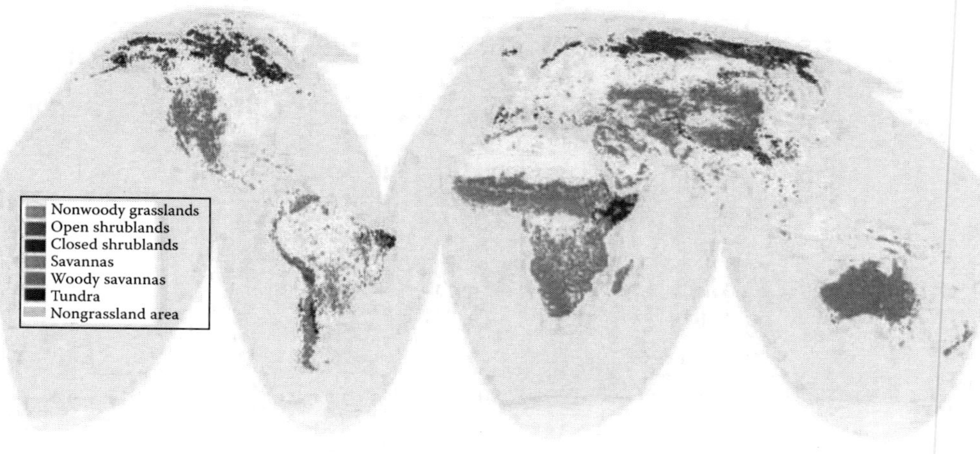

FIGURE 9.1 Global extent of grassland. (From White, R. et al., *Pilot Analysis of Global Ecosystems: Grassland Ecosystems*, World Resources Institute, Washington, DC, 2000, p. 12.)

Grasslands comprise approximately 40% of the global land surface area, excluding Greenland and Antarctica (White et al. 2000, p. 12; see Figure 9.1), and are typically areas of low and seasonal rainfall (see Figure 9.2).

For many centuries, grasslands have been degrading primarily through destructive grazing practices and poorly managed agricultural encroachment of permanent grasslands. At least one billion rural and urban people depend on grasslands for their livelihoods, often through livestock production or for ecosystem services (Ragab and Prudhomme 2002; Neely et al. 2009). There are huge economic and social costs associated with this degradation apart from the diminished role they are able to play in sequestering carbon.

Restoration and sustainable use of billions of hectares of grasslands worldwide requires low-input technology and adaptive management procedures that use a suitably flexible framework to restore ecosystem function (Neely et al. 2009). An objective of this chapter is to outline how such restoration is being accomplished in many locations around the world by changing how livestock managers make decisions to achieve ecological restoration and to enhance their livelihoods and quality of life. A direct consequence of restoring ecological function to these ecosystems is that net carbon sequestration and other essential ecosystem services are significantly enhanced, using an approach that has developed in the natural world over millions of years and has minimal likelihood of unintended consequences.

Need for Sustainable Solutions

Future challenges to humankind are considerable, with climate change at the top of the list. Food production will have to increase by 50% by 2050 to keep pace with global population needs. Our ability to produce food has increased dramatically through technological and biological innovations driven by cheap fossil fuels, but that has not, and will not, freed us from the fact that global resources are finite (Janzen 2011). Ironically, overuse of agrochemicals is gradually decreasing our ability to produce food. While their application dramatically increased yields during the green revolution, their continued overuse is making farms less productive by exhausting the soils, and overreliance on chemical inputs has put the global agriculture industry at major risk as these inputs are running out.

Large urban populations and the intensive agriculture they depend on are severely impacting fresh water supplies (Carpenter and Biggs 2010). We use water extravagantly—from personal use at home to irrigating lawns and crops. In the future, we will have to use water resources much more carefully and efficiently while managing our water catchments more effectively.

A considerable part of our ability to produce food now and in the future depends on having affordable energy from sustainable sources. Cheap energy is a thing of the past because alternative sources of

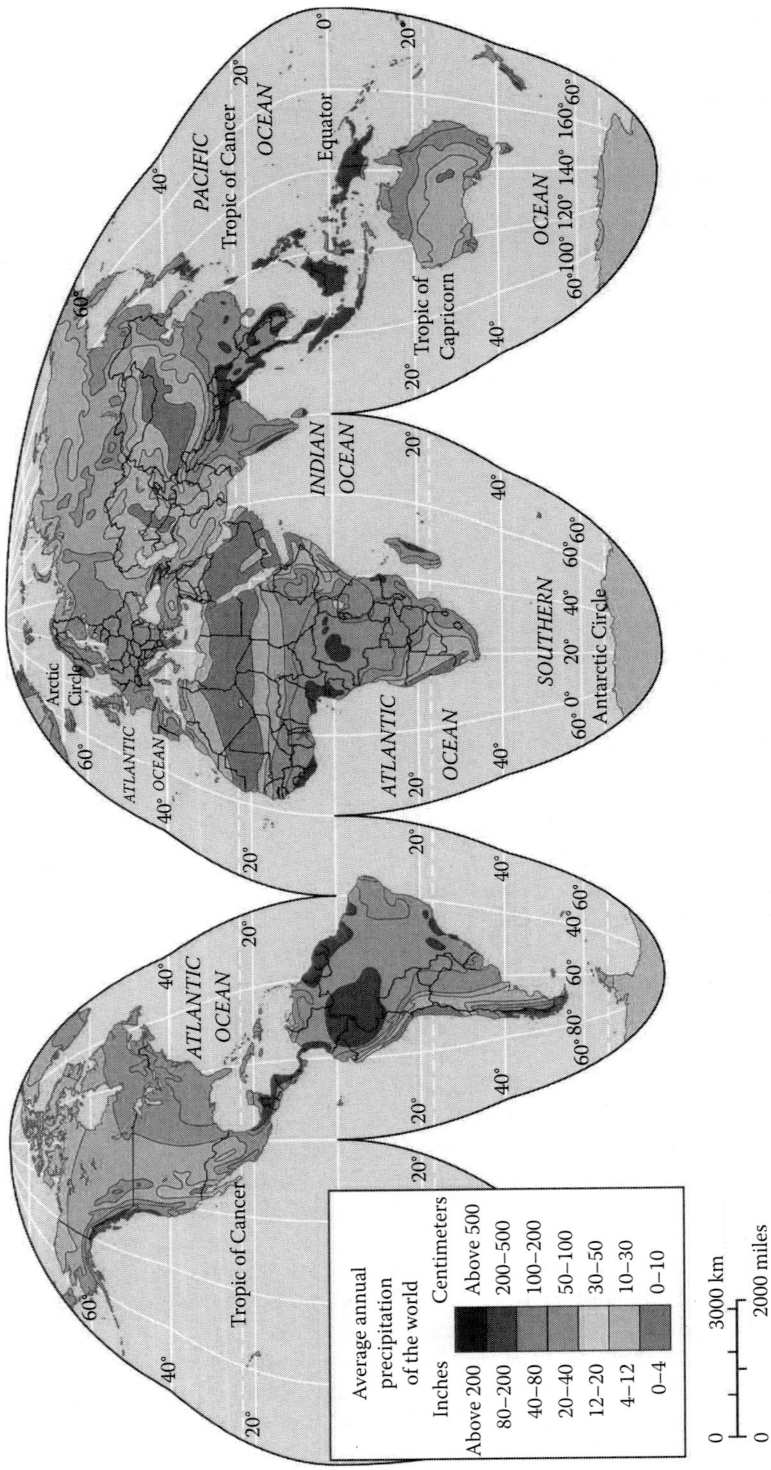

FIGURE 9.2 Average annual precipitation of the world. (From Planetolog, n.d., Map of average annual precipitation of the world, http://planetolog.com/map-world-detail.php?type=RES&id=2.)

energy—solar, wind, water, biofuels, and nuclear—are not abundant enough or are more expensive or more difficult to use (Pimentel and Pimentel 2008). To become more sustainable, we will have to transition to renewable, non-fossil-fuel-based systems as quickly as possible (MacKenzie 1996; MEA 2005). Global industrial production and development will have to restructure significantly to reduce emissions and pollution and use material inputs more efficiently (Hawken et al. 1999). Whether humanity can proceed with resource consumption at anywhere near the rate of the exuberant past several hundred years is open to serious question.

Role of Rangelands in Ecosystem Health

It is in the interests of humankind to increase food production by linking ecology with economics to produce foods as nature does, with minimal inputs (Ikerd 2005; Provenza 2008). As energy becomes more expensive, we will have to manage for healthier ecosystems with fewer purchased and synthetic supplements. We will have to change our land use philosophy from a maximum production to a regenerative and minimum-cost approach. We will have to learn to work harmoniously with the landscapes we inhabit and breed plants and livestock able to thrive locally and sustainably (Provenza 2008). Better management expertise will enable us to manage land more effectively with less labor, less capital, and fewer inputs.

Much of the shrinking habitat available to conserve wild species is in rangelands. These ecosystems are too dry or have topographic limitations that dictate they remain dominated by native vegetation, not food crops. As seminatural communities, they have high floral and faunal diversity, which, if managed in harmony with nature, will avoid dependence on mechanical interventions, synthetic fertilizers, pesticides, and other expensive inputs that threaten farmer and rancher income and exacerbate problems such as soil erosion and depletion, pollution, and loss of biodiversity (MEA 2005). Since they cover such large areas, rangelands are extremely important in managing water catchments and sequestering soil carbon.

Unsustainable technologies of large-scale industrialized farming have become common on rangelands, increasing inputs and costs. The high production philosophy of the industrial agriculture paradigm causes many concerns regarding health, pollution, and ecological degradation (Provenza 2008; Teague et al. 2010). These include

- Replacing locally adapted livestock with genetic lines that are productive but dependent on high inputs of feed and medication
- Stocking with excessive numbers and single species of livestock
- Use of performance enhancing drugs and hormones
- Feeding purchased supplements to maintain animal numbers in times of low productivity
- High mechanization and transport costs
- Relying on heavy machinery and equipment for land management
- Replacing diverse plant communities with few or single species forage sources that often require high inputs of water, fertilizer, and herbicides

Some rangeland managers reject expensive and ecologically unsound practices, and low-input managers are practicing more sustainable and economically viable management in rangeland ecosystems around the world (Savory and Butterfield 1999; Gerrish 2004; Provenza 2008; Teague et al. 2011). These examples provide an excellent knowledge base on which to build more sustainable management in all rangeland ecosystems.

Soil Health and Ecosystem Function

Although many rangelands have been badly degraded (Figure 9.3 and Table 9.3), it is possible to manage them to reverse degradation. Well-managed rangelands have a very important role to play globally as providers of livelihoods, water catchments, and biodiverse habitat for a multitude of plants and animals,

FIGURE 9.3 Soil erosion, New Mexico. Shows almost 5 ft of soil loss. The mound that remains is held together by an isolated stand of native bluestem grass roots; the rest of the soil was dispersed by the winds. (From *National Geographic Magazine*, September 2008:150, http://ngm.nationalgeographic.com/flashback/2008#/2008-09-FB.jpg; Photo by USDA.)

and many are managed to provide good wildlife habitat as well (Milchunas and Lauenroth 1993; Savory and Butterfield 1999). They hold a large reserve of soil C, which, released when rangelands degrade, adds to CO_2 emissions. However, under restorative management, such rangelands can significantly enhance soil C sequestration (Derner et al. 2006; Allard et al. 2007; Soussana et al. 2010; Teague et al. 2011).

Successful conservation-minded rangeland managers enhance the health of the ecosystems upon which they depend by using soil, water, and plant resources efficiently and sustainably (Walters 1986; Holling and Meffe 1996; Walker et al. 2002). By combining scientific principles and local knowledge to manage animals adaptively, they influence four key ecosystem processes (Stinner et al. 1997; Reed et al. 1999; Savory and Butterfield 1999; Gerrish 2004; Barnes et al. 2008; Provenza 2008; Diaz-Solis et al. 2009; Teague et al. 2009a):

- Efficient sequestration of solar energy by plants
- Interception and retention of precipitation in the soil
- Optimal cycling of nutrients
- Promotion of high ecosystem biodiversity with more complex mixtures and combinations of desirable plant species

Scientific Evidence of Grazing Management to Improve Ecosystem Health

Rangelands with healthy soil and diverse plants and animals are more productive, stable, and resilient than those in poor condition and provide greater earnings and ecosystem services. They may represent the most benign use of natural resources. *Contrary to common belief, they regenerate with disturbance and they deteriorate in the absence of periodic disturbance in the form of proper grazing or appropriate and limited use of fire* (Savory and Butterfield 1999, p. 430). They retain diverse ecological forms and function under infrequent and light to moderate defoliation and the impact of grazing herds (Knapp 1985; Hulbert 1988; Oesterheld et al. 1992; Milchunas and Lauenroth 1993; Seastedt 1995).

Large herbivores play a key role in ecosystem function by increasing forage concentration, grazing efficiency, forage nutrient concentration, and aboveground plant production (Frank and Groffman 1998;

TABLE 9.3

Dramatic Decline in Central North American Prairies since 1830

Prairie Type/Location	Past Area (km²)[a]	Current Area (km²)[a]	Decline (Percent)
Tall-grass prairie	677,300	21,548	96.8
Manitoba	6,000	3	99.9
Illinois	85,000	9	99.9
Indiana	28,000	4	99.9
Iowa	120,000	121	99.9
Kansas	69,000	12,000	82.6
Minnesota	73,000	450[b]	99.3[b]
Missouri	60,000	320	99.5
Nebraska	61,000	1,230	98.0
North Dakota	1,300	1	99.9
Oklahoma	52,000	N/a	N/a
South Dakota	26,000	200	99.2
Texas	72,000	7,200	90.0
Wisconsin	24,000	10	99.9
Mixed-grass prairie	628,000	225,803	64.0
Alberta	87,000	34,000	60.9
Manitoba	6,000	3	99.9
Saskatchewan	134,000	25,000	81.3
Nebraska	77,000	19,000	75.3
North Dakota	142,000	45,000	68.3
Oklahoma	25,000	N/a	N/a
South Dakota	16,000	4,800	70.0
Texas	141,000	98,000	30.5
Colorado	N/a	N/a	N/a
Kansas	N/a	N/a	N/a
Montana	N/a	N/a	N/a
Wyorring	N/a	N/a	N/a
Short-grass prairie	181,790	62,115	65.8
Saskatchewan	59,000	8,400	85.8
Oklahoma	13,000	N/a	N/a
New Mexico	N/a	12,552	N/a
South Dakota	1,790	1,163	35.0
Texas	78,000	16,000	79.5
Wyorring	30,000	24,000	20.0
Colorado	N/a	N/a	N/a
Kansas	N/a	N/a	N/a
Montana	N/a	N/a	N/a
Nebraska	N/a	N/a	N/a
Total	1,487,090	309,467	79.2

Source: White, R. et al., *Pilot Analysis of Global Ecosystems: Grassland Ecosystems,* World Resources Institute, Washington, DC, 2000, p. 22.

[a] Estimates of past and current area are based on information from The Nature Conservancy's Natural Heritage Data Center Network; the provinces of Alberta, Manitoba, and Saskatchewan; universities, and state conservation organizations.

[b] Original data reported as a range: 300–600 km² at 99.2–99.6%.

Frank et al. 1998). Their digestive systems are designed to provide humidity and microorganisms to produce biological decomposition of huge volumes of plant material.

Nutrients return to the soil in the form of urine and dung, accelerating cycling of nutrients from aboveground vegetation biomass. They improve mineral availability by enhancing soil microbial nutrient enrichment and soil microbiological processes that augment plant nutrition and photosynthesis (Hamilton and Frank 2001) in addition to increasing nutrient cycling within patches of their urine and excrement (Holland et al. 1992). *Consequently, grazing is an optimization function*: excessively low or high levels of herbivory (eating plants without killing them) yield low levels of primary production, while intermediate levels of herbivory yield maximum productivity (McNaughton 1985; Dyer et al. 1993; Turner et al. 1993).

The grasslands of the world evolved under large herds of migratory herbivores that supported more biomass and sustained considerably higher levels of herbivory than any other terrestrial habitats (Frank and McNaughton 1993). The relatively recent replacement of free-ranging herbivores and herding pastoralists with sedentary livestock has resulted in widespread overgrazing (Oesterheld et al. 1992; Milchunas et al. 1993; Weber and Shannon 2011), which often has degraded vegetation and soils, reduced biodiversity, and led to the loss of the capacity of soils to sequester carbon.

Domesticated livestock, without pressure from predators, do not graze uniformly over a landscape but instead apply constant grazing pressure and repeatedly consume preferred plants and patches of vegetation. Such excessive herbivory removes biomass and litter that causes soil exposure and degradation in heavily used areas.

Soil, plants, and animals benefit when herbivores roam. These are natural behaviors in response to changes in forage quality and availability and in response to predators. Satiety mechanisms ensure herbivores eat a variety of foods and forage in a variety of places (Provenza 1996; Bailey and Provenza 2008). Variety stimulates appetite and enhances nutrition, production, and the health of soil, plants, herbivores, and human beings (Provenza et al. 2007; Provenza 2008).

Managing animals sustainably requires long-term planning horizons, conserving primary resources, appropriate management goals, and continually adapting to dynamic ecological, social, and economic conditions (Savory and Butterfield 1999). Unless sufficiently sensitive indicators of change are continually monitored, landowners who focus on short-term profit maximization may not realize that the ecosystems upon which their production systems depend are being systematically degraded (Kothmann 1971; Whitson et al. 1982; Knight et al. 1990; Teague et al. 2009b). Narrowly focusing on *maximizing* livestock production from rangelands is an unsustainable goal ecologically and economically (Workman 1986).

Life of the Soil

Healthy soils store carbon in complex biomolecules, which can remove carbon from global cycles for hundreds or even thousands of years. Australian soil scientist Christine Jones notes (Jones 2009b):

> [i]f soil is of high ecological integrity, soil microbes, especially fungi, resynthesise and polymerise labile carbon (mostly exuded from plant roots) into high molecular weight stable complexes, referred to collectively as humic substances. Humus, a gel-like substance that forms an integral component of the soil matrix, is the best known of the long-lived stable organic fractions.
>
> Humified carbon differs physically, chemically and biologically from the labile pool of organic carbon that typically forms in agricultural soils. Labile organic carbon arises principally from biomass inputs (such as crop residues) which are readily decomposed. Conversely, most humified carbon derives from direct exudation or transfer of soluble carbon from plant roots to mycorrhizal fungi and other symbiotic or associative microflora.
>
> Humus is composed of large, complex molecules made up of carbon, nitrogen, soil minerals and soil aggregates. It is an inseparable part of the soil matrix that can remain intact for hundreds, sometimes thousands, of years.
>
> Once carbon is sequestered as humus it has high resistance to microbial and oxidative decomposition.

Plants in a healthy biodiverse soil release much of their photosynthetic sugar (perhaps 40%) to symbiotic mycorrhizal fungi. In return, the fungal mycelia distribute energy in the form of sugars to microbial communities deep in the soil, which find and extract minerals for the plant. The mycelia are active in finding water in pores inaccessible to plant roots. Mycorrhizae produce a carbon-rich glycoprotein called glomalin, which comprises a large percentage of organic matter in healthy soils. These are sticky substances that bind soil particles together, providing air spaces and structure for the movement of water and soil organisms and holding many times their own weight in water (Comis 2002; USDA 2008).

Christine Jones has studied this *liquid carbon pathway* and proposes the following:

> In addition to reducing levels of atmospheric carbon dioxide, the activation of the soil sequestration pathway results in the release of plant nutrients from the theoretically insoluble mineral fraction, which comprises by far the largest proportion (96–98%) of the soil mass. This increased mineral availability improves the health of pastures, crops, livestock and the people consuming agricultural produce. Everyone benefits when food is more nourishing.
>
> Mineral availabilities are determined more by the rate of carbon flow from plants than by the stock of carbon in the soil. The "key" to mineral management is appropriate groundcover management.

Jones has been studying pasture cropping on the Seis Ranch in New South Wales, Australia, for over 20 years. From 2008 to 2010, the sequestration rate was 33 t of CO_2 per hectare per year, the equivalent of 9 t of SOC. Stable, nonlabile compounds comprise 78% of the carbon gain. Jones is finding that over time, the biggest gains are made deeper in the soil. "That is, over time, fertile carbon-rich topsoil will continue to build downwards into the subsoil." (Jones 2011, p. 3). Recent evidence from boreal forests shows that healthy fungal networks are critical to soil carbon stores (Clemmensen et al. 2013).

Conventionally managed soils using synthetic fertilizers and pesticides destroy essential soil biota and lead to widespread soil degradation (Neely et al. 2011). Mycorrhizal fungi and associated bacteria are strongly inhibited by excessive soil disturbance such as plowing and by high levels of dissolved phosphorus and nitrogen commonly used as fertilizers in agriculture (Killham 1994; Leake et al. 2004). Such soils become *addicted* to artificial inputs, requiring ever larger *fixes* over time while yielding diminishing outputs (Khan et al. 2007). Farmers compensate by using even more fertilizer, creating a vicious cycle of overfarming and chemical abuse.

The energy supplied by green plants is necessary to keep fungi functioning effectively; the fungi reciprocate by providing plants with essential minerals. Mineral depletion and structural degradation of grassland soils have largely been the result of human mismanagement, diminishing the effective function of fungi. When soil carbon supply is limited by loss of primary plant production (sugars and other carbon compounds created via photosynthesis and metabolic processes), carbon sequestration is markedly reduced (Jones 2008).

To maintain soil health in rangeland and more intensive pastures means moderate defoliation during the growing season and adequate recovery before once again grazing them (Gerrish 2004; Teague et al. 2011). By doing so, it is possible to keep plants photosynthesizing as long as soil moisture permits. This increases the growing season, animal diet quality, and energy for the plants, fungi, and other soil organisms that obtain energy from them.

Restoring Grasslands

Synergy is at the heart of effective ecorestoration. To restore grasslands as healthy ecosystems and effective carbon sinks, we must reestablish evolutionary relationships between grazing animals and their habitats. Grassland habitats have suffered worldwide since the advent of agriculture and pastoralism by human use of wild lands for food and other products (Suttie et al. 2005). Recently, degradation has accelerated (Weber and Shannon 2001, p. 2):

Some pastoral cultures (e.g., the Herero of Namibia and the Samburu of Northern Kenya) have degraded their environments to the point where temporary abandonment is required. Still, numerous pastoral cultures (e.g., Rashayada Bedouin of the Sudan, Mongolian and Chinese herdsman, and Pyrenean herders) have subsisted on rangelands for thousands of years despite various hardships and challenges. Indeed, the degraded landscapes observed today are considered a relatively recent phenomenon that accelerated during the late twentieth and early twenty-first centuries.

This has culminated in the modern industrial approach to animal husbandry that has so distorted our raising of goats, sheep, and cattle that we do not understand how essential they are to healthy landscapes.

Role of Livestock in Grassland Ecosystem Destruction

Livestock have a large and growing impact on global ecosystems. In biomass, they exceed humans and other mammals, and because of the growing demand for livestock products, this impact is likely to increase markedly (Janzen 2011). Increased demand for livestock products is a major driver in changing land use to expand livestock production, such as clearing of forests in Brazil. Serious loss of wildlife habitat, biodiversity, and terrestrial carbon stocks are resulting (Steinfeld and Wassenaaar 2007). Approximately a third of cultivated land is used to produce animal feed, decreasing the amount available for human consumption. The amount of water livestock required, either directly or in the production of animal feed, competes with human and agricultural uses. Improperly managed livestock may also be a threat to water quality through their impact on watersheds and directly through fecal contamination (UN FAO 2006; de Vries and de Boer 2010).

Where more intensive animal agriculture is conducted, there are major problems with livestock excrement disposal and pollution (Galloway et al. 2008), heavy dependence on nonrenewable energy (Bartlett 1978), and competition for human food, with up to 40% of cereal production globally being used for livestock feed (Garnett 2010). Intensively reared livestock are significant contributors to GHG emissions (Steinfeld et al. 2006). These issues have resulted in numerous calls to reduce livestock agriculture to control these many and widespread problems (Popp et al. 2010).

Role of Livestock in Grassland Ecosystem Restoration

Livestock are major determinants in maintaining and restoring ecosystem function and services while providing products that humans need without being in competition with them. In contrast to cereal grains that humans can digest, herbivores transform pasture and rangeland plants (grasses, forbs, and shrubs high in cellulose, which humans cannot digest) into animal products (meat, milk, and fiber) humans can use.

Rangelands occupy approximately one-third of the Earth's land area not covered by ice (White et al. 2000). At least a billion rural and urban people depend on them for livelihoods through livestock products and ecosystem services (Ragab and Prudhomme 2002; Neely et al. 2009). Rangelands are not suitable for cultivation and should not be cultivated as this would virtually eliminate their value in maintaining human and biodiverse populations and supplying ecosystem services.

Grazing animals are *essential* for rapid restoration of degraded rangelands (Savory and Butterfield 1999, Gerrish 2004). Properly managed grazing by herbivores can create diverse plant species with different chemical characteristics, enhance soil organic matter (SOM) and nutrients, moderate soil temperature, and increase water infiltration rates (Figure 9.4). Proper grazing management also can increase the fungi to bacteria ratio by more than four times over land not grazed. Fungi access and transport nutrients, extend root volume and depth, and increase drought resistance (Killham 1994; Leake et al. 2004).

Standard practice prescribes long-term, indefinite *rest* for overgrazed grasslands, that is, removal of animals for years or decades. However, such rest does not restore grasslands in *brittle* environments, that is, environments with highly seasonal or erratic precipitation patterns and often long dry periods (see, e.g., West et al. 1984). Brittle environments comprise 75% of all grasslands on the planet (grasslands with year-round rainfall are far more resilient). Despite clear evidence that overrest is destructive, not restorative,

FIGURE 9.4 Land in the semiarid Karoo region in South Africa. Healthy land on the left was managed with HPG; degraded land on the right is overrested, with no animal impact for decades. Average annual rainfall here is 250–450 mm/year. (From Palmer, T. and Ainslie, A., n.d., United Nations Food and Agricultural Organization, Country Pasture/Forage Resource Profiles: South Africa, http://www.fao.org/ag/AGP/AGPC/doc/Counprof/southafrica/SouthAfrica.htm.)

the application of overrest remains firmly entrenched in the dominant paradigm (Beschta et al. 2012). There is a vast difference between land partially rested with plants either grazed or overgrazed and land where rest has been replaced by periodic high impact of grazing animals and plant recovery.

An appropriate undisturbed recovery period, punctuated by well-planned grazing, is essential for maintaining plants and ecosystem processes in brittle environments. The purpose of the recovery period is, after proper grazing pressure, to allow bitten plants to regrow. Recovery periods may be anywhere from 30 days to 2 years depending on factors such as the climate, time of year, animal density, precipitation, and many other factors (Savory and Butterfield 1999, p. 501 ff.) This mimics the way nature cycles animal impact in wild herds. An appropriate recovery period is never as long as several years or decades, over which time deleterious effects of overrest become increasingly apparent.

Introducing Holistic Management

Rangeland ecosystems are complex and variable so to achieve desirable sustainable goals a flexible management framework is essential. As each ranch or management area is unique, such a framework must be able to incorporate a great deal of complexity and monitoring to achieve multiple goals while maintaining a flexible format.

Holistic planning protocols (Savory and Butterfield 1999) provide a comprehensive management framework based on maintaining or improving ecological function and biological resources upon which productivity is based, while simultaneously addressing social and economic complexity. Traditional teachings on management of rangeland, exemplified by Briske et al. (2008), concentrate on achieving the highest productivity without examining long-term consequences on ecosystem function or ranch profitability. This management approach has been a major cause of rangeland deterioration.

The holistic framework ensures that all objectives are consistent and all actions are socially, environmentally, and economically sound, both short and long term. HPG (Savory and Butterfield 1999) is achieved by monitoring and making adjustments as conditions change to continually improve efficient sequestration of solar energy by plants, interception and retention of precipitation in the soil, optimal cycling of nutrient, and promotion of high ecosystem biodiversity of desirable plant species.

The complexity that ranchers and pastoralists have to contend with is too great for any prescribed rotational or other management system and is therefore addressed through a planning process involving a planning tool (referred to as an *aide memoire* by Savory) that breaks the many factors into small easy-to-manage segments. Such segments build upon one another until the plan is complete. Because different dimensions such as land areas, time, numbers of animals, and seasonal variations are involved, the actual plan is done on a chart with symbols and figures where planned moves of the animals are plotted.

Key principles employed are as follows:

- Partial or total rest of the land is detrimental and thus the herd effect of the animals is used to ensure repeated and periodic high disturbance to ensure annual rapid biological decay of dead plant material and soil cover. This is put into practice through moving livestock in tight packs (Figure 9.5) and in regular movements of the corrals (every 7–10 days) to maximize the cover of dung and plant litter (Figure 9.6).

FIGURE 9.5 Dense packs. Livestock, of mixed species, are moved in tight and mobile packs, replicating the coevolutionary and beneficial behavior of wild herds. (Photo by Africa Center for Holistic Management.)

FIGURE 9.6 Mobile corrals. Corrals are moved every 7–10 days to maximize the area covered in dung and plant litter, providing cover for bare or highly degraded ground and helping to ensure better water retention the during rains. (Photo by Seth J. Itzkan.)

- Only plants (not land) can be overgrazed or overbrowsed. Overuse is a function of time in a paddock, not animal numbers. Through planned grazing, plants are assured adequate recovery between grazings for full root reestablishment—planning is based on plant recovery periods, not grazing periods.
- Drier than average years is a regular feature in low rainfall environments; accordingly, every year is planned as a drought year, but in terms of months of reserve of forage and not areas of reserve forage. This is done to keep production of both animals and plants high in all years including those not drier than average.
- Planning always includes dovetailing livestock moves with all other land uses, including preventing conflict with wildlife.

The holistic management process provides the manager with protocols and forward planning to implement timely adjustments to counter overgrazing and overrest as well as to plan ahead to achieve biodiversity and wildlife goals. The process allows for such goals to be tailor-made to specific situations in different portions of a management area and to coordinate disparate requirements of different areas (Savory and Butterfield 1999).

Long-term ranch-based research and theoretical analyses indicate that reducing livestock numbers when forage availability declines is insufficient to maintain rangeland health and productivity and may in fact prevent maintaining health (Müller et al. 2007; Teague et al. 2004; 2011).

Adequate postgrazing recovery during the growing season is also necessary. Subdividing grazing units into smaller pastures (*paddocks*) effectively increases the area and biomass of plants animals graze. In the process, animals can learn to eat mixtures of plant species, including the so-called unpalatable plants, in ways that can increase forage intake and improve animal health (Provenza 2003a,b; Provenza et al. 2003; Villalba et al. 2004; Shaw et al. 2006; Provenza and Villalba 2010). This increases available forage, a significant advantage over continuous grazing (Teague et al. 2011). Under such management, biodiversity and ecosystem function are considerably enhanced. When combined with a decision-making process that responds to frequent changes in conditions, it can also proactively mitigate damage caused by recurring drought periods.

Much evidence supports HPG as a highly effective means of restoring soil health. This includes articles from the mainstream literature (Savory Institute 2013), empirical data from ranchers and various agricultural services, reasonable inference from conventional soil carbon measurements, and historical data on SOM and SOC loss. The same is true for management-intensive grazing (MIG) (Gerrish 2004). While the former (HPG) has focused more on rangelands than pastures and the latter (MIG) more on pastures than rangelands, the principles and practices with regard to grazing are the same. Collectively, these approaches have led to planned and managed grazing at sites throughout the world.

Rangeland and pasture restoration can play a central role in addressing climate change, since restoring soil health *always* results in storing significant quantities of carbon: soil health, biodiversity, and SOC go hand in hand. The potential for carbon sequestration in soils is massive, likely equal to the task of storing over 250 years of industrial carbon emissions should we choose to implement it in earnest.

How Pasture and Rangeland Soils Come Back to Life

Holistic management is successful because it mimics nature's approach to a sustainable ecosystem. A typical illustration of the essential process of soil restoration, beginning with mostly bare, dry ground and using any of a wide variety of animals as grazers, from cattle to goats to sheep to bison to antelope is as follows:

First a grazing plan is created in order to manage livestock properly in their particular and unique habitat. Even on severely degraded lands, there is usually some forage available—the more diverse the better (Provenza et al. 2007)—to get the process started. Infrequently, it may be necessary to provide hay or other feed, for example, in an area where the soil has been degraded by materials such as mine tailings (Dagget and Bean 2005). The animals are run in tight groups, confined to relatively small paddocks, or herded in sequences that optimize use of all plants on landscapes (Meuret 2010; Meuret and

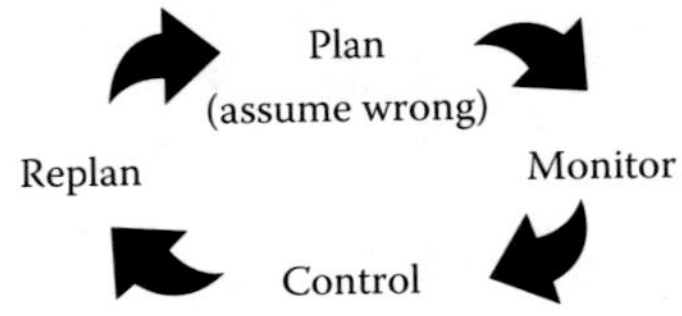

FIGURE 9.7 Holistic management feedback loop.

Provenza 2013), having intense but brief impact (several hours to a few days) on the land. Planning is done on the basis of ensuring adequate recovery of grazed plants. During grazing, constant monitoring is used to ensure that plants are not being overgrazed. When a new situation is experienced, the manager assumes the decisions are wrong and monitoring focuses on the earliest detectable changes to ensure that management is moving the system toward the desired state. If not, further changes are made to ensure the desired state is reached via proactive management. These are essential elements of holistic management—plan (and assume that original plan may have been wrong), monitor, control, and replan (Savory and Butterfield 1999; Figure 9.7).

During their first impact on degraded soils, the animals break the soil cap with their hooves, fertilize it with urine and dung rich in gut bacteria, and trample plant matter into the soil surface, including fresh seeds and dead biomass that interferes with new growth. This disturbance stimulates biotic activity by facilitating circulation of oxygen, carbon dioxide, and other gases, by providing nutrients, allowing penetration of water (Weber and Gokhale 2011), and providing land cover to minimize or eliminate bare ground (Naeth et al. 1991). Covering the ground with dung and trampled grass increases the effectiveness of rainfall, as the increase in organic matter helps to hold water in the soil (McGinty et al. 1995). Biodiversity flourishes as the area of bare soil diminishes. Reclamation begins with vegetation-covered ground to make the available rainfall more effective through increasing infiltration of water into the soil, reducing soil surface water evaporation and runoff during high-rainfall events.

Over as few as 3 years, many long-disabled processes come back to life. Insects such as dung beetles retrieve excreta and store it more than 18 in. beneath the surface (Richardson and Richardson 2000). Worms and small mammals such as moles and prairie dogs churn the soil, while deep-rooted perennial grasses regrow and create channels for water and gases. Mycorrhizal fungi, with literally thousands of miles of hyphae in a small patch, transport nutrients that they have the unique ability to obtain from soil minerals and exchange them for carbohydrates from photosynthetic plants. The fungi synthesize a stable glycoprotein, glomalin, which holds 4–20 times its weight in water (Reicosky 2005). Microorganisms join the elaborate fray, accessing minerals that they supply to fungi that in turn supply them to the green plants, and in the process create complex carbon molecules that store carbon deep in the soils for a long period of time (Jones 2009a). This is the embodiment of carbon sequestration.

When grazing animals eat perennial plants, their root systems die back proportional to the amount of the plant that is eaten and become food for communities of bacteria (Baskin 2005), leaving porous passages and complex carbon molecules that are aggregated into humus (Pucheta et al. 2004). This is why healthy soils are dark in color (like elemental carbon), relatively low in density, and clump, not crumble, when handled.

Following defoliation, perennial plants, after first sacrificing root and root hairs to mobilize energy for leaf growth, then regrow new roots that season. The process repeats itself, increasing soil porosity and water and carbon content annually. Through each cycle, every component of this complex system nourishes all the others, resulting in rich soils that, storing vast quantities of water and carbon, remain moist even during periods of drought, raise the water table to restore and maintain perennial streams and ponds, and create habitats for richly diverse microbial, plant, insect, and animal life.

These are the healthy soils we strive to recreate, capturing carbon, providing food, reestablishing balanced hydrological and nutrient cycles, and imparting beauty to the land. Herbivores are organs as essential to the form and function of these processes as stomachs and hearts are to the form and function of mammals.

Livestock Ecorestoration Case Studies: Lessons, Finances, Topsoil, Pasture, and Rangeland Creatures

Holistic Planned Grazing and Management-Intensive Grazing

Holistic Management: A New Framework for Decision Making (Savory and Butterfield 1999) examines the nature of the decision-making process that leads to destruction and desertification of terrestrial ecosystems and an alternative decision-making process that makes possible reestablishing healthy, whole, functioning ecosystems while also meeting human needs. Savory gives examples from arid regions worldwide—Zimbabwe, South Africa, Namibia, and the Southwest United States—where the reversal of desertification was achieved through a changed decision-making framework and modifications of livestock management (see Figures 9.8 and 9.9).

Management-Intensive Grazing (Gerrish 2004) is based on 30 years of research and filled with examples of how to use understanding of ecological principles and processes to implement grazing practices, especially pasture lands.

Scientists in academic institutions have less readily embraced HPG than MIG (e.g., Briske et al. 2008), but managers have readily embraced both HPG and MIG (Teague et al. 2011). The discrepancy has to do mainly with how academics study grazing (analysis of parts, control of variables) and how

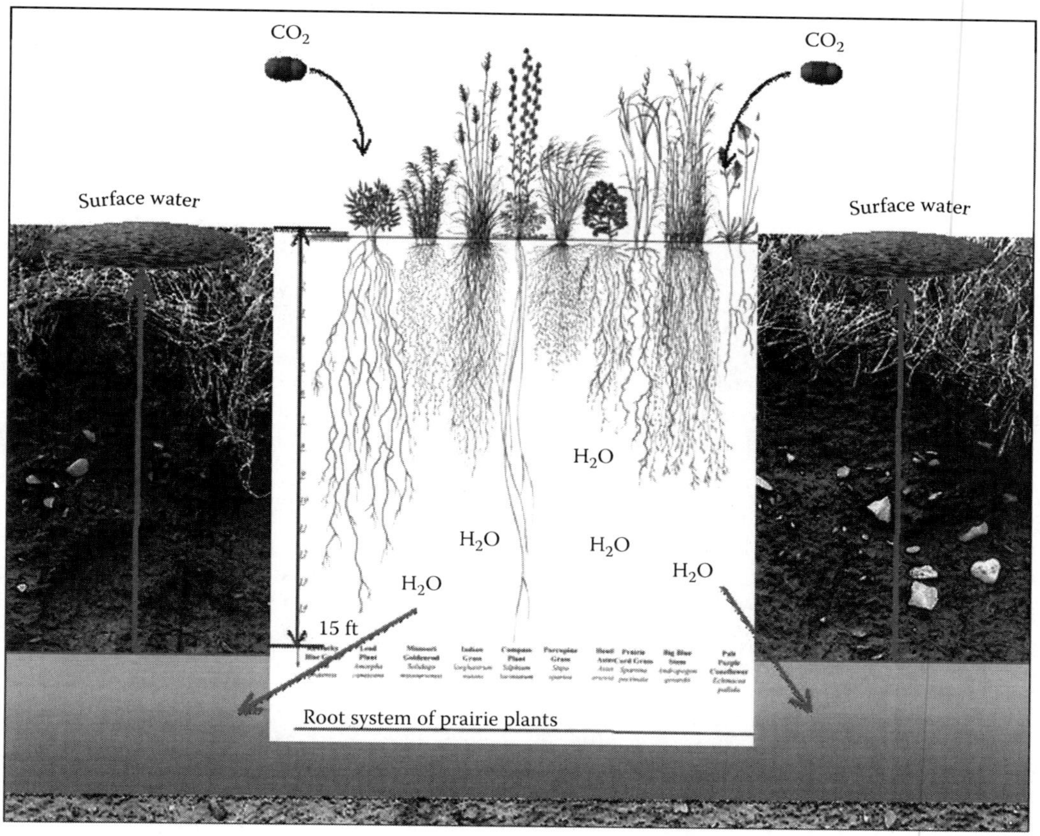

FIGURE 9.8 Restoration of water table and surface water in grasslands. Animal impact stimulates deep-rooted prairie plant growth creating porous, absorbent soils. Water sinks into the ground rather than getting lost to evaporation and runoff. The increased water recharges and raises the water table, which feeds aboveground streams and ponds. Holistic management practitioners routinely report new surface water, often within a few years, where it had not appeared for as long as anyone could remember. (Graphic by Seth J. Itzkan; USDA/Agricultural Research Service, Glomalin is key to locking up soil carbon, *Science Daily*, June 16, 2008, p. 5, http://www.ars.usda.gov/is/pr/2008/080617.htm.)

FIGURE 9.9 Reversal of desertification in Zimbabwe. Average rainfall 600–700 mm. In the left picture (2006), wooden paddock fences can be seen in the background. Animals were corralled in the paddocks overnight for several consecutive nights to maximize the cover of dung and plant litter then were moved to new sites in need of intense impact. After the rains, new plant growth (2009) is incorporated into the holistic grazing plan. Arrow points to the same tree in both pictures. (Photo Credits by Africa Center for Holistic Management.)

practitioners use grazing to accomplish ecological, economic, and social goals (synthesis of wholes, flexibility) (Van Der Ploeg et al. 2006).

Both practices have been adapted by practitioners at modest rates (Stinner 1997), but there is increasing interest in managing grazing as an integral component of meeting ecological, economic, and social goals. This interest is being driven by ever-increasing costs of fossil fuels, combined with growing awareness of the high ecological, economic, and social costs of management that relies on fertilizer and pesticide inputs that damage ecological systems. It is being accelerated by increasing consumer awareness of the relationship between diet and health and the desire to eat whole foods raised under healthful conditions, including pasture-reared animals treated humanely.

Dan Dagget and Examples from North America

Ecologist and rangelands activist Dan Dagget (Dagget and Bean 2005) has collected several illustrated examples of ecorestoration across the Southwestern United States and Northern Mexico using livestock and HPG. He emphasizes that degraded and desertified rangelands, most in drought-prone environments with irregular rainfall, require animals for systemic health. Restoration examples with livestock include barren gold-mining land in Nevada contaminated with cyanide, riparian embankment recovery along the Gila River in New Mexico, and native grassland recovery on the Audubon National Wildlife Refuge in North Dakota.

Jim Howell and the Role of Grazers

Rangeland expert Jim Howell has been practicing HPG for over 20 years and has working experience with ranchers throughout the American West, Mexico, Argentina, Australia, New Zealand, and Africa to improve the health and productivity of their lands in a wide variety of climates and conditions. He summarizes the importance of grazing animals to grasslands as follows (Howell 2008; emphases in the original manuscript):

> Perennial grass plants and large grazing mammals (in association with their pack hunting predators) have been living together for millions of years. Grasses are adapted to periodic defoliation. Most grasses grow from growth points that lie at the base of the plant, beyond the reach of the grazing animal's muzzle…
>
> Given sufficient **time**, the plant can completely recover from the severe grazing…. and many grass species in fact rely on properly timed grazing…. Overgrazing takes place when a plant is severely regrazed before it has had the chance to completely recover both its carbohydrate reserves and above ground leaf area.
>
> … [T]he bulk of the world's grazing animals live in areas with either highly seasonal or highly erratic precipitation patterns. The soil surface condition is therefore dry and arid for much of the

year, which means that microbes and insects in charge of decomposing plant material into the soil die off or become dormant during the dry, non-growing season. ***The only place that these decomposers survive during these long dry spells is in the gut of the grazing animal.*** In these seasonal environments, the grazing animals are therefore critical in maintaining the flow of carbon from the atmosphere to the plants and into the soil.

The Maddox Family: Saving the Farm while Improving the Land

In 1986, Joe and Peggy Maddox were managing an 8800 ha ranch near Colorado City, Texas, and like many ranchers at the time, they were financially desperate. They had been following the recommendations of the local extension service regarding livestock stocking rates. That year they had a $200,000 bill for synthetic chemicals, mostly to fight mesquite, but the mesquite persisted in abundance.

They began HPG and financial planning and eliminated a major share of their expenses and most of the chemicals. They ran three to four times as many animals on the land but concentrated them in one herd and moved them weekly. It was 6 months before the herd came back to any paddock after it had been grazed. Within 3 years, the land was much better able to capture rainfall and the perennial grasses were coming back. Their income greatly improved and wildlife began to show up to take advantage of the ponds and streams emerging from the land. The Maddox family often had open houses and shared what they were learning (Laurie 1997).

Joel Salatin and Restoring Topsoil

Joel Salatin's family has been grazing in Virginia's Shenandoah Valley since 1960. Estimates of soil lost after plowing began range from 3 to 8 ft. When Salatin's father began grazing, there were areas of bare shale rock as extensive as 100 ft in diameter. In 2000, after 40 years of grazing, the largest of these rocky galls had been reduced to a few feet in diameter. By 2010, he cannot find any of these areas with less than 8 in. of new soil. He is making a case for 8 in. of soil created within a decade using a grazing plan with high-density herd impact followed by ample recovery time (Salatin 2010).

Walt Davis and Dung Beetles

Walt Davis is a rancher in southern Oklahoma. When he stopped using insecticides in 1975, he noticed within a year that dung beetles were burying his cow pads. Over the years, his dung beetle populations have increased, and by 2000, there were five species of dung beetles on the ranch.

Patricia and Dick Richardson visited the Davis Ranch in the 1990s and discovered that dung beetle activity increased the soil's water infiltration by over 120%. They also estimated that dung beetles were capable of burying 2.5 t of wet manure per hectare per day. Over 1000 dung beetles were seen on single dung piles. Their potential to transport tons of organic matter into the soil is significant.

In 1975, Walt could find no earthworms in his pasture soils. Over the next decade, in areas where he observed concentrated dung beetle activity, he also began to see earthworm castings. At a 1996 field day at the ranch, he dug large core soil samples from a number of these pastures and counted 12–30 earthworms per cubic foot. Today, at several sites on the ranch, harvester ants have built their mounds out of earthworm castings (Richardson and Richardson 2000).

It may be that dung beetle activity was necessary to bring organic matter into the soil before earthworms were viable on the Davis Ranch. Water is also affected:

> We have compared water infiltration rates where a cow pad has been buried by dung beetles vs. no cow pad on a variety of pasture and soil conditions. We find on average a 129% increase in infiltration rate over control (Table 9.1). An extra inch (25 mm) of rainfall absorbed per acre means the addition of 27,225 gallons/acre (254,530 l/ha) of water in the soil, reducing the detrimental effects of either drought or flood years (Richardson and Richardson 2000).

The many benefits that dung beetles bring to an ecosystem (and to human enterprise) are substantial: improved soil health and reduced runoff, increased pasture productivity, reduced infection of livestock by parasitic worms, reduced fly pests and human disease, reduced GHGs, and economic benefits (extensive references at http://dungbeetle.org.nz/benefits/).

Greg Judy and Earthworms

Greg Judy owns and leases several farms in central Missouri. He has written extensively about grazing cattle, sheep, goats, and pigs, yet he considers his most important *livestock* resource to be earthworms. Judy's goal is 25 earthworms per square foot, which he estimates would yield 250 t/ha/year of worm castings at no expense. Many of Judy's fields now have worm densities of $17/ft^2$.

Earthworms are of key importance to healthy soils. Their burrows, which may be mainly horizontal or vertical depending on species, can be extensive. As they work their way through the soil, they disperse litter from the surface and may move biomass up to several meters deep, spreading nutrients and providing easy passage for root growth (Brady 2002).

Judy's grazing strategy, built on HPG, is very-high-density animal herds grazing in a paddock for 1 day and then moved to the next paddock. The goal is for animals to consume a third to half the grass in the paddock and to trample the rest into the soil. This feeds his earthworms and soil microbes (Judy 2008).

A critical point is that he does not use deworming agents and other chemicals that will poison soil biota such as dung beetles, whose beneficial impact is significant. Dung beetle holes help create paths for earthworm activity and other soil creatures, and the dung dried out by beetle activity and taken belowground reduces the fly populations.

Based on soil depth measurements, Judy estimates that his grazing practice has built 7.5 cm of soil in many fields in 4 years. The trampled grass, animal manure, and worm casting mix become a very rich layer of topsoil. The dry matter of worm castings is typically 70% organic matter and 30% minerals. Estimating the SOM of this new soil at a very conservative 8%, the SOC would be at least 4.5% or over 8.4 t/ha (Table 9.4).

In addition to earthworms, the activities of dung beetles carrying animal dung balls down as much as 1 m and the increased flow of sugars from healthy plants to support microbial communities will also increase SOC. Conventional soil science, which does not generally assess capacity of new biologically created soils, considers 2.5 t of carbon sequestered per hectare each year to be significant, but the true soil carbon-sequestering potential may be much larger (Judy 2011).

Abe Collins: The Keyline Plow and Prairie Dogs

Abe Collins is a grazier in Vermont who runs dense herds and moves them once or more daily using electric fencing. He observes directly how this builds healthy soils. He says:

> As the cattle graze, they cream - eat the top third of the plants - and trample. The large amounts of plant material left behind form a mulch layer, which encourages fast regrowth, preserves soil moisture and helps build soil organic matter.

Collins has seen several inches of organic matter develop in his soils within a few years, and he has been communicating with several universities to study this soil building process. There is interest, but he has been asked to provide several million dollars to fund the research, a sum not readily available to an

TABLE 9.4
Greg Judy's Example of Soil Creation: Calculating Soil Organic Carbon (SOC)

If we assume a soil bulk density of 1.0 g/cm³:

1 g/cm³ × 108 cm²/ha × 7.5 cm
= 7.5 × 108 g soil/ha
= 750 t soil/ha × 0.045 SOC
= 33.75 t SOC/ha

Dividing by 4 years yields a SOC annual increase of 8.4 t/ha/year. For further explanation of these calculations, see "Calculating Soil Carbon Sequestration Potential."

innovative small farmer. That agribusiness interests are now doing the preponderance of soil research is a serious barrier to soil improvement, but beneficial to the corporate bottom line: poor soils will require ever more chemical inputs to produce crops.

Collins has used the Keyline system, an approach to subsoil contour plowing that can rapidly increase the depths at which soil biota are active (Yeomans Plow n.d.; Keyline Vermont n.d.). From a whole-system perspective, however, this technology is a proxy for the essential impact of burrowing mammals, many of which have been eliminated as a result of modern agricultural and rangeland practice.

The digging and churning activities of these animals enable the capture of far more rainwater on capped soils and begin ecorestoration in areas where it is difficult to bring livestock on a regular basis. Prairie dogs and moles were once numerous in North America when soils were much deeper. In Australia, there were once many more small marsupials such as potoroos and bettongs, which broke up the waxy surface caps of soils in their search for truffles. They may be critical to the hydrology of Australian land and to the reduction of wildfires in a warming climate (Maser et al. 2008).

At Janos Preserve in the Mexican Chihuahuan Desert, there are thousands of prairie dogs and very few mesquite or juniper trees. Healthy stands of grass predominate and some grazing of livestock and buffalo occurs:

> The black-tailed prairie dogs are vital to this grassland system as ecosystem engineers that create habitats for other plants and animals through their burrowing activities. They are herbivores and also serve as prey for other animals creating an important link in the food chain. The threatened status of these prairie dogs causes a cascade of effects throughout the grassland ecosystem (Nature Conservancy 2010).

In Texas, many millions of dollars are spent to control mesquite and to eradicate prairie dogs as enemies of the grass. Prairie dogs at Janos chew mesquite roots, constantly impeding their growth. They want a better view of the landscape looking for predators like hawks and coyotes. Prairie dogs may be critical to the return of grasslands in relatively dry and brittle landscapes. A single prairie dog town in Texas was the size of Maine and contained an estimated 400 million animals. That was when grasslands were much healthier than now. One hundred forty species of wildlife benefit directly from prairie dog towns, including large grazing animals, predators, small birds, mammals, reptiles, and insects (Williams 2008). Prairie dog dens are dug as deep as 3–4 m, making pathways easier for all the other soil biota from worms and beetles to fungal mycelia seeking minerals (Outwater 1996).

Pasture Cropping: A Regenerative Solution from Down Under

Since the late 1990s, Australian farmer Colin Seis has been successfully planting a cereal crop into perennial pasture on his sheep farm during the dormant season using no-till drilling, a method that uses a drill to sow seeds instead of the traditional plow. He calls it pasture cropping and he gains two crops this way from the land—a cereal crop for food or forage and wool or lamb meat from his pastures.

Today, over 2000 farms practice pasture cropping across Australia and many more worldwide. The practice is spreading for several reasons: high crop yields; sustained high pasture and animal production from cropped land; increased fodder for livestock; high rates of carbon sequestration; marked improvement in soil water-holding capacity; improved nutrient cycling; increased biodiversity and resilience, even in drought; greatly reduced input costs and risks; improved economic return from the vertical stacking of enterprises; and improved happiness on the farm.

Capture and Storage of Atmospheric Carbon in Grassland Soils

Factors in Soil Carbon Storage

SOC compounds, constituting approximately 60% of SOM, have beneficial effects on the chemical, physical, and biological functions of soil quality (Bardgett 2005). SOC increases water-holding capacity and contributes to soil structural stability. Organic matter increases adsorption of nutrients, cations, and trace elements that are of importance to plant growth; prevents nutrient leaching; and includes the organic acids that make minerals available to plants. It buffers soil from strong changes in pH. Consequently, it is widely accepted that the C content of soil is a major factor in overall soil health, plant production, the health of water catchments, and, central to this discussion, a sink for atmospheric C to offset climate change (Charman and Murphy 2000; Lal 2008).

The manner in which land is used and managed affects the soil's ability to sequester and retain organic C. Practices that increase plant productivity and C inputs to the soil and those that decrease soil exposure to erosion and exposure to sunlight allow higher levels of C to accumulate in soil (Parton et al. 1987).

Soil carbon, whether roots, associated microbes, or SOM, is sensitive to climatic conditions such as rainfall and temperature. Temperate grasslands store more carbon than tropical grasslands, although all grasslands show improvement with managed grazing. There will be benefits for all such habitats, but some areas will be more effective than others at storing carbon. When tropical savanna soils are compared to temperate grassland soils, the cooler soils are found to have much higher SOM and higher soil carbon to plant biomass carbon ratios and store carbon for longer than warmer grasslands (Baldock and Bruce 2012). High nighttime respiration and decomposition is thought to be the cause of inefficient carbon storage in warmer soils.

In rangeland ecosystems, maintaining normal soil function and ecosystem health is only possible if adequate plant and litter cover is present to provide protection from soil loss and allow soil microorganisms to perform optimally (Thurow 1991; Rietkerk et al. 2000; Bardgett 2005). Plant and litter cover enhances infiltration, buffers soil temperatures, and decreases evaporation so that soil moisture is retained longer after each precipitation event. This enhances soil microbial activity, promotes soil aggregate stability, sustains nutrient availability and plant nutrient status, improves plant growing conditions, and results in incorporation of more organic matter into the soil.

The amount of bare ground is a good indicator of soil function and erosion risk (Thurow 1991). Bare ground is not protected from the sun and gets much hotter than covered soil, decreasing microbial activity, accelerating loss of organic matter, and increasing erosion risk if there is insufficient cover to dissipate the energy of raindrops before they strike the soil (Blackburn 1975; Blackburn et al. 1986). Elevated soil temperature and soil loss have a direct negative effect on infiltration rates, water evaporation from soil, nutrient retention, and biological functions that contribute to ecosystem function (Wright and Bailey 1982; Neary et al. 1999).

Mycorrhizal fungi play an important part in facilitating sequestration of soil C. They access the mass of soil volume and increase availability of water and nutrients, thereby increasing plant productivity (Bardgett 2005). Consequently, management of the soil to achieve C sequestration goals requires carefully managing plants to reach their full potential.

Carbon Measurements in Soils

Lal (2004) estimates that the total carbon in the soil measured to a depth of 1 m is 2500 gigatons (Gt), comprising 1550 Gt of soil organic carbon (SOC) and 950 Gt of soil inorganic carbon (SIC). (i.e., largely limestone concretions in dry soils). Hiederer and Kochy (2011) estimate global SOC at 2469 Gt (Figures 9.10 and 9.11).

In estimating SOC, it is important to note the depth of the soil samples because reports of SOC measurements vary considerably, as do estimates of carbon-storing potential, based on depth. In many studies, soils are only measured to depths of 30 cm, which underestimates SOC capacity by a large margin. For example,

> The magnitude of SOC is estimated at 684 to 724 Pg [1 Petagram = 1 Gt] in the 0- to 30-cm layer, 1468 to 1548 Pg in the 0- to 100-cm layer and 2376 to 2456 Pg in the 0- to 200-cm layer. (Lal 2001, citing Batjes 1996)

The topsoil layer (0–10 or 0–30 cm) SOC stock comes to 967.3 Gt C and the subsoil layer to 1502.2 Gt C.

The subsoil layer thus contains about 1.5 times the amount of OC of the topsoil (Hiederer and Kochy 2011).

It is crucial to note that in our measurements we may not have explored soils nearly deep enough to have an adequate understanding of the extent of carbon soil sequestration possibilities. As much as

FIGURE 9.10 Reversal of desertification in Mexico: Las Pilas Ranch in Coahuila, Mexico. Average rainfall 350–450 mm. The arrows mark the same point on the horizon. Over a 25-year period, from 1978 to 2003, the barren landscape was completely revived (although the first picture is from 1963, the restoration with holistic management did not start until 1978). During the restoration period, the livestock population was doubled and grazing was done according to a plan that paid close attention to grass health. Although there appears to be more water in the left photo, the man-made pond shown is merely runoff captured by a constructed dirt dam. The dry, sandy land surrounding it retains no water. The restored terrain in the photo on the right is estimated to hold six times as much water as the depleted terrain on the left, but now the water is held in the soil and vegetative matter in a state referred to as *green water*. Previously, a 1 in. rainfall would fill the trough. Today, even a 6 in. rainfall is all absorbed (as it should be). The trough has been overgrown and is no longer needed to water the animals, as formerly dried-up springs in the vicinity have begun flowing year-round once again. (Photos by Guillermo Osuna.)

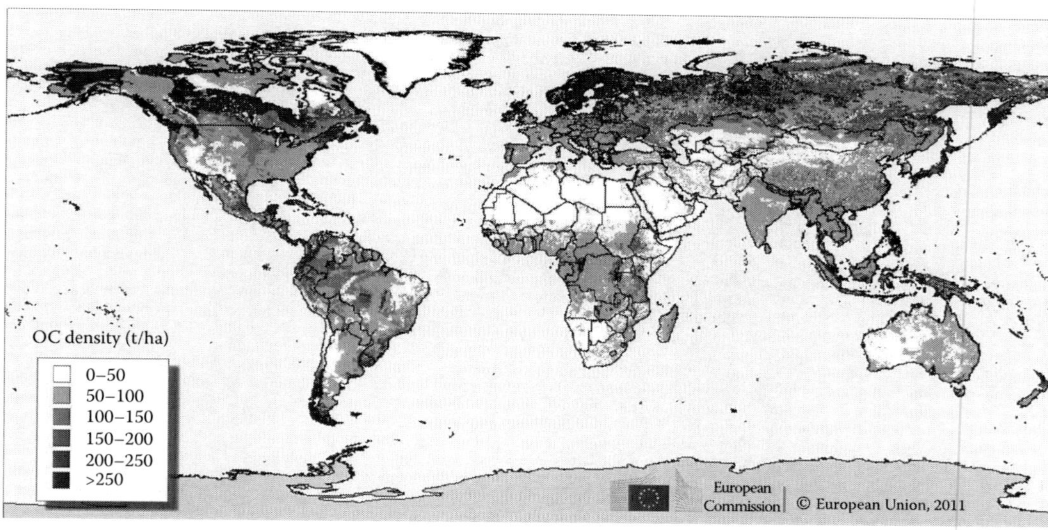

FIGURE 9.11 SOC density. (From Hiederer, R. and Köchy, M., *Global Soil Organic Carbon Estimates and the Harmonized World Soil Database*, EUR 25225 EN, Publications Office of the European Union, 2011.)

three-quarters of the mass of plants is below ground. Roots grow to well over 2 meters in depth, with interesting and complex interrelationships among different life forms—grasses, forbs, and shrubs (e.g., Weaver 1950; Richards and Caldwell 1987). Thus, assumptions about the ability of soil to store carbon may vary by greater than threefold depending on soil depths sampled in a study, especially considering that deeper stores of carbon are more stable and removed from the cycling carbon pool for hundreds or thousands of years. These differences take on even greater significance in restoring degraded and desertified rangelands over a period of years, since deep-rooted grassland plants may store carbon to depths of greater than 2.5 meters (Nippert et al. 2012; Brady 2002) to almost 5 meters in depth (USDA 2008) (Figure 9.12).

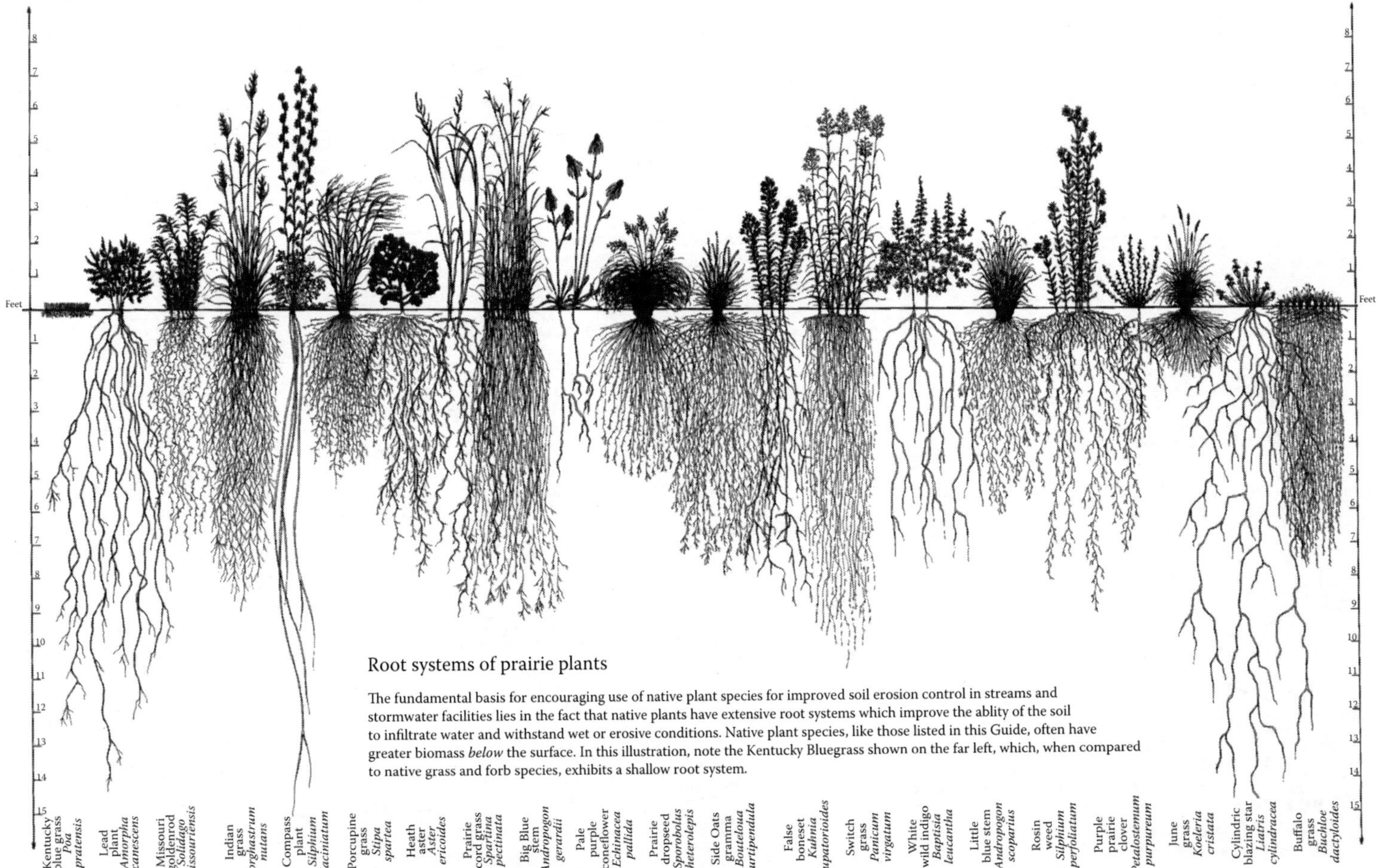

FIGURE 9.12 Illustrating the deep root systems of many prairie plants; they create porous soils and interact in complex ways with soil biota, including creating long-lived stores of carbon in complex biomolecules. (From Heidi Natura and Conservation Research Institute)

Case for Biological Soils

Another conventional assumption that is open to question leads to a potentially dramatic underestimation of SOC storage capacity in assessing sequestration potential with respect to global warming. That is, that the only soils available for restoration are those that already exist; any additional soil will be of geological origin—which accumulates only slowly, over thousands of years.

Some investigators assume that soils have a limited carbon sequestration capacity. Follett et al. (2001) assert that "turnover time of 25 years is necessary for achieving a practical upper limit for C sequestration in soil." Jones and Donnelly (2004) estimate that "… time to saturation range from 10 year (Janzen et al. 1998) to 100 year, but other models of SOC dynamics predict that soil C stocks can, in theory, be increased without limit" (Jones and Donnelly 2004). Upper-limit estimates, however, effectively remove new soil creation from the assessment and thereby underestimate soil sequestration capacity by an unknown but potentially significant magnitude.

Thus, soil creation of biological origin, which can be rapid, remains unaccounted for in these analyses. Christine Jones notes that

> the rates of soil formation provided in the scientific literature usually refer to the weathering of parent material and the differentiation of soil profiles. These are extremely slow processes, sometimes taking thousands of years.
>
> Topsoil formation is a separate process to rock weathering and can occur quite rapidly under appropriate conditions. In fact, soil building occurs naturally in most terrestrial habitats unless reversed by inappropriate human activities… (Jones 2002)

Jones cites examples of rapid soil accumulation:

> The late P.A. Yeomans, developer of the Keyline system of land management,… was able to produce 10 cm of friable black soil within three years, on what was previously bare weathered red shale on his North Richmond farm (Hill 2002).
>
> Bennett (1939) calculated a rate of topsoil formation of just over 11 t/ha/yr for soils in which organic material was intermixed into surface layers. In situations where plant root mass is high, rates of topsoil formation of 15–20 t/ha have been indicated (Brady 1984). Healthy groundcover, high root biomass and high levels of associated microbial activity, are fundamental to the success of any technique for building new topsoil.
>
> … If the land management is appropriate, evidence of new topsoil formation can be seen within 12 months, with quite dramatic effects often observed within three years (Jones 2002).

Historical Data on Loss of Soil Organic Matter

The extent of historic loss of SOC since the advent of agriculture ranges from 40 to 537 Gt, the wide variations being "attributed to differences in methods used, variability in the available data, and the baseline used as a reference point" (Lal et al. 2007). On the high end, Buringh (1984) estimated that the amount of organic carbon in soils—carbon in complex molecules formed by living things—is approximately 1477 Gt. This is roughly 537 Gt less than total organic carbon in soils in prehistoric times "or 27 per cent of the amount present prior to the spread of civilization in the last two millennia" (Buringh 1984).

Rattan Lal's estimates are lower. He states that the "maximum sink capacity of the world soils, equal to the historic carbon loss (78 +/− 12 Pg C), is substantial and must be included among the possible strategies of stabilizing atmospheric CO_2 at a desired level" (Lal 2008). Note that Lal's assertion of limited carbon sequestration capacity seems to assume that there is no biological accumulation of soils. Elsewhere, Lal estimates that there is considerably more soil carbon lost (456 +/− 5 Gt), but does not seem to consider the major part of it replaceable (Lal et al. 2004).

Discrepancies in estimates of lost SOC are of concern when proposing a strategy for sequestering over 200 Gt of carbon in soils and will be addressed in "Calculating Soil Carbon Sequestration Potential."

If we accept Buringh's 537 Gt number for the historic loss of carbon from soils, it is the equivalent of 218 ppm that was once safely stored in the ground instead of in the atmosphere and is roughly twice the excess carbon that we have injected into the atmosphere since 1750 (half of which was absorbed by the ocean and other carbon sinks). We need only put 224 Gt (112 ppm) back into the ground, even though a percentage of the excess did not come from soil, but from burning of fossil fuels. To emphasize, *replacing just half of the soil carbon we have lost in the past ten thousand years has the realistic potential for reducing atmospheric carbon to a preindustrial 280 ppm*. Note that in so doing, we would also sequester *all* past and current emissions from fossil fuels.

Changes in land use change soil carbon:

> Soil carbon stocks decline after land use changes from pasture to plantation (−10%), native forest to plantation (−13%), native forest to crop (−42%), and pasture to crop (−59%). (Guo and Gifford 2002)

Practices such as HPG, MIG, and pasture cropping reverse the most significant of these land-use changes: they can return degraded croplands to pasture, and even more significantly, semidesert to pasture, thereby storing significant quantities of carbon in the ground, while restoring communities, economies, and production.

Calculating Soil Carbon Sequestration Potential

Soils are created through geological and biological processes. Geological soil generation is primarily the weathering of rock that removes carbon dioxide from the atmosphere by wearing down rocks and limestones over millennia. Biological soil generation is the combining of plant litter and animal dung and the action of myriad soil life forms to create soil. Under proper management, biological processes can add an inch or more of soil per year and turn 2.5 t or more of atmospheric carbon into stable complex biological compounds per hectare, compounds that stay in the ground for centuries or millennia.

A 5-year study of grasslands planted with switchgrass for use as biofuels demonstrated that soil carbon could increase by more than 1.7 t/acre (4.25 t/ha) (Liebig et al. 2008). This was on conventional farmlands *without* the added benefit of grazers and the diverse assortment of animals, insects, and fungi that capture significant additional quantities of carbon dioxide.

The Rodale Institute in Pennsylvania has conducted the longest running US study comparing organic with conventional farming. Its results are as follows: "organic systems have shown an increase of almost 30 percent in soil carbon over 27 years," and "during the 1990s, results from the Compost Utilization Trial (CUT) at Rodale Institute—a 10 year study comparing the use of composts, manures and synthetic chemical fertilizer—show that the use of composted manure with crop rotations in organic systems can result in carbon sequestration of up to 2000 lb/acre/year. By contrast, fields under standard tillage relying on chemical fertilizers lost almost 750 lb of carbon per hectare per year. Storing—or sequestering—up to 2000 lb/acre/year of carbon means that more than 7000 lb of carbon dioxide are taken from the air and trapped in that field soil" (LaSalle et al. 2008). The Rodale study shows sequestration of up to 1 t of carbon per acre, without employing the relationship between grazing animals and grasslands.

A third example shows that perennial pasture on sandy soils in Western Australia can sequester 5–10 t CO_2 per hectare per year, or approximately 0.5–1 t C/acre (Wiley et al. 2013). The site has mixed grazing with nongrazing and there was below average rain during the trial period.

With respect to soil characteristics (Jones 2002),

> Mineral soil has a higher bulk density (is more compact) than living soil, and is far more easily eroded. Soil loss figures usually assume an average bulk density (weight per unit volume) of around 1.4 g/cm^3. If one millimetre of soil is eroded that represents about 14 t/ha soil loss.
>
> When new topsoil is forming, it will have better structure and will contain more air and more pore spaces than degraded soil, so the bulk density will be less. That is, a given volume of new topsoil will weigh less than an equal volume of non-living mineral soil.

> The bulk density of healthy topsoil may be as low as 0.5 g/cm^3. In practical terms, a one millimetre increase in the height of new soil would equate to the formation of around 5 to 10 t/ha of organically enriched topsoil.

For calculation purposes, using a midpoint of soil densities, we can reasonably assume bulk soil density in healthy, newly created biological soils of 1 g/cm^3.

Since there are 1×10^8 cm^2/ha, to a depth of 1 cm, we have 1×10^8 g of soil per hectare, or 100 t/ha. SOM will vary according the characteristics of the soil. SOC is generally estimated at 58% of SOM (Lal 2001); Jones gives a range of 50%–62%. If we assume that 1% of total topsoil weight is composed of SOM (Troeh 2005), then the weight of SOC will be as follows (Tables 9.5 and 9.6):

To calculate the quantity of carbon captured from the atmosphere and stored in organic molecules in the soil in terms equivalent to those used by climate advocates, the equivalent formula is expressed in Table 9.6.

That is, at a conventional soil measurement depth of 30 cm, we would have a total soil carbon weight of 17.4 Gt C captured per billion hectares or, in terms of atmospheric carbon dioxide, a value of 8.7 ppm. How long it would take to store 8.7 ppm depends on how quickly we are able to restore degraded soils and create new soils, and that will vary across climate and soil types. Keep in mind that there are an estimated 5.25 billion hectares of grasslands worldwide (Suttie et al. 2005) that could be candidates for restoration.

These calculations may underestimate soil carbon storage potentials because they do not account for the greater and more stable carbon accumulation in deeper soils, up to 5 m deep, created by interactions of plant roots, mycorrhizal fungi, bacteria, small mammals, and insects as the health of the land returns.

Another example is a soil that, at 100 t/ha, with 2% increase SOM of which 58% is SOC, at a density of 1.2 g/cm^3, 60 cm deep over 3 billion hectares of grasslands, would yield storage of approximately 250 GtC (Table 9.7):

TABLE 9.5

Calculating Weight of Soil Organic Carbon per Centimeter of Soil Depth

Total Weight of Soil	×	% Soil Organic Matter (SOM)	×	% Soil Organic Carbon (SOC)	=	Weight of SOC per cm of Soil Depth
100 t/ha	×	0.01 (1%)	×	0.58 (58%)	=	0.58 t C/ha/cm

TABLE 9.6

Calculating Carbon Captured in Soils from Atmospheric Carbon Dioxide, in Gigatons per Billion Hectares, for the First 30 cm of Soil Depth

Total Weight of Soil	SOM as a Percent of Total Soil Weight	×	SOC58% of SOM in This Example	×	Soil Density in g/cm^3	×	Depth in cm	×	Billions of Hectares	=	Total Weight of Carbon in Soil in Billions of Tons (Gigatons, Gt) per Billion Hectares
100 t/ha	0.01 (1%)	×	0.58 (58%)	×	1	×	30	×	1	=	17.4 Gt C

TABLE 9.7

Example of Calculating Atmospheric Carbon Sequestered in Soils with 2% Soil Organic Matter to a Depth of 60 cm Over 3 Billion Hectares

Total Weight of Soil	SOM as a Percent of Total Soil Weight	×	SOC58% of SOM in this Example	×	Soil Density in g/cm^3	×	Depth in cm	×	Billions of Hectares	=	Total Weight of Carbon in Soil in Billions of Tons (Gigatons, Gt) per Billion Hectares
100 t/ha	0.02 (2%)	×	0.58 (58%)	×	1.2	×	60	×	3	=	~250 Gt C

Note that sequestration of 250 Gt C is equivalent to 125 ppm, or approximately 62.5 times the carbon currently added to the atmosphere annually.

5-Billion-Hectare Potential

If we were to capture 2.5 t of carbon per hectare per year on roughly 5 billion hectares of grasslands worldwide, we would remove 12 Gt of C from the atmosphere per year. If gross soil sequestration is approximately 6 ppm/year, after subtracting current net annual carbon emissions of 2.5 ppm/year net sequestration would be 3.5 ppm/year.

In principle, this returns us to preindustrial atmospheric CO_2 levels in less than 40 years. How realistic these numbers are is unknown, but it is apparent that the potential may indeed be promising.

Ruminant Methane

Methane from grazing ruminant livestock has been estimated to be approximately 22% of anthropogenic methane emissions in 2005, or 67.6 Mt out of a total 313 Mt (Hoglund-Isaksson 2012). Historical emissions from wild herds before European settlement in the United States are estimated at 86% of current emissions (Hristov 2012). Preindustrial atmospheric methane did not exceed 788 parts per billion by volume (ppbv) for 650,000 years and was 600 ppbv or less for most of that period (Spahni et al. 2005), so it seems that ruminant methane emissions were balanced by natural methane decomposition processes.

Livestock's Long Shadow (UN FAO 2006) points out that

The respiration of livestock makes up only a very small part of the net release of carbon that can be attributed to the livestock sector. Much more is released indirectly by other channels including the following:

- Burning fossil fuel to produce mineral fertilizers used in feed production
- Methane release from the breakdown of fertilizers and from animal manure
- Land-use changes for feed production and for grazing
- Land degradation
- Fossil fuel use during feed and animal production
- Fossil fuel use in production and transport of processed and refrigerated animal products

Except for use of fossil fuels in final product processing and transport, HPG and MIG incur none of those burdens and in fact reverse most of them. Furthermore, there is evidence that better animal management will reduce methane production (DeRamus et al. 2003):

> Projected CH_4 annual emissions in cows reflect a 22% reduction [resulting] from BMP [Best Management Practices] when compared with continuous grazing in this study. With the BMP application of MIG [Management-Intensive Grazing], less methane was produced per kilogram of beef gain.

Microorganisms that break down methane (methanotrophs) live in healthy soils but they are inhibited by artificial nitrogen fertilizers and pesticides.

Carbon Calculation Summary

Our soil carbon sequestration calculations are based on the following assumptions, which may be speculative at the current time:

- Biological restoration of crop, pasture, and rangeland soils will sequester carbon deep in the ground every year worldwide.
- The average carbon sequestration over a wide variety of soils and climate worldwide may approximate 2.5 t/ha.

We assume that it will be strategically, politically, and economically possible to mount a massive effort to restore billions of hectares of grasslands in a relatively short period of time. Whether global political will can overcome longstanding preconceptions about land management, even when conventional approaches carry great risk, remains to be seen. Current emissions-reduction efforts have proved fruitless to date and to hold limited promise, and grassland restoration has the potential to contribute significantly to reversing anthropogenic climate change. There may be no better way to store carbon, at so little expense and with so many potential benefits.

Reconciling Perspectives of Advocates for Climate Action and Practitioners of Planned Grazing

Many climate researchers discount use of grazing animals, particularly livestock, to sequester carbon in soils. There are several reasons for this: misunderstanding of the capacity of soils to sequester carbon, lack of awareness of the extent of carbon emissions from soil loss through human activity over millennia (Ruddiman 2003), concerns about the damage inflicted by poorly managed livestock (United Nations Food and Agricultural Organization 2006), and the almost exclusive focus on emissions reductions.

Given the diversity of climatic, edaphic (soil-related), and topographic features that structure plant communities on landscapes and the influence that has on fixing carbon, no one knows for sure how much carbon can be captured through managing cropping, pastures, and rangelands, but people are actively working to determine those amounts (Donovan n.d.). Regardless of the exact amounts of carbon that can be fixed in soil through managed grazing, changing farming and grazing practices is valuable ecologically, economically, and socially. It is important to build the health of soil, the diversity of plants and animals, and the welfare of the ecological communities we inhabit and upon which we depend.

Soil carbon measurements in holistically managed lands have been limited because what is important to ranch managers is the visible and functional health of the soil that they restore and maintain, and there are few direct measurement of soil carbon changes. Ranchers want to know the following:

- Is the plant life vigorous and diverse, and are perennial grasses reappearing with roots that penetrate up to 5 m deep into the ground (USDA 2008)?
- Is there good soil moisture content, and do streams and ponds have water even during a drought?
- Are the dung beetles, earthworms, and other soil organisms present and functioning in a productive manner?
- Is the diversity of wildlife returning?
- Is the soil turning black (the color of carbon), soft, and sweet from biotic activity?
- Is there less need for winter hay because healthy land in combination with planning provides more options for dormant season sustenance of the livestock?
- Are the animals healthy, less prone to disease and parasites?
- Are people more productive with less work, and are our lives happier and more secure?

Grazing animals, including goats, sheep, and cattle, are "essential" to healthy lands and climate restoration. They are part of the solution when raised in accordance with nature's rules, not part of the problem the way they are when raised on factory farms or grazed continuously on large tracts of land.

In nature, everything is connected to everything else. Global warming is more than a carbon or other GHG problem; it is a problem of how humans are living on Earth—and we can learn to live well differently. There are no simplistic, linear, reductionist solutions, only complex dynamic systems ever transforming. Only by understanding wholes and transformation with the systems we inhabit can we address the loss of ecological integrity in systems upon which we depend for life.

Obstacles to reestablishing the evolutionary grassland–grazer relationship for long-term sequestration of carbon in soils and restoring atmospheric carbon dioxide to preindustrial levels include assumptions based on traditional livestock mismanagement, such as that

- Grazing animals chronically overgraze rangelands and destroy soils and plants that must be *rested* to be restored
- Climate action should focus on reducing emissions; livestock are destructive and contribute to emissions
- Soils are a limited carbon sink, and new biologically generated soils are not part of the equation
- Soil sequestration of carbon is only significant in the first 60 cm
- "The turnover time of 25 years is the time for achieving a practical upper limit for C sequestration in soil" (Follett et al. 2001)

They can be addressed readily by dedicated measurements of carbon in soil roots, mycorrhizae, and bacteria and as soil humin through time as a function of management. Unfortunately, the needed data are still scarce. Yet there is good evidence that proper management can restore biomass at much faster rates than have been assumed.

The number of scientific publications and observations cited in this chapter suggest an alternate grazing/soil carbon paradigm is credible. We maintain that the current paradigm is preventing the implementation of management that shows widespread evidence of being able to more effectively restore ecosystem function on grazing lands. Furthermore, such ecosystem restoration will positively affect the fundamental biogeodynamic of carbon and water cycles and mitigate, quite possibly significantly, the extremely destructive anthropogenically generated atmospheric carbon burden.

While there are other potential methods for restoring carbon in soils, as outlined in this book, none have so many associated benefits, not the least of which is that managed grazing, by producing marketable animal and other products, generates income and food for local communities worldwide, requires minimal short-term investment after which it is self-supporting, needs no new or complex technology (little beyond optional portable electric fencing), and is sustainable in the long-term. Moreover, it is culturally relevant throughout virtually all grassland contexts (Figure 9.13).

FIGURE 9.13 Young woman teaching HPG in an African village.

Conclusions

Since the advent of agriculture, humanity has been moving carbon out of the soils and into the atmosphere. We plow and leave agricultural land bare for much of the year, exposing soil life and humus to sunlight, desiccation, and oxidation. On bare ground, solar energy necessary for plants to enable humification is not harvested. Today we make matters worse by using nitrogen fertilizers, seriously inhibiting the flow of energy to fungi that make stable, carbon-sequestering molecules, and by stimulating growth in bacterial populations destructive to soil fungi.

In some circumstances, unnecessary use of fire also exposes bare ground, destroys organic soil matter, and emits massive amounts of carbon into the atmosphere. And of course, we emit vast amounts of geological carbon from burning fossil fuels, amounts unprecedented since the beginning of life on earth (DePaolo 2012). While these practices seemed like good ideas at the time, now we know that they are leading us to soil and climate catastrophe.

Today, we are in the early stages of understanding the extent to which soils can sequester atmospheric carbon, and it is evident that far more information is needed for a comprehensive understanding of the conditions that maximize storage of soil carbon. But, given the exigencies of climate change, it is necessary to act on removing carbon from the atmosphere immediately.

Therefore, we must rely on qualitative empirical evidence, which is unequivocal with respect to ecorestoration of grazing lands, as the basis for moving forward. Furthermore, outcomes of improper management of livestock and inadequate soils models have led us to significantly underestimate the potential of soil carbon capture and storage for reversing climate change. Restoring the evolutionary relationship between grazing animals and their grassland habitat is an essential part of the equation.

As strange as it may seem from an experimental perspective, we know when carbon is in the ground and when it is not *just by looking*. A green biodiverse landscape is brimming with carbon dioxide transformed into the molecules of life, in trees, grasses, flowers, and many other life forms. When the ecosystem is intact, much of this carbon is stored deep in the soils. A bare landscape is doing none of this, and when left alone in a brittle (seasonal rainfall) environment, it only gets worse.

To the extent the global store of carbon is in the ground, it is not in the atmosphere, and *vice versa*. Grazing animals can be used to create healthy soil and plants as much as healthy soil and plants create grazing animals. They are part of a complex system that includes microbes, insects, fungi, grasses and forbs, birds, and mammals, large and small. They all feed each other. They created the deep soil that we humans continue to diminish and destroy. We can and need to restore them using the most effective management we are capable of.

We have lost the huge wild herds that used to roam the plains, steppes, and savannas of the world, but livestock are herbivores too, and their function in ecosystems is the same as their wild relatives, *when they graze the way nature designed it.*

We can create those soils again. Restoring the hectares we have destroyed can sequester the carbon that we have lost, potentially all of it. When we aim for the health and productivity of our life support systems, as wholes, the numbers will take care of themselves.

Obstacles to the extraordinary effort that will be necessary for successful soil carbon sequestration will include changing the cultural mindset about the role of herbivores and soils in reversing global warming, the learning curve that new practitioners will have to master, the current shortage of expert teachers, gaining access to land, adequate numbers of animals, and equitable distribution of benefits.

Yet, despite daunting challenges and the uncertainties regarding precise quantities of attainable soil carbon levels, there are so many ecological, social, and economic benefits that accrue from proper grassland management that we are well advised to pursue such ecorestoration with all due dispatch:

- Planned grazing can restore soils, and HPG has been the most broadly successful and widely implemented planned grazing approach to date.
- Soils are a prodigious carbon sink that are inadequately assessed, and building biological soils is an important mechanism for storing carbon.

- Healthy soils are complex collections of interdependent life forms, and as a synergistic system, soil activity may become seriously impaired when any of its elements are compromised or destroyed, as currently occurs worldwide with chemical agriculture, mismanagement of grazing animals, and overrest of grasslands.
- Effective ecorestoration, of which HPG, MIG, and pasture cropping are all examples, can restore geophysical cycles in grasslands and savannas and has the potential to stabilize carbon in the atmosphere at around 280 ppm in time to significantly mitigate the damaging effects of global warming.

ACKNOWLEDGMENTS

The authors would like to express their gratitude for the suggestions and encouragement of several reviewers: Andrea Malmberg, Tony Malmberg, Jim Howell, Allan Savory, Don Nelson, Karl Thidemann, Jane Hammer, and Peter Donovan.

REFERENCES

Allard, V., Soussana, J.-F., Falcimagne, R., Berbigier, P., Bonnefond, J.M., Ceschia, E., D'hour, P. et al., 2007. The role of grazing management for the net biome productivity and greenhouse gas budget (CO_2, N_2O and CH_4) of semi-natural grassland. *Agriculture Ecosystems & Environment* 121, 47–58.

American Meteorological Society, 2013. *Geoengineering the Climate System: A Policy Statement of the American Meteorological Society*, Boston, MA, http://www.ametsoc.org/policy/2013geoengineeringclimate_amsstatement.html.

Archer, D. et al., May 2009. Atmospheric lifetime of fossil fuel carbon dioxide. *Annual Review of Earth and Planetary Sciences* 37, 117–134.

Bailey, D.W. and Provenza, F.D., 2008. Mechanisms determining large-herbivore distribution. In: H.T.T. Prins and F. van Langevelde (eds.), *Resource Ecology, Spatial and Temporal Dynamics of Foraging*. Springer, Dordrecht, the Netherlands, pp. 7–28.

Baldock, J.A. and Broos, K., 2012. Soil organic matter. In: P.M. Huang, Y. Li, and M.E. Sumner (eds.), *Handbook of Soil Sciences: Properties and Processes*, 2nd edn., CRC Press, Boca Raton, FL, pp. 11-1–11-52.

Bardgett, R.D., 2005. *The Biology of Soil: A Community and Ecosystem Approach*. Oxford University Press, New York.

Barnes, M.K., Norton, B.E., Maeno, M., and Malechek, J.C., 2008. Paddock size and stocking density affect spatial heterogeneity of grazing. *Rangeland Ecology and Management* 61, 380–388.

Bartlett, A.A., 1978. Forgotten fundamentals of the energy crisis. *American Journal of Physics* 46, 876–888.

Baskin, Y., 2005. *Under Ground: How Creatures of Mud and Dirt Shape Our World*, Island Press, Washington, DC.

Batjes, N.H. 1996. Total carbon and nitrogen in soils of the world. *European Journal of Soil Science* 47, 151–163.

Bennett, H.H., 1939. *Soil Conservation*, McGraw-Hill, New York (cited by Jones 2002).

Beschta, R.L., Donahue, D.L., DellaSala, D.A., Rhodes, J.J., Karr, J.R., O'Brien, M.H., Fleischner, T.L., and Deacon Williams, C., 2012. Adapting to climate change on western public lands: Addressing the ecological effects of domestic, wild, and feral ungulates. *Environmental Management*, 51(2), 474–491.

Blackburn, W.H., 1975. Factors influencing infiltration and sediment production of semiarid rangelands in Nevada. *Water Resources Research* 11, 929–937.

Blackburn, W.H., Thurow, T.L., and Taylor, C.A., 1986. Soil erosion on rangeland. In: *Proceedings Use of Cover, Soils and Weather Data in Rangeland Monitoring Symposium, Society for Range Management*, Denver, CO, pp. 31–39.

Brady, N.C., 1984. *The Nature and Properties of Soils*, 9th edn., Macmillan (cited by Jones 2002); also, with Weil, R.R., 2002. 13th edn., Prentice-Hall.

Briske, D.D. et al., 2008. Rotational grazing on rangelands: Reconciliation of perception and experimental evidence. *Rangeland Ecology & Management* 61(1), 3–17.

Brysse, K. et al., February 2013. Climate change prediction: Erring on the side of least drama? *Global Environmental Change* 23(1), 327–333. Summarized in "Climate Scientists Erring on the Side of Least Drama," *Skeptical Science*, January 30, 2013, http://www.skepticalscience.com/climate-scientists-esld.html.

Buringh, P., 1984. Organic carbon in soils of the world. In: G.M. Woodwell (ed.), *The Role of Terrestrial Vegetation in the Global Carbon Cycle: Measurement by Remote Sensing*, John Wiley & Sons Ltd., http://globalecology.stanford.edu/SCOPE/SCOPE_23/SCOPE_23_3.1_chapter3_91-109.pdf.

Carey, J., November 2012. Global warming: Faster than expected? *Scientific American*, 307, 50–55.

Carpenter, S.R. and Biggs, R., 2010. Freshwaters: Managing across scales in space and time. In: F.S. Chapin, G.P. Kofinas, and C. Folke (eds.), *Principles of Ecosystem Stewardship*, Springer Science + Business Media, New York, pp. 197–220.

Charman, P.E.V. and Murphy, B.W., 2000. *Soils, their Properties and Management*, 2nd edn, Oxford University Press, Melbourne, Victoria, Australia.

Clemmensen, K.E. et al., March 2013. Roots and associated fungi drive long-term carbon sequestration in boreal forest. *Science* 339(29), 1615–1618.

Comis, D., September 2002. Glomalin: Hiding place for a third of the world's stored soil carbon. *Agricultural Research* (United States Department of Agriculture Agricultural Research Service), 50, 4–7.

Conant, R.T., 2010. *Challenges and Opportunities for Carbon Sequestration in Grassland Systems: A Technical Report on Grassland Management and Climate Change Mitigation*, Plant Production and Protection Division, Food and Agriculture Organization of the United Nations (FAO), Rome, Italy.

Dagget, D. and Bean, T., 2005. *Gardeners of Eden: Rediscovering Our Importance to Nature*, Thatcher Charitable Trust/EcoResults! Press, Santa Barbara, CA.

de Vries, M. and de Boer, I.J.M., 2010. Comparing environmental impacts for livestock products: A review of life cycle assessments. *Livestock Science* 128, 1–11.

DePaolo, D., September 24, 2012. Lecture: Global carbon cycle change: A geological perspective on carbon and climate, Lecture at Stanford Woods Institute for the Environment, http://energyseminar.stanford.edu/node/461.

DeRamus, H.A., Clement, T.C., Giampola, D.D., and Dickison, P.C., 2003. Methane emissions of beef cattle on forages: Efficiency of grazing management systems. *Journal of Environmental Quality* 32, 269–277.

Derner, J.D., Boutton, T.W., and Briske, D.W., 2006. Grazing and ecosystem carbon storage in the North American Great Plains. *Plant and Soil* 280, 77–90.

Diaz-Solis, H., Grant, W.E., Kothmann, M.M., Teague, W.R., and Díaz-García, J.A., 2009. Adaptive management of stocking rates to reduce effects of drought on cow-calf production systems in semi-arid grazinglands. *Agricultural Systems* 100, 43–50.

Donovan, P., n.d., http://soilcarboncoalition.org/.

Dyer, M.I., Turner, C.L., and Seastedt, T.R., 1993. Herbivory and its consequences. *Ecological Applications* 3, 10–16.

Follett, R.F. et al., 2001. Potential of U.S. grazing lands to sequester soil carbon. In: R.F. Follett, J.M. Kimble, R. Lal (eds.), *The Potential of U.S. Grazing Lands to Sequester Carbon and Mitigate the Greenhouse Effect*, CRC Press, Boca Raton, FL.

Frank, D.A. and Groffman, P.A., 1998. Ungulate vs. landscape control of soil C and N processes in grasslands of Yellowstone National Park. *Ecology* 79, 2229–2241.

Frank, D.A. and McNaughton, S.J., 1993. Evidence for the promotion of aboveground grassland production by native large herbivores in Yellowstone National Park. *Oecologia* 96, 157–161.

Frank, D.A., McNaughton, S.J., and Tracy, B.F., 1998. The ecology of the Earth's grazing ecosystems. *BioScience* 48, 513–521.

Galloway, J.N., Townsend, A.R., Erisman, J.W., Bekunda, M., Cai, Z., Freney, J.R., Martinelli, L.A., Seitzinger, S.P., and Sutton, M.A., 2008. Transformation of the nitrogen cycle: Recent trends, questions, and potential solutions. *Science* 320, 889–892.

Garnett, T., 2010. Livestock and climate change. In: J. D'Silva and J. Webster (eds.), *The Meat Crisis: Developing More Sustainable Production and Consumption*, Earthscan, London, U.K., pp. 34–56.

Gerrish, J., 2004. *Management-Intensive Grazing: The Grassroots of Grass Farming*, Green Park Press, Ridgeland, MO.

Goreau, T.J., 1992. Control of atmospheric carbon dioxide. *Natural Resources Forum*, 2, 5–11.

Guo, L.B. and Gifford, R.M., April 2002. Soil carbon stocks and land use change: A meta analysis. *Global Change Biology* 8(4), 345–360, http://onlinelibrary.wiley.com/doi/10.1046/j.1354-1013.2002.00486.x/abstract;jsessionid=60D26918F786756A58B3365D3DE18B3A.d03t03.

Hamilton, E.W. and Frank, D.A., 2001. Can plants stimulate soil microbes and their own nutrient supply? Evidence from a grazing tolerant grass. *Ecology* 82, 2397–2402.

Hawken, P., Lovins, A., and Lovins, H., 1999. *Natural Capitalism: Creating the Next Industrial Revolution*, Little Brown and Co., New York.

Hiederer, R. and Köchy, M., 2011. *Global Soil Organic Carbon Estimates and the Harmonized World Soil Database,* EUR 25225 EN, Publications Office of the European Union.

Hill, S.B., April 6, 2002. 'Redesign' for soil, habitat and biodiversity conservation: Lessons from Ecological Agriculture and Social Ecology'. In: *Proceedings 'Sustaining Our Future: Through Healthy Soils, Habitats and Biological Diversity' Launch of the 'Healthy Soils Campaign'*, Nature Conservation Council of NSW, Sydney, New South Wales, Australia (cited by Jones 2002).

Hoglund-Isaksson, L., 2012. Global anthropogenic methane emissions 2005–2030: Technical mitigation potentials and costs. *Atmospheric Chemistry and Physics* 12, 9079–9096, http://www.atmos-chem-phys.net/12/9079/2012/acp-12-9079-2012.pdf.

Holland, E.A., Parton, W.J., Detling, J.K., and Coppock, D.L., 1992. Physiological responses of plant populations to herbivory and other consequences of ecosystem nutrient flow. *American Naturalist* 140, 685–706.

Holling, C.S. and Meffe, G.K., 1996. Command and control and the pathology of natural resource management. *Conservation Biology* 10, 328–337.

Howell, J., 2008. *For the Love of Land: Global Case Studies of Grazing in Nature's Image*, BookSurge Publishing, Charleston, South Carolina.

Hristov, A.N., April 2012. Historic, pre-European settlement, and present-day contribution of wild ruminants to enteric methane emissions in the United States. *Journal of Animal Science* 90, 1371–1375, http://www.journalofanimalscience.org/content/90/4/1371 (pay to view).

Hulbert, L.C., 1988. Causes of fire effects in tallgrass prairie. *Ecology* 69, 46–58.

Humphreys, D., June 2011. Smoke and mirrors: Some reflections on the science and politics of geoengineering. *The Journal of Environment Development* 20(2), 99–120, http://jed.sagepub.com/content/20/2/99.abstract.

Ikerd, J.E., 2005. *Sustainable Capitalism: A Matter of Common Sense*, Kumarian Press, Inc., Bloomfield, CT.

International Energy Agency, May 24, 2012. Global carbon-dioxide emissions increase by 1.0 Gt in 2011 to record high, http://www.iea.org/newsroomandevents/news/2012/may/name,27216,en.html.

Janzen, H.H., 2011. What place for livestock on a re-greening earth? *Animal Feed Science and Technology* 166–167, 783–796.

Janzen, H.H., Campbell, C.A., Izaurralde, R.C., Ellert, B.H., Juma, N., McGill, W.B., Zentner, R.P., 1998. Management effects on soil C storage on the Canadian prairies. *Soil and Tillage Research* 47, 181–195.

Jones, C., May 3, 2002a. Building new topsoil. In: *Stipa Native Grasses "Changing Landscapes" Forum*, Armidale.

Jones, C., 2002b, Recognise Relate Innovate, Department of Land and Water Conservation, New South Wales Government "Ecological and Technical Support for Landcare on Rangelands" Final Report, http://www.amazingcarbon.com/PDF/JONES-RecogniseRelateInnovate.pdf.

Jones, C., July 2008. Liquid carbon pathway unrecognised. *Australian Farm Journal,* http://amazingcarbon.com/PDF/x07afj_0128.ZIP.

Jones, C., September 2009a. Mycorrhizal fungi—Powerhouse of the soil. *Evergreen Farming Newsletter*, http://amazingcarbon.com/PDF/JONES-MycorrhizalFungiEVERGREEN(Sept09).pdf.

Jones, C., December 12, 2009b. *Inquiry into Soil Sequestration in Victoria*, Submission to Environment and Natural Resources Committee (Australia), Melbourne, Victoria, Australia, http://www.amazingcarbon.com/PDF/JONES-SoilSequestrationInquiry(17Dec09).pdf.

Jones, C., March 2011. *Carbon that Counts*, New England and North West Landcare Adventure, Guyra, New South Wales, Australia, http://soilcarboncoalition.org/files/JONES-Carbon-that-counts-20Mar11.pdf.

Jones, M.B. and Donnelly, A., 2004. Tansley review: Carbon sequestration in temperate grassland ecosystems and the influence of management, climate and elevated CO_2. *New Phytologist* 164, 423–439, http://onlinelibrary.wiley.com/doi/10.1111/j.1469-8137.2004.01201.x/pdf.

Judy, G., 2008. *Comeback Farms: Rejuvenating Soils, Pastures and Profits with Livestock Grazing Management*, Green Park Press.

Judy, G., February 2011. The healing effects of holistic high density grazing on land, livestock and people's lives, *12th Annual Virginia Biological Farming Conference*, http://www.youtube.com/watch?v=W6HGKSvjk5Q.

Kerr, R.A., 2010. Ocean acidification unprecedented, unsettling. *Science* 328, 1500–1501.

Keyline, Vermont, n.d., http://www.keylinevermont.com/Keyline_Subsoil_Plowing.html.

Khan, S.A., Mulvaney, R.L., Ellsworth, T.R., and Boast C.W., November 2007. The myth of nitrogen fertilization for soil carbon sequestration. *Journal of Environmental Quality* 36(6), 1821–1832.

Killham, K., 194. *Soil Ecology*, Cambridge University Press, Cambridge, U.K.

Knapp, A.K., 1985. Effect of fire and drought on the ecophysiology of Andropogon gerardii and Panicum virgatum in a tallgrass prairie. *Ecology* 66, 1309–1320.

Knight, J.C., Kothmann, M.M., Mathis, G.W., and Hinnant, R.T., 1990. Cow-calf production with alternative grazing systems. *Journal of Production Agriculture* 3, 407–414.

Kothmann, M.M., Mathis, G.W., and Waldrip, W.M.J., 1971. Cow-calf response to stocking rates and grazing systems on native range. *Journal of Range Management* 24, 100–105.

Kravcík, M., Pokorný, J., Kohutiar, J., Kovác, M., and Tóth, E., 2007. Water for the Recovery of the Climate—A New Water Paradigm, TypoPress, Kosice, Slovakia, Available at http://www.vodnaparadigma.sk/indexen.php?web =./home/homeen.html.

Kravcík, M. et al., 2009. Water, Vegetation and Climate Change—Košice Civic Protocol (addressed to COP15—Copenhagen), Available at http://www.ourclimate.eu/ourclimate/newsdetail.aspx?itemid=9f3fc0e3-d5de-47e4-b505-848047195781.

Lal, R., 2001. Soils and the greenhouse effect. In: R. Lal (ed.), *Soil Carbon Sequestration and the Greenhouse Effect*, Soil Science Society of America, Inc., Madison, WI.

Lal, R. et al., 2004. Soil carbon sequestration impacts on climate change and food security. *Science* 304, 1623–1626.

Lal, R., July/August 2008. Promise and limitations of soils to minimize climate change. *Journal of Soil and Water Conservation* 63(4).

Lal, R., Follett, R.F. et al., December 2007. Soil carbon sequestration to mitigate climate change and advance food security. *Soil Science*, 172(12).

LaSalle, T.J. and Paul, H., 2008. *Regenerative Organic Farming: A Solution to Global Warming*, Rodale Institute, Kutztown, PA, Slideshow of paper available at http://www.princeton.edu/morefoodlesscarbon/speakers/timothy-lasalle/Timothy_LaSalle.pdf; full paper no longer available online.

Laurie, J.H., 1997. Clear lake: Biological and educational tools for reversing the loss of biodiversity in the 21st Century, Master's Thesis, University of Houston, Houston, TX.

Laurie, J.H., 2009. Personal conversations with Joe and Peggy Maddox.

Leake, J.R. et al., 2004. Networks of power and influence: The role of mycorrhizal mycelium in controlling plant communities and agroecosystem functioning. *Canadian Journal of Botany* 82, 1016–1045.

Liebig, M.A. et al., September 2008. Soil carbon storage by switchgrass grown for bioenergy. *Bioenergy Research* 1, 215–222, http://naldc.nal.usda.gov/download/28132/PDF.

Lyall, S., January 10, 2013. *Heat, Flood or Icy Cold, Extreme Weather Rages Worldwide*, New York Times, http://www.nytimes.com/2013/01/11/science/earth/extreme-weather-grows-in-frequency-and-intensity-around-world.htm.

MacKenzie, J.J., 1996. *Oil as a Finite Resource: When will Global Production Peak*? World Resources Institute, Washington, DC.

Maser, C. et al., 2008. *Trees, Truffles and Beasts: How Forests Function*, Rutgers University Press, New Brunswick, NJ.

McGinty, A., Thurow, T.L., and Taylor, C.A. Jr., 1995. *Improving Rainfall Effectiveness on Rangeland*, Texas A&M University, College Station, TX.

McNaughton, S.J., 1985. Ecology of a grazing ecosystem: The Serengeti. *Ecological Monographs* 55, 259–294.

Meuret, M., 2010. *Un Savoir-Faire de Bergers*, Versailles & Dijon, Paris, France.

Meuret, M. and Provenza, F.D., 2013. *Shepherds' Know-How*, ACRES U.S.A., Austin, TX.

Milchunas, D.G. and Lauenroth, W.K., 1993. Quantitative effects of grazing on vegetation and soils over a global range of environments. *Ecology* 63, 327–366.

Müller, B., Frank, K., and Wissel, C., 2007. Relevance of rest periods in non-equilibrium rangeland systems—A modeling analysis. *Agricultural Systems* 92, 295–317.

Naeth, M.A., Bailey, A.W., Pluth, D.J., Chanasyk, D.S., and Hardin, R.T., 1991. Grazing impacts on litter and soil organic matter in mixed prairie and fescue grassland ecosystems of Alberta. *Journal of Range Management* 44, 7–12.

National Climate Assessment and Development Advisory Committee, Draft, January 2013, http://ncadac.globalchange.gov/download/NCAJan11-2013-publicreviewdraft-fulldraft.pdf *National Geographic Magazine*, September 2008. http://ngm.nationalgeographic.com/flashback/2008#/2008-09-FB.jpg.

National Wildlife Federation, February 2013. Wildlife in a Warming World, http://www.nwf.org/~/media/PDFs/Global-Warming/Reports/NWF_Wildlife-Warming-World_Report_web.pdf.

Nature Conservancy, 2010. Janos Valley Grasslands, http://www.nature.org/ourinitiatives/regions/northamerica/mexico/placesweprotect/janos-valley-grasslands.xml.

Neary, D.G., Klopatek, C.C., DeBano, L.F., and Folliott, P.F., 1999. Fire effects on belowground sustainability: A review and synthesis. *Forest Ecology and Management* 122, 51–71.

Neely, C., Bunning, S., and Wiles, A., 2009. *Review of Evidence on Drylands Pastroral Systems and Climate Change: Implication and Opportunities for Mitigation and Adaptation*, Food and Agriculture Organization of the United Nations (FAO).

Neely, C. and Fynn, A., 2011. Critical choices for crop and livestock production systems that enhance productivity and build ecosystem resilience, SOLAW (The State of the World's Land and Water Resources for Food and Agriculture) Background Thematic Report—TR11, FAO, n.d., http://www.fao.org/fileadmin/templates/solaw/files/thematic_reports/TR_11_web.pdf.

Nippert, J.B., Wieme, R.A., Ocheltree, T.W., and Craine, J.M., January 19, 2012. Root characteristics of C_4 grasses limit reliance on deep soil water in tallgrass prairie. *Plant Soil*, http://www.k-state.edu/ecophyslab/pdf's/NippertPlantSoilRoots2012.pdf.

NOAA, January 7, 2013. Carbon dioxide annual means, ftp://ftp.cmdl.noaa.gov/ccg/co2/trends/co2_annmean_mlo.txt.

Oesterhceld, M., Sala, O.E., and McNaughton, S.J., 1992. Effect of animal husbandry on herbivore carrying capacity at the regional scale. *Nature* 356, 234–236.

Outwater, A., 1996. *Water: A Natural History*, Basic Books, New York.

Pacala, S.W. et al., 2001. Consistent land- and atmosphere-based U.S. carbon sink estimates. *Science* 22, 2316–2320.

Palmer, T. and Ainslie, A., n.d. United Nations Food and Agricultural Organization. Country Pasture/Forage Resource Profiles: South Africa, http://www.fao.org/ag/AGP/AGPC/doc/Counprof/southafrica/SouthAfrica.htm.

Parton, W.J., Schimel, D.S., Cole, C.V., and Ojima, D.S., 1987. Analysis of factors controlling organic matter levels in grasslands. *Soil Science Society of America Journal* 51, 1173–1179.

Pimentel, D. and Pimentel, M.H., 2008. *Food, Energy, and Society*, Taylor & Francis, Boca Raton, FL.

Planetolog, n.d. Map of average annual precipitation of the world, http://planetolog.com/map-world-detail.php?type=RES&id=2.

Popp, A., Lotze-Campen, H., and Bodirsky, B., 2010. Food consumption, diet shifts and associated non-CO_2 greenhouse gases from agricultural production. *Global Environmental Change* 20, 451–462.

Prentice, I.C. et al., 2001. The carbon cycle and atmospheric CO_2. In: J.T. Houghton and D. Yihui (eds.), *Climate Change: The Scientific Basis, the Contribution of WGI of the IPCC to the IPCC Third Assessment Report (TAR)*, Cambridge University Press, Cambridge, U.K., pp. 183–237.

Provenza, F.D., 1996. Acquired aversions as the basis for varied diets of ruminants foraging on rangelands. *Journal of Animal Science* 74, 2010–2020.

Provenza, F.D., 2003a. *Foraging Behavior: Managing to Survive in a World of Change*, Utah State University Press, Logan, UT.

Provenza, F.D., 2003b. Twenty-five years of paradox in plant-herbivore interactions and "sustainable" grazing management. *Rangelands* 25, 4–15.

Provenza, F.D., 2008. What does it mean to be locally adapted and who cares anyway? *Journal of Animal Science* 86, E271–E284.

Provenza, F.D. and Villalba, J.J., 2010. The role of natural plant products in modulating the immune system: An adaptable approach for combating disease in grazing animals. *Small Ruminant Research* 89, 131–139.

Provenza, F.D., Villalba, J.J., Dziba, L.E., Atwood, S.B., and Banner, R.E., 2003. Linking herbivore experience, varied diets, and plant biochemical diversity. *Small Ruminant Research* 49, 257–274.

Provenza, F.D., Villalba, J.J., Haskell, J.H., MacAdam, J.A., Griggs, T.C., and Wiedmeier, R.D., 2007. The value to herbivores of plant physical and chemical diversity in time and space. *Crop Science* 47, 382–398.

Pucheta, E., Bonamici, I., Cabido, M., and Diaz, S., 2004. Below-ground biomass and productivity of a grazed site and a neighbouring ungrazed exclosure in a grassland in central Argentina. *Austral Ecology* 29, 201–208.

Ragab, R. and Prudhomme, C., 2002. SW—Soil and water: Climate change and water resources management in arid and semi-arid regions: Prospective and challenges for the 21st century. *Biosystems Engineering*, 81, 3–34.

Reed, F., Roath, R., and Bradford, D., 1999. The grazing response index: A simple and effective method to evaluate grazing impacts. *Rangelands* 4, 3–6.

Reicosky, D.C., de Maio. 2005. Alternatives to mitigate the greenhouse effect: Emission control by carbon sequestration, Simpósio sobre Plantio direto e Meio ambiente; Seqüestro de carbono equalidade da agua, pp. 20–28, Anais, Foz do Iguaçu.

Richards, J.H. and Caldwell, M.M., 1987. Hydraulic lift: Substantial nocturnal water transport between soil layers by *Artemisia tridentata* roots. *Oecologia* 73, 486–489.

Richardson, P.Q. and Richardson, R.H., 2000. Dung beetles and their effects on soil. *Ecological Restoration* 18, 116–117, http://managingwholes.com/dung-beetles.htm.

Rietkerk, M., Ketner, P., Burger, J., Hoorens, B., and Olff, H., 2000. Multiscale soil and vegetation patchiness along a gradient of herbivore impact in a semi-arid grazing system in West Africa. *Plant Ecology* 148, 207–224.

Ruddiman, W.F., 2003. The anthropogenic greenhouse era began thousands of years ago. *Climatic Change* 61, 261–293.

Salatin, J., 2010. *The Sheer Ecstasy of Being a Lunatic Farmer*, Chelsea Green, Swoope, VA.

Savory, A., 2009. *Keeping Cattle: Cause or Cure for Climate Crisis*? Feasta Lecture, Ireland, http://vimeo.com/8239427.

Savory, A., 2013. *How to Fight Desertification and Reverse Climate Change*, TED Talk, http://www.ted.com/talks/allan_savory_how_to_green_the_world_s_deserts_and_reverse_climate_change.html.

Savory, A. and Butterfield, J., 1999. *Holistic Management: A New Framework for Decision Making*, Island Press.

Savory Institute, 2013, http://www.savoryinstitute.com/wp-content/uploads/2013/03/Savory_Inst_HM_Research_Portfolio_March2013.pdf.

Schimel, D.S. et al., 2001. Recent patterns and mechanisms of carbon exchange by terrestrial ecosystems. *Nature* 414, 169–172.

Schwartz, J.D., 2013. *Cows Save the Planet,* Chelsea Green, White River Junction, VT.

Seastedt, T.R., 1995. Soil systems and nutrient cycles of the North American Prairie. In: A. Joern, and K.H. Keeler (eds.), *The Changing Prairie: North American Grasslands*, Oxford University Press, New York, pp. 157–176.

Shaw, R.A., Villalba, J.J., and Provenza, F.D., 2006. Resource availability and quality influence patterns of diet mixing by sheep. *Journal of Chemical Ecology* 32, 1267–1278.

Soussana, J.F., Tallec, T., and Blanfort, V., 2010. Mitigating the greenhouse gas balance of ruminant production systems through carbon sequestration in grasslands. *Animal* 4, 334–350.

Spahni, R. et al., 2005. Atmospheric methane and nitrous oxide of the late Pleistocene from Antarctic ice cores. *Science* 310, 1317–1321. Cited in Spahni, Renato, Methane concentration in ice cores: A tool to reveal firn-ice properties and past climate changes, Dissertation, 2006, p. 55, http://www.climate.unibe.ch/~spahni/papers/spahni06phd.pdf.

Steinfeld, H., Gerber, P., Wassenaar, T., Castel, V., Rosales, M., and de Haan, C., 2006. *Livestock's Long Shadow: Environmental Issues and Options*, FAO, Rome, Italy.

Stinner, D.H. et al., 1997. Biodiversity as an organizing principle in agroecosystem management: Case studies of holistic resource management practitioners in the USA. *Agriculture, Ecosystems & Environment* 62, 199–213, http://www.sciencedirect.com/science/article/pii/S0167880996011358.

Suttie, J.M., Reynolds, S.G., and Batello, C. eds., 2005. *Grasslands of the World,* Food and Agriculture Organization of the United Nations, Rome, Italy; Chapter 1, Available at http://www.fao.org/docrep/008/y8344e/y8344e05.htm-bm05.

Tainton, N.M., Aucamp, A.J., and Danckwerts, J.E., 1999. Principles of managing veld. In: N.M. Tainton (ed.), *Veld Management in South Africa*, University of Natal Press, Pietermaritzburg, South Africa, pp. 169–193.

Teague, W.R., Dowhower, S.L., Baker, S.A., Haile, N., DeLaune, P.B., and Conover, D.M., 2011. Grazing management impacts on vegetation, soil biota and soil chemical, physical and hydrological properties in tall grass prairie. *Agriculture Ecosystems & Environment* 141, 310–322.

Teague, W.R., Dowhower, S.L., and Waggoner, J.A., 2004. Drought and grazing patch dynamics under different grazing management. *Journal of Arid Environments* 58, 97–117.

Teague, W.R., Provenza, F.D., Norton, B.E., Steffens, T., Barnes, M.K., Kothmann, M.M., and Roath, R.L., 2009a. Benefits of multi-paddock grazing management on rangelands: Limitations of experimental grazing research and knowledge gaps. In: H.G. Schroder (ed.), *Grasslands: Ecology, Management and Restoration*, Nova Science Publishers, New York, pp. 41–80.

Teague, W.R., Kreuter, U.P., Grant, W.E., Diaz-Solis, H., and Kothmann, M.M., 2009b. Economic implications of maintaining rangeland ecosystem health in a semi-arid savanna. *Ecological Economics* 68, 1417–1429.

Teague, W.R., Kreuter, U.P., and Fox, W.E., 2010. Economically efficient rangeland management to sustain ecosystem function and livelihoods. In: V. Squires (ed.), *Range Livestock and Resource Management*, In Encyclopedia of Life Support Systems (EOLSS), Developed under the Auspices of the UNESCO, EOLSS Publishers Co. Ltd., Oxford, U.K., http://www.eolss.net.

Thurow, T.L., 1991. Hydrology and erosion. In: R.K. Heitschmidt and J.W. Stuth (eds.), *Grazing Management: An Ecological Perspective*, Timber Press, Portland, OR, pp. 141–159.

Troeh, F.R. and Thompson, L.M., 2005. *Soils and Soil Fertility*, 6th edn., Blackwell Publisher.

Turner, C.L., Seastedt, T.R., and Dyer, M.I., 1993. Maximization of aboveground grassland production: The role of defoliation frequency, intensity and history. *Ecological Applications* 3, 175–186.

United Nations Food and Agricultural Organization, 2006. *Livestock's Long Shadow: Environmental Issues and Options*, Rome, Italy, http://www.fao.org/docrep/010/a0701e/a0701e00.htm.

USDA/Agricultural Research Service, June 16, 2008. Glomalin is key to locking up soil carbon. *Science Daily*, http://www.ars.usda.gov/is/pr/2008/080617.htm.

Van Der Ploeg, J.D., Verschuren, P., Verhoeven, F., and Pepels, J., September 2006. Dealing with novelties: A grassland experiment reconsidered. *Journal of Environmental Policy & Planning* 8(3), 199–218.

Villalba, J.J., Provenza, F.D., and GouDong, H., 2004. Experience influences diet mixing by herbivores: Implications for plant biochemical diversity. *Oikos* 107, 100–109.

Walker, B.H., Carpenter, S., Anderies, J., Abel, N., Cumming, G., Janssen, M., Lebel, L., Norberg, J., Peterson, G.D., and Pritchard, R., 2002. Resilience management in social-ecological systems: A working hypothesis for a participatory approach. *Conservation Ecology* 6(1), 14, URL: http://www.consecol.org/vol6/iss1/art14.

Walters, C.J., 1986. *Adaptive Management of Renewable Resources*, Collier Macmillan, New York.

Weaver, J.E., 1950. Effects of different intensities of grazing on depth and quantity of roots of grasses. *Journal of Range Management* 3, 100–113.

Weber, K. and Horst, S., 2011. Desertification and livestock grazing: The roles of sedentarization, mobility and rest. *Pastoralism: Research, Policy and Practice* 1, 19, http://www.pastoralismjournal.com/content/1/1/19.

Weber, K.T. and Gokhale, B.S., 2011. Effect of grazing on soil-water content in semiarid rangelands of southeast Idaho. *Journal of Arid Environments* 75(2011), 464–470.

West, N.E., Provenza, F.D., Johnson, P.S., and Owens, M.K., 1984. Vegetation change after 13 years of livestock grazing exclusion on sagebrush semi-desert in central Utah. *Journal of Range Management* 37, 262–264.

White, R., Murray, S., and Rohweder, M., 2000. *Pilot Analysis of Global Ecosystems: Grassland Ecosystems*, World Resources Institute, Washington, DC.

Whitson, R.E., Heitschmidt, R.K., Kothmann, M.M., and Lundgren, G.K., 1982. The impact of grazing systems on the magnitude and stability of ranch income in the Rolling Plains of Texas. *Journal of Range Management* 35, 526–532.

Wiley, T., Hasson, A., and Woolston, G., 2013. Whole soil sampling" to compare carbon sequestration under perennial pastures of Western Australia. *Nature Environment and Pollution Technology* 12, 203–208.

Willliams, T.T., 2008. *Finding Beauty in a Broken World*, Pantheon Books.
Workman, J., 1986. *Range Economics.* Macmillan Publ. Co., New York, 217p.
World Economic Forum, 2013. *Global Risks Report 2013*, World Economic Forum, Cologny/Geneva, Switzerland, http://reports.weforum.org/global-risks-2013/section-five/x-factors/.
Wright, H.A. and Bailey, A.W., 1982. *Fire Ecology*. John Wiley & Sons, New York.
Xu, J., 2011. China's new forests aren't as green as they seem. *Nature* 477, 371, http://www.nature.com/news/2011/110921/full/477371a.html.
Yeomans, Plow n.d. *Yeomans Concepts—How to End Global Warming & Stop Climate Change*, http://www.yeomansplow.com.au/.

10

Geology into Biology: Carbon, Minerals, and Microbes—Tools to Remineralize Soil, Sequester Carbon, and Restore the Earth

David Yarrow

CONTENTS

A Sense of Humus

It is no joke to say 50 years ago soil science had no sense of humus. Soil was seen as inert dirt—a structural media to support plants and hold water and dissolved nutrients. No one investigated soil residues of dead fungal bodies. No one studied the unique roles of charred carbon in soils. No one measured soil microbe respiration.

A consequence of this purely chemical view is that in the twentieth century, soil carbon declined steadily and, in far too many soils, approached zero. On the Great Plains, soil organic matter (SOM) has decreased from a historic 8% down to a current 2%. Similarly, biological activity in soil dropped, as soils became increasingly sterile and lifeless—inert dirt. This reduction in soil carbon and life is unhealthy for plants but also had a big impact on human carbon emissions.

Humus, a complex organic substance, occurs naturally in soil and compost, the result of plant and animal matter that rots, decays, and decomposes. Like air, humus is abundant, renewable, essential to life, but far more complex. After hundreds of years of research, no one knows exactly what humus is.

Soil science uncertainty about humus is mirrored by an ambiguous definition. *Humus* originated as a scientific word in late eighteenth century from Latin for *earth or ground*. Like *organic*, *humus* has multiple meanings defined with varied precision. Most often, humus is a generic term for all kinds of carbon. In agriculture, humus is mature compost, or natural compost from forest or prairie soil. Humus also is a topsoil horizon that contains organic matter (OM).

Technically, precisely, humus is the final product of decaying plant and animal residues. Scientists say if it is not stable, it is not humus. Stable humus complexes survive thousands of years. Compost is able to decompose further and thus is classified as *active humus*. *Passive* humus is humic acids and substances so tightly bound to clay and hydroxides; microbes cannot penetrate them, and thus they resist further decay. So, stable humus adds few available nutrients to soil but does supply essential physical structure.

But humus is never static. It is dynamic, always changing, in constant digestion. So, it is hard to refer to it as *end product*. This confusion reveals vast unknowns about humus chemistry.

Humus begins when plant and animal residues are eaten by microbes. Carbon molecules synthesized by plant and animal are now food for bacteria, fungi, algae, actinomycetes, yeasts, and all microbes of decay. Microbes dismantle sugar, starch, protein, cellulose, and other carbon-based molecules. Nutrients and energy released by this microbial digestion are used by microbes and plants. Some revert to CO_2, water, and other gas. Some mineralize back into plant foods. Some resist rot and accumulate as indigestible residue.

As nature nibbles at complex, large biomolecules, something remains. Every biomass leaves some sort of indigestible residue. This inert, inedible, black residue is true *humus*—leftovers from microbial feasts.

This residual volume varies with type of biomass, climate, and environmental conditions but is in the range of 5%–15%. The decay-resistant molecules have been altered by microbial digestion. They consist mostly of carbon in the form of extremely large molecules, usually with multiple carbon rings, embedded in long, branching carbon chains (Figure 10.1).

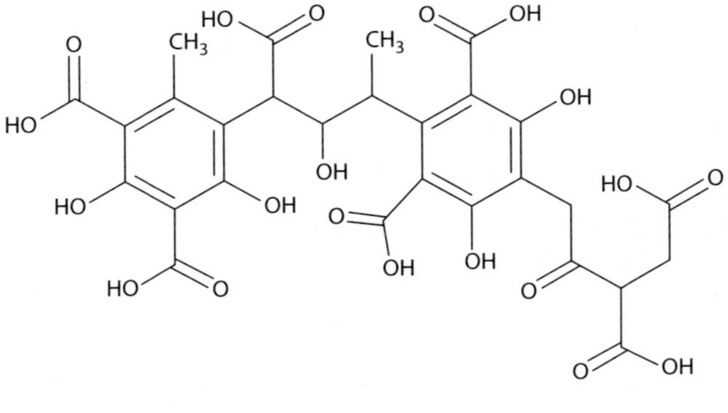

FIGURE 10.1 Humic acid—small water-soluble molecule 28 carbons—2 rings.

Information on humus identifies specific chemicals and certain components. Its nature and properties are well known. Factors that control it are common knowledge.

Yet, an accurate way to extract humus from soil has yet to be devised. This itself limits the study of humus. Further, humus in one soil differs structurally, chemically, and visibly from humus in another soil. So, it is simplistic to call them all the same thing.

Until recently, humus molecules were too large and complicated to study. Only in the last decade did science acquire tools to accurately map and measure these complex carbon structures. Now, with advanced high-tech tools such as x-ray crystallography, scanning electron microphotographs (SEMs), and supercomputer 3D imaging, scientists can peer at these massive molecules and begin to decipher their structures and functions.

Humus, while very stable, continues to decompose at a very slow rate, dependent largely on external climate factors such as temperature and moisture but also the character of the source biomass. The lifetime of humus in soil is variable, but easily several decades, and in some circumstance, over a century. In some environments, humus may remain for a few centuries.

Soil Organic Matter

Plants create OM by photosynthesis and metabolism. In death, their carbon-based bodies become debris and residue. Microbes eat this dead biomass and turn it back into water, CO_2, plant food, and humus.

As OM's complex biomolecules break down by weathering and digestion into simpler substances, they release energy and nutrients to feed other organisms. Easily digested residues are reused over and over by organisms and never become humus.

OM, then, is organic debris in the process of becoming humus. And humus is the most stable, persistent form of OM.

In the 1980s, we drafted *certified organic standards* for food, not just in the United States, but worldwide. Agreement was universal that organic farming's foundation is use of natural methods and materials to maintain soil fertility. Our complete consensus was that soil must have a minimum of carbon as *OM*. Certified farms must show soil tests with 4%–5% soil carbon. But carbon's complexity was lumped under one broad, vague term: *soil organic matter* (*SOM*).

To most minds, *SOM* is decayed and decaying biomass—rotting bodies of dead organisms—mostly plants but also animals and their manures. SOM is not a stable, fixed substance, but a complex concoction of biochemicals in the process of digestion and transformation. In most environments, SOM is in a state of rapid flux as it changes and degrades into simpler substances. Most SOM—85%–95%—eventually reverts back to CO_2 and water—the *CHO trinity*.

SOM contains more than carbon, or only organic elements. Dry and burn it and a whitish powder remains: ash—the oxidized minerals. So, embedded in SOM's carbon matrix are atoms and ions of metals and nonmetals—major minerals, trace elements, and all the least elements. These embedded ions support SOM structures and clusters and create electric charges on surfaces of huge SOM molecules.

SOM holds other mineral ions on its surfaces by *adsorption*, a weak electric attraction between atoms with opposite charge. The adsorbed minerals enrich the SOM and allow soil to acquire and hold electric charges. The ions are not bonded, but lightly held in loose association, so they are easily available to soil organisms and plant roots. Once held by SOM, ions are less likely to dissolve and flow through soil with water from rain or irrigation. Thus, nutrients stay in soil near the top, where they are available to microbes and rootlets.

SOM improves soil structure and increases water absorption and retention. SOM's large, complex carbon molecules insert themselves between soil particles to separate them, create open spaces, and form supramolecular superstructures that allow soil to breathe. Looser soil structure also allows water to easily enter and pass into soil. Open spaces in and around clusters of SOM molecules store water in soil, keep it wetter.

SOM is the essence of *recycling*—the breakdown, release, and reuse of nutrients to grow new organisms. In gardens, the icon of this microbial recycling is *compost*—a deliberate, sometimes carefully constructed, heap of OM that turns plant wastes and weeds into black, crumbly, lightweight soil ingredient. It is hardly hyperbole to say compost is legendary to enhance fertility and renew plant growth.

Soil Organic Carbon

Unfortunately, SOM is a simplistic idea to describe one of the planet's most complex structures: *soil organic carbon* (SOC). SOM is not just dead, decaying biomass. Missing from this SOM concept are living organisms—the microbes and larger organisms—tiny life forms that actively digest and disassemble OM's massive molecules. Complex communities of microbes inhabit the biomass, converting rotting carbon skeletons into food and energy.

So, OM is not just rotting debris. OM also includes the living organisms actively digesting the inert biomass. OM contains and shelters this microbial community—an interactive assembly of diversity of Nature's smallest, simplest, most ancient organisms.

A significant fraction of soil organic carbon exists as cells and bodies of living organisms. One cup of living soil is estimated to contain a few billion bacteria, a few million fungi, a few thousand pinhead-size creatures, and a few earthworm cocoons. The interaction of these microbial communities takes our quest for carbon's complexity into a deeper dimension of diversity.

Agriculture has just begun to assess this living biomass as a critical to sustainable soil fertility and able to convert currently infertile land into productive soil. The new twenty-first century agriculture studies this soil biology, applies probiotic methods, mimics microbial interactions, and harnesses their symbiotic functions to grow stronger, healthier crops. We need to learn much more about this living soil ecology and employ them as primary allies in soil fertility.

Currently, the only simple, economical way to evaluate the presence of soil organisms is a Solvita respiration test. A soil sample is sealed in a jar, and the conversion of O_2 to CO_2 is measured. This provides a rough measure of biological activity in the soil sample. However, we need better, simpler tools to evaluate soil biology—not only quantity, but diversity.

Dozens of types of organisms are the denizens of soil, each with families and species, all working together to sustain soil's fertility and stability. This world of tiny, invisible life forms is easily as complex and diverse as our large-scale realm of biology and ecology. But microbes are among the Earth's oldest communities of life—among the Earth's earliest inhabitants. Soil is not just inert geological matter, but a living tissue—Gaia's thin skin, covering the surfaces of the planet.

Soil Food Web

In recent years, science began to study the complex biochemistry and ecology of soil biology, and thus SOC. My own encounters with USDA and NRCS directors and board members convinced me this new biological paradigm is firmly in the minds of most agricultural leaders, and they are receptive to new probiotics strategies to work with soil life.

Dr. Elaine Ingham at Oregon State University, a leader in this research, developed the concept of the *soil food web* to identify and describe this complexity of cooperating organisms. Dr. Ingham created ways to assay and quantitatively evaluate the many different categories of soil organisms. Dr. Ingham trained hundreds in soil food web technology, who are now soil consultants all over the world. Dr. Ingham also developed ways to propagate and promote this living soil biology, most notably, by the technique of compost tea.

So, peering into carbon's roles in soil opens windows into a tremendous complex biochemistry of tiny cells, microbiology, and molecular life. We are still learning about these invisible-to-the-eye soil communities. Science is discovering nature knows and has used nanotechnology and quantum physics for millions of millennia.

The Humus Society, a new international association of scientists and researchers, meets every year to review new scientific papers on the complexities of soil carbon. In 2010, for the first time, the society was addressed by a biocarbon expert on *biochar*, Dr. Hugh McLaughlin.

UNIQUENESS OF CHARCOAL

Charcoal is the most unique adsorbent known to man because of variable size and shape micropores *combined* with constant electric surface charges. Far beyond empty space, the walls of char maintain many surface charge sites to trap and hold much smaller *stuff* than pore size and configuration would indicate.

Plants know how to go into the char and deenergize or reverse polarize the plates to overpower attractive forces. Root hairs and gels coordinate to become a microscopic organic ion vacuum cleaner.

Charcoal is a catalyst, releasing its goodies as conditions dictate. The analogy that works for me is electrostatic filters. Relatively open flow, but catching the small stuff by using surface attraction.

Black powder makers use very specific charcoal recipes to tailor powders for fast or slow "pop." To this day, biomass-based char from distilled mountain alder consistently provides the best "pop"—highest instant energy release of any char available.

Doug Brethower
Freedom Biomass Resilience Movement, Missouri

Charred Carbon

SOC's newest, most intriguing disguise is as charcoal, made by charring biomass in a smoldering fire. Charcoal is a common, well-known substance worldwide—preferred fuel for cookout and barbecue and preferred media for water filtration. Yet, 10 years ago, adding charcoal to soil was nearly unknown in western agriculture and ecology.

But, a 2009 international conference in Birmingham, England, adopted the term *biochar* to describe special charcoal made to add to soil. And since that soil will grow food to feed humans, special, higher standards must apply to this charcoal.

Biochar is a carbon-rich solid made by *carbonization* when biomass is heated with little or no oxygen (O_2) at low temperatures (<700°C). Instead of complete combustion to ash, an oxygen-starved fire smolders by *gasification*—releasing gas, vapors, and smoke, leaving behind charred carbon, or charcoal. Biomass can be wood, manure, leaves, straw, cornstalks, rice hulls, or any burnable organic waste.

Biochar added to soil improves soil fertility and health, filters nutrients, keeps them from leaching, provides carbon storage, cuts greenhouse gas emissions, and a host of other services. Biochar's promise and promotion is a way to sequester carbon that is measurable and verifiable for carbon offset protocols.

A side effect of adding biochar to soil is it is then easy to grow nutrient-dense crops.

In 2010, the USDA Sustainable Agriculture Research and Education (SARE) funded Iowa farmer John Topic to build charcoal kilns and add biochar to his heavy clay soils (FNC10-807). In fall, he tilled a few tons per acre into a test plot, with cow manure compost. Next spring, he wrote in his report:

> Untreated soil was gooey and waterlogged. Soil with only composted manure was also sticky. Soil with char only was more friable, less gooey. But the char and manure mix was like potting soil, requiring only a twist of the wrist to loosen the surface.

Yet, until very recently, hardly anyone in agriculture, botany, or soil science ever thought to use charred carbon in soil. But what is new to science and farming has 6000 years of success in South America (Figure 10.2).

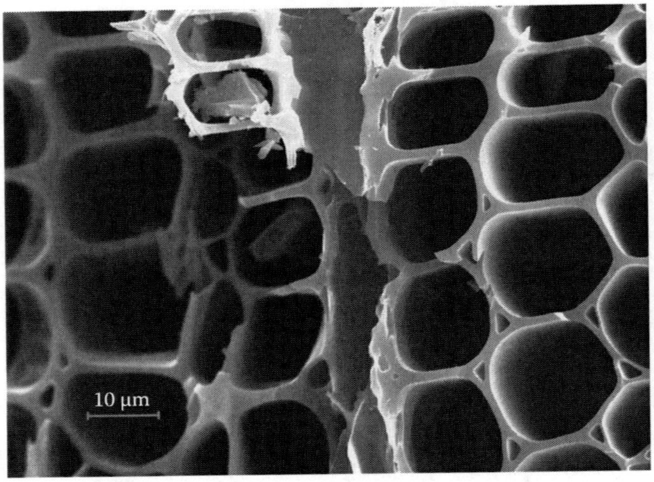

FIGURE 10.2 Cedar chip biochar.

Terra Preta: Amazon Dark Earths

While putting charcoal in soil is new to Western science, this method began 6000 years ago, in western Amazon rain forests, where indigenous tribes used charcoal to convert heavy, acid clays into fertile, stable topsoil. In 2000 years, this spread east to the mouth of the Amazon. Enough land was converted to feed millions of people.

WHAT IS BIOCHAR?

Biochar is a fine-grained charcoal, high in organic carbon, largely resistant to decomposition, produced from pyrolysis of plant and waste feedstocks. As soil amendment, *biochar* creates a recalcitrant soil carbon pool that is *carbon negative*—a net withdrawal of atmospheric carbon. The enhanced nutrient retention capacity of biochar-amended soil reduces total fertilizer requirements and also the climate and environmental impact of croplands.

International Biochar Initiative
www.biochar-international.org

The first Spanish conquistadores to travel the Amazon saw many large towns of people living densely along the river. Yet, 70 years later, when more Europeans visited the Amazon interior, no sign of huge populations was evident. The indigenous tribes vanished without a trace, leaving only a few scattered villages. Reports of densely populated Amazon cities became a myth named *El Dorado.*

Early Portuguese settlers, surprised by these remarkable, productive soils, named them *terra preta*, meaning *dark earth*, due to their near-black color. They were preferred for agriculture, especially to grow sugarcane, so highly prized, they were dug to sell as potting soil. Even in intensive cultivation, *terra preta* sustained strong, healthy plant growth. In contrast, surrounding rain forest clays are poor, acidic, and weak, have virtually no carbon, have low ion storage capacity, and grow inferior, stunted crops.

At first, explorers assumed these desirable, dark soils were unusual geologic deposits, or river sediments.

Then, in 1966, Dutch soil scientist Wim Sombroek published *Amazon Soils*, in which he made a case that *terra preta* soils were man-made by indigenous tribes and are black due to large amounts of added charcoal. At first, his idea was rated preposterous, but Wim made a solid scientific case.

Since then, further studies documented that *terra preta* is man-made. Archaeologists from several countries made dozens of digs to document *terra preta*, verify their human origin, and assay their characteristics. Aerial surveys with canopy-penetrating radar detected thousands more hectares of carbon-rich soils. We now know indigenous tribes created enough *terra preta* to feed a population of 5 million, perhaps 25 million.

BENEFITS OF BIOCHAR

Replenished soil depleted by overfarming
Crops growing larger and more bountiful
Improved soil structure and tilth
Improved water penetration and absorption
Increased water holding capacity of soil
Reduction and prevention of runoff and erosion
Decreased mineral leaching
Retention of nutrients in soil better
Increased soil pH/Lower soil pH
Aiding roots to adsorb nutrients
Boosting cation exchange capacity (CEC)
Creating anion exchange capacity (AEC)
Greater Ca, Mg, K bioavailability
Greater micronutrient bioavailability
Greater P retention and bioavailability
Greater nitrogen bioavailability
Increased nitrogen fixation
Increased nitrogen retention and cycling
Accelerated compost processes
Reduced greenhouse gas emission
Absorption of gas and odor of compost and manure
Lessened aluminum toxicity
Proliferation of beneficial microbes
Increased activity of mycorrhizal fungi
Promotion of earthworm activity
Dissolved SOM energy for microbes
Sheltering beneficial microbes from predators
Higher sorption of microorganisms
Better formations of biofilms
Sorption of inhibitory compounds
Sequestering carbon for 1500+ years

Carbon Sequestration

In 1992, Wim Sombroek published a book that suggested *terra preta* may be a way to sequester carbon, reduce greenhouse gas levels, and thus reverse global warming and mitigate climate change. If we convert plant biomass to char and add it to soil, significant CO_2 can be moved out of the atmosphere into a safe, solid form. If done globally on farmland, using crop and forestry wastes, perhaps enough carbon can be sucked out of air to make a difference for our planetary future.

So, plants are our primary allies in this sequestration strategy. By photosynthesis, plants fix CO_2 into carbohydrates and build their cellulose skeletons from this sweetness. By pyrolysis, we then convert carbohydrate into char—a superstable kind of carbon that remains for centuries. Thus, we sequester carbon, to reduce the impact of climate change—perhaps reverse the excess greenhouse gas in the Earth's atmosphere.

Once biochar is properly put in soil, it stays for centuries. The USDA soil scientist David Laird believes char's half-life in soil is over 1600 years. Controlled biochar decomposition experiments reveal a mean residence time in soil between 1300 and 4000 years. Thus, farmers who add biochar to soil invest in several lifetimes of sustained fertility. To sequester carbon, science calculates it must be removed from the atmosphere 100 years. Thus, biochar sequesters carbon 16 times longer than this minimum.

By comparison, raw OM has a lifetime in soil measured in years. In healthy, living soil, grasses, weeds, and leaves decay and disappear within a few years. Large woody limbs rot away in a decade or two. Even stable humus has a soil life cycle of a century. But biochar is superstable in soil, resists microbial digestion, and persists with barely measurable annual decay.

This, then, is our best, most natural strategy to quickly remove carbon from the Earth's atmosphere. The potent advantage is that the biochar is not a new form of waste, but a resource that boosts soil fertility so it grows even more biomass each year to sequester more carbon as plants and soil biology. Thus, soil regeneration sets in motion multiple positive feedback loops that accelerate carbon capture and sequestration.

Currently, Australia and Japan are the only nations to officially admit soil can be a carbon sink! This needs to be more widely known, and more importantly, followed.

However, while sequestering carbon in soil is a significant way to remove carbon from the Earth's atmosphere, this is a simple, superficial, single element view of a far greater reality. It is crucial to realize that it was microorganisms in seas and soils that—by their respiration of breathing out—created the Earth's atmosphere. And the continuing activities of these least of all life forms has sustained and stabilized the gases composing the Earth's air so that larger, more complex organisms could survive, thrive, and evolve.

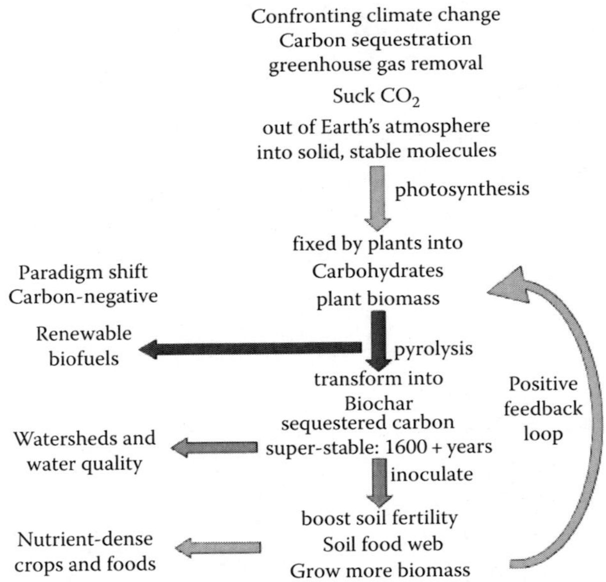

FIGURE 10.3 Carbon sequestration.

Today, we know from scientific studies and news reports that the microbial life in the sea is not faring well. Dead zones, red tides, and coral bleaching are serious signs warning that life in the ocean is under stress and declining. But little attention is given to the reality that soils are becoming sterile due to deforestation, acid rains, harsh agricultural fertilizers, and toxic treatments. These ancient microbial communities are collapsing due to effects of industrial civilization.

Ten years after Wim Sombroek published his sequestration suggestion, an international association formed to advance his idea. By 2012, networks and associations had formed in countries worldwide involving thousands of scientists, investigators, experimenters, inventors, and enthusiasts, guided by the International Biochar Initiative, headed by Dr. Johannes Lehmann at Cornell University. Already, in America, thousands of backyard burners have made biochar to test in gardens, farms, and greenhouses.

Thus, in a single decade, biochar rapidly went from unknown idea to mainstream strategy in global efforts to mitigate climate change, restore depleted soils, create new arable land, jumpstart a new green revolution to transform agriculture into carbon-smart farming, and restore a solvent foundation to community economies.

And we can grow high-quality, nutrient-dense foods to restore human health (Figure 10.3).

Is Biochar a Fertilizer?

In a word, "no." Biochar is not a nutrient. Nor is it a food source. Bacteria and fungi do not eat it. Neither will earthworms. Water will not dissolve it. It does not chemically degrade or weather. In fact, archaeology commonly does radiocarbon dating with charcoal because it resists decay so well.

Fresh char added to soil loses 3%–5% of its mass in 5 years. After that, annual mass loss drops to a fraction of a percent. Most of this loss is tar and resin residues held in micropores—tiny drops and thin films of oily hydrocarbons left from cooked cellular substances. They are eaten by bacteria and simple organisms. As char ages, a tiny fraction of lightweight, water-soluble carbon molecules are also lost. A smaller fraction of carbon escapes as volatile gas, such as ethylene, a plant growth stimulant.

But over 95% of well-made char is superstable and stays in soil for centuries—truly an inert ingredient. Biochar continues to sustain soil fertility in the Amazon, apparently for centuries.

So, if biochar is not a fertilizer, what does it do?

Micropores: Nature's Nanotechnology

Plants are mostly water. Thus, their physical structure is mostly plumbing. Seen in a microscope, plant biomass looks like bundles of pipes, tubes, and tunnels to move water around. Amid all these stacks of water channels are larger cavities, each occupied by a plant cell (Figure 10.4).

Thus, char is hollow inside. Charcoal is lightweight because it is empty. It floats because air is trapped in those tiny tubes and pipes. Those bundles are like a sponge and have a huge capacity to soak up water and other substances. And even as char is empty, open and porous inside, it causes soil to become lighter, looser, more open to air and water.

A bit of char has limited external surface area, but this is tiny compared to char's vastly greater inner surfaces. The walls of all the inner chambers are many times greater than outer surfaces. The added internal capacity of char is estimated between a few thousand, to maybe a million, times more than external surface.

Nature makes multiple uses of this emptiness. Char's profusion of microscopic pores is nature's nanotechnology. Water, oil, sugars, proteins, and other substances are drawn in and stretched into thin films that support biological organisms. Char's empty inner spaces give it tremendous storage capacity for water, ions, electrons, and nutrients—even whole organisms take up residence in the char.

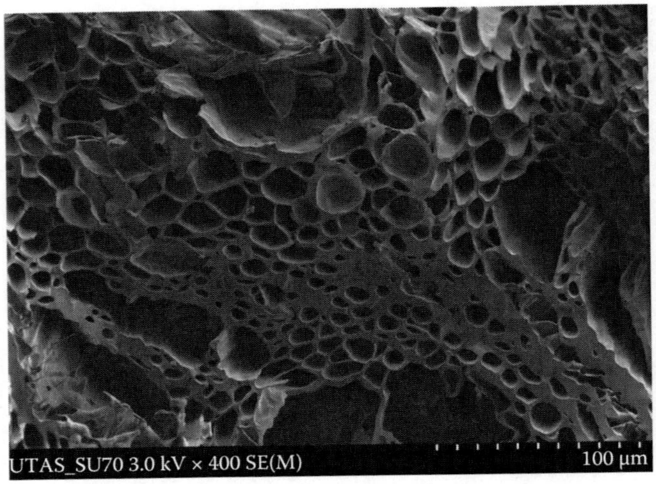

FIGURE 10.4 Red cedar biochar scanning electron microphoto. UTAS_SU70 3.0 kV × 400 SE(M).

MICROPORES IN BIOCHAR

Biochar pores are great refuges for microbes, but none will live—not even microbes—in the most beautiful palace without a nicely filled fridge, well-stocked pantry, and fresh water. Thus, a key condition for microbes to colonize biochar is nutrients available on and in the char. Micropores must contain nutrients and be recharged continually.

Char can be inoculated by blending with compost. Another approach is feeding it to livestock. But once in soil, nutrients of SOM go to equilibrium with biochar and recharge them continually. Nutrient flows must be steady to keep microbes active in pores and surfaces. Biochar enables more efficient nutrient flow between soil, root, and microbe. Biochar decreases nutrient leaching from soil, as the biochar matrix becomes a nutrient reservoir and relay.

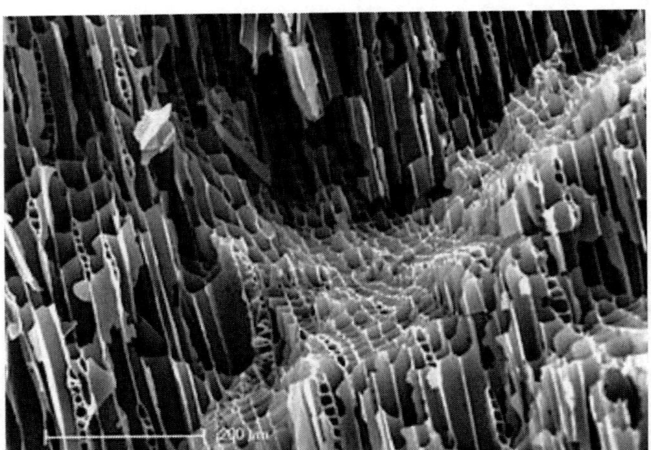

FIGURE 10.5 Biochar micropores: bundles of tubes, scanning electron microphoto.

So, in soil, biochar's first service is to soak up and hold water and thereby keep soil wetter. Like a dry sponge, char's micropores draw in water by capillary suction. Then, gradually, micropores meter moisture back out into soil for microbes and roots. Thus, a small amount of char greatly increases soil's water holding capacity and improves its moisture management ability.

The combination of micropore sponge with soil microbiology in active symbiosis with roots enhances drought tolerance of plants, from annual crops to trees. In 2011, the US Forest Service released preliminary results of adding char to soils after forest fires in the West. Clearly, soils with char stay wetter, so soil biology, including shrubs and trees, regenerates faster (Figure 10.5).

Adsorption: Why Carbon Is Black

Carbon's valence electrons make four covalent bonds in tetrahedral symmetry, nature's simplest 3D geometry. This four-arm arrangement attracts and holds many kinds of energy and chemistry, from photons of light to neutrons in nuclear reaction. Carbon's perfect symmetry captures photons, atoms, and ions and thus is black, the color of complete absorption and zero radiation.

But char is not just carbon. Held in the carbons are minerals. If you burn charcoal, the final product is whitish ash—mostly metal oxide minerals. These embedded mineral ions create sites in the char with electric charge. Also, char has numerous oxygen atoms (O, OH, OOH, etc.) that create other electric charge sites.

These embedded electric charges attract atoms and ions in molecules of opposite polarity. This charge is not strong enough to form bonds in which atoms exchange or share electrons, but able to draw and hold atoms near the char. This loose association is based on a subtle mutual attraction, like a hydrogen bond in water. Nature makes many uses of these subtle electric fields.

Adsorption is a technical term for this electric attraction between atoms and molecules. Char's *absorption* of water operates largely by capillary action from differential pressures. But *adsorption* occurs due to electric charges that cause atoms and molecules to be attracted, form clusters, and align in groups.

Charcoal has abundant charged sites that give it tremendous capacity to pull molecules out of solution and hold onto atoms on—and even in—the char. Charcoal's very high adsorption potential is what makes it an ideal media for water filtration. In water purification, ions in solution are considered *pollutants*. But in soil, the important ions are *nutrients*. Biochar is a sponge that also soaks up and holds ions of elements and biomolecules.

Biochar has an added adsorption capacity, far beyond other soil particles like sand and clay. Like any particle, a bit of biochar has a fixed, measurable external surface area to attract and adsorb ions. But biochar's limited external surface is augmented by complex internal surfaces of its micropores. Those hollow, inner chambers provide a far greater ion adsorption capacity than almost all other natural materials. Surface area calculations vary, as do biomass characteristics, but biochar easily has a few thousand to over a million times more internal than external surfaces.

Char draws ions to itself, separates them from solution, and holds them tightly at char surfaces by weak electric attraction, not fixed atom-to-atom bond. This loose association is easy to affect or alter, so adsorbed ions can become mobile, or jump around, such as from char to root, or microbial membrane. This easy access ion exchange makes adsorbed nutrients bioavailable.

Adsorbed charges can also stack up in thin layers and form rudimentary thin films—a special structural strategy of cells and microbes. Thin films allow nutrients and electrons to move around a cell in remarkable ways that are orderly and efficient.

Like most soil particles, char has negative charge sites that attract and adsorb positive ions—the cations. Adding char to almost any soil will boost *CEC*. Since metal cations are soil's primary electron donors and charge sources, CEC is a valuable numerical measure of soil's potential energy to support and sustain growth. The higher the CEC, the faster and better plants are likely to grow. Typically, biochar added to clay or sandy soils sees CEC rise by 10 or more points.

After char was accepted as a soil additive, scientists began a global search to reexamine other soils. Ecologists discovered ancient prairie soils had significant fractions of charred carbon—something no one had studied before. Apparently, significant charred carbon was created by slow-moving, low-temperature prairie fires, especially fires with green, moist biomass. Centuries of recurring fires accumulated charred carbon in soils and changed those soils to improve their fertility.

In July 2012, American Chemical Society's *Environmental Science and Technology* journal published six scientists' paper that the Midwest's highly productive, grassland-derived soils (Mollisols) contain char from presettlement fires that structurally compare to *terra preta*. These soils are much more abundant than previously thought (40%–50% organic C). Oxidized char residues are particularly stable, abundant, fertility-enhancing forms of SOM. Further, if just 40% of SOC is char (low estimate for Mollisols), essentially the entire soil CEC is attributed to char.

What was good for Amazon forest soils found a natural use in North American prairies.

BIOCHAR AND CLIMATE CHANGE

Soil has enough capacity to store more carbon than is currently in the atmosphere, plus all the green growing biomass on the planet. Biochar is produced by pyrolysis or gasification—processes that heat biomass in absence or shortage of oxygen. Biochar resists decay to stay in soil hundreds of years, likely thousands. So, buried in soil, biochar is *carbon negative*—permanently removes carbon from air into superstable soil. This simple, yet powerful process can store 2.2 Gt of carbon annually.

Biochar production can coproduce liquid and gas byproducts to process into biofuels, providing clean, renewable energy. This twin win of biochar and biofuel confronts global climate change, replaces fossil fuels, sequesters carbon in stable soil, improves fertilizer efficiency, and reduces nitrous oxide emissions.

It is inexpensive, widely applicable, and fully scalable—a strategy we cannot afford not to implement worldwide.

Fertilizer Efficiency and Water Quality

However, biochar has a further feature to dramatically boost its ion adsorption properties. Amazon studies of *terra preta* document that char also has positive charges embedded in the carbon matrix. This positive polarity attracts negative ions and molecules: the anions. These atoms are electron acceptors that serve in cells as charge carriers to move energy around. So, unlike most soil particles, char has *AEC*. The two most critical soil anions are nitrogen and phosphorus—N and P of NPK fertilizers.

Anion adsorption is a potent tool in soil fertility. By gathering and holding anions out of the soil solution, biochar immediately curbs leaching and loss of these nutrients. Instead of washing out with rain or irrigation, nitrogen, phosphorus, and other anions are held on and in the bits of char. On the supply side, this holds critical nutrients in the root zone and delivers more fertilizer to the plants, sharply increasing overall useful fertilizer efficiency.

Multiple research studies in Japan, Australia, United States, and other countries found that biochar added to soil cuts nitrate migration out of the root zone into groundwater by 50%–80%. Since nitrate and phosphate from farm fertilizers is a primary source of nonpoint water pollution, this facility of biochar has great significance for watershed management. Biochar's adsorption power is also useful in stormwater purification and management.

Federal-funded research at University of Vermont is testing biochar use as a phosphorus trap on dairy farms. Phosphorus pollution is a recurring problem fouling water in the Great Lakes. Protecting Lake Champlain, the sixth Great Lake, from phosphorus pollution from farmland runoff is an environmental priority for two states and two nations. Early efforts at Shelburne Farms near Burlington, VT documented over 50% reduction in phosphorus leaving the farm to enter the lake.

Research in Japan, Australia, America, and Germany has documented that biochar added to soils curtails outgassing of greenhouse gases by 37%–90%. Biochar seems to alter microbial activity in soil, reducing soil respiration (CO_2) and conversion of Nitrogen fertilizer to nitrous oxide (NO_x) that otherwise goes into the air. Nitrous oxide has a greenhouse gas effect over 300 times that of CO_2. This is partly biochar's adsorption power and capacity but also microbial activity to manage the nitrogen cycle and keep these volatile molecules in soil. Methane emissions are also reduced.

The ability of biochar to curtail leaching, outgassing, and pollution and improve fertilizer effectiveness occurs at two levels. First is the physical chemistry of adsorption—char's immense capacity to capture and hold nutrient ions. The second level is biological, involving soil microbes that process nutrients into biomolecules and protoplasm.

Habitat, Housing, and Shopping Malls

The final step is for biochar to be colonized by microbes. With water and nutrients stored in char micropores, bacteria, fungi, and other soil biology move into the char and claim its empty chambers as residences. We do not eat our houses, and microbes do not eat char. They live in it.

Studies of Amazonian *terra preta* describe char-enriched soils as full of microbial life—so much so, the term *soil reef* was used to highlight how the soils are bursting with microflora. Even as coral reefs provide calcium carbonate structures to house simple cellular organisms on shallow sea floors, char provides carbohydrate shelters for communities of soil organisms.

GLOBAL WARMING TURNAROUND

Replacing slash-and-burn agriculture with slash-and-char and use of agricultural and forestry wastes for biochar production could provide CO_2 drawdown of about 8 ppm or more in half a century. Carbon sequestration in soil has significant potential. Biochar, produced in pyrolysis of crop residues, forestry, and animal wastes, can restore soil fertility while storing carbon for centuries to millennia. Biochar helps soil retain nutrients and fertilizers, reducing emissions of greenhouse gases such as nitrous oxides.

Dr. James Hansen, Director
NASA Goddard Institute for Space Studies
America's leading climate scientist

Bacteria are the tiniest life forms, with diameters of 0.2–70 μm (average), 3 μm (so tiny), over a million fit on a pinhead. Bacteria have no trouble populating biochar. In fact, bacteria and creatures their size build the basic structures of all higher organisms. Their natural home is spaces like biochar micropores—skeletons of structures built by their own kind—like squatters living in abandoned buildings.

Fungi are hundreds to thousands of times larger than bacteria. While bacteria are small, independent cells, fungi form extensive networks by growing long, thin threads—*mycelium*—that scavenge their neighborhood for food and water. We have all seen the dense, whitish mats of fungal mycelium that grow like fuzzy fur on moldy food or leafy litter. These threads can travel 10 or 100 ft through soil searching out nutrients.

Rice-grain-size char has enough micropore capacity to easily house a million bacteria and a thousand fungi. A grain of char is a microcondominium to provide residences for complex communities of soft-body microbes. And with water and nutrients stored in micropores, char also provides microbial shopping malls and supermarkets.

Char is a refuge for soil organisms. Walled chambers and tunnels are safe shelter from predators roaming the soil solution. Char's inner recesses are safe space to germinate spores and nurse new bacteria, fungi, and their helper networks. Living in char provides microbes protection to survive, thrive, interact, and evolve to optimum density and diversity.

And because biochar is superstable for centuries, microbes do not get short-term leases for a few months. Instead, microbial communities can become established in full-scale symbiosis.

Substrate for Microbes

In 25 years of research, Japanese charcoal scientist Dr. Makoto Ogawa showed repeatedly that *mycorrhizae* fungi have a distinct preference for char. Their rate of spore formation is greater with biochar, and a greater percent of spores germinate. From cozy quarters in char, these fungi more rapidly grow their feeding networks of *hyphae.*

These symbiotic fungi in char micropores grow dense mats of whisker-thin white fungal threads that extend throughout surrounding soil. Each thread is a tiny tube with an acid-secreting mouth at its far end to scavenge water and nutrients. The *mouth* dissolves ions such as phosphorus from soil and rock to pipe to fungi living in char. Thus, mycorrhizal networks actively, efficiently gather and concentrate water and nutrients in bits of char. In this adsorbed state, nutrient ions are readily bioavailable to microbes and rootlets. Thus, *mycorrhizae*'s networks of hyphae become extensions of a plant root system, improving the uptake of nutrients and thus tolerance for drought and other stresses. Recognition of these benefits has quickly made inoculation with *mycorrhizae* a standard practice in production agriculture, both conventional and organic.

I have seen plants quickly become intimate friends with char. The very finest root hairs easily penetrate biochar's micropores in search of water and nutrient ions that are *sorbed* inside. Roots will grow into bits of char and tenaciously hold onto the biochar, even after being pulled from soil. Char and fungi thus become extensions of plant roots, upgrading its ability to find and get water and minerals from soil, even in dry or deficit conditions.

Growth and proliferation of *mycorrhizae* sets the stage for a further transformation of soil structure and function. Dense mats of fungal hyphae scavenging through the soil eventually die and wither, leaving behind a tiny trace of sticky protein. These fungal residues accumulate over months to become a significant SOM fraction. In 1996, in a quest to find the *missing carbon* in soils, USDA research scientist Sara F. Wright discovered glomalin, named for the *Glomales* order of fungi (Figure 10.6).

FIGURE 10.6 *Fungal mycelium* cover leaves, searching for food.

Missing Carbon

Glomalin glycoprotein produced by arbuscular *mycorrhizae* in soil and roots accounts for 27% of soil carbon, a major component of SOM. Humic acid is only about 8% of soil carbon. Glomalin is 30%–40% carbon, stored as both protein and carbohydrate. Fungi use carbon from plants to grow glomalin as hairlike filaments, called *hyphae*, that function as pipes to funnel water and nutrients—particularly phosphorus—to plants.

Glomalin permeates SOM, binding to silt, sand, and clay particles. Because it is sticky, glomalin glues soil granules into clumps called *aggregates* that enhance soil structure and give soil *tilth*—a texture growers judge by flowing granules through their fingers—stable, to resist wind and water erosion, but porous, to let air, water, and roots flow through, harbor beneficial microbes, hold more water, and resist soil surface crusting.

Glomalin is causing a rethinking of SOM, including strategies for carbon storage and soil quality. Glomalin weighs up to 24 times more than humic acids, once believed to be the main source of stable soil carbon. Glomalin also keeps other soil carbon from escaping, thus boosting carbon stores.

Glomalin is unique among soil components for its strength and stability. Other soil components with carbon and nitrogen do not last long. Microbes quickly break them into by-products, and plant proteins are degraded quickly in soil. Research estimates hyphae have a life of days to weeks, but glomalin lasts up to 50 years, depending on conditions, longer than most soil carbon.

Science has only begun to peer into the microscopic world of the Earth's smallest organisms and most ancient communities. The sodium hydroxide used to separate humic acids from soil misses most glomalin, and it was thrown away with insoluble humus and minerals—soil's *missing carbon*. Glomalin and other benefits of mycorrhizae in soil are just early glimpses into the intelligence and function of these least of all life forms.

Soil Reef: Microbial Communities

Microbes are not solitary creatures. They prefer to live in communities with other microbe species and larger organisms. Microbial associations are interdependent and interactive and share and recycle nutrients, providing mutual services. They collaborate to build infrastructures to gather resources, distribute nutrients, and share information. These microbial communities can adapt to changing environments to sustain a stable habitat.

Many organisms provide special services that benefit the larger community. Superstable biochar offers microbes long-term residence to develop full diversity and supporting services. Fully functional cultures of microbes with enough diversity can adapt and respond to changing food sources, climate, geology, and other symbiotic organisms.

Consider the lowly earthworm—plowman of native soil. Recent research discovered earthworms are farmers. They eat raw plant wastes then regurgitate partially digested pulp to line their tunnels. Fungal spores sprout, encase the plant biomass, and further digest starch, cellulose, and protein. Earthworms return later to graze on the fungi.

Studies of infertile Amazon clays show that adding tons of raw biochar to poor soil actually retards plant growth for 1 or 2 years. But after that, plants grow remarkably better each year, with less fertilizer. It seems biochar must undergo a gradual transformation before it is able to sustain strong plant growth.

The primary reason it takes several months for new char to boost fertility is because first, microbes must colonize the char. This microbial population explosion consumes most immediately available nutrients. But after microbial communities and networks are established, they gather and make available surplus nutrients directed to roots and plant growth.

GLOBAL COOLING AND FOREST REGENERATION

EUROPE'S LITTLE ICE AGE TRIGGERED BY AMAZON FOREST REGROWTH

September 2010, six scientists in *Annals of the Association of American Geographers* reported that Europe's Little Ice Age was due largely to post-Columbian collapse of Amazon population (*Columbian Encounter and the Little Ice Age: Abrupt Land Use Change, Fire, and Greenhouse Forcing*; Robert A. Dull Richard J. Nevle, William I. Woods, Dennis K. Bird, Shiri Avnery, and William M. Denevan).

Pre-Columbian farmers in Amazon lowlands—estimated at 25 million by 1492, 80+ percent living in forests—burned and cleared forests for agriculture. Population pressure on forest resources increased steadily, peaking after European arrival. European diseases caused epidemics, resulting in a population crash in the Amazon. An estimated 95% of indigenous inhabitants perished by 1650.

Fire history, high-resolution charcoal records, and demographic estimates show Amazon lowlands went from being CO_2 source before Columbus, to carbon sink for centuries after Columbus. Amazon forest regrowth after Columbus led to carbon sequestration of up to 5 Pg, contributing to well-documented CO_2 decreases seen in Antarctic ice cores from 1500 to 1750.

Post-Columbian carbon sequestration was a global cooling process that caused Europe's Little Ice Age. This was previously attributed entirely to decreased solar irradiance and increased global volcanism. But new evidence supports the theory that depopulation of the Amazon resulted in rapid forest regeneration, particularly on *terra preta* and *terra mulatta*. Added carbon fixed from the Earth's atmosphere by this tropical forest regeneration was sufficient to lower the Earth's thermostat.

Thus, the evidence of Earth history, read to us by science, confirms that global cooling is not only possible, but is achievable.

Tragically, in 2013, man-made carbon emissions are still climbing, and the rate of rain forest loss is still increasing. Humans not only are making climate change troubles greater but have yet to resolve to take the actions required to turn around our direction and certain destination.

Microbial Nitrogen Cycle

Of special interest to soil science are nitrogen-fixing (N-fixing) bacteria that use a *nitrogenase* enzyme to create usable nitrogen from inert nitrogen gas in the air. In the 1800s, French chemist Louis Pasteur, the father of microbiology, discovered the *Rhizobium* bacteria that live in pink nodules on legume roots, where they convert nitrogen that is passed to the plant. These microbes are well known and widely used by farmers and gardeners as seed inoculants. But over 200 other bacteria convert nitrogen gas into plant food. Each species adapts to specialized environments, often using unique enzymes and metabolic strategy. Most are symbiotic with particular plant species. All depend on certain *helper* bacteria.

High nitrogen levels in *terra preta* suggest that certain nitrogen-fixing bacteria take up residence in char. Ongoing research is peering into biochar's microbial ecology to learn which microbes prefer char as a host environment. If biochar hosts special nitrate-producing microbes, farmers have added pathways to add natural nitrogen to soils and decrease their dependence on buying imported synthetic, fossil-fueled nitrogen.

Not just nitrogen-fixing bacteria, but all the organisms that the drive the nitrogen cycle seem to reside in char and work together to support full function fertility in a stable, balanced soil. Some bacteria convert nitrate to ammonia. Other organisms flip ammonia back to nitrate. They seem to coexist in char and keep the form of nitrogen in soil balanced. Research must identify microbial strains, understand their interactions, and culture colonies for specific soils, crops, and climates.

Feedstocks: Weedy or Woody?

Photos (Figure 10.7) at the lower left show microscopic slices of eight hardwoods. It is obvious the microscopic structure of each wood is different, unique, and complex. Each type of biomass is so specialized wood species can be identified by their different cell and cellulose structures in a microscope. Similarly, each biomass type has different chemical composition. This feedstock variability adds another confounding layer of complexity to this carbon sequestration strategy.

Thus, biochar characteristics vary greatly with each biomass type. Even given the same species, many factors alter biochar quality: growing conditions, soil minerals, harvest method, postharvest processing, and more. Thus, biochars differ in particle size, porosity, surface area, pH, and biologically available compounds. Most biochar pores are at least 5 μm in diameter, and some are 20 μm. Production methods vary temperature during pyrolysis, time of heating, pressures, oxygen content, moisture content, and other parameters that affect the final char. Because of structural and chemical differences, some biochars break down more quickly in soil than others.

To understand and assess this complexity, and develop uniform, reliable standards, is an immediate technical challenge. IBI has invested 3 years to address this complexity and draft professional, universal standards to characterize biochar for commercial certification.

A fundamental question is: "woody or weedy?" By reflex, we believe char is made from wood—in particular, dense hardwood, such as oak—to burn for fuel. Like charcoal and campfires, we expect biochar to be made from woody trees, maybe brush.

But for soil use farm biomass is very likely more desirable than forestry wastes. Annual plant biomass such as straw, hay, cornstalks, seed husks, weeds, and other grassy, weedy debris may prove more suitable in soil than dense, tough, chunky wood.

Oak exemplifies this principle. It is very mineralized, and char from this dense hardwood is lumpy and hard as rocks. Much energy and machinery is needed to crush oak char to soil particle sizes. Even as dust, oak char is dense, heavy, with smaller micropores and tighter internal spaces.

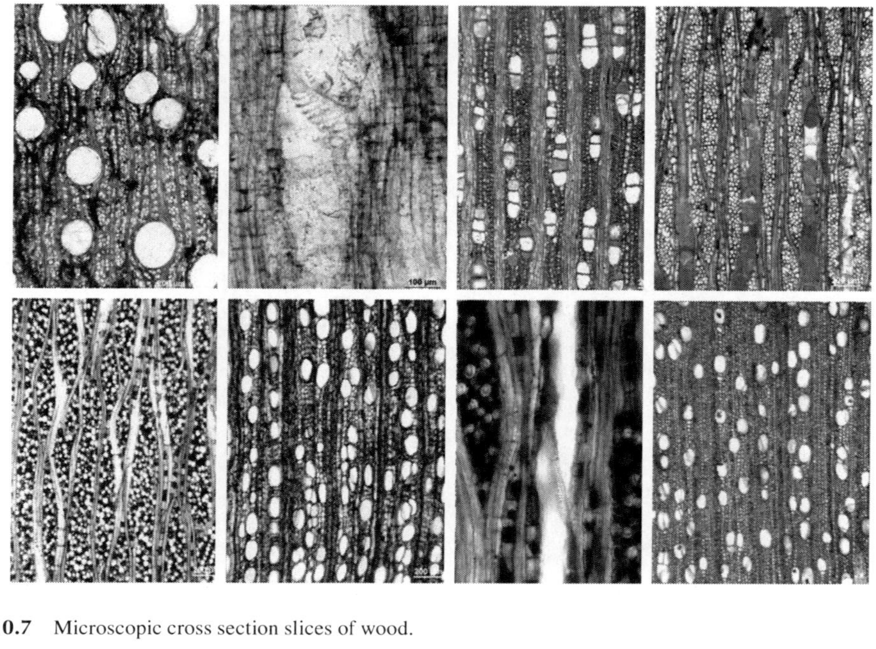

FIGURE 10.7 Microscopic cross section slices of wood.

On the other hand, char made from weedy biomass is quite light—as fluffy as downy feathers. Weedy char will crush in your hand to a fine powder that will disperse and vanish into soil. Such soft, fluffy, fine textured char has better effects on soil structure and adsorption capacity.

Preparing Biochar for Soil

In November 2007, scientists at the USDA National Laboratory for Agriculture and the Environment (NLAE) in Ames, IA, began multiyear field trials to assess biochar effects on crop productivity and soil quality. Scientists amended almost 8 acres with biochar made from hardwood. Twelve plots got 4 t/acre; 12 got 8 t/acre.

BLENDING BIOCHAR WITH COMPOST

We have no hard numbers on biochar-to-compost ratio. Farmers I work with use five parts compost to one part biochar (crushed, average size 3 mm, like coarse sand) by volume with good anecdotal results. We do not have replicated, statistically valid trials at this point.

Wayne S. Teel
James Madison University

Most compost we see is done with wood char, but biochar from grass and manure will work. We do not see much char from grass except energy crops like switchgrass. In Germany, bioactivation is done primarily by wood char with manure.

Two operations find poultry litter char composted with dairy manure is a great combination. Carbonized poultry litter provides compost with nutrients removed in the dairy flush.

Pyrolysis is in the absence air. Gasifiers are slightly oxidizing, as is staged combustion (e.g., high carbon boiler ash).

Particle size is highly variable. Lots of theories; no rules. It depends on porosity of other materials in compost and what is the bulking agent. Aeration is an important attribute of char in compost.

In Germany, Spain, and Japan, compost is up to 50% by weight (w/w), mostly 5%–25% by volume (v/v). We blend 1:2 (v/v) biochar/organics (municipal yard waste) to end with 1:1 compost/biochar blend to use in stormwater bioretention trials.

When composting grasses, your hands get black at 15% biochar: 85% organics by volume (1:6 v/v). If organics lose half their volume, the end is 1:3, 25% v/v (about 12% biochar by weight). That works well for tree nursery growing media.

As little as 5% biochar by volume is beneficial in compost. We put that much in kitchen compost (food scraps), and worms love it!

Tom Miles

They found no significant difference in the 3-year average grain yield from either treatment. Other USDA field and laboratory studies in Idaho, KY, Minnesota, South Carolina, and Texas showed hardwood biochar can improve soil structure and increase sandy soils ability to retain water. But soil fertility response was more variable.

USDA scientists violated four key principles for biochar use:

1. Bulk char, in one large load
2. Raw, uncharged char
3. Sterile, uninoculated char, with only a tad of microbial life
4. Synthetic salt fertilizer, tillage, and other antibiotic practices

After all, soil may get 25 or more inches of rain a year, but not all at once in a single event. Biochar, like water, is best added in a series of small doses so soil has adequate time to distribute and digest it. We already know from Amazon research that dumping 5, 10, even 20 t of raw char all at once into poor soil retards plant growth for 1 year, maybe 2. But after that, plants erupt in impressive, vigorous growth.

But a dip in yield is not acceptable for production agriculture. Farmers cannot wait a year to harvest a profitable crop. Professional growers need fast response and strong stimulus to growth. Economics and handling logistics require convenience and low cost, with vigorous growth from minimal applied material.

Terra preta was not made with just char or created in a single fortnight. Clearly, char initiates a process that transforms soil, and this requires an extended period of months. Other ingredients are needed to create *terra preta*. According to scientists studying Amazon village "middens" know indigenous people added all kinds of *cultural debris* to soil, including garden debris, kitchen wastes, building material, and clothing—even urine, feces, and kid toys. In a culture with no metal, plastic or, synthetic substances, everything is organic and fully composts.

Ultimately, *terra preta* is highly fertile, not due to carbon, but microbes. So, while microbial populations initially grow, diversify, and become well established in fresh char, nutrients are delivered to microbes and are not available to plant roots. But once char is fully inhabited and soil biology is fully alive, productivity rises each year to soon surpass chemical fertilizers. Multitudes of symbiosis in fully functional microbial culture take over the feeding of plants.

Fortunately, we are learning how to prepare char for optimum effects in soil and on crops. Biochar research in America is hardly 10 years old, but solid research shows that properly prepared, intelligently applied biochar has dramatic effects on soil structure and plant growth at as little as 500 lb/acre.

To prepare biochar for optimum effective use in soil, there are four fundamental steps—*the 4 Ms*: *M*oisten, *M*ineralize, *M*icronize, and *M*icrobial inoculation.

Moisten: First Ingredient of Life

The first step is to add enough water to moisten char without becoming waterlogged.

Fresh from a production burner, char is bone dry. It is heated to over 500°C and hardly has a molecule of water in it.

But water is the first ingredient to cook up biological life. Without water, even earthworms avoid char. But properly moist, worms like char mixed with their food and microbes rapidly move in to colonize the char.

Fresh char is not just dry, it is hydrophobic, it actually resists water penetration. Residues of tar and resin left in the char are oily hydrocarbons and repel water. Until thin films and beads of tar are etched out of char, it will not easily accept water.

However, char left lying on soil a few months loses those black beads of resin, and its color shifts from sparkly black to lusterless gray. Microbes inhabiting char see those hydrocarbon residues as food. Carbon chains and rings contain electrons and energy, so bacteria and other organisms eat it like candy.

Without water, char is very dusty. Fresh char is weak, brittle, and shatters easily. So, dry char easily sheds fine black dust that hovers around like a dark cloud. This dust is hard to handle, easily airborne, not healthy to inhale, and blows away in the wind. And yet, that very fine dust is the most precious portion to add to soil and transform its structure because it most widely and intimately inserts itself between soil particles.

In normal production, water is used to kill the fire that makes the char—to cool it down and stop the charcoal fire. People beginning to make char are surprised how much water is needed to extinguish a charcoal fire. A lot of heat is held in char, and char's micropores soak up lots of water. Often engineers quench fresh, hot char by dumping it in water. However, too much water yields char that is soggy, sticky, and heavy, which makes handling messy, and screening for particle size difficult. Water-logged char is anaerobic and poor habitat for beneficial microbes.

But, with careful attention and proper protocol, a minimum of water will put out a fire, yielding lightweight char that is easy to process. With the right moisture, char is not dusty, but cohesive enough to hold together in soft clumps that are easy and safe to handle and do not disappear in the wind. And with just enough water, char is suitable media for strong colonies of microbes, and earthworms are attracted to it.

Micronize: Intimate Relations

The second step is to reduce particle size.

Smaller particles disappear into soil quicker, mixing more thoroughly and intimately with soil particles and organisms. Thus, crushing, grinding, and screening char are valuable to increase char's dispersal throughout soil and optimize its effects on soil structure, ion adsorption, and microbial colonization. On the other hand, if the particles are ground too fine they can clog fine soil pores and impede movement of water and oxygen.

The first benefit of smaller particle size is increased surface area. For water, ions, and microbes to penetrate char, they must enter at an exterior surface. Smaller bits have more total surface available for absorption and adsorption. A 1 in. chunk has a surface area of—at best—6^2 in. The same chunk shattered in a thousand fragments has thousands times more surface area. Due to extremely fine microporosity, 1 g of biochar has over 4000 m^2 of surface area, and 12,000 is achievable. Water, nutrients, and microbes quickly get inside smaller particles and access interior spaces.

Smaller particle sizes also distribute in soil more widely, more intimately. Dust—the smallest particles, smaller than most soil particles—inserts itself between soil particles. Carbon isolates soil granules, insulating their electric charges. Thus, clay is less sticky, while sand has more cohesive body.

Smaller particles hold water better, because water penetrates more easily and quickly into char's sponge-like micropores. Large chunk of char have difficulty drawing water into its deepest recesses and do so slowly.

Similarly, smaller particles allow ions better penetration into the char's sponge-like internal micropore matrix. A large chunk of char has difficulty drawing ions into its deepest interior spaces.

Ultimately, think like a microbe. What size micropores are fit for bacteria? What size will satisfy a fungus? Rice grain kernels of char are large enough to house thousands of microbes. A 1 in. chunk of char is a microbial metropolis—millions of denizens inhabit and share such a charred carbon matrix.

Because char performs an assortment of services to soil, a variety of particle sizes seem best. Rice-grain-size char is large enough for large microbial communities. Powdered char provides condominiums for microbes. Fine dust is most effective to separate soil particles and shift soil structure and tilth.

One advantage of weedy biomass is its char easily crushes to dust in your hand. Minimal effort and machinery is needed to create extra fine, fluffy char, and such char seems to further enhance soil structure and boost its CEC and AEC. After a few years of field trials, we may decide weedy char is better than woody for many agricultural soils and crops.

One fascinating facet is water-soluble complex carbons. Tiny bits of char—the finest dust, with up to 100 carbons—are small and light enough to suspend in water. These extremely small carbon molecules can be harvested by rinsing fresh, dusty char with water. They disappear in water, making it slightly dark. The microparticles are useful in foliar and other sprays, where minerals ions and nutrients are packaged in these ultralight carbons. Such nutrients are more easily and efficiently assimilated through leaf pores. And inside a plant, the carbons strengthen plant structure and energy.

Mineralize: Charge the Soil Battery

The third step is to add minerals to biochar.

Soil is a battery that stores electric charge. Electrons and ions are electric charges that adsorb onto soil particles, especially SOM and biochar. The ability of soils to capture and hold these charges—both positive (CEC) and negative (AEC)—creates a fundamental electric storage capacity—like the electric potential, or ampere-hours, of a battery. Beyond simple quantity, it is important to also examine how soils can easily and quickly make these electric charges available to organisms.

Both SOM and biochar have remarkable capacity to gather and store both negative (electron) and positive (proton = H+) charges and are highly efficient to hold and deliver ionized minerals and their electrons to cells. SOC's large, complex molecules, with their embedded minerals and multiple surface charges, are especially effective to capture and hold free electrons. The more electric charges char can store, the more energy soil has to deliver to growing microbes and plants.

Charging this biological battery begins by adding ionized and ionizing minerals. Cations, although they have a positive charge, contribute electrons to the soil battery by giving up their valence electrons

FIGURE 10.8 Carbon super battery. In 1962, science discovered *graphene*—pure carbon atoms in a regular hexagonal pattern as a one-atom thick monolayer sheet. Atoms are densely packed in flat honeycomb lattice, best visualized as atom-scale chicken wire. Graphene is very light; a 1 m^2 sheet weighs only 0.77 mg.

and thus deliver these fundamental mobile charges to power cell biology. Meanwhile, anions are electron receptors (or acceptors) that hold and safely transport electrons around soil or inside a cell and deliver them to metabolic reaction sites (Figure 10.8).

Research now makes graphene batteries that charge 100–1000 times faster, are inexpensive, nontoxic, and surpass current battery technology in efficiency and performance. Graphene nanotechnology deposits an atom-thin layer of carbon on an electrode. This super thin coating quickly adsorbs huge numbers of electrons, stores that charge, and slowly meters electrons out for use.

An iPhone graphene battery can charge in 5 s. MacBook graphene battery can charge in 30 s. Electric cars with a graphene battery can charge as quickly as filling a car with gas. Besides superfast charging, the battery has few negative environmental impacts and is biodegradable, even compostable.

Biochar's high adsorption capacity makes it an ideal delivery system for minerals and their electric charges. Electrons and ions adsorbed onto and into biochar are safely, efficiently placed in the root zone and kept there, ready for ion exchanges with plant roots. Char-adsorbed minerals are removed from the soil solution and thus have minimal mobility to leach and outgas. So, any electrons in char are kept in the root zone, in locations that attract plant roots.

Micronized minerals that are finely powdered are more able to blend into intimate contact with bits of char and thus deliver electric charges where they are needed. Like micronized char, smaller particles mean greater surface area and faster, easier digestion by microbes. Stone meal or rock dusts are natural, insoluble forms of minerals that can be mixed with char to charge the soil battery. Synthetic, soluble commercial fertilizers can also be micronized or solubilized to blend into char to supply electric charge.

Most farm soils have deficits and/or imbalances of major minerals and thus require amendments. Similarly, char made from biomass grown in mineral-deficient soil will be short of essential elements. Any soil test quickly, cheaply reveals what minerals are needed by soil and in what amounts. Adding these minerals to soil directly can create losses and lower effective use of the minerals. But blending these minerals with biochar adds needed nutrients in a high-efficiency, targeted delivery system.

Full Spectrum Fertility

Nature's best source for trace elements is sea minerals. For over a billion years, minerals from the land washed into the sea. Every element that dissolves in water is in the sea—including nano- and picoelements. Seawater is a full spectrum source for every and all elements that are essential for life.

Sea minerals are in ratios that are ideal to sustain cells and organisms. This mix of minerals was blended, sorted, stirred, precipitated, titrated, and energized by geological and biological processes of Earth evolution. For example, early in planetary history, bacteria in the sea ate iron and precipitated this magnetic mineral out of solution. Later, cells evolved that fixed calcium and phosphorus to create crystal shells to shield their soft bodies. Many biological and geological processes have created the mix of minerals in the sea. Thus, this mix is no accident, but a full menu of all minerals needed by biology in a deliberate, balanced blend, created by intelligent evolution.

Given the weak, depleted, deficit state of most farm soils, I recommend charging biochar with sea minerals as a way to deliver the full spectrum menu of elements needed by biology. Immediately after making char, douse the char with sea minerals dissolved in water. Adding this solution to fresh, hot char accomplishes five tasks in one operation:

1. *Extinguish*: Sea mineral solution puts out the charcoal fire by cooling and by steam.
2. *Moisten*: Hot char flushed with water generates steam with enough pressure to force itself into char micropores.
3. *Fracture*: Steam pressure strong enough cracks open char to expose inner area as external surface, boosting adsorption.
4. *Scour*: Alkali ions of sea minerals loosen and remove tar and resin residues from char's micropores.
5. *Mineralize*: Sea mineral ions adsorbed onto char's electric charge sites load carbon matrix with a full menu of elements.

Then, when microbes and roots invade char micropores, they find fundamental food all cells require, already in solution, ready to feed them. This elemental feast assures soil a foundation of essential nutrients to support rapid, vigorous, healthy growth.

Biochar's AEC also allows it to adsorb and supply negative ions, mostly nitrates and phosphates. Nitrogen and phosphorus were #1 and #2 fertilizers in twentieth-century farming. Sea minerals do not supply much of either, so N and P must be added in various forms. Biochar is a high-efficiency delivery system for anions, with high bioavailability and minimal loss by leaching. In properly prepared biochar, N and P are present not only as nutrients but also as organic biomolecules of living soil biology. Microbes living in biochar are an ideal medium to retain and recycle these anions.

Microbial Inoculation

The fourth step to prepare biochar for soil is to add life to it. With water, nutrient ions, and vast, empty micropores in the char, microbes move in. We do not eat our houses, and microbes do not eat char. They live in it.

In the paradigm shift from twentieth-century chemistry to twenty-first-century biology, the culture and care of symbiotic organisms is crucial for soil fertility. For endless eons of geological evolution, microbes managed and improved soil to sustain fertility. Creating fertile soil is a fundamental job description for these *no-see-ums* of soil. These least and smallest of all life forms are among the Earth's most ancient communities. Yet, we ignore what we cannot see—microbes—and focus on obvious, visible bulk ingredients—OM, compost, and mineral fertilizers.

Bacteria are Earth's oldest life forms. While a single bacteria is simple, as a collective community or culture, they are complex and more intelligent than plants. One expression of this is that it is primarily bacteria that consume trace elements and build them into complex biomolecules that perform key, often fundamental, metabolic and regulatory functions.

If minerals are the foundation of biology, microbes are the sill plate—where foundation meets superstructure. Microbes transform minerals into protoplasm in living cells. Bacteria are the primary consumers of mineral nutrients. They also synthesize critical biomolecules that more complex organisms require. For example, all B vitamins are synthesized by microbes, who then supply them to plants and animals.

GRANDFATHER'S CHARCOAL GARDEN

I have been using biochar in my organic nursery over 50 years.

My grandfather was a charcoal burner in Sussex, England. He brought home smaller pieces, which he fed to his dogs (and me). He kept his charcoal in a hessian sack next to his compost. One day rain ran off compost to soak the charcoal. Grandfather was so mad he threw the charcoal on his garden and got the best yields ever. This was his *secret* for years, passed on to my father, then to me.

A few years ago, I learned he was not the first to discover charcoal's benefits. Amazonians did a few thousand years ago. He did not know charcoal is changed to biochar by inoculation with beneficial bacteria and microbes. I realized growing organic with biochar gives better yields and nutrient-rich crops.

Adding inoculated charcoal to poor soil and using chemical fertilizer has no benefit, as chemicals kill bacteria that extract nutrients from soil. The greatest benefit is adding biochar to soil amended by OM with healthy bacteria. Biochar absorbs water, filters runoff, and sequesters carbon.

A farmer must see immediate results and learn why these results duplicate themselves. He must also learn to add rockdust, bonemeal, and other OM. It is cheaper and better for his crops, family, and profits.

Ken Bourne, 10-03-2012

For example, vitamin B_{12}—the so-called "vegetarian vitamin"—is not made by animals but only by a bacterium. B_{12}'s mineral cofactor cobalt has six valence electrons in outer orbitals and thus makes six bonds to other atoms. B_{12} bacteria build a complex structure of carbon rings to enclose an atom of cobalt. Unlike chlorophyll and heme, whose carbon rings are in a 2D, flat disk, B_{12}'s more complex carbon rings form a 3D structure. Thus cobalt's magnetic energy is focused and harnessed to perform certain critical, universal energy exchange functions needed for DNA replication, hormone synthesis, and red blood cell formation.

Thus, these simplest and least of all life forms digest trace elements to transform them into key biomolecules that are metabolic catalysts and regulators in larger organisms. Similarly, bacteria assemble minerals and biomolecules into primary units of protoplasm, which are fed to cells of larger, more complex life forms, including plants and animals.

Studies show char is an ideal substrate to culture many beneficial microbes. Features that make char ideal for water filtration media also make it optimum habitat for the smallest microbes. Larger particles soak up nutrients and water. With essential nutrients abundant and bioavailable, microbes take up residence inside char. Each bit of biochar becomes a host for colonies of bacteria and fungi, the primary cycle of the soil food web, and their presence anchors the rest of the soil biology. Larger char particles provide enough space for large-scale, fully diversified microbial communities to become established and nurse the growth of their food-seeking networks.

For example, mycorrhizal fungi take up residence in char, proliferate by abundant sporulation, send whisker-thin hyphae out into soil to search for water and nutrients, and pump them to fungi living in char. Thus, char becomes an active storehouse to stockpile essential nutrients. Fungi interact with other organisms in symbiotic networks to create complex feeding webs. This interactive community helps soil breathe, absorbs water, mobilizes nutrients, and supports larger life forms. Together, this living community forms intelligent, adaptable, and resilient systems.

Thus, char preparation's final step is inoculation with a wide diversity of microbes to jumpstart their presence and function.

Biodiversity of Digestion

As the emerging twenty-first century biology paradigm spreads, it is spawning businesses and services that offer a variety of new inoculants to enliven soils. Many new probiotic products are becoming available on the market.

Most often, simply and easily, char is inoculated by blending it with compost. Properly mixed, in just a few days, char becomes colonized by beneficial microbes. These microbial colonies form a *culture* of interactive organisms that dramatically improve char's value as a soil enhancement. Compost not only provides microbes biochar to inhabit but also supplies partially digested OM to feed microbes as they get settled into their new homes.

Compost and biochar are not competitors contesting for available raw materials, but complementary. Wet biomass is best made into compost, while dry biomass is ideal to convert to char. After production, char and compost are perfect partners for soil. Fresh char is lifeless and needs compost to populate its empty micropores. Char is an ideal media to transfer microbes into new biomass and kick-start compost's digestive processes. Compost benefits from char's superstable habitat for microbes.

Unfortunately, quality compost teeming with a full diversity of microbes is hard to find. Compost is often treated as inert OM, and a way to dispose of *wastes*. Much compost is made from high carbon monomaterials, like wood chips or leaves, and lack fully diverse microbial cultures. Inadequate attention is given to assure strong, diverse, multifunction micro-organisms for soil digestion.

Thus, for biochar inoculation, it is crucial to select compost that delivers a full diversity of microbes. One solution is to use compost made with animal manures, especially from herbivores. Animals with ruminant digestive systems have far more diverse and active microbial cultures. It also helps to include a diversity sources of raw biomass that break down into a variety of organic substances. The highest-quality composts even include herbs, mineral supplements, and even biochar.

In Europe, newly developed EU protocols recommend most biochar is fed to animals. This accomplishes microbial inoculation by passage through animal intestines. Indigestible char moves through an animal gut to be spread with the manure on soil, fully colonized by a full range of digestive organisms.

Currently, several microbial inoculants, inoculation strategies, and microbial starter foods are available for purchase.

Rudolf Steiner, in early twentieth-century Germany, was among the first to envision holistic soil microbiology. Steiner taught biodynamic agriculture, an advanced farming method that uses eight preparations (BD preps). Some Steiner preparations are microbial cultures and stimulants to inoculate soils with microbes and nurse these cultures into optimum activity. Most relevant to soil microbes is Steiner's cow horn prep.

Dr. Teruo Higa, in 30 years of research in Japan, developed effective microorganisms (EMs)—cultures of *Lactobacillus*, photosynthetic bacteria, and algae. EM is not just a product but a complete technology that includes techniques to incubate and propagate cultures, application methods for many situations and environments, and extensive scientific testing.

Dr. Michael Melendrez, founder of Soil Secrets in New Mexico, first coined the phrase *soil food web* in a 1976 research paper. He developed high-quality humus-based inoculants and microbial foods to transform extreme soils of the Southwest. His passion is to create North America's largest collection of oaks, but he first had to transform New Mexico's alkaline, arid soils.

Dr. Elaine Ingham popularized *soil food web*, and perfected compost tea as a way to propagate and distribute soil microbes. Compost tea involves adding a little compost, with all its digestive organisms, to a lot of water, plus essential nutrients—minerals, sugar, and OM. Air bubbled through the water (or whirled in a vortex) for 24 h adds abundant oxygen to nurse a microbial population explosion. This brewed tea is sprayed on plants, soil, and biochar as inoculant. Thus, microbes in a small amount is compost proliferate to cover a large land area or biomass volume.

Dr. Michael Amaranthus at Oregon State University authored over 70 scientific papers on mycorrhizal fungi and their use in soil restoration and maintaining productivity of plants. He created a full array of mycorrhizal inoculants and had two mycorrhizal genera and many species named in his honor. His company Mycorrhizal Applications produces the world's largest selection of mycorrhizal inoculum used by landscapers, farms, nurseries, soil restoration and erosion control companies, and distributors.

Many new probiotic microbial products are appearing each year. For example, SCD Probiotics, a new company in Kansas City, offers a wide range of microbial cultures packaged for varied applications, including agriculture, livestock, household, compost, and personal health. Their BioAg™ product, with lactic acid bacteria, yeast, and phototrophic bacteria, is used as a liquid biochar inoculant by soaking char in a solution for 48 h.

A special concern is nitrogen-fixing bacteria, since nitrogen, the fourth organic element of life, is a major growth-limiting nutrient. A National Science Foundation report cataloged over 250 species of nitrogen-fixing bacteria. Each inhabits its own specialized environment, including strains that live in symbiosis on the roots of specific host plants, such as *Rhizobia* on legumes. Of particular interest for biochar are free-living N-fixing bacteria.

High nitrate levels found in Amazonian *terra preta* suggest that biochar hosts unique strains of free-living, nitrogen-fixing bacteria. Thus, it seems properly inoculated char can provide additional ways to fix nitrogen into nitrates. The more the needs and nutrients of these specialized bacteria are readily available, the faster they will proliferate and spread in soil. Biochar plus trace elements can meet these requirements.

So, a key issue is to find these strains of free-living bacteria and be sure they are inoculated into new biochar.

Carbon-Smart Farming

In 2006, Virginia Tech University began field trials adding biochar to soil growing tomatoes, potatoes, and sweet corn. Virginia Tech research demonstrated and highlights that biochar properly prepared by being charged with minerals and inoculated with microbes can achieve remarkable results with very low rates per acre, instead of several tons per acre—instead of several tons per acre, only a few hundred pounds acre, even less, gave significantly greater yields. These results are immediate, not in the second or third year.

Biochar was supplied by Carbon Char, a new company producing inoculated biochar. Jon Nilsson, an east coast soil scientist and compost expert, guided Carbon Char to create a biochar-based inoculant with beneficial microbes, substrates, and microbial food. In initial research with sweet corn, Nilsson found that biochar-based inoculant applied in the planting row at rates as low as 7.5 lb/acre saw significant increases in yield and mycorrhizal colonization of roots.

When a biochar-based inoculant was used at 2.5% by volume in a transplant growing mix, it had the following results:

- 30 lb/acre savings in nitrogen for potatoes (2006).
- 10% increase in sweet corn yield (2006–2007).
- 22% increase in tomato yield (2007).
- 51% increase in tomato yield at first pick (4 year average).
- Tomato yield was 1–2 weeks earlier than untreated plots.

Tomato yields were achieved with 2 cups of char plus inoculant in 5 gal of transplant potting mix. In 2010 trials, 4 cups per 5 gal of mix was used, resulting in even higher yields.

Jon Nilsson reports, "Virginia Tech research was replicated results over 4 years. Trials were not based on biochar alone, but also included changes in cultural practices:

- Soil had decent fertility when we started.
- Low salt fertilizer used at transplant time (fish emulsion).
- Cover crop turned in prior to planting.
- Low salt fertilizer at low amounts in growing season."

"We saw changes in the field through soil penetrometer readings. If my stuff works, who cares what I call it, or what your beliefs are. Bottom line is beneficial microbes are salt intolerant, and much easier to kill than a germinating seed. Farmers using high-salt fertilizers won't benefit from this technology."

Jon Nilsson tells farmers, "Microbes are more efficient than any product you ever bought. They work 24/7, and are lots cheaper than fertilizer. Inoculated products lower costs, yet maintain or increase yield. That's money in a farmer's pocket."

Jon is emphatic: "Agriculture will turn on a dime when bio-inoculants show farmers permanent changes to soil fertility and cost reductions. My job is to cut costs and increase yields." (For further information, see Chapter 29).

This is our twenty-first century challenge: to transform farming from antibiotic to probiotic. The future of farming is not methods that just increase yield, but rather create a full menu of soil minerals, regenerate microbes and the entire soil food web, and grow nutrient-dense foods. And the future of humanity hinges on farming that restores carbon to soil and reduces global warming due to the greenhouse effect.

Key to Agriculture Transformation

To change agriculture, a key principle is that farmers grow food for a market. Farming is not a hobby. It is a business to produce profit. Before buying seed and fertilizer, farmers need markets to buy the food, feed, and fiber they grow. If the buyer is ADM or Cargill, farmers will grow GMO grain. In recent decades, consumers demanded organic, so farmers adapted to grow foods for that special market.

If consumers demand carbon-sequestering food, then farmers will change how they farm to supply that demand. If shoppers buy carbon-smart food in preference to other food, farmers have a market-based reason to adopt carbon-conserving methods. And if consumers pay premium prices for carbon-smart food, this is a potent financial incentive for a carbon farming revolution.

Currently, no market exists for *climate-conscious* or *carbon-conservative* foods. Thus, farmers lack a monetary incentive to apply biochar to soils. Without markets for carbon-smart foods, only altruism will motivate change to farming practices. Without financial reward, agricultural change will be too slow to alter our oncoming climate calamity.

For consumers to buy carbon-smart foods, they need a way to identify food grown by methods that remove carbon from the Earth's atmosphere. Shoppers have no clue what foods decrease our planet's CO_2 burden. Consumers have no way to recognize such foods, and select them. Nobody educates them about carbon-smart shopping or assures that products are authentic and effective to reduce our carbon footprint.

The first key issue is consumer choice. To encourage shift to a green economy, consumer products are tagged to indicate their energy-saving value. Appliances have Energy Star tags for efficiency. Cars are rated by miles per gallon. Even buildings are assessed by a LEEDS rating. And a battle is underway for consumers' right to know if foods contain GMOs.

So too, consumers need a label to identify foods that reverse our carbon footprint and sequester carbon. A simple, uniform system must define and mark carbon-sequestering foods to track them from farm to market and assure point-of-sale authenticity. Marketplace visibility, product identity, and quality distinction are primary purposes for trademarks and brand names.

This is why 25 years ago, *certified organic* was created as a market label for foods grown without synthetic chemicals. A USDA insignia assures consumers of authentic no-chemical food. Without scientific support, with active resistance by corporate industry, and despite a higher price, *certified organic* is now the fastest growing sector of America's food system.

Now, we need a simpler, hopefully easier way to label carbon-smart foods that reverse our farm and food carbon footprint. Consumers need clear choices about the carbon consequence of foods they buy. It is imperative to create a marketplace identity for carbon-smart foods grown by carbon-sequestering methods, produced by earth-sensible, soil-regenerating, climate-mitigating farming, by farmers pledged to learn and use sustainable, global cooling farming and marketing.

Cool Food

A plan has been proposed for a *Cool Food* label—a new trademark to identify foods grown and marketed with materials and methods that remove carbon from the Earth's atmosphere and restore it to soil. First qualification to use the *Cool Food* insignia is to add a minimum of biochar per acre to soil. Also, growers must adopt probiotic methods to increase SOM, SOC, and soil biology. Currently, there will be seven criteria, principally the following:

1. Add a minimum annual amount of biochar to soil
2. Sign a carbon-smart farmer's pledge
3. Soil test of minerals
4. Soil test of organic carbon
5. Soil biology assessment
6. Soil carbon management plan
7. Eco-local, carbon-smart marketing

To issue licenses to use *Cool Food* labels and marketing materials, *carbon-smart standards* must be drafted and published. This must be led by a panel of growers and marketers with experience and insight. The *Cool Food* board must also prepare a simple, practical *Growers Manual of Carbon-Smart Farming* to detail *best practices* to prepare, apply, and evaluate biochar use in soils, information sources, and research.

The immediate task is to create a simple model to show how to link a market label with carbon-sequestering farming. We need to implement a Cool Food label and marketing service to illustrate how to direct food dollars into agricultural transformation and ecological regeneration. More than just a label, this must link eco-conscious consumers with eco-responsible farmers.

Carbon-conservative agriculture does not exist yet as full function farming systems. Carbon-smart farming is not distinct, and well-articulated, with field-tested guidelines. While many basics are known, much must be done to define an agriculture that captures carbon in soil and minimizes emissions. We must create carbon-smart farming systems in the next few years and teach the methods to a new generation of growers.

We need this *Cool Food* label fast, because a melting arctic ice cap and massive methane hydrate release warn we are at a tipping point of global climate calamity. We do not have another decade to fiddle with short-sighted solutions. Failure of America's 2012 corn crop from historic extreme drought in the Midwest previews what lies ahead if we do not curb our greenhouse gas emissions, start sucking CO_2 out of the air, regenerate soils, and repair our planet's damaged climate engine.

Planetary Healing, Not Geoengineering

Today, carbon is the cause of a global ecological crisis called climate change, and this may precipitate another mass extinction on Earth. Conventional science says accumulating CO_2, CH_4 (methane), and NO_x (nitrous oxide) in the Earth's air is driving global warming, overheating atmosphere and oceans, unbalancing the planet's thermal engine, fueling extreme weather, and warming climate. But carbon is merely a scapegoat.

The honest truth is we, the human species, are the cause of this problem. We deforested much of the temperate zone and now clear-cut tropical and boreal forests, too. We plow and poison soil and plant monocultures so soil has less than 1% carbon and is largely devoid of microbes or other life. We burn fossil carbon fuels and dump our combustion wastes in the Earth's atmosphere. We add acids to atmosphere and ocean, stressing the remaining trees, bleaching coral reefs. We dump and pump our wastes in waters, creating red tides and dead zones.

But carbon is not the problem. We are the problem. It is humans that must change.

If we restore to soil the elemental foundation of life—carbon and all the minerals and trace elements needed by biology—then Earth and its ecosystems can heal themselves and restore the balance and biodiversity of life. Restoring carbon to soils in its full diversity and complexity will reestablish the matrix for life's full, complete, complex diversity at every scale, every landscape. Putting carbon back in soils will restore the soil food web, and all the web of life—living organisms and systems that collaborated for millions of millennia to assure a stable, habitable biosphere.

And it may be that returning carbon to soils will initiate changes that support functional, successful, peaceful, healthy human communities. Carbon creates a matrix of structure to support essentially endless complexity and capacity. If the next generations of humans dedicate themselves to the humble, difficult work that regenerates soils from whence almost all life on Earth grows, we may also transform ourselves and our society.

Carbon Creates Community

Carbon is the element of life—the backbone of biomolecules. So, carbon is special among all the elements and is a signature element needed in abundant amounts, with qualities that support and sustain the vast diversity of forms and functions of biology.

At atom scale, carbon has a high bonding energy to link to other atoms. So, carbon is slow to react, but when it does, it forms bonds that are strong, stable, and durable. Carbon's stable bonds give persistent, fixed form to molecules to build structure for living organisms. This also means carbon burns at high temperature. So, in arc welding, carbon's high combustion heat melts most metals and is preferred fuel for forges and smelters. In an electric arc, carbon emits brilliant white light—preferred light for early Hollywood movies.

At molecule scale, carbon forms four bonds with other atoms in perfect tetrahedral, four-direction, triangular symmetry. Thus, carbon is *the great connector* that links atoms into complex structures. As a semiconductor, carbon has low charge polarity and prefers to share valence electrons with other atoms and form covalent bonds. Thus, carbon brings other elements together in structures that carry complex pattern and hold memory, information, and intelligence. An ultimate expression of this potential is DNA, the spiral helix of genetic code.

At cell scale, carbon creates shape. Carbon forms a vast assortment of molecules to build structures and functional units as containers and the machinery for life. A cell is a bubble of water enclosed in thin film of hydrocarbon chains (lipids) that form a cell membrane, reinforced with carbohydrates and proteins. Inside that bubble of fluid, carbon's complex molecules impart rigid structure to organize the space in a cell and direct flows of fluids, electrons, and charge.

At organism scale, carbon is form and food. Carbon—mostly complex carbohydrates and proteins—forms a mesh and matrix to surround cells with shells, shields, and suprastructure. Thus, carbon supplies safe, stable habitat within which cells can aggregate into multicell colonies. Colonies expand, evolve, and collaborate together, sharing space and energy, forming tissue and organ in larger and larger organisms. Complex carbon molecules also are food to supply substance to build more suprastructure and charge to carry energy into and within organisms.

At ecosystem scale, organisms modify the environment and create habitat to support greater diversity and stability for life. Early in evolution, primitive bacteria in the sea learned to eat iron and extract energy for metabolism. The iron became insoluble and precipitated to the sea floor. Over eons, these sediments became iron ore, which today is mined to make steel. Then a new bacterium evolved to capture sunshine to combine carbon with water to make oxygen and carbon rings of sugar. They became blue-green algae, the first plants. This created conditions able to support the emergence of the first single-celled animals. This began the carbon cycle, which transformed life on Earth.

And at planet scale, carbon creates community. Organisms interact to form colonies of independent, yet interdependent individuals. Blue-green algae grew together in thick colonies and enclosed themselves in a membrane to become stromatolites, the first multicell organisms. Stromatolites transformed the Earth's anaerobic atmosphere with free oxygen to cause the first mass extinction. Plants eventually evolved into trees that grew together in forests that supplied shade and shelter for hosts of other organisms. It was in the forests that humans evolved to join this growing Earth community. All this is formed from carbon in ever-evolving complexity and diversity. This web of relationships created *culture*.

Appendix 10.A: Soil Carbon: Geometry, Chemistry, and Ecology

We Are the Whirled

The twentieth century saw science rise to solar prominence in the Western society as men analyzed the physical and chemical nature of biology, ecology, and industry—of the whole cosmos. A prime paradigm of this previous century was that life is chemistry. This mindset assumed living organisms can be reduced to a menu of atoms and molecules.

Organic chemistry—the chemistry of carbon—grew, as science made rapid advances to spawn a technology revolution and give birth to new industries and global economy. Food was defined as nutritional chemistry. Food processing refined complex living organisms into pure crystals and simple chemical formulas. Farming became a chemical-dependent industry. Medical science used chemical therapy with *miracle drugs* to give birth to pharmaceutical industry.

Now, a more profound and astounding shift to a twenty-first-century paradigm is underway. The true nature and full scope of this sea change of consciousness is not well recognized or understood yet. Not only is the planet's climate changing but multiple geologic, ecologic, economic, sociopolitical systems are simultaneously shifting to reach *limits*. Even theology is undergoing rebirth. Our concepts and images of the universe, physics, nature, human identity, and spiritual destiny are in accelerating transformation. Our ways and means to inhabit Earth is also in rapid reformation.

Most mystifying is that not only are our ideas of space–time reality changing, our brains are evolving, too. Our very neurological process of awareness is gradually being altered, awakened, and fully remembered.

Biological Revolution

The first phase of this revolution in evolution is to realize that life is *biology*. Emerging insights see that life is more than a matter of physical chemistry. Atoms and molecules are organized in discrete, complex, functioning units called *cells*. These tiny, fragile units aggregate into organized, specialized, unified organs and organisms. Our bodies are at least a trillion cells, sharing space inside our skin, all guided by a single, unitary identity.

The earth's diversity of living organisms use specific life cycles, architectures, and interactive functions to associate into family, community, and ecology. These very orderly, functional structures of living organisms are a form of biological intelligence encoded in the dance of physical matter. The form, context, and organization of nutrients is as important as their chemical formulas. To confront this complexity of biology's interactive interrelationships requires a higher-order resource called *culture*.

Thus, twenty-first century farming is probiotic, not antibiotic. This reversal of relationship respects microbes as allies, not enemies. Rather than eradicate them, enlightened growers encourage population explosions of beneficial organisms, starting with microorganisms. Biological farmers minimize disruptive practices such as tillage, herbicide, fungicide, toxic, or harsh chemicals. Instead, carbon-smart farmers inoculate to spread special microbial *cultures*. Rather than killing pests, the focus is to create stable environments and favorable conditions for microbes to proliferate and differentiate into stable, complex communities.

Life Energy

The second phase of this revolution is that life is *energy*. We now understand life's bits of matter also hold and transmit energy. Life's atoms and molecules carry specific energy as polarized charge, kinetic motion, harmonic vibration, and bandwidth.

Charge

Bits of quanta-fied matter share, store, and circulate energy as charge, which is fully polarized force. Beginning with pH (acid/alkaline balance), living cells and organisms are electric—and magnetic.

Each biological molecule is an ion, with specific electric charges arranged in specific geometries to sustain specific shape, structure, and function. The molecular motions of metabolism is this shared kinetic energy of life.

Frequency

Energy is not just polarity and charge. Energy is dynamic oscillation, or vibration. As energy, life is also frequency, to transmit not only power, but information as geometric order and synchronous action. Many biomolecules ring at radio frequency—they emit and react to specific frequencies. Orderly, rhythmic, kinetic movement of molecules in cells allows them to coordinate function, hold memory, carry information, and transmit intelligence (Figure 10.9).

Unifying Principle

Information

Energy carries pattern and rhythm encoded as frequency—as in-*form*-ation. Cells communicate by exchanging electric charge but also by wireless vibrational harmonics. Science has just begun to grasp the electronic functions of living cells, organisms, and eco-communities.

Yet, biological cells are crystalline electromagnetic units operating at radio frequencies. Like our smart electronic devices, cells use radio frequencies to operate and communicate at specific scales and bandwidth. Contemporary scientists describing *bioplasmic body*, *solid-state biology*, and *quantum coherence* have begun to hear the symphony and chorus of life.

The third phase of this paradigm shift is *spirit*. An emerging holistic perspective implicitly understands the universe is one. This unifying view appreciates that everything is connected—a single, unitary field. All living beings form a single community. Modern culture and consciousness is evolving toward renewed respect for the oneness of creation and our human responsibility as stewards to support the planetary web of life, to protect biodiversity and the sanctity of all its participants.

But reverting to the twentieth-century paradigm of physical chemistry, life is exquisitely simple.

Elements in biology

Chemical composition of life

Organic	95%
Hydrogen	45%
Oxygen	30%
Carbon	15%
Nitrogen	5%
Major	4%
Trace	1%

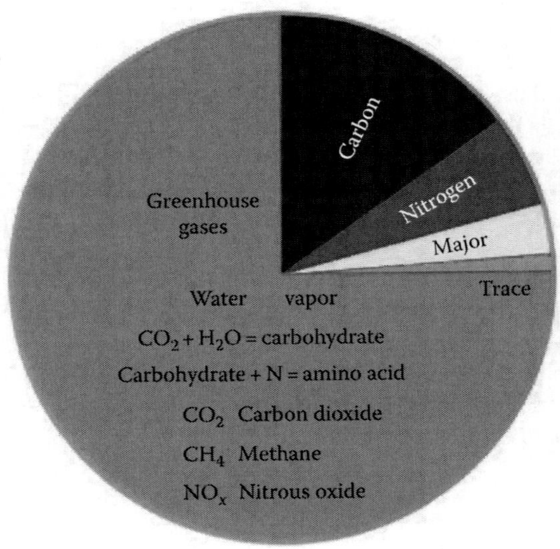

FIGURE 10.9 Organic elements. Major minerals.

Life Elements

Most matter in living cells is four of the lightest, simplest elements. In organisms, 95% of the atoms are, in order of amount: *hydrogen, oxygen, carbon*, and *nitrogen*. These four *organic elements* are most of life's physical substance and structure.

Of these, the first two together are *water* (H_2O). Most living organisms are 75% or more water. A living cell is a tiny bubble of water held inside a thin-film, oil membrane.

Water's gift to life is, as a fluid, to allow atoms, molecules, cells, and organisms to move around. Movement is life.

Water in an organism is much more orderly than in a cloud, stream, or pipe. Water inside a cell membrane is precisely organized into specific structures yet is still a freely flowing fluid. Cell water is not formless, shapeless, or motionless, yet still a liquid. Such water is a *liquid crystal*—a dual-phase state, when atoms share unifying energy fields and move and circulate in orderly, synchronous patterns. This harmonic organic unity of discrete objects is *quantum coherent*.

Water's mystery is this capacity to use the implosion power of vortex geometry to infold pattern through fractal scales to contain, carry, and compress spin energy as information, memory, and intelligence.

Carbon: Backbone of Biochemistry

Carbon is #3 of the organic elements. Carbon is most of the rest of the atoms in a cell. Life on Earth consists of carbon-based organisms. Complex biological molecules such as sugar, oil, fat, and protein have a backbone of carbon atoms.

Carbon makes four bonds with tetrahedral (four-direction) symmetry. Unlike water's fluid nature, carbon builds fixed, rigid structure. Carbon's four-arm common connector links to two other carbon atoms, to thus create long chains, or *hydrocarbons*. Carbon chains loop to form closed rings. Thus, a small, lightweight atom—carbon—creates complex structures (Figure 10.10).

Carbon, thus, is the backbone atom of biochemistry. Nature builds complex biomolecules out of carbon chains and rings. Biomolecules are composed of hundreds, even thousands, of carbon atoms interlinked in chains, rings, sheets, spirals, and tubes—even bubbles (*buckyballs*). Among nature's most complex carbons are humus particles in *SOM*—indigestible residue of bacterial decay.

Supreme expression of carbon's creative complexity is DNA—twin helix spiral stairway, with treads and risers made of carbon—key code to memory and instruction in every cell's heart.

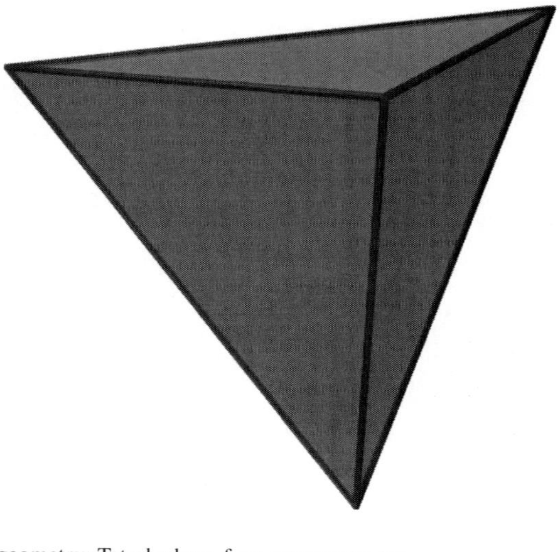

FIGURE 10.10 Carbon bond geometry: Tetrahedron, four-arm symmetry.

Biocarbon

Carbohydrate is chemistry's name for the sweetness of life: sugar. Sweet is made with the first three organic elements. This *CHO trinity* of biochemistry forms various five- and six-carbon ring molecules. Carbon itself holds a lot of energy. But these large loops of carbon hold extra electrons to deliver extra energy in an organism.

But those carbon rings are also crystals that carry an extra quality of energy. Those extra electrons vibrate at higher harmonic frequencies. Yes, carbohydrates ring in resonance!

Sugar is also structure. Plants build their bodies out of sugar spun in spirals. Linked one way, sugars create cellulose fiber, the primary skeletal substance of living plants. Linked another way, sugars weave pleated sheets of starch—a compact, efficient way to stack and store fuel for energy.

Biocarbon forms complex, massive molecules. Tens of thousands of carbon atoms interlink as cellulose fiber, cell membrane, enzyme, hormone, and nuclear DNA. Ribose, a five-carbon sugar, sits at opposite ends of DNA's spiral stairway, linked by amino acid *treads* of each rising step in the helix. Complex biocarbon structures are made by organisms as diverse as whisker-thin, white fungal threads deep in soil or photosynthetic green needles 300 ft high in a redwood—as varied as mosquito or whale.

Nitrate, Ammonia, and Amino Acid

Nitrogen, the odd member of the organic elements, makes three bonds to other atoms, also in tetrahedral, four-arm symmetry.

Nitrogen is ubiquitous on Earth. Nitrogen gas (N_2) is 78% of the Earth's atmosphere—an inert, stable form plants cannot use. Not until nitrogen combines with oxygen to form nitrate (NO_3)—by nitrogen-fixing bacteria or by synthetic fertilizer industry—can crops grow productively, lushly, and rapidly.

Nitrogen is a reversible ion. It binds with oxygen to form anion *nitrate* (NO_{3-}). It bonds with hydrogen to make cation *ammonia* (NH_{4+}).

Nitrogen bonds with two carbons to form *amino acid*: C–C–N. *Amino acids* link together in longer –C–C–N–C–C–N– chains to become *proteins*. Some proteins are hundreds, even thousands, of amino acids, joined in long chains that are folded, twisted, and bundled into orderly shapes and structures.

Nature uses the *nitrogen cycle* to change this three-arm atom from nitrate (–) to ammonia (+), and back to nitrate (–). This cycle is fundamental to almost all biology, especially animals, whose growth depends on amino acids and proteins. Nitrogen fixation in soil, and the entire nitrogen cycle, is run by bacteria. Microbes control this primary engine of biosynthesis.

These two organic elements must be in proper proportion. *Carbon/nitrogen ratio* is a key fundamental balance in any biological system, whether soil, compost, microbe, food, or tissue. This ratio also appears as carbohydrate/protein, the two basic nutrients in every organism's diet.

Plants build their bodies from *carbohydrates* that are spun into chains of *cellulose* fibers.

Animals grow their tissues with *protein* formed by chains of *amino acids* (Figure 10.11).

Geology into Biology

After organic elements, the 5% of atoms left in a living body are loosely labeled *minerals*. These physical elements are the simplest matter—tiny bundles of protons, electrons, and neutrons that are called *atoms*.

Atom is a physical science term. *Mineral* is a word from geological science. The two are similar, but different, often confused. In geology, *mineral* is a molecule formed when two or more physical *atoms* react, bond, and associate together. In short, minerals are atoms that got married. An element atom is one of many in a mineral molecule (Figure 10.12).

These discrete bits of matter are not synthesized by plants, animals, or humans. These elements are the cold stardust of the cosmos, collected, condensed, and compressed into the bedrock and basalt of a planet. We cannot make or manufacture these centers of density. We get them direct from nature.

To see the majestic Rocky Mountains, or peer into the deep valleys like the Grand Canyon, is a glimpse at ancient layers of the planet's bedrock of minerals. Those dense, hard rocks are the source and

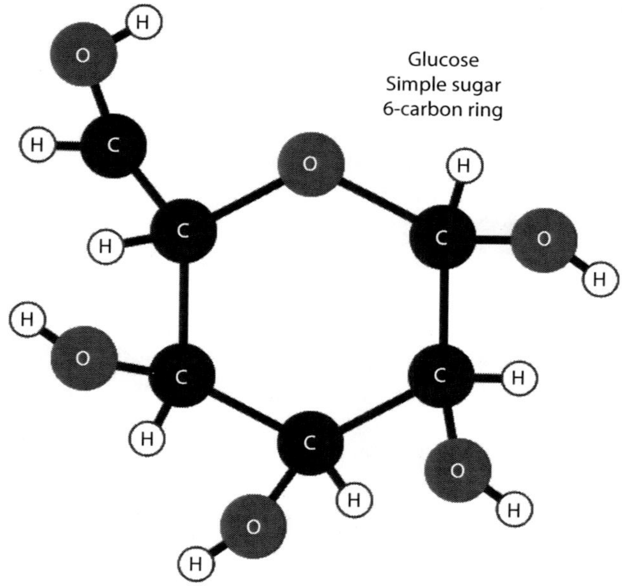

FIGURE 10.11 Glucose: Let Carbon Ring!

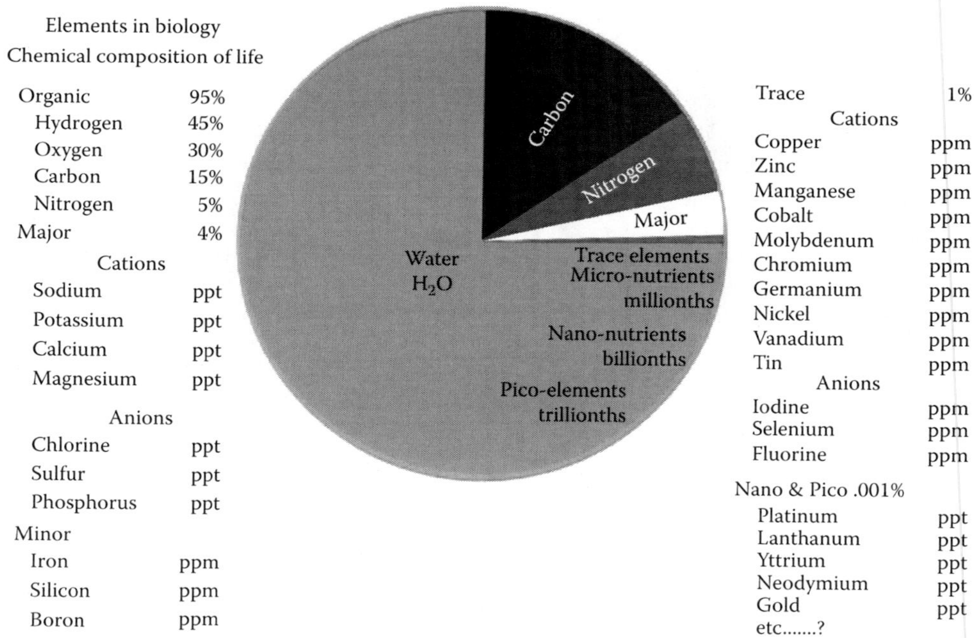

FIGURE 10.12 Trace elements.

substance of the basic building blocks of life. From crystal bedrock density, nature dissolves, selects, and organizes mineral elements to build cells and organisms—atom by atom, molecule by molecule.

Elements and minerals are fundamental for life. My analogy is building a house. First, a foundation is laid of natural or synthetic stone. Similarly, to build biology, after water, next comes the minerals—the geology. Then comes sills, walls, floors, doors, windows, rafters, roof, closets, counters, shelves, and appliances. Biological molecules—carbohydrate, protein, oil, and all—are like the house framing, flooring, fittings, and furniture. Even wooden members need little metal mineral nails or screws to join together and hold their shape.

This is the mineral imperative: geology underlies biology. Minerals are the stable, solid foundation to support the biological infrastructure of cells and organisms.

This is a key insight into how nature works—how geology becomes biology. Specific, special processes, partners, and relations govern how raw, elemental minerals transform into dynamic, organic, living, cellular protoplasm. Gradually, stones and bones of the land are ground down to dust and dissolved into watery solution, to become life's elemental building blocks.

IRON FORTIFICATION

Iron fortification of food is recommended when dietary iron is insufficient or of poor bioavailability. This is reality for most developing nations and vulnerable populations in developed nations. Iron deficiency and anemia in vegetarians and developing nations dependent on cereals or tubers is higher than in omnivore populations. Iron is present in foods in two forms:

Heme: from flesh foods (meats, poultry, fish); highly absorbed (20%–30%); bioavailability relatively unaffected by dietary factors.

Nonheme: Inorganic form in plant foods (beans, grains, nuts, vegetables); lowers absorption rates (2%–10%); depends on balance between dietary iron absorption inhibitors (phytates, polyphenols, calcium, phosphate) and enhancers (ascorbic acid, citric acid, cysteine peptides, ethanol, fermented products).

Staple foods worldwide provide mostly nonheme iron of low bioavailability. Traditional staple foods are excellent for iron fortification: wheat flour, corn (maize) flour, rice, salt, sugar, cookies, curry powder, fish sauce, soy sauce.

Benefits of consuming iron absorption enhancers are extensively proven and should be promoted (i.e., consume vitamin-C-rich food together with nonheme iron source).

United Nations World Health Organization (WHO)
Food and Agriculture Organization (FAO)
Human Vitamin and Mineral Needs, 2005

Bread from Stones

In the Christian *New Testament*, when Jesus was fasting for 40 days, Satan tempts him to turn stones into bread. This gospel parable of rocks and food is a profound and true teaching about minerals and grain. The miracle of life transforms bedrock into biology—turns stones into food. In eons of evolution, nature experimented with countless ways to transform bedrock elements into edible nutrients, then into cell protoplasm in living organisms.

And of all foods, seeds are the key, source, and secret of plant life—and thus, of all animals.

Plants, through their roots, take in water and minerals from soil to transform into protoplasm and cells. The mineral elements must be in the soil for plants to get them. And the minerals must be in plants for animals to get them. If any essential element is not in soil, then it must be imported from elsewhere and added to the soil or else biology suffers a deficit.

So, these elements and minerals are gotten from the rocks of the planet—geology of the Earth. Varied chemical and biological processes transform inert rock into living biology. Most minerals are digested by microbes, which nurse the roots of plants, which become food for animals. Thus, stone becomes bone.

Most of biology depends on this disassembly, dissolution, and decay of rocks into soil. Most of all, it is bacteria that digest minerals in cellular protoplasm and thus build the fundamental molecules of life. It begins with water, heat, and cold etching and weathering the bare rocks. Eventually, lichen, moss, and tiny, simple life forms encrust the stones to slowing dissolve and digest the minerals into replicas of themselves. In their death, their bodies become food for the next cycle of the soil food web that sustains most of the life on Earth.

So, soil is a living matrix to turn minerals into organisms. Microbes as intermediaries turning rock minerals into primary cell metabolism.

There are three kinds of rocks in the Earth's crust: igneous, sedimentary, and metamorphic.

Igneous bedrock generally forms from very hot, molten magma inside the Earth. Volcanic lava is magma that reaches the Earth's surface, but many igneous rocks form under the sea or inside the Earth. Igneous rocks are very dense, highly crystalline, with heavier elements. Elements in these rocks are not reacted or weathered into complex mineral compounds. Electrons in their outer orbits are available and ready to form bonds with other atoms to begin building complex molecules.

Primary igneous rock is weathered, worn away, and washed into solution and carried by rivers to the sea. Sediments from these eroded rocks were deposited on coastal sea floors. Other sediments formed from biological activity and bodies of living organisms. Then, over tens of millions of years, these deposits were hardened again into new bedrock—layers of sedimentary stone—mostly shale, limestone, and sandstone.

In planetary geology, certain processes periodically renew the minerals in soil. Annual spring flooding of rivers is a common way primary soil mineral are replenished, which is why many civilizations arose in river valleys. Volcanic eruptions distribute new minerals on the Earth's surface as ash and lava. At longer intervals, glaciers in ice ages grind rocks to powder and distribute them by ice and water. Mineral springs are another process that brings deep minerals to the surface to renew soils.

In a more limited, local way, deep-rooted plants mine minerals from subsoil and scatter them on the land surface to replenish primary minerals in the root zone of soils.

Today, we face an urgent crisis of soil, water, and climate. This man-made calamity begs for a man-made remedy. Or we can wait centuries for natural processes to restore minerals, renew soil, and regenerate microbiology. Creating sustainable, enduring soil fertility is the foundation of any effort at Earth restoration and nature regeneration.

Major Minerals

Most of this, 5% of essential elements are the very lightest atoms. The seven *major minerals* are commonly present in organisms at parts per thousand.

> Four are cations, with positive, alkaline charge: sodium, potassium, calcium, and magnesium.
>
> Three anions have negative, acid-forming charge: phosphorus, sulfur, and chlorine.

These small, simple atoms have few electrons in their outer orbitals, and their orbit diameters are very short. Thus, they easily give or take one, or two, or three electrons to become ions with a strong electric charge. Thus, they are very reactive. In cell biology, minerals are the centers of charge: negative and positive electric ions. In solutions, their electric charge exerts strong attraction on other atoms to organize their arrangement in space. This electric charge is the *fire in the water* of biological life—the spark of chemical fire that allows atoms to react and bond.

Major elements have specific functional relationships and exist in cells and living tissue in fundamental, broadly universal proportions. In general, these seven exist in easy-to-remember ratios—about 1:7.

A prime example of this proportion principle is calcium and phosphorus. This cation–anion pair is essential for energy exchange and transport, especially sugar synthesis and metabolism. Calcium and phosphorus are the two main minerals in your bones and are also essential in soil. Ideally, calcium is 65%–75% of base saturation, and phosphorus is 1/7 (9%–12%).

Trace Elements

Beyond four organic elements and seven major minerals, other elements are known to be necessary for life and health. But these other elements are needed at much less than parts per thousand. A few lesser elements are minor minerals (iron, boron, silicon). The rest are *trace elements*, required at extremely minute amounts—commonly, a few parts per million.

Trace elements are new to science and medicine, due to ongoing research discoveries in recent decades. At present, over 20 trace elements are identified as essential for healthy biology. Most were discovered only in the last century, and our understanding of their roles in cell biology is a still-unfolding story.

Yet, because of solid science about these minerals, it is public policy since 1941 that all *refined* flours are *enriched* with iron. The medical reality of one trace element nutrient is so well known that all table salt (except sea salt) is *iodized*.

Iodine is added to commercial salt because it is documented science that iodine deficiency results in birth defects affecting the brain, mental development, and the thyroid. Inadequate iodine is a proven cause of thyroid hormone deficiency. Iodine deficiency disrupts neurologic and endocrine hormones, while excess is still a common medical disinfectant.

Cobalt is the trace element cofactor in vitamin B_{12} (cyanocobalamin), its only known use in humans. Cobalt is only needed at parts per million—a tiny speck that fits on a pinhead.

Yet, without cobalt in B_{12}, your body cannot make adequate red blood cells, and red cells it makes are swollen, enlarged, and weakly attract oxygen. Without B_{12}, nerves have less ability to transmit energy, causing numbness. Without a microgram of cobalt, DNA replication into messenger RNA slows, and key protein synthesis can slow, even halt. Without a speck of one element, key pineal and pituitary hormones are not made.

Without a few micrograms of cobalt, you are a corpse.

IODINE FORTIFICATION

Iodine is sparsely distributed on Earth. Foods grown in soil with little or no iodine will lack adequate amounts. Only marine-origin foods are naturally iodine rich. Thus, iodine deficiency disorders were prevalent in many countries before salt iodization.

Salt is used by most people worldwide, so salt iodization is the best way to eliminate iodine deficiency. A well-implemented salt iodization can eradicate iodine deficiency disorders.

Iodine must be added within safe, effective ranges, usually 25–50 mg/kg salt. Actual amount is set by each country's salt intake level and deficit magnitude.

Salt iodization is not just a legislated mandate. Each nation must determine the best fortification technique, coordinate implementation at all salt production sites, establish monitoring and quality control, and measure iodine fortification periodically.

Difficulty implementing salt iodization arises mostly when salt industry is dispersed among many small producers. A monitoring plan must assess the amount at consumer level. United Nations agencies provide technical support to implement, monitor, and evaluate to ensure sustainability.

United Nations World Health Organization (WHO)
Food and Agriculture Organization (FAO)
Human Vitamin and Mineral Needs, 2005

The Most from the Least

The powerful systemic effects of iodine and cobalt illustrate the power of trace elements. These atoms are needed in very tiny amounts, yet have tremendous effects on biological organisms. This miracle of the microdose is because these least of all the elements are key cofactors in catalysts, enzymes, hormones, and other biomolecules that accelerate and regulate metabolism. Iodine and cobalt are key trace elements for endocrine hormones. Their miniscule presence is enhanced and amplified by their special uses as critical metabolic regulators.

Modern electronics is based on a similar microdose miracle. Diodes, transistors, and solid-state electronic devices are made by growing highly purified crystals of semiconductors such as germanium or silicon. These crystalline materials have their atoms arranged in orderly, prefect arrays, with precise geometry and symmetry. These superpure, perfectly orderly crystals then have another element added at extremely low levels of a few parts per million. Adding a tiny trace of another element distort the crystal structure and alters electron distribution, thus affecting their electric characteristics.

Such exotic electric effects of crystals carefully contaminated with parts per million of another element allow a tiny electric voltage to control a large flow of electrons. These semiconductor crystals then become microscopic, superfast, supersensitive switches and gates in sensor, amplifier, and logic circuits of modern electronics.

Similarly, in biology, tiny amounts of certain elements have dramatic effects on energy flow in and between living cells. A trace element at parts per million embedded in endocrine hormone molecules can turn on or off key biological systems for growth, immunity, awareness, memory, reproduction, or detoxication. And often, trace element deficits do not appear as a disease, but as less than optimum biological function (Figure 10.13).

Nanonutrients and Picoelements

In recent decades, accumulating evidence in biology and medicine indicates that some elements are needed by biology at *beyond trace* levels—at parts per billion, even less. These *nanonutrients* use complex geometries and dense energies of the heavy elements to build nature's most complex biomolecules. Greater complexity stores higher intelligence.

But lab data hints that a few elements are needed at parts *beyond billionths*—parts per trillion—millionth of a millionth. These *picoelements* are present in cells at the current threshold of detection by laboratory equipment and methods. Even if biochemical assays detect these rare elements at these beyond-micro levels, biology has little insight into their roles in cell structure and function. But clearly, they are keys and catalysts to enhance biochemical reactions and metabolic pathways.

These dense, heavy atoms are very large, with huge, complex clouds of orbiting electrons. This multitude of spinning electrons provides these elements with multiple valence energy levels to bond with other atoms. These heavy atoms use unique and complex geometries to coordinate and organize their electron bonds to other atoms—not four bonds, like light and simple carbon, or 6, like cobalt—but 12 bonds, even as many as 20 links to other atoms.

These more elaborate geometries are used to build nature's most complex biomolecules. This molecular complexity encodes the most intelligent functions of cells and organisms: immunity, reproduction, neurology, endocrine hormones, and high-level metabolic processes. If we appreciate the critical and essential roles of trace elements to regulate and accelerate metabolism, we can begin to imagine the synergistic effects of picoelements.

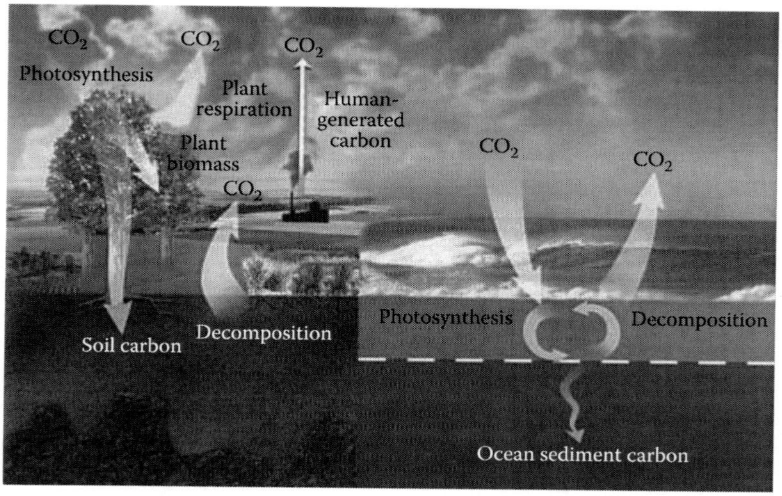

FIGURE 10.13 The carbon cycle (simplified).

Taken together, a gathering mass of scientific data and detail suggests human physiology does not need a few elements. Our bodies do not need a mere dozen elements, or even 25, or only 33. Increasing insights from research encourage the assumption that human biology needs nearly every element in the periodic table—as many as 76, likely more.

Consider the platinum series, at the very bottom of the periodic table of elements. These heaviest elements have 72–80 electrons in orbitals organized by the geometry of 5-sided symmetry. These elements have over 12 valence electrons, allowing them to create gradual, controlled cascades of electron energy. These rare elements provide very special services to cells needed by nature's most complex, exotic, potent, and intelligent biomolecules, including DNA, endocrine hormones and neuroactivators.

Thus, intravenous solutions for medical injection should contain—not just sodium chloride, or even only the major minerals—but nearly every water-soluble element. And these elements must be in specific ratios or ranges suited to human physiology. And the same is true for fertilizers to supply primary minerals to assure optimum, healthy plant growth.

To get the best, the most growth, we must provide these least of all the elements—not just three (NPK) or merely a dozen (Figures 10.14 and 10.15).

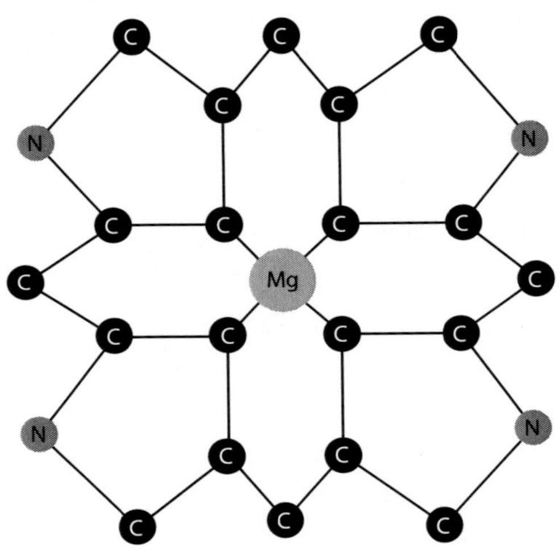

FIGURE 10.14 Chlorophyll.

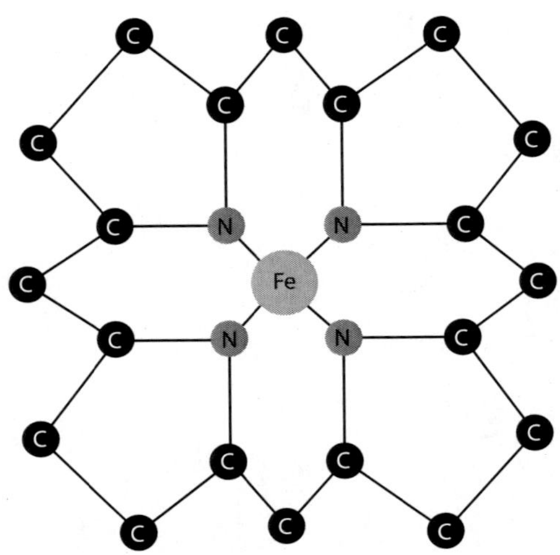

FIGURE 10.15 Heme.

Chelation: Pack and Stack with Carbon

As ions, atoms deliver power to chemical reactions of cell metabolism. The strong electric charge of these mineral atoms—mostly metals and nonmetals—attracts other atoms and arranges them to form orderly structures with specific geometries and symmetries. The ion charge of highly polar atoms is too strong to simply float in solution. Pure sodium, for example, is so reactive, it explodes with a "pop" when it touches water.

So, Nature separates these strong ion charges, to carefully isolate their electric power. An ion's powerful charge must be buffered, contained, and controlled, lest it disrupt the careful order inside a cell. An ion's strong electric field must be organized and focused to form stable structures and to deliver useful charge to enzymes in as cell's chemical reactions.

When I studied basic electricity in high school, carbon was a *resistor* in our circuits. Carbon is neither a good conductor, nor a good insulator. It is a *semiconductor*, like germanium and silicon crystals in solid-state electronics. Thus, carbon allows strong electric charges to be isolated, separated, and contained. In biology, individual ions are almost always enclosed in carbons.

So, in a cell, nature embeds an ion amid carbon atoms or other molecules to insulate and focus the charge. A cell organizes an energy field around an ion to create a structure to package and harness a strong electric charge. In biochemistry, ions are enclosed in carbon rings or a cloud of biomolecules.

In cell protoplasm and body fluids, mineral ions are held in a matrix of carbon atoms. Thus, a cell can hold a strong electric charge in a stable, sustaining shape. The ability to hold charge in a stable shape is what creates cellular memory. At molecular, cellular, and organism scales, shape creates identity.

In biochemistry, embedding ions is called *chelation*. In modern nutrition, mineral supplements are *chelated* with biomolecules to improve their absorption in intestines. Calcium, for example, as raw mineral element, is only 15% absorbed. But *chelation* with carbon biomolecules can boost calcium assimilation to 80%. Modern medicine uses *chelation* to transfer specific minerals or biomolecules in the body or to target specific cells.

Sunshine into Sugar: Freedom Rings

One key biomolecule is a perfect portrait of this marriage of mineral with carbon: *chlorophyll*. This large, complex molecule gives plants their green color and is the antenna of *photosynthesis* to capture sunshine to create sugar. Ultimately, plants use sunshine to combine carbon dioxide (CO_2) with water, converting them to carbohydrates and oxygen.

Chlorophyll is a large biomolecule formed from a few dozen atoms. A single magnesium atom sits at the center, surrounded by rings of carbons and four nitrogens. When a photon of light strikes the magnesium in the heart of this orderly carbon array, this creates a molecular vibration to advance a five-step cascade that converts high frequency light into electric charge.

So, photosynthesis begins when a magnesium antenna intercepts a photon of sunlight. High-frequency light transmits an impulse to the magnesium atom. This pulse triggers ringing in carbons around it—a molecular vibration that is transferred to other molecules in the plant's photoreaction center.

Eventually, a five-step water splitting clock will pry a hydrogen off a water molecule nearby, liberating an electron and a proton. Plant photosynthesis captures four photons to pry four hydrogens off two water molecules, yielding one oxygen molecule, four protons, and four electrons. Now the plant has the strong electric charge needed to fix carbon into carbohydrate.

Chlorophyll's remarkable ion–carbon geometry, symmetry, and synergy gets evermore extraordinary.

Heme: Flip Side of Photosynthesis

Animal blood is red due to cells packed with *iron* in *heme* molecules of *hemoglobin* (globin = protein). Thousands of heme molecules stack inside each red blood cell to attract oxygen and carry it to a cell to burn sugar and release energy for metabolism. Heme has nearly the same geometry as chlorophyll, except four nitrogens flip inward and peripheral carbons and oxygens are rearranged.

Besides their remarkable symmetry of structure and function, chlorophyll and heme illustrate how strong electric charge of ions is buffered by being embedded in carbons and thus organized into a functional bioenergy field. Carbon allows cells to safely acquire, move, and use the electric charge of mineral ions and harness their energy in cell metabolism.

Carbons distribute the electric charge of an ion to shape it into a focused field. Thus, electric charge can be harnessed to make or break bonds, to build specific geometry and structure, and to create energy flows, including electron cascades and radio frequency ringing.

Carbon Cycle: Complexity and Diversity

Carbon is one of the Earth's most mobile, labile, and facile atoms. It is in constant change, moving around the planet, making molecules in new organisms, and assuming seemingly endless forms and complex structure. Carbon's most amazing disguises are as carbohydrate and amino acid. As sugar and protein, carbon forms the biological bodies of plants and animals. As living organisms, carbon achieves its greatest complexity and diversity, from microscopic fungal mycelia to giant redwoods and whales.

In soil, diversity is as crucial as quantity and proportion.

Today, billions of extra tons of CO_2 sit in the atmosphere, changing climate around the world. Global warming makes us aware of carbon's movement as CO_2 and methane into the Earth's atmosphere. Carbonates (CO_3) are carbon bedrock, mostly limestones—fossil skeletal shells of cells that lived in the Earth's ancient sea—buried as bedrock for millions of years. Coal is fossil carbon from the cellulose bodies of ancient trees that we mine today to burn for energy.

Soil is the largest carbon storage on land: three-quarters of land-based carbon resides in soil. Plants take in carbon during growth. When they die, this carbon enters soil. Soil carbon is like a bank: it can either remain in soil or quickly return to the atmosphere. Modern soil management drains the soil carbon bank and depletes our capital, rather than gaining new interest.

Carbon is accumulating in the Earth's atmosphere, fed by fossil fuels, deforestation, soil destruction, and industrialization. Carbon movement from biosphere to atmosphere is causing a thermal imbalance known as *greenhouse effect*. Our carbon emissions are now known as a pollutant that is a menace to life on Earth. The carbon cycle that kept the biosphere in a remarkable state of stability is now falling out of balance, threatening rapid-onset, runaway climate change.

Food production releases at least 30% of man-made greenhouse gas emissions. Deforestation to remove trees for farmland is 20% of carbon emissions. Carbon and nitrogen outgas from fields subject to routine tillage, monoculture, and synthetic fertilizers. Half of US-produced hydrogen is used to make nitrogen fertilizer—most from methane, a greenhouse gas.

Carbon in soil comes in an incredible array of chemical forms. A fertile, functional soil is alive, and needs not only lots of carbon but also carbon in a wide variety of forms. However, until lately, soil science interest in soil carbon was minimal. Science has just begun to recognize and sort out the diversity and complexity of soil carbon.

11

Biochar: The Field Experience

Kurt Spokas and Jeff Novak

CONTENTS

Introduction

Despite the relatively new concept of biochar, which utilizes chemical–thermal alterations of biomass to achieve carbon sequestration (Goldberg 1985, Kuhlbusch and Crutzen 1995), the purposeful addition of biochar-like materials to soils can be traced back to the beginning of our modern science record (Kirwan 1793, Ogawa and Okimori 2010). In addition, there is archaeological evidence of biomass pyrolysis residues and products being used in earlier civilizations (Rapp 2009). The process of pyrolysis converts the biomass into a carbon form that is more resistant to degradation than the parent biomass (Okimori et al. 2003, Lehmann 2007, Belanger and Pinno 2008).

Even though the amendment was not called biochar, we have been utilizing the solid residue from biomass pyrolysis as a soil fertilizer agent for at least the 300 years of our modern science record. The use of charcoal amendments in the past has been limited by the economics, which hampered scientific investigations into the mechanisms of charcoal's improvement of soils. Despite this, there are many reports of charcoal improving soil physical properties, water holding capacity, and the action of other fertilizers after application. In the 1900s, new scientific developments expanded the potential uses of charcoal, thereby removing some of the emphasis on agricultural use. However, charcoal continued to be investigated as a filtration media and uses in nontarget agrochemical protection. Now in the early 2000s, biochar has entered the public's vocabulary, as seen in the Google™ Trends graph (Figure 11.1). The emphasis is on production temperatures and feedstock type to describe the biochar, but past research has also had demonstrated that the rates of heating and cooling can be more important than the maximum pyrolysis temperature reached in regard to the influence on surface chemistries (Sahu et al. 1988, Kashiwaya and Ishii 1991). These factors stress the importance of fully documenting the source and production conditions of biochar, so that future efforts will not be hindered by the lack of adequate characterization data or varying product physical and chemical properties.

Even though we have been using charred amendments for a long time, there is a significant amount of hype, skepticism, and uncertainty surrounding how soil–biochar additions have performed historically and what guidance this provides for the future. This chapter will synthesize what we know from looking back at the past and current field experiments investigating biochar additions.

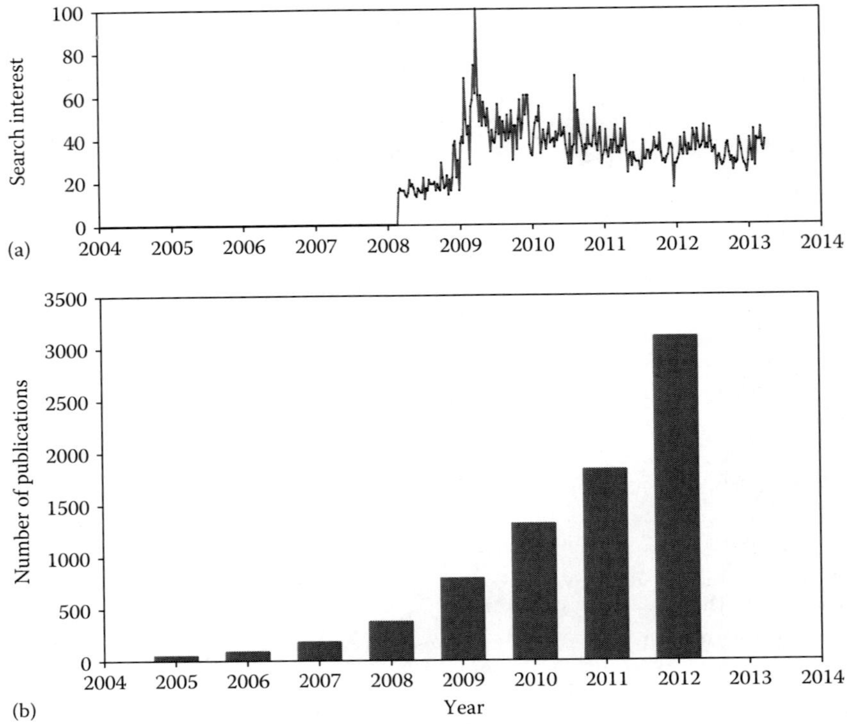

FIGURE 11.1 (a) Google™ Trends report on the search frequency of *biochar* over the last 8 years. The search volume has been normalized to the maximum search volume. (b) Google Scholar™ reports on the number of publications (i.e., reports and manuscripts), which include the term *biochar* over the last 8 years.

Biochar Field Application History: 1700s to Current

1700s

There were limited agricultural field research efforts in the early 1700s as modern science was still in its infancy. Despite this lack of field data, there were a few hypotheses that are noteworthy during this time period as to charcoal's interactions with soil and plants. During the 1700s, there was a surge of research interest into the properties of charcoal, primarily focused on electrical conductivity and gunpowder production (Napier 1788, Proust 1811). During this time, there was the realization that even charcoals made from the same wood feedstock under the exact same production conditions did not always produce charcoal with identical properties (Mitchell 1748, Priestley 1770). This was illustrated by significant differences in electrical conductivity between similar pieces of wood charred within the same batch (Priestley 1770, Becket 1773). Most of the hypotheses behind charcoal's benefits to soil at this time were centered on supplying *carbonic acid* or carbon to the soil for the plant to grow (Bradley 1727). This was due to the fact that it was not until after Priestley (1775) that the scientific foundation of photosynthesis was known. Despite the false scientific hypothesis that the charcoal was supplying carbon directly to the plant, there were still indications of its beneficial use in agriculture (Kirwan 1793). However, the profitable use of charcoal in agriculture during the late 1700s was limited by the high charcoal price that was the result of additional uses for charcoal ranging from iron production, energy source, and gunpowder manufacturing (Tucker 1756, Kirwan 1793, Dundonald 1795, Sneyd et al. 1798). Incidentally, just prior to the American Revolutionary War, there were efforts by England to acquire cheaper charcoal from the more resource-abundant US colonies (Tucker 1756).

Near the end of the 1700s, there was a more concentrated scientific effort on the use of charcoal as an *artificial manure* (fertilizer) for agriculture (Dundonald 1795). Some of the first documented field trials were those conducted at the Rothamsted farm in the United Kingdom, which were conducted just prior to the formal establishment of the Rothamsted Research Center (Hall 1917). For those who are not aware, Rothamsted is the world's oldest operating agricultural experimental site. A number of major advancements in the long-term cycling of carbon, guidelines for crop rotations, and impact of tillage have resulted from this location's research (i.e., Jenkinson and Rayner 1977). At this time, these research plots were established to evaluate yield responses to various artificial manures (fertilizers), since it was viewed that this type of data was required to guide the developing agricultural fertilizer industry. Among the various treatments, they did include a peat charcoal application (Rafn 1799). The addition of the peat charcoal (unknown application rate) was reported to increase turnip yields, compared to the control plots (Rafn 1799).

1800s

In the 1800s, there were a significantly higher number of reported field trials with biochar materials. A number of agricultural crops were observed to have beneficial impacts following biochar (charcoal) additions. There were observations of doubling yields for some crops, including oats, turnips, and grasses following charcoal incorporation (Raynbird 1847, Durden 1849, Davy 1856). One report stated that a farmer observed 90 bu/ac (5649 kg/ha) corn grain yield following guano and charcoal additions (it was even stated that the yield was measured by his neighbors) at a time when corn yields were averaging 45 bu/ac (2825 kg/ha) (Robinson 1853, Voelker 1864). Furthermore, there were numerous observations of yield improvements following charcoal additions in home gardens for flowers (increased growth and blooms), cucumbers, peach and apple trees, potatoes, tomatoes, and squash (Kirwan 1793, Otto 1804, Davy 1818, Morfit 1848, Browne 1852). However, these reports are often for a single unreplicated field plot or a single growing year, with no indication if the effect was long term.

One of the first observations regarding the mechanisms of charcoal's improvement in field plots was the increase in the *permeability of the soil to water* following charcoal additions (Davy 1818). There were also observations of increased seed germination, which was initially attributed to the charcoal improving the aeration of clay soils (Davy 1818). However, these types of yield increases were not predictable, and the scientific conclusion reached up through that time was that "using charcoal as a fertilizer depends on circumstances" (Holbrook 1849). As an aside, charcoal additions at this time were also viewed as the sole manner in which to modify field soil temperatures, as it was noticed that charcoal-amended soils (particularly top dressed) possessed a higher temperature than nonamended soils, which was attributed to the sorption of the solar radiation by the charcoal (Browne 1852, Nash 1895).

There were also observations during this time that charcoal coapplication improves the efficiency of manure-based fertilizers. Some of the initial observations of this phenomenon were observed in improved yields from the combined guano manure and charcoal application (Clark 1845). There were specialized mechanical planters developed to simultaneously place the guano, charcoal, and seed in proper position and ratios (three parts guano to one part charcoal/wood ash) (Robinson 1853). Charcoal also was used to facilitate the reuse of slaughter house wastes as soil fertilizers, most with positive yield improvements (Terne 1882). Lefroy (1883) presented guidelines for producing charcoal for agricultural use through pyrolysis, which was quoted from a 1833 report. He stated that one should initially pile biomass (wood or kelp was assumed at the time), ignite the biomass, and then cover with enough soil as to allow just enough oxygen to enter the pile to keep the fire smoldering. After production, this charcoal was to be immediately applied at a rate of 5000 kg/ha (0.5 kg/m^2; 1 lb/yd^2) (Lefroy 1883).

Although not fully elucidated during this time period, there was a growing set of observations that charcoal did continue to react after production and while present in the field. Lefroy (1883) stressed that the charcoal "lost nearly all of its value" as an agricultural amendment, if it was stored for a long period after production. However, there was no indication of what a long period was or what type of reaction was occurring. Others believed that there was a delayed reaction of the charcoal that could be initiated with water. This was the basis of *immediate watering* following soil application to ensure proper action in the soil (Clark 1845). There were also suggestions that charcoal should be applied and incorporated in late fall and to allow the charcoal to weather during the winter period in the soil to improve its use (Browne 1852). The mixing of charcoal with various manures and other organic wastes was also common practice prior to soil application (e.g., Raynbird 1847, Davy 1856, Davis and Copeland 1874, Terne 1882). It was known a priori that the charcoal undergoes chemical alterations with storage (i.e., contact with water and oxygen), but was not fully documented until nearly 100 years later (Puri et al. 1958).

Not only were there improvements to the physical and nutrient state, there were also reductions in the number of crop pathogen infections when charcoal was applied to soil. Potatoes and peach trees grown in the presence of charcoal had reduced pathogen attacks compared to companion field controls (Raynbird 1847, Walden 1860). Problems with the fungal powdery mildew infection of gooseberry and strawberries could be reduced with charcoal amendments (Walden 1860). These observations lead to a US patent in 1874 for an antiseptic fertilizer, based on charcoal and sorbed hydrocarbons from the pyrolysis of wood (Chesebough 1874). In addition, there were other advancements that utilized charcoal for improving the action of night soil (sewage) as an agricultural fertilizer (Pitts 1863, Sewell 1872). Charcoal also was a key ingredient in several of the initial commercial agricultural fertilizers (Amies 1872). These agricultural developments for charcoal's use cite the reduction in odors and ammonia volatilization losses, but indirectly they claimed that charcoal improved the soil fertilization potential and duration of the fertilizer action. Along these same lines, there was a conceptual outline for the possible use of city sewage as a feedstock for pyrolysis, and then the solid charcoal could be applied to agricultural fields as a fertilizer (Universal Charcoal and Sewage Company Ltd 1875). Charcoal also was used in seedling nurseries at the time, due to improved soil physical and chemical characteristics coupled with the observations of reduced plant pathogen infestations on charcoal-amended soils (Retan 1915).

In spite of these promising observations for charcoal use in agriculture during the 1800s, there were unforeseen hurdles. There was a summary published by Durden (1849) that stated "the results of trials with peat charcoal as a manure are very conflicting and very difficult to reconcile; serving to show us that we have yet very much to learn with regard to the action of artificial manures." This was primarily due to the fact that the soil fertility improvements were not always consistent after charcoal additions. As stated earlier, there were some studies reporting double yields and others with no yield improvements. There were hypotheses that charcoal was sorbing needed and required soil nutrients (Davy 1856), as well as reacting with oxygen (Davy 1840). Incidentally, charcoal at this time was known to sorb significant quantities of ammonia nitrogen as well as sulfur (Blake 1893). Charcoal also was known to sorb noxious (inhibitory) chemicals in the soil as well as improve soil water holding characteristics by retaining a good balance of moisture around plant roots (Walden 1860). However, just like the earlier century, efforts to spur charcoal use in the 1800s in agricultural operations were thwarted by the lack of consistent yield improvements, which resulted in charcoal being regarded as a very expensive amendment and typically not worth the cost to farmers (Durden 1849, Davis and Copeland 1874).

1900s

At the start of the 1900s, Schreiner and Brown (1912) observed that the *stable organic fraction* of soils also contained a series of higher-molecular-weight, complex organic compounds resembling coal, charcoal, and/or resins. This was one of the very first observations that charcoal (biochar) is naturally present in soils, which has been supported by more recent assessments (Haumaier and Zech 1995, Glaser et al. 1998, Schmidt et al. 1999, Schmidt and Noack 2000). Skinner (1908) observed a yield increase in peas and lettuce grown in a media with charcoal present in isolated tubes (Figure 11.2). The most significant increases (70%) were observed in poor-quality clay loam soils. As shown in Figure 11.2, direct mixing of the charcoal into the soil did not result in significant increases; in fact, nonsignificant yield decreases were noticed (Skinner 1908). On the other hand, Vantsis and Bond (1950) observed an increase in the dry weight of peas grown in charcoal-amended media and attributed this response to a combination of multiple factors: (1) the sorption by the charcoal of inhibitory excretions from plant roots or microorganisms, (2) charcoal sorbing excess nutrients, and/or (3) pH modification of the rooting medium. Turner (1955) hypothesized that the improved growth and nodulation for clover grown in charcoal-amended soils was due to the charcoal sorbing compounds that were inhibitory to clover's growth/nodulation. Tryon (1948) observed that the impact on water holding improvements is not only charcoal, but also soil

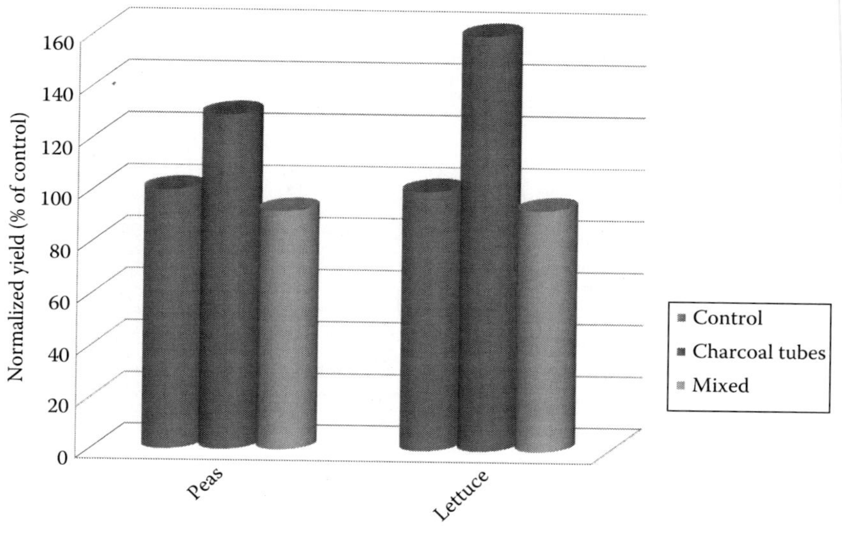

FIGURE 11.2 Yield improvements in lettuce and peas growth in the presence of charcoal isolated in gas-permeable tubes, but no significant yield impacts observed when charcoal was mixed into the soil. (Data from Skinner, J.J., *Plant World*, 11, 249, 1908.) The cause of this is most likely due to the sorption of nitrogen species (i.e., nitrate and ammonia) that has been observed for some types of charcoals. (From Asada, T. et al., *J. Health Sci*., 52, 585, 2006; Rivera-Utrilla, J. et al., *J. Hazard Mater*., 187, 1, 2011; Harper, H., *Ind. Eng. Chem*., 16, 180, 1924.)

texture dependent. Coincidentally, the hypothesis of charcoal sorbing *noxious* or inhibitory compounds and improving the water holding capacity of soils was suggested almost 100 years prior (Walden 1860).

It should be pointed out that during the early 1900s, there was a large emphasis on the role of other organic compounds present in soils, which was largely linked to the observations of Lathrop (1917) that "...certain soils may be infertile, not on account of the lack of any essential elements, such as potash, phosphorus, or nitrogen, but because of the presence of organic compounds in the soils, detrimental to the normal growth of crops." Research on the sorptive nature of charcoal combined with the emphasis on chemical signaling led to a corresponding shift in charcoal research during the 1900s. During this period, much of the work was focused on charcoal use as a sorptive media for naturally produced inhibitory chemicals and as a protection strategy for exposure to nontarget plant species from emerging agrochemicals (Hitz et al. 1953, Linscott and Hagin 1967, Bovey and Miller 1969, Kratky et al. 1970, Chandler et al. 1978). Some of these studies included charcoal-amended and control field plots, which provide a glimpse into the continued observations of occasionally increased yield performance from charcoal additions, particularly on weathered tropical soils (Figure 11.3). However, as baseline, agronomy yields have improved from farming intensification (higher seed planting densities), soil fertility management, and seed genetics; the relative yield boost from charcoal amendments has correspondingly decreased.

The role of soil chemical signaling compounds is still an active area of soil research (Insam and Seewald 2010) and one of the potential interests to the biochar research area. A current hypothesis is that there are chemical compounds that are plant stimulants present on charred materials (Nelson et al. 2012). These signaling compounds are produced during the thermal pyrolysis of biomass, which can correspondingly trigger/stimulate seed germination (Keeley et al. 1985). There are thousands of different chemical compounds that can be produced during the pyrolysis of biomass, with some of these compounds sorbing/condensing to the solid residual (biochar) (Smith et al. 2003, Hansson et al. 2004, Spokas et al. 2011). Therefore, some of the produced compounds can also be inhibitory compounds to both plant and microbial activity (Asita and Campbell 1990, Brown and van Staden 1997, Guillén and Manzanos 2002, Olsson et al. 2004). Nevertheless, these chemical signaling compounds are not the sole mechanism behind the actions of charcoal in soils as other soil chemical, physical, and microbial alterations also occur, which are specific to the particular soil type, biochar, crop, and environment (Merritt et al. 2007, Lehmann et al. 2011).

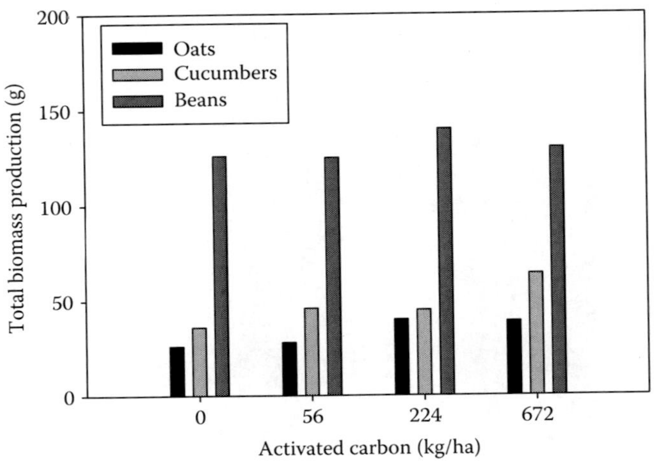

FIGURE 11.3 Impacts of activated charcoal additions on oats, beans, and cucumber yields in field plots on a tropical soil (Tao silty clay; Puerto Rico). (Data modified from Bovey, R.W. and Miller, F.R., *Weed Sci.*, 17, 189, 1969.)

The start of the twentieth century also brought an explosion of new analytical instrumentation. In the early 1920s, charcoal was being utilized for the chemical cleanup for various plant and soil extracts that were being analyzed on these newly developed tools. However, during this method development, several researchers noted occasional disappearance of nitrogen-containing compounds (i.e., nitrite, nitrate, and ammonium) when charcoal was used for purification of soil/plant extracts (Harper 1924, Burrell and Phillips 1925, Emmert 1930). They also noted that this disappearance was not consistent across all charcoal types (Harper 1924). These accidental observations created the theoretical foundation for the development of specialized charcoal activations to optimize this interaction of carbon and nitrogen compounds (Holmes and Beebe 1957, Shevlin et al. 1983, Jansen and van Bekkum 1994, Biniak et al. 1997, Petit and Bandosz 2007, Seredych and Bandosz 2007), which incidentally can be catalyzed by the presence of metal cations (Baker 1989, Hansen et al. 1996, Samarkin et al. 2010). Therefore, the production and postproduction processes determine if biochar can be a source of sorbed nitrogen compounds or if the biochar can react with nitrogen to form other organic compounds, thereby reducing nitrogen's availability (Spokas et al. 2012). Simultaneously, we were gaining knowledge on various chemical alterations of the charcoal's surface to improve its sorptive characteristics and eventually developed *activated charcoal* (Sheldon 1920, El-Shobaky and Youssef 1978). Additional studies provided insights on charcoal's interaction with ambient oxygen and moisture (Puri et al. 1958, Puri 1966), which was shown to cause significant alterations in the sorption capacity of charcoal (Rideal and Wright 1926).

Charcoal use in allelopathic research was also expanding during the 1900s (Fuerst and Putnam 1983, Keeley et al. 1985, Keeley and Pizzorno 1986). These studies showed occasional suppression of seed germination in activated charcoal treatments (Keeley et al. 1985, Keeley and Pizzorno 1986), suggesting the potential presence of germination inhibitors on the charcoal. Another complicating factor was the occasional sorption of soil nutrients by the charcoal, which has been one of the difficulties in separating out charcoal effects from nutrient deficiencies using charcoal (Keeley et al. 1985, Inderjit and Callaway 2003). However, there was the suggestion at the time of adding additional fertilizer to the charcoal treatments to compensate for this effect (Putnam and Duke 1978). All of these results emphasize the conclusion that not all charcoals behave similarly.

During the 1900s, charcoal was also investigated as a carrier for mycorrhizal fungi. The stimulatory actions of charcoal on fungi germination were known from preliminary work in Japan (Saito 1990, Ogawa and Okimori 2010). Similar to the work on the inhibitory soil compounds impacting seed germination, the work on the mycorrhizal fungi also showed mixed effects, with occasional sorbed stimulatory and inhibitory compounds on the charcoal (Fries 1978). Again, this research also supported the conclusion that not all charcoal possesses the same chemical species or capacity to sorb inhibitors.

In addition to specific biochar properties, there will be differences in the generation of secondary compounds that could condense on the charcoal product. In fact, it has been known that pyrolysis of biomass materials produces various chemical species that can be plant/microbe stimulants (e.g., increased seedling vigor, seed germination, and root development) as well as chemicals that are plant inhibitors (Nelson et al. 2012). Keeley et al. (1985) performed some of the very first experiments that confirmed the stimulation of seed germination from the presence of wood-derived chars. This stimulation effect was later observed in the water extract of various charcoals, indicating that the actual stimulant chemical was also water soluble (Keeley and Pizzorno 1986). However, not all charcoals possessed these compounds. From this work, they postulated that the stimulant compound was derived from hemicellulose through heating above 175°C. Further work has expanded our knowledge now to a wider range of potential chemical stimulants (e.g., acetone, cyanohydrin, glycolonitrile, ketones, aldehydes, karrikins, and mandelonitrile) (Keeley et al. 1985, Footitt and Cohn 2001, van Staden et al. 2004, Nelson et al. 2012). On the other hand, other compounds sorbed on charcoal have also been shown to be inhibitory to plant and microbe growth, particularly naphthalene, and some extractable biomass compounds like vegetation-derived abscisic acid (ABA), terpenes 1,8-cineole, bornane-2,5-dione, camphor, and β-thujaplicin (Brown and van Staden 1997, Krock et al. 2002, Nelson et al. 2012). However, these chemical signaling compounds are not the sole mechanism as other chemical and physical alterations also occur, which are specific to the particular soil, biochar, and plant types as well as the environment (Merritt et al. 2007).

2000s: The Recent Biochar Renaissance

As highlighted in the other chapters of this book, the driving facets of the current biochar renaissance have been closely tied to the discovery of the *terra preta* soils in the Amazonia forest area (Sombroek 1966, Sombroek et al. 2004). As shown in the aforementioned historical accounts, the mechanisms and knowledge of charcoal's potential role in soils were known earlier. The unique aspect to the current biochar research is the additional purpose for the creation of the carbon-rich residue as an agent for carbon sequestration. The chemical composition of biochar is highly variable and spans the entire black carbon spectrum (Spokas 2010).

Following the *terra preta* concept for use in modern agricultural practices, there have been numerous published field studies of biochar's potential role since the early 2000s (Figure 11.1b), which have been summarized by two recent meta-analyses (Jeffery et al. 2011, Biederman and Harpole 2013). Both of these studies have concluded a relatively small anticipated yield increase of 10%–20% following biochar use based on data contained from these field studies (around 30 studies). Furthermore, as highlighted by both of these meta-analyses and other literature reviews (Atkinson et al. 2010, Sohi et al. 2010, Manyà 2012, Spokas et al. 2012), there is an urgent need for long-term research plots on biochar's impact, particularly for different soils and climates.

To fill this void, scientists at the US Department of Agriculture–Agricultural Research Service (USDA-ARS) established collaborative research among locations in different climatic zones with soils formed in different pedogenic processes. The objectives of the study were finalized in 2007, and by 2008, a hardwood biochar was distributed to six locations: Prosser, WA; Kimberly, ID; St. Paul, MN; Ames, IA; Bowling Green, KY; and Big Spring, TX (Figure 11.4). The agreed experimental design was that each location would use the same type of hardwood biochar to examine its impact on crop yield responses under soil and climate variability.

This project was initially conceived by Dr. David Laird formerly with the USDA-ARS Ames, IA, location (currently located at Iowa State University). This is believed to be the first multilocation experiment that utilized the same source of biochar and is planned for a multiyear effort (>5 years is planned). The experimental protocol at each site was a complete randomized design, with three field replicates comparing the effect of biochar, manure, and biochar + manure to control (unamended) plots (total of 12 plots at each location) (Table 11.1). The numbers and the size of field plots were limited (triplicate ~25 m^2 plots), due to the restricted sources of biochar. At this time (2007–2008), one of the few options for field-scale quantities of biochar was Dynamotive Energy Systems (Vancouver, British Columbia, Canada). This particular biochar was produced from the fast pyrolysis of hardwood sawdust and leftover hardwood flooring and shipped to the locations by means of sealed barrels.

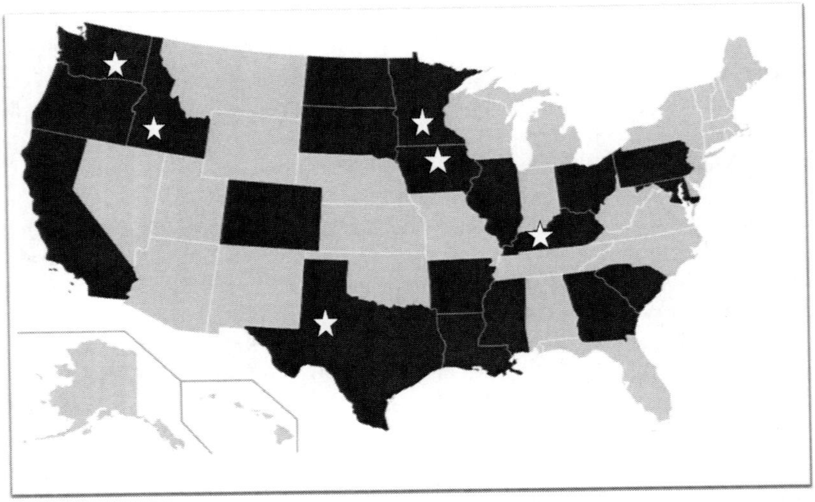

FIGURE 11.4 States with USDA-ARS locations involved in the USDA-ARS biochar and pyrolysis initiative (dark shading) and those specific locations involved in the biochar multilocation research effort (indicated by star on map). Additional details of other ongoing USDA-ARS biochar research efforts can be located through the USDA-ARS website (http://www.ars.usda.gov; USDA-ARS home page) by searching for *biochar.*

TABLE 11.1
Description of Plot Treatments

Treatment Identifier	Description of Treatment
Control	No additions
Biochar	22,000 kg/ha
Biochar + manure	Biochar 22,000 kg/ha Manure 2,268 kg/ha
Manure	2,268 kg/ha

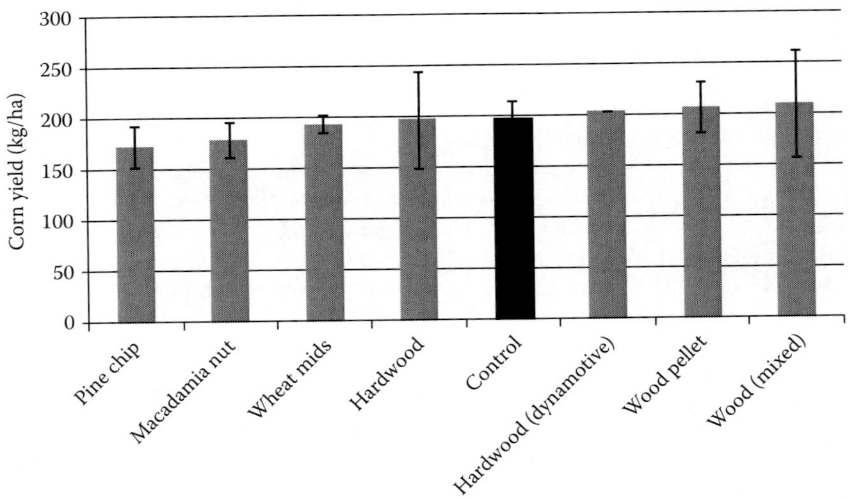

FIGURE 11.5 Average corn grain yield from the biochar research plots in Rosemount, MN, for 2011 (3 years after application). The various biochar feedstocks are shown on the *x*-axis with the control (no amendment) shown with solid black fill. The standard deviation of the three replicate plots is shown. There were no statistically significant differences in corn yield as a function of the biochar additions in the Upper Midwest soil type (Waukegan silt loam).

This project is still ongoing, and the manuscripts describing the outcomes of this project are expected near the end of 2013. Annual corn grain yields and soil properties were measured at each location and will be summarized in those upcoming publications. At the St. Paul location, additional types of biochar are being evaluated along with the Dynamotive biochar (all at the same application rate of 22,000 kg/ha; Figure 11.5). Preliminary data published from the Kimberly site (Lentz and Ippolito 2012) and data from the St. Paul, MN, location (Figure 11.5) do not suggest any universal yield improvements resulting from the biochar additions. These data also echo the importance of understanding the interactions between the soil, biochar, and plant locations and selecting a biochar based on the soil deficiency that you would like to remediate with *designer biochar* (Novak et al. 2009).

Conclusions

So what can we conclude from looking over the past and current biochar field researches? Are we any further along in our understanding from the field plot research of charcoal's use as a soil fertilizer than "using charcoal as a fertilizer depends on circumstances?" (Holbrook 1849)

From all the available studies, we know that *biochar is a heterogeneous material*, which even under similar production conditions can result in chemical and physical variability that impacts how that particular biochar will interact with the soil–plant–microbial system. As indicated in the preceding quote, this conclusion is analogous to those observations since the dawn of modern science. This is further complicated by the fact that a number of current studies do not adequately report the production, feedstock, postproduction handling, and actual biochar properties to fully comprehend where the biochar fits on the black carbon continuum to advance the state of this knowledge.

As explored in this chapter, biochar's interactions are due to several different potential mechanisms, which intimately depend on the soil, particular type and amount of biochar used, crop, and climate at any particular location. The interactions and driving forces of these interrelated dependencies are not fully understood, despite the fact that there has been over 250 years of investigation into the interactions of charcoal and soil. It is clear that there will not be one single mechanism for biochar's impact and the importance of any particular mechanism will depend on the unique soil–biochar–microbe–plant–climate combination. These data do emphasize the need for *designer biochar*. The use of biochar should be targeted to the specific soil deficiency one is trying to remedy.

Biochar's use as a soil fertilizer for productive soils has not consistently produced yields benefits to justify the cost of the material solely as a soil fertilization aid. This has been known since the 1700–1800s and has been further illustrated in the previously identified meta-analyses of recent biochar field studies with relatively low yield increases. However, unlike in the 1700–1800s, yield is not the primary motivation behind the creation of biochar. In other words, for good soils, the addition of biochar has not created significant yield declines, except when the biochar contains a high amount of sorbed organic compounds (Deenik et al. 2010, Rogovska et al. 2012). This demonstrates its beneficial use as a carbon sequestration agent.

More importantly, biochar's true value might not be in the routine use in agricultural production. Biochar's ability to increase crop yields appears to be directly linked to degraded and low productivity soils. Consequently, biochar could be part of the solution to (1) sequester atmospheric carbon dioxide and (2) as a management tool for creating additional productive land to provide food for the growing global population.

REFERENCES

Amies, W.S. 1872. Improvement in artificial manures. Patent #128578, US Patent and Trademark Office.

Asada, T., T. Ohkubo, K. Kawata, and K. Oikawa. 2006. Ammonia adsorption on bamboo charcoal with acid treatment. *Journal of Health Science* 52(5):585–589.

Asita, A.O. and I.A. Campbell. 1990. Anti-microbial activity of smoke from different woods. *Letters in Applied Microbiology* 10(2):93–95.

Atkinson, C., J. Fitzgerald, and N. Hipps. 2010. Potential mechanisms for achieving agricultural benefits from biochar application to temperate soils: A review. *Plant and Soil* 337(1):1–18. doi: 10.1007/s11104-010-0464-5.

Baker, R.T.K. 1989. Catalytic growth of carbon filaments. *Carbon* 27(3):315–323. doi: 10.1016/0008-6223(89)90062-6.

Becket, J.B. 1773. *An Essay on Electricity, Containing a Series of Experiments Introductory to the Study of that Science.* Bristol, U.K.: JB Becket.

Belanger, N. and B.D. Pinno. 2008. Carbon sequestration, vegetation dynamics and soil development in the Boreal Transition ecoregion of Saskatchewan during the Holocene. *Catena* 74(1):65–72. doi: 10.1016/j.catena.2008.03.005.

Biederman, L.A. and W.S. Harpole. 2013. Biochar and its effects on plant productivity and nutrient cycling: A meta-analysis. *GCB Bioenergy* 5(3):202–214.

Biniak, S., G. Szymanski, J. Siedlewski, and A. Swiatkowski. 1997. The characterization of activated carbons with oxygen and nitrogen surface groups. *Carbon* 35(12):1799–1810. doi: 10.1016/s0008-6223(97)00096-1.

Blake, W.P. 1893. Note upon the absorption of sulphur by charcoal. *Science* 21(541):326–327. doi: 10.2307/1765283.

Bovey, R.W. and F.R. Miller. 1969. Effect of activated carbon on the phytotoxicity of herbicides in a tropical soil. *Weed Science* 17(2):189–192.

Bradley, R. 1727. *Ten Practical Discourses: Concerning Earth and Water, Fire and Air, as they Relate to the Growth of Plants. With a Collection of New Discoveries for the Improvement of Land, Either in the Farm or Garden.* Westminster, U.K.: Printed by J. Cluer and A. Campbell, for B. Creake.

Brown, N.A.C. and J. van Staden. 1997. Smoke as a germination cue: A review. *Plant Growth Regulation* 22(2):115–124. doi: 10.1023/A:1005852018644.

Browne, D.J. 1852. *The American Muck Book: Treating of the Nature, Properties, Sources, History, and Operations of All the Principal Fertilisers and Manures in Common Use, with Specific Directions for their Preparation, Preservation, and Application to the Soil and to Crops; as Combined with the Leading Principles of Practical and Scientific Agriculture; Drawn from Authentic Sources, Actual Experience, and Personal Observation.* New York: CM Saxton (Agricultural Book Publisher).

Burrell, R.C and T.G. Phillips. 1925. The determination of nitrate nitrogen in plants. *Journal of Biological Chemistry* 65(1):229–234.

Chandler, J.M., O.B. Wooten, and F.E. Fulgham. 1978. Influence of placement of charcoal on protection of cotton (*Gossypium hirsutum*) from diuron. *Weed Science* 26(3):239–244. doi: 10.2307/4042706.

Chesebough, E.A. 1874. Improvement in antiseptic fertilizers. Patent #152725, US Patent and Trademark Office.

Clark, J. 1845. *Practical Instructions for Using Guano as a Manure*, 6th edn. London, U.K.: James Clark.

Davis, H.V. and H. Copeland. 1874. *Davis' Superphosphate and Animal Charcoal Fertilizer: Manufactured by H.V. Davis, New Bedford, Mass. Office, 40 North Water Street. Manufactory, Court Street, West of the Jail.* New Bedford, MA: E. Anthony & Sons.

Davy, E.W. 1856. XXII. On some experiments made with a view to determine the comparative value of peat and peat-charcoal for agricultural purposes. *The London, Edinburgh, and Dublin Philosophical Magazine and Journal of Science* 11(71):172–178.

Davy, S.H. 1818. *A Treatise on Soils and Manures, as Founded on Actual Experiences, and as Combined with the Leading Principles of Agriculture: In Which the Theory and Doctrines of Sir Humphry Davy, and Other Agricultural Chemists, are Rendered Familiar to the Experienced Farmer*, Vol. 8. London, U.K.: Printed for T. Cadell and W. Davies.

Davy, S.H. 1840. *The Collected Works of Sir Humphry Davy: Agricultural Lectures, pt. 2, and Other Lectures*, Vol. 8. London, U.K.: Johnson Reprint Corporation.

Deenik, J.L., T. McClellan, G. Uehara, M.J. Antal, and S. Campbell. 2010. Charcoal volatile matter content influences plant growth and soil nitrogen transformations. *Soil Science Society of America Journal* 74(4):1259–1270. doi: 10.2136/sssaj2009.0115.

Dundonald, A.C. 1795. *A Treatise, Showing the Intimate Connection that Exists between Agriculture and Chemistry: Addressed to the Cultivators of the Soil, to the Proprietors of Fens and Mosses, in Great Britain and Ireland, and to the Proprietors of West India Estates.* London, U.K.: Printed by the author, and sold by R. Edwards.

Durden, E.H. 1849. On the application of peat and its products, to manufacturing, agricultural, and sanitary purposes. *Proceedings of the Geological and Polytechnic Society of the West Riding of Yorkshire* 3:339–366. doi: 10.1144/pygs.3.339a.

El-Shobaky, G.A. and A.M. Youssef. 1978. Chemical activation of charcoals. *Surface Technology* 7(3):209–216. doi: 10.1016/0376-4583(78)90051-1.

Emmert, E.M. 1930. A method for the rapid determination of phosphate in fresh plant tissues. *Plant Physiology* 5(3):413.

Footitt, S. and M.A. Cohn. 2001. Developmental arrest: From sea urchins to seeds. *Seed Science Research* 11(01):3–16. doi: 10.1079/SSR200055.

Fries, N. 1978. Basidiospore germination in some mycorrhiza-forming Hymenomycetes. *Transactions of the British Mycological Society* 70(3):319–324.

Fuerst, E.P. and A.R. Putnam. 1983. Separating the competitive and allelopathic components of interference. *Journal of Chemical Ecology* 9(8):937–944.

Glaser, B., L. Haumaier, G. Guggenberger, and W. Zech. 1998. Black carbon in soils: The use of benzenecarboxylic acids as specific markers. *Organic Geochemistry* 29(4):811–819.

Goldberg, E.D. 1985. *Black Carbon in the Environment: Properties and Distribution*. New York: John Wiley & Sons Inc.

Guillén, M.D. and M.J. Manzanos. 2002. Study of the volatile composition of an aqueous oak smoke preparation. *Food Chemistry* 79(3):283–292. doi: 10.1016/s0308-8146(02)00141-3.

Hall, S.D. 1917. *The Book of the Rothamsted Experiments*. New York: Dutton.

Hansen, H.C.B., C.B. Koch, H. Nancke-Krogh, O.K. Borggaard, and J. Sørensen. 1996. Abiotic nitrate reduction to ammonium: Key role of green rust. *Environmental Science & Technology* 30(6):2053–2056. doi: 10.1021/es950844w.

Hansson, K.-M., J. Samuelsson, C. Tullin, and L.-E. Åmand. 2004. Formation of HNCO, HCN, and NH_3 from the pyrolysis of bark and nitrogen-containing model compounds. *Combustion and Flame* 137(3):265–277. doi: 10.1016/j.combustflame.2004.01.005.

Harper, H.J. 1924. The accurate determination of nitrates in soils. *Industrial and Engineering Chemistry* 16(2):180–183.

Haumaier, L. and W. Zech. 1995. Black carbon—Possible source of highly aromatic components of soil humic acids. *Organic Geochemistry* 23(3):191–196.

Hitz, C.W., K.H. Pak, and E.M. Rahn. 1953. Use of a crop protectant with certain herbicides on strawberry plants. *Proceedings of 7th Conference on Northeastern Weed Control*, New Brunswick, NJ, pp. 139–141.

Holbrook, F. 1849. Improvements of lands by green manuring. *The Cultivator* (Albany, New York*: New York State Agricultural Society*) 6(7):201–232.

Holmes, J.M. and R.A. Beebe. 1957. An example of desorption hysteresis at low relative pressures on a non-porous adsorbent: Ammonia on graphitized carbon black. *The Journal of Physical Chemistry* 61(12):1684–1686. doi: 10.1021/j150558a040.

Inderjit and R.M. Callaway. 2003. Experimental designs for the study of allelopathy. *Plant and Soil* 256(1):1–11. doi: 10.1023/A:1026242418333.

Insam, H. and M. Seewald. 2010. Volatile organic compounds (VOCs) in soils. *Biology and Fertility of Soils* 46(3):199–213. doi: 10.1007/s00374-010-0442-3.

Jansen, R.J.J. and H. van Bekkum. 1994. Amination and ammoxidation of activated carbons. *Carbon* 32(8):1507–1516. doi: 10.1016/0008-6223(94)90146-5.

Jeffery, S., F.G.A. Verheijen, M. van der Velde, and A.C. Bastos. 2011. A quantitative review of the effects of biochar application to soils on crop productivity using meta-analysis. *Agriculture, Ecosystems and Environment* 144(1):175–187. doi: 10.1016/j.agee.2011.08.015.

Jenkinson, D. and J. Rayner. 1977. The turnover of soil organic matter in some of the Rothamsted classical experiments. *Soil Science* 123:298–305.

Kashiwaya, Y. and K. Ishii. 1991. Kinetic analysis of coke gasification based on non-crystal/crystal ratio of carbon. *ISIJ International* 31:440–448.

Keeley, J.E., B.A. Morton, A. Pedrosa, and P. Trotter. 1985. Role of allelopathy, heat and charred wood in the germination of chaparral herbs and suffrutescents. *Journal of Ecology* 73(2):445–458. doi: 10.2307/2260486.

Keeley, S.C. and M. Pizzorno. 1986. Charred wood stimulated germination of two fire-following herbs of the California Chaparral and the role of hemicellulose. *American Journal of Botany* 73(9):1289–1297. doi: 10.2307/2444063.

Kirwan, R. 1793. What are the manures most advantageously applicable to the various sorts of soils, and what are the causes of their beneficial effect in each particular instance? *The Transactions of the Royal Irish Academy* 5:129–198 (Article Type: Research-article/Full publication date: 1793/1794/).

Kratky, B.A., D.L. Coffey, and G.F. Warren. 1970. Activated carbon root dips on transplanted strawberries. *Weed Science* 18(5):577–580.

Krock, B., S. Schmidt, C. Hertweck, and I.T. Baldwin. 2002. Vegetation-derived abscisic acid and four terpenes enforce dormancy in seeds of the post-fire annual, *Nicotiana attenuata*. *Seed Science Research* 12(4):239–252. doi: 10.1079/ssr2002117.

Kuhlbusch, T.A.J. and P.J. Crutzen. 1995. Toward a global estimate of black carbon in residues of vegetation fires representing a sink of atmospheric CO_2 and a source of O_2. *Global Biogeochemical Cycles* 9(4):491–501. doi: 10.1029/95gb02742.

Lathrop, E.C. 1917. The organic nitrogen compounds of soils and fertilizers. *Journal of the Franklin Institute* 183(2):169–206. doi: http://dx.doi.org/10.1016/S0016-0032(17)90622-2.

Lefroy, J.H. 1883. *Remarks on the Chemical Analyses of Samples of Soil From Bermuda*. Royal Gazette, Hamilton, UK: Foreign and Commonwealth Office Collection, pp. 42.

Lehmann, J. 2007. Bio-energy in the black. *Frontiers in Ecology and the Environment* 5(7):381–387.

Lehmann, J., M. Rillig, J. Thies, C.A. Masiello, W.C. Hockaday, and D. Crowley. 2011. Biochar effects on soil biota—A review. *Soil Biology and Biochemistry* 43:1812–1836. doi: 10.1016/j.soilbio.2011.04.022.

Lentz, R.D. and J.A. Ippolito. 2012. Biochar and manure affect calcareous soil and corn silage nutrient concentrations and uptake. *Journal of Environmental Quality* 41(4):1033–1043.

Linscott, D.L. and R.D. Hagin. 1967. Protecting alfalfa seedlings from a triazine with activated charcoal. *Weeds* 15(4):304–306.

Manyà, J.J. 2012. Pyrolysis for biochar purposes: A review to establish current knowledge gaps and research needs. *Environmental Science & Technology* 46(15):7939–7954. doi: 10.1021/es301029g.

Merritt, D.J., S.R. Turner, S. Clarke, and K.W. Dixon. 2007. Seed dormancy and germination stimulation syndromes for Australian temperate species. *Australian Journal of Botany* 55(3):336–344. doi: 10.1071/bt06106.

Mitchell, J. 1748. An account of the preparation and uses of the various kinds of pot-ash. *Philosophical Transactions* 45(485–490):541–563.

Morfit, C. 1848. *Manures, Their Composition, Preparation, and Action upon Soils: With the Quantities to be Applied. Being a Field Companion for the Farmer. From the French of Standard Authorities*. Philadelphia, PA: Lindsay and Blakiston.

Napier, G. 1788. Observations on gun-powder. *The Transactions of the Royal Irish Academy* 2:97–117. doi: 10.2307/30079234.

Nash, G.V. 1895. Notes on some Florida plants. *Bulletin of the Torrey Botanical Club* 22(4):141–161.

Nelson, D.C., G.R. Flematti, E.L. Ghisalberti, K.W. Dixon, and S.M. Smith. 2012. Regulation of seed germination and seedling growth by chemical signals from burning vegetation. *Plant Biology* 63:107–130.

Novak, J.M., I. Lima, B. Xing, J.W. Gaskin, C.Steiner, K.C. Das, M. Ahmedna et al. 2009. Characterization of designer biochar produced at different temperatures and their effects on a loamy sand. *Annals of Environmental Science* 3:195–206.

Ogawa, M. and Y. Okimori. 2010. Pioneering works in biochar research, Japan. *Soil Research* 48(7):489–500. doi: 10.1071/SR10006.

Okimori, Y., M. Ogawa, and F. Takahashi. 2003. Potential of CO_2 emission reductions by carbonizing biomass waste from industrial tree plantation in South Sumatra, Indonesia. *Mitigation and Adaptation Strategies for Global Change* 8(3):261–280. doi: 10.1023/B:MITI.0000005643.79908.5a.

Olsson, M., O. Ramnäs, and G. Petersson. 2004. Specific volatile hydrocarbons in smoke from oxidative pyrolysis of softwood pellets. *Journal of Analytical and Applied Pyrolysis* 71(2):847–854. doi: 10.1016/j.jaap.2003.11.003.

Otto, M. 1804. On the influence of the component parts of the soil on vegetation. *The Philosophical Magazine: Comprehending the Various Branches of Science, the Liberal and Fine Arts, Agriculture, Manufactures, and Commerce* 18(69):75–83.

Petit, C. and T.J. Bandosz. 2007. Role of aluminum oxycations in retention of ammonia on modified activated carbons. *The Journal of Physical Chemistry C* 111(44):16445–16452. doi: 10.1021/jp074118e.

Pitts, E.B. 1863. Improvement in preparing night-soil for manure. Patent #37685, US Patent and Trademark Office.

Priestley, J. 1770. Experiments and observations on charcoal. *Philosophical Transactions (1683–1775)* 60:211–227. doi: 10.2307/105888.

Priestley, J. 1775. *Experiments and Observations on Different Kinds of Air*, Vol. 1. St. Paul's Church-Yard, U.K.: J. Johnson.

Proust, M. 1811. Some account of experiments on different kinds of charcoal, for improving the manufacture of gunpowder, and of the slowness of combustion of chestnut wood. *The Belfast Monthly Magazine* 58–61.

Puri, B.R. 1966. Chemisorbed oxygen evolved as carbon dioxide and its influence on surface reactivity of carbons. *Carbon* 4(3):391–400. doi: http://dx.doi.org/10.1016/0008-6223(66)90052-2.

Puri, B., D.D. Singh, J. Nath, and L. Sharma. 1958. Chemisorption of oxygen on activated charcoal and sorption of acids and bases. *Industrial & Engineering Chemistry* 50(7):1071–1074. doi: 10.1021/ie50583a046.

Putnam, A.R. and W.B. Duke. 1978. Allelopathy in agroecosystems. *Annual Review of Phytopathology* 16(1):431–451.

Rafn, M. 1799. IV. Extract of a memoir, and experiments on the nutrition of plants. *Philosophical Magazine Series 1* 5(19):233–236. doi: 10.1080/14786449908677145.

Rapp, G. 2009. Pigments and colorants. In *Archaeomineralogy*, pp. 201–221. Berlin, Germany: Springer Science & Business Media.

Raynbird, H. 1847. On peat charcoal as a manure for turnips and other crops. *Journal of the Royal Agricultural Society of England* 7:539–548.

Retan, G.A. 1915. Charcoal as a means of solving some nursery problems. *Journal of Forestry* 13(1):25–30.

Rideal, E.K. and W.M. Wright. 1926. CCXL.-Low temperature oxidation at charcoal surfaces. Part II. The behaviour of charcoal in the presence of promoters. *Journal of the Chemical Society (Resumed)* 129: 1813–1821.

Rivera-Utrilla, J., M. Sánchez-Polo, V. Gómez-Serrano, P.M. Álvarez, M.C.M. Alvim-Ferraz, and J.M. Dias. 2011. Activated carbon modifications to enhance its water treatment applications. An overview. *Journal of Hazardous Materials* 187(1–3):1–23. doi: 10.1016/j.jhazmat.2011.01.033.

Robinson, S. 1853. *Guano: A Treatise of Practical Information for Farmers: Containing Plain Directions How to Apply Peruvian Guano to the Various Crops and Soils of America*. New York, NY: M.F. Barreda and Brother and T.W. Riley.

Rogovska, N., D. Laird, R.M. Cruse, S. Trabue, and E. Heaton. 2012. Germination tests for assessing biochar quality. *Journal of Environmental Quality* 41(4):1014–1022. doi: 10.2134/jeq2011.0103.

Sahu, R., Y.A. Levendis, R.C. Flagan, and G. R. Gavalas. 1988. Physical properties and oxidation rates of chars from three bituminous coals. *Fuel* 67:275–283.

Saito, M. 1990. Charcoal as a micro-habitat for VA mycorrhizal fungi, and its practical implication. *Agriculture, Ecosystems & Environment* 29(1–4):341–344. doi: http://dx.doi.org/10.1016/0167-8809(90)90298-R.

Samarkin, V.A, M.T. Madigan, M.W. Bowles, K.L. Casciotti, J.C. Priscu, C.P. McKay, and S.B. Joye. 2010. Abiotic nitrous oxide emission from the hypersaline Don Juan Pond in Antarctica. *Nature Geoscience* 3(5):341–344.

Schmidt, M.W.I. and A.G. Noack. 2000. Black carbon in soils and sediments: Analysis, distribution, implications, and current challenges. *Global Biogeochemical Cycles* 14(3):777–793.

Schmidt, M.W.I., J.O. Skjemstad, E. Gehrt, and I. Kogel-Knabner. 1999. Charred organic carbon in German chernozemic soils. *European Journal of Soil Science* 50(2):351–365.

Schreiner, O. and B.E. Brown. 1912. Occurrence and nature of carbonized material in soils. *US Department of Agriculture Bureau of Soil Bulletin* 90.

Seredych, M. and T.J. Bandosz. 2007. Mechanism of ammonia retention on graphite oxides: Role of surface chemistry and structure. *The Journal of Physical Chemistry C* 111(43):15596–15604. doi: 10.1021/jp0735785.

Sewell, T. 1872. Improvement in compositions for deodorizing and preparing fertilizers. Patent #126418, US Patent and Trademark Office.

Sheldon, H.H. 1920. Charcoal activation. *Physical Review* 16(8):165–172.

Shevlin, P.B., D.W. McPherson, and P. Melius. 1983. Reaction of atomic carbon with ammonia. The mechanism of formation of amino acid precursors. *Journal of the American Chemical Society* 105(3):488–491. doi: 10.1021/ja00341a031.

Skinner, J.J. 1908. Water culture method for experimenting with potatoes. *Plant World* 11(11):249–254.

Smith, C.J., T.A. Perfetti, R. Garg, and C. Hansch. 2003. IARC carcinogens reported in cigarette mainstream smoke and their calculated log P values. *Food and Chemical Toxicology* 41(6):807–817. doi: http://dx.doi.org/10.1016/S0278-6915(03)00021-8.

Sneyd, J., W. Carlisle, T. Davis, J. Phillips, S. Holt, H. Heathcote, J. Holt et al. 1798. Papers in agriculture. *Transactions of the Society, Instituted at London, for the Encouragement of Arts, Manufactures, and Commerce* 16:115–261. doi: 10.2307/41322028.

Sohi, S.P., E. Krull, E. Lopez-Capel, and R. Bol. 2010. A review of biochar and its use and function in soil. *Advances in Agronomy* 105:47–82.

Sombroek, W.G. 1966. *Amazon Soils*. Wageningen, the Netherlands: Centre for Agricultural Publications and Documentation (PUDOC).

Sombroek, W., M. Ruivo, P. Fearnside, B. Glaser, and J. Lehmann. 2004. Amazonian dark earths as carbon stores and sinks amazonian dark earths. In J. Lehmann, D. Kern, B.Glaser, and W.Wodos, eds., pp. 125–139. Dordrecht, the Netherlands: Springer.

Spokas, K.A. 2010. Review of the stability of biochar in soils: predictability of O:C molar ratios. *Carbon Management* 1:289–303.

Spokas, K.A., K.B. Cantrell, J.M. Novak, D.W. Archer, J.A. Ippolito, H.P. Collins, A.A. Boateng et al. 2012. Biochar: A synthesis of its agronomic impact beyond carbon sequestration. *Journal of Environmental Quality* 41:973–989.

Spokas, K.A., J.M. Novak, C.E. Stewart, K.B. Cantrell, M. Uchimiya, M.G. duSaire, and K.S. Ro. 2011. Qualitative analysis of volatile organic compounds on biochar. *Chemosphere* 85(5):869–882.

Spokas, K., J. Novak, and R. Venterea. 2012. Biochar's role as an alternative N-fertilizer: Ammonia capture. *Plant and Soil* 350(1):35–42. doi: 10.1007/s11104-011-0930-8.

Terne, B. 1882. Utilizing tank-waters of slaughter-houses. Patent #269487, US Patent Office.

Tryon, E.H. 1948. Effect of charcoal on certain physical, chemical, and biological properties of forest soils. *Ecological Monographs* 18(1):81–115.

Tucker, J. 1756. *The Case of the Importation of Bar-Iron, from Our Own Colonies of North America*. London, U.K.: Thomas Trye.

Turner, E.R. 1955. The effect of certain adsorbents on the nodulation of clover plants. *Annals of Botany* 19(1):149–160.

Universal Charcoal and Sewage Company Limited. 1875. *The Profitable Utilization of Towns' Refuse: As Carried on by the Universal Charcoal And Sewage Company Limited*, Ireland and Co. Printers, Manchester, UK. pp 1–28.

van Staden, J., A.K. Jager, M.E. Light, and B.V. Burger. 2004. Isolation of the major germination cue from plant-derived smoke. *South African Journal of Botany* 70(4):654–659.

Vantsis, J.T. and G. Bond. 1950. The effect of charcoal on the growth of leguminous plants in sand culture. *Annals of Applied Biology* 37(2):159–168.

Voelker, A. 1864. On Peruvian guano and the means of increasing its efficiency as a manure. *Journal of Royal Agriculture Society* 25:186–209.

Walden, J.H. 1860. *Soil Culture: Containing a Comprehensive View of Agriculture, Horticulture, Pomology, Domestic Animals, Rural Economy, and Agricultural Literature.* New York: CM Saxton.

12

Survey of Biochar Field Trials

Erich J. Knight

CONTENTS

Terra Preta Nova

The "founding father" of terra preta soil study is Dr. Wim Sombroek. Before Sombroek died in 2003, he called together a team of internationally renowned soil scientists and gave them the challenge of creating a new terra preta capable of solving some of the world's most pressing problems. He called this new terra preta, *terra preta nova*, created using modern thermal conversion technologies: pyrolysis, gasification, and hydrothermal carbonization. The question now is whether it is possible to meet this challenge Sombroek laid out before us. Are we capable of creating rich, self-sustaining soils without doing any further harm to the Earth's already fragile ecosystems? I believe the answer to this question is yes, yet the real work lies in ferreting out the mechanisms, the soil physical property changes, how to improve soils' sustainably on a bioregional basis, and doing it at a scale that will make a difference.

Historical Notes: A Brief History of Agricultural Time

Our farming for over 10,000 years has been responsible for two-thirds of our excess greenhouse gases (GHGs). This soil carbon, converted to carbon dioxide, methane, and nitrous oxide, began a slow stable warming that is now greatly accelerated with burning of fossil fuel. The unintended consequence has been the flowering of our civilization. Our science has now realized the consequences and developed a more encompassing wisdom [1].

Modern agriculture has evolved in the ability to remove the limitations to plant growth, from burning forest for ash fertilizers, to bison bones, to Guano islands, then in 1913, to Germans figuring out how to suck nitrogen from the air to make fertilizers, now using natural gas energy. These chemical fertilizers have overcome nutrient limits to growth for 100 years.

NPK and the *Green Revolution* in genetics have brought us to where we are, all made possible by basically mining soil carbon stocks. So we have now hit a carbon limit in two distinct ways. The first is continued loss of soil carbon, and the second is fossil carbon energy cost. The present farming system spends ten cents of fossil energy delivering one cent of food energy. We cannot go back, but we can go forward with our newly acquired wisdom. Wise land management such as conservation agriculture and afforestation can build back our soil carbon. Biochar allows the soil food web to build much more recalcitrant organic carbon (living biomass and glomalins) in addition to the elemental carbon in the biochar.

We can rectify the carbon cycle and, beyond that, heal the nitrogen and phosphorous cycles and toxicity in soils and sediments, and as a feed additive cut the carbon hoof and paw prints of livestock. At some 250 farms and vineyards in Europe, the health metrics of both fauna and flora demonstrate the closed loop now achieved. Managing nutrients and feed in a cascade of higher value utilization and allowing the wee-beasties, the microbiology, to do the heavy lifting both in the soil and in the enteric digestion of farm animals [2].

Long-Term Biochar Field Trials

If you enter *biochar field trials* into the search engine Google Scholar, it will bring up thousands of peer-reviewed studies, for the last 2 years growing exponentially, like biological systems. The heavy metal remediation studies show great promise for in situ binding of these toxins from the food chain. Coal power plants utilizing pyrolytic co-combustion would then have a char by-product for mine scarred land reclamation and remediation of arsenic, selenium, uranium, mercury, etc., into both air and ash piles. The recent American Chemical Society's Soil Science Conference had over 100 biochar presentations. The volume of published, peer-reviewed research on biochar is doubling every year; biochar continues to increase exponentially, with 60 papers published in 2008, 129 in 2009, 140 in 2010, 171 in 2011, and to date 235 in 2012. Peer-review studies are maintained and expanded at the International Biochar Initiative's (IBI) directory of research institutions worldwide and continue to be maintained and updated at IBI's searchable electronic biochar bibliography that contains 1000 published citations on biochar [3].

My focus here will be on the longer-duration biochar trials and those showing unique results by manipulating plant growth variables. Early trials tended to use raw biochar, unamended and not formulated with compost or other nutrient charging, with treatment plots receiving various rates of biochar application from 0.5 to 10 tons per acre. The several meta-analytical surveys, incorporating these early trials, may incorporate a negative bias in their results. Future surveys incorporating trials more attuned to the soil microbiology will favor increasingly positive outcomes [4].

Internationally, Japan, Australia, and the EU are 10–15 years ahead of the United States in developing biochar soil technologies. The United States has been catching up fast, particularly with the development of pyro/catalytic thermal conversion of biomass directly into transport fuels [5]. Much of the work in Japan can be reviewed at the Japanese Biochar Association website and other regional research networks. There is much work that is not yet translated [6]. The US total number of acres under biochar protocols is in the hundreds, while the totals probably number in the thousands of hectares across Asia, Australia, and Europe. A new discovery from the Advanced Light Source at Berkeley about fungal potassium salts being the primary nucleating catalyst for clouds and rain brings new importance to soil carbon solutions. That vision of how life itself calls the rain is another unaccounted for ecological service provided by a healthy soil. Several other findings this year concerning soil microbiology and soil carbon in forest and fields are extremely supportive of Dr. Jim Hansen's 100 GtC of new forests to avoid dangerous climate change, the national Carbon Farming Initiatives & Soil Carbon Standards [7].

The accounting of soil carbon as the base measurement of sustainability and aligning incentives to get a farmer paid for their good works is where carbon markets should all grow from. The farmer will always have the lowest cost system for sequestration of carbon, and it is about time that the carbon markets recognize that as its very foundations, a foundation far more secure than any other market. All political persuasions agree. Building soil carbon is good. All LCAs are hollow unless grounded in soils. Soil is the bank everything else dividend; if soil carbon is the rule, biochar will be the game [8].

There is a dearth in trials comparing the effects of plant-based and manure biochars in the field; combined biochar/bokashi compost trials and the effects of biochar on the production of fruits and vegetables are lacking in the literature. Currently, there are only a few published large-scale field trials spanning several growing seasons, but there are some studies in the pipeline that will likely have results published within the next year.

Project Rainbow Bee Eater. Broad field biochar studies for 9 years in a host of common field crops, utilizing a banding application right in the planting furrow, have shown consistent yield increases and increases in water holding capacity and are providing an economic model for inclusion in Australia's carbon farming initiative.*

* Australian and New Zealand Biochar Research Network. http://www.anzbiochar.org/projects.html#nine.

Virginia Tech is in the sixth year of field trials with the Carbon Char Group's *CharGrow*-formulated bagged product. These studies have consistently shown that just 5% of biochar placed in the growing medium of tomato transplants, when planted out to the field, has an average 40%–50% increase in yield and fruit quality.*

The BlueLeaf Inc. Four years of study, using several different biochars in different cropping systems, has exciting results given how far north the site is at 45° and the low application rates. I suspect, as we saw with the Imperial College test. Other parameters showing increases with biochar included earthworms, nematode, and mycorrhizal root colonization, supporting the hypothesis that biochar may serve as a refuge for soil microbes. Surface soil water infiltration was also greater in biochar-amended soil. These studies are the most thoroughly measured, including GHG soil emissions, leaf temperature, and protein content, in soybean and forage crops. Generally, biochar amendments increased crop productivity 10%–15%. Given the low metabolic turnover in these Canadian soils, fueled increases are significant.†

My own studies covered only a 2-year period: first, as a sister study with the Rodale Institute in 2009 and, with no additional biochar amendments, the measured yield from a commercial soybean crop in 2010. In both years, the yields increased roughly 20% over the controls in our temperate Virginia clay loam soils. Also of note were increased soil bacteria and rates of fungal infection, earthworks increased 25%, and SPAD readings (a proxy for tissue nitrogen) increased proportionally to yield increases.

Alterna Biocarbon and Cowboy Charcoal Virginia field trials 2009 field testing for the 09 corn season with JMU and consultation with Dr. Hepperly at Rodale Institute.‡

The NC Farm Center has large-scale field application trials encompassing 40 acres on two farms in southeastern North Carolina.§

Imperial College test. This work in temperate soils gives data from which one can calculate savings on fertilizer use, which is expected to be ongoing with no additional soil amending.¶

Nikolaus Foidl's work with Aspirin & Char is Amazing in Maize, demonstrating plant chemical signaling with resultant epigenetic effects on corns' metabolism, 250% yield gains, and 15 cobs per plant. While the *biochar-only* treatment plots showed increases in yield, the large increases reported were in conjunction with a foliar feeding of ascorbic acid, a known growth stimulant. This additional application had little effect on the crops without biochar. Nikolaus explains this result as a combination plant chemical signaling: first, the plants sense the pyrolytic substances from the biochar giving them the message that a forest fire occurred, an opportunity to shine in the sun, and then, when the aspirin is applied at the 4-leaf growth stage, this caused the corn to reach deep into its own genetic library, to selectively express long dormant genetic capabilities that produced these startling results.**

Bartlett Tree Experts and The Morton Arboretum, assisted by the Chicago Bureau of Forestry and Department of Environment, studied biochar applications on about 60 trees on Milwaukee Ave., in Chicago's Wicker Park/Bucktown community. Bartlett's research indicates improvements in soil fertility and water retention, but another unexpected benefit has also started to emerge. Experimental trees in soil with biochar are demonstrating improved resistance to certain insects and disease. Premium Landscape Biochar is a new soil amendment offered by Bartlett Tree Experts. Both the organic and agricultural chemical schools of soil science recognize biochar as a powerful tool to foster biodiversity and nitrogen efficiency in soils. My hope is that biochar can form the basis for an uncomfortable marriage between commercial agriculture and organic practice—that is, improved synthetic nitrogen efficiency and livestock manure management and increasingly biodiverse soil microbiology, forming ever more recalcitrant forms of humic substances, respectively. Work by

* http://www.carbonchar.com/plant-performance.

† http://biomassmagazine.com/articles/3501/dynamotive-blueleaf-complete-two-year-biochar-trial.

‡ http://biochar.bioenergylists.org/node/1408.

§ http://www.biochar-international.org/profiles/northcarolinafarmcenter.

¶ http://www.iop.org/EJ/article/1755-1315/6/37/372052/ees9_6_372052.pdf?request-id=22fb1902-1c23-4db8-8801-2be7e2f3ce1b.

** http://terrapreta.bioenergylists.org/content/trials-maize-reactivating-dormant-genes-using-high-doses-salicylic-acid-and-charcoal.

Christoph Steiner, then at U of Georgia, showing a 52% reduction of NH_3 loss when char is used as a composting accelerator. This will have profound value-added consequences for the commercial composting industry by reduction of their GHG emissions and the sale of compost as an organic nitrogen fertilizer.*

REFERENCES AND NOTES

1. The opening presentation at the fourth USBI Biochar Conference in Sonoma California: http://2012.biochar.us.com/299/2012-us-biochar-conference-presentations
   Sonoma Biochar Conference Report: http://tech.groups.yahoo.com/group/biochar-policy/message/3921
   The Anthropogenic Greenhouse Era Began Thousands of Years Ago: http://www.springerlink.com/content/h328n0425378u736/
   The Columbian encounter led to terrestrial biospheric carbon sequestration on the order of 2 to 5 GtC. Dull, R. A., R. J. Nevle, and W. I. Woods, 2010, The columbian encounter and the little ice age: Abrupt land use change, fire, and greenhouse forcing. *Annals of the Association of American Geographers*, 100: 775–771, Special Issue: Climate Change, DOI: 10.1080/00045608.2010.502432.
2. Ninety percent of the biochar produced in Europe is used in livestock farming. Whether mixed with feed, added to litter, or used in the treatment of slurry, the positive effect of biochar very quickly becomes apparent. The health—and consequently the well-being—of the livestock improve within just a short space of time. As regards nasty smells and nutrient losses, the use of biochar could even herald a new age of livestock farming, closing agricultural cycles of organic matter. Biochar – a key technology for the planet:
   http://www.ithaka-journal.net/pflanzenkohle-eine-schlusseltechnologie-zur-schliesung-der-stoffkreislaufe?lang=en
   In a study with field trials on six different grape-growing sites around Europe (Switzerland, Spain, France, and Italy), the authors found a wide spectrum of results with different patterns of reactions in biochar-amended soils. The major nutritional elements, for example, phosphorus (P) and potassium (K), show a clear positive trend in the nutrient uptake by the vine in biochar-amended soils compared to the control. In some cases, they observed decreased growth of grape leaves in biochar-amended soils compared to the control plot, which may have been due to nutrient blockages associated with the use of non-nutrient inoculated biochar in the experiments: C. Niggli, and H. P. Schmidt, 2012, Biochar in European viticulture: Results of the 2011 Season. *Ithaka Journal* (2012): n. pag. Ithaka Journal Delinat Institut www.ithaka-journal.net/druckversionen/e022012-bc-viticulture.pdf.
3. International Biochar Initiative (IBI), 2012, A Year in Review for IBI http://www.biochar-international.org/node/3847
   IBI Biochar Standards can serve as a guide to characterizing biochar for use in field trials and other research trials as well as for commercial applications: http://www.biochar-international.org/characterizationstandard.
4. Spokas, K. et al., 2012; Biochar: A Synthesis of Its Agronomic Impact beyond Carbon Sequestration. Fifty percent of the reviewed studies reported yield increases after black carbon or biochar additions, with the remainder of the studies reporting alarming decreases to no significant differences: https://www.agronomy.org/publications/jeq/abstracts/41/4/973.
   Another meta-analysis of biochar field trial results has been published to date by Jeffery et al. This analysis included both pot and field trials and used all available published literature with quantitative results up to March 2010. Overall, the mean results for each analysis performed within the meta-analysis covered a wide range (from −28% to 39%) but found that when compared to a control, biochar application to soils showed a small but statistically significant benefit to crop productivity, with a grand mean increase of 10%.
   Jeffery, S., F. G. A. Verheijen, M. van der Velde, and A. C. Bastos, 2011, A quantitative review of the effects of biochar application to soils on crop productivity using meta-analysis. *Agriculture, Ecosystems and Environment*, 144: 175–187.

* http://www.ibi2010.org/wp-content/uploads/BiocharPoultrySteiner.pdf.

Biederman and Harpole performed a meta-analysis of 371 independent studies culled from 114 published manuscripts finding an average increase of aboveground productivity, crop yield, soil microbial biomass, rhizobia nodulation, plant K tissue concentration, soil phosphorus, soil potassium, total soil nitrogen, and total soil carbon.

Biederman, L. and W. S. Harpole, 2013, Biochar and its effects on plant productivity and nutrient cycling: A meta-analysis, *Global Change Biology–Bioenergy*, Special Issue: Biochar. 5: 202–214. http://onlinelibrary.wiley.com/doi/10.1111/gcbb.12037/abstract

5. CoolPlanet Biofuels, guided by Google's Ethos and funding, along with GE, BP, and Conoco, are now building the reactors that convert 1 ton of biomass to 75 gallons of Biogasoline and one-third ton biochar for soil carbon sequestration: http://www.coolplanet.com/.
6. Japan Biochar Association; http://www.geocities.jp/yasizato/pioneer.htm eBRN European Biochar Research Network http://cost.european-biochar.org/en

   UK Biochar Research Centre http://www.geos.ed.ac.uk/sccs/biochar/

   The Australian and New Zealand Biochar Researchers Network http://www.anzbiochar.org/index.html
7. The demonstration, using quantitative 13C nuclear magnetic resonance (NMR) spectroscopy measurements, concludes that both terra preta soils and Midwest dark soils contain 40% to 50%+ of their organic carbon (SOC) as pyrolytic carbon char and that this pyrolytic carbon can account for all cation-exchange capacity (CEC). Mao, J.-D., R. L. Johnson, J. Lehmann, D. C. Olk, E. G. Neves, M. L. Thompson, and K. Schmidt-Rohr, 2012, Abundant and stable char residues in soils: Implications for soil fertility and carbon sequestration. *Environmenal Science and Technolology*, Article ASAP. DOI: 10.1021/es301107c American Chemical Society http://pubs.acs.org/doi/abs/10.1021/es301107c

   (Potassium) Salt Seeds Clouds in the Amazon Rainforest; http://newscenter.lbl.gov/feature-stories/2012/09/10/amazon-aerosols/

   Fertile soil does not fall from the sky. The contribution of bacterial remnants to soil fertility has been underestimated until now.

   http://www.alphagalileo.org/ViewItem.aspx?ItemId=126987&CultureCode=en

   Biologists Unlock 'Black Box' to Underground World: How Tiny Microbes Make Life Easier for Humans, http://www.sciencedaily.com/releases/2013/01/130103092030.htm?goback=.gde_4767237_member_201276911.

   Cross-biome metagenomic analyses of soil microbial communities and their functional attributes, http://www.pnas.org/content/109/52/21390

   Afforestation Soil-C Support: A review of 46 research papers published between 1957 and 2010 as well as an analysis of 409 soil profiles from the National Soil Carbon Network database. Soils in newly forested areas store substantial carbon that could help offset climate change http://www.eurekalert.org/pub_releases/2013-04/uom-sin040113.php. International Soil Carbon Network: http://soilcarb.net

   Boreal Forest Deep C Storage; New Data for modeling boreal forest root exudate deep carbon storage. Identifying the young carbon deep in the soil transported by fungi fed by the sweet root sugars they love adds a new refinement to carbon flows of our northern forest. This should add significantly to Dr. Jim Hansen's 100 GtC of new forests to avoid dangerous climate change.

   Fungi pull carbon into northern forest soils. http://www.sciencenews.org/view/generic/id/349265/description/Fungi_pull_carbon_into_northern_forest_soils

   Clemmensen, K. E. et al., 2013, Roots and associated fungi drive long-term carbon sequestration in boreal forest. *Science*, 339: 1618. DOI:10.1126/science.1231923.
8. FAO on Conservation Agriculture: "In general, soil carbon sequestration during the first decade of adoption of best conservation agricultural practices is 1.8 tons CO_2 per hectare per year. On 5 billion hectares of agricultural land, this could represent one-third of the current annual global emission of CO_2 from the burning of fossil fuels (i.e., 27 Pg CO_2 per year)." http://www.fao.org/ag/ca/doc/CA_SSC_Overview.pdf Adding just 1 ton of biochar per hectare (800 lbs/acre) would cover 100% current annual fossil CO_2 emissions. "Greenhouse Gas Mitigation Potential of Agricultural Land Management in the United States: A Synthesis of the Literature" An extensive scientific literature review providing a side-by-side comparison of the biophysical greenhouse gas (GHG) mitigation potential of more than 40 agricultural land management activities in the United States:

   http://nicholasinstitute.duke.edu/ecosystem/land/TAGGDLitRev

   Re-Building the World's Soil: The Role of Soil Carbon Methodology for U.S. and Global Carbon Offset Projects, http://www.prweb.com/releases/2012/12/prweb10185341.htm

The first methodology approved by the UNFCCC that will allow carbon credits to be generated from the biochar arena. The approved CDM methodology will help to clean up charcoal production in developing countries and—for existing biochar projects that are using traditional technologies—will also help the biochar arena

The Final Frontier for Biochar. Carbon Credits;http://www.britishbiocharfoundation.org/?p=725

A copy of the methodology can be found;

http://www.britishbiocharfoundation.org/wp-content/uploads/EB67_repan08_ACM0021_NM0341.pdf

NASA's EO-1 hyperspectral imagery data have been used to discern Amazonian black earth or terra preta soils.

The full complement of earth-sensing satellites will soon provide multiple proxy measurements for soil carbon itself. Using direct moisture measurements to 3 feet depth, temperature, and density, even reading GHG emissions and dead and alive biomass from the tree tops down, at 1 hectare resolutions, when the Orbital Carbon Observer 2,(OCO-2), is aloft in 2014.

NASA's Space Archaeology; $364K Terra Preta Program http://archaeologyexcavations.blogspot.com/2010/08/time-traveling-via-satellite.html

Sowing Seeds With New Agricultural Carbon Accounting Tool. Carbon dioxide emissions from agricultural activity in the United States can now be tracked with unprecedented resolution because of a method developed at Oak Ridge National Laboratory team led by Tristram West. http://www.sciencedaily.com/releases/2010/06/100602131436.htm

13

Hydrological, Ecological and Economic Advantages of Aliquot Biochar Dosing for Soil, Climate, and Ecosystem Remediation

D. Nathaniel Mulcahy

CONTENTS

Introduction

With rising CO_2 levels and the very real prospect of peak water and the loss of cultivable lands, the ability of biochar to sequester carbon, economically and safely, combined with biochar's water and nutrient holding capacity, are creating a growing demand for this product. As with any new product, a sudden increase in demand requires a careful analysis of costs and benefits of both product and application techniques to achieve the greatest possible benefit while at the same time reducing, or eliminating, environmental or economic risks. Over the past several years, our work has shown that biochar's full potential is viable even with application rates as little as 1 metric ton per hectare (mt/ha). Surprisingly, this approach has shown that biochar's potential may be even greater than originally anticipated, providing specific measures are taken when applying it to the field. In this chapter, we will review a specific application process we have developed, discuss both laboratory and field experiments, and suggest future directions that could be advantageous.

Where Is It Applicable?

As knowledge of biochar has grown and begun to enter into accepted agricultural practices, current schools of thought for its application, while varying in the recommended rates, can be divided into two methods, field-wide or concentrated (side dressing or banded) (see Blackwell et al., 2009). An alternative approach to these, described here, is the aliquot dosing of biochar. This allows for a more economical and environmentally sustainable approach to biochar application, provides the same documented benefits of field-wide and side dressing approaches, and may in fact provide additional advantages. In this chapter, we will describe the theory, technique, and benefits of this third method, which we have called *aliquot dosing* application.

While biochar has long been shown to be applicable in many locations, its greatest impact has been seen in sandy or desertified soils (Tryon, 1948). In our own work, we have seen positive results following the application of biochar in coastal zones, desertified areas, and, in contrast to many studies, even rich loamy soils in areas with abundant rainfall. While the production method and feedstocks are critical parameters for any biochar study, in this chapter, we limit our discussion to the application method we have developed; most importantly, this effective method has surprising additional advantages and is applicable to both subsistence farm level and industrial scale agriculture.

Biochar and Ecosystem Remediation

The need for ecosystem remediation is nowhere greater than in the arid and semiarid ecosystems of the world. These cover 45% of the terrestrial surface (Schimel, 2010) and are expanding (Brown, 2006). Deserts and arid zones exhibit sparse vegetation, and they lack soil organic matter that would increase the retention of water. In the Kalahari Desert of Botswana, for example, Silberbauer (1981) wrote that "… even fleeting showers sank quickly out of reach in the sandy soils. …the sun dried, and the wind blew most leaf debris away before it could decompose into rich organic humus." (See also Workman, 2002.) Thus, both water and nutrient elements (Novak et al., 2009) are lost without vegetation. Introduced vegetation often fails to survive for lack of these resources, and desertification becomes self-reinforcing. Breaking this destructive cycle requires that introduced vegetation persists long enough to improve both nutrient content and soil structure, allowing sparse rainfall to be retained. Our studies have shown that biochar can help seedlings survive long enough to break this cycle (Mulcahy et al., 2013).

Economic Aspects of Aliquot Dosing

While it is becoming ever more evident that there are significant advantages to using biochar in both small- and large-scale agriculture, some current practices may pose serious economic and environmental concerns. Novak and Watts (2013) describe a study utilizing 2% by weight field-wide application rate for biochar. At this recommended application rate, they note that 40 t/ha would be needed and estimate the cost of required biochar to be $6000 per ha. While it is very likely that the eventual cost of biochar will diminish as worldwide biochar production increases, the environmental implications of this requirement deserve careful consideration. Assuming a biochar to feedstock ratio of one to five, a 2% per weight application rate would require 200 t/ha of biomass.

Assuming the production of biochar is done in an environmentally responsible manner and that the net CO_2 calculations are shown to be ecologically advantageous, there still remains the matter of feedstock requirement of 200 t/ha. This would be unattainable for farms in developing nations and on the continent of Africa where over 90% of food production comes from farms that are around 1 ha in size (Spencer, 2002; Wiggins, 2009). Other researchers tested application rates as high as 200 t/ha (Kammann et al., 2011); this however would require 1000 t/ha of biomass for feedstock, given that an average yield per year of six varieties of switchgrass is approximately 8.6 t/ha.

According to Sladden et al. (1991), application rates 200 t/ha of biochar would require 1000 t/ha of dry biomass, which would take the average African small farm the entire yield of their hectare 116.27 years to produce sufficient biomass to meet the suggested application rates. Even if we were to apply the more attainable recommendation of 40 t/ha (Novak and Watts, 2013), it would still require 23.25 years of dedicated biomass growth and harvesting to meet recommended application rates.

Given this daunting requirement for biochar, and fully aware that too little biochar would have little or no effect, we began a program of greenhouse experiments and field trials using biochar made of differing feedstocks and in differing soil types and climates. The original objective was to determine how small an amount of biochar would have measurable positive effects. As a baseline for our research, we began with greenhouse experiments to confirm the results from accepted application rates. Clearly, the costs of field-wide tests were not advantageous, and given that we were also testing our biochar, it was important to first understand that our specific biochar was safe to apply to soil before any field studies were even contemplated; pot trials in the greenhouse provide an economic way to do both.

Preliminary studies indicated that half-liter containers would be appropriate, equivalent to the contents of a 10 × 10 cm square pot or the half-liter plastic water bottles, universally available in the 13 different countries in which we work. As a side note, plastic water bottles are an environmentally serious concern, and by using them as our standard test pot, we not only could guarantee that all our partners could find affordable sources of planting containers but we would have the additional benefit of helping the environment by removing some of the bottles from the waste streams. In the event that plastic water bottles were not available or that the quantity needed was greater than the available supply, we were able to use, with equal efficacy, 1.5 mm polyfilm tubes of comparable dimensions and volume (Figure 13.1).

While initial experiments were done with ryegrass, we have also worked with tomatoes (Mulcahy et al., 2013) and Greveria, a rapidly growing tree in the Proteaceae family, widely planted for erosion control, firewood, animal fodder, and lumber (Greveria is a local name for *Grevillea robusta* A. Cunn. ex R. Br. (Buyinza et al., 2008)). In an effort to provide reproducible procedures even in areas without modern scientific equipment and digital scales, we chose to do percentage comparisons based on volume rather than weight. Converting from weight to volume, we have found that 320 g of biochar is 1 L of biochar based on a measured pyrolysis temperature of over 550°C and a feedstock of hardwood pellets. We based our experiments on accepted peer-review application rates of 0%, 2%, 4%, 8%, 15%, 20%, 30%, 50%, and 100% biochar per volume. In this discussion, we focus on 30% v/v.

Initial attempts to premix larger volumes of biochar/soil mixtures proved to be unsuccessful. Even when soaked, the biochar density varies so greatly from that of soils, sands, and clays that with even minimal vibrations, the mixtures would separate. To compensate for this, each container was incrementally

FIGURE 13.1 Seedling nursery in Gisenyi, Rwanda. Polyfilm, 1.5 mm thick was used as a seedling container holing a mixture of 30% (v/v) biochar and local soil. Total volume of container was 500 mL.

filled with the required mixture. For example, a 20% per volume uniform mixture was achieved by adding small doses at a time, as little as 5 mL; a uniform mixture throughout the aliquot is easily achieved. One 5 mL dose of biochar was followed by four 5 mL doses of soil and the process repeated until the desired fill level of the container was reached. Additions of doses smaller than 5 mL can be used but with containers of 500 mL or over; the resulting difference of smaller increments did not result in a statistically significant difference. Adding in smaller amounts is helpful only in the case of smaller containers, and smaller grain size, but with much smaller containers, the effect of biochar becomes difficult to distinguish from the effects of the container itself.

We feel it is important not to powder or crush biochar prior to application. While crushed biochar may appear to be easier to mix, there are three reasons for not doing so. The first reason, as well documented by Husk and Major (2010), is that even during periods of little wind, as much as 30% may be lost. Loss would be reduced with pelleted biochar. Pelleted biochar, which can come from pyrolyzing pelleted feedstocks or pelleting pryrolyzed material, is larger and sufficiently porous that the roots from seedlings will rapidly surround, and even penetrate, the moisture holding pellets, keeping them in place and therefore not only assist in preventing erosion but secure the biochar's location to where it will be most beneficial to the plant. In addition, if biochar is crushed too fine, it will clog soil pores and impede flow of water and oxygen.

A second reason to avoid crushed biochar concerns health risks. Many in the biochar world came through clean cookstove programs. The Clean Cookstove movement came into being in an effort to reduce fuel consumption and respiratory illnesses associated with smoke inhalation during traditional cooking. Each year, between 1.9 and 3.0 million premature deaths are attributed to illness caused by traditional cooking options (WHO, 2007). Given that crushed biochar is fundamentally the same as crushed coal, encouraging the use of crushed biochar would spur a new resurgence of the equivalent of miner's lung for the field workers and those preparing biochar and distributing it.

The third reason for not crushing biochar is that the larger size of pelleted biochar (typically 5–8 mm in diameter with lengths that vary between 5 and 15 mm) is the microbiological benefit. "Evidence suggests that bacteria are found in preferentially colonized patches in soil located close to pores in the subsoil" (Nunan et al., 2006). Pore size in biochar is controllable and stable; therefore, the establishment of microbial populations can depend on these factors. Surprisingly, particle size, and in the case of biochar pore size, is more influential on microbial communities than external treatments such as fertilizers, green manure, animal manure, sewage sludge, and peat (Sessitch et al., 2001).

Both laboratory and greenhouse experiments seemed to confirm that the recommended application rates were well founded. But it was at this point that we came to a new hypothesis. Wishing to see how the plants would do once transplanted, the original potted plants were transferred to the field. Special care was taken when removing the root balls from the containers so as to not disturb the soil/biochar mixtures. Remarkably, the same drought tolerance observed in the greenhouse continued in the field. This suggested that while the recommended application rates were accurate, the same results could be archived by limiting the same application rate to the root zone of plants.

Field Planting Methods

Since 2009, we have been testing, evaluating, and refining this technique. The procedure has proven effective in greenhouse, nursery, and field settings. While the technique for both nursery and greenhouse is the same, for direct field planting and application, the process was modified. With both greenhouse and nursery settings even with differing containers, the process was the same. First, determine the type of soil, biochar, and needed ratios and then add the mixtures following the 5 mL loading method described earlier. Once the seedlings are well established, the roots hold the mix in the same ratio and allow it to remain undisturbed when transplanting to the field. For direct field planting, we found that the most effective approach was to produce half-liter cylindrical containers of paper or other biodegradable material. Holes were then dug into the field and once the container was placed in the hole, the same filling procedure was repeated. In the event that the paper cylinders were prefilled, seeds should not be sown in them until the cylinder has been buried up to its fill line. We found that placing seeds in the mix filled cylinders prior to placing the cylinders in the ground may cause the seed to sink below optimum planting depth and will therefore result in poorer germination rates.

With each successive application of this procedure, it became more and more evident that aliquot dosing allowed for the same results as full-field biochar application while dramatically reducing the amount of biochar needed. Additionally, while it is noted that biochar is most effective in sandy or poor soils (Tryon, 1948), it is also known that in some of these, biochar may have a negative effect for the first few years. It has been suggested that this is because biochar can absorb limiting nutrients in these poor soils (Chan and Xu, 2009), thereby competing with plant roots for limited resources. Our own experience confirms that in nutrient poor soils biochar can indeed reduce growth significantly.

Grevaria Studies

Grevaria was used in one of our reforestations programs in Rwanda. Of 58,000 Grevaria plants, 8,000 were planted in 500 mL containers containing local soil, and 25,000 were planted in 500 mL containers containing a 30% mixture of biochar applied as described earlier to local soil. The remaining 25,000 were planted in 500 mL containers with a 30% mixture of biochar also applied in the same process, but in this case, the biochar was soaked in a mixture of compost and manure tea for a period of 2 weeks. For this third application, the biochar was permitted to soak until sinking. Biochar pellets that were still floating on the water's surface were skimmed off and disposed of prior to application of the remaining saturated pellets. The saturated pellets were then drained and permitted to dry on a tarp for three days under full sun.

Half the 58,000 seedlings were planted on a hillside with an average slope of 12% and half were planted on a level field. Our study site was characterized by rich soil and average rainfall of 1600 mm/year, so the conditions could not be called arid nor the soils poor.

After the first year, the results began to show a pattern. Because of a rich soil and abundant rainfall, seedling mortality was negligible. On the level field, the group with soaked biochar was slightly taller than the other two, the control (no biochar) and with unsoaked biochar. There was no appreciable difference between the control group and the plants with unsoaked 30% biochar.

The appreciable difference came with the plants on the 12% slope, where both groups with biochar were 20%–25% taller than the control. There was no appreciable difference between the two biochar groups. For an answer as to why this might be the case, we turn to one of our programs in a desert climate.

In the desertified areas between Senegal and Mauritania, one finds funnel-shaped disks in the soil. The disks measured between 500 and 1500 mm in diameter and have a central bore measuring 80–110 mm in diameter and having a depth of 180–300 mm below the base of the cone (Figure 13.2). The soil in the cone is smoothed by hand when wet and dries hard. The desertified area between the two countries was once a lush jungle. Years of deforestation and poor agricultural practices resulted in the barren landscape we find there today, and the annual, alternating rainy season and dry windy seasons contributed to the

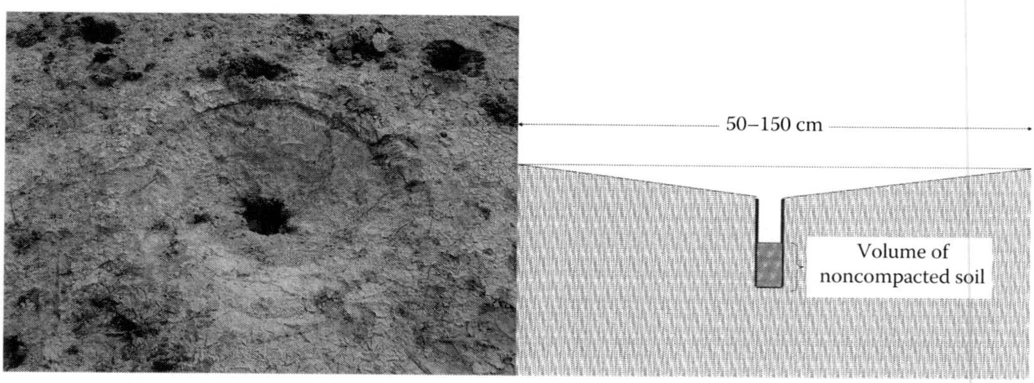

FIGURE 13.2 Fulani water catchment system and cross-sectional diagram of same (Senegal).

FIGURE 13.3 The Fulani are a nomadic group located between Senegal and Mauritania. Once a lush forest, the area is now mostly desert. Water catchment systems (including biochar aliquots) are now vital to survival.

exponential loss of topsoil. Now with the exception of the occasional thorny acacia, little can be found for tens of kilometers at a time. People crossing these areas are often faced with long journey with little or no sources for water. Local members of the Fulani people (Figure 13.3), nomadic herdsmen, have taken it upon themselves to dig these shallow reservoirs so that people or animals might find the occasional sip of water and thereby make it across the open areas. Dew and the rare rainfall are gathered by these excavations and directed to the central reservoir. These trap some of the water out of the sun and, with a reduced surface area in the central reservoir, are less vulnerable to evaporation. While these ingenious structures work remarkably well, the significance is not that they store water but that they permit what little water that might arrive to penetrate the surface.

While it may come as a surprise to some, the reality is that there are more people who drown each year in deserts than those are who die there of thirst (USGS, 1997). One reason for this is that with exposed soils, crusts form as fine particles, moved by wind or water, and clog surface pore spaces. Water thus tends to run across the surface rather than penetrate. This in turn leads to flashfloods and the surprising number of people drowning. Agriculturally speaking, this means very little water from the rare rainfall is able to penetrate the soil. Biochar however facilitates penetration of water into soils. The result then is that if biochar is applied to a field, when it does rain, the water, rather than running off and stripping nutrients and topsoil, is able to penetrate to where plants might be able to use it. Once the surface is penetrated, however, in sandy soils, water is able to percolate, which decreases any residence time near the plant roots. Biochar, however, has the double advantage that it not only facilitates penetration of water into soil but it captures the water, increasing the residence time in the area where biochar is present. If the biochar is concentrated in the areas where the plant roots are, then the water will be available to the plants; conversely, if the layer of biochar is not where the plant roots are, the water will collect in an area where it will not be accessible except when potential differences produce a wicking effect from the surrounding soils.

With regard to our approach to biochar application, the implications of this are quite significant. With field-wide applications of biochar in arid areas, what little rainfall there might be would be uniformly absorbed over the entire area of application. With the aliquot dosing method, however, the biochar is concentrated in the areas of greatest root density while the surrounding areas remain unchanged. The net result then is more water for each plant. Water falling on untreated areas would tend not to penetrate and

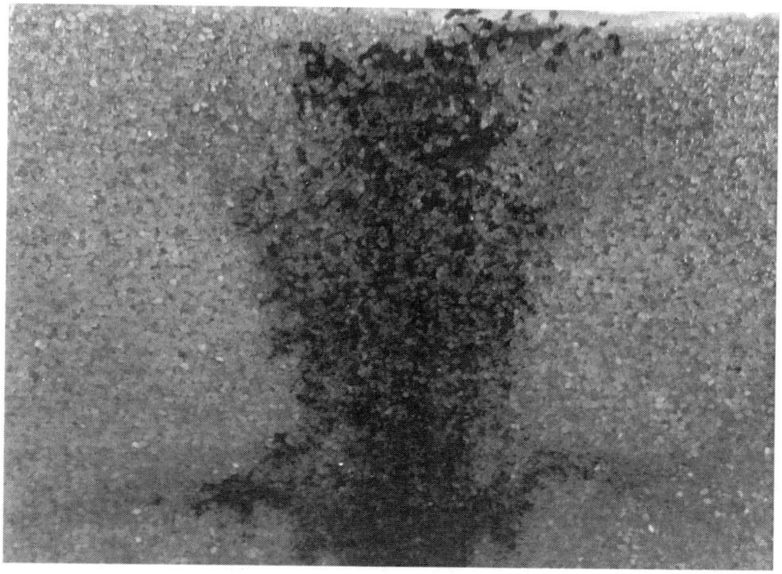

FIGURE 13.4 Cross section of a 30% (v/v) biochar sand aliquot flanked on either side by sand showing the increased soil penetration and retention of water and nutrients facilitated by the use of biochar aliquots.

run along the surface and favor flow toward areas with increased permeability. The areas containing the biochar aliquot then function in much the same way as the desert catchments made by the Fulani, collecting not only the water in the central bore (aliquot) but the water of the surrounding circumference as well.

Laboratory experiments allowed us to both verify and measure this funneling effect of a central aliquot even when the surrounding soil surface is perfectly level (Figure 13.4). The hydrological significance of this is that for arid zones where rain is scarce, the aliquot dosing method allows more water per plant rather than uniformly distributing the water over an entire field. Surprisingly, this funneling effect also has implications for more tropical zones.

In our Rwanda field study of Grevaria trees, the funneling effect could explain why there was significantly more growth in the two biochar aliquots versus the soil-filled control aliquot. Rainfall is not only a source of water but also a major factor in leaching and erosion. Normally, nutrients wash down the mountains, and if there is not sufficient organic matter to capture replacements, the eventual result is less fertile mountains and more fertile valleys. If however we take into account the funneling effect of the biochar aliquots, then not only is the rainfall being funneled into the root areas of the seedlings but the nutrients as well. Given the length of the rainy seasons and the presumed amount of nutrient runoff, this would give an advantage to any plant whose aliquot would facilitate soil permeability and would explain not only why the two biochar groups were taller than the soil-only control group but also why there was no noticeable difference between plain biochar and that which was preloaded with nutrients. Over the length of the rainy season, it is likely that the amount of nutrient provided during runoff exceeded that which was originally present in the biochar.

But if we have penetration without retention, there is little or no advantage for the plants. Fortunately, the advantages of biochar are, at least twofold, it facilitates percolation and it increases water-holding capacity.

Biochar Increases Water-Holding Capacity of Some Soils

The porous structure and resistance to decay make biochar an obvious candidate for amending soils that are deficient in water-holding capacity. The quantification of how water interacts with biochar, and the other components of soil, is complex, and these interactions will determine the availability of water and the ability of plants to survive drought. Later, we review recent papers on this issue.

Biochar and Estimates of Water Availability

The following terms are approximately synonymous (or often used so): water-holding capacity, field capacity (FC), moisture equivalent, and pot holding capacity. Each refers to the amount of water a soil will retain after having been first saturated and then drained by gravity, centrifugation, or suction. Suction plates have become widely used because of the speed and versatility (indicating also wilting points [WPs]), but Novak et al. (2013) point out that fine particles of biochar can plug the plates and lead to spurious data. For that reason, pot holding capacity may be often preferred. Nevertheless, relying on centrifugation, Tryon (1948) provided an early quantitative analysis of the water-holding capacity (moisture equivalent) of biochar. In that study, 0%, 15%, 30%, and 45% (v/v) biochar was mixed with clay, loam, or sandy soils. That study revealed a strictly linear relationship between volume of biochar and his estimate of moisture equivalent.

The slopes (±b) of the three soil/biochar mixtures show how FC increases (+b) or decreases (−b) with increasing volumetric proportion of biochar in the soil/biochar mixtures. Slope values were, respectively, −0.111, for clay; +0.015, for loam; and +0.065, for sand, indicating that biochar reduces the FC in clay, increases it somewhat in loam, and strongly so in sand. The strict linearity and precision of the slopes allow estimating the FC of biochar by extrapolating each slope to the point of 100% biochar. The three intercepts are 20.9 for clay, 17.5 for loam, and 16.5 for sand, converging, as an average, on a moisture equivalent (FC) of 18.3% (Figure 13.5). To the extent that this estimate is correct, it indicates that the FC of pure biochar in Tryon's study was between that of loamy sand (14%) and sandy loam (20%), according to the IAEA (2005) and sandy loam (18%), according to Saxton and Rawls (2006). Data provided by the IAEA suggest that, considering only FC, biochar employed by Tryon will *increase* FC of loamy sand (which has an FC of 12%) or sand (FC of 10%) and *decrease* FC of sandy clay loam (FC 27%), loam (FC 28%), and silt loam (31%). Knowing the FC of biochar gives a guide to predict the effect of biochar amendments in other soils, and we use this baseline comparison to estimate the FC and, compared with local soil, to regulate the volume (v/v) needed for specific soils and crops.

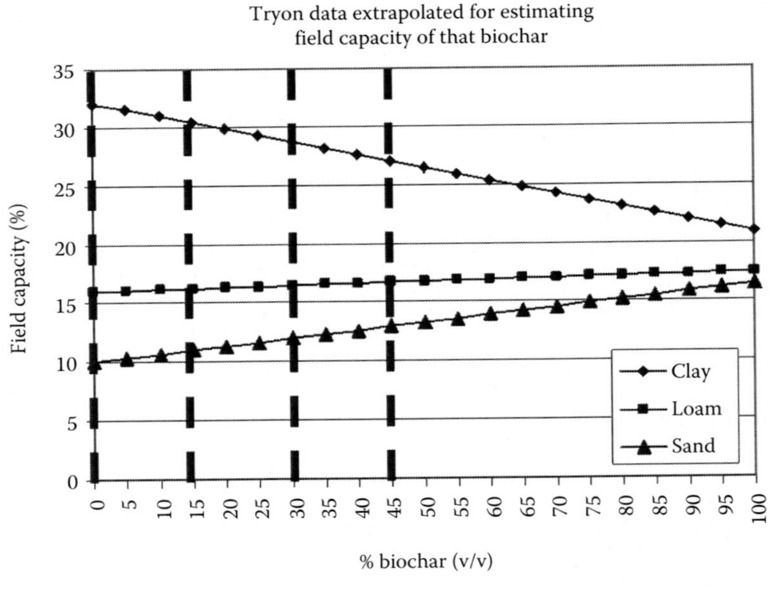

FIGURE 13.5 Influence of biochar on FC in three different soil types. (From Tryon, E.H., *Ecological Monographs*, 18, 81, 1948.) In this figure, his 12 data points have been extrapolated in order to estimate FC of 100% (v/v) biochar.

Some Relevant Interactions between Plants, Water, and Soil

Biogeochemical cycling is a complex topic, perceptively discussed by Major (2009), but here, for purposes of discussion, the following diagram illustrates some interactions between biochar, selected components of an ecosystem, and water, as we have observed in tropical, temperate, and desert climates (Figure 13.6).

Some aspects of the diagram need explanation. For example, Bormann (2012) points out that climatic models predict reduced precipitation in midlatitudes of important crop zones in Europe. This will cause drier soils and reduced evaporative cooling, leading to extreme summer heat waves. (From this, we presume that transpiration of crops will also accelerate.)

Ayodele et al. (2009) compared soils from charcoal kiln sites with those from adjacent, unamended soils and observed an 88% increase in saturated hydraulic conductivity (percolation) and, with a 5% slope and with simulated rainfall of 200 mm/h, a reduction of 37% in runoff volume.

Laird et al. (2010) investigated the impact of biochar amendments (0, 5, 10, and 20 g biochar/kg soil in columns) and found that soil bulk density increased with time for all columns but was significantly lower for biochar-amended soils relative to the unamended soils. The biochar-amended soils retained more water at gravity drained equilibrium (up to 15%) and had greater water retention at −1 and −5 bars soil water matric potential (13% and 10% greater, respectively). These effects were accomplished with smaller quantities than were used in other studies, perhaps due to inputs of organic nutrients. Karhu et al. (2010) found that incorporating 9 mt/ha biochar on an agricultural soil in Southern Finland increased soil water-holding capacity by 11%.

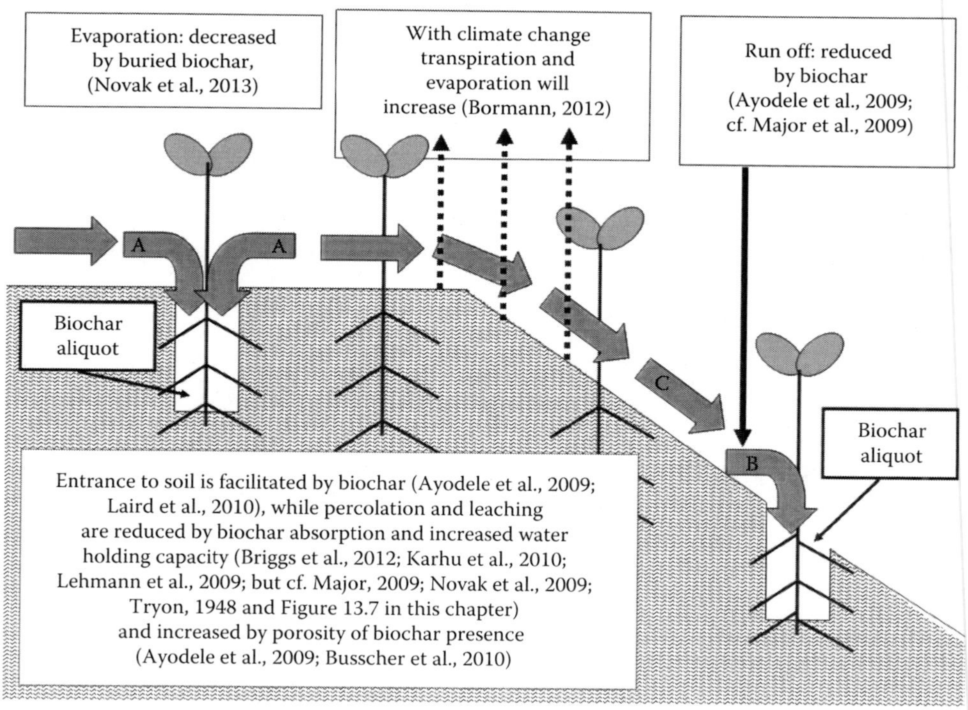

FIGURE 13.6 Schematic diagram of major interactions between biochar amendments to soil as biochar aliquot deposits. Arrow A indicates that biochar in the aliquot enhances entrance of water from surrounding areas to the root zone of seedlings. Arrow B shows that runoff will be captured by biochar, and arrow C indicates that runoff will be reduced by biochar.

Some Measures of How Biochar Influences Water Loss to Gravity

Our experiments and the literature have demonstrated that biochar facilitates percolation of water into soil. To measure the dynamics of this process, we mixed biochar with Windsor sandy loam (a widespread glacial outwash soil: 93.1% sand, 5.2% silt, and 2.6% clay, obtained from Warner Bros. Corp. Sunderland, Massachusetts, United States). The sand was sieved to remove stones over 3 mm in diameter, oven dried, and then mixed with 0%, 15%, and 30% (v/v) biochar pellets. (Biochar, pH 7.0–7.5, was produced in a WorldStove LuciaStove, by pyrolyzing wood pellets from West Oregon Wood Products, Columbia City, Oregon. Feedstock was primarily Douglas fir, *Pseudotsuga menziesii*. Approximately 20% of pellet weight is converted to biochar in this stove, and the original 8.0 mm pellet diameter shrinks to 5.0 mm. A liter of the charred pellets weighs 292 g and contains almost no fine particles.)

We used the different biochar/soil mixtures to fill 250 mL graduated glass cylinders. Filling was done in small aliquots of about 5 mL each, rotating the cylinder a quarter turn after each addition in order to reduce separation of different textures and to favor the production of fairly horizontal strata. Ten mL of water was quickly and evenly dispensed from a 10 mL wide-bore pipet onto each filled cylinder. This was the equivalent of 11 mm of precipitation falling on a soil surface since the cross section of the cylinder was 9.079 cm^2. The front of water penetrating the soil cylinder was then recorded each hour. Since the water front was often not strictly horizontal, we recorded both the most and the least advanced points of the front and averaged these two for data analysis (Figure 13.7).

Analysis of covariance indicated a significant interaction between time and treatments.

For the 0%, 15%, and 30% biochar/soil mixtures, 15% slows the water front advance more than does the 0% biochar ($F = 158.2$, df = 3, 28, $p < 0.001$) as does the 30% treatment ($F = 275.1$, df = 3.28, $p < 0.001$). There was no significant difference between 15% and 30% treatments ($F = 41.72$, df = 3, 28, $p = 0.035$).

Even 15% (v/v) slowed the percolation of water in this sandy soil. Biochar slowed the early entrance of water into these columns, but this may be a hydrophobic effect or an artifact of the unnatural conditions in laboratory constructed soil cylinders.

Another aspect of interactions between biochar and rapidly percolating water is how quickly water enters the biochar. To determine how quickly biochar absorbs water, packets of oven-dried biochar were wrapped in a fine stainless steel mesh and immersed in water. At intervals of 5 min, these packets, replicated three times, were drained briefly and then weighed. The averaged data are presented in Figure 13.8, with time in minutes plotted as natural logarithms to expand the scale.

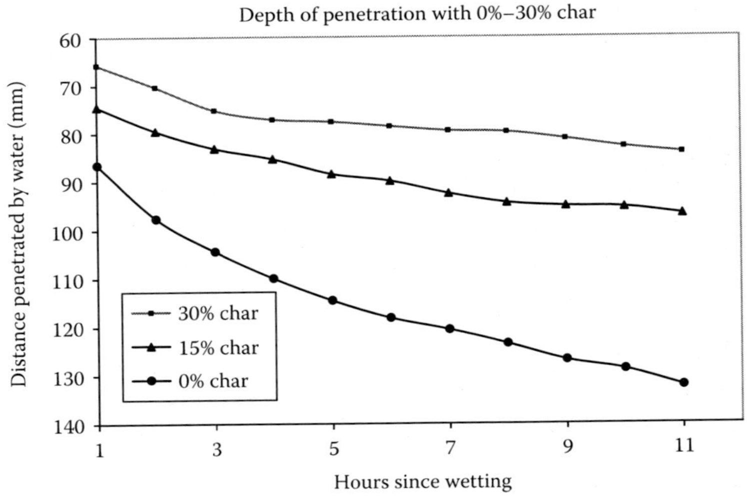

FIGURE 13.7 Effect of 0%, 15%, and 30% biochar on water drainage in this sandy soil.

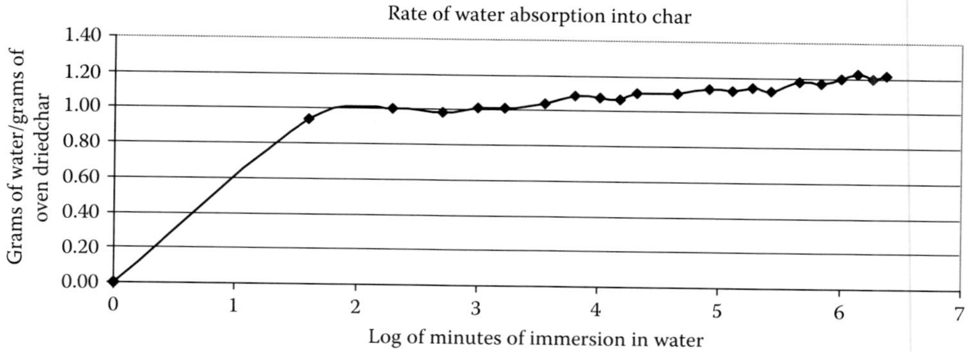

FIGURE 13.8 Speed with which water enters biochar.

The first reading, after 5 min of immersion (nlog = 1.6), indicated that each gram of biochar had absorbed 0.94 g of water. At 10 min (nlog = 2.3), 1.01 g water, and after nearly 10 h (585 min), each gram of char held 1.22 g water. The rapidity of absorption may be overestimated because the biochar had been oven dried. Nonetheless, the data suggest that even water from a brief shower may be captured by biochar.

The data presented in Figures 13.7 and 13.8 describe and quantify some aspects of how biochar influences the availability of water in ecosystems. Neither of these aspects is new but demonstrated applications of many tons per hectare, even though large quantities of biochar are not yet available. Here, because the seedling phase of growth is one of the most vulnerable of developments, we suggest that biochar aliquots could provide significant benefits in arid and semiarid zones with as little as 1 t of biochar per hectare. Surviving beyond the seedling phase, plant roots would have access to deeper and larger volumes of soil held resources for continued growth. With the aliquot system, a single ton of biochar, mixed with 70% (v/v) soil, would provide over 20,000 half-liter biochar aliquots, well within the planting densities for a hectare of maize (*Zea mays*) in arid zones (Esechie, 1992). Furthermore, a family cooking with a small pyrolyzing cookstove, in 1 year of normal cooking, could produce over 4000 half-liter aliquots of 30% biochar planting mix (Mulcahy et al., 2013). Char produced from cooking fuels to make plant fertilizer would provide statistically significant protection for seedlings, the most sensitive stage in development, should an episode of drought occur.

How Much Biochar Is Needed to Increase Drought Resistance

The ease and importance of measuring water-holding capacity, that is, FC, has resulted in many studies that take into consideration how biochar amendments to soil effect water-holding capacity (Atkinson et al., 2010; Briggs et al., 2012; Chan et al., 2007; Kammann et al., 2011; Liu et al., 2012). Useful as these data are, they do not indicate how much water is available for plants use. The water availability (WA) in well-drained soil is less than the water-holding capacity, because plants are unable to extract water held most tightly in soil. For example, loam can have an FC of 27%, WP of 12%, and thus a WA of 15%, whereas clay with FC of 40% and WP of 25% would also have WA of 15% (IAEA, 2005). In other words, few studies have demonstrated how much biochar is needed to increase wilting resistance. Exceptions include Tryon (1948), Kammann et al. (2011), Ippolito et al. (2011), and Mulcahy et al. (2013).

Ippolito et al. (2011) reported that 2% biochar addition, by weight, increased the moisture content of two aridisols, by 3%–7% relative to the control, and when relevant evapotranspiration rates were considered, it was concluded that this could lead to an additional 0.4–2.5 days of available water for crop growth. But 2% biochar in soil, by weight, is the equivalent of 40–44 mt/ha (Novak et al., 2009). Kammann et al. (2011) found that increased wilting resistance required 50 mt/ha. Mulcahy et al. (2013) concluded that adding 30% (v/v) of biochar in the top 8 cm of soil would require over 78 mt/ha.

Aliquot Dosing Is Applicable to Small Farms But Is It Scalable?

Over the past several years, our studies have shown that there are significant advantages to aliquot dosing of biochar. The advantages of this approach are particularly clear with regard to the needs of subsistence farmers who may not have sufficient funds or land to provide the needed biomass for full-field applications. This, however, does not mean that this approach and related advantages are not applicable to industrial scale agriculture. There exists an Australian company with an approach that could very well be used with aliquot dosing of biochar. Transplant Systems (63 Enterprise Avenue, Berwick, Victoria, 3806) has developed field transplanters, nursery machinery, and forestry automated transplanters that are all easily adapted to aliquot dosing. Seedlings of vegetables or trees could be started in greenhouses and, when sufficient roots have grown to hold the aliquot together, the automated transplanters could then be used to plant vegetable fields or assist in reforestation programs. The GPS guided systems can transplant at a rate of 30,000 seedlings per hour.

This not only means that industrial scale agriculture could benefit from the advantages of biochar aliquot dosing but would sustainably increase the use of biochar thereby increasing the CO_2 sequestration potential of this method.

Summary

Biochar has been shown to have positive effects increasing both percolation and retention of nutrients and water in soils thereby also reducing nutrient runoff. For these effects to be achieved, current literature recommends application rates of 40–100 mt/ha. These recommendations would require so much biomass as to question the economic and environmental feasibility of field-wide biochar application approaches. Aliquot dosing allows for the same application rates but requires significantly less biochar and as a consequence significantly less biomass for the same positive effects to be realized. Seedlings are the most vulnerable stage in a plant's life; aliquot dosing is ideally suited for protecting seedlings. We have discovered additional benefits of aliquot dosing in that it funnels water and nutrient from larger areas and concentrates them near plant root balls. Top-dressing full-field applications have been shown to increase soil temperature and possibly accelerate water loss; aliquot dosing also helps avoid this specific problem. Finally, while initially intended for subsistence farming, new agricultural equipment means aliquot dosing is also applicable to industrial scale agriculture.

ACKNOWLEDGMENTS

We thank Gabriella Bergamini Mulcahy for editing and for the use of unpublished data shown in Figure 13.7 and also Chi-Yun Kuo, Department of Organismal and Evolutionary Biology, for ANOVA and CONOVA analyses of Figure 13.8.

REFERENCES

Atkinson, C., J. Fitzgerald, and N. Hipps, 2010. Potential mechanisms for achieving agricultural benefits from biochar application to temperate soils: A review. *Plant and Soil* 337: 1–18.

Ayodele, A., P. Oguntunde, A. Joseph et al., 2009. Numerical analysis of the impact of charcoal production on soil hydrological behavior, runoff response and erosion susceptibility. *Revista Brasileira de Ciência do Solo* 33: 137–145.

Blackwell, P., G. Riethmulle, and M. Collins, 2009. Biochar application to soil. In *Biochar Environmental Management—Science and Technology*, pp. 207–226. (eds.) J. Lehmann and S. Joseph. Earthscan, London, U.K.

Bormann, H., 2012. Assessing the soil texture-specific sensitivity of simulated soil moisture to projected climate change by SVAT modeling. *Geoderma* 185–186: 73–83.

Briggs, C., J.M. Breiner, and R. Graham, 2012. Physical and chemical properties of *Pinus ponderosa* charcoal: Implications for soil modification. *Soil Science* 177: 263–268.

Brown, L., 2006. *Plan B 2.0 Rescuing a Planet under Stress and a Civilization in Trouble*. W.W. Norton Press, New York.

Busscher, W., J. Novak, D. Evans et al., 2010. Influence of pecan biochar on physical properties of a Norfolk loamy sand. *Soil Science* 175: 10–14.

Buyinza, M., A. Banana, G. Nabanoga et al., 2008. Farmer's adoption of rotational woodlot technology in Kigorobya sub-county of Hoima district, Western Uganda. *Ethnobotany Research and Applications*, North America, vol. 6, pp. 107–115, August 6, 2008. Available at:_http://lib-ojs3.lib.sfu.ca:8114/index.php/era/article/view/202. Accessed March 26, 2013.

Chan, K. and Z. Xu, 2009. Biochar: Nutrient properties and their enhancement. In *Biochar Environmental Management—Science and Technology*, pp. 67–84. (eds.) J. Lehmann and S. Joseph. Earthscan, London, U.K.

Chan, K.Y., L. Van Zwieten, I. Meszaros et al., 2007. Agronomic values of greenwaste biochar as a soil amendment. *Australian Journal of Soil Research* 45: 629–634.

Esechie, H., 1992. Effect of planting density on growth and yield of irrigated maize Zea mays in the Batinah Coast region of Oman. *The Journal of Agricultural Science* 119: 165–169.

Husk, B. and J. Major, 2010. Commercial scale agricultural biochar field trial in Québec, Canada over two years: Effects of biochar on soil fertility, biology and crop productivity and quality. http://xa.yimg.com/kq/groups/22438052/740714074/name/BlueLeaf+Biochar+Field+Trial+08-09+Report.pdf. Accessed March 29, 2013.

IAEA, 2005. *Field Estimation of Soil Water Content: A Practical Guide to Methods, Instrumentation and Sensor Technology*. IAEA, Vienna, Austria. http://www.iaea.org/inis/collection/NCLCollectionStore/_Public/39/079/39079658.pdf. Accessed March 29, 2013.

Ippolito, J.A., J. Novak, W. Busscher et al., 2011. Switchgrass biochar affects two aridisols. *Journal of Environmental Quality* 41: 1123–1130.

Kammann, C., S. Linsel, J. Gößling et al., 2011. Influence of biochar on drought tolerance of *Chenopodium quinoa* Willd and on soil–plant relations. *Plant Soil* 345: 195–210. doi 10.1007/s11104-011-0771-5.

Karhu, K., T. Mattila, I. Bergström, and K. Regina, 2011. Biochar addition to agricultural soil increased CH4 uptake and water holding capacity—Results from a short-term pilot field study. *Agriculture, Ecosystems and Environment* 140: 309–313.

Laird, D.A., P. Fleming, D. Davis et al., 2010. Impact of biochar amendments on the quality of a typical Midwestern agricultural soil. *Geoderma* 158: 443–449.

Liu, J., H. Schulz, S. Brandl, B. Miehtke et al., 2012. Short-term effect of biochar and compost on soil fertility and water status of a Dystric Cambisol in NE Germany under field conditions. *Journal of Plant Nutrition and Soil Science* 175: 698–707.

Major, J., 2009. Biochar effects on nutrient leaching. In *Biochar for Environmental Management: Science and Technology*, pp. 271–287. (eds.) J. Lehmann and S. Joseph. Earthscan, London, U.K.

Mulcahy, D.N., D. Mulcahy, and D. Dietz, 2013. Biochar soil amendment increases tomato seedling resistance to drought in sandy soils. *Journal of Arid Environments* 88: 222–225.

Novak, J.M., W. Busscher, D. Laird et al., 2009. Impact of biochar amendment on fertility of a southeastern coastal plain soil. *Soil Science* 174: 105–112.

Novak, J.M. and D. Watts, 2013. Augmenting soil water storage using uncharred switchgrass and pyrolyzed biochar. *Soil Use and Management* 29: 98–104. doi: 10.1111/sum.12026.

Nunan, N. et al., 2006. Spatial distribution of bacterial communities and their relationships with the microarchitecture of soil. *FEMS Microbiology Ecology* 44.2: 203–215.

Saxton, K. and W. Rawls, 2006. Soil water characteristic estimates by texture and organic matter for hydrologic solutions. *Soil Science Society of America Journal* 70: 1569–1578.

Schimel, D., 2010. Drylands in the Earth system. *Science* 327: 418–419.

Sessitch, A., Weilharter, A., Gerzabek, M.H. et al., 2001. Microbial population structures in soil particle size fractions of a long-term fertilizer field experiment. *Applied and Environmental Microbiology* 67: 4215–4424.

Silberbauer, G., 1981. *Hunter and Habitat in the Central Kalahari Desert*. Cambridge University Press, Cambridge, U.K.

Sladden, S.E., D. Bransby, and G. Aiken, 1991. Biomass yield, composition and production costs for eight switchgrass varieties in Alabama. *Biomass and Bioenergy* 1: 119–122.

Spencer, D., 2002. The future of agriculture in Sub-Saharan Africa and South Asia: Whither the small farm? In *Sustainable Food Security for All by 2020. Proceedings of an International Conference*, September 4–6, 2001, Bonn, Germany. International Food Policy Research Institute, Washington, DC.

Tryon, E.H., 1948. Effect of charcoal on certain physical, chemical, and biological properties of forest soils. *Ecological Monographs* 18: 81–115.

USGS, 1997. Desert features. http://pubs.usgs.gov/gip/deserts/features/. Accessed October 29, 1997.

Wiggins, S., 2009. Can the small holder model deliver poverty reduction and food security for a rapidly growing population in Africa? FAC Working Paper No. 08 July. http://www.future-agricultures.org/Documents/Smallholder_S-Wiggins_Jul-09.pdf. Accessed March 29, 2013.

WHO, 2007. New country-by-country data show in detail the impact of environmental factors on health. http://www.who.int/mediacentre/news/releases/2007/pr30/en/index.html. Accessed March 29, 2013.

Workman, J., 2009. *Heart of Dryness—How the last Bushmen Can Help us Endure the Coming Age of Permanent Drought.* Walker and Co., New York.

14

Mobilizing Biochar: A Multistakeholder Scheme for Climate-Friendly Foods and Rural Sustainable Development

Steven R. McGreevy and Akira Shibata

CONTENTS

Introduction

The story of the relationship between charcoal and agriculture in Japan is long and complex. There are a number of connections between the history of charcoal production, natural resource management, and agriculture that span centuries. Awareness of this history is prerequisite to understanding how charcoal has reemerged, in the form of biochar, as a tool for rural sustainable development and carbon sequestration in Japan.

The first written record of charcoal use in agriculture in Japan is in the *Nogyo Zensho*, an agricultural encyclopedia written in 1697 during the Edo period (1603–1867) by the wandering samurai-turned-peasant Yasusada Miyazaki, although there is reason to believe the practice has a much longer history throughout Asia. The emergence of the practice itself is less likely the result of planned experimentation and more likely the serendipitous result of the cyclical nature of traditional Japanese agricultural in which organic wastes and byproducts from daily living are meticulously recycled, composted, and applied to fields and gardens. Of course, throughout Japanese history, charcoal's primary use was as a source of fuel, where it was used for heating and cooking along with firewood or straw. Ash and cinders from household hearths and fireplaces were mixed with compost and used on agricultural lands. It was common for farmers to spend their winters making charcoal from hard and soft woods, the feedstock garnered from carefully managed forestland from nearby mountains (referred to as *satoyama*) and then set off down to cities for sale. Subsequently, the ash and charred wood produced in urban dwellings were purchased by farmers or *ash traders* and brought back to the countryside to be composted and amended to fields (Ishikawa 2000).

With this story in mind, a pattern of resource use and circulation in rural Japan becomes clear, connecting forest and agricultural land management, charcoal production and use as fuel, and its eventual utilization as a soil fertility agent. Charcoal also linked rural producers and urban consumers in a mutually beneficial relationship, providing a means of income for farmers and a valuable service as heat for urban homes. Incredibly, this pattern of use continued from the Edo period until the 1960s, when post-WWII

Japan embraced fossil fuels as a both a source of fertilizer and heat. Domestic charcoal production for fuel use decreased from over 2 million tons in 1950 to a low of 150,000 tons in 2009 (MAFF-FA 2013).

Despite the move to petroleum-based fertilizers and decline in charcoal use as a fuel, research on alternative uses for charcoal flourished during the 1970–1980s. The areas of agricultural and forestry-related deployment were particularly fruitful, leading the Japanese government's Ministry of Agriculture, Forestry, and Fisheries (MAFF) to officially recognize and classify charcoal as a material authorized for soil amendment in 1984.

Research associations and organizations were created, including the Technical Research Association for Multiuse of Carbonized Material established in 1986, to contribute to the study of the effects of charcoal and wood vinegar as soil amendments. Various university and governmental research institutes have continued experimentation to the present day.

Japanese research on biochar is extensive and not only covers agricultural and forestry applications but also studies on use in livestock and poultry operations, pyrolysis technique and charcoal production, and the chemical properties of charcoal. We have summarized an excellent, open-source review paper by two leading Japanese researchers, Dr. Makoto Ogawa and Mr. Yasuyuki Okimori, in Table 14.1, and offer a

TABLE 14.1

Brief Summary of Japanese Biochar Research

Research Target	General Findings	References
Ag, AMF, PG	Bark charcoal from hardwood trees was mixed with a series of fertilizer media, prepared in soils, and planted with soybean. Yields for control plots prepared with inorganic fertilizers and the charcoal plots were comparable, indicating that the required amount of fertilizer when accompanied with charcoal was far less. AMF increased in general. Straight charcoal plots with no fertilizer treatment did not see yield increases, highlighting the importance of AMF for crop growth	Ogawa and Yambe (1986)
Ag, AMF	Hardwood-derived charcoal increased AMF in soybeans	Saito (1990)
Ag, AMF	Rice husk charcoal increased AMF formation in citrus seedlings	Ishii and Kadoya (1994)
Ag, AMF, SC	Rice husk charcoal mixed with uncharred rice husks improved soil conditions and AMF formation in marigold (*Tagetes patula* L.)	Ezawa et al. (2002)
Ag, SC	Coconut shell charcoal inoculated with AMF suppressed *Fusarium* sp. (soil-borne pathogen)	Matsubara et al. (2002)
Ag, SC, AMF, PG	Bark charcoal from Acacia *mangium* increased yields of maize and peanuts, increased AMF growth (particularly in maize), neutralized soil pH, increased total N, increased CEC, and decreased exchangeable Al ions.	Yamato et al. (2006)
F, SC, AMF, PG	Bark charcoal and a small amount of chemical fertilizer was applied to *Pinus thunbergii*. Root growth, AMF development (including *Rhizopogon* sp.) increased. Further experiments with applications at 30cm depth around *Pinus* roots showed significant growth of new roots and AMF nine months and one year afterwards. The overall quality of tree morphology increased after treatment. Water content was high in the charcoal-rich soil even during the dry season.	Ogawa (1983)
F, AMF, PG	Rice husk charcoal was an effective inoculation medium for Dipterocarp seedlings. Excessive application of rice husk charcoal showed no negative plant growth effects.	Mori and Marjenah (2000)
F, AMF, PG	Dipterocarpaceae (rain forest tree species) seedlings benefited from charcoal application, stimulating AMF development.	Ogawa (2006)
F, AMF, PG	*Pinus* sp. seedlings were planted with charcoal powder, phosphate, and AMF fungi in a sand dune environment. Survival rates of charcoal and AMF treated seedlings were consistently better than the untreated controls.	Ogawa (2007)

Source: Adapted from Ogawa, M. and Okimori, Y., *Aust. J. Soil Res.*, 48, 489, 2010.
Note: Ag, agriculture; F, forestry; SC, soil conditions; AMF, arbuscular mycorrhizae fungi; PG, plant growth.

brief overview of some of the studies pertaining to plant growth, soil effects, and arbuscular mycorrhiza fungal (AMF) development as applicable in agriculture and forestry (Ogawa and Okimori 2010). We stress that the studies presented here are only a small slice of the overall wealth of research on charcoal in Japan.

In general, experimentation covering a range of soils and treatments saw plant growth increase or remain constant when charcoal and fertilizer were applied. Ogawa and Yambe observed that the amount of fertilizer necessary when applied in conjunction with charcoal was much less than typical application rates in order to achieve equivalent yields in soybeans (1986). Charcoal and fertilizer were also effective in promoting growth for plants in soils with low fertility (Ogawa 2007). Plant growth was also positively influenced by increased development of AMF infection in roots due to charcoal's porous structure, attractive for microbial and fungal colonization. In addition, Matsubara et al. inoculated coconut shell charcoal with AMF and saw a decrease in *Fusarium* sp., a soilborne pathogen that can cause root rot and wilting in crops (2002). Various soil conditions were seen to improve under charcoal application, including pH neutralization, increased total N, increased cation exchange capacity (CEC), decreased toxicity such as Al ions, and increased water retention (Ogawa 1983; Yamato et al. 2006). Japanese researchers have also led the way in charcoal applications in forestry, as in the rehabilitation of pine forests suffering from pine wilt, weakened decorative garden pines, and in rain forest reforestation efforts with species such as those in the Dipterocarpaceae family. As a general conclusion, charcoal derived from a variety of feedstocks, including hardwoods and rice husks, *showed no negative impacts* on plant growth or soil conditions and in many cases saw positive growth impacts.

The history of use, official recognition by the government, and body of applicable research give charcoal (and its modern incarnation as *biochar*) a certain level of social acceptance in Japan that may not exist in countries without such a history. In fact, the average Japanese citizen encounters charcoal on a daily basis. It is common to find a variety of products in Japanese stores and homes today that are charcoal based or include charcoal as an ingredient, including toiletries (toothpaste, soap, shampoo, towels), food (sweets, noodles, powder), water purification/filtration products, interior decorations, odor control and removal goods (bedding, storage bins), and other indoor air quality and humidity control products for use in construction. Bags of biochar are sold in home and garden stores. Smoking mounds of rice husk charcoal being produced are readily observable when driving through rural areas after the fall rice harvest. Wood vinegar, also known as pyroligneous acid and collected from the condensation of smoke emitted during the carbonizing process, is commonly used in agriculture to promote plant growth, repel insect and fungal pests, and deodorize livestock pens. Needless to say, Japan's charcoal literacy and level of acceptance is quite high.

It is partly because of this awareness of charcoal in agriculture that we have not seen the explosion of interest in biochar among Japanese farmers and researchers that has taken place in other countries (e.g., Australia, the United Kingdom, the European Union, and the United States), a boom fueled equally by its potential as soil ameliorator and climate change mitigation tool. It is also likely that a number of structural and socioeconomic factors have had greater influence tempering enthusiasm for biochar. First of all, Japanese agricultural soils are quite fertile, which lessens the impact biochar can have on crop yields. In addition, small farm size, 1.8 ha average per household in 2006, minimizes any potential economic benefit derived from selling carbon credits via carbon sequestration with biochar (MAFF 2008).

But in reality, serious challenges currently face the very prospect of continued agricultural production and farming livelihoods in rural Japan. Much of this socioeconomic instability stems from increasingly liberalized trade agreements that have opened Japan to global agricultural markets and a flood of cheap food. Small farm size, coupled with the rising cost of agricultural inputs and falling returns, has made gaining a livable income from farming very difficult for the typical, small-scale farmer (MAFF 2008). Many farmers, who are aging rapidly, do not have a successor and encourage their children to take up more viable professions. This has led to a massive depopulation of rural areas—9.27 million farming households in 1970 to 2.85 million in 2000 (Teruoka 2008)—and contributed to an increase in abandoned farmland, up 194% from 131,000 ha in 1975 to 386,000 ha 2005 (MAFF 2008). At the same time, the long history of forest resource management that was essential to the cyclical flow of resources from rural to urban areas and critical to maintaining animal habitats before the introduction of fossil fuels has all but ceased. Consequently, we see a spike in wildlife damage to crops of some 19.6 billion yen, or US$21 million, in 2006 (ibid.).

Before biochar can contribute fully to carbon sequestration efforts in Japan, it must first help to solve the puzzle of rural socioeconomic and agricultural decline that is kicking farmers off the land and devaluing forest resources—a similar challenge exists in many countries that are experiencing the loss of family farms and rural livelihoods. From a historical perspective, the old cycle of resource use from forest and agricultural land management, charcoal production and its eventual utilization as soil fertility agent, and finally the cultivation of agricultural produce that bridged rural producers and urban consumers offers intriguing possibilities. There is little doubt that biochar is an innovative technology able to address some of our most pressing and wicked modern-day problems, such as climate change and soil fertility maintenance, but unless both rural and urban stakeholders *mobilize* biochar throughout society, its impact will remain minimal. This challenge is made more difficult by the rural social and agricultural hardships described earlier. However, Japanese urban stakeholders are in a position to assist the mobilization process. As awareness of environmental problems grows, urban citizens are seeking for ways to contribute to environmental conservation, the reduction of green house gases, and *living green* so long as the costs and time spent associated with such activities are kept at a minimum. At the same time, companies from all industries face mounting pressure to address climate change, decrease carbon emissions, and do their part for the environment.

The remainder of the chapter will focus on a multistakeholder scheme, Carbon Minus Projects, to mobilize biochar throughout Japanese society (McGreevy and Shibata 2010). Its goals are the economic and ecological revitalization of Japanese farming, rural society, and landscape while using biochar as the driving, organizing force for local natural resource management, carbon sequestration, and an eco-branding marketing campaign. The scheme links rural and urban areas and a variety of stakeholders to increase farmer incomes and reduce greenhouse gas emissions. We believe that this scheme can serve as a model for rural and urban communities in international settings and, if widely distributed, contribute significantly to the fight against climate change and enhancing farmer livelihoods.

Carbon Minus Projects

Carbon Minus Projects are organized around four simple ideas: (1) carbon capture and storage via biochar (hereinafter referred to as biochar CCS) is an effective and convenient method for reducing atmospheric carbon and mitigating climate change; (2) biochar CCS should be carried out on agricultural lands in rural areas in need of agricultural revival and regional development; (3) biochar CCS activities should be monitored and measured to generate carbon credits when possible; and (4) concurrently, the eco-brand COOL VEGE™ should be established for agricultural products derived from biochar CCS lands to increase their commercial value and sales appeal to consumers. The model scheme that emerges from these ideas is shown in Figure 14.1. Multiple stakeholders, including farmers and farmer cooperatives, land managers, retailers, consumers, companies, and local and state governments create a system that attaches value to previously devalued rural resources—unused or undermanaged local biomass as *biochar CCS feedstock*, agricultural products as *climate-friendly vegetables*, and agricultural lands as *climate sinks*—ultimately moving capital from urban centers (companies and consumers) to rural communities.

In the realization and undertaking of these schemes, there are a number of structural and potentially difficult points of interest to note.

Life-Cycle Assessment for CO_2, Economic Efficiency, and Carbonization Technology

Any plan that is truly serious about reducing the impact of climate change must, at a minimum, be *net carbon negative* in its operation, which can be measured by a life-cycle assessment for CO_2 (LCA-CO_2). The scope of the range of a typical Carbon Minus Project LCA-CO_2 is outlined in Figure 14.2.

In the Japanese countryside, rice husks, forestry trimmings, wood waste from construction, and bamboo from thinned bamboo stands are common sources of local biomass. This biomass is seasonal and sparsely dispersed, which can make it difficult to collect, store, and carbonize with the lowest possible amount of carbon emitted. Reducing the LCA-CO_2 below zero by sequestering carbon in agricultural lands via biochar is critical. Generally speaking, this requires the amount of carbon sequestered

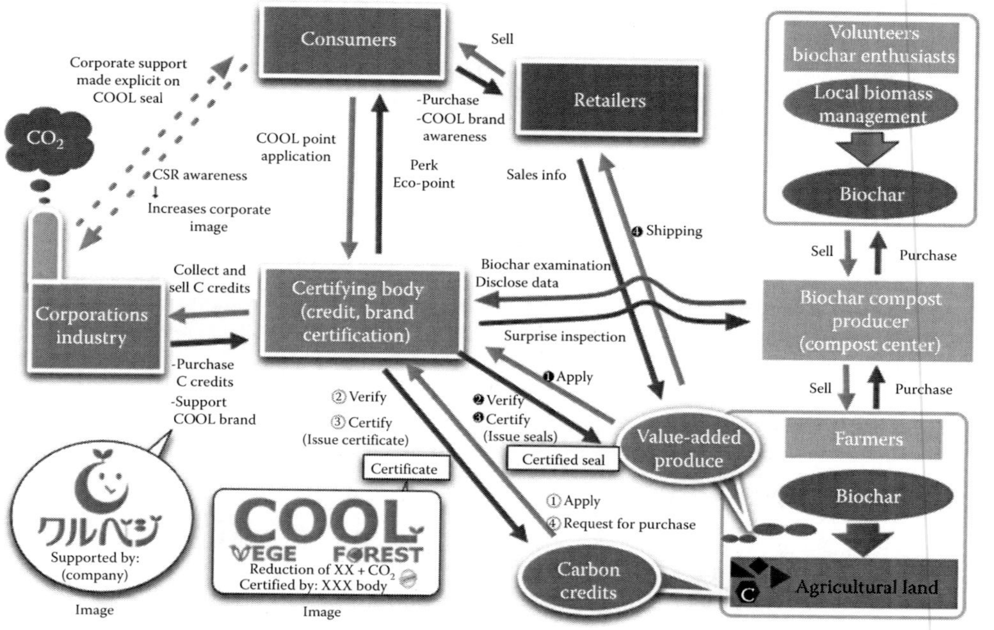

FIGURE 14.1 Carbon Minus Project model scheme.

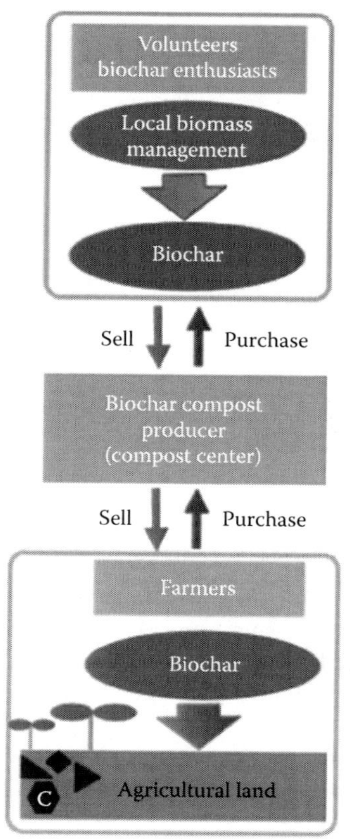

FIGURE 14.2 Range of Carbon Minus Project LCA-CO_2.

in agricultural lands to exceed that being emitted throughout the range of operations being measured. The processes by which biomass collection, storage, and carbonization take place, as well as the way in which the biochar is processed and applied to agricultural lands, should also be done in a way to minimize their carbon emissions. As a general rule, it is important to use local biomass generated within a region, carbonize it there without long-distance transportation, and then utilize the biochar in agricultural lands nearby. Reducing the time and space from the site of generation to the place of ultimate use makes sense not only from an LCA viewpoint but also from an economical one.

Additional consideration is important when deciding on carbonization technology. Generally speaking, machinery capable of carbonizing large volumes of biomass have high investment and maintenance costs and are constructed from large quantities of metal parts produced by consuming fossil fuels. To efficiently use these machines, a great deal of biomass processing is needed and a constant supply of local biomass must be stored in a designated place. As a result, the LCA-CO_2 rises and the economic efficiency (cost per amount of carbon sequestered) decreases significantly. Under these circumstances, dispersible and flexible carbonization technology able to be moved to the places where biomass is being generated is appropriate for regional biochar production.

Carbon Credits and Reasonable Biochar Production and Pricing

Another factor that could possibly influence the price of biochar production is cost reductions incurred by the utilization of carbon dioxide emission offsetting credits, or commonly known as *carbon credits*. Unfortunately, the price for carbon credits on the international market is quite low (1400–1600 JPY per ton as of December 2010 on the EU voluntary carbon market) when we consider the fixed amount of carbon contained in biochar. The total carbon amount of carbonized biomass is around 40%–85%, although the figure may differ very much according to feedstock and carbonization method. The total carbon weight in 1 ton of biochar (absolutely dry) is 400–850 kg. When this is converted to CO_2, the weight of the biochar increases 3.666 times (molar mass ratio: $((12 + 16 \times 2)/12))$ to 1467–3117 kg. At an average price of 1500 JPY per ton for CO_2, the estimated maximum income via carbon credits for the biochar is only 2200–4675 JPY per ton, or 2.2–4.7 JPY per kg, not including distribution costs. When compared with the price for biochar sold in gardening stores or around 90–200 JPY per kg, any economic benefit claimed from carbon credits would only result in covering a few percentage points of the overall cost. Of course, these calculations may change in the future if the price for carbon credits increases.

In the meantime, we maintain that biochar should be produced in a *reasonable* manner, meaning that the entire collection, storage, and production process, including the carbonization technology, is within the capacity of local stakeholders. In Japan, this often means that feedstock is harvested and collected voluntarily by neighborhood and municipal groups or derived from local waste streams for free or at low cost. These same groups or the farmers or farming cooperatives (who will eventually use the biochar) perform the carbonization themselves. With this in mind, the carbonizing technology must be easy to use. A simple and commonly used piece of equipment in Japan for producing biochar is shown in Figure 14.3.

This *smokeless carbonizing kiln*, produced by Nagano-based MOKI Manufacturing Co. Ltd., is a simple stainless steel ring (dia. 150 cm) and weighs about 17 kg, making it easily transportable using a light truck. Biochar production is straightforward—place the feedstock within the ring, ignite it, and extinguish when everything has been carbonized. The biochar is usually extinguished with water, but sand or dirt can also be used. Using this method, the charcoal yield (biochar weight/feedstock weight) is about 39% (wet/wet) or about 24% (dry/dry) for bamboo. The approximate production cost is estimated at about 79 JPY per kg (wet) or about 245 JPY per kg (dry), and the carbon ratio is about 80% by weight (Shibata et al. 2010b). It is inevitable that extinguishing the biochar with water increases the moisture content (around 60% on average). However, the average charcoal yield of 24% (dry/dry) compares favorably with other carbonizing techniques and technology that typically range between 5% and 30% for carbon yield. If labor, equipment, and maintenance costs are factored in, however, this method is quite efficient.

Also essential to this effort is a common understanding among local stakeholders (both rural and urban) that they are engaging in activities aimed at revitalizing their local communities, economies, and ecosystems through the establishment of a regional eco-brand for agricultural products, by managing lands in a sustainable manner, and through a growing sense of identity.

FIGURE 14.3 Moki kiln.

Growing a Regional Eco-Brand, Engaging the Public and Private Sectors, and Building Trust

Since carbon credits cannot be considered a viable source of cost offsetting at the moment, biochar production schemes need to focus on adding value to the economic bottom line, even under the very low biochar production costs seen with the techniques described earlier. Carbon Minus Projects add value not to the biochar itself, but to the agricultural product derived from agricultural lands applied with biochar in the form of a regional, eco-brand called COOL VEGE (see Figure 14.4). The brand

FIGURE 14.4 COOL VEGE label and in-store display.

itself displays not only the COOL VEGE logo but also the name of the town, city, or region in which the vegetables were cultivated. Branding vegetables and processed foods in this way (1) raises awareness of the need for climate change action for consumers and offers an opportunity to act by *voting with their pocketbook* and (2) builds a sense of identity for rural communities as an emphasis is placed on both the sustainable management of local natural resources and the region itself as a site for climate-friendly food and activities that resonate internally and with consumers outside the region.

Consumers, especially in urban areas, are currently without many options when it comes to acting on the need to do something to protect the environment or combat climate change. Taking action for the environment is difficult for the general public because of the broadness of the issue itself and the diluted sense of benefit derived from such actions despite the reality being that everyone ultimately benefits from a healthy, livable environment. Eco-branding of food not only educates consumers on the need to act to counter global environmental problems such as climate change but also gives them the chance to do so in a way that is simple and functional since we must buy food regularly. Also, money spent on COOL VEGE flows back to rural communities and farmers, strengthening local economies and agricultural livelihoods (see Figure 14.5).

That said, market investigations of COOL VEGE in Japan revealed that consumers do have limits when it comes to price. A consumer survey showed that 47% of respondents said that they would buy COOL VEGE products if they were the same price as other similar products, while 15% replied that they would buy them even if they were more expensive than similar products because they were combatting climate change (Shibata et al. 2010a). This would suggest that, at the moment, consumers with high environmental awareness should be targeted for biochar eco-branding schemes. Plans are being made to integrate an eco-point system to consumers who collect and redeem COOL VEGE labels within existing government and green-business eco-point systems (see Figure 14.5).

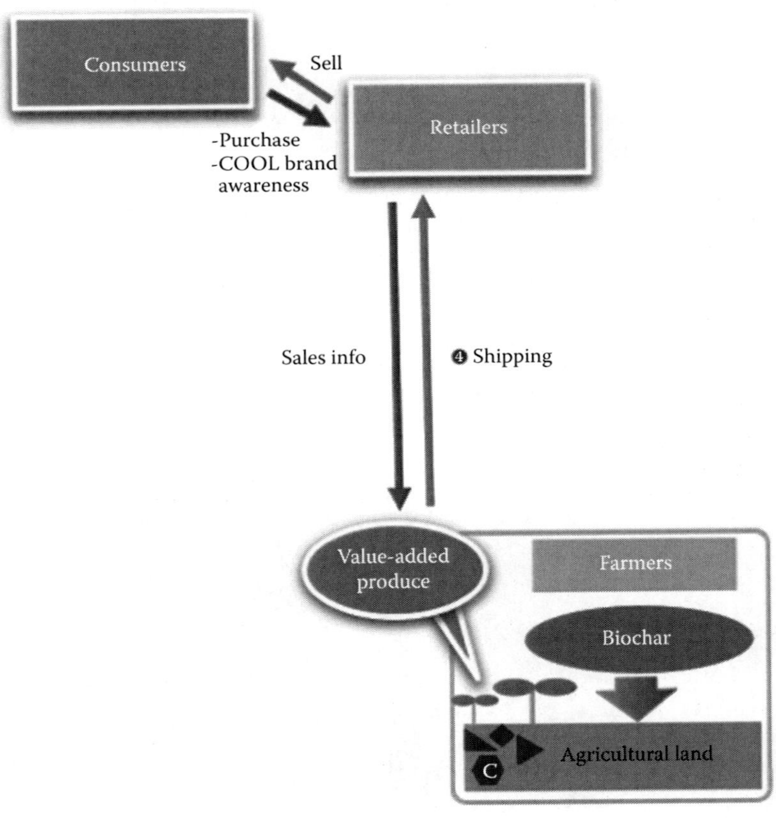

FIGURE 14.5 Farmers to consumers.

An additional measure to increase the economic feasibility of the overall scheme is involvement of business through corporate social responsibility (CSR) sponsorship. CSR is a way for corporations to be cognizant of the way their business practices impact society and the environment and to be active in supporting ethical and positive projects in the public sphere. Within the scheme, farmers are presented with additional costs for purchasing biochar or biochar compost, the COOL VEGE labels themselves, as well as for paying costs related to carbon credit certification (if possible). As was mentioned earlier, the current market for carbon credits does not provide sufficient remuneration for the costs associated with biochar production or purchase. While selling COOL VEGE products at a slight premium may result in a small economic benefit, a certain degree of financial burden will exist for farmers until the COOL VEGE brand becomes recognized and established in the market. In order to supplement the costs for farmers, a space on the COOL VEGE label itself has been left for advertisements for corporations conducting CSR activities. For corporations, advertising on eco-brand labeling enables them to appeal directly to consumers and communicate that they support environmental protection through carbon sequestration in agricultural lands and they support regional agricultural development. Since food is so commonplace in our daily lives, advertising on food labeling is quite visible for consumers, making the acquirement of corporate sponsors an easy sell. Corporations engaging in CSR activities are also supportive of the idea of purchasing carbon credits to offset their carbon emissions, which are funds that can also reach farming communities (see Figure 14.6).

Another important element in the establishment of any brand is a sense of trust between all stakeholders that the activities elicited by brand labeling are actually being carried out, as well as to ensure the safety of the product and that the production processes themselves are done in a sustainable manner. This brings about the need for certification by some kind of authoritative body.

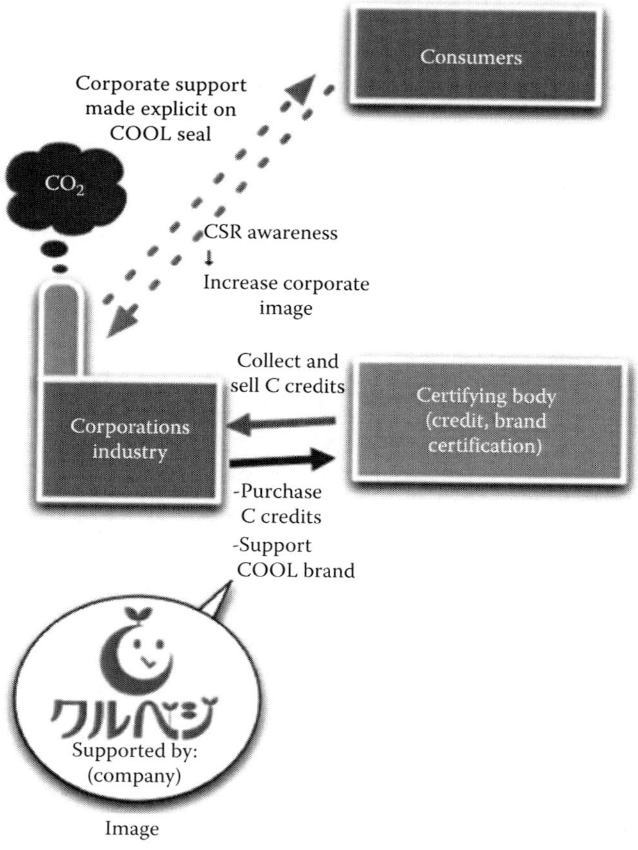

FIGURE 14.6 Corporate involvement in Carbon Minus Projects.

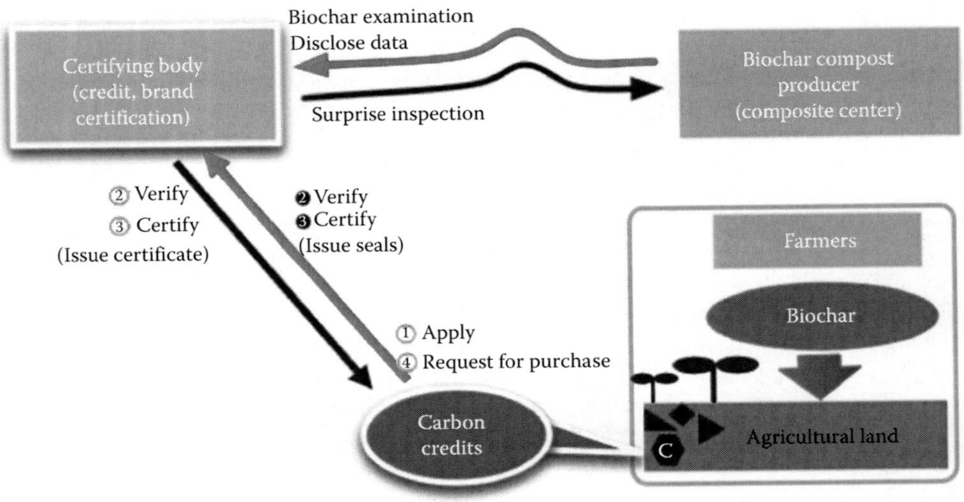

FIGURE 14.7 Biochar carbon credit certification system.

In the case of the Carbon Minus Project scheme, specialists' skills are required for quantitative and qualitative analyses of the biochar produced and amended to agricultural land. To secure objectivity, a certification system should be established that, at a minimum, verifies the amount of carbon in the biochar to expedite carbon credit integration. Verification of the sustainable management of feedstocks and the ecosystems within which they are found and, at least until competence is demonstrated, the actual production of the biochar are other targets to include. The actual carrying out of a certification system will involve costs, labor, and time that may represent an additional burden for rural stakeholders involved in the process, which can be problematic.

One possible solution, which has been tested on the ground in Japan, includes the involvement of a regional composting facility, or some similar business, as a key actor (see Figure 14.7). For the purpose of reducing costs associated with experimentation, handling the biochar, and the administration of carbon credits, the biochar produced in a region is collected and stored in bulk in the regional composting facility, which commonly operate with large volumes of composting materials. The composting facility, then, takes the lead in applying for qualitative and quantitative analyses of large lots of biochar to certifying bodies. The certified biochar is then mixed with compost at ratios appropriate for agricultural use and farmers' preference. After that, the compost facility stores, sells, and applies the certified biochar compost to fields, taking data on the overall amounts applied and periodically visiting the fields to perform checks. Additional data, including soil samples, cultivation records, and other field amendment records will also be collected either by the facility or by the certifying body. At that point, the certifying body performs the experiments, notifies the parties of the results, and certifies carbon credits. In principle, the certifying body should inspect the biochar compost facility and witness the compost manufacturing and application.

As the overall goal of a Carbon Minus Project is regional development and revitalization, forests, in addition to agricultural lands, should be considered as places for suitable for biochar CCS. Similar to COOL VEGE products derived from agricultural lands, COOL FOREST products, such as lumber or wood pulp, are conceivable, where biochar is buried during seedling planting or sustainable management of forests. Taking this possibility in mind, a logo has been proposed in Figure 14.8 to be allowed for use after obtaining certification. If the effectiveness and safety of biochar CCS in rivers or oceans are verified, similar certification systems could be enacted in the future.

FIGURE 14.8 COOL VEGE FOREST.

Areas Needing Further Work

We see two areas that require further development before the scheme outlined here can operate effectively—(1) improvement of the technical analysis of carbon in soil and (2) the establishment of a carbon credit brokerage network.

Despite the growth in biochar-related research, technical shortcomings still exist. For example, simple, cost-effective techniques for performing quantitative analyses measuring the labile and long-term stable fraction of carbon sequestered in soils via biochar have yet to be developed. This is particularly relevant when calculating LCA-CO_2. There are also gaps in our understanding of the agronomic properties of biochar, produced under various conditions and from various feedstocks, in diverse soils. Further research remains to determine the effective functions of biochar and biochar composts and the duration of those functions. One possibility, a *biochar assessment toolkit* being developed at the UK Biochar Research Centre, will directly address these technical needs (UK Biochar Research Centre 2013).

Another area still evolving is a carbon credit brokerage network. A system to collect the credits certified by the certifying body and sell them to businesses and corporations is required. Furthermore, to raise sales efficiency, when carbon credits are certified, they should be immediately sold to the corporations involved in the aforementioned COOL VEGE label sponsorship.

Carbon Minus Projects and COOL VEGE in Action

The pilot Carbon Minus Project, located in Kameoka City, Kyoto Prefecture, has been in operation since 2008, but there are plans for enacting Carbon Minus Projects in five additional sites throughout Japan including Kashiwa and Yotsukaidou Cities in Chiba Prefecture beginning in the fall of 2013. The organizing groups involved in setting up the projects are usually a collection of NPO, universities, farming cooperatives, and local governments who gravitate toward the idea of revitalizing agriculture and combatting climate change together. In the case of Kameoka City, the Japan Biochar Association (NPO), Ritsumeikan University, Ryukoku University, Kyoto Gakuen University, the Hozu area farming cooperative, and Kameoka City municipal government are working in partnership. An additional partner, the Kyoto Prefectural government, has emerged as a certifying body capable of approving the project as a carbon credit offsetting activity and including it on the *menu* of approved projects available to companies looking to offset their emissions. As of this writing, the Carbon Minus Project in Kameoka has been approved as an *appropriate eco-friendly activity* and is in the process of being approved as a carbon credit offsetting activity.

The pilot project in Kameoka City involves the management of bamboo forests that have largely been left to overgrow, causing damage to roads and buildings. Bamboo is thinned by volunteer groups and carbonized in the steel ring kilns shown in Figure 14.3. Bamboo makes an excellent feedstock because it is fast growing and easy to manipulate by hand. The bamboo biochar is collected by a local composting facility, mixed with manure, spread over agricultural fields at a minimum rate of 100 kg carbon per 0.1 ha per year, and planted by the Hozu area farming cooperative with vegetables such as cabbage, carrots, onions, and potatoes. The harvested vegetables are then sold to local retailers as COOL VEGE and either sold directly or processed into salads to be sold in the deli. Currently, three supermarkets are selling COOL VEGE in Kameoka City and the average yearly sales from 2011 to 2012 are around 12–15 million JPY, or about US$125,000. The proposed activities in Chiba Prefecture are involving a large supermarket chain and larger-scale farming cooperative on the order of ten times the size of Kameoka.

Four Japanese companies are also involved in supporting the Kameoka Carbon Minus Project as part of their CSR activities. Daiwa House, Suntory, Toyopet, and Kyoto Bank have given money to the project in exchange for exposure on COOL VEGE labeling and/or on billboards near biochar CCS farmland.

The COOL VEGE brand is also a way to educate the public on the importance of climate change action and carbon sequestration through biochar. In-store displays near COOL VEGE products have information on greenhouse gas emissions and eco-friendly lifestyles. An elementary and junior high school level curriculum including a *Dr. COOL VEGE* picture book has been developed and used throughout Kameoka City and students have participated in planting and harvesting of COOL VEGE. A COOL

VEGE manga and animation contest was held in 2012 in Kochi Prefecture by Kochi University as a way to get young people interested in agriculture and farming lifestyles and to promote ecologically friendly food choices. COOL VEGE and Carbon Minus Projects have been featured in both local and national newspaper articles and on television news programs. Finally, the Kameoka City Carbon Minus Project activities were awarded the Gold Prize at the *Japan Low Carbon Cup 2013*, a contest sponsored by the Japanese Ministry of Environment for proposals from companies, NGO, and NPO that have been tackling environmental problems.

Conclusion

It is well established that the potential of biochar to address climate change is high, but mobilizing it among various stakeholders and throughout society has proven challenging. Furthermore, the diverse socioeconomic contexts within which biochar projects find themselves influence the extent to which projects can be effective and can make it difficult for stakeholders to participate. Carefully positioning biochar projects within these contexts and paying special attention to the needs of producers, farmers, consumers, companies, and governments is required. The Carbon Minus Project scheme presented here is one approach to meeting these various needs in a holistic fashion—communities are able to manage local resources, farmers are given more opportunity for income, consumers can participate in climate change mitigation in a simple way, companies can respond to pressure to reduce or offset carbon emissions, and governments can demonstrate their authority toward proenvironmental causes.

Of course, in the case of Japan, a long history of charcoal use and culture, the socioeconomic struggles in agriculture, and rural demographics are a unique set of conditions that have shaped the scheme and its structure—not all Carbon Minus Projects are alike or will remain the same in the future. The Carbon Minus Project in Kameoka City has evolved as better understanding of the problems facing stakeholders has surfaced and as more stakeholders have expressed an interest in participating. The evolution process is ongoing since issues such as the feasibility of carbon markets, involvement of corporate support, and consumer response will surely have an effect on the interworking of the overall system. However, we do feel strongly that the general scheme itself is applicable in other locations throughout the world by tailoring certain aspects to meet local conditions and we invite the reader to start a Carbon Minus Project in their own communities.

Like the organic agriculture movement, a worldwide biochar agriculture movement is taking shape and we can only hope that someday *COOL VEGE* will be just as recognizable as the *organic* label has become—cooling the Earth from the dinner table and turning *carbon minus* into a *global plus*.

ACKNOWLEDGMENTS

In conducting this study, the authors have received financial support from the Toyota Foundation Research Grant Foundation (FY 2008) and the Japanese Ministry of Education, Culture, Sports, Science and Technology (MEXT)/Japan Society for the Promotion of Science (JSPS) KAKENHI grant #23310034. The authors are deeply grateful for the cooperation and assistance of many people in Kyoto Prefecture, Kameoka City, and the Hozu area of Kameoka City. We are also very appreciative of the guidance received from executive officers of the Japan Biochar Association, Prof. Hidehiko Kanegae (Graduate School of Policy Science at Ritsumeikan University), Dr. Yusuke Tanabiki and Mr. Ryo Sekiya (Kinugasa Research Organization at Ritsumeikan University), and Dr. Terukazu Kumazawa (Research Institute for Humanity and Nature).

REFERENCES

Ezawa, T., Yamamoto, K., and Yoshida, S. (2002) Enhancement of the effectiveness of indigenous arbuscular mycorrhizal fungi by inorganic soil amendments. *Soil Science and Plant Nutrition*, 48(6): 897–900.

Ishii, T. and Kadoya, K. (1994) Effects of charcoal as a soil conditioner on citrus growth and vesicular-arbuscular mycorrhizal development. *Journal of the Japanese Society for Horticultural Science*, 63(3): 529–535 (in Japanese).

Ishikawa, E. (2000) *O-edo ecology jijo (Japan in the Edo Period - An Ecologically-Conscious Society).* Kodansha Publishing Company: Tokyo, Japan. (in Japanese).

Matsubara, Y., Hasegawa, N., and Fukui, H. (2002) Incidence of *Fusarium* root rot in asparagus seedlings infected with arbuscular mycorrhizal fungus as affected by several soil amendments. *Journal of the Japanese Society for Horticultural Science*, 71(3): 370–374 (in Japanese).

McGreevy, S. and Shibata A. (2010) A rural revitalization scheme in Japan utilizing biochar and eco-branding: The carbon minus project, Kameoka city. *Annals of Environmental Science*, 4: 11–22.

Ministry of Agriculture, Forestry, and Fisheries (MAFF). (2008) Annual report on food, fisheries, and rural areas in Japan FY 2007 (Summary provisional translation). MAFF-Information and Policy Evaluation Division. Online resource: http://www.maff.go.jp/e/annual_report/2007/index.html (Accessed February 21, 2013).

Ministry of Agriculture, Forestry, and Fisheries, Forestry Agency (MAFF-FA). (2013) "Mokutan Kankei Shiryo" (Data on Charcoal). MAFF-FA. Online resource: http://www.rinya.maff.go.jp/j/tokuyou/tokusan/megurujoukyou/pdf/3mokutan.pdf. (Accessed February 21, 2013) (in Japanese).

Mori, S. and Marjenah. (2000) A convenient method for inoculating dipterocarp seedlings with the ectomycorrhizal fungus, *Scleroderma columnare*. In *Rainforest Ecosystems of East Kalimantan: El Niño, Drought, Fire and Human Impacts*, (eds.) E. Guhardja, M. Fatawi, M. Sutisna, T. Mori, and S. Ohta. Springer-Verlag: Tokyo, Japan. Series: Ecological Studies, Vol. 140, pp. 251–255.

Ogawa, M. (1983) The mushroom *Rhizopogon rubescens* and charcoal. *Forestry and Forest Products Research Institute News*, 233(2): 1–3.

Ogawa, M. (2006) Inoculation method of *Scleroderma columnare* onto dipterocarps. In *Plantation Technology in Tropical Forest Science*, (eds.) K. Suzuki, S. Sakuarai, K. Ishii, and S. Sasaki. Springer-Verlag: Tokyo, Japan, pp. 185–197.

Ogawa, M. (2007) *Sumi to kinkon de yomigaeru matsu (Reviving pine trees with charcoal and mycorrhizae).* Tsukiji Shokan: Tokyo, Japan (in Japanese).

Ogawa, M. and Okimori, Y. (2010) Pioneering works in biochar research, Japan. *Australian Journal of Soil Research*, 48: 489–500.

Ogawa, M. and Yambe, Y. (1986) Effects of charcoal on VA mycorrhizae and root nodule formation of soybean: Studies on nodule formation and nitrogen fixation in legume crops. *Bulletin of Green Energy Program Group II 8, Ministry of Agriculture, Forestry, and Fisheries*, Japan, pp. 108–134 (in Japanese).

Saito, M. (1990) Charcoal as a micro-habitat for VA mycorrhizal fungi, and its practical implication. *Agriculture Ecosystems and Environment*, 29(1–4): 341–344.

Shibata, A., McGreevy, S., Kumazawa, T., Sekiya, R., and Kanegae, H. (2010a) Toward diffusing "Cool Vegetables"—Reconstructing rural socio-economic systems in Japan based on an eco-branding strategy biochar cultivated vegetables. Presented at *3rd International Biochar Conference*, Rio de Janeiro, Brazil, September 14, 2010. Available online at: http://www.ibi2010.org/agendaagenda.

Shibata, A., Sekiya, R., Kumazawa, T., McGreevy, S., and Kanegae, H. (2010b) Analyzing a simple biochar production process and the cultivation and assessment of "cool" cabbages in Kameoka city, Japan. *3rd International Biochar Conference*, Rio de Janeiro, Brazil, September 13, 2010. Available online at: http://www.ibi2010.org/agendaagenda.

Teruoka, S. (ed). (2008) *Agriculture in the Modernization of Japan 1850–2000.* Manohar: New Dehli, India.

UK Biochar Research Centre. (2013) Biochar assessment. Online resource: http://www.biochar.org.uk/toolkits.php. (Accessed July 3, 2013).

Yamato, M., Okimori, Y., Wibowo, I.F., Anshori, S., and Ogawa, M. (2006) Effects of the application of charred bark of *Acacia mangium* on the yield of maize, cowpea and peanut, and soil chemical properties in South Sumatra, Indonesia. *Soil Science and Plant Nutrition*, 52(4): 489–495.

15

Role of Biochar in Farming Systems Producing Food and Energy from Biomass

T.R. Preston

CONTENTS

The long-term survival of planet earth requires that air temperatures do not rise more than 4°C–5°C over present levels. To achieve this goal, it is not only necessary to stabilize and—even better—reduce atmospheric concentrations of greenhouse gases but also more importantly to reduce anthropogenic heat production (Chaisson 2008; Nickolaenko 2009; Nordell and Gervet 2009). This in turn requires the replacement of fossil fuels with those derived from renewable resources, the source of which can only be the sun. Biochar both has the potential to be a major player in this endeavor and at the same time acts as a catalyst for the promotion of activities that will lead to the achievement of the required goals.

Biochar is the product of the carbonization of fibrous biomass in a restricted flow of air and at high temperatures (500°C–1000°C). These are the conditions found in gasifiers be they updraft (TLUD) stoves or downdraft reactors producing gas for internal combustion engines, turbines, or simply process heat (as in brickmaking). Thus from the outset, by using fibrous waste as feedstock, the gasification system, and the associated production of biochar, avoids the conflict of using natural resources of soils and water for biofuels instead of food. Biochar should therefore always be a by-product of the process of producing energy from nonedible biomass. Preston and Leng (2004) drew attention to the future effects of using food grains for ethanol production referring to it as an insane approach. They warned that the development of the ethanol industry would lead to competition for resources used for food and feed production and that prices of these commodities would rise with disastrous consequences particularly for the poor. The recent statement by the chairman of Nestle (see Box 15.1) suggests that 10 years later,

BOX 15.1 2012 WORLD WATER WEEK IN STOCKHOLM

The chairman of Nestlé, at a recent international conference in Sweden, stated that "the high price of food (which has intensified following the severe drought in the U.S. in the summer of 2012) was the result of biofuels that are grown in soil and water otherwise used for crops intended for human or animal consumption" and that "Biofuels compete with food production. This is unacceptable and cannot be justified."

the situation is potentially becoming even more calamitous without action to curb biofuel production that conflicts with that for food and feed.

Biochar as a Component of Integrated Farming Systems

A major challenge facing agricultural planners is in designing farming systems that combine production of food and energy with an overall negative carbon footprint. Growing of crops solely to produce biochar, as a major tool for sequestration of atmospheric carbon dioxide, is not encouraged. Such proposals are rightly criticized by environmentalists as being both undesirable and unnecessary.

The aforementioned criticisms can be largely avoided by ensuring that the farming system is geared to the needs and opportunities presented by small-scale farms and rural communities that are best suited to utilize efficiently local resources and to recycle wastes.

In tropical latitudes, the cropping systems best suited to integrated production of food, energy, and biochar are rice, sugar cane, and tree crops such as coffee, cocoa, oil palm, and coconut. In all cases, the primary objective is production of food or animal feed, while fibrous residues can be the feedstock for gasification. Maize is commonly grown in the tropics but should not be encouraged as the latter third of the growth cycle is concerned with transfer of nutrients from the stalk to the grain, rather than photosynthesis, as in this stage the leaves die and no longer utilize solar energy, which is thus wasted.

Sugar Cane for Energy, Feed, and Biochar

A model based on sugar cane, developed and used in Colombia over the past 8 years, is indicative of the potential of the small to medium scale of operation that can now be adapted to the needs of rural communities. Sugar cane stalks are crushed in a traditional three-roll mill (as used for production of gur and panela) to yield sugar cane juice that is the basis of the feeding system for livestock (and as a sweetener for the family). The crushed stalks (bagasse) are gasified to produce a gaseous fuel (for an engine-generator-producing electricity) and biochar (Figure 15.1).

The model shown in Figure 15.1 was designed for, and has been operating in, a small-scale farm (3 ha) in the Santander province in Colombia (Rodriguez 2012). However, it can be adapted to the needs of rural communities of 50–100 families. At this larger scale, the sugar cane juice is concentrated to *syrup* (about 65% sugars) to ensure a stable product that can be stored and transported for use as replacement for cereal grains in the same community.

Energy, Food, and Biochar from the Rice Crop

Presently, the production of biochar from the rice crop is almost exclusively from the use of rice husks as feedstock for downdraft industrial-scale gasifiers. Cambodia is at the forefront in application of this technology with an estimated 40–50 medium- to large-scale units (250–500 kVA) already operating in rice mills and rural communities (Khieu Borin, personal communication). Increasingly, the biochar produced in these gasifiers, considered originally as a waste product, is being used as a soil conditioner for growing

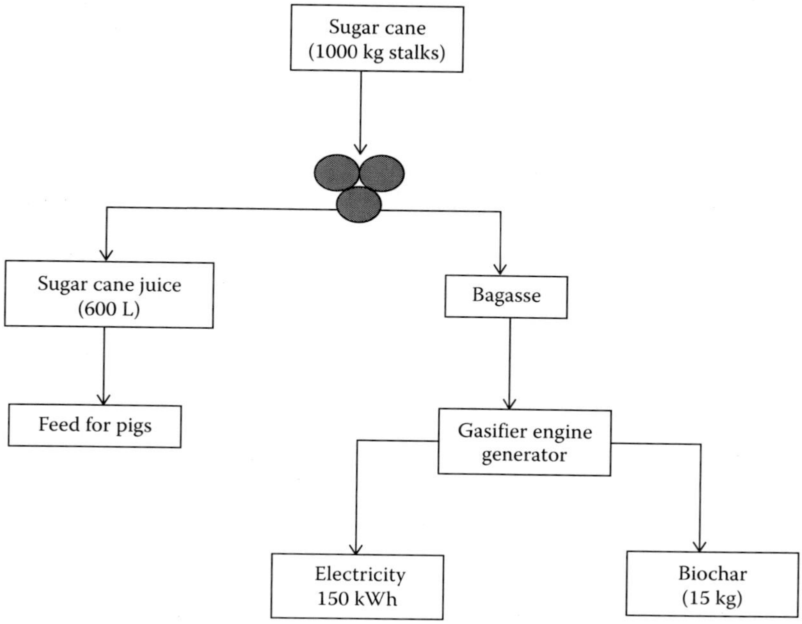

FIGURE 15.1 Integrated use of sugar cane for food/feed, biochar, and electricity. (From Rodriguez, L., Integrated farming systems for food and energy in a warming, resource-depleting world, 2012, http://edoc.hu-berlin.de/dissertationen/rodriguez-lylian-2010-10-12/PDF/rodriguez.pdf.)

vegetables and flowers. Research is underway to proceed to the next step, which is to compact the rice straw into briquettes or pellets, a form compatible with the operating conditions in a downdraft gasifier.

It can be estimated that if the 40–50 million tonnes of rice straw produced annually in Vietnam were gasified, it could produce the balance of electrical energy needed to complete that from hydro and make the country a 100% clean energy producer. The associated production of biochar would be of the order of 5–6 million tonnes. The potential is enormous. Similar calculations can be applied to Cambodia, Lao PDR, and other rice-growing countries.

There has been much interest in, and development of, stoves that use the gasifier principle to produce a clean gas for cooking. The main aim of these initiatives was to improve the conditions under which women, in particular those in poor households, cooked the family meals, a process traditionally carried out over open fires. With the increasing awareness of the potential role of biochar in the farming system, the focus on cook stoves now includes their dual role, with production of biochar becoming as important as their use for cooking. The gasifier stove that was designed and is being constructed in Vietnam was developed originally in the Philippines (Belonia 2005) with the special purpose to use rice husks as fuel. However, it has already served a valuable purpose as the means to facilitate production of biochar for research in universities and research centers throughout Cambodia, Laos, and Vietnam. Already more than 25 research papers have been produced in the region as a result of this initiative.

Role of Biochar in Farming Systems

To date, most of the emphasis given to biochar has been on its role as an amendment in soils and its potential to sequester atmospheric carbon (Lehmann and Joseph 2009). However, recent research shows that its role is much wider with a potential application in all components of farming systems that involve microbial activities. This review will therefore relate primarily to research findings where biochar has been evaluated as a component of integrated farming systems.

Synergism between Biochar and Biodigester Effluent

The first experiment in this series was carried out on a farm in the Colombian foothills where a downdraft gasifier had been installed and where effluents from tubular plastic plug-flow biodigesters were the principal sources of fertilizer (Rodriguez et al. 2009). Biochar derived from sugar cane bagasse, used as feedstock in a downdraft gasifier, was added to a fertile soil from a coffee plantation and to subsoil taken from 1 m below the surface of an unfertilized pasture. In each case, the soil was fertilized with the effluent from a biodigester charged with pig manure. Maize was used as the indicator plant over a growth period of 40 days (Figures 15.2 and 15.3).

There was an obvious synergism between the biochar and the availability of plant nutrients. On the subsoil, application of biochar had no effect until it was combined with biodigester effluent, when the combination of the two amendments supported the same growth of maize as that achieved on the fertile soil. The impact of the combined treatment in modifying the structure and nutrient supply in the subsoil was even more impressive (Photos 15.1 and 15.2).

Biochar versus Charcoal

It is generally considered that the difference between biochar and charcoal traditionally produced from wood in simple earth-covered kilns arises from the fact that in the former the carbonization takes place at much higher temperatures. It has long been known that application of charcoal to soils can be beneficial for plant growth. However, the results summarized in Figures 15.4 and 15.5 (and in Photo 15.3) show clearly that biochar (produced in an updraft gasifier stove from rice hulls) was a more effective soil

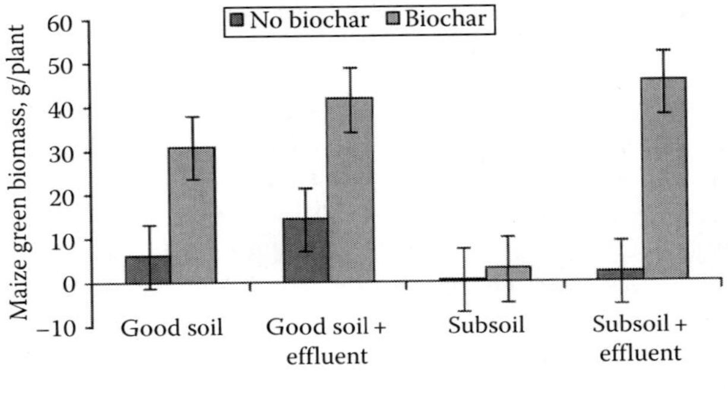

FIGURE 15.2 Effect of biochar and effluent added to fertile soil and subsoil on fresh weight of aerial part of maize (40 days of growth). (From Rodriguez, L. et al., *Livestock Res. Rural Dev.*, 21, Article #10, 2009, http://www.lrrd.org/lrrd21/7/rodr21110.htm.)

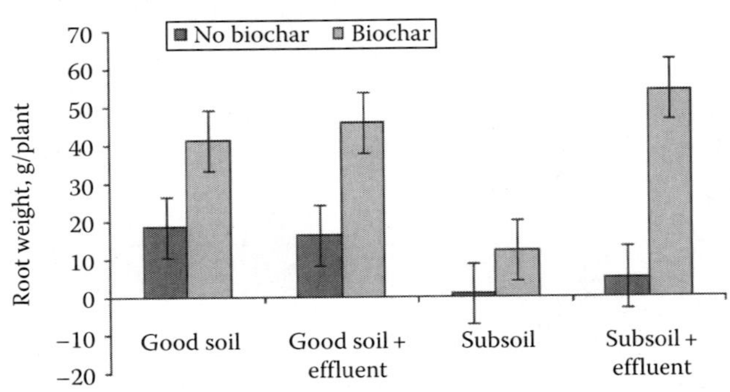

FIGURE 15.3 Effect of biochar and effluent added to fertile soil and subsoil on fresh weight of the roots of maize (40 days of growth). (From Rodriguez, L. et al., *Livestock Res. Rural Dev.*, 21, Article #10, 2009, http://www.lrrd.org/lrrd21/7/rodr21110.htm.)

PHOTO 15.1 The subsoil with added biochar but no effluent, at the end of the trial. The maize germinated but subsequently died. (From Rodriguez, L. et al., *Livestock Res. Rural Dev.*, 21, Article #10, 2009, http://www.lrrd.org/lrrd21/7/rodr21110.htm.)

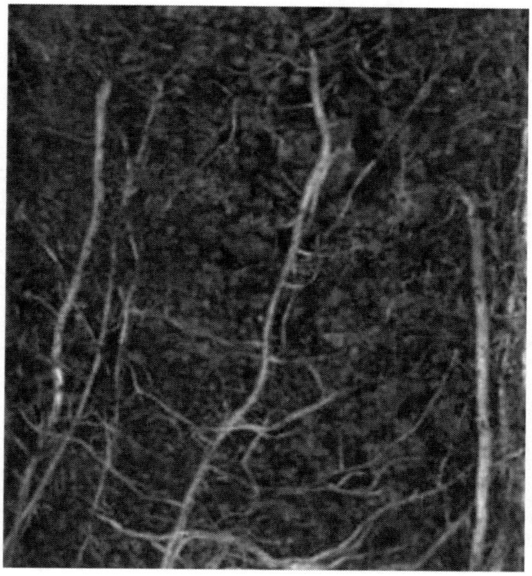

PHOTO 15.2 The subsoil at the end of the trial after amendment with biochar and effluent. (From Rodriguez, L. et al., *Livestock Res. Rural Dev.*, 21, Article #10, 2009, http://www.lrrd.org/lrrd21/7/rodr21110.htm.)

amendment especially when the regrowth of the indicator plant (in this case, water spinach) was measured over a second harvest with no further addition of the two soil amenders.

Residual Effects of Biochar in Soils

The belief that biochar may have mean residence times in soil, ranging from 1,000 to 10,000 years (Skjemstad et al. 1998; Swift 2001), implies that there should be long-term benefits in soils from

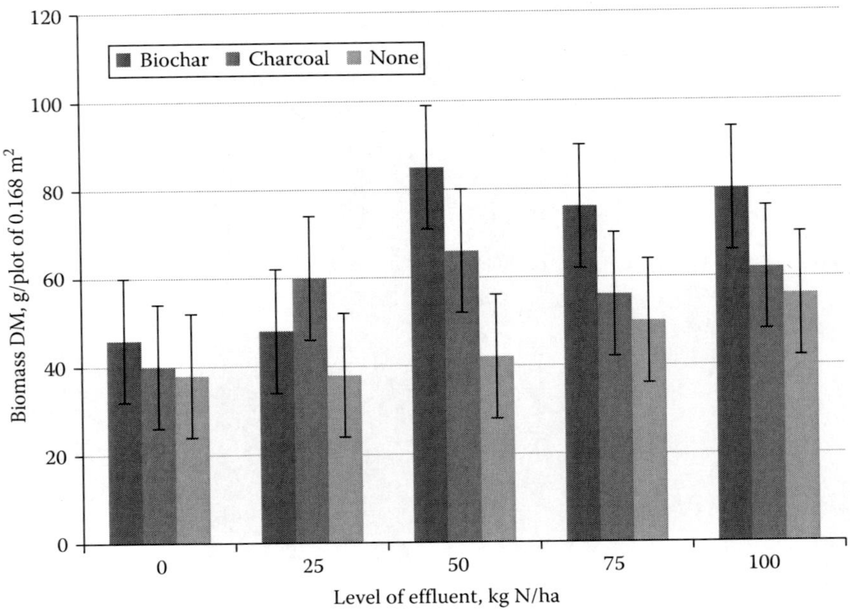

FIGURE 15.4 Effect of biochar, charcoal, and biodigester effluent on biomass yield of water spinach (first harvest). (From Southavong, S. et al., *Livestock Res. Rural Dev.*, 24, Article #26, 2012b, http://www.lrrd.org/lrrd24/2/siso24026.htm.)

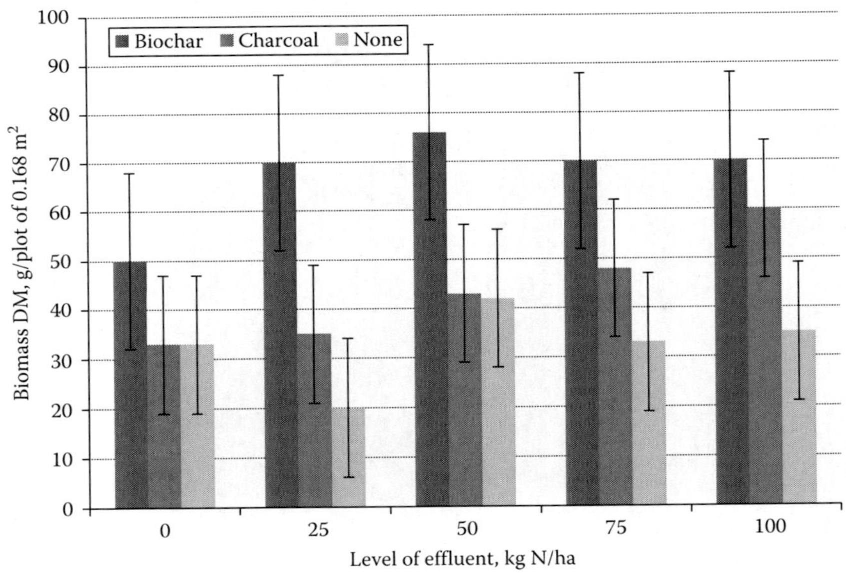

FIGURE 15.5 Effect of biochar, charcoal, and biodigester effluent on biomass yield of water spinach (second harvest). (From Southavong, S. et al., *Livestock Res. Rural Dev.*, 24, Article #26, 2012b, http://www.lrrd.org/lrrd24/2/siso24026.htm.)

application of biochar. This is in line with the observations of maintained fertility in *terra preta* black soils in Amazonia (Glaser 2007). Yet, surprisingly, little attention appears to have been given to evaluating such residual soil fertility aspects in a more modern context.

Results from an experiment in an acid soil (pH 4.7) in Pakse province, Lao PDR, planted with taro (*Colocasia esculenta*), give some indication of positive carryover effects of biochar applied once at the

PHOTO 15.3 Contrasting growth of water spinach in soils amended or not with charcoal or biochar. (From Southavong, S. et al., *Livestock Res. Rural Dev.*, 24, Article #26, 2012b, http://www.lrrd.org/lrrd24/2/siso24026.htm.)

start of the trial. Increased yields relative to control plots were recorded consistently over consecutive harvests at 84-day intervals (yields for the first and third harvests are shown in Figures 15.6 and 15.7).

Recent observations on farmer field plots in Kenya indicate that the benefits from biochar on yield of grain sorghum were greater in the second year with no further addition of biochar (http://www.re-char.com/2013/01/24/the-longterm-impact-of-biochar-in-soil-season-2/).

Enriching Biochar with Biodigester Effluent

Biochar is reported to remove nutrients such as phosphorus (http://news.ifas.ufl.edu/2011/05/12/uf-researchers-develop-method-to-remove-phosphate-from-water-using-biochar/) and ammonia from wastewater. However, there appear to be no reports on the effects on plant growth of biochar after it has been used to absorb nutrients from biodigester effluent. It is possible that the combination of effluent-treated biochar with biodigester effluent would have synergistic effects on soil improvement as manifested in increased crop growth and/or reduced requirements for exogenous fertilizer.

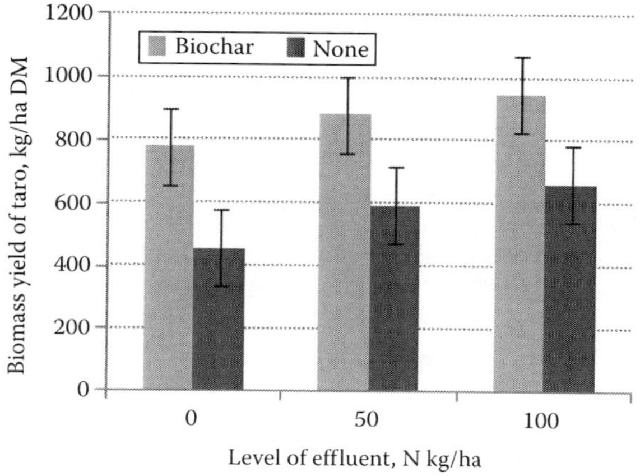

FIGURE 15.6 Effect of biochar on total biomass yield of taro for the first harvest (84 days after planting). (From Vivasane, S. et al., *Livestock Res. Rural Dev.*, 24, Article #107, 2012, http://www.lrrd.org/lrrd24/6/viva24107.htm.)

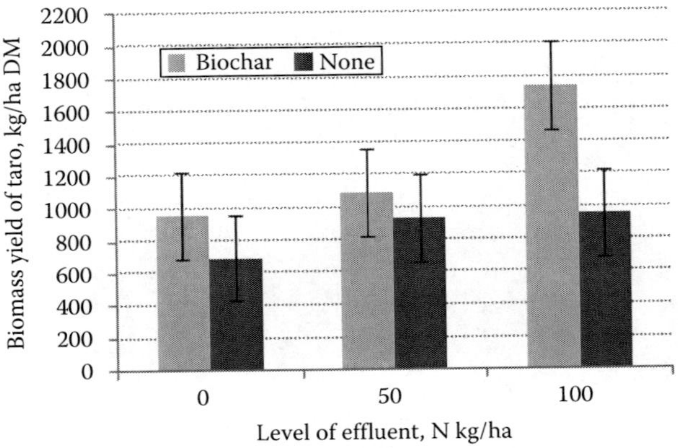

FIGURE 15.7 Effect of biochar on total biomass yield of taro for the third harvest (140 days after planting). (From Vivasane, S. et al., *Livestock Res. Rural Dev.*, 24, Article #107, 2012, http://www.lrrd.org/lrrd24/6/viva24107.htm.)

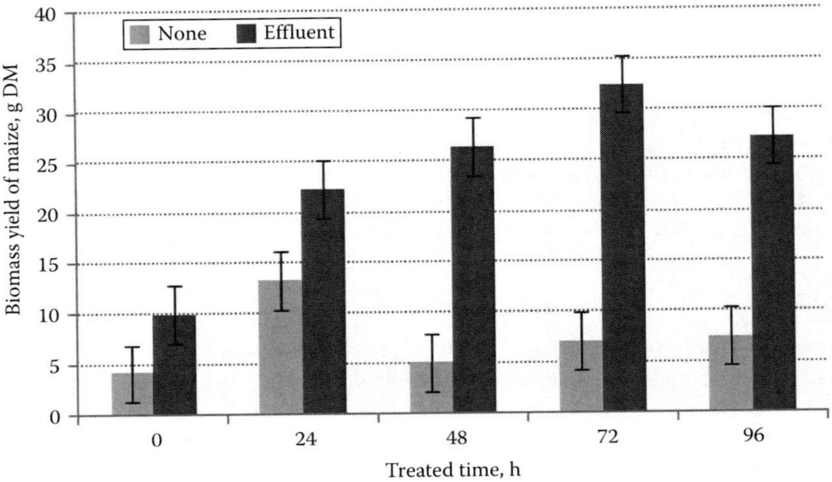

FIGURE 15.8 Effect of biodigester effluent and length of time the biochar was suspended in biodigester effluent on aboveground biomass yield of maize after 50 days growth. (From Southavong, S. et al., *Livestock Res. Rural Dev.*, 24, Article #104, 2012a, http://www.lrrd.org/lrrd24/6/siso24104.htm.)

Results of an experiment to test this hypothesis are summarized in Figures 15.8 and 15.9. Biochar was soaked in effluent from a biodigester for periods of 24, 48, 72, or 96 h or was not treated. The different biochars produced by this process were then added (4 kg/m^2) to an acid soil (pH 4.7) with maize as the indicator plant. An additional treatment in a factorial arrangement was the application or not of the same biodigester effluent as source of plant nutrients (at a level equivalent to 50 kg N/ha).

Soaking the biochar in the effluent for 72 h gave a threefold increase in aboveground growth of maize and of the maize roots when effluent was also applied as fertilizer. In the absence of effluent as fertilizer, there was no response to prior soaking of the biochar in effluent. These responses support the concept of the biochar acting as a habitat where microbial biofilm facilitates the action of consortia of microorganisms in releasing minerals located in close proximity to their substrates.

Biochar and Nutritive Value of Vegetables

We earlier described the positive effects of biochar on biomass yield of water spinach (Figures 15.4 and 15.5). More recent experiments showed linear increases in dry matter (DM) yield of celery cabbage (Figure 15.10) and green mustard (Figure 15.11). These yield increases were accompanied by positive changes in nutritive

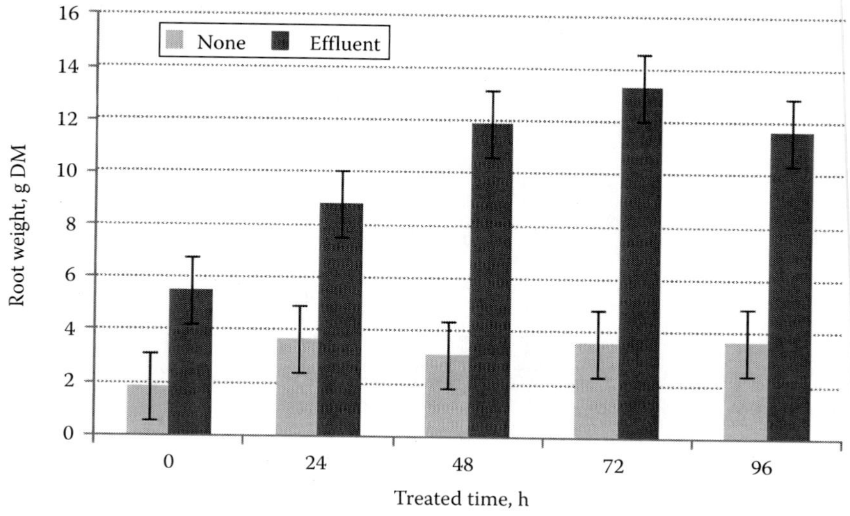

FIGURE 15.9 Effect of biodigester effluent and length of time the biochar was suspended in biodigester effluent on root weight of maize after 50 days growth. (From Southavong, S. et al., *Livestock Res. Rural Dev.*, 24, Article #104, 2012a, http://www.lrrd.org/lrrd24/6/siso24104.htm.)

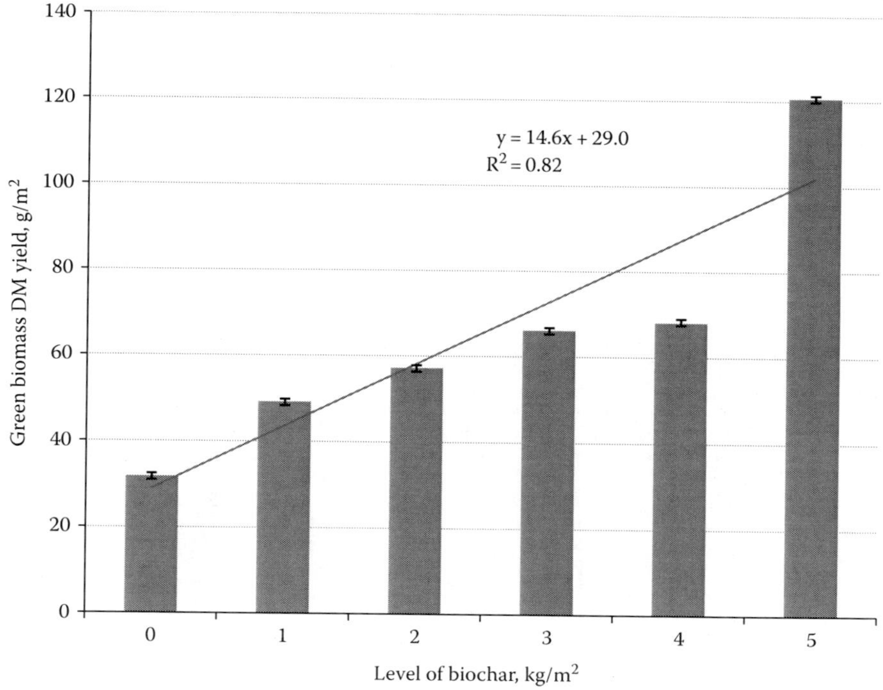

FIGURE 15.10 Relationship between level of biochar and green biomass DM yield of celery cabbage. (From Chhay, T. et al., *Livestock Res. Rural Dev.*, 25, Article #8, 2013, http://www.lrrd.org/lrrd25/1/chha25008.htm.)

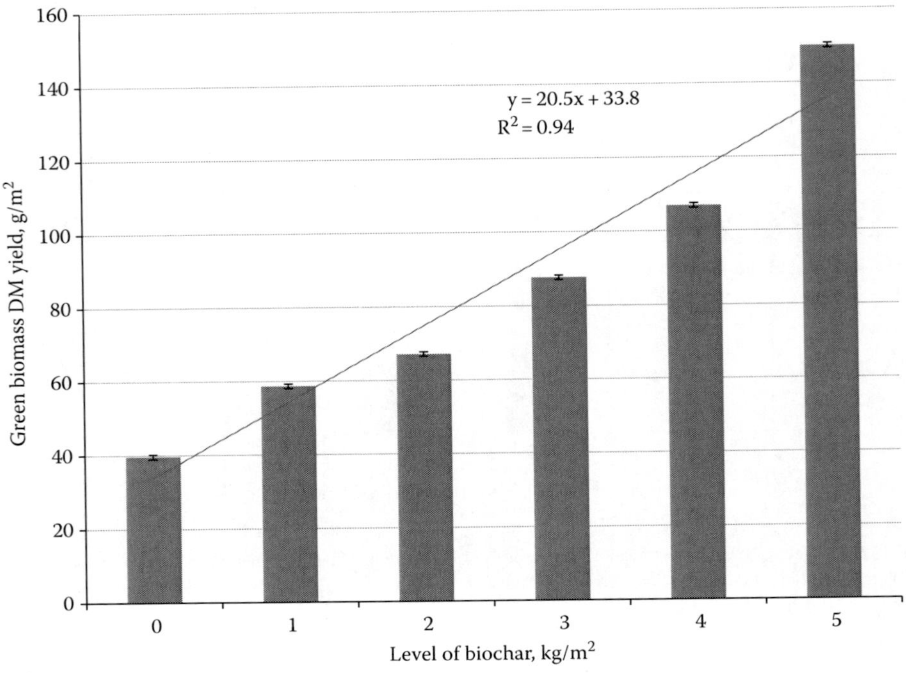

FIGURE 15.11 Relationship between level of biochar and green biomass DM yield of mustard green. (From Chhay, T. et al., *Livestock Res. Rural Dev.*, 25, Article #8, 2013, http://www.lrrd.org/lrrd25/1/chha25008.htm.)

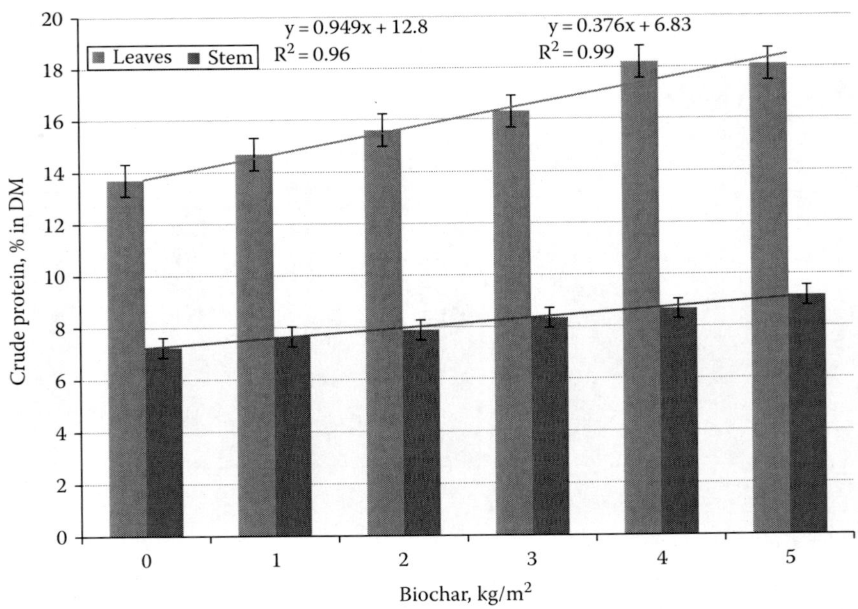

FIGURE 15.12 Relationship between level of biochar and crude protein content of leaves and stems of four vegetables (mean values of celery cabbage, Chinese cabbage, mustard green, and water spinach). (From Chhay, T. et al., *Livestock Res. Rural Dev.*, 25, Article #8, 2013, http://www.lrrd.org/lrrd25/1/chha25008.htm.)

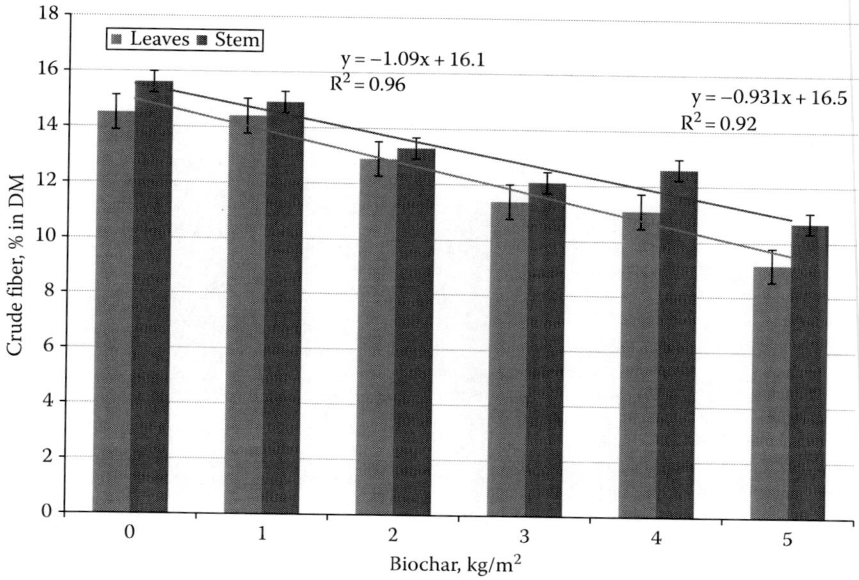

FIGURE 15.13 Relationship between level of biochar and crude fiber content of leaves and stems of four vegetables (mean values of celery cabbage, Chinese cabbage, mustard green, and water spinach). (From Chhay, T. et al., *Livestock Res. Rural Dev.*, 25, Article #8, 2013, http://www.lrrd.org/lrrd25/1/chha25008.htm.)

value with increases in content of crude protein (Figure 15.12) and decreases in crude fiber (Figure 15.13) as biochar applications to soil were increased.

Biochar as an Additive in Diets of Ruminant Animals

Results of preliminary trials in Lao PDR (Leng et al. 2012a,b,c) indicate that biochar reduced methane production in an in vitro incubation with rumen fluid taken from an unadapted buffalo (Figure 15.14). When the biochar was fed at 1% of the diet to local cattle, this was reflected in improved growth rates and concomitant reduction in enteric methane emissions (Figures 15.15 and 15.16).

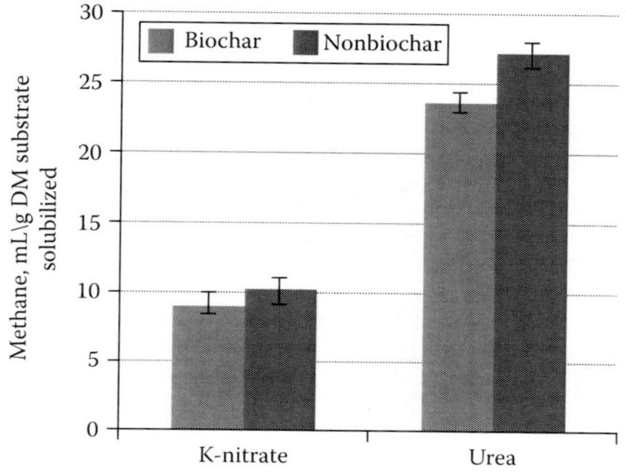

FIGURE 15.14 Effect of biochar and potassium nitrate or urea as NPN source on methane production per unit DM solubilized after 24 h of fermentation of a substrate of cassava root meal with cassava leaf meal and urea.

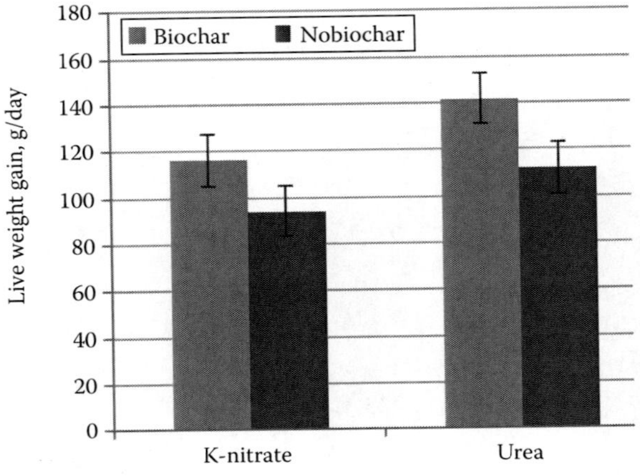

FIGURE 15.15 Effect of biochar and NPN source on ratio of methane to carbon dioxide in mixed air and eructed rumen gases for *yellow* cattle fed with cassava root and cassava foliage.

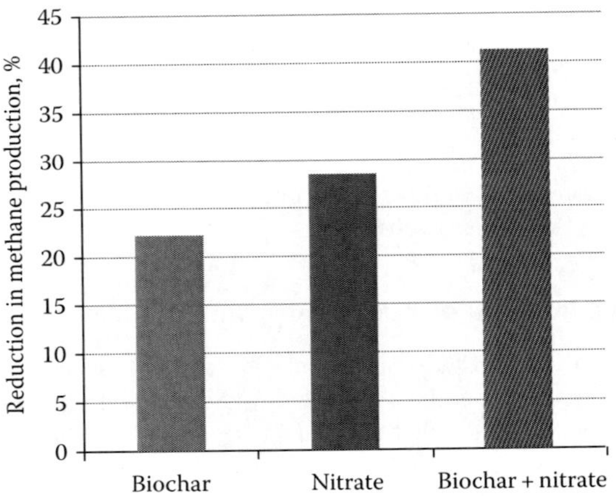

FIGURE 15.16 Reduction in methane due to biochar and nitrate in local *yellow* cattle fed with cassava root and cassava foliage supplemented or not with biochar and with urea or potassium nitrate as NPN source.

Further evidence for the effect of biochar on rumen microbial activity was that rumen fluid, taken from the cattle previously fed with biochar, supported lower levels of methane production compared with rumen fluid from unadapted cattle (Figure 15.17). When the in vitro system combined rumen fluid from adapted cattle plus additional biochar, the reduction in methane production was threefold that observed in the treatment with rumen fluid from unadapted cattle (Leng et al. 2012a).

Biochar as a Source of Biofilm in Biodigesters

On the basis of the results from in vitro systems with rumen fluid taken from buffaloes and cattle and the practice of adding activated charcoal to anaerobic biodigesters to enhance the stability and efficiency of conversion of organic wastes to methane (Aktaş and Çeçen 2007), it was hypothesized that biochar should also influence the process of fermentation in biodigesters charged with cattle excreta. After an

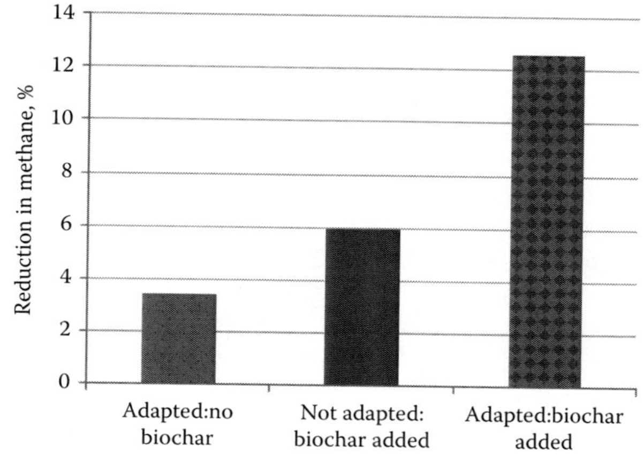

FIGURE 15.17 Effect of adaptation to biochar in the diet, and of addition of biochar to the substrate, on percent reduction in methane production (mL methane/g DM solubilized) in an in vitro incubation of cassava root meal with cassava leaf meal and urea. (From Leng, R.A. et al., *Livestock Res. Rural Dev.*, 24, Article #211, 2012a, http://www.lrrd.org/lrrd24/11/sang24211.htm.)

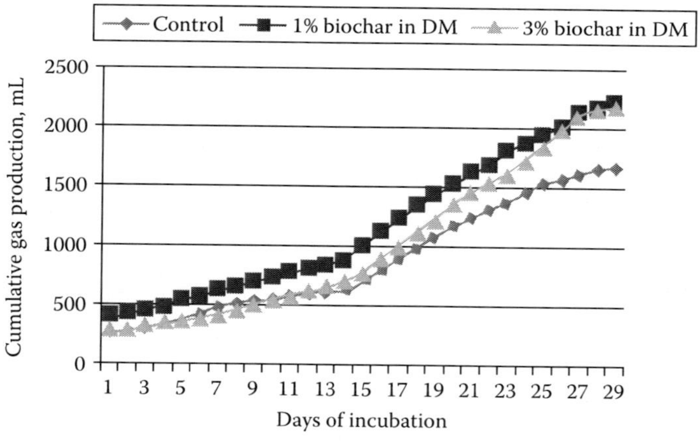

FIGURE 15.18 Development of gas production in a batch biodigester charged with cattle manure to which was added 1% or 3% of biochar (DM basis). (From Inthapanya, S. et al., *Livestock Res. Rural Dev.*, 24, Article #212, 2012, http://www.lrrd.org/lrrd24/12/sang24212.htm.)

initial lag period, incorporation of 1% of biochar in the cattle manure introduced into a batch biodigester led to a consistent increase in production of biogas with an overall improvement of 31% over a 30-day period of incubation (Figure 15.18). There was no further improvement with 3% of biochar in the influent. The methane content of the gas was unchanged at 43% (Figure 15.19).

By contrast, the effect of the biochar was quite different when added to biodigesters managed as a continuous process with daily additions of cattle manure containing 1% (in DM) of biochar (Figures 15.20 and 15.21). Gas production was increased but by a much smaller degree (5%) than in the batch biodigester (30% increase), and the methane content was decreased.

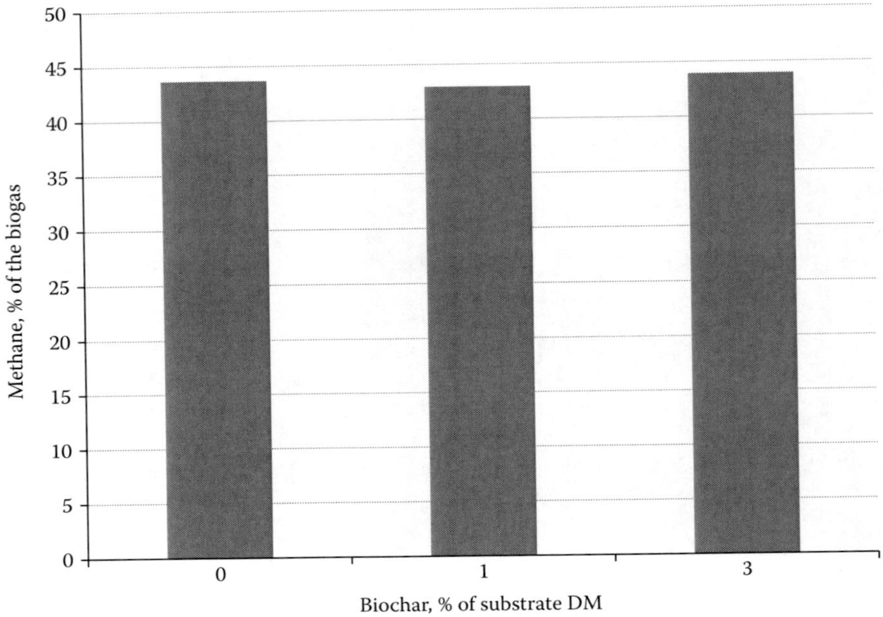

FIGURE 15.19 Methane content of the biogas after 30 days of fermentation of cattle manure in a batch biodigester to which was added 1% or 3% of biochar (DM basis). (From Inthapanya, S. et al., *Livestock Res. Rural Dev.*, 24, Article #212, 2012, http://www.lrrd.org/lrrd24/12/sang24212.htm.)

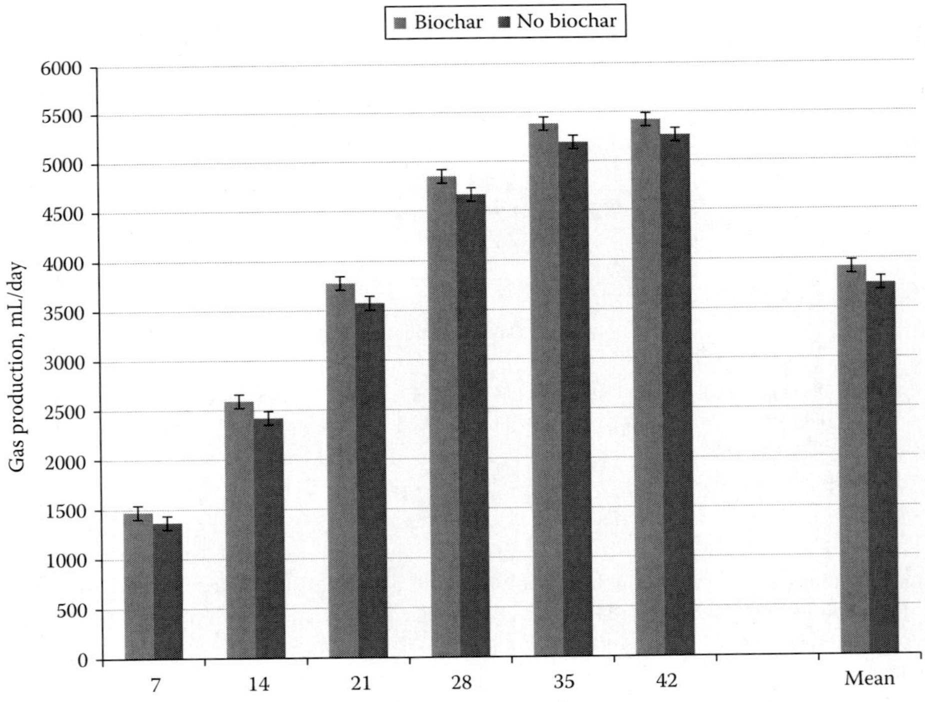

FIGURE 15.20 Daily addition of biochar (1% on DM basis) increased the gas production in a continuous plug-flow biodigester charged daily with cattle manure. (From Inthapanya, S. et al., *Livestock Res. Rural Dev.*, 24, Article #212, 2012, http://www.lrrd.org/lrrd24/12/sang24212.htm.)

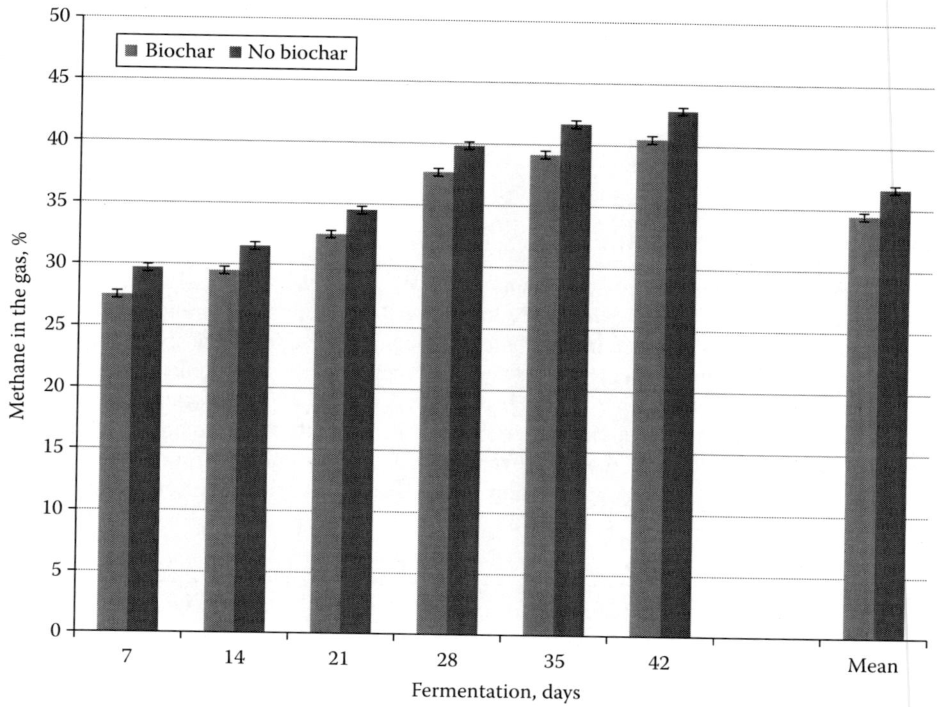

FIGURE 15.21 Daily addition of biochar (1% on DM basis) reduced the methane content of the gas in a continuous plug-flow biodigester charged daily with cattle manure. (From Inthapanya, S. et al., *Livestock Res. Rural Dev.*, Unpublished data.)

Research Priorities

Source of the Biochar

In almost all the experiments reported in this chapter, the biochar was produced from rice husks carbonized in an updraft gasifier stove (Olivier 2010). The exception was in the research of Rodriguez et al. (2010), where the biochar was from sugar cane bagasse carbonized in a downdraft gasifier. The experiments were done on farm (Rodriguez et al. 2009) or in laboratories in newly established universities in Laos where analytical facilities were minimal; thus, there are no data on the composition of the biochar.

Gasification versus Pyrolysis

Competing claims are made for the virtues of biochar produced by pyrolysis compared with gasification. However, there is a lack of information on the respective merits of the biochar produced from the two processes as a component of farming systems. Comparative data from use of biochar produced by different technologies are urgently needed. From the systems standpoint, gasification appears to have the advantage as the products can be used efficiently on farm at all scales of operation. By contrast, the bio-oil that is a major product from pyrolysis will require refinement at some central facility before it can be used to replace liquid fuels. Use of electricity produced as the output of gasification is facilitated by the ease of transporting it over existing transmission lines.

Long-Term Effects of Biochar

The belief that biochar has mean residence times in soil, ranging from 1,000 to 10,000 years (Skjemstad et al. 1998; Swift 2001), implies that there should be long-term benefits in soils from the application of biochar. This is in line with the observations of maintained fertility in *terra preta* black soils in Amazonia (Glaser 2007). There is also support from recent findings from Kenya (http://www.re-char.com/2013/01/24/the-longterm-impact-of-biochar-in-soil-season-2/) that indicated that the benefits from biochar on yield of grain sorghum were greater in the second year with no further addition of biochar. This feature of long-term impact on soil fertility of the use of biochar has so far received insufficient attention from researchers.

Biochar in the Farming System

The indications that incorporation of biochar in livestock diets has beneficial effects on productivity with reduced emissions of enteric methane need to be verified under a wide range of farming systems. Similarly, the potential role of biochar in enhancing biogas production from livestock wastes, and in purification (recovery of plant nutrients) of wastewaters, merits increasing attention.

Most reports on biochar as soil amender refer to effects on yield of biomass and its components. The finding that in green mustard vegetable the crude protein percentage was increased, and crude fiber reduced (Chhay et al. 2013), implies that improvement in nutritive value of certain crops may be another virtue of soil application of biochar. Validation of these findings with other crops and vegetables is another urgent research need.

Bottom Line

The potential role of biochar as a means of sequestering atmospheric carbon has so far been the main focus of attention. However, its virtues are much wider with potential impacts on soil fertility, plant yield, and quality and as an additive in livestock feed, biodigesters, and other liquid waste disposal systems. More important is the emphasis that biochar has given to the overall concept of the systems approach to utilization of renewable resources and recycling of wastes with maintenance or improvement in soil fertility. Biochar makes most sense when it is derived from a process that aims to optimize the varied characteristics of biomass to satisfy the major needs of humanity, which are food, energy, shelter, and a healthy environment. It needs to be produced and used locally for maximum benefits.

REFERENCES

Aktaş O and Çeçen F. 2007. Bioregeneration of activated carbon: A review. *International Biodeterioration and Biodegradation*, 59, 257–272.

Belonio A T. 2005. *Rice Husk Gas Stove Handbook*. Appropriate Technology Center, Department of Agricultural Engineering and Environmental Management, College of Agriculture, Central Philippine University, Iloilo, Philippines.

Chaisson E. 2008. Long-term global heating from energy usage. *Transactions American Geophysical Union*, 89(28), 253. doi:10.1029/2008EO280001.

Chhay T, Vor S, Borin K, and Preston T R. 2013. Effect of different levels of biochar on the yield and nutritive value of Celery cabbage (*Brassica chinensis* var), Chinese cabbage (*Brassica pekinensis*), Mustard green (*Brassica juncea*) and Water spinach (*Ipomoea aquatica*). *Livestock Research for Rural Development*, 25, Article #8. http://www.lrrd.org/lrrd25/1/chha25008.htm.

Glaser B. 2007. Prehistorically modified soils of central Amazonia: A model for sustainable agriculture in the twenty-first century. *Philosophical Transactions of the Royal Society B*, 362, 187–196.

Inthapanya S, Preston T R, and Leng R A. 2012. Biochar increases biogas production in a batch digester charged with cattle manure. *Livestock Research for Rural Development*, 24, Article #212. http://www.lrrd.org/lrrd24/12/sang24212.htm.

Inthapanya S, Preston T R, and Leng R A. 2013. Biochar increases biogas production but decreases the methane concentration in the gas in a continuous plug-flow biodigester charged with cattle manure. *Livestock Research for Rural Development*, Unpublished data.

Lehmann J and Joseph S. (eds.) 2009. *Biochar for Environmental Management; Science and Technology.* Earthscan Publishers Ltd, Sterling, VA.

Leng R A, Inthapanya S, and Preston T R. 2012a. Methane production is reduced in an in vitro incubation when the rumen fluid is taken from cattle that previously received biochar in their diet. *Livestock Research for Rural Development*, 24, Article #211. http://www.lrrd.org/lrrd24/11/sang24211.htm.

Leng R A, Preston T R, and Inthapanya S. 2012b. Biochar lowers net methane production from rumen fluid in vitro. *Livestock Research for Rural Development*, 24, Article #103. http://www.lrrd.org/lrrd24/6/sang24103.htm.

Leng R A, Preston T R, and Inthapanya S. 2012c. Biochar reduces enteric methane and improves growth and feed conversion in local "Yellow" cattle fed cassava root chips and fresh cassava foliage. *Livestock Research for Rural Development*, 24, Article #199. http://www.lrrd.org/lrrd24/11/leng24199.htm.

Nickolaenko A P. 2009. Concept of planetary thermal balance and global warming. *Journal of Geophysics Research*, 114, A04310, doi:10.1029/2008JA013753.

Nordell B and Gervet B. 2009. Global energy accumulation and net heat emission. *International Journal of Global Warming*, 1, 378–391.

Olivier P. 2010. *The Small-Scale Production of Food, Fuel, Feed and Fertilizer: A Strategy for the Sustainable Management of Biodegradable Waste*. http://www.mekarn.org/workshops/pakse/html/olivier.docx.

Preston T R and Leng R A. 2004. Global decline in oil and natural resources; implications for the scope and content of papers for publication in LRRD. *Livestock Research for Rural Development,* 16, Art. No. 87. http://www.lrrd.org/lrrd16/11/pres16087.htm.

Rodriguez L. 2012. Integrated farming systems for food and energy in a warming, resource-depleting world. PhD thesis dissertation. http://edoc.hu-berlin.de/dissertationen/rodriguez-lylian-2010-10-12/PDF/rodriguez.pdf.

Rodríguez L, Salazar P, and Preston T R. 2009. Effect of biochar and biodigester effluent on growth of maize in acid soils. *Livestock Research for Rural Development,* 21, Article #110. http://www.lrrd.org/lrrd21/7/rodr21110.htm.

Skjemstad J O, Janik L J, and Taylor J A. 1998. Non-living soil organic matter: What do we know about it? *Australian Journal of Experimental Agriculture*, 38, 667–680.

Southavong S, Khammingsavath K, Vyraphet P, and Preston T R. 2012a. Effect of effluent-treated biochar and biodigester effluent on growth of maize (*Zea mays*) and on soil physical properties. *Livestock Research for Rural Development*, 24, Article #104. http://www.lrrd.org/lrrd24/6/siso24104.htm.

Southavong S, Preston T R, and Man N V. 2012b. Effect of soil amender (biochar or charcoal) and biodigester effluent on growth of water spinach (*Ipomoea aquatica*). *Livestock Research for Rural Development*, 24, Article #26. http://www.lrrd.org/lrrd24/2/siso24026.htm.

Swift R S. 2001. Sequestration of carbon by soil. *Soil Science*, 166, 858–871.

Vivasane S, Southavong S, Vyraphet P, and Preston T R. 2012. Effect of biochar and biodigester effluent on growth performance of taro (*Colocasia esculenta*). *Livestock Research for Rural Development*, 24, Article #107. http://www.lrrd.org/lrrd24/6/viva24107.htm.

16

Vetiver System: Reversing Degradation On and Off Farm to Keep Soil Carbon In Place, Build Up Root Biomass, and Turn Degraded Areas into Biofuel Sources

Elise Pinners

CONTENTS

Introduction

This chapter focuses on the remarkable capacity of vetiver grass (Chrysopogon zizanioides) for atmospheric carbon sequestration and soil carbon capture and storage.

Vetiver system (VS) semipermeable hedges reduce loss of soil, moisture, and fertility and are directly important for farmers and food production (Truong et al., 2008).

Vetiver, due to high biomass production, is great for production of biofuel (dried grass bunches, briquettes, or pellets for home cooking and feed for ethanol plants and gasifier installations feeding the grid and for residual biomass from essential oil plantations for furnaces).

Vetiver's promise of soil carbon capture and carbon sequestration is immense, given vetiver's enormous capacity to produce biomass and its impressive deep root system that can possibly capture more carbon than any other grass. Most plant carbon biomass research focuses on grassland grasses or trees, but vetiver is not that kind of grass nor a tree. Yet it has to fit into databases on carbon sequestration and storage that are currently developed.

Technologies, methodologies, and algorithms need to be improved to allow measurement of vetiver carbon sequestration, capture, and storage. This requires observations under different climate and soil conditions, as well as different applications of the VS. India and China, having good research institutions and a wide range of growing conditions, are well placed for this (Pinners, 2011).

Biofuel from Degraded Areas

Biofuel from Vetiver Grown as a Field Crop

In nearly all applications of vetiver, as *VS*, the grass is planted as a dense hedge, typically placed along contour lines or (when farms are small) on boundaries of fields, embankments, or edges of water courses (Truong et al., 2008). This applies for on-farm applications, bioengineering for protection of slopes (infrastructure), and artificial wetlands. Perhaps the odd exceptions are just two: floating vetiver for treatment of wastewater in ponds (see photo below) and the planting of vetiver as a field crop, for biofuel.

As a field crop, producing about 100,000 plants/ha (0.3 × 0.3 m spacing), it provides potentially up to 120 t of dry biomass per year per hectare on soils of reasonable depth and fertility. On saline soils in the Dominican Republic, dry vetiver biomass production is reported at 70 t/ha. For modeling it is fair to use the 100 t/ha when we have deep and good soils, and 80 t/ha for poorer conditions. (From Truong, P., Personal Communication, 2013.)

Not Displacing Food Crops for Fuel Production

Vetiver thrives in places and supports revegetation under difficult conditions where nothing else will grow or replace conventional nonvegetative engineering. For smaller scale use, it could be obtained from road cut-and-fill batter and drains (Photos 1–3), landslides, gully rehabilitation work, and decommissioned waste dumps (where toxic leachate is recycled back on the dump [Photos 4–6], on highly erodible canal banks made of coastal acid sulfate soils, shifting dune sand, and degraded laterite soils [Photos 7–9]).

Vetiver is well placed as a potential furnace feedstock and has the added advantage in that it could be grown on marginal lands. Vetiver could be grown on a large scale, producing high yields with very little supplementary irrigation to be used for biofuel in ethanol plants, in conventional coal-fired furnaces (as in the Dominican Republic example), or in the latest state-of-the art gasification installations that have recently been developed for woodchips.

In the following example, from the Dominican Republic, vetiver is grown on saline soils. In northern and northwest India and Pakistan, there are huge areas of saline irrigated land with high water tables (marginal to agriculture because of poor drainage that has led to saline soils and saline groundwater). Additionally, vetiver would help rehabilitate saline land so that it might again be used for profitable purposes. This was done in the 1950s on Ussar lands in Uttar Pradesh where vetiver was used to rehabilitate saline land where little else grew (The National Research Council, 1993).

In some parts of India, it is traditional to move hand-cut sugarcane and sugar grass to nearby cane and paper processing plants. There would be little difference in doing the same for vetiver.

Cellulosic Ethanol

In the United States, research is being carried out on plants such as *Miscanthus*, switchgrass, and corn as potential feedstocks for cellulosic ethanol plants. Table 16.1 shows the characteristics for an ideal biomass fuel.

TABLE 16.1
Characteristics of an Ideal Biomass Energy Crop Present (+) in Corn, Short-Rotation Coppice, and *Miscanthus*, Developed in Part from Long (1994) and Khosla (2006)

Crop Characteristic	Corn	Short-Rotation Coppice	*Miscanthus*
C4 photosynthesis	+		+
Long canopy duration		+	+
Perennial (no need for annual tillage or planting)		+	+
No known pests or diseases			+
Rapid growth in spring to outcompete weeds		+	+
Sterile; prevent *escape*			+
Stores carbon in soil (soil restoration and carbon sequestration tool)		+	+
Partitions nutrients back to roots in fall (low fertilizer requirement)			+
Low nutrient content, i.e., <200 mg/MJ nitrogen and sulfur (clean burning)		+	+
High water use efficiency	+		+
Dry down in field (zero drying costs)			+
Good winter standing (harvest when needed; zero storage costs)		+	+
Utilizes existing farm equipment	+		+
Alternative markets (high-quality paper, building materials, and fermentation)	+	+	+

Vetiver grass meets all the characteristics of *Miscanthus sp.* and more. Most importantly, it is a drought-tolerant, long-term perennial crop that needs only moderate inputs to be productive (Grimshaw, 2008).

The dry matter yield of *Miscanthus* is—after 2 years of stand development—about 25 t/ha (University_of_Florida, 2012), whereas that of vetiver, if planted as a crop, has a potential of up to 100 t/ha . But even so, when used in lesser conditions, on degraded land, one should expect lower yields (Pinners, 2011). Vetiver productivity, in particular belowground, is superior to *Miscanthus* (in photos: (1) *C. zizanioides*, or vetiver as we refer to here, in the background, compared to *C. nemoralis* in front; (2) vetiver, 6 months old; (3) vetiver plant on the right, 9 months old; (4) in China, an effort to uproot a plant).

Coal-Fired Furnaces for Electricity

When a medium-sized power plant adds biomass to its mix, its global warming emission reductions are equivalent to taking 17,000 cars off the road (ALSTOM, 2011).

Vetiver has an energy value of 16.3 GJ/t compared to petroleum 41.9 GJ/t, coal 12–30.3 GJ/t, dry wood 19.8 GJ/t, and sugarcane bagasse 9.3 GJ/t. These nonvetiver biomass sources are used as feedstock to generate electricity.

The advantages of vetiver feedstock over other fuels are as follows:

1. It is entirely renewable and in the burning process does not add net carbon amounts to the atmosphere.
2. CO_2 reduction from coal power plants can be expensive (\$100–\$200 t^{-1} for some methods), but cofiring biomass with coal (coal in quantities as small as 15% of the biomass) gives 95% cost reduction compared to using coal only.
3. Biomass has much lower sulfur content than even low-sulfur coal (which is costlier than regular coal), so there will be less SO_2 release into the atmosphere.
4. Nitrous oxide emissions from biomass combustion are also expected to be less than from coal firing processes.
5. Coal mines release gases for a long period; for a vetiver farm, this is only temporary, when the field is established (the crop remains permanently, so no annual plowing is required).

Boucard's Biofuel Case (a 50 MW Power Plant in Barahona, Dominican Republic)

- Under tropical conditions, under irrigation or good annual rainfall 70 up to 80 t/ha year, dry vetiver leaves are produced.
- As it is never a good idea to burn wet fuel (energy is used to evaporate the water, and steam reduces furnace temperature), vetiver is dried prior to burning, to reach a moisture content of about 10%, which takes 3–4 days of sun drying.
- Planting vetiver for fuel is one of the easiest and safest farming operations: unlike sugarcane, vetiver plantations need no replanting, it can be harvested annually at any time of the year and has no diseases or weather damage, and production can go on for more than 20 years, by just adding fertilizer.
- Owning land and farming operation, once the plantation is established, Boucard estimates the production of vetiver biomass costs to be below US \$16.5 t^{-1}.
- Baling is done with a standard hay baler; after 3–4 days of sun drying, the moisture content becomes 10%.

- Dry grass calorific value is 16.3 GJ/t (about half that of coal).
- Modern thermal power plants need about 900 t/day of vetiver bales to produce 50 MW (or 450 t/day coal import from Colombia, providing 30.3 GJ/t).
- 900 t of vetiver bales more than equal to 450 t coal, to be harvested from 12.5 ha in 1 day; so, *to run a 50 MW plant 365 days/year requires 4563 ha, surrounding the plant*; unlike sugarcane, vetiver does not need replanting as it simply grows back, with natural rainfall, and can be harvested any time of the year for 20 years or more (just add some fertilizer). The raw cost of producing vetiver biomass fuel is no more than US \$16.5 t^{-1}, once the plantation is established.

The cost of producing 1 GJ (in 2009):

- Vetiver gives 16.3 GJ/t dry leaf, so we need 64.4 kg; it costs *US \$1.06.*
- Cost if coal, 32.6 GJ/t, as the cheapest fuel costs US \$40 t^{-1} → *US \$1.60.*
- Cost if crude petroleum, US \$655 t^{-1} crude petroleum, which has 42 GJ/t → *US \$16.*

Vetiver is the most lucrative use for tropical farmland (Boucard, 2009). If third world countries realize this, one of the biggest problems will be to ensure that food crops are not displaced for fuel production!

New Technology Gasification for Electricity

In Kenya, an initiative is in preparation to use vetiver (grass bales) as an alternative biofuel in a wood chip to electricity plant (biomass to electricity [BTE]). The initiative is from the African American Environmentalist Association (AAEA), in cooperation with National Clean Fuels (NACF) and Dyson Engineering and Technical (DET) Services. Vetiver grass would be gasified with the methane burned in a turbine engine to produce electricity. A comparison of heat content shows the following:

- Petroleum has on the average 42 GJ/t.
- Coal has 28 GJ/t.
- Dry wood has 20 GJ/t.
- Sugarcane bagasse has 9.3 GJ/t.
- Dry vetiver grass has 16.3 GJ/t.

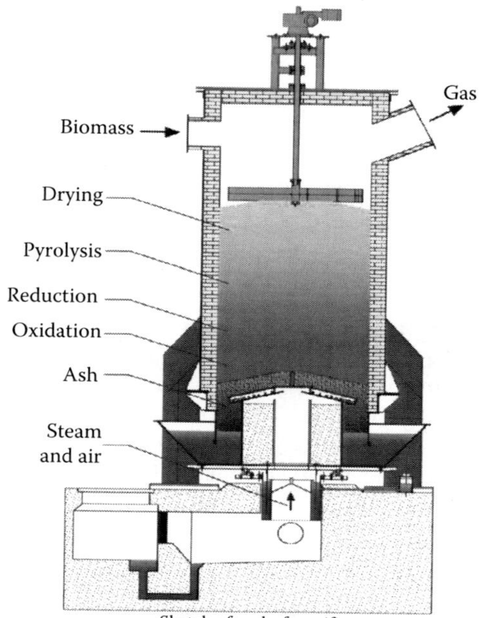

Sketch of updraft gasifier.
(Source: Volund Systems,
Waste and Energy Technologies)

The source of vetiver grass bales is to be a combination of nearby farming of vetiver and delivery from other farms; a combination with sugarcane bagasse may also be tried. The idea is still in a very early stage and a feasibility study is being prepared.

Gasifiers generally run on a portion of the supplied feedstock as well as converting the balance of it to a fuel for other applications. Biofuel-generated gas is a syngas that is produced by thermal gasification of biomass. It is the result of two high-temperature reactions (1292°F): an exothermic reaction where carbon burns to CO_2 but is then reduced partially back to CO where carbon reacts with steam, producing carbon monoxide (CO), molecular hydrogen (H_2), and carbon dioxide (CO_2). In several gasifiers, the actual gasification process is preceded by pyrolysis ("a thermochemical decomposition of organic material at elevated temperatures in the absence of oxygen; pyrolysis typically occurs under pressure and at operating temperatures above 800°F"), where the biomass or coal turns into char, releasing methane (CH_4) and tar rich in polycyclic aromatic hydrocarbons (PAHs). Other gasifiers are fed with previously pyrolysed char. Biofuel-generated gas is flammable because of the carbon monoxide, hydrogen, and methane content. A gasifier takes biomass as fuel and is burned incompletely in a firebox, producing solid ashes and soot (which have to be removed periodically from the gasifier). The gas can then be filtered for tars and soot/ash particles, cooled, and directed to an engine. Most of these engines have severe purity requirements of the gas, so the gas often has to pass through extensive gas cleaning in order to remove or convert (i.e., to *crack*) tars and particles. The removal of tar is often accomplished by using a water scrubber. Exhaust gas emission levels from an internal combustion engine are significantly lower on biofuel gas than on petrol. Especially low are HC emissions. A normal catalytic converter works well with biofuel gas, but even without it, emission levels less than 20 ppm HC and 0.2% CO can be easily achieved by most automobile engines (Wikipedia); see also technology brief (Lunds_University, 2003).

Other fuel ideas were discussed in Forti's 2013 comment on *flash pyrolysis* (Forti) (in Italian).

R&D on Combustion

More research on perspectives and constraints is required. Some estimates indicate that if income from biomass for fuel is combined with credits for carbon offsetting, economic rates of return would be in the order of 15%. These estimates are made a few years ago when oil prices were lower, so the return rate would be even higher today.

There are two challenges that need further research:

1. Vetiver's high nitrate absorption does not all go into the air in negative way, if primary N combustion product is elemental nitrogen (N_2, which is the bulk of the atmosphere). But poor combustion can lead to all sorts of noxious compounds. Much of knowledge about combustion is directly applicable to vetiver, but some practical research may be needed to check noxious compound emission. The factors that affect results of combustion are the following:
   a. N content of foliage: In vetiver, this is not particularly high.
   b. How the grass is *cured* and how it is burned in coal-fired units.
   c. Impact of silica on furnaces.
   d. Silica in foliage: Vetiver has higher silica levels that are higher than in many other grasses, increasing with age and variable per genotype; it is consolidated in leaf as mineralized *phytoliths* (whose shape is unique to each species). Silica (and other minerals) can create ash disposal burden in combustion as well as aerial fly ash deposition, plus silica itself can serve as a flux in combination with other elements than can wreak havoc in furnaces. So silica is a downside to burning vetiver (compared to *Miscanthus*), but how serious is that?

   And possibly complex silicate salts capture and precipitate some of the heavy metals, sulfur compounds, and nitrous oxides in mixed combustion reaction, preventing them from escaping through the stack. This is relevant for large furnaces and combustion systems, where there is great concern for emissions. Another plus for vetiver?
2. Silica affecting the furnace: Vetiver forms a glassy type of *klinker* that stays in furnace with ash (complex oxides and salts). Vetiver contains so much silicates (more than in *Miscanthus*); it tends to flux firebrick by lowering the melting point of brick. This is why furnaces in Texas (where spent wood from extraction process was burnt at very high temperatures >2000°F) had to go to a brick with 70% alumina content, to prevent *fluxing*. Cheap bricks with only 50% alumina melt like butter. High alumina bricks are costly and there are thousands of bricks in a large furnace.

Carbon Sequestration, Capture, and Storage

There are two areas describing carbon sequestration:

1. *Sequestration of carbon* (from the air) that is—simply put—a way to offset carbon emission elsewhere. Natural vegetation when in a balanced, normal life cycle does sequester carbon as well as emit carbon—so that would be neutral. But VS has the ability to reverse degradation, to transform a biomass-poor situation (degraded land, gullies, roadsides) into something with much more biomass.
2. *Carbon capture and storage* (from atmosphere via the vegetation into the soil).

Carbon Sequestration

Vetiver is one of the world's most unique plants in its range of applications and also has to be one of the world's best carbon-sequestering plants. Vetiver's advantage over other candidates is best summarized when referring to its high potential for biomass production in combination with its other characteristics that support a wide range of applications.

From a spatial perspective, the carbon sequestration can be classified as follows:

1. *Increasing biomass in situ* (3–9 kg/m depending on soil and rainfall) where vetiver is planted, it increases biomass on mostly barren land and water surface, for example,
   a. Planted on degraded lands.
   b. Vegetating mining sites.

c. Decontamination of water in artificial wetlands or hydroponic (in wetlands, vetiver doubles kikuyu grass effectiveness in biomass production and N uptake (Vieritz, Truong, Gardner, and Smeal)).

d. Planted on infrastructure slopes, for example, road and railway embankments and gullies that would otherwise be poorly vegetated or paved with any combination of stones and concrete (sand, cement, steel), or around houses, on compounds (around L. Victoria compounds, landing sites, footpaths, and unpaved roads are 2.2% of the study area but contribute 85% to soil loss) (De Meyer and Jean Poesen, 2011). *Photos: Machakos, March 2013, and Nyanza, 2011 (biogardening innovations).*

e. Replacing the inadequate (outdated) engineering structures for soil conservation (e.g., the constructed contours that typically are not so well vegetated). In Ethiopia, crop productivity increased by as much as 30% after replacing structures by semipermeable vetiver hedgerows.

To illustrate, the following pictures show the *before–after* situation on degraded lands (first three projects were done by Habtamu Wubshet, in connection with the Ethiopian NGO *Sustainable Land Use Forum* (SLUF) and with GIZ funding).

May 2003–October 2004 (Ethiopia, Magera).

February 2004–October 2005 (Ethiopia, Kanat).

June 2005–October 2005 (Ethiopia, Kanat).

February 2010–May 2010 (Madagascar).

February 2010–May 2010 (Madagascar).

2. *Increasing biomass and boosting a more biodiverse organic productivity in its vicinity*, as it supports rehabilitation of vegetation through reduced runoff and improved water retention, for example,
    a. Supporting revegetation of mining sites, polluted or otherwise degraded areas
    b. Supporting vegetation (or even farming) on infrastructure slopes, for example, roadsides and gullies
    c. Boosting production (biomass) on farm
3. *Off-site impact*: With a more biodiverse organic productivity and its extensive, deeply penetrating root system (which also penetrates hard layers), there are also these benefits:
    a. Increased groundwater recharge along vetiver's copious and deep roots, which supports increasing organic productivity further down in a catchment; this happens when groundwater resurfaces in wetlands, or is used for irrigation, as well as when better groundwater storage reduces farm crop and infrastructure flood damage (we need to intensify crop production to avoid expansion of farm areas into natural vegetation); let us call this *catchment carbon benefits*.

b. By preventing soil erosion, vetiver prevents soil organic carbon (SOC) from being eroded and oxidized downstream, preventing its addition to the atmosphere.

In Ethiopia, trials with contour hedges on farmland showed that in the third year, runoff (in comparison with that of bare land) was reduced by 88% and soil loss by 97%. In that year (2003), soil loss was reduced to 3.66 t/ha, whereas in the previous 2 years (when the hedge was not yet fully effective), soil losses were, respectively, 86.3 and 48.1 t/ha year. So the difference is some 80 t/ha year even when comparing coffee farms with or without vetiver. Farmers say that when they have 20 year-old hedges, there is virtually no soil loss or water runoff (Kebede, 2004).

c. Improving infiltration and reducing soil loss also mitigate storm and flood damage to infrastructure.

In a country like Kenya, the *mainstream* extension message is still that farmers should use the rigid engineered structures to reduce runoff; these structures ("fanya juu") indeed slow down water and increase infiltration above the structure, but with heavy rains, they accumulate too much water, and the excess water is diverted to the side of the farm, where it runs off in a more concentrated (and devastating) manner. Anyway, after so many years of promoting only this *mainstream* extension message, adoption of the labor-intensive structures is poor, maintenance even poorer. Yet these are the problems (ICRAF, 2004):

- A total of 3.2 million t soil was washed into L. Victoria since 1963, equivalent to one million truckloads in 40 years
- In Nyando river basin, US $42.7 millions worth of soil is lost every year (based on US $12 t^{-1})
- >50% of land was abandoned due to soil nutrient depletion.
- In Kenya, the annually lost soil value is 3–4 times higher than the annual income from tourism!

Add to this the impact that floods and landslides have, every year, on hundreds of thousands in Kenya alone (and that includes dozens of human casualties and loss of livestock and homes).

Some of this could certainly be avoided and had the extension system looked more critically at the content of its message as well as the reasons why adoption is so poor. The writer's own field study on the matter, in Machakos, Busia, Samia, Kerio Valley, Kisii, around L. Elmenteita and L. Naivasha, and other areas, reveals two things: (1) when farmers are given a well-informed choice, they choose vetiver; but (2) most of the formal extension apparatus poorly informs farmers about VS, and their own lack of understanding perpetuates two critical misconceptions, that is, (1) *animals do not eat vetiver* (they do very well but it needs to be known that they only eat fresh leaves; vetiver is not suitable for making hay) and (2) *vetiver is planted on top of the structure* (in that way, it is wasted as it does not form a semipermeable hedge); the hedge needs to be semipermeable to allow water to follow its natural course down a slope—rather than diverting water to the side of the farm where a drainage system is typically absent.

Farmer Participatory Research by CIAT in SE Asia (Vonkasem, Klakhaeng, Watananonta, and H. Howeler), involving more than 850 farmers, revealed that cassava producers preferred the combination of contour ridging with vetiver, among half a dozen of alternative choices. In the first year, the impact of vetiver is negligible, but in the second year, the use of vetiver contour hedges did better, reducing soil loss between 0.4 and 18 t/ha, compared to alternative practices of soil and water conservation.

Other sources report reduced soil loss of between 1.3 and 143 t/ha (Greenfueltech).

There are a number of examples of crops in which the belowground biomass does contribute significantly to SOC (sequestration), including plants such as *Andropogon gayanus* (Fisher, 1994), *Miscanthus* × giganteus (Clifton-Brown et al., 2007; Heaton et al., 2008; Dohleman and Long, 2009; Dondini et al., 2009, 2010; Hillier et al., 2009a,b), *Panicum virgatum* (switchgrass) (Ma et al., 2000; Ma 2001; Liebig et al., 2005; Al-Kaisi and Grote, 2007; Anderson-Teixeira et al., 2009; Collins et al., 2010), and vetiver (*C. zizanioides* L.) (Grimshaw, 2008; Lavania and Lavania, 2009) grasses and even sugarcane (Otto et al., 2009; Galdos et al., 2010).

Once this more-carbon-retaining vegetation finds its new balance, there seem to be no further gains, but taking a closer look at the preceding description, we see further potential:

Ad 1c: Apart from the biomass, the water decontamination is itself highly beneficial for reducing greenhouse gases: it prevents emission of CH_4 and NH_3 and N_2O.

Ad 1d: Replacing conventional engineering (using mined materials) with bioengineering implies a clear reduction of carbon emissions through transport (replacing faraway resources by locally grown resources) and by avoiding mining of materials, requiring vegetation removal.

Ad 1e: The digging up of soil, for example, for contour bunds, is having its own negative impact, as with plowing a field: it causes release of soil carbon and more greenhouse gases that would otherwise be retained in the soil. *Hedge establishment does not require much digging, and being permanent, trapped sediments are kept buried* so they do not otherwise oxidize (similar to minimum tillage).

Ad 2a: Degraded areas have potential for planting vetiver for biofuel, replacing nonrenewable sources that emit CO_2 and more harmful greenhouse gases (CFCs, SO_2, and nitrous oxides).

Ad 2b: On-farm biomass production can be further enhanced considering that the following applies:

- Vetiver mulch can be used to reduce soil temperatures, allowing higher levels of carbon storage (carbon sink) and preventing production of other greenhouse gases.
- Permanent vetiver hedges maintain a better (predator) insect balance, which supports practices of conservation farming or minimum tillage that are favorable to keeping soil carbon locked up.
- Vetiver's deep-rooting system absorbs nitrates from lower levels in a field (which would otherwise end up in oceans, bringing the nitrates (and potassium and phosphates) back into the biomass).

Some examples of projects with a more comprehensive approach are from poor rural and mountainous areas (e.g., in Bali, Indonesia (Booth), where the focus is on poverty reduction and vetiver is taught to school children to protect vegetable gardens on steep slopes and gradually involve the entire community in protecting roadsides, water sources, and homes, and in Ethiopia (Bottenberg, 2012)). More recently, a more futuristic concept of using vetiver in urban areas was presented (Aglietti): it opens the floodgates to possible urban installations where vertical fields of vetiver can play a role cleaning the polluted city environment (see picture).

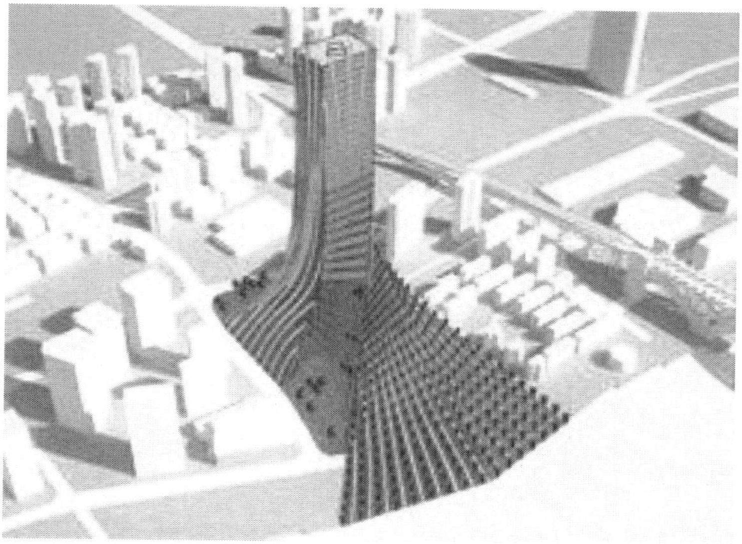

From Sequestration to Carbon Capture and Storage

Carbon capture and storage is a step further than recycling CO_2 as renewable biomass fuels, in which carbon-retaining vegetation is captured in the soil and retained in a more stable, mineral form: SOC or biochar (elemental carbon).

A good overview of the different pools of carbon is provided in "Biological Labile Carbon" (Hoyle, 2006).

Singh in his book "*A Strategy for Sustainable Carbon Sequestration Using Vetiver*" (2011) shows vetiver producing 30.18 t/ha year dry matter in comparison with lemongrass (11.07 t/ha year) and palmarosa (11.76 t/ha year). The carbon sequestrations in biomass for each plant were 15.24, 5.38, and 6.14 t C/ha year, respectively.

Carbon sequestration of vetiver was found to be 15.42 t C/ha year (p. 9, Table 1: 14.46 in the shoot and 0.78 in the root, with dry matter production of 28.62 t/ha year in shoot and 1.56 t/ha year in root) (Singh et al., 2011).

Fisher (1994) states that 100–507 million t C probably is stored in grasslands with deep-rooted grasses. They estimate that there are 35 million ha of these grasslands in South America. According to these data, this stores 3–14 t C/ha.

A. gayanus (vetiver close relative, also a deep-rooting tropical grass) stores 53 t/ha year CO_2 as organic matter (Chomchalov, 2003); he cites this from Noel Vietmeyer who picked this up from a CIAT publication in *Nature* in 1995 on deep-rooted pasture grasses as carbon sinks. This is equivalent to 5.3 kg/m^2. Comparing this *Andropogon* with vetiver "whose roots are much more extensive and deeper," it is estimated that a full-grown clump of vetiver, taking up 0.5 m^2, would absorb at least 2.5 kg CO_2/year. But it could be more, as vetiver has a more massive and deeper root system than *Andropogon* and huge biomass above the ground; the reason for this is its mycorrhizal association.

Taranet et al. (2010) observe that vetiver total biomass yield (above- *and underground*) for four ecotypes is between 84.4 and 114.7 t C/ha. Compared to nonvetiver grass cultivation, vetiver-cultivated areas on average increase soil carbon stock by 16.35 t C/ha, whereas in nonvetiver grass cultivation area, this parameter decreased by 12.3 t C/ha. So vetiver considerably increases carbon sequestration by enhancement of SOC with subsequent reduction of CO_2 in the atmosphere.

Linking to the Kew grass database (Clayton et al., 2006) and vetiver publications in Google Scholar since 2009 (*C. zizanioides*) provides some very interesting information. Lavania and Lavania (2009) conjecture that a vetiver hedge could sequester up to 1 kg C/m year (with a hedge of 0.8 m that would be 12.5 t C/ha year). The figure may be already high, and it still does not include carbon retained in trapped sediments (vetiver's *hidden contribution*: a feature not shared by other plants under consideration—in the section "Biofuel from Vetiver Grown as a Field Crop" a similar estimate is made).

More research is needed to refine these for specific soil types, climates, and slopes, but it is still clear that vetiver's potential is extraordinary.

Currently, there is no simple way to measure carbon uptake or conversion of vetiver:

- Lack of site uniformity.
- Even in well-documented cases, *hard numbers* can be specious.

Only counting aerial biomass gives a serious underestimate because of vetiver's very high underground biomass to aboveground biomass ratio:

- Vetiver has a regular foliar turnover (harvested or not).
- Not trimming plants is (in most applications) also poor management.

There are specific research questions related to different applications:

1. Vetiver wetlands: how much ammonia (NH_3) can be recovered?
   a. Treatment of raw sewage (black water) and/or septic tank overflow, compared to conventional water treatment method(s)

b. In slums, rural areas, and refugee camps where the method replaces the *no-sanitation* option
c. Treating waste dump leachate
2. Vetiver ponds (along boundaries and on pontoons): idem?
3. *Reinforcing carbon capture in reforestation* on poor (mostly degraded, inferior quality) soils: how much is the carbon capture potential?

VS contribution to carbon capture and storage in soils requires assessment of different pools of carbon in the soil and different ways in which VS contributes to increasing those pools and avoiding reduction in others. Literature indicates that root-derived soil carbon accumulation is being estimated by scientific studies across the globe under both grassland and forests, either in the tropics or in temperate areas. In most cases, VS is not applied in grassland or forestry settings, but on-farm, for infrastructure protection and for decontamination of polluted soils and wastewater.

For carbon to be truly sequestered, that is, captured and stored in the soil, it must be transformed to mineralized carbon aided by the microbial activity in the soil associated with the root zones. One of the reasons for high vetiver plant vigor is a result of its mycorrhizal association. This association is one of the reasons why vetiver hedgerows produce such high amounts of biomass. Some of that biomass can result in high amounts of organic carbon added to the sequestered soil carbon pool, both from decomposition of root biomass and dissolved organic carbon exudates from roots. In addition, aboveground biomass can be converted into biochar (Grimshaw, 2011) (VS Biomass Fuel, 2011) (Institutional cook stoves) (Alternative fuel for Haiti). Both produce energy for cooking or power generation and an elemental carbon residue, which when buried in soils lasts many times longer than SOC and acts to retain soil water and nutrients, increasing plant productivity (Vetiver/biochar_international_team, 2013).

Under cultivation systems, a very significant part of the labile carbon fraction is lost to oxidation (removed with the crop). So vetiver hedgerows, unless used for mulching as well, are not going to have much direct impact on the other 90%–92% of the field within a cropping system. Considering the *VS* within the context of good management of soil fertility and increasing organic matter in the rest of the field, the hedge is only compensating for a fraction of the much larger fraction being lost soil carbon oxidation in agricultural soils between the hedges. Practices such as no-till conservation farming as being introduced to southern African countries should overcome some of the latter problem and underscore the need for other appropriate farm practices to be associated with VS.

We need to learn a lot more about vetiver's ability to sequester carbon under different climate and soil conditions. Scientists in China, India, and elsewhere would be advised to investigate vetiver's potential. US scientists are working on *Miscanthus sp.* and switchgrass because they are known to be high-biomass North American grasses. We should be doing the same for vetiver throughout the tropics and subtropics.

Biological nitrogen fixation can greatly increase in soils with high organic carbon, along with P and K availability, thus reducing the need for synthetic (fossil fuel requiring) fertilizer and all the fossil carbon its production generates. The ability of vetiver to retain and enhance high-carbon soils in a stabilized agricultural system is a qualitative and quantitative advantage, regardless of what farming system is used. If biochar is incorporated, it is best utilized on-site.

Biochar

Biochar (old charcoal, a highly concentrated carbon compound) has recently received a lot of attention. The economist (Biochar Could Enrich Soils and Cut Greenhouse Gases As Well, 2009) reported about the virtue of various ways of making biochar from different raw materials and how big the benefits would be. Two major issues came to light:

1. *Beneficial effects*, when biochar is properly incorporated into the soil, would include the following:
   a. Remains in the soil for a long time, improving soil fertility and reducing leaching of nitrate, phosphate, and potassium to groundwater.

b. Biochar-containing soils release less methane and nitrous oxide than untreated soils, probably due to physical adsorption and outcompeting plant roots for ammonium, inhibiting nitrification. As methane and nitrous oxide are more potent greenhouse gases than CO_2, this effect should help in combating global warming considerably.
c. The biochar-making process creates beneficial by-products: heat, syngas or fuel, and heavy oil, which can be used as an energy source.

2. *Major concerns* were also raised:
   a. Use of existing farmland or virgin land for biochar crops.
   b. Tillage would release CO_2 and methane.
   c. Growing switchgrass for biochar may do more harm than good; garden waste and forestry offcuts are better.

More recent research on biochar is also highlighting some concerns, see http://www.wageningenur.nl/en/Expertise-Services/Research-Institutes/alterra/Facilities-Products/Laboratories-Environmental-Sciences-Group/Soil-Physics-Laboratory/Research/Biochar.htm, it would be worthwhile to await results from that research.

Vetiver could be the best option, having all the benefits and none of the aforementioned concerns:

1. Producing very high biomass, up to 130 t/ha year under subtropical conditions and probably higher in the wet tropics, with a high carbon content of 45%.
2. Tolerating a wide range of climatic conditions: tropical (55°C) to subtemperate (−14°C as long as the ground is not frozen), drought, flood, and prolonged submergence (3 months)
3. Growing vetiver grass does not compete for farmed land as it can be grown on wasteland or highly degraded agricultural lands or (tolerating heavy metals) in mine rehabilitation areas or (tolerating pollutants) in artificial wetlands and ponds to treat and dispose of industrial, municipal, and domestic wastewater.
4. As a perennial crop, there is only minimal release of greenhouse gases due to tillage.

Capturing Dissolved Ammonium from Wastewater

As with biogas, also the use of vetiver for treatment of wastewater deserves more serious attention. When capturing dissolved ammonium from wastewater (to outcompete nitrifying bacteria that are a major source of N_2O) and pumping oxygen from roots into the water, it creates a habitat for methane-oxidizing bacteria. For a reference on this, see Singh and Kalha (2009).

Conclusion: More Research Is Needed

Much more fieldwork is needed to fully quantify the benefits of vetiver. Technologies, methodologies, and algorithms need to be improved to allow measurement of carbon sequestration and carbon capture and storage by vetiver. This requires observations under different climate and soil conditions, as well as different applications of VSs. India and China are well placed for this, having some good research institutions and a wide range of growing conditions.

Here are more general questions for research:

1. Plug in vetiver in overall data structure developing on grasses; for example, there is an increasing amount of work done on carbon sequestration, but the focus is on *Miscanthus*. Comparative research could be an efficient way to establish the relative contribution of vetiver:
   a. Broader research on grasses (versus trees)
   b. Observational and experimental trial comparisons among candidate species and vetiver (make practical judgments on utility of vetiver, using physiological and morphological

TABLE 16.2

CDM Technology for Which VS Can Be Relevant

Refs.	Methodologies Title	Options with VS, Research
AMS- I.A.	Electricity generation by the user	Vetiver biomass for biofuel: small furnaces?
AMS-4I.D.	Grid connected renewable electricity generation	Idem—large scale (see Boucard's example)
AMS-III.A.	Agriculture: keeping the soil, keeping carbon in place	SWC, mulching, reducing oxidation in soil
AMS-III.H.	Nutrient recovery in wastewater treatment	VS wetlands as secondary treatment of household wastewater
AMS-III.I.	Avoidance of methane production in wastewater treatment through replacement of anaerobic lagoons by aerobic systems	Idem?

indicators, e.g., energy efficiency, nutrient partitioning, biomass production and turnover, and water infiltration).

2. Adding vetiver to (comparative) research scenarios already underway with other grasses (fairly inexpensive, short time frame, practical):
   a. What are CH_4, N_2O, NH_3, and SO_2 absorption quantities for vetiver (compared to other plants), in different conditions (using vetiver for different methods of (1) wastewater treatment; (2) infrastructure, e.g., embankments of roads, railways, rivers, and canals; and (3) on-farm uses including mulch and fodder)?
   b. Root nutrient absorption and prevention of eutrophication as a result of improved soil microbiology.
   c. CO_2 mineralization, for different conditions: quantities for roots and leaves.

Under current-day advanced remote monitoring procedures, it might also be feasible to provide carbon credit income to other users of vetiver planting it for land protection, slope stabilization, etc.

Two new methodology elements are being assessed under the Verified Carbon Standard (VCS) methodology approval process:

1. Methodology for Sustainable Grassland Management (SGM) (Li, 2011)
2. Methodology for Soil Carbon (The_Earth_Partners, 2012).

Also see "Methodologies Under Development" (Vetiver/biochar_international_team, 2013) and "What is a Methodology?" about the VCS approval process (Carbon_Verified_Standards) (Table 16.2).

Jim Smyle (TVNI chairman, personal communication) suggests it would be very useful if we had actual figures for vetiver and proposes a (relatively) simple research protocol for sampling SOC in the top 25–50 cm of areas with long-term (>5–10 years) presence of vetiver hedgerows and adjacent sites for purposes of comparison. You basically only need a good methodology for selecting sites, a laboratory-grade scale, and an oven to cook organic matter out of the samples. With a simple protocol and awards for researchers, data from around the world (different conditions) could be collected.

Everything counts, but this is crucial:

- *Emphasizing the rooting depth*, with the assumption that (bioavailable) carbon from roots *below plow-zone depth* (where most outgassing occurs, regardless whether you plow or not) became immobilized and perhaps eventually mineralized. This gives vetiver an enormous *direct* edge over other grasses that have been proposed for mass plantings. Taking an average plow zone of 20 cm deep, comparing that with a depth of some 1 m (80% of the roots of vetiver are in the first meter, still another 20% beyond that depth), one can see the significance of vetiver's rooting system.

- The ability for new humus to form in stabilized debris behind hedges.
- Vetiver grass has high biomass both below- and aboveground. There has been virtually no research work carried out to quantify how much carbon is sequestered as root and mycorrhizal biomass and under what growth and climatic conditions.
- There are many genotypes of *C. zizanioides*. Research in Taiwan (Wang) indicates differences as to where different genotypes store their carbon; some store more in the roots, while others in the leaves and stem. Plant mass varies too. Fresh weight of plants grown from individual seeds in 1 year varied from 77 to 8.3 kg/plant. Other results show that vetiver is an effective CO_2 absorber in twice the CO_2 concentration level.
- *Comparisons with other plants*: If the vetiver hedge *retains 80 t/ha year of topsoil* that would otherwise be lost, this is perhaps the most important bonus of VS, sequestration-wise, keeping carbon in its place!

Several of the calculations are speculative, but we hope that research will prove that this ubiquitous plant will once again come up with trumps in this important area.

Taranet et al. (2010) (cited earlier) show a difference of 28.5 t C/ha between vetiver-cultivated areas and noncultivated (degraded) vetiver grass areas, but it appears the study considers a pure stand of vetiver and the typical few hedges in a field of annual food crops.

Suppose that in the more typical context of on-farm soil and weather conservation only 5% of the field is used for vetiver hedges, then 1.5 t C/ha would be prevented from being lost (or added to soil carbon), equivalent to sequestering 5.5 t CO_2/ha year *by the vegetation of vetiver hedges alone*.

Taking another angle, preventing soil loss of 80 t/ha (Kebede, 2004) and taking this as equivalent to 5% of the topsoil carbon (from the soil top 20 cm), 5% of 2400 t/ha (Lieu, 2007) gives 96 t SOC/ha year kept in place, preventing emission of 352 CO_2/ha year. But with this estimation, it is assumed that all the lost soil is ending up in a way that it releases all its carbon (ignoring any deposit of lost soil in a way that it keeps its carbon), and in not all areas is the soil loss as high as 80 t/ha. And preventing soil loss with VS is rarely attributed to only the planting of vetiver hedges: it goes hand in hand with other measures (as part of a comprehensive zero-till package or organic farming methods).

If combined with other practices, it can achieve 90% reduction of soil loss (Doanh et al., n.d.), but taking together a number of studies from across the world, the soil loss reduction is between 20% and 95%, and it matters of course whether the study reaches the third year after planting; in some of these studies, also reduction of water runoff is mentioned, between 40% and 80% (TVNI) (Abate and Simane, 2010). In cases where indeed the study measures soil loss reduction in the third year, it is typically found to be over 50%.

In Kenya, soil loss on farms is about 30 t ha^{-1} in a farmed area of roughly 10 million ha (Cohen et al., 2005). If all was adequately protected against erosion and assuming this would prevent soil loss of 30 t CO_2/ha year, Kenya would prevent soil loss of 300 million t CO_2 ha^{-1} $year^{-1}$ (compared with that of the United States, its agricultural sector emits annually 700 million t CO_2).

The correct farming techniques can sequester carbon into the soil and reverse the 25% of Australia's greenhouse gases created by agriculture. The processes to increase soil carbon can be divided into three steps:

1. Use plants to grow soil carbon.
2. Use microorganisms to convert soil carbon into stable forms.
3. Avoid farming techniques that destroy soil carbon (Lieu, 2007).

VS contributes to that in two ways: growing soil carbon and avoiding soil carbon loss.

Nyando sediment plume in Lake Victoria, Feb 2000

Nyando river basin: US \$42.7 millions worth of soil is lost every year (based on US \$12 t^{-1}, it is 3.6 million t soil/year from 365,000 ha, or an average of 10 t/ha) (World Agroforestry Centre, 2006).

Vetiver's carbon contributions are just bonus features to its already enormous suite of positive benefits, providing individuals with flexible tools and mobilizing local resources to protect themselves from climatic fluctuations and other natural disasters and in addition its other benefits (for pest control, fodder or materials for handicraft). This help-yourself quality is more important than sequestration, as this is the quality that convinces the end user.

Appendix: Energy Units

Quantities

- 1.0 joule (J) = 1 Newton applied over a distance of 1 m (= 1kg m^2/s^2)
- 1.0 J = 0.239 calories (cal)
- 1.0 cal = 4.187 J
- 1.0 gigajoule (GJ) = 10^9 J = 0.948 million Btu = 239 million cal = 278 kWh
- 1.0 British thermal unit (Btu) = 1055 J (1.055 kJ)
- 1.0 Quad = 1 quadrillion Btu (10^{15} Btu) = 1.055 exajoules (EJ), or approximately 172 million barrels of oil equivalent (BOE)
- 1000 Btu/lb = 2.33 gigajoules per tonne (GJ/t)
- 1000 Btu/US gallon = 0.279 megajoules per liter (MJ/L)

Power

- 1.0 W = 1.0 J/s = 3.413 Btu/h
- 1.0 kilowatt (kW) = 3413 Btu/h = 1.341 horsepower
- 1.0 kilowatt-hour (kWh) = 3.6 MJ = 3413 Btu
- 1.0 horsepower (hp) = 550 foot-pounds per second = 2545 Btu/h = 745.7 W= 0.746 kW

Energy Costs

- $1.00 per million Btu = $0.948/GJ
- $1.00/GJ = $1.055 per million Btu

Some Common Units of Measure

- 1.0 US ton (short ton) = 2000 pounds (lb)
- 1.0 imperial ton (long ton or shipping ton) = 2240 lb
- 1.0 metric tonne (tonne) = 1000 kilograms = 2205 lb
- 1.0 US gallon = 3.79 L = 0.833 imperial gallon
- 1.0 imperial gallon = 4.55 L = 1.20 US gallon
- 1.0 L = 0.264 US gallon = 0.220 imperial gallon
- 1.0 US bushel = 0.0352 m^3 = 0.97 UK bushel = 56 lb, 25 kg (corn or sorghum) = 60 lb, 27 kg (wheat or soybeans) = 40 lb, 18 kg (barley)

Areas and Crop Yields

- 1.0 ha = 10,000 m^2 (an area 100 m × 100 m, or 328 × 328 ft) = 2.47 acres
- 1.0 km^2 = 100 ha = 247 acres
- 1.0 acre = 0.405 ha
- 1.0 US ton/acre = 2.24 t/ha
- 1 metric tonne/ha = 0.446 ton/acre
- 100 g/m^2 = 1.0 tonne/ha= 892 lb/acre
  - For example, a *target* bioenergy crop yield might be 5.0 US tons/acre (10,000 lb/acre) = 11.2 tonne/ha (1120 g/m^2).

Biomass Energy

- *Cord*: A stack of wood 128 cubic feet (3.62 m^3); its standard dimensions are 4 × 4 × 8 ft, including air space and bark. One cord contains approx. 1.2 US tons (oven-dry) = 2400 lb = 1089 kg
  - 1.0 metric tonne *wood* = 1.4 m^3 (solid wood, not stacked)
  - Energy content of *wood fuel* (high [gross] heating value [HHV], bone dry) = 18–22 GJ/t (7600–9600 Btu/lb)
  - Energy content of *wood fuel* (air-dry, 20% moisture) = about 15 GJ/t (6400 Btu/lb)
- Energy content of *agricultural residues* (range due to moisture content) = 10–17 GJ/t (4300–7300 Btu/lb)
- Metric tonne *charcoal* = 30 GJ (=12,800 Btu/lb) (but is usually derived from 6 to 12 t air-dry wood, i.e., 90–180 GJ of original energy content)

- Metric tonne *ethanol* = 7.94 petroleum barrels = 1262 L
  - Ethanol energy content (lower (net) heating value [LHV]) = 11,500 Btu/lb = 75,700 Btu/gallon = 26.7 GJ/t = 21.1 MJ/L. HHV for ethanol = 84,000 Btu/gallon = 89 MJ/gallon = 23.4 MJ/L
  - Ethanol density (average) = 0.79 g/mL (=metric tonnes/m^3)
- Metric tonne *biodiesel* = 37.8 GJ (33.3–35.7 MJ/L)
  - Biodiesel density (average) = 0.88 g/mL (=metric tonnes/m^3)

Fossil Fuels

- *Barrel of oil* equivalent (BOE) = approx. 6.1 GJ (5.8 million Btu), equivalent to 1700 kWh. *Petroleum barrel* is a liquid measure equal to 42 US gallons (35 imperial gallons or 159 L); about 7.2 barrels of oil is equivalent to 1 tonne of oil (metric) = 42–45 GJ.
- *Gasoline*: US gallon = 115,000 Btu = 121 MJ = 32 MJ/L (LHV). HHV = 125,000 Btu/gallon = 132 MJ/gallon = 35 MJ/L
  - Metric tonne gasoline = 8.53 barrels = 1356 L = 43.5 GJ/t (LHV); 47.3 GJ/t (HHV)
  - Gasoline density (average) = 0.73 g/mL (=metric tonnes/m^3)
- *Petro-diesel* = 130,500 Btu/gallon (36.4 MJ/L or 42.8 GJ/t)
  - Petro-diesel density (average) = 0.84 g/mL (=metric tonnes/m^3)
- Note that the energy content (heating value) of petroleum products per unit mass is fairly constant, but their density differs significantly—hence, the energy content of a liter, gallon, etc. varies between gasoline, diesel, and kerosene.
- Metric tonne *coal* = 27–30 GJ (bituminous/anthracite); 15–19 GJ (lignite/subbituminous) (the preceding ranges are equivalent to 11,500–13,000 Btu/lb and 6,500–8,200 Btu/lb).
  - Note that the energy content (heating value) per unit mass varies greatly between different *ranks* of coal. *Typical* coal (rank not specified) usually means bituminous coal, the most common fuel for power plants (27 GJ/t).
- *Natural gas*: HHV = 1027 Btu/ft^3 = 38.3 MJ/m^3; LHV = 930 Btu/ft^3 = 34.6 MJ/m^3
  - Therm (used for natural gas, methane) = 100,000 Btu (=105.5 MJ)

Carbon Content of Fossil Fuels and Bioenergy Feedstocks

- *Coal* (average) = 25.4 metric tonnes carbon per terajoule (TJ)
  - 1.0 metric tonne *coal* = 746 kg carbon
- Oil (average) = 19.9 metric tonnes carbon/TJ
- US gallon *gasoline* (0.833 imperial gallon, 3.79 L) = 2.42 kg carbon
- 1.0 US gallon *diesel/fuel oil* (0.833 imperial gallon, 3.79 L) = 2.77 kg carbon
- 1.0 *Natural gas (methane)* = 14.4 metric tonnes carbon/TJ
- 1.0 cubic meter *natural gas (methane)* = 0.49 kg carbon
- Carbon content of *bioenergy feedstocks*: approx. 50% for woody crops or wood waste; approx. 45% for graminaceous (grass) crops or agricultural residues.

ACKNOWLEDGMENTS

Many thanks to Jim Smyle, Mark Dafforn, Umesh Lavania, Guéric Boucard, Paul Truong, Richard Grimshaw, M. P. Singh, Thomas Goreau, and many others who have assisted the author in the preparation of this article. Also a hug for my family who endured my absence from family life in weekends and evenings.

REFERENCES

Abate, H. and Simane, B. (2010). Multiple benefits of the Vetiver System and its environmental application in Ethiopia. Retrieved August 3, 2014 from TVNI: http://www.vetiver.org/ETH_ERO.pdf.

Aglietti, E. (n.d.). VetiVertical City. Retrieved April 21, 2013 from Architect, the magazine of the American Institute of Architects: http://www.architectmagazine.com/projects/view/vetivertical-city/1659/.

ALSTOM. (2011). ALSTOM announces successful results of mountaineer carbon capture and storage (CCS) project. Retrieved March 31, 2013 from NRDC: http://www.alstom.com/press-centre/2011/5/press-releases-xn-52/.

Alternative fuel for Haiti. (n.d.). Retrieved March 27, 2013 from http://haitireconstruction.ning.com/page/alternative-fuel-for-haiti.

Al-Kaisi, M. and J.B. Grote. (2007). Cropping systems effects on improving soil carbon stocks of exposed subsoil. *SSSAJ*, DOI: 10.2136/sssaj2006.0200.

Anderson-Teixeira, K.J., Davis, S.C., Masters, M.D., and Delucia, E.H. (2009). Changes in soil organic carbon under biofuel crops. *GCB Bioenergy*, 1, 75–96.

Biochar Could Enrich Soils and Cut Greenhouse Gases As Well. (August 29, 2009). *The Economist Magazine*, p. 69 (Science and Technology section).

Booth, D. (n.d.). Sustainable Agricultural Development. Retrieved April 21, 2013 from East Bali Poverty Project: http://eastbalipovertyproject.org/what-do-we-do/projects/sustainable-agricultural-development/.

Bottenberg, C. (2012). Constructed Vetiver Wetlands in Jimma, Ethiopia. (University of Colorado at Boulder). Retrieved April 21, 2013 from http://www.vetiver.org/ET_coffee%20ethiopia.pdf.

Boucard, G. (2009). Coal-fired furnaces for electricity. Personal communication with Elise Pinners.

Carbon_Verified_Standards. (n.d.). What is a Methodology? (Carbon_Verified_Standards). Retrieved March 27, 2013 from http://v-c-s.org/methodologies/what-methodology.

Chomchalov, N. (January 2003). The role of vetiver in controlling water quantity and treating water quality: An overview with special reference to Thailand. *Australian Journal of Technology*, 6(3), 145–161 from http://vetiver.com/THN_vetiver_water.pdf.

Chrysopogon zizanioides. (n.d.). Retrieved May 10, 2012 from Google Scholar: http://scholar.google.com/scholar?hl=en&q=Chrysopogon+zizanioides&as_sdt=0%2C9&as_ylo=2009&as_vis=0.

Clifton-Brown, J.C., J. Breuer, M.B. Jones. (2007). Carbon mitigation by the energy crop, Miscanthus. DOI: 10.1111/j.1365-2486.2007.01438.x.

Cohen, M. J., Brown, M. T., and Shepherd, K. D. (December, 2005). Estimating the environmental costs of soil erosion at multiple scales in Kenya using emergy synthesis. Retrieved April 22, 2013 from http://www.worldagroforestry.org/downloads/publications/PDFS/ja06026.pdf#Rbib90.

Collins, H.P., Smith, J.L., Fransen, S., et al. (2010). Carbon sequestration under irrigated switchgrass (*Panicum virgatum L.*) production. *SSSAJ*, Doi: 10.2136/sssaj2010.0020.

Clayton, W. D., Vorontsova, M., Harman, K., and Williamson, H. (2006). GrassBase—The Online World Grass Flora. (Kew Royal Botanic Gardens) Retrieved March 31, 2013 from http://www.kew.org/data/grasses-db/www/imp02311.htm.

De Meyer, A. and Jean Poesen, M. I. (2011). Soil erosion loss in tropical villages. *Catena*, 84, 89–98. Retrieved April 20, 2013 from https://lirias.kuleuven.be/handle/123456789/298749.

Doanh, L. Q., Loan, T., Ngoan, T., Tuan, H., Hung, N., Chien, D. et al. (n.d.). Integration of Northern Upland Program with Farmers' Participatory Research for Sustainable Cassava Production in North Vietnam. (CIAT, Ed.).

Dohleman, F.G. and S.P. Long. (2009). American Society of Plant Biologists. More productive than maize in the midwest: How does Miscanthus do it? Retrieved August 3, 2014 from http://www.plantphysiol.org/content/150/4/2104.full.

Dondini, M., van Groeningen, K., Del Galdo, I., and Jones, M.B. (2009). Carbon sequestration under *Miscanthus*: A study of ^{13}C distribution in soil aggregates. DOI: 10.1111/j.1757-1707.2009.01025.x.

Dondini, M., Hastings, A., Saiz, G., Jones, M.B., and Smith, P. (2010). The potential of *Miscanthus* to sequester carbon in soils: Comparing field measurements in Carlow, Ireland to model predictions. DOI: 10.1111/j.1757-1707.2010.01033.x.

Fisher, M. J., Rao, I. M., Ayarza, M. A., et al. (1994). Carbon storage by introduced deep-rooted grasses in the South American savannas. *Nature*, 371, 236–238.

Forti, M. (n.d.). *Biomasse*. Retrieved March 27, 2013 from Vetiver Italia: http://www.vetiveritalia.it/biomasse/.

Galdos, M.V., Cerri, C.C., Lal, R., Bernoux, M., Feigl, B., and Cerri, C.E.P. (2010). Net greenhouse gas fluxes in Brazilian ethanol production systems. *GCB Bioenergy*, 2, 37–44.

Greenfueltech. (n.d.). Vetiver Technology. Retrieved April 20, 2013 from Greenfueltech Fuel for the future: http://www.greenfueltech.net/vetiver.htm.

Grimshaw, R. (2008). From The Vetiver Network International. Retrieved August 3, 2014 from http://www.vetiver.org/TVN_INDIA_1stWORKSHOP_PROCEEDINGS/Chapter 1-1.pdf.

Grimshaw, R. (December 13, 2011). Vetiver stoves and biofuel. Retrieved March 27, 2013 from The Vetiver Network International: http://vetivernetinternational.blogspot.com/2011/12/vetiver-stoves-and-biofuel.html.

Grimshaw, R. and Truong, P. (2007). *Impact of the Vetiver System on Climate Change.* TVNI. TVNI.

Haiti Reconstruction International combats land desertification with vetiver grass hedgerows. (n.d.). Retrieved March 27, 2013 from http://haitireconstruction.ning.com/.

Heaton, E.A., F.G. Dohleman, S.P. Long. (2008). Meeting US biofuel goals with less land: the potential of Miscanthus. *Global Change Biology* 14, 1–15.

Hillier, J., Whittaker, C., Dailey, G. et al. (2009a). Greenhouse gas emissions from four bioenergy crops in England and Wales: Integrating spatial estimates of yield and soil carbon balance in life cycle analyses. *Global Change Biology Bioenergy*, 1, 267–281.

Hillier, J., Hawes, C., Squire, G., et al. (2009b). The carbon footprints of food crop production. *International Journal of Agricultural Sustainability*, 7, 107–118.

Hoyle, F. D. (2006). Biological labile carbon. Retrieved March 27, 2013 from www.soilquality.org.au: https://s3.amazonaws.com/soilquality-production/fact_sheets/4/original/Biol_-_Labile_Carbon.pdf.

Kebede, T. T. (2004). Retrieved April 2013, 2013 from http://www.vetiver.org/ETH_WORKSHOP_09/ETH_A7a.pdf.

Kell, D. (2011). Breeding crop plants with deep roots: Their role in sustainable carbon, nutrient and water sequestration. *Annals of Botany*, 108, 407–418.

La Plantation Bemasoandro. (2010). Lavaka stabilization for railway protection. (Y. Coppin, Ed.) Madagascar.

Lavania, U. and Lavania, S. (2009). Sequestration of atmospheric carbon into subsoil horizons through deep-rooted grasses—Vetiver grass model. *Current Science*, 97, 618–619.

Li, Y. E. (June 20, 2011). Methodology for Sustainable Grassland Management. (C. Institute of Environment and Sustainable Development in Agriculture, Producer). Retrieved March 27, 2013 from Verified Carbon Standard: http://v-c-s.org/methodologies/methodology-sustainable-grassland-management-sgm.

Liebig, M.A., Johnson, H.A., Hanson, J.D., and Frank, A.B. (2005). Soil carbon under switch grass stands and cultivated cropland. *Biomass and Bioenergy*, 28, 347–354.

Lieu, A. (March 21–22, 2007). Organics and soil carbon: Increasing soil carbon, crop productivity and farm profitability. Retrieved April 22, 2013 from http://www.amazingcarbon.com/Leu%20-%20Organics%20and%20soil%20carbon.pdf.

Long (1994). From http://www.aces.uiuc.edu/DSI/MASGC.pdf.

Lunds_University. (2003). Toxicity of wastewater generated from gasification of woodchips. J.nr. 511 66/0 1 -00 08, 51.

Ma, Z., Wood, C.W., and Bransby, D.I. (2000). Soil management impacts on soil carbon sequestration by switchgrass. *Biomass and Bioenergy*, 18, 469–477.

Ma, Z. (2001). Switchgrass (*Panicum virgatum*) for biofuel production. *Farm Energy*, Retrieved August 3, 2014 from https://www.extension.org/pages/26635/switchgrass-panicum-virgatum-for-biofuel-production#.U95UDUjdj7s.

Otto, R., Trivelin, P.C.O., Franco, H.C.J., Faroni, C.E., and Vitti, A.C. (2009). Root system distribution of sugar cane as related to nitrogen fertilization, evaluated by two methods: Monolith and probes. *RevistaBrasileira de Ciência do Solo*, 33, 601–611.

Pinners, E. (2011). Some thoughts on the potential contributions of Vetiver System in mitigating Climate Change. *Proceedings of the 5th International Conference on Vetiver (ICV-5).* Lucknow, India: The Vetiver Network International/CIMAP.

Singh, M. (July 2011). A strategy for sustainable carbon sequestration using Vetiver. Retrieved August 3, 2014 from www.cmmacs.ernet.in/cmmacs/Publications/proj_docs/PDCM1101.pdf.

Singh, M., Guleria, N., Rao, E. P., and Goswami, P. (2011). A strategy for sustainable carbon sequestration using Vetiver (*Vetiveria zizanioides* (L.)): A quantitative assessment over India, Project Document under the CSIR Network Project Integrated Analysis for Impact, Mitigation and Sustainability (IAIMS), 97, 618–619.

Singh, M. and Kalha, G. (August 2009). Evaluating and quantifying Carbon Sequestering, GHG Emission Reduction, through a community development project in Punjab, India. *Earthizenz.*

Standards, V. C. (n.d.). Methodologies under development. Retrieved March 27, 2013 from http://v-c-s.org/methodologies/in-development.

Taranet, P., Wattanaprapat, K., and Meesing, I. (2010). Carbon sequestration and carbon dioxide emission. Retrieved March 28, 2013 from https://docs.google.com/file/d/0B3E8MMCy36wZMTAxMGYzZTEtNmQ5ZC00OWZlLWE3OTctYzdkZmU4ZTlkYWEy/edit?hl=en_US.

The National Research Council (NRC). (1993). *Vetiver Grass—A Thin Green Line Against Erosion.* Washington, DC: The National Academic Press.

The_Earth_Partners. (November 16, 2012). Soil carbon quantification methodology. Retrieved March 27, 2013 from http://v-c-s.org/methodologies/VM0021.

Truong, P., Van, T. T., and Pinners, E. (2008). *Vetiver System Applications: Technical Reference Manual* (2nd edn. (color) ed.). Leesburg, VA: The Vetiver Network International (TVNI).

TVNI. (n.d.). Vetiver grass hedgerow technology. (TVNI, Producer) Retrieved April 7, 2013 from http://www.ipcp.org.br/IPCP-MA/Projects/CNPqSolos/VetiverDocs/TecStuff/VETIVER%20GRASS%20HEDGE%20ROW%20TECHNOLOGY.htm.

University_of_Florida. (January 18, 2012). *Miscanthus* yields high dry matter across range of conditions. Retrieved May 10, 2013 from The Crop Site: http://www.thecropsite.com/articles/1153/miscanthus-yields-high-dry-matter-across-range-of-conditions.

Vetiver publications in Google scholar since 2009. (n.d.). From http://scholar.google.com/scholar?hl=en&q=Chrysopogon+zizanioides&as_sdt=0%2C9&as_ylo=2009&as_vis=0.

Vetiver System biomass fuel. (November 21, 2011). Retrieved March 27, 2013 from The Vetiver Network International: http://vetivernetinternational.blogspot.com/2011/11/vetiver-system-biomass-fuel.html.

Vetiver/biochar_international_team. (January 28, 2013). Vetiver/biochar solution. Retrieved March 27, 2013 from http://www.vetiver.org/TVN_Vetiver_biochar_finalL.pdf.

Vieritz, A., Truong, P., Gardner, T., and Smeal, C. (n.d.). Modelling Monto vetiver growth and nutrient uptake for effluent irrigation schemes. Retrieved April 20, 2013 from http://prvn.rdpb.go.th/files/2–03t.pdf.

Vonkasem, W., Klakhaeng, K., Watananonta, W., and H.Howeler, R. (n.d.). The use of vetiver for soil erosion prevention in cassava fields in Thailand. Retrieved April 20, 2013 from CIAT library: http://ciat-library.ciat.cgiar.org/articulos_ciat/THE_USE_OF_VETIVER_FOR_SOIL.pdf.

Wang, Y.-W. (n.d.). (Department of Agronomy, National Taiwan University) Retrieved April 21, 2013 from The root extension rate of vetiver under different temperature treatments.

Wikipedia. (n.d.). Wood gas. Retrieved March 27, 2013 from http://en.wikipedia.org/wiki/Wood_gas.

World Agroforestry Centre. (2006). Improved land management in the Lake Victoria Basin. Retrieved April 22, 2013 from http://www.worldagroforestry.org/downloads/publications/PDFs/OP14439.PDF#page=89.

17

Basalt Powder Restores Soil Fertility and Greatly Accelerates Tree Growth on Impoverished Tropical Soils in Panama

Thomas J. Goreau, Felix Lufkin, Carlos A. Arango, Gabriel Despaigne-Matchett, Gabriel Despaigne-Ceballos, Roque Solis, Marina Goreau, and Joanna Campe

CONTENTS

Introduction

The major cause of low productivity of agricultural crops, tree plantations, pastures, and natural habitats in the tropics is the extremely low fertility of most tropical soils. Low fertility results from the extremely rapid soil weathering, decomposition, erosion, and leaching of nutrients caused by high temperatures and high rainfall. This results in extremely acidic soils in which plant growth rates are far below their maximum potential due to lack of major soil nutrients (nitrogen, phosphorus, potassium), minor essential soil nutrients (magnesium, calcium, and trace essential metals such as zinc, manganese, copper, and molybdenum), high acidity, aluminum toxicity, and insufficient organic matter to retain nutrients and water in the soil. Addressing these problems through addition of chemical fertilizers is often too expensive to be practical, and unbalanced nutrient ratios in most chemical fertilizers frequently result in poor plant responses due to extremely inefficient uptake of fertilizers. Chemical fertilizers typically contain only N, P, and K, so if the other essential elements are missing, plants will be unable to utilize them, and most will be wasted.

A simple and cost-effective way to increase soil fertility and plant growth rates on impoverished tropical soils should have revolutionary impacts by increasing productivity and reducing the need for further deforestation in the warm climate zone where most of the world's people live. In this report, we present the results of a method that is capable of producing precisely such results by increasing growth rates of trees, grass, shrubs, and crops on severely impoverished soils by more than five times using locally available basalt rock powder as a natural mineral supplement.

Addition of rock powder contains the entire spectrum of elements in slow-release form, although the concentration range of elements and their release rates vary, depending on chemical weathering rates of the mineral phases containing them. Rock powders mimic the natural release of minerals from weathering of basement rock but speed it up by presenting fresh unweathered mineral surfaces and by greatly increasing their surface area for weathering by water and dissolved carbonic acid (CO_2). Their release rate is more closely matched to plant uptake needs, being highest when it is wet and warm, they are far less likely to be leached out of the soils as happens when large amounts of soluble chemical minerals are used, and in general, less fossil fuel energy is used to produce them. While much work has been done on the effects of rock powders on seedling and crop growth (Chapters 4 and 22), there is little or no data available on the long-term growth of trees in rock powders under field conditions, especially in tropical conditions. This chapter provides what are believed to be the first such data.

Materials and Methods

This project was conducted on the grounds of Cantera Pedrera, a basalt rock quarry located near the border of the provinces of Cocle and Panama in the Republic of Panama, Central America. This region is regarded as having some of the poorest soils in the country, being classified as nonarable with very severe limitations due to extreme infertility and high erosion risk, and suitable only for pastures, forest, and land reserves. Around 83% of the country's soils are regarded as nonarable due to various limitations (see map of Panamanian soil fertility in Atlas Nacional de la Republica de Panama, 1988). The basalt is quarried from an open pit dug into a subsurface basalt intrusion. The basalt is overlain by a red lateritic soil derived from weathering of the basalt, in a tropical area with strong seasonal rainfall (mean temperature around 26°C, annual monthly average range 2.5°C–2.9°C, average daily range 7°C–8°C, mean annual rainfall 2500 mm, monthly average range 10–260 mm with a unimodal distribution peaking in October, all data from Atlas Nacional de la Republica de Panama). The elevation is around 600 m above sea level.

The soil covering the surrounding region is an extremely poor lateritic soil derived from intensive leaching and weathering of old basalts and andesites of Miocene age (5–23 million years), overlain by pale pyroclastic volcanic ash and tuff deposits, which have been largely weathered to quartz sand. These are derived from eruptive volcanism in the Pleistocene period (age approximately 1–3 million years ago), during the intensive subduction that resulted in the closing of the ocean passage linking the Atlantic and Pacific Oceans that had previously separated North from South America. During this time, Panama was a chain of active volcanic islands stretching between the two continents (Coates, 2001).

This region provides clear visual evidence for extremely low soil fertility:

1. The undisturbed native vegetation found in remnant patches in the region consists of low biomass and low-diversity forest dominated by short (roughly 3–5 m tall), twisted, widely spaced trees dominated by two native fruit tree species that predominate in the poorest soils in Panama, namely, nance (*Byrsonima crassifolia*, Malpighiaceae) and marañon or cashew (*Anacardium occidentale*, Anacardiaceae), separated by low biomass scrub vegetation. In contrast, Panamanian forests growing in rich alluvial valley bottom soils in areas with similar climate have some of the highest biomass and productivity recorded (Leigh, 1999).
2. Areas of cleared vegetation are almost worthless for agriculture or pasture due to exceptionally low productivity and have largely been abandoned. Abandoned cleared areas have failed to regain more than a tiny fraction of the biomass of the original vegetation and consist of low scrub bushes and clumping grasses separated by bare areas of deeply eroded soils. This gives the area the appearance of semiarid conditions despite the high rainfall, and gullies and badland topography quickly develop due to inadequate vegetation cover.

In order to assess the relative increases in biomass, several comparative approaches were used:

1. Seedlings of *Acacia mangium*, a leguminous nitrogen-fixing tree, were planted during September 1997 in soils derived from pale weathered volcanic ash, red lateritic soils derived from underlying weathered basalt, and black basalt powder and gravel derived from wastes from the stone crushing plant at the quarry. The seedlings were derived from a uniformly sized and aged batch of nursery seedlings approximately 20 cm tall and were planted at the same time. No chemical fertilizers, organic matter, or irrigation were provided. Tree heights were measured by photographing the trees 19 and 59 months later, with a pole of known length held by a person of known height standing next to the tree. Tree heights were then measured as the

ratio of pixels spanning tree height compared to the scale length. The relative leaf abundance, size, and color were estimated from digital video images.

2. Grasses, bushes, and weeds growing in the three different soil types were qualitatively examined for their height and ground cover using digital images.
3. A crop of maize (*Zea mays*) sown in fresh basalt powder and gravel with no additional fertilizer or irrigation produced a bumper crop. Maize sown in adjacent soils would produce little or nothing.
4. Around 11 soil samples (5 from the basalt powder soil and 3 each from the lateritic and pale sandy soils) were analyzed by the Soil Science Lab of the National Agricultural Research Institute (IPIAT), and pH, phosphorus, potassium, calcium, magnesium, aluminum, organic matter, manganese, iron, zinc, and copper were measured.

Results

Soon after planting, the *A. mangium* seedlings showed dramatic growth differences in different soil types (Figures 17.1 through 17.6). After 19 months, seedlings planted in surrounding pale soil were 3.58 ± 1.07 m tall, and after 59 months, they were 6.25 ± 1.03 m. Seedlings planted in red lateritic soil

FIGURE 17.1 Trees on low-fertility site at 19 months.

FIGURE 17.2 Trees on medium-fertility site at 19 months.

FIGURE 17.3 Trees on high-fertility site at 19 months.

FIGURE 17.4 Trees on low-fertility site at 59 months.

FIGURE 17.5 Trees on medium-fertility site at 59 months.

FIGURE 17.6 Trees on high-fertility site at 59 months.

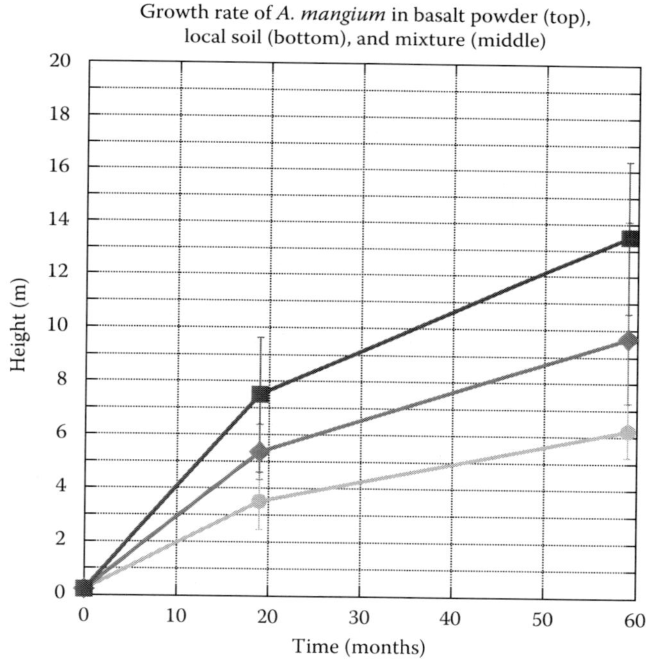

FIGURE 17.7 Tree heights as a function of time since planting.

were 5.36 ± 1.02 m after 19 months and 9.72 ± 4.42 m after 59 months. Seedlings planted in black basalt gravel and powder were 7.50 ± 2.17 m after 19 months and 13.54 ± 2.85 m after 59 months (Figures 17.7 and 17.8). Regression of tree biomass versus height (Figure 17.9) was used to calculate biomass of various treatments (Figure 17.10). Tree survival was proportional to tree growth, with low survival of the slowest-growing trees (Figure 17.11).

The chemical analyses showed that there was a gradient in phosphorus, potassium, calcium, magnesium, manganese, iron, and zinc with highest values in basalt, medium values in laterite, and lowest values in sand but that pH, organic matter, aluminum, and copper showed little variation with soil type (Figures 17.12 through 17.22).

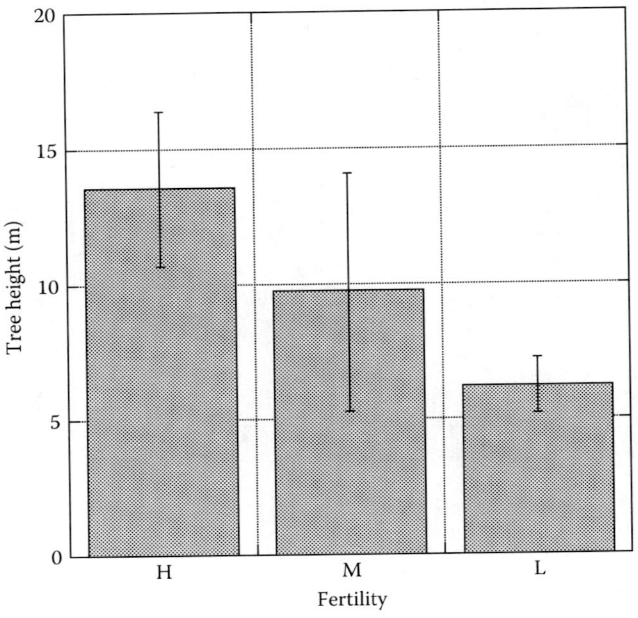

FIGURE 17.8 Tree height as a function of soil fertility at 59 months.

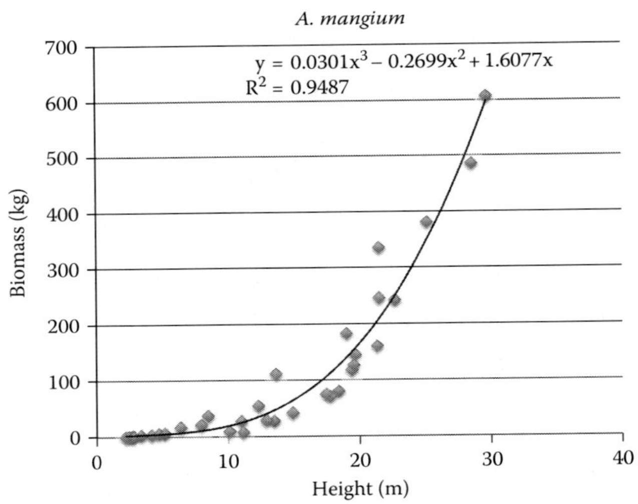

FIGURE 17.9 Biomass versus height regression for *A. mangium*. (Data kindly provided by Torres-Velez, D.A. and del Valle-Arango, J.I., *New Forests*, 34, 293, 2007.)

Dramatic differences were also seen in trunk thickness, leaf abundance, size, and color in seedlings planted in different soil types (Figures 17.1 through 17.6). Seedlings planted in white ash had thin trunks, sparse leaf cover, and small leaf size, and the leaves were chlorotic, with numerous brown spots typical of severely nutrient-deficient plants. Seedlings planted in red laterite had much thicker trunks and had high abundance of uniformly light green leaves with larger size. Seedlings planted in black basalt had the thickest trunks and largest leaves and were much more densely covered in dark green leaves free of blemishes.

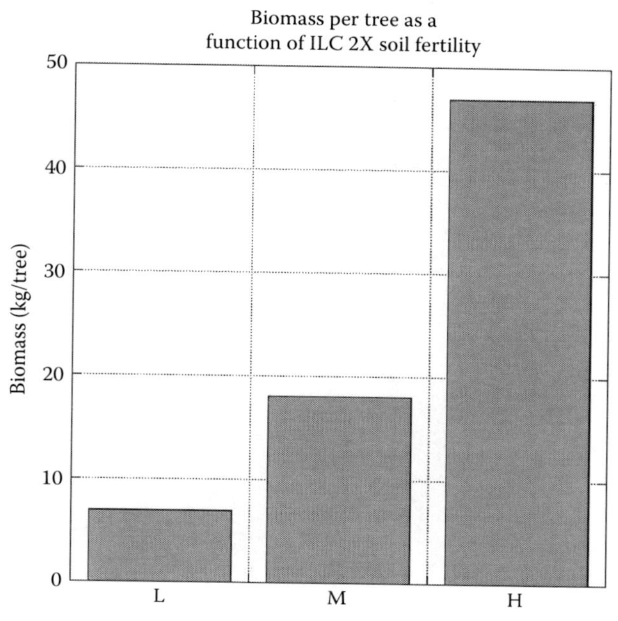

FIGURE 17.10 Biomass versus soil fertility at 59 months, using the regression equation fit to the data in Figure 17.9.

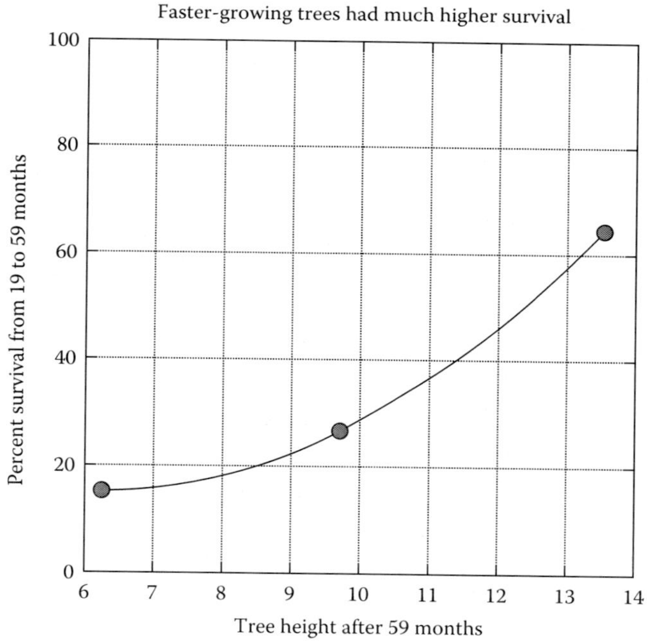

FIGURE 17.11 Survival greatly increased with tree growth rate.

Comparison of plant size and vegetative ground cover of shrubs, herbs, and grasses also showed the same dramatic differences between soil types. For each species, the height, plant density, size of leaves, and dark green leaf color were by far the highest on basalt, lowest on ash, and intermediate on laterite, indicating that both native and exotic plants showed similar relative growth responses.

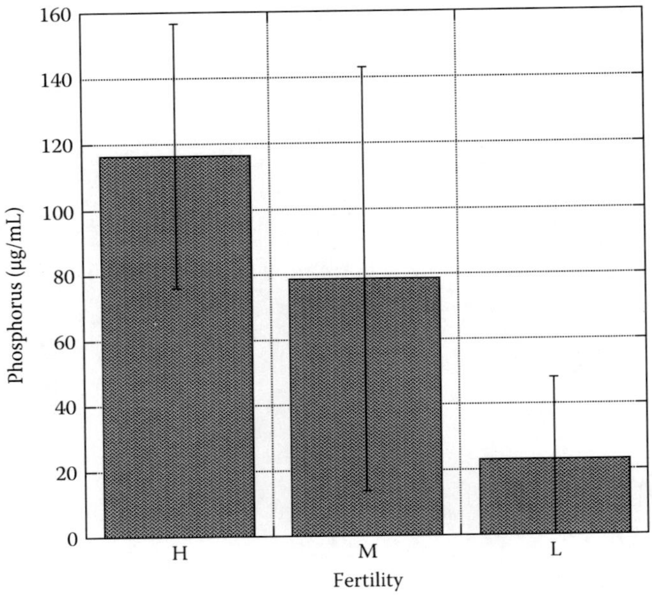

FIGURE 17.12 Phosphorus content of various soil classes.

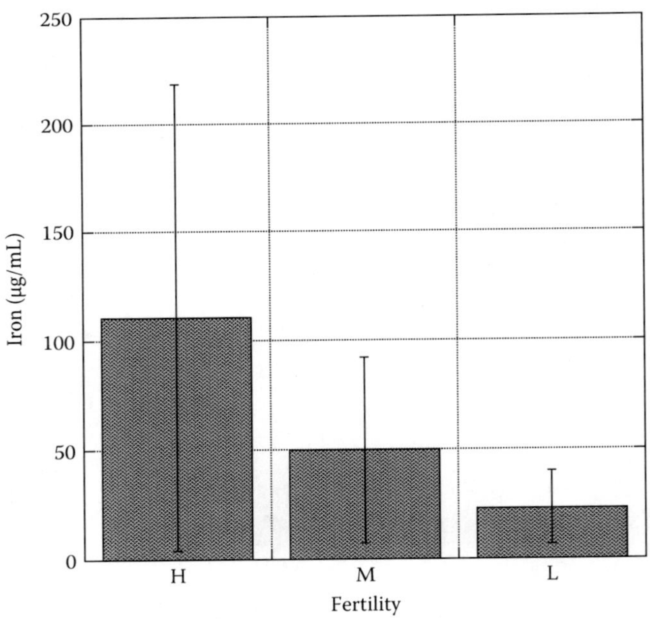

FIGURE 17.13 Iron content of various soil classes.

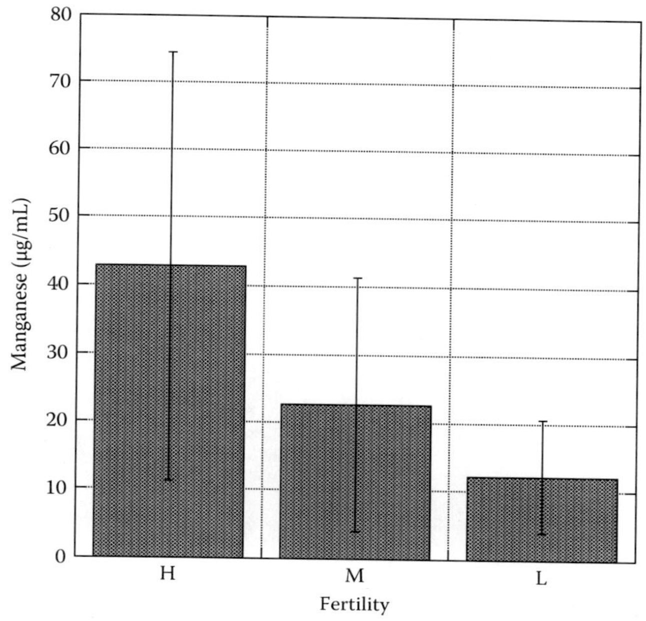

FIGURE 17.14 Manganese content of various soil classes.

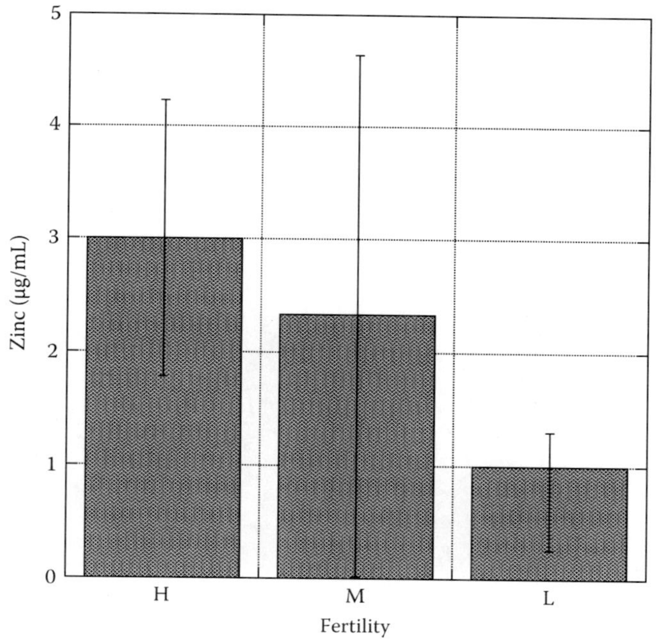

FIGURE 17.15 Zinc content of various soil classes.

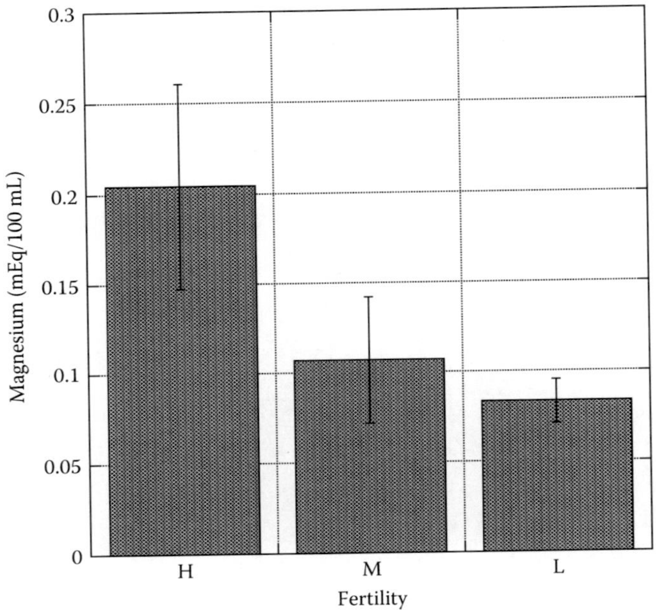

FIGURE 17.16 Magnesium content of various soil classes.

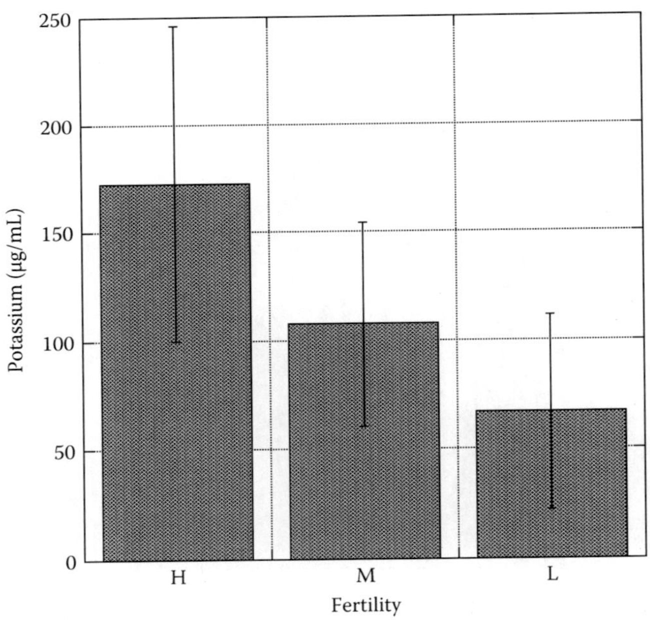

FIGURE 17.17 Potassium content of various soil classes.

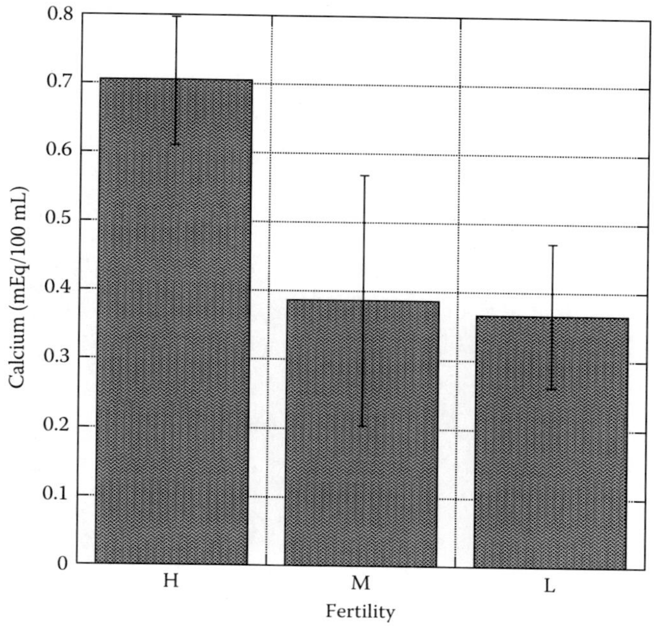

FIGURE 17.18 Calcium content of various soil classes.

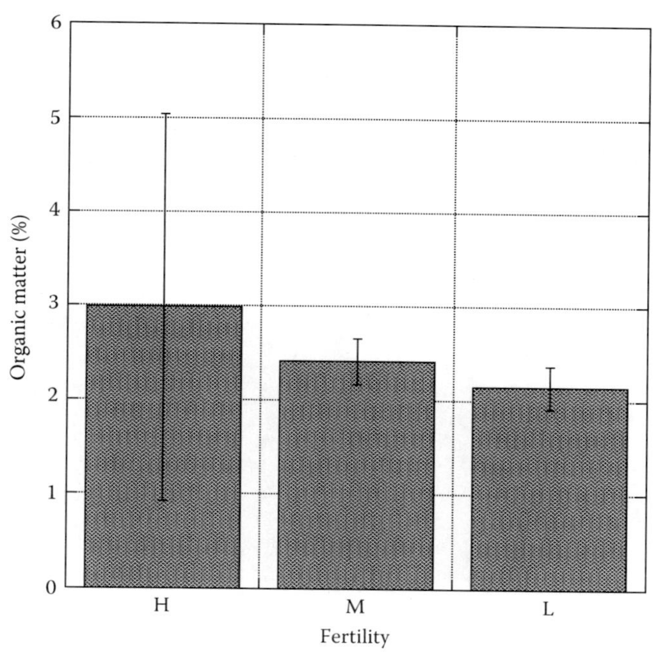

FIGURE 17.19 Organic matter content of various soil classes.

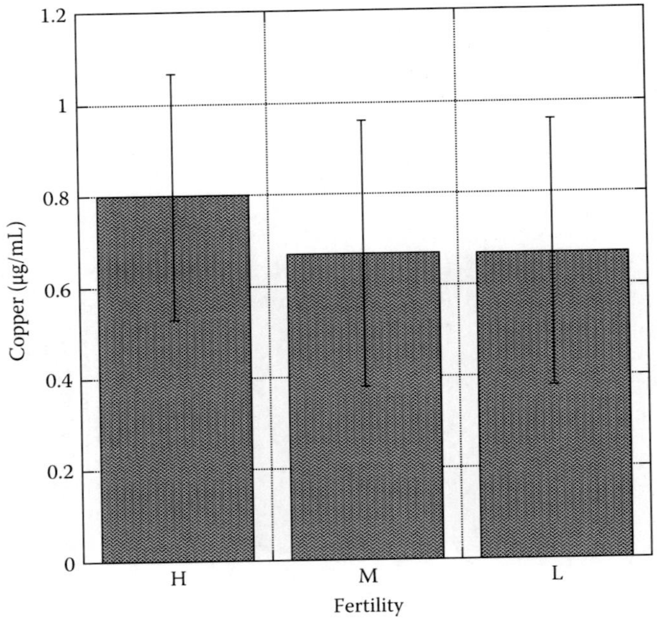

FIGURE 17.20 Copper content of various soil classes.

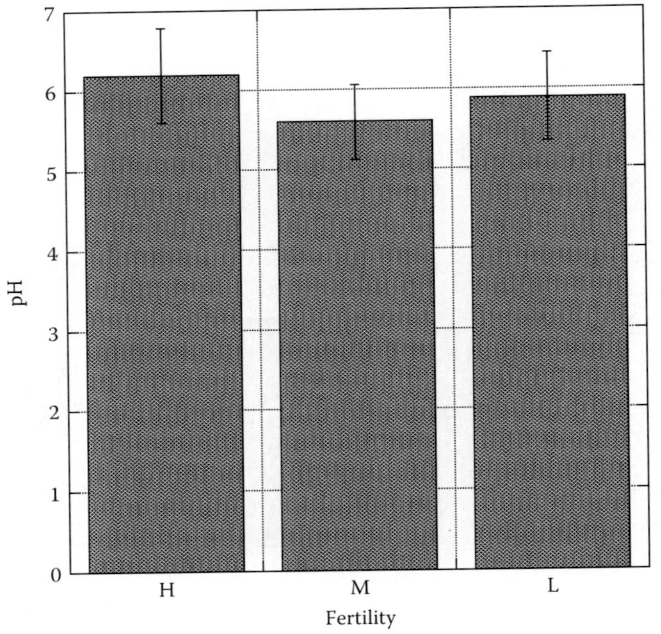

FIGURE 17.21 pH of various soil classes.

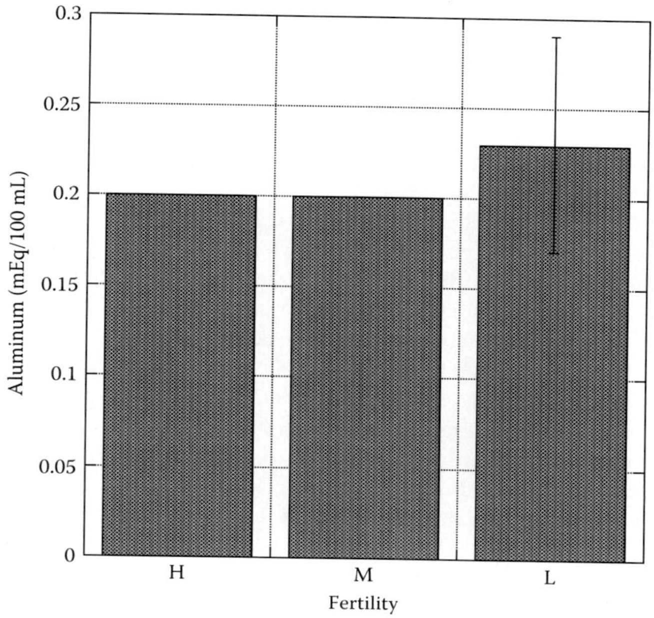

FIGURE 17.22 Aluminum content of various soil classes.

Discussion

Growth rate differences expressed in terms of biomass increase would be much higher than the roughly 2.17-fold increase in height found between local soil and fresh basalt due to the thicker trunks and stems, more abundant leaves, larger leaves, improved nutritional status, and hence leaf weight per unit area in soils supplemented with fresh basalt powder. This ratio underestimates the tremendous increase in leafy material growth that would have been found had the trees been cut down, split into component parts, and weighed directly. Photosynthesis should be proportional to surface area and the trees on basalt had higher leaf densities and were darker green with chlorophyll. Biomass should be proportional to volume, but trees on basalt had more branches and thicker stems. Simple allometric scaling of plant geometry, assuming that all portions of the plant scale similarly to length, suggests that a 2.17 times increase in height corresponds to 4.71 times higher surface area and 10.22 times higher volume of trees on basalt compared to local soils. Using the biomass to tree height regression (Figure 17.9) of Torres and Del Valle (2007), measured on the same tree species under similar conditions in the neighboring country, Colombia, we get an increase of biomass of 6.71 times (Figure 17.10).

Igneous rocks have long been proposed as an effective natural mineral fertilizer (Chapter 4), but this method has not been widely applied, in part due to lack of long-term data on trees under field conditions. Our results strongly confirm the applicability of previous work to trees on infertile tropical soils. The specific productivity increases that result are likely to be highest in tropical soils for several reasons. First, tropical soils tend to be much more dominated by extremely low-fertility soils than temperate climate zones because of the much greater rates of soil weathering, erosion, and nutrient leaching caused by the higher temperatures and rainfall. Second, tropical soils are generally less able to retain nutrients due to the lower organic matter contents resulting from more rapid rates of litter and humus decomposition by microbes and soil microfauna. Third, the rates of plant nutrient uptake can be much higher in the tropics due to the much longer growing season and the much higher potential productivity under wet and hot climatic conditions. Fine-grained basic igneous rocks such as basalt should be much more effective than coarse-grained acidic igneous rocks like granites due to higher concentrations of nutrient elements, more rapid release due to higher surface areas and greater mineral weathering caused by their lower thermodynamic stability at ambient temperatures, and reduced acidity of mineral surfaces.

Callahan has proposed that paramagnetic field strength is a measure of soil fertility, and experiments show that increasing amounts of ground paramagnetic rock increases plant growth (Dykstra, 2000). While Callahan proposes that those experimental results are due to the paramagnetic field itself, an alternative suggested by the present data is that paramagnetic field strength could be the result of crystals of minerals with high amounts of essential elements and so correlate with soil fertility for biogeochemical reasons rather than physical ones.

Despite the dramatic increases in growth rates reported here, there are good reasons to believe that considerably higher increases can be obtained with further experimentation. Optimal growth rates will take place when all the nutrients required for plant growth are provided in precisely the ratios needed for optimal uptake by plants. Any imbalance in these ratios will result in plant growth rates limited by the least available nutrient, and additional amounts of the other nonlimiting elements will not be taken up by plants. Unbalanced nutrient ratios are the major cause of wasteful and economically inefficient use of chemical fertilizers. Applications of standard NPK chemical fertilizers formulated for nontropical conditions very rarely produce the growth increases that are predicted and factored into cost–benefit analyses. This failure is often caused by lack of base cations and trace metals in most fertilizer mixes, whose lack prevents the plant from taking up the major nutrients in fertilizers such as nitrogen. Much of the fertilizer nitrogen is subsequently leached from the soil, polluting groundwaters and causing eutrophication of rivers, lakes, estuaries, and coastal coral reefs by overfertilizing algae growth. Determining the optimal mineral supplementation response requires adjusting the amount of mineral supplementation per unit area to an extent that depends on the ratios of all major and minor nutrients in the mineral supplements, the nutrient ratios available in the soil, and the ratios needed by the plant. Thus, the optimal conditions would be expected to differ for each plant type, each soil, and each mineral supplement.

The dramatic results we obtained are probably due to a combination of greatly increased supply of major base cations, especially calcium (needed to complex lignins with ionic bridges and replace that lost in litter fall), magnesium (needed for chlorophyll formation and to stabilize nucleic acids), potassium (needed for cell osmotic balance), phosphorus (needed for ATP and nucleic acid synthesis), and trace metals (needed as essential cofactors to optimize enzyme activities).

Basalts are known to be high in all these elements, which are readily leached in tropical soils or made unavailable through complexation by iron and aluminum oxides and hydroxides. Phosphorus is provided by apatite mineral, a significant component of basalt and other basic igneous rocks, while sulfur (needed for protein stability) and trace metals are likely to be provided in the form of sulfide minerals. Further contributions to increased productivity come from reduced soil acidity and reduced aluminum toxicity provided by adding basic cations contained in the calcium and magnesium minerals in basalt, such as pyroxenes, amphiboles, and feldspars. The large increase in phosphorus, potassium, calcium, magnesium, manganese, iron, and zinc in basalt over local soils indicates that these were the primary nutrient limitations in this region because there was little difference in pH, organic matter, aluminum, or copper. Nitrogen values were not measured, but in any case were not relevant in this case as (1) basalt contains little nitrogen and (2) *A. mangium* is a nitrogen-fixing plant.

Although mineral supplementation with fresh basalt powders is likely to resolve many or most nutrient deficiencies in tropical soils, it is unlikely to make any significant contribution of nitrogen (needed by non-nitrogen-fixing plants in large amounts for protein and nucleic acid synthesis) due to the very low levels of ammonium ions in basaltic rocks. Nor does it supply any significant amount of organic carbon, needed to complex and retain nutrients, which would otherwise be leached and lost from the soil, and to absorb and retain water, making it more available during the dry season.

We believe that it is likely that our results were by far suboptimal and that further increases can be obtained with further research. We therefore expect that greatly improved results would be found if organic carbon and nitrogen are added, especially if it is in the form of manures or green mulches, which release the nitrogen at a slower and more efficient rate than chemical nitrogen fertilizers. Further research is needed to optimize the manure to basalt ratios to ensure the most efficient use of nutrients, which will also depend on the nutrient composition of the manure or mulch. In addition, we believe that the productivity increases found here could be further increased by use of biochar (*terra preta*), a form of elemental carbon similar to finely divided charcoal, which retains water, carbon, and nutrients in soils and greatly enhances their fertility (Sombroek et al., 2000; Steiner, 2007; Lehmann and Joseph, 2009).

Even though we are confident that further and more sophisticated research programs can greatly increase the response of plants grown on tropical soils, the remarkable results obtained here indicate that this approach is highly effective even without any effort to optimize it. Basalts are readily and cheaply available in Panama and many other parts of the world, and the increases in growth rates found here could dramatically increase the profitability of tree plantations by greatly reducing harvest times, perhaps by up to an order of magnitude on very infertile soils. Such productivity increases on already cleared and degraded tropical soils would greatly reduce the demand for clearing fresh forest and would greatly reduce the negative impacts of soil erosion on soil fertility and on downstream fresh and salt waters that receive them, especially for coral reefs.

Basalt supplementation is likely to be highly cost effective. The production cost of basalt powder and gravel depends on the grain size, since the finer it is, the more extensive crushing is required. Existing rock-crushing plants are largely designed for the production of coarse gravel for construction of roads, building foundations, seawalls, and similar uses. The material we used is an incidental by-product of rock crushing for gravel and aggregates to be used in concrete. However, there is an increasing demand for finer grades to be added to concrete mixtures, and all available supplies are immediately sold. Further demand and increased production of finer grades could decrease unit costs. Transportation costs increase with distance between the source and the end-use site since basalt is heavy and bulky and must be transported by truck.

A typical retail cost for basalt powder delivered in Panama City is approximately \$95.85 per tonne, which compares with a retail price for urea of \$451.94 per tonne and \$634.34 per tonne for standard NPK (12-24-12) fertilizer. It is clear that farmers can save immensely by using basalt powders that cost only 21.2% as much as urea and only 7.2% as much as NPK by weight while providing many more minerals for much longer. Much of this difference in cost is due to the fact that chemical fertilizers are imported from abroad and require much larger amounts of fuel energy to produce while basalt powder is a local product. Assuming that the productivity increases are similar, the much greater cost-effectiveness of basalt is readily seen. Because most of the nutrients in basalt powders are slowly released by weathering, the effects are likely to be much longer lasting than those of quickly released chemical fertilizers and to be a function of grain size.

We believe that further research to improve and apply this approach may be one of the most important priorities for tropical agriculture, ranching, and silviculture and will also provide important benefits for the health of aquatic and coastal marine ecosystems. In addition, increasing productivity of deforested areas will reduce the rate of new deforestation and increase the rate of net carbon dioxide absorption from the atmosphere, with important consequences for the chemistry of the atmosphere, earth's radiation, and heat balances and hence for reducing rates of global climate change. Large-scale, long-term research programs are needed for a wide range of rock powders, soil types, climate regimes, and plant types to determine the optimal and most cost-effective amendments of rock powders, mulches, and biochar needed for the increases in agricultural and soil fertility needed to feed a growing population while preserving forest, soil, and water resources and sequestering carbon to stabilize atmospheric CO_2 and global temperatures at safe levels.

Conclusions

1. Trees showed a dramatic growth increase in fresh basaltic soil over local soils as a result of greatly improved nutritional status.
2. The specific increase in growth rate is a function of the nutrient levels in local soils, their levels in basalt, and the specific uptake rates of each different type of plant.
3. Basalt powder is readily available and cheaply transported in Panama and could be used to greatly accelerate the growth rate of tree plantations, crops, and pastures on poor tropical soils.

4. These results are likely to be applicable to many other parts of the tropics, with the greatest benefits occurring on the poorest soils.
5. Optimizing the response to basalt mineral fertilization will take further work on a larger range of soil types, rock powder compositions, plant species, and climates.
6. Growth rate stimulation is likely to be increased further by amendments of organic carbon and nitrogen and of elemental carbon (biochar), and further work is required to determine the most cost-effective ratio of basalt to organic fertilizer and biochar supplementation in a wide range of soil and plant types.
7. The increases in growth rates reported here indicate that many tropical soils can undergo dramatic increases in soil fertility with cost-effective additions of natural mineral fertilizers supplemented with manures.
8. This could result in tremendous reductions in harvest intervals for tree plantations, increases in food productivity, and increases in pasture and livestock production on poor tropical soils, decreasing pressure to cut native forest while sequestering CO_2 from the atmosphere.
9. Further research to optimize these methods is one of the most important priorities for tropical agriculture, tree plantations, livestock production, reduction of erosion, protection of aquatic and marine ecosystems, and reduction of rates of global climate change.
10. Greatly increased funding for such research to *remineralize the earth* is a vital and urgent global necessity to feed the people of the tropics and reverse global environmental degradation of soils, waters, ecosystems, and the atmosphere.

Summary

A. mangium seedlings were planted in three soil types spanning a gradient between highly infertile soils and freshly crushed basalt rock powder. Basalt provides slow release of minerals essential for plant growth, including essential elements not found in local soils or included in most chemical fertilizers. The growth rate of the trees was followed for 5 years. Seedlings growing in basalt powder grew 2.17 times taller than in local soils. This corresponds to a tree allometric surface area 4.71 times higher and a volume (or biomass) 10.22 times higher than on local soils. Trees on basalt had noticeably more leaves and branches and darker-colored leaves. Basalt powder was enriched in phosphorus, potassium, calcium, magnesium, manganese, iron, and zinc, but its pH, organic matter, copper, aluminum, or nitrogen was not appreciably different from local soils. These results indicate that highly infertile tropical soils can be made productive with rock powder minerals alone. Benefits depend on local soil fertility and nutrient element deficiencies, the specific mineral needs of each plant, and climate. Even more dramatic growth enhancement would be expected if biochar, organic carbon, or nitrogen were also added. A wide variety of rock powders of many kinds are available in most locations needing environmental restoration and soil fertility enhancement for forestry or agriculture. Studies with a wide range of plants, soils, rock powders, biochar, nitrogen, and climate regimes are needed to optimize the productivity and carbon storage benefits and minimize the cost for large-scale applications.

ACKNOWLEDGMENTS

We gratefully acknowledge the support of the staff of Cantera Pedrera, the Pew Fellowship Program for the travel and equipment, and the Leslie Jones Sauer Trust for the Global Coral Reef Alliance for environmental restoration projects. We thank the IPIAT Soil Science Laboratory for the soil analyses. We thank Foster Derrick, Richard Ruschman, Leslie Jones, and Gill Hill for helpful discussions on rock powder fertilization. We thank Jorge Ignacio del Valle Arango and Danny Alexander Torres Velez for providing the biomass–height regression data.

REFERENCES

Atlas Nacional de la Republica de Panama, 1988, Instituto Geografico Nacional Tommy Guardia, Panama, RP.

Coates, A., 2001, En la historia geologica, Panama ha cambiado el mundo, in S. Heckadon-Moreno (ed.), *Panama*: *Puente Biologico*, pp. 18–25, Instituto Smithsonian de Investigaciones Tropicales, Panama, RP.

Dykstra, T. M., 2000, Paramagnetic effects on plant growth, http://www.dirtdoctor.com/Paragmagnetism-Effects-on-Plant-Growth_vq127.htm.

Lehmann, J. and S. Joseph, 2009, *Biochar for Environmental Management*: *Science and Technology*, 416pp., Earthscan, London, U.K.

Leigh, E. G., 1999, *Tropical Forest Ecology*: *A View from Barro Colorado Island*, 248pp., Oxford University Press, Oxford, U.K.

Sombroek, W. G., P. M. Fearnside, and M. Cravo, 2000, Geographic assessment of carbon stored in Amazonian terrestrial ecosystems and their soils in particular, in R. Lal, J. M. Kimble, and B. A. Stewart (eds.), *Global Climate Change and Tropical Ecosystems*, 483pp., CRC Press, Boca Raton, FL.

Steiner, C., 2007, *Slash and Char as an Alternative to Slash and Burn: Soil Charcoal Amendments Maintain Soil Fertility and Establish a Carbon Sink*, 185pp., Cuvillier Verlag, Gottingen, Germany.

Torres-Velez, D. A. and J. I. del Valle-Arango, 2007, Growth and yield modelling of *Acacia mangium* in Colombia, *New Forests*, 34:293–305.

18

Basalt Dust and Biochar Interactions at New Harmony Farm, Massachusetts

Thomas J. Goreau, Erin Stack, Elaine Senechal, Jianwu Tang, Rebecca Ryals, Tom Vanacore, and Joanna Campe

CONTENTS

Introduction

Biochar and rock dusts such as basalt are widely regarded as excellent slow-release, long-lasting, natural fertilizers by themselves, which should work even better in combination. Many field studies showing dramatic increases in productivity with rock powders or with biochar alone are shown in this book. But there are many reasons to believe that combining the two methods would synergistically increase both the magnitude and duration of biological responses to both of these natural plant growth enhancers.

Rock powders are sources of essentially all the biological elements except (in general) for nitrogen, while biochar is less a fertilizer per se since it largely consists of elemental carbon. However, biochar acts physically and chemically as a sponge that absorbs and retains nutrients and water, making them available to plant roots via interaction with symbiotic fungi and bacteria (Ladygina and Rineau, 2013). So biochar and rock powders should be highly complementary, especially when combined with nitrogen-rich composts.

Nevertheless, there are few studies of their interactions or optimal combinations. We have started a long-term study of biochar and basalt dust interactions in 32 plots using combinations of 3 levels of biochar and 4 levels of basalt dust, with replicates and controls.

Site Location and History

The site, New Harmony Farm (Figure 18.1), in West Newbury, Massachusetts, is a fertile soil that was the riverbed of the Merrimack River until the river suddenly cut a new channel during a catastrophic hurricane. Local historians claim this catastrophic event was the Colonial Hurricane of 1635 that caused devastation to the newly formed Massachusetts Bay Colony. A map of the area published in Woods (1634) (Figure 18.2a) shows an island at the site. New Harmony farm was located in the southern river channel that was plugged up by hurricane-transported sediments (Figure 18.2b). This occurred before any deforestation for agriculture had taken place in the Merrimack watershed as all English settlement at that time was along the coast. All descriptions of the hurricane come from coastal sites (http://en.wikipedia.org/wiki/Great_Colonial_Hurricane_of_1635; Foster et al., 2004).

The Merrimack River watershed includes most of the state of New Hampshire, including the eastern slopes of the White Mountains, and much of northeastern Massachusetts (Figure 18.3). The watershed was essentially entirely deforested for agriculture during the colonization of New England by European settlers in the late 1600s, 1700s, and early 1800s. Searches for a small watershed with primary (uncut) forest in the White Mountains for scientific research purposes found there were none at all; the best site was the Hubbard Brook watershed, which had been logged for yellow birch in 1918. This site was used to set up the Hubbard Brook Experimental Forest, in the uppermost reaches of the Merrimack watershed, now the longest monitored forest watershed in the world and site of many fundamental forest ecology and biogeochemistry studies (Bormann and Likens, 1979).

Cores in the Plum Island salt marsh, at the mouth of the river, show that the marsh grew very rapidly by trapping sediment from eroded soils after deforestation (Kirwan et al., 2011), connecting the mainland to Plum Island, which had previously been a separate island (Woods, 1634). This erosion decreased after most of the farmland in the hills was abandoned in mid-1800s and reverted to secondary forest (Bormann and Likens, 1979), and the sediment transport was blocked as the river was dammed upstream of the farm site for industrial purposes (Cumbler, 2001).

FIGURE 18.1 New Harmony Farm. The plot of the biochar/rock dust experiments is shown in the white rectangle. The Merrimack River is at lower left.

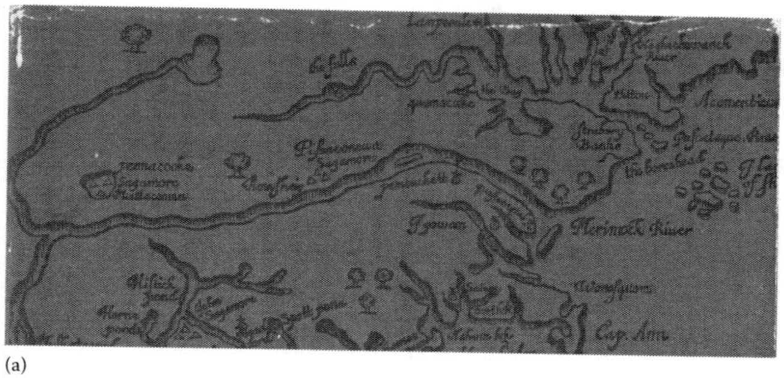

(a)

(b)

FIGURE 18.2 (a) A portion of William Wood's 1634 map showing an island in the Merrimack River. New Harmony Farm is located on the southern channel, which was plugged up by the 1635 hurricane. (b) A perspective view of the New Harmony Farm site. The white dot marks the project site. The approximate boundaries of the Merrimack River channel before the hurricane filled it are shown in the dashed stars, based on the topographic contours of the site.

As a result of this history, the soils are rich alluvial deposits with few fertility limitations and so would not be expected to respond strongly to the addition of plant nutrients. Due to the low and nearly flat elevation, the chemical quality of the soil is quite uniform, but the physical properties are controlled by very small elevation differences of only around 10 cm that determine the depth to the groundwater table. The lowest points are permanently wet and are left as conservation areas in wetland vegetation that cannot be farmed. The highest areas are ploughed and farmed, and intermediate areas are prone to spring snowmelt flooding and prone to flooding during very rainy summers. Russ Barry and Bonnie Silva purchased the land in 2007, planted the field in timothy grass and clover for hay production, and had the land certified organic. Prior to this, the soil had lain fallow for several years after many years of conventional corn production. In July 2011, Erin Stack purchased the land for a certified organic Community Supported Agriculture (CSA) farm with a mission to support research and educational efforts.

Because the benefits of both basalt dust and biochar become greatest after several years, long-term changes in crop yields, elemental composition of soil and crops, carbon storage, and greenhouse gas emissions will be measured. The hope is to continue measurements for at least 3–5 years with two crops per year. This chapter describes the setting up and the first 6 months of results from the study.

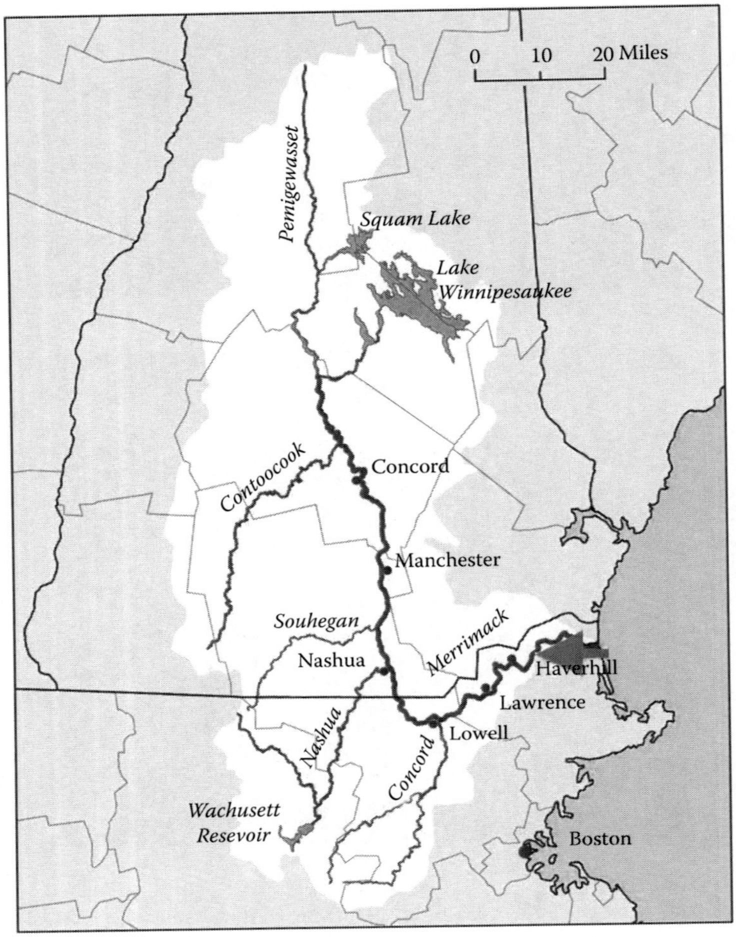

FIGURE 18.3 Merrimack River watershed, with the project site indicated by the arrow.

Materials and Methods

Soil Types and Fertility

Soils were classified by the USDA (Figures 18.4 and Table 18.1). Approximately 50 soil samples from all over the farm were pooled for an average chemical analysis (Figure 18.5).

Fertilization

The experimental plot was ploughed twice, once in each direction. The plot was then marked with stakes and strings into 32 subplots each measuring 5 ft × 10 ft, with replicates of each treatment and controls (Figure 18.6). Each subplot was then fertilized with various combinations of biochar or rock dust. Biochar plots received 0, 20, or 40 lb of biochar per subplot. Rock dust plots received 0, 10, 20, or 30 lb of rock dust per subplot. These were spread across each subplot and raked by hand into the soil three times to a depth of about 5 cm.

FIGURE 18.4 New Harmony Farm showing the boundaries of New Harmony Farm and of soil types surveyed by the US Department of Agriculture. Soil types are classified in Table 18.1. Note that the image was taken before the greenhouse (long white structure in Figure 18.1) was built. At the time this image was taken, the land had been ploughed for hay.

TABLE 18.1
Soil Types Shown in Figure 18.4

Map Unit Legend			
Essex County, Massachusetts, Northern Part (MA605)			
Map Unit Symbol	**Map Unit Name**	**Acres in AOI**	**Percent of AOI**
98A	Wincoski very fine sandy loam, 0%–3% slopes	1.7	24.6
275A	Agawarn fine sandy loam, 0%–3% slopes	0.1	1.1
405D	Charlton fine sandy loam, 15%–25% slopes	0.6	8.5
713A	Limerick and Rumney soils, 0%–3% slopes	4.0	58.0
718A	Saco variant silt loam, 0%–3% slopes	0.5	7.9
Totals for area of interest		6.9	100.0

Biochar was obtained from New England Biochar in Cape Cod, Massachusetts. The biochar had been made the year before, largely from oak and pine wood undergoing pyrolysis at a temperature of 450°C, and had been mixed 50:50 with leaf compost, again predominantly oak and pine, and matured for a year prior to bagging and application. It should be noted that this compost is fairly acidic and poor in nitrogen.

Basalt rock powder was kindly donated for the project by Tom Vanacore of Rock Dust Local, in Vermont. The igneous basalt used at the New Harmony farm is from an extrusive igneous rock of geochemistry typical of *a silica-oversaturated basalt* from the Holyoke region. The geologic environment of its formation is primary magmatic lava flows. Typical mineralogy for the basalt: Essentials: plagioclase (labradorite-bytownite, anorthite content greater than 50%), pyroxene (augite, pigeonite, hypersthene). Accessories: magnetite, hematite, apatite, quartz, olivine, glass. Accidentals: amphibole (titaniferous hornblende), biotite.

Soil Report

Job Name Erin Stack — Date 11/30/2011

Company New Harm ony Farm — Submitted By

	Sample Location		Field		
	Sample ID		1		
	Lab Number		84		
	Sample Depth in inches *Total Exchange Capacity (M. E.)*		6 5.38		
	pH of Soil Sample *Organic Matter, Percent*		6.40 4.84		
ANIONS	SULFUR:	p.p.m.	17		
	Mehlic hll l Phosphorous:	ppm	151		
EXCHANGEABLE CATIONS	CALCIUM: ppm	Desired Value Value Found Deficit	731 777		
	MAGNESIUM: ppm	Desired Value Value Found Deficit	100 57 –43		
	POTASSIUM: ppm	Desired Value Value Found Deficit	100 71 –29		
	SODIUM:	ppm	20		
BASE SATURATION %	Calcium (60 to 70%)		72.23		
	Magnesium (10 to 20%)		8.83		
	Potassium (2 to 5%)		3.38		
	Sodium (5 to 3%)		1.60		
	Other Bases (Variable)		5.00		
	Exchangable Hydrogen (10 to 15%)		9.00		
TRACE ELEMENTS	Boron (p.p.m.)		<0.2		
	Iron (p.p.m.)		218		
	Manganese (p.p.m.)		12		
	Copper (p.p.m.)		1.48		
	Zinc (p.p.m.)		1.14		
	Aluminum (p.p.m.)		1028		
OTHER	Cobalt		0.17		
	Molybdenum		0.09		
	Selenium		0.7		
	Silicon ppm		8.8		
	EC				

Logan Labs, LLC

FIGURE 18.5 Soil chemical analysis for New Harmony Farm. Note the generally fertile conditions, with high organic matter and calcium, but deficiency in magnesium, potassium, and boron. These measurements were made on a composite sample made from mixing about 50 samples from all over the farm.

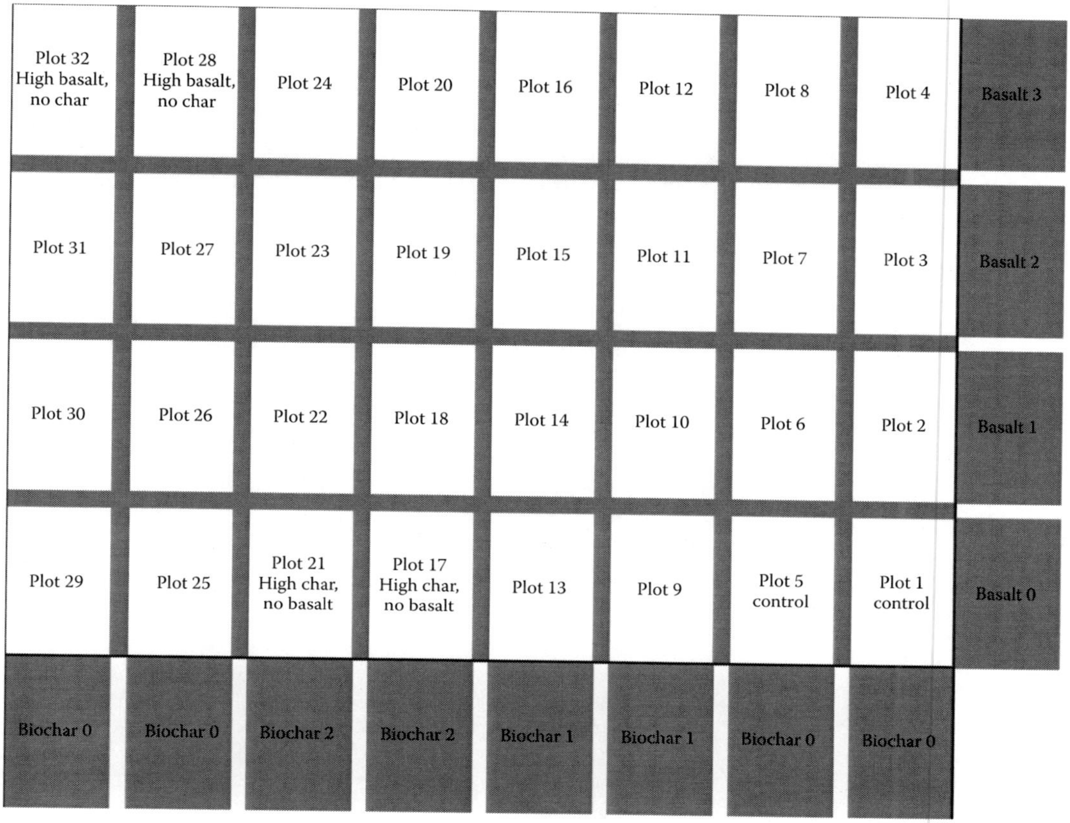

FIGURE 18.6 Arrangement of subplots used in this experiment. The legends at the top and bottom show the relative amounts of biochar or rock dust in each column and row, respectively. Plots 1, 5, 17, 21, 28, and 32 were used for the initial greenhouse gas flux measurements.

The quarried stone is extracted from hard rock ledge in central MA, crushed and pulverized without water. The screened and air-floated gradations are recombined to produce *BrixBlend*. Gradation of the basalt *BrixBlend*: 100% passing 2 mm and 60%+ passing the 200 mesh screen, of which at least 50% of the total is highly micronized air floated fraction (average estimated at 37 μm or 400 mesh). This material is *virgin rock dust*, unweathered.

Although the complex silicate crystalline matrix is technically insoluble, the minerals appear to be highly reactive in the dynamic environment of agricultural soils that include soil biota, aqueous solutions, and fluctuations in pH, which present a distinctly diverse environment from geologic stasis. Harvest weight in the first season field crop indicates rapid assimilation against the controls.

According to the Phil Callahan Soil Meter (PCSM, Pike Agri) measurements of magnetic field strength in CGS units, the basalt *BrixBlend* generates a reading of 2400 (CGS $\times 10^{-6}$). Although the PCSM readings are skewed by iron, the basalt measured against other rock powders generates a strong steady attractive force in comparison to other rock powders with equal or greater proportion of Fe producing either low magnetic readings or low momentary readings. The steady elevated field strength in the and ferromagnetic elements or molecules. Paramagnetic atomic basalt indicates the presence of paramagnetic structures having one or more unpaired electrons in their outer shells and in totality producing a net attractive force in a magnetic field.

A chemical analysis of this basalt is listed in Table 18.2.

Site Management

After the biochar and rock powder had been raked into the soil, there were five soil samples taken from each subplot, one in the center and four more, each taken two-thirds of the way from the center toward

TABLE 18.2
Chemical Composition of Basalt

Element	Weight Fraction
SiO_2	49.38%
Al_2O_3	13.32%
Fe_2O_3	13.78%
MnO	0.2%
MgO	5.77%
CaO	9.21%
Na_2O	3.14%
K_2O	0.96%
TiO_2	0.916%
P_2O_5	0.15%
Loss on Ignition	4.02% (probably carbon, hydrogen, oxygen, and nitrogen)
C-total	0.47%
S-total	0.04%
B	297 ppm
Au	<1 ppb
Ag	<0.5 ppm
As	5 ppm
Ba	137 ppm
Be	<1 ppm
Bi	<2 ppm
Br	1.1 ppm
Cd	1 ppm
Co	43.1 ppm
Cr	25.3 ppm
Cs	<0.2 ppm
Cu	81 ppm
Hf	2.2 ppm
Hg	<1 ppm
Ir	<1 ppb
Mo	0.6 ppm
Ni	35 ppm
Pb	12 ppm
Rb	20 ppm
S	0.054%
Sb	<0.1 ppm
Sc	39.3 ppm
Se	<0.5 ppm
Sr	164 ppm
Ta	2.3 ppm
Th	2.6 ppm
U	0.9 ppm
V	324 ppm
W	<1 ppm
Zn	96 ppm
Zr	88 ppm
La	10.7 ppm
Ce	22 ppm
Nd	14 ppm
Sm	3.07 ppm
Eu	0.86 ppm
Tb	0.9 ppm
Yb	2.67 ppm
Lu	0.21 ppm

each corner, and stored in a dark cool basement for later chemical analysis. Results of those analyses are not available at the time of writing and will be reported elsewhere. Each plot was then seeded. The first crop of beet and radishes were seeded with a hand-drawn seeder. As it was noted that seeding was not regular, with gaps and with several seeds (two to five) being inserted into the same holes, where they competed with each other, the second seeding of radishes was done by hand. The plot was then regularly hand weeded by a team of dedicated volunteers.

Crop Yields

At crop time each subplot had all beets and all radishes pulled up by hand, soil brushed off, and counted and weighed to provide total fresh weight for each crop in each subplot. Because of irregular seeding success, the data were presented as mean weight per plant for each crop in each subplot. Because the first crop was sown late in July and harvested in September, the second crop of radishes was started very late. They were harvested in November but had been affected by early frost.

Greenhouse Gas Fluxes

Greenhouse gas fluxes were measured as a pilot study on September 28, 2012, from selected bare field plots, between crops, using a mobile gas analyzer system. Duplicate control plots with no biochar or rock dust, plots with the highest level of biochar and no rock dust, and plots with the highest level of rock dust and no biochar were monitored, as shown in Figure 18.6. Gas fluxes into a cylindrical chamber with a circular soil surface area of 0.0639 m^2 were analyzed instantaneously with a cutting-edge technology in measuring trace gases, that is, the new generation of cavity ringdown spectroscopy techniques. During the measurement, the chamber was attached to a preinstalled collar in the soil. The air inside the chamber was pumped through the CO_2 analyzer and N_2O analyzer and then back to the chamber in a closed loop. The gas flux was calculated based on the increase in concentration during an approximately 5 min measurement period (depending on the gas flux rate) when the chamber was attached with the soil collar. The slope of a linear curve of concentration against time is used to calculate the flux.

Results

Crop Yields

The first crop of beets and radishes showed very strong stimulation of plant productivity (measured as total fresh weight per plant at harvest in pounds) by basalt powder, up to 2.5 times greater for beet and 2.2 times for radish (Figure 18.7). We found inhibition of beet yields, but not of radishes, by biochar in the first crop (Figure 18.8), probably due to nutrient immobilization caused by the high C/N ratio. The maximum yield per plant for both crops was attained at 20 lb of basalt per plot, with a small reduction at the highest loading. This suggests that some factor was in excess and inhibitory at the highest loading.

The full first crop data showing results from the interaction plots are shown in Figures 18.9 through 18.14.

The inhibitory effect of biochar and the stimulatory effect of the basalt largely disappeared by the second crop of radishes, which was sown late and affected by early frost (Figures 18.13 and 18.14). This is presumably because the microbial flora associated with the biochar and rock powder became more mature and because base cations from basalt weathering were taken up on biochar cation exchange sites.

Weed Growth

A strong gradient in weed growth was noted along the paths between the plots. The biochar made little obvious difference in the amount and type of weeds, but weeds grew much faster next to the higher basalt dust plots. In addition, there was a very marked difference in the types of weeds as a function of the amount of basalt added. In the areas with no basalt, the weeds were dominated by sedge, but with

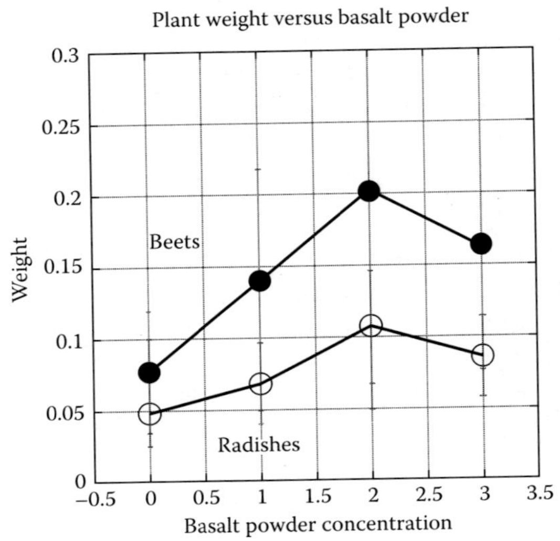

FIGURE 18.7 First crop yields in pounds per plant for beets and radishes as a function of basalt powder concentration. Note an increase with basalt but an inhibition at the highest levels, indicating that that there is an optimal concentration above which benefits decrease. All graphs show means plus or minus standard deviations of replicate plots.

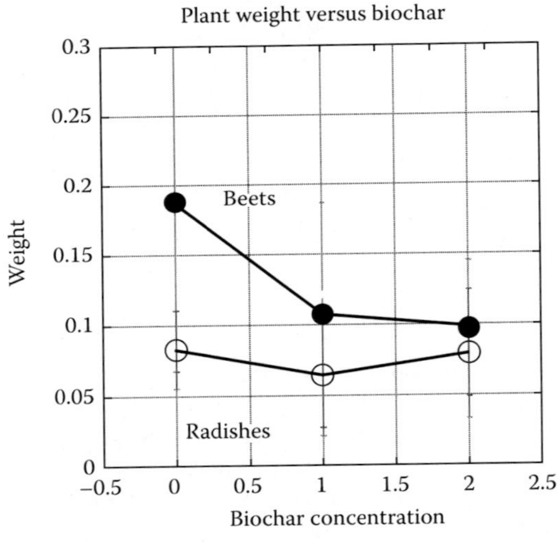

FIGURE 18.8 First crop yields in pounds per plant for beets and radishes as a function of biochar concentration. Note that the fresh biochar inhibited growth of beets but not radishes.

increasing basalt, the weeds were dominated by large grasses with runners and weeds with deep roots. Note that neither basalt dust nor biochar was applied to the paths or raked into them, so this effect could represent windblown material from the plots, translocation by earthworm castings, or that the roots of the weeds in the path grew into the adjacent soil plots. Weeds were indeed concentrated along the edges of the plots and not in the centers of the grass strips between them.

Earthworm Activity

Basalt powder was observed to strongly increase earthworm activity. Between reaping the first crop and sowing the second crop, a period of heavy rain followed by a few dry days resulted in earthworm

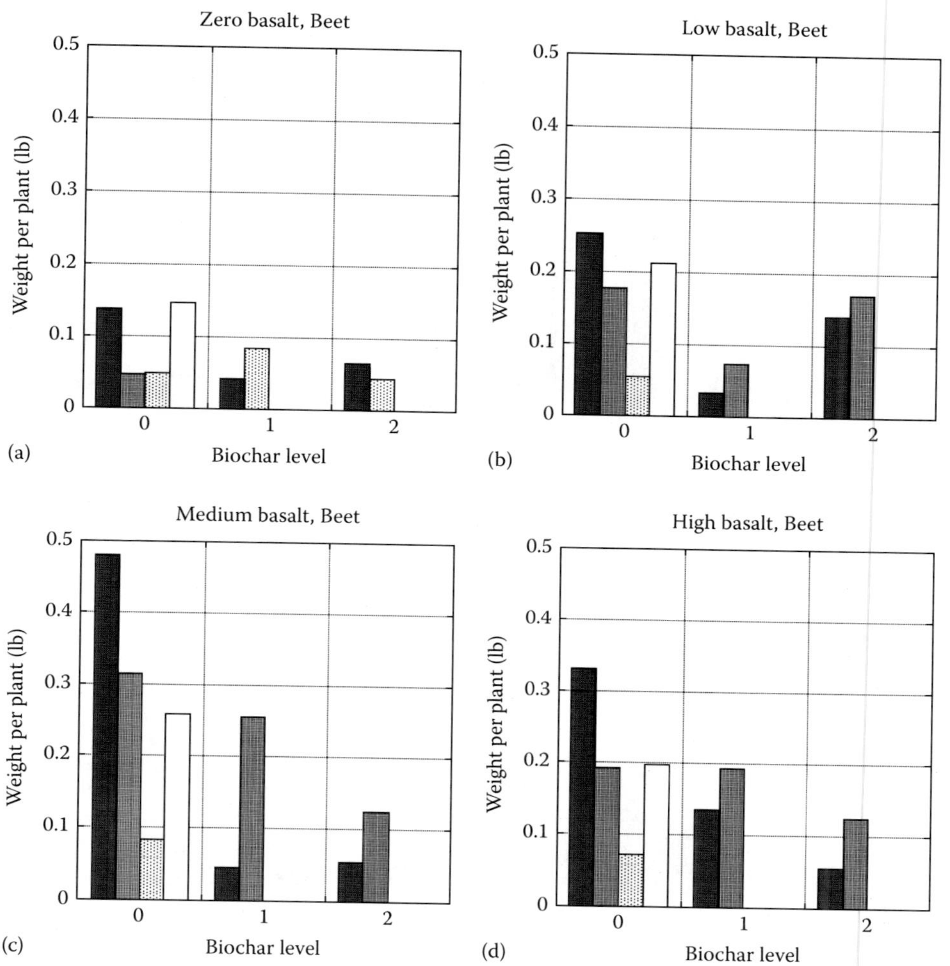

FIGURE 18.9 First beet crop as a function of biochar (a) with zero basalt, (b) with low basalt. (c) with medium basalt, and (d) with high basalt.

castings being clearly visible as dry light-colored clods on top of dark, wet soil. We photographed the areas where the chambers were used to measure greenhouse gas fluxes in control plots, high biochar only plots, and high basalt dust only plots. The number of earthworm castings was counted in each. These numbers are underestimates, especially for the basalt plots, where the castings were so numerous that they made large lumps joined to each other, which were generally counted as single castings although they were likely from 2 to 10 separate lumps combined. The control plots had 3.5 ± 2.12 castings, the high biochar only plots had 8.0 ± 4.24 castings, and the high basalt dust only plots had 86 ± 16.97 castings. Addition of biochar more than doubled the number of castings above the controls, and addition of basalt powder increased them more than 20 times higher than controls. Results are shown in Figure 18.15. Increased earthworm activity is associated with many benefits for soils (Darwin, 1881) including better soil granulation, larger soil pores, increased aeration, and increased microbial and fungal activity.

Greenhouse Gas Emissions

Preliminary gas flux measurements suggest that the high-basalt and high-biochar plots release less carbon dioxide to the atmosphere than control plots. The basalt plots had lower releases of nitrous

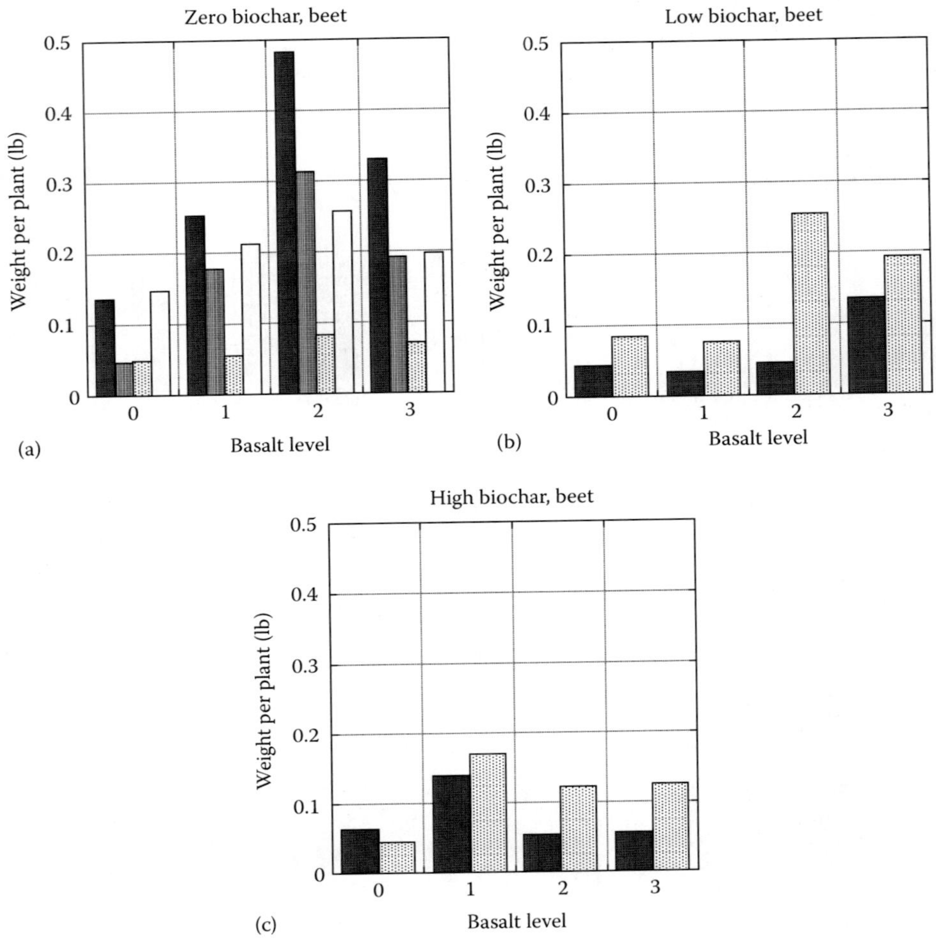

FIGURE 18.10 First beet crop as a function of basalt (a) with zero biochar, (b) with low biochar, and (c) with high biochar.

oxide but biochar plots had higher emissions, and both basalt and biochar reduced absorption of nitric oxide from the atmosphere (Figure 18.16). As these measurements were made at one time, on a wet day, and since methane fluxes were not measured, measurements need to be made throughout the season for a better understanding of the patterns of gas emissions between soil and the atmosphere. While these preliminary patterns are interesting, since there were only two measurements per treatment, more measurements are needed to determine whether these differences are statistically significant.

A recent paper (Lubbers et al., 2013) predicts that CO_2 emissions should increase by 33% and N_2O emissions by 42% with earthworm activity due to the increased exchange of gases through earthworm burrows and because worm castings and worm burrow linings may be sites of enhanced microbial activity that could affect greenhouse gas production and consumption. That prediction was largely based on lab experiments rather than field data, because it is difficult to manipulate earthworm density in the field.

Our experiment provides a natural experiment that has increased earthworm activity about 20-fold, but there seems to be no corresponding increase in fluxes except for nitrous oxide. Our preliminary results suggest that basalt dust with high earthworm activity appears to reduce both CO_2 and N_2O emissions,

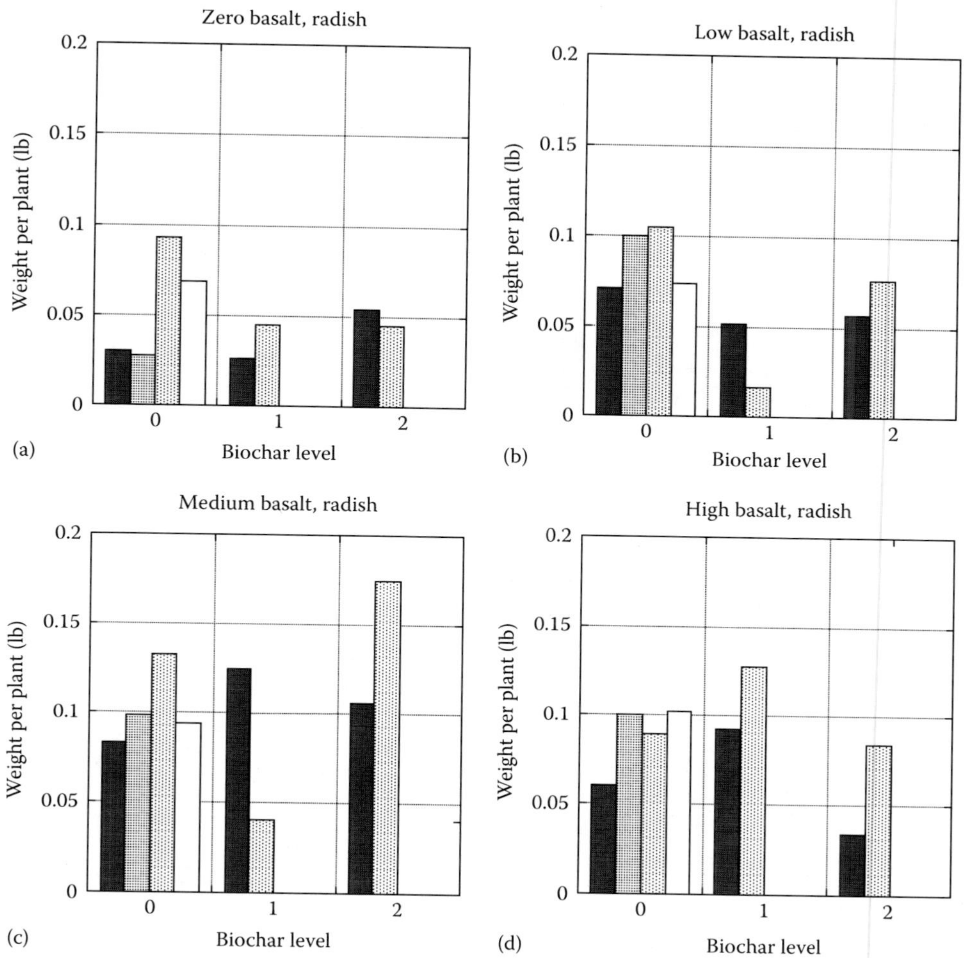

FIGURE 18.11 First radish crop as a function of biochar (a) with zero basalt, (b) with low basalt, (c) with medium basalt, and (d) with high basalt.

in contrast to predictions. This could result from basalt powder acting as a CO_2 sink through chemical weathering.

Some papers have suggested that biochar should reduce N_2O and CO_2 emissions (Spokas and Reicosky, 2009; van Zwieten et al., 2010; Scheer et al., 2011, Case et al., 2013). While we confirm the CO_2 decrease, our results seem to show an increase in N_2O emissions with biochar. The patterns appear to be complex and highly variable, depending in part on the available compost nitrogen and the level of soil oxygenation. Much further work is needed to resolve these patterns.

In addition, rock mineral weathering should serve as a CO_2 sink by converting it into bicarbonate ions. Measurements of soil alkalinity changes are needed to clarify this. It is clear that further work will be needed to better understand the long-term impacts of biochar and rock powder on soil greenhouse gas emissions.

In summary, the effect of basalt and biochar treatments on greenhouse gas emissions is complicated and involves biological (microbial and invertebrate communities), chemical (pH values, redox conditions, carbon to nitrogen ratios), and physical (soil properties) processes. Short-term effects may be different from the long-term. To evaluate the contribution of basalt and biochar to greenhouse gas emissions and

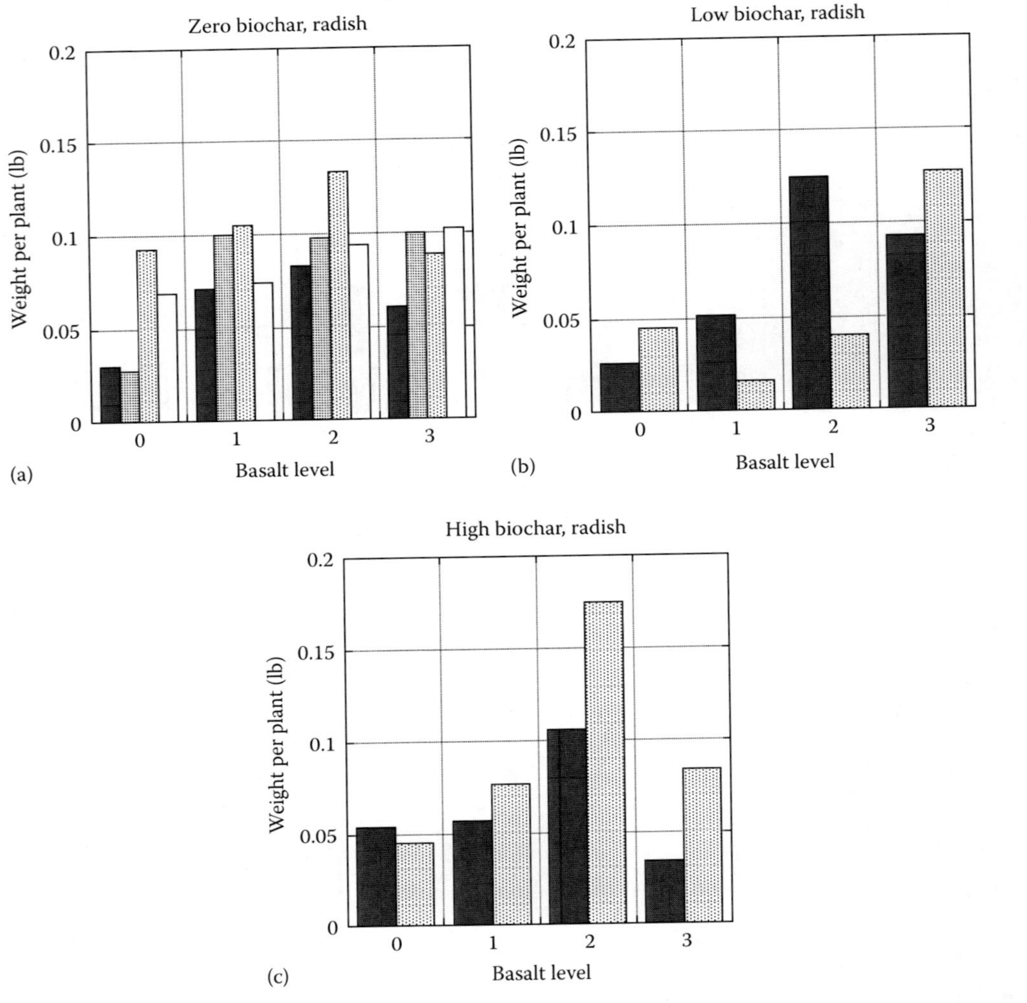

FIGURE 18.12 First radish crop as a function of basalt (a) with zero biochar, (b) with low biochar, and (c) with high biochar.

carbon sequestration, long-term measurements that cover the seasonal variation and periodic events (rain, snow, flooding) are needed.

Conclusions and Future Prospects

The strong immediate response to basalt powder was a surprise, as we had thought soils at this site were highly fertile, but the results indicate significant mineral deficiencies in the soil that were remedied by the added materials. We had initially thought that the soils at the site were derived from river channel deposits that were composed of farm and forest soils, which were eroded from upstream in the watershed following deforestation and agriculture and which would have been among the richest soils in the watershed. But the fact that the hurricane that plugged the channel took place before regional deforestation means that New Harmony soils represent natural river sediment before it was

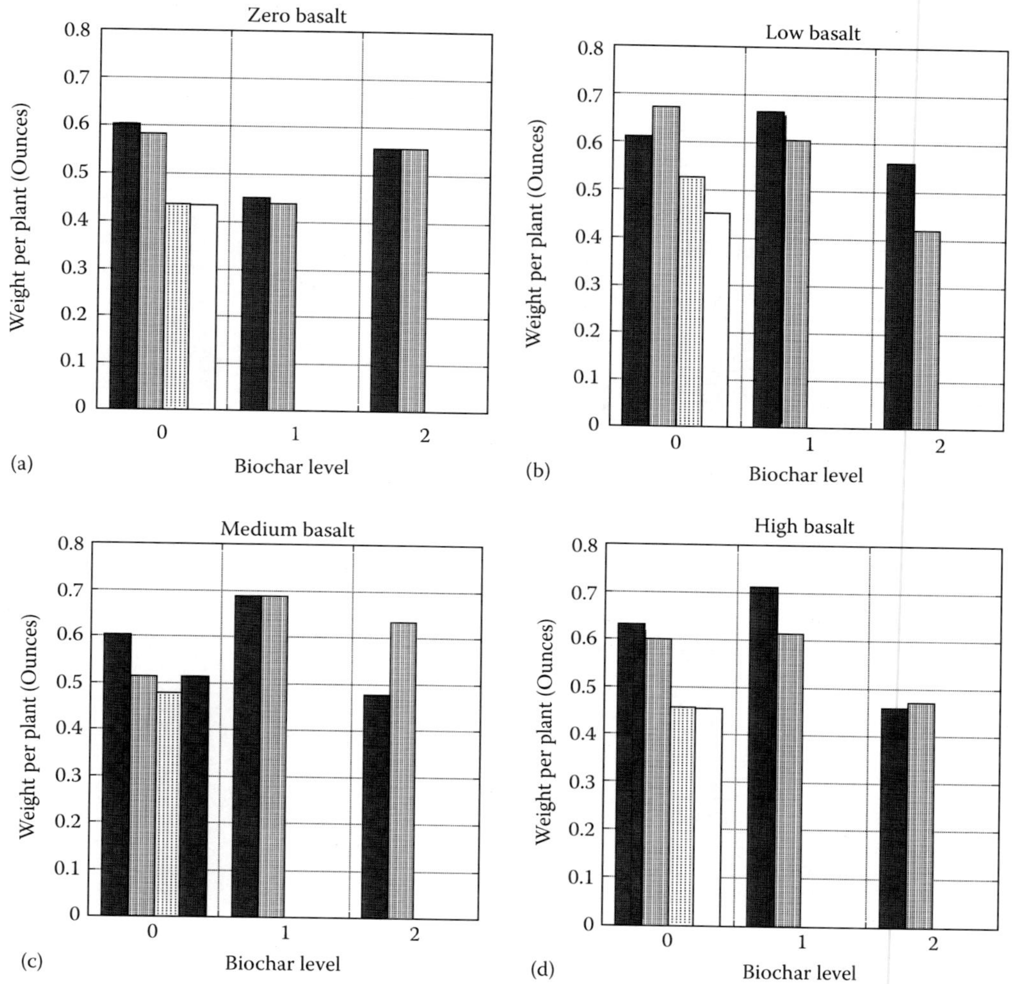

FIGURE 18.13 Second radish crop as a function of biochar (a) with zero basalt, (b) with low basalt, (c) with medium basalt, and (d) with high basalt.

enriched by erosion of forest and farm soils, which may explain the unexpectedly strong response to basalt powder fertilization.

Interactions between biochar and rock powder, and with soil biology, all of which lead to effects on plant nutrient uptake, growth, and nutritional quality, will be smallest at the very start of this long-term experiment and should change in the subsequent years as the materials mature and interact.

Our results to date represent only the very first steps on a long journey. Already they have revealed surprises, and we expect more to come. We intend to apply a much larger suite of measurements, including detailed soil elemental analysis, carbon and nitrogen isotope analysis, measurements of a larger suite of climatically active gases that are involved in soil biochemical metabolism, and measurements of crop nutritional quality in future work.

We expect that the agricultural productivity, crop nutritional quality, soil carbon storage, and greenhouse gas emissions will evolve as biochar and basalt powder continue to mature and interact. The long-term results will provide useful guidelines for managing biochar/rock dust combinations to optimize their economic, environmental, ecological, and climate change benefits.

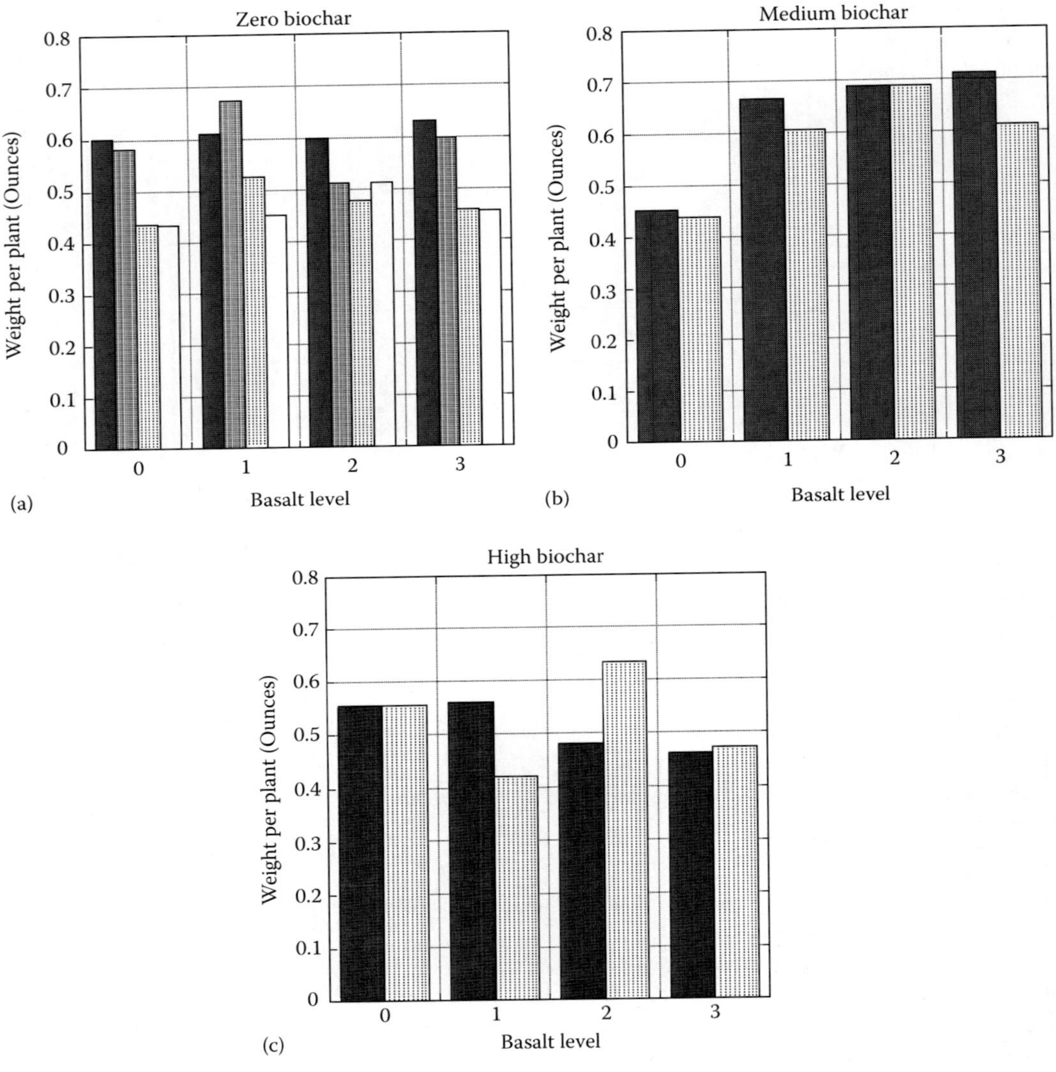

FIGURE 18.14 Second radish crop as a function of basalt (a) with zero biochar, (b) with low biochar, and (c) with high biochar.

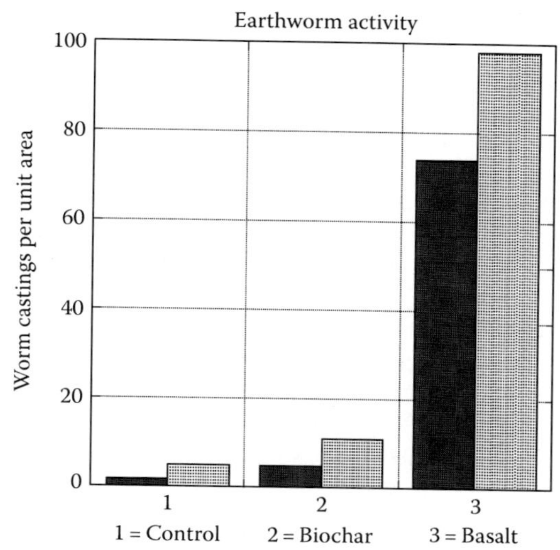

FIGURE 18.15 Earthworm castings per unit area.

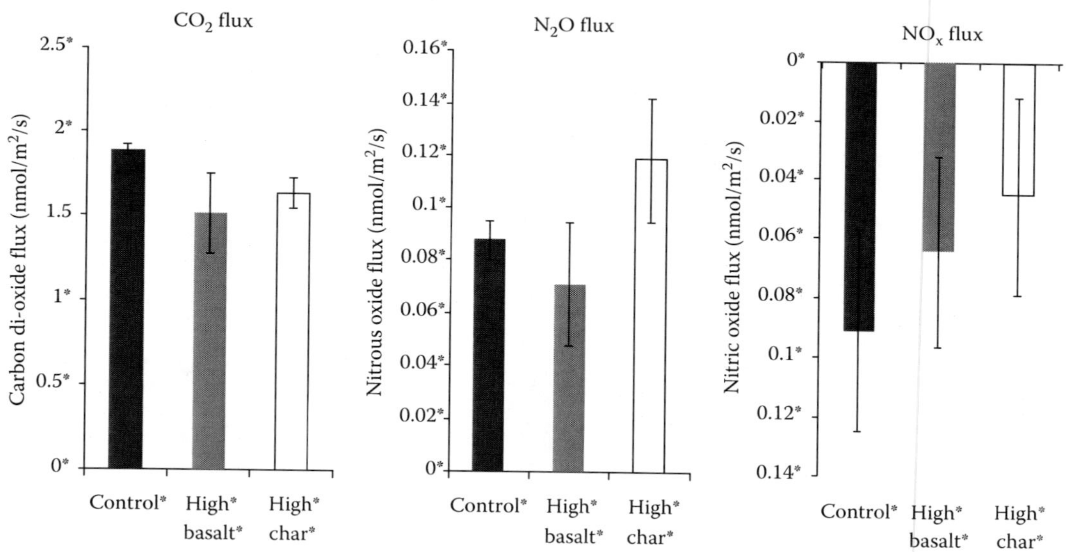

FIGURE 18.16 Gas emissions from plots measured between crops.

ACKNOWLEDGMENTS

We thank Rock Dust Local for donating the basalt powder for this experiment, Bob Wells of New England Biochar for providing the biochar, and all the dedicated volunteers who kept the plots weeded.

REFERENCES

Bormann, F. H. and G. E. Likens. 1979. *Pattern and Process in a Forested Ecosystem: Disturbance, Development, and the Steady State Based on the Hubbard Brook Ecosystem Study*. Springer, New York.

Case, S. D. C., N. P. McNamara, D. S. Reay, and J. Whitaker. 2013. Can biochar reduce greenhouse gas emissions from a Miscanthus bioenergy crop? *Global Change Biology Bioenergy* 6:76–89.

Cumbler, J. T. 2001. *Reasonable Use: The People, the Environment, and the State, New England 1790–1930*. Oxford, New York.

Darwin, C. 1881. *The Formation of Vegetable Mould, through the Action of Worms, with Observations on Their Habits*. John Murray, London, U.K.

Foster, D., S. Cooper-Ellis, A. B. Plotkin, G. Carlton, R. Bowden, A. Magill, and J. Aber. 2004. Simulating a catastrophic hurricane. In D. R. Foster and J. D. Aber (eds.). *Forests in Time: The Environmental Consequences of 1,000 Years of Change in New England*, pp. 235–258. Yale, New Haven, CT.

Kirwan, M. L., A. B. Murray, J. P. Donnelly, and D. R. Corbett. 2011. Rapid wetland expansion during European settlement and its implication for marsh survival under modern sediment delivery rates. *Geology* 39:507–510.

Ladygina, N. and F. Rineau (eds.). 2013. *Biochar and Soil Biota*. CRC Press, Boca Raton, FL.

Lubbers, I. M., K. J. van Groenigen, S. J. Fonte, J. Six, L. Brussard, and J. W. van Groenigen. 2013. Greenhouse gas emissions from soils increased by earthworms. *Nature Climate Change* 3:187–194.

Scheer, C., P. R. Grace, D. W. Rowlings, S. Kimber, and L. Van Zwieten. 2011. Effect of biochar amendment on the soil-atmosphere exchange of greenhouse gases from an intensive subtropical pasture in northern New South Wales, Australia. *Plant Soil* 345:47–58.

Spokas, K. A. and D. C. Reicosky. 2009. Impacts of 16 different biochars on soil greenhouse gas production. *Annals of Environmental Science* 3:179–193.

Woods, W. 1634. *New England's Prospect*. Reprinted 1977. University of Massachusetts Press, Amherst, MA.

Van Zwieten, L., S. Kimber, S. Morris, A. Downie, E. Berger, J. Rust, and C. Scheer. 2010. Influence of biochars on flux of N_2O and CO_2 from Ferrosol. *Australian Journal of Soil Research* 48:555–568.

19

Soil Remineralization Trial: Preliminary Effects of Montserrat Volcanic Ash on Barbuda Limestone Soils

John Mussington

CONTENTS

Introduction

Barbuda is the 62 mile^2 sister island of the two island states of Antigua and Barbuda. It is a flat coral limestone island with its highest point being just about 39 m. The nature of its limestone terrain meant that it could not be used for intensive-type agriculture to produce sugar cane as many islands in the Caribbean were. Despite this, Barbuda had a traditional history of agricultural production based on shifting cultivation involving slash-and-burn techniques that benefited from the advantages of biochar. This production was so successful that Barbudan produce was used to sustain the Codrington plantations which existed on the other islands. Agriculture though still remains a challenge since many of the older members of the community with traditional knowledge are passing away.

Montserrat lies about 32 miles south of Antigua and Barbuda. Its challenge has to do with trying to live with an active volcano that has erupted in recent times making a significant proportion of its land space unlivable, being buried in volcanic ash. Finding ways to make the Montserrat community sustainable is a major concern.

This scenario makes for an interesting research question. Could the abundant ash on Montserrat be utilized in Barbuda to improve soil fertility thereby improving agriculture production? Montserrat would find a useful market for its ash and Barbuda would gain better earnings from its agriculture. This would benefit both communities (Figure 19.1).

Volcanic ash was collected by Dr. Thomas J. Goreau from the bed of the major river draining the ash-affected area, who hand delivered the bag of ash by air to Antigua and by boat to Barbuda. This trial involved 25 students in the agriculture science and biology programs of Sir McChesney George Secondary School (MGSS) and had the advantage of teaching basic science while at the same time finding solutions for challenges facing two island communities.

Because harvesting was still underway at the time of submission of this chapter, we show key results in the following in terms of photographs. Full data will be reported on elsewhere.

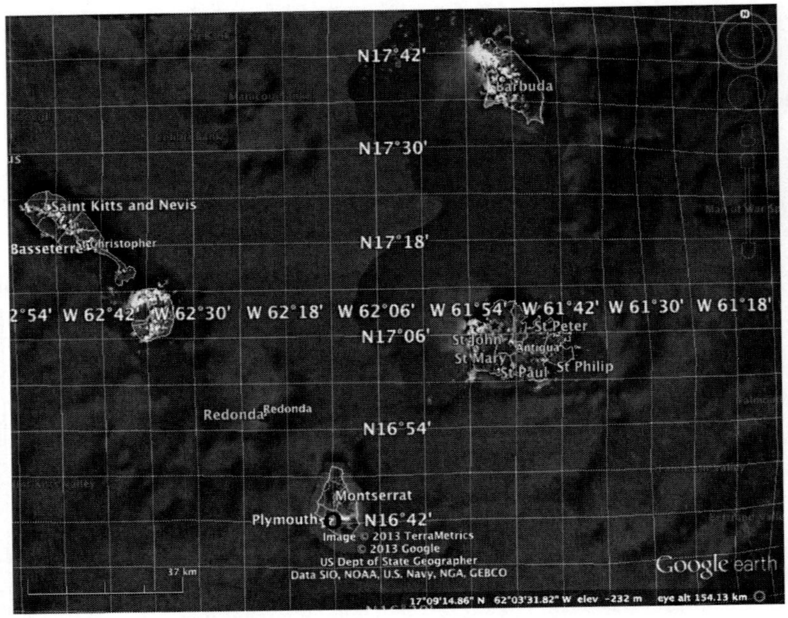

FIGURE 19.1 Barbuda (top) is a flat, dry limestone island. Montserrat (bottom) is a high, wet island with an active volcano.

Basic Research Question

Do seedlings of selected plants grow better and produce more when planted in Barbuda limestone soil that has been mixed with Montserrat volcanic ash than similar seedlings grown in Barbuda limestone soil that has not been mixed with ash?

Experimental Design

Seeds of okra, spinach, hot pepper, season pepper, and eggplant were collected from the school's garden in August. At the start of the academic year, the students were supervised in planting the seeds into individual seedboxes. Once the seeds germinated, 20 of the strongest of each type were selected for the trials. They were transplanted into individual pots: 10 into pots that contained only Barbuda limestone soil and 10 into pots containing Barbuda limestone soil mixed with ash in the ratio 1:5 or 1:9. The latter ratio was used for most of the sets since it was realized after preparing one set that enough ash would not be available for all the sets if we continued to use the 1:5 ratio.

Data Recorded

Over the first 6 weeks after transplanting, daily measurements of seedling height were taken to investigate growth rates. The next phase of data collection commenced after the potted seedlings were transplanted out onto prepared planting beds. Observations were made of qualitative parameters such as plant vigor, time of first flowering, time of fruit setting and ripening and finally, fruit yield by mass. The height of the seedlings were measured daily during the initial growth period.

Potting soil and seed boxes were prepared for germinating the seeds that would produce seedlings for use in the trials. The seeds were collected from plants in the school's garden a few weeks before: eggplant, hot peppers, season peppers, okra (Figure 19.2).

FIGURE 19.2 Students sowing seeds to produce seedlings per the trials.

FIGURE 19.3 Students care for the seedlings after germination.

Seeds germinate and students care for them until they are ready for transplant into the test media (Figure 19.3).

After a week, okra seedlings were moved into full sun for hardening in preparation for transplanting into test media (Figure 19.4).

Students review the experiment design before moving to the next stage where two batches of potting soil will be prepared: one containing no ash and the other containing volcanic ash at a ratio of 1:9 (ash to soil) (Figure 19.5).

Barbudan topsoil was collected to be used as the growing medium for the trials. For each kind of seedling, half the plants were transplanted into normal soil and half into soil mixed with volcanic ash (Figures 19.6 and 19.7).

The soil was mixed and seedlings transplanted into marked pots (Figure 19.8).

Okra seedlings were selected and transplanted into the marked pots—half in normal soil and half into soil mixed with volcanic ash (Figure 19.9).

The trial setup is as follows: there are 20 plants in each set, 10 plants in pots with normal soil and 10 plants in pots wits soil mixed with volcanic ash (Figure 19.9).

FIGURE 19.4 Okra seedlings placed in full sunlight for hardening before transplant.

FIGURE 19.5 Experiment design being reviewed before implementing next phase.

FIGURE 19.6 Collecting Burbudan limestone soil.

FIGURE 19.7 Burbudan limestone soil being prepared for use in the trials.

FIGURE 19.8 Seedlings being transplanted into either mixed or unmixed limestone soil.

FIGURE 19.9 Trial setup: 10 plants in pots with unmixed soil and 10 plants in pots with limestone soil mixed with volcanic ash.

FIGURE 19.10 Measuring seedling height.

FIGURE 19.11 Season peppers after 3 monts. The larger plant was from the group growing in limestone soil mixed with volcanic ash.

Measurements were taken daily of seedling height to investigate whether there were differences in growth rate (Figure 19.10).

Season peppers: After 3 months, the differences between the plants in the trial groups can be seen (Figure 19.11).

In the second phase of the trials, seedlings were planted out into the field to enable production of fruits. Here, eggplant fruits are weighed to investigate differences in yield and quality (Figures 19.12 and 19.13).

Students investigate differences in production by recording weight and quality of harvested produce (Figures 19.14 and 19.15).

FIGURE 19.12 Eggplant fruits being weighed to investigate differences in yield.

FIGURE 19.13 Eggplant fruits being examined to determine if there are differences between plants grown in soil mixed with volcanic ash and plants in unmixed soil.

FIGURE 19.14 Hot pepper plants and fruits being examined for differences in fruit yield and quality.

FIGURE 19.15 Hot pepper plants and fruits being examined for differences.

Hot pepper plants were inspected for number of flowers, number of fruits produced and the quality of the peppers with respect to size and color (Figure 19.15).

Overall health and vigor of plants were compared.

Summary of Main Findings

The differences in growth rate with respect to height increase were inconclusive. Plants becoming pot bound was an issue. This was particularly the case with okra that did not produce any fruits and most did not survive.

The most striking differences were recorded in the yield of the eggplant set. One group began flowering earlier, produced larger and more fruits, and generally were more vigorous in growth.

Hot peppers showed significant increase in fruit yield and fruit size. The effects in season peppers were also significant with one set showing more growth by the transplant stage than the other. Differences were also recorded with respect to time of first flowering and total yield.

Conclusions

The trials were a good opportunity to engage students in basic science and problem solving. There were measurable results that had real-life applications and that could allow Barbuda to produce more food.

The availability of ash was a problem since it was not sufficient to apply to the field beyond these initial experiments. The preliminary results were such that further research is very promising, and they need to be replicated on a larger scale, with more crops and with fruit trees, using mixtures of composts as well as rock powder and local soil. This time more attention to rigorous data collection for the purpose of comparative data analysis should be paid. Student participation in these efforts will give them a stake in helping develop a more productive future for their island.

20

Building Soil Where There is None: Feasibility of Using Recycled Glass in Growing Media

Fred Riger

CONTENTS

Introduction

The government of the Bahamas has long been concerned about food security as about 80% of foodstuffs are imported [1]. The island nation is literally days away from shortages if for some reason cargo is suspended. This has already happened due to weather and war events. It is a recurrent discussion that production should be increased, which seems obvious due to the warm and sunny climate [2]. Agricultural development has not followed desire or expectations. There are problems.

The soil is quite thin, where it exists; much of the land is limestone rock outcrop that makes preparing land very expensive (Figure 20.1) [3].

The surface needs to be crushed into a small enough size to act as a medium for growth, which is an expensive proposition and requires return to justify. This crushed limestone gravel needs application of expensive fertilizer to allow for growth [3]. During the summer season, these growing areas can get 5 in. or more of rain each afternoon [4], which washes the majority of the fertilizer into the groundwater and directly to the coral reef that fringes all the islands [5]. The farmer needs to add fertilizer in excess of what would be needed in other areas, and the nearby marine environment suffers.

The warm climate of the Bahamas is actually a problem. Most modern agricultural production actually depends upon frost; cold winter temperatures reduce plant disease and pests and allow crops to get a head start. Many crops require winter cold. In the subtropical climate of the Bahamas, we find ourselves a bit too cool for many full tropical plants and too warm for almost all temperate crops, especially perennials.

This also means that plant disease and pests operate year round. Without periodic suppression of these dangers to the crops, scheduled applications of pesticide, fungicide, and bactericide become problematic as there is a dearth of research into the special needs of our area.

Another problem is that our country is made up of deposited calcium carbonate, chemically quite pure when deposited. Dust storms in Northern Africa add small amounts of iron to the surface [6], which rain water washes down into the substrate, but plants capture some of it before it is taken too deep and becomes unavailable. Seaweed and flotsam wash up on shore, which brings small

FIGURE 20.1 Typical rock outcrop demonstrating the lack of soil. Taken near the garden plots in Freeport, Grand Bahama. (Photo by F. Riger.)

amounts of phosphorus and other micronutrients to the island. Many of the plants that grow on the shoreline and dunes are nitrogen fixing [7].

A very serious issue is calcium carbonate is quite alkaline, relative to most crop preferences. The chemical dictionary lists calcium carbonate as having a pH of 9 [8], although most of my measurements of the pH of soil and our hard water are 8.2–8.4. Most garden vegetables are listed as preferring 5–6, with some requiring 4–5 and a few that will *tolerate* 7–8 [9]. All of my experience prior to this experiment also showed that root-knot nematode is endemic to our soil [10].

The heavy rains, already mentioned, also serve to bind the calcium carbonate into a sort of cement that the locals call *beach rock* [11].

These difficulties have slowed commercial agricultural development and left local food production mostly in the hands of backyard gardeners.

In response to the aforementioned difficulties with the soil, local gardeners have to either use raised bed plots or containers or purchase or create their garden soil.

Background: Methods

Among the factors that led to this experiment was a desire to have a garden that produced useful food for the table, in the process to find a way to recycle available materials, and see if gardening costs could be lowered to make gardening feasible as more than just a hobby.

My experience with gardening in the Bahamas started with gardening in containers. I acquired some soil, scraped from an area of mangroves that had been targeted for mining. This became the base of my gardening medium. I was not terribly happy with this soil. It tended to harden too much, and growth of all plants was generally poor. As part of my business, I have a supply of activated charcoal and thought to start adding it as a soil amendment. Second guessing myself caused me to search the Internet for information concerning adding charcoal to garden soil. I quickly found a wealth of information of work being done on terra preta, making charcoal, and research on biochar [12].

FIGURE 20.2 One of the steel retorts used to make charcoal for the garden. A perforated pipe is welded to the bottom of the barrel. Gas pressure pushes pyrolysis products out a series of holes. The gases ignite and provide additional heat for the process. (Photo by F. Riger.)

Retorts were easily built using an oxyacetylene torch, each being made from a steel drum and a section of steel pipe.

Basalt powder used in these experiments was provided by Dr. Thomas J. Goreau and is the same material as described in Chapter 18 (Figure 20.2).

This retort purpose was to turn woody waste into char and charcoal. The retorts are fired from below. As pyrolysis commences, smoke and gases from the reaction are routed back under the retort and ignited to help heat and continue the reaction. Watching the smoke from the retort's burners can give a good idea of the temperature in the retort. White smoke signals the water is boiling out of the woody material. As the smoke turns brown, it ignites and pyrolysis continues until the burner flames stop. This is the signal the reaction has gone to completion (Figures 20.3 and 20.4).

Previous experience with adding this charcoal to garden soils improved the growth of plants. I did not quantify this improvement because previous to the addition of carbon to the soil, tomato and sweet pepper plants died before ripening fruit. After adding carbon at levels of 10%–20% by volume, the plants survived long enough to ripen some fruit.

I also had a project concerned with recycling glass or more specifically repurposing used bottle glass and decided to look into using crushed glass cullet as a base for garden soil. Waste glass is easily available.

Waste glass is stable, neutral, nontoxic, and heavy. It does not bond together or clump.

I had already been using it as hydroponic grow medium in my aquaponics unit, and it works quite well (Figure 20.5).

Charcoal is extremely light. It was not milled to powder before applying it; it was smashed to a mix of powder to granule and applied. Dry charcoal will float, and during heavy rains, this is one problem. Another problem with charcoal's lack of density is the soil needs to be heavy to allow the plants to survive the high winds that periodically visit the Bahamas. The local calcium carbonate sand is very porous and may not add enough mass to the soil, especially with the charcoal preventing it from solidifying as beach rock. I thought glass might solve this issue of weight.

FIGURE 20.3 Three retorts filled with woody material stacked in a frame awaiting ignition. Steel panels help to keep the heat directed at the retorts. (Photo by F. Riger.)

FIGURE 20.4 The fire is burning, and pyrolysis has started. Fuel will be added to the fire until no more flame comes from the retort burners. The fire will then be allowed to burn out, and the retorts opened after they are cool to touch. (Photo by F. Riger.)

FIGURE 20.5 Vegetables growing in a hydroponic bed using crushed bottle glass as the media, a demonstration of the viability of using glass as an inorganic component of garden soil. (Photo by F. Riger.)

Experimental Field Results

This experiment was done in two parts. The first part was a comparison of two plots, one based on the local calcium carbonate sand and the other on recycled glass cullet. Each was a 10 cm raised bed plot to compare the two garden soils (Figure 20.6).

In the second part, some basalt powder was acquired, thanks to Dr. Thomas Goreau, and expanded the comparison to four plots, calcium versus glass and calcium with basalt versus glass with basalt.

In the first part, two plots were prepared, side by side raised 10 cm, using media as follows:

One plot used local sand, collected from landscape runoff, 113 L by volume, and abbreviated Ca for the calcium carbonate that makes up the local sand.

Other plot used crushed bottle glass, all screened to 10 mm or smaller, 113 L of it, and abbreviated Si for the silicon of glass.

In total, 113 L of fresh biochar was added to both plots. The biochar was made using a retort method. Woody material was contained in a steel barrel and heated using fire fueled by both wood and the gasses produced by pyrolysis of the material.

Both plots got 56 L of fresh grass clippings.

A half kilogram of 6-4-6 fertilizer with Bo, Cu, Fe, Md, Mo, and Zn (South Florida formulation) was added and worked in.

The plots were turned and worked for 3 weeks before planting.

August 29, 2012

Final mix and smooth both plots from north row to south.

Planted; ea. Plot:

Seven corn seeds (Ferry-Morse Golden Cross Bantam Hybrid)

Ten okra (Ferry-Morse Emerald)

Four golden zucchini (Livingston's Golden)

Five watermelon (Ferry-Morse Jubilee)

Twelve eggplant (Ferry-Morse Black Beauty)

September 7, 2012

Corn and okra have sprouted, and most of the squash and watermelon have spouted. Some eggplant is showing through the Ca side, none on the Si yet.

(a) (b) (c)

FIGURE 20.6 (a) Preparing the garden beds and mixing the components. The glass bed is in the foreground; the limestone sand bed is in the background. Picture is taken from east to west from a height of about a meter. (b) A closeup picture of the soil mix made with the local calcium carbonate sand. The picture was taken from under a meter away with a wide angle lens. (c) A closeup picture of the soil mix made with crushed bottle glass. The glass was screened to a size not larger than 10 mm. The picture was taken from under a meter away with a wide angle lens. (Photos by F. Riger.)

It has been watered (or rained on) daily. When irrigated, care was taken to measure the same amount of water for each plot from the same rain barrels.

September 21, 2012

It has been added a top feed of fertilizer, about a half kilogram, 6-4-6 citrus formulated for South Florida with minerals to each plot.

Okra on the Ca plot shows noticeable chloridosis (Figure 20.7).

Corn seems to be less green as well but not as noticeable. This side also shows more insect damage on the squash.

November 9, 2012

Hurricane Sandy did extensive damage to the plantings on October 26. Plants on the Si plot were less damaged than the plants on the Ca side and showed signs of recovering. Virtually all plants in the Ca died or were so damaged as to preclude any hope of them continuing to set a crop.

The plants on the Si side have started to recover, but as there was no longer hope of a comparison, all the plants on the Si plot were transplanted to an aquaponics system. As a side note, the eggplant and okra recovered and have gone on to develop and produce, but the watermelon did not survive transplanting.

FIGURE 20.7 (a) Okra grown on calcium sand mixture showing stunted growth and chloridosis. (b) Okra grown on the glass mixture showing increased growth and healthy color. Both pictures were taken from east to west from about a half meter in height. (Photos by F. Riger.)

TABLE 20.1

Crop Comparison of the First Trial Between Silica-Based Soil and Calcium-Based Soil

Crop	Number of Plants	Condition	Length/Height (in)	Average Size
Ca:				
Corn	6	Nearly dead	10,10,9,5,10,8	8.67
Okra	0	Dead		
Squash	2	Sickly	10	10
Eggplant	1	Single leaf	5	
Watermelon	1	Weak	14	
Si:				
Corn	3	Nearly dead	8,8,12	9.33
Okra	6	Healthy	8,12,9,11,9,10	9.8
Squash	2	Flowering	24	12
Eggplant	2	5 and 3 leaves	4 and 3	
Watermelon	2	One flowering	24	24

Table 20.1 shows the number of plants, their size, and condition at the time that a hurricane ended this phase of the experiment.

At this point, it was decided to expand the experiment by adding two new plots to create a four-way comparison of Ca versus Si and each with the addition of basalt rock powder as a mineralization agent.

Each new plot was made with the same basic formula of charcoal, substrate, and lawn clippings but had the addition of 1.4 kg of basalt powder. Each of the four plots was fertilized with 45 kg of 6-4-6 fertilizer with Bo, Cu, Fe, Md, Mo, and Zn (South Florida formulation).

November 17, 2012

Each plot was planted in the same pattern: one sunflower, four corn, four okra, one cucumber, four broccoli, three golden zucchini, five beefsteak tomato, row of basil, and row of Simpson lettuce.

The Ca basalt had the strongest germination and growth. The zucchini in that plot flowered a week earlier, December 23, 2012, than any of the other plots and had two fruits harvested a week before any other plot (Figure 20.8).

(a) (b)

FIGURE 20.8 Taken on the same day, from about half a meter and toward the north, (a) zucchini grown on the calcium soil enriched with basalt, (b) zucchini grown on the calcium soil. (Photos by F. Riger.)

The Ca plot was the only plot to have a sunflower blossom on the first planting; they did not germinate on either of the basalt plots until the second planting.

All the corn died early.

First harvest of all four plots was on January 15, 2013 as follows:

Ca basalt—2.75 kg zucchini, 1 kg lettuce, and 1 kg purslane (*Portulaca oleracea*), as well as a half cup of basil and a sprig of dill (the dill and purslane were volunteer planting)

Si basalt—1.75 kg zucchini

Si—1.75 kg zucchini

Ca—0 fruit (in spite of three plants)

January 22, 2013

The zucchini on both Ca plots succumbed to a leaf fungus (Figures 20.9 and 20.10).

The zucchini on the Si plots were still alive and have small fruits (Figures 20.11 and 20.12).

On the Ca plot, the only one to start a cucumber, the cucumber plant died suddenly, apparently of the same leaf fungus that looks like powdery mildew.

January 25, 2013

I picked three leaves of chard from Ca plot.

February 1, 2013

The zucchini on the Ca basalt plot succumbed to the leaf fungus. I harvested one more zucchini from the Si basalt plant.

I harvested a colander of basil tips from the Ca basalt plot as well as a similar amount of lettuce.

Time ran out before any of the other crops matured enough to quantify. At the deadline, there were tomato plants with green fruit and bad cases of late blight. The cabbage on all four plots is setting heads. The best view of the cabbage heads is on the Si basalt plot. One broccoli matured on each of the Ca and Si plots. There was no difference in size of the broccoli heads; each was approximately a kilogram. I replanted sunflower, and while only the plant on the calcium soil is ready for harvest, the leaves on the Si plot are significantly larger than on the Ca plots. The sunflower leaves on the Si plot average 17 cm long × 16 cm at the widest versus 14 cm long × 11 cm wide on the Ca plots (Figure 20.13).

FIGURE 20.9 The zucchini has developed that looks like powdery mildew. On the basalt-enriched calcium soil, it is almost growing ahead of the disease.

FIGURE 20.10 On the calcium soil not enriched with basalt, the zucchini has succumbed to the fungal disease. (Photo by F. Riger.)

FIGURE 20.11 The zucchini plant on the glass-based soil enriched with basalt is resisting the powdery mildew.

FIGURE 20.12 On glass-based soil not enriched with basalt, the zucchini is doing about as well as the basalt-enriched calcium soil. (Photo by F. Riger.)

FIGURE 20.13 Typical sunflower leaves, on the right from the Si plot and on the left the Ca plot. (Photo by F. Riger.)

FIGURE 20.14 Sunflower heads, right side is from the Si plot and Ca is on the left. The larger leaf size is reflected in the larger head size. (Photo by F. Riger.)

Here are the sunflower heads from the Ca and Si plots. The sunflower head on the Si plot was not ready for harvest, but the size differential was significant (Figure 20.14).

Lettuce was distinctly different on the different plots. The biggest differences were between the calcium plots and the glass plots. A surprise difference was that the lettuce on the calcium plots grew leaves off a stem. The lettuce on the glass plots grew as heads (Figures 20.15 through 20.17).

The same thing happened with the cabbage. The cabbage plants on the Ca plots had long stems and small heads, while the cabbage on the Si basalt plot had shorter stems and larger denser heads. As the deadline was extended, it gave time to quantify the cabbage (Figures 20.18 through 20.21).

Table 20.2 shows the length of stem versus overall length and weight of stem versus overall weight demonstrates this effect and the relative efficiency of the two growing media. It also shows the quantity of cabbage produced in each plot and demonstrates how soil makeup can influence percentage of useable product.

FIGURE 20.15 Lettuce, grown on the Ca plot to the left and grown on the Si plot to the right. (Photo by F. Riger.)

FIGURE 20.16 Lettuce, grown on the Ca with basalt on the left and grown on the Si with basalt on the right. (Photo by F. Riger.)

FIGURE 20.17 A side-by-side comparison, Si with basalt on left and Si alone on right. (Photo by F. Riger.)

FIGURE 20.18 All the cabbage from the calcium plot. The stems were clipped off at ground level. (Photo by F. Riger.)

FIGURE 20.19 The cabbage from the calcium plot with basalt. (Photo by F. Riger.)

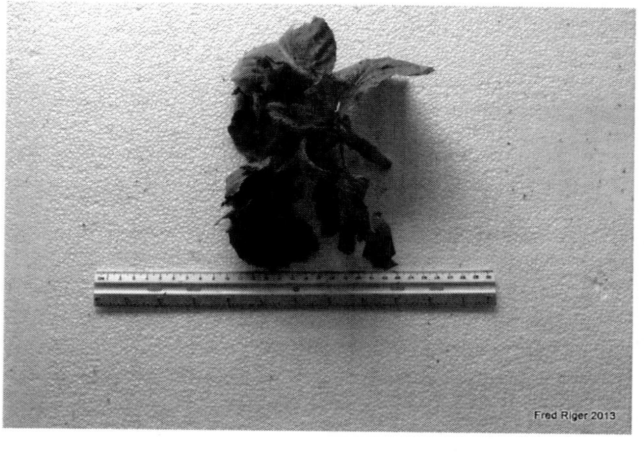

PICTURE 20.20 Only one cabbage plant survived on the glass plot. (Photo by F. Figer.)

FIGURE 20.21 Cabbage grown on the glass plot with the addition of basalt powder outperformed the other three test beds. (Photo by F. Riger.)

TABLE 20.2
Cabbage Results

Length Overall	Length of Stem	Stem (%)	Weight Overall	Weight Stem	Stem Weight (%)
Si Basalt					
20	6	30	153.5	3.1	2
22	4	18	398.8	7.9	2
30	6	20	219.8	8.8	4
28	4	14	652.7	1.4	0
Si					
11	4	36	27	2.7	10
Ca					
16	10	63	18.4	5.3	29
15	12	80	16.1	4.9	30
27	15	56	100	18.7	19
45	40	89	309.6	39.6	13
Ca B					
31	20	65	43.5	10.4	24
42	17	40	598.3	25.2	4

Discussion

Local gardeners have noticed that very few plants will produce well on the local soil. Most plants succumb to disease or insects before harvest, regardless of fertilizer or pesticide applications. Most gardeners import *soil* or build garden soil that eventually has little resemblance to the local soil. I had already done experiments that convinced me that adding charcoal would improve the local soil to the point where gardening might be worthwhile.

My original thought was to do a comparison of calcium/carbon-based soil and silicon/carbon-based soil. It was my impression from years of gardening that the high amount of calcium in any local soil mix caused many plants to grow poorly, either because of the high pH or calcium's tendency to bind vital micronutrients.

The first planting showed this tendency quite strongly. I was very careful to treat both plots the same, even to the point of measuring irrigation water and always using the same water source on both plots (my three basic water sources, tap water, rain water collected in plastic barrels, and rain water collected in steel barrels). Plants on the Si plot grew better and started to produce earlier. When hurricane Sandy caused me to replant, many of the plants on the Si plot survived to be moved to my aquaponics grow beds. The plants on the Ca side either did not survive the storm or succumbed shortly afterward.

It was when I decided to add two more plots to do a cross comparison with basalt dust that the results became less clear-cut. In the 2 × 2 array, the Ca plot that had the addition of basalt powder by far and away out produced the other plots at first. But it was not the case across the board for all crops. Two plants, chard and sunflower, started out doing better in the non-enriched calcium soil. The sunflower was later surpassed by the planting on the Si soil. Broccoli and cabbage started out doing better in the non-enriched silicon soil. Lettuce, basil, and zucchini started out doing much better in the basalt-enriched calcium soil. The tomatoes have not set fruit yet, so it is too soon to tell, but the zucchini showed the typical tendency of early die-off on the calcium soils. The basalt-enriched Si soil seems poised to have the highest yield over time as the plants have remained healthier. On the basalt-enriched Si soil, the zucchini plant has powdery mildew but has survived 1 month after all the other zucchini plants succumbed. It has immature fruits on it, and I may get another harvest after the deadline for this paper.

Conclusions

Using crushed glass as an inorganic base for anthropogenic soil is a viable option for building soil in poor soil areas like the Bahamas. Crushed glass mixes well with biochar, compost, and other garden media. It is a way to increase depth of the soil, so instead of taking up space in a landfill or becoming litter, it can be used to produce food that is in demand. While the use of glass instead of sand or crushed limestone is not a simple case of *better or worse*, it does allow for the production of a wider variety of garden vegetables and improves the quality of some.

Some vegetable crops will grow well on alkaline soil, and farmers in the Bahamas might do well to target growing these crops if all they have access to is alkaline soil. Many more types of vegetables will do much better if the soil is neutral to slightly acid. Using glass instead of limestone or calcium carbonate sand will serve to allow the backyard or small farmer to grow these crops for which there is a fairly high demand. In the case of crops that head, like cabbage and lettuce, growing in a glass-based media allows more efficiency of water, space, and fertilizer as more usable product is grown with less stem.

The addition of basalt powder as a mineralizing agent enhanced the growth of plants on either base, calcium or glass, leading to earlier and greater production. The relatively small amount used in this experiment had a significant effect on the growth and health of the plants and is something to continue to look into. Further work needs to be done on various mixes of glass and calcium sand to find if there is an optimum mix and will be forthcoming.

REFERENCES

1. The CIA World Factbook. https://www.cia.gov/library/publications/the-world-factbook/geos/bf.html.
2. http://www.baic.gov.bs/Agriculture%20Department.html.
3. Sealey, N, April 2010. Soil and land resources of the Bahamas. http://www.iica.int/Eng/regiones/caribe/bahamas/ATT/Documents/Soil%20and%20Land%20Resources.pdf.
4. Geerts B. Convective and stratiform rainfall in the tropics. http://www-das.uwyo.edu/%7Egeerts/cwx/notes/chap10/con_str.html. Retrieved on November 27, 2007.
5. Goreau, T.J. Waste nutrients: Impacts on coastal coral reefs and fisheries, and abatement via land recycling united nations expert meeting on waste management in small island developing states, October 27, 2003.
6. Mahowald, N.M. et al. 2009. Atmospheric iron deposition: Global distribution, variability, and human perturbations. *Annual Review of Marine Science* 1, 245–278.
7. Lonard, R.I., Judd, F.W., and Stalter, R. 2012. The biological flora of coastal dunes and wetlands: Salicornia bigelovii J. Torrey. *Journal of Coastal Research* 28(3)719–725.
8. http://www.engineeringtoolbox.com/bases-ph-d_402.html.
9. http://pss.uvm.edu/ppp/pubs/oh34.htm.

21

Soil Remineralization in Scotland

Jennifer A. Brodie

CONTENTS

Introduction

In Scotland, no account of the efficacy of adding mineral-rich dust, produced purely from crushing and screening mineral-rich rock to a fine power and adding to compost, soil, and wormeries, would be complete without reference to SEER Rockdust and its co-founders, Cameron and Moira Thomson. This inimitable Scottish couple devoted their lives to demonstrating and developing a brand of powdered rock, which they called SEER® Rockdust® and I first visited them in November 2004.

Starting out as professional art teachers, they were setting up their home, near Dundee in Scotland, in the 1980s with a young family, striving to grow their own food organically when one day they heard a Radio 4 broadcast reviewing a book called *The Survival of Civilization* by John Hamaker and Don Weaver. A recording of Cameron and Moira Thomson telling their own story can still be heard live on Binn Soil Nutrients website: http://www.binnsoilnutrients.com/seer-centre/.

In the late 1990s, the couple moved to a 6 acre site, 1000 ft above sea level in the Scottish Highlands and established the SEER Centre Trust. SEER is an acronym for Sustainable Ecological Earth Regeneration. Here, Cameron and Moira developed what professor David Bellamy called the best example of *soil remineralization in the northern hemisphere*. Such was the climate and condition of the soil when they arrived, that local farmers told them they would never even grow potatoes. Undaunted, they built terraces made of local authority green waste compost mixed with quarry dust from a local basalt quarry. They also had areas where they used only SEER Rockdust and natural soil; this to quash any claims that results were due to compost and not SEER Rockdust. SEER achieved significant grant support and the gardens the Thomsons developed were nothing short of phenomenal in production of exceptionally large, healthy, tasty vegetables, fruit, and flowers. During the dry summers, the crops required no extra watering, and in the wet summers, the crops thrived equally well. Once harvested, the crops proved to have exceptionally long shelf lift. In the early days, they received extensive media coverage including Scottish, UK, and, in one instance, Swedish television.

Market Penetration

In 2004, after the SEER Centre launched SEER Rockdust onto the U.K. market, I was recommended by colleague John Ferguson to visit the SEER Centre. I was simply blown away by what Cameron and Moira Thomson had achieved and what they had to say. Since then, working to get SEER Rockdust to the market has basically taken over my life. At an early stage, SEER had the product approved by four UK organic accreditation bodies and there is now a significant and growing market niche among the expanding *grow your own* market. Market is UK wide and involved export in larger and smaller amounts, to Slovenia, Sweden, Holland, Denmark, Majorca, and Austria. The export of the dust from the UK is viewed very much as a precursor to encouraging companies to work with us to source suitable mineral supplies, as a by-product of quarrying, in their own countries. Within the United Kingdom, we started with privately owned Garden Centres and whilst the sale of the first pallet was easy enough repeat business was slow. Newspaper and magazine articles featuring SEER Rockdust promoted sales. Cracking the farming and horticulture market however is going to require more professional research and trials. We received a boost in recent years when the DIY chain, B & Q, took it on in their own bespoke bags of 10 kg each.

Users Feedback

The SEER Rockdust sales leaflet lists the benefits that the Thomsons have noted in growing crops at the SEER Centre. To date, the most interesting aspect of my work has been receiving and collating user reports from very wide range of SEER Rockdust users that confirm, anecdotally, the Thomsons' finding. A good cross section is included on SEER website (http://www.seercentre.org.uk) and I could fill this book with them. Instead, I will pick out a few to illustrate this chapter.

Health and Vigor

Raspberry Farmer (Photos 21.1 and 21.2)

My first task, after first visiting the SEER Centre, was to organize "The Secrets of Rockdust Conference 2005" at Pitlochry Festival Theatre. A nonorganic raspberry farmer Grant Williamson, who attended the conference, was inspired to take in a bulk load of SEER Rockdust shortly thereafter in early spring of 2005. As I did not hear back from the farmer by late autumn, I phoned him and was taken aback to hear him proclaim, "SEER Rockdust is a con, just a pure con." I thought all was lost but I need not have worried.

PHOTO 21.1 Raspberry canes with no SEER Rockdust as seen the following Spring.

PHOTO 21.2 Raspberry canes that had been fed SEER Rockdust as seen the following Spring after application.

The following spring the tide had turned. Grant phoned me and, in a different tone, confessed that *he was wrong, just plain wrong*, and he invited me to visit his farm. When I arrived, he met me at the car and was able to point out the difference that I was about to examine in more detail, from quite some distance. Grant fed his crops by fertigation, that is, pumping water around his raspberries via seep hoses and putting the feed into his barrel, from where the water was pumped prior to filling the hoses. What was now evident were the 12 rows of raspberries that had had the SEER Rockdust added were bolt upright, leaves were significantly stronger and richer, and the young fruit far better developed.

As shown by the photographs, the rows that had not had the SEER Rockdust were significantly less developed and healthy.

A key point to this was that the previous year Grant had also taken a bulk load of PAS100 accredited green waste compost. While in year 1 he had the best crop of raspberries he had ever had, he had seen no difference with the Rockdust. It was not until the following spring when the advantage of the SEER Rockdust was to show itself loud and clear—as Grant said, "to the row."

Regatta Lane Allotments (Photo 21.3)

Right around the United Kingdom, the rise in demand for allotments has been dramatic. Sites with plots that have laid empty through the 1980s and 1990s have suddenly been taken up and have waiting lists as more folk become eager to grow their own. Via the United Kingdom's largest allotment associations' parent body, the National Allotments and Gardeners Association, www.nslag.org.uk, we have been able to access this market with many associations having their own shop. In 2012, I had an enthusiastic association on the phone asking to order another pallet. Ever eager to hear in detail of users experiences, I asked the chairman if he would round up his members views.

Here they are

1. Best *bedding plants* and display I have ever had.
2. My *onions, shallots, and garlic* are much bigger than other years and there was a reduction in white rot.
3. *Runner beans* have cropped really well this year.
4. It's the first year my *cauliflowers* have not had stem rot and the size is great.
5. I potted half my *sweet corn* in a mixture of compost and Rockdust and half just in compost. The former plants were twice the size of the latter.
6. I mixed Rockdust in my *leek* seed bed; results were fantastic—roots were so numerous and large I had trouble transplanting them into their final holes.
7. My *brassicas* plants are tremendous; the plants look very vigorous and strong plus the overall size of the plants is great.
8. Most economic product I have ever purchased!

PHOTO 21.3 Regatta Lane Allotments shop with a 20 kg bag of SEER Rockdust on the counter.

Dundee Flower Show

The co-founders of SEER Rockdust, Cammi and Moira Thomson, have strong connections with Dundee and no where is the appreciation of SEER Rockdust more evident than at the annual Dundee Flower Show. In 2011, I rounded up the following:

1. *Cabbages and Gladioli (Photo 21.4)*: Before the show had even started, Keith Brand came on the stand and proclaimed: "That stuff is magic! I don't know how it works, but it fair makes things grow. It is just amazing. I used in on my cabbages and I could not believe it, they doubled in size in a week. Come and see what it has done for my gladioli! (He was overall Gladioli champion for the show!) I do a lot of hybridization and SEER Rockdust improves quality and over wintering of the corns."
2. *Parsnips (Photo 21.5)*: Tom Caldwell, a private gardener told us: "I could never grow parsnips in my garden but I got a couple of bags from you at the Glasgow Show two years back and I have had a good crop of parsnips each year since."
3. *Alpine Plants (Photo 21.6)*: Graeme and Hilary Butler are fellow stand holders and I have visited their *Rumbling Bridge Alpine Nursery*. From the show, with his usual tinge of humor, Graeme gave me this statement: "In some cases SEER Rockdust is a problem because the alpines grow too fast – they forget they are meant to be Alpine! I found however if I added a small amount to the New Zealand Daisy, as a top dressing, it really helped get this tricky Alpine established. Without SEER Rockdust I am not sure if I could grow it."
4. *Cabbages and Roses (Photos 21.7 and 21.8)*: Keith Pool, then head gardener of a Fife estate, told us: "Not so much in 1st year, but by years 2 and 3, wow! This cabbage was 9.5 lbs and our roses where we applied SEER Rockdust excelled while those we left without got Tar Spot were spindly and had few flowers." The latter comment strays into the next category.

PHOTO 21.4 Keith Brand with his winning gladioli at Dundee fruit and flower show.

PHOTO 21.5 Gardener visiting Dundee flower show who could not grow parsnips without his SEER Rockdust.

PHOTO 21.6 Graham and Hilary Butler who used SEER Rockdust to grow their New Zealand Daisy.

Pest Resistance

It can be startling when you find users trying to educate you what you have been used to trying to advise them. A good example of this was when Mrs. Shirley Watt announced "Normal plant food make the plants soft and sappy and much more vulnerable to greenfly infection. This does not happen when you grow your plants with SEER Rockdust as the greenfly cannot get a bite into the stronger plants." She went on to tell me her garden was nominated *the best garden in Banchory* (Photo 21.9).

PHOTO 21.7 Keith Poole, Fife gardener who grew bumper cabbages using SEER Rockdust.

PHOTO 21.8 Keith Poole whose roses with SEER Rockdust did not catch Tar Spot.

Shirley is a real fan of SEER Rockdust and later told me that a LABUR NUM TREE she had had for years had never flowered and she was contemplating on removing it. Instead, she gave it a good dose of SEER Rockdust and the next spring it exploded into bloom—certainly saving its skin (Photo 21.10).

With a lady on the island of Orkney who had a large greenhouse that was suffering from a dreadful dose of greenfly, I advised that she use a Thomson tip and add SEER Rockdust to as good a shake as possible,

PHOTO 21.9 Shirley Watt's garden named "The best garden in Banchory" after using SEER Rockdust.

PHOTO 21.10 Shirely Watt's Laburnum Tree bloomed for first time after application of SEER Rockdust.

and then overwater her plants with it. She contacted me a short time later to say all the greenfly had gone—could it have been the SEER Rockdust? Indeed, from where I am standing, we have only scraped the surface of what SEER Rockdust can achieve.

Flavor

Feedback is often unexpected. A lady told me "Oh yes, we bought PEAS from the SEER Centre a few years back and the flavor was amazing. My husband and I thought we have lost the ability to taste, but one

PHOTO 21.11 Worms on ECOWORM's worm farm showing they relished the SEER Rockdust.

bite into the SEER Centre PEAS and it all came right back!" I know our supermarket apples and strawberries have lost their flavor over the years—please rest assured it is not our own loss of the ability to taste that is happening here. Our crops have lost their flavor. We may have the answer with SEER Rockdust.

Worms

The quantity and quality of worms, in recent years, in the SEER gardens have been absolutely phenomenal. When we had the TV cameras there after "The Secrets of Rockdust Conference," it was a wet morning and you could not move but you would stand on a worm. Independently, a gardener in Inverness had just moved house and complained of no worms. We supplied her with green waste compost and SEER Rockdust. The following year, she reported that every spade of soil had at least one healthy worm in it and they got the biggest tastiest strawberries they had ever grown.

Photo 21.11: Bringing us right up to date at The Edible Garden Show last year, our stand was next to a new UK company *Ecoworm*. Their worm farms are in Lithuania, and over there, they have since trialed SEER Rockdust in their wormeries. Within a couple of days, the worms congregated around the SEER Rockdust and showed increased activity. I have observed this in my own wormery. Mention has to be made in the category of Ron Gilchrist, Scotland's leading worm-man and running Greenway Consulting (www.green-way.org.uk). He has used SEER Rockdust in his wormeries for years and seen the benefits to the worm. This is well featured in his new book (Gilchrist et al. 2011).

Flower Quality

Photo 21.12: There are many I could mention here, but to save space, I have picked out one. We have just started producing SEER Rockdust in 450 g packs and I supplied a lady last autumn and this his spring, she told me she tried it right away on a *peace lily* that she had had for 5 years. This plant had never flowered since she got it. Within a week, it started its first flower—and 5 months later, it remains in flower looking healthier than it had ever looked.

Export

And so to Animals

Photos 21.13 and 21.14: Export to Slovenia is thanks to Sonja Cernic Langerwall. She imported a lorry load and saw such a difference with her vegetables and begonias she persuaded gardening chain Kalia, part of seed company, Semenarna to meet me. We translated bags and fliers to Slovenian and several

PHOTO 21.12 Peace Lily that flowered for first time in 5 years, 2 weeks after application of SEER Rockdust.

PHOTO 21.13 Slovenia: Remineralized Begonias flourish outside Agent's house.

lorry loads have been exported. Export to Majorca is thanks to Englishman Robert Heron who bought a 24 ton load. Such was his success with tomatoes, bush beans, onions, and strawberries he now sells via his company http://www.terranova-balear.com.

Not If, but How

After seeing the Thomsons results, listening as they recount their knowledge and experience, then seeing for myself evidence, albeit *unscientific* from around the United Kingdom and now Europe, for me, it is not a case of IF SEER Rockdust works but *HOW*. If we can work out how, I suggest we can better

PHOTO 21.14 Majorca: Agent's stockist holds families remineralized aubergines.

hone our research to understand, prove, and so benefit from what is happening when we apply SEER Rockdust. I do buy Cameron Thomson's line that, generally speaking, scientists are specialist and do not take a step back and objectively view the full picture, outwith their own discipline.

Five Components of Soil (Diagram 21.1)

While my honors degree is in zoology, it took a local geologist, Dr. Peter Craig who gardens with SEER Rockdust and who guides us in finding potentially good sources for SEER Rockdust, to show where it fitted into soil. In Peter's words, and as echoed by him, in our online video

> Soil is made up of 5 things, (i) AIR, (ii) WATER, (iii) DECAYED ORGANIC MATERIAL (basically what we see as soil), (iv) MICROFAUNA (bacteria to worms) and (v) MINERALS. When it comes to SEER Rockdust what we are dealing with is this fifth component, minerals.

(v) MINERALS: *Basically* a tiny but critical percentage of every crop we harvest is minerals. If we do not replace these minerals, or only replace a small percentage of some of them with plant food, the soil is bound to be running short of minerals. It is 10,000 years since the last ice age when glaciers crushed and transported and deposited the minerals to help make up Europe's soil. The analysis of SEER Rockdust shows it contains all of the macronutrients and all but one (iodine) of the micronutrients. Furthermore, a full analysis shows that it contains two-thirds of all the nonsynthetic elements in the periodic table, only not necessarily in the elementary form. The most informative book to understand the elements and their human, food, medical, economic, environmental, chemical, and surprise aspects is (Emsley 2003). The complexities involved here, certainly not yet fully understood by today's science, are outwith the scope of this chapter, but this book helps to unravel the full picture.

There is an entire vitamin industry built up trying to replace the vitamins and minerals now lacking in our diets. The explosion of obesity, heart disease, cancer, etc., all have their roots in our diet and a big factor in this is, I think, that our diets now lack a full quota of minerals. This is because our soil, where our food comes from, is running out of them. We get minerals from our food, most of which is ultimately supplied from our soil, seafood being the exception. Again SEER Rockdust has, we think, an as yet not understood role to play in improving our health. As John Muir noted, the crushed rock in the mineral-rich

DIAGRAM 21.1 Back to basics: Role of minerals in soil fertility.

moraines left by the glaciers in Yosemite National Park grow the biggest plants on the planet (see section Wisdom of the Ancients). Do we have now the wisdom to learn from nature and emulate these procedures for ourselves—this within our own lifetime—rather than waiting for the next ice age?

Liebig's Barrel (Diagram 21.2)

The best theory I have ever come across when considering how SEER Rockdust works was introduced to me by a senior and well-respected Scottish agronomist, Bill Petrie, who calls his consultancy *Liebig Technology*. I learned from Bill that a German chemist, Justus von Liebig, born 1803, given the name *father of the fertilizer industry* stated that "The YIELD of the crop is LIMITED by the NUTRIENT in SHORTEST SUPPLY." This is illustrated by what is now known as the Liebig Barrel, with the shortest stave of the barrel dictating the maximum that the barrel can hold. This could explain a couple more of our User Reports, that is, why a vegetable grower and a couple trying to establish a new garden on *builders* soils had tried for several years to make their garden grow only achieved success when they applied SEER Rockdust. Both gardens were probably primed with the best the shops could provide in the way of manure, lime, growmore, etc—but it took one of the multitudes of minerals in SEER Rockdust to peg up the last stave of the barrel so that their respective gardens could flourish at last.

(iv) MICROFAUNA: Having an Honours Zoology Degree, I have most affinity with the fourth component, microfauna. Soil science is of course a well-established and respected discipline to which

DIAGRAM 21.2 Liebig Barrel illustration showing how shortage of one mineral can impede soil fertility.

this *out-of-the-box* thinking will be little more than an irritation. However, in support of my back to basics approach, I found the "Loss of Biodiversity" section in a 2006 paper entitled (Towers et al. 2006) http://www.scotland.gov.uk/Publications/2006/09/21115639/1 on the Scottish Government's website particularly interesting. It makes the following key statement: "The diversity of life (invertebrates and micro-organisms) in soils is vast and un-explored. *Soil biodiversity is therefore a true scientific frontier*." As it is the microfauna that is making the minerals in the SEER Rockdust available to the plants, this paper dated 2006 puts into perspective how little is really known about soil.

A breakthrough in thinking has come with the new book (Lowenfels and Lewis 2010). This book is a joy to read and it clearly illustrates how this discipline has so far been largely ignored by our soil scientists. It takes you up through the food web, from bacteria to worms. We learn that once soil has passed through a worm, it has *10 times available potash, 5 times nitrogen, 3 times useable magnesium, and one and a half times calcium*. The pH of soil of course has a big part to play in how readily plants can uptake nutrients from the soil—soil microfauna has an as yet often unacknowledged part to play in this. To date, growers and suppliers of their plant food have concentrated on feeding the plant, not the soil. Not until we look after our soil, and its five components, will we reverse what is now happening to our soils and our food.

Charcoal

It has occurred to me to wonder where charcoal (trade names for example, carbon gold, among others) fits into the five component theory. Firstly, charcoal is not a natural ingredient of soil and it is not adding nutrients. We think it is the porosity of charcoal that is making the difference. It serves as a surface that benefits the reproductive rate of the microfauna, actually a feature that has also been attributed to freshly crushed rockdust, and it also helps to hold the nutrients in the soil and prevents them being washed away.

Chemical Industry

One of the speakers at our 2005 "The Secrets of Rockdust Conference" was Graham Harvey. He went on to write a seminal book (Harvey 2006) that has never received the credit it deserved. I have kept a stash of these books and given them to various contacts from farmers to casual observers over the years, and many of them rave about it. Graham explains how the chemical industry has the vast majority of farmers and other growers of our food, totally dependent on nitrogen-based fertilizers that are damaging our soils. Not to demonize the chemical industry in its entirety, I do think SEER Rockdust has a role to play in weaning us, at least in part, off dependency on substances that, while boosting plant growth in the short term, can damage our soil in the long term. All the other benefits outlined in this chapter will then also be at our disposal!

Organic Movement

In the United Kingdom, the debate continues as to whether organically produced food is more nutritional than nonorganically produced food. The organic movement has proven that food can be produced without chemical fertilizers, herbicides, and pesticides—but the case for organically grown food being nutritionally better only very recently secured scientific evidence of it's superior nutritional value as shown here: http://www.soilassociation.org/organicisdifferent. That is the question that SEER Rockdust and the soil remineralization movement have the potential to resoundingly answer positively. If it can be shown that food grown with SEER Rockdust, and other forms of mineral-rich dust, is more mineral rich than food grown without it, then history will indeed be made from which the whole soil remineralization movement and society can significantly benefit.

Wisdom of the Ancients

Before touching on the *scientific* side, I would like to visit what I call *the wisdom of the ancients*. Readers of contemporary literature on the benefits of organic gardening will be well versed with profound quotations from the ancients. In 350 BC, Aristotle is claimed to have said "Worms are the intestines of the soil." Yes, it may look as if we are no further forward when considered what is written about soil biodiversity, as noted earlier, on the Scottish Government's website. We know that Darwin *got it* with regard to the role of worms in our soils. Another one, not so well known, is John Muir, the Scot that emigrated to America and became a key figure in the establishment of national parks in America. There are several places in his books (Muir 1992) where he shows he has picked up on the role of soils being made dramatically more productive by the addition of mineral deposits left by *God's glacial mills*, that is, the icebergs that have crushed rocks and left a trail of nutrients in the form of glacial moraines. In California, it is these moraines that form the nutrient beds for the largest plants on the planet, the giant sequoia or redwood trees. John Muir tells us that the location of these magnificent trees follows the paths of the glaciers. A point it is imperative I make here—when we are using mineral-rich dust ground from Scotland's mountains, it is always, and will always be, a by-product of established quarrying operations. With so much available, it is unlikely to forecast that it could ever reach a stage where commercial pressures would demand fresh stone to be quarried for SEER Rockdust alone.

Scientific Research

The lack of scientific trials to prove results achieved at the SEER Centre, as shown in our testimonials thanks to SEER Rockdust, remains the missing link required to convince professional growers to take SEER Rockdust seriously. One trail partly funded by the Scottish Government and run by Glasgow University adjacent to the SEER Centre was thought to be fatally flawed as the dust used was not fit for purpose as the dust used at the SEER Centre itself and the outcome was not deemed significant.

A leading figure in the development of Waste Regulation in the Scottish Environment Protection Agency, in the early days at the SEER Centre, is my already mentioned colleague John Ferguson. John continues to take an active role in the development of SEER Rockdust. In 2004, John co-wrote a leading research review with Dr Robin Szmidt, (Ferguson and Szmidt 2004), as included in the research page http://www.seercentre.org.uk.

While receiving reports back from SEER Rockdust users is the most interesting part of my work, the most frustrating is not having the resources to get more scientific trails underway. So saying, the following four trials are underway in Scotland in 2013: (1) production of oil seed rape, (2) potato propagation, (3) organic vegetable growing, and (4) bee foraging.

If there is room, I will provide the background of the first two. At £400 per ton, the high value of oil seed rape seed is making it an ever more popular crop among Scottish farmers. In Germany however, they can produce up to 8 tons/ha, while in Scotland, the average production is about half this, despite the investment in the optimum fertilizer additives. If a means can be found to increase the seed production per hectare, bringing it up closer to the German harvest level, we would be making an early breakthrough proving the value of SEER Rockdust—regardless of how it works—money talks! (Photo 21.15): The presence or absence of a trail is underway on a Scottish farm with application at time of sowing in the autumn of 2012. It should be noted that 2013 opened with the coldest spring for 30 years, and at the time of going to press, it is stated the UK's OSR production is already recognized as being 30% lower because it simply has not become established in time to produce a crop.

Seed potatoes can only be produced for seven annual cycles before they begin to weaken and get so prone to disease they are deemed unsatisfactory. For this reason, there are four locations in Scotland where seed potatoes/tubers are produced by bringing them on from the cuttings of young plants. One of these, TLC Potatoes, has trialed SEER Rockdust on two different years and found a significant increase in the number and size of tubers when SEER Rockdust is added to the growing medium (Photos 21.16 and 21.17). At the time of going to press, I am arranging for SEER Rockdust to be provided to be trialled again at TLC Potatoes to grow their tubers in 2013.

PHOTO 21.15 Oil seed rape seedling showing good root bed with SEER Rockdust.

PHOTO 21.16 Trialling SEER Rockdust with potato tuber propagation.

Carbon Sequestration

The ability of mineral dusts to help stabilize climate change by sequestering carbon into soil was the principal theme of Don Weaver and John Hamakers's book. The organic movement now throws up papers showing that organically approved soil holds more carbon than the nonapproved soil. Another paper on the Scottish Government's website, entitled "(ECOSSE 2010), tells us that while the organic soils of Scotland contain 2735 million tonnes of carbon, the vegetation in the whole of the United Kingdom contains only 114 million tonnes of carbon. The claim can be made that soil remineralization has potentially a far bigger role to play in carbon sequestration than "planting a tree." Preoccupation with emissions

PHOTO 21.17 Close up showing surface application of SEER Rockdust in potato propagation trial.

reduction could be well overturned if we can prove that we can put the carbon that has formed a balaclava around the globe back to where it came from—our soils.

Conclusion

The foregoing touches on how soil remineralization can benefit crops; improve food's nutrient value; reduce dependency on less environmentally friendly, soil-damaging synthetic fertilizers; and play a part in stabilizing climate change. Both SEER Rockdust and REMIN volcanic rock dust are now marketed in Scotland. In conclusion, I ask the question: Is it the multitude of potential benefits of soil remineralization that is holding up any single benefit being recognized, researched, and embraced as a mainstream soil and compost improver by twenty-first century gardeners and professional growers, or do we revert to conspiracy theories about fertilizer/drug companies making more money if the public remains poorly informed?!

REFERENCES

ECOSSE: Estimating Carbon in Organic Soils–Sequestration and Emissions: University of Aberdeen, Centre of Ecology and Hydrology, Macaulay Institute, Cranfield University, Rothamsted Research, Harpenden, U.K., 2010. ISBN 978 0 7559 1498 2 (Web only publication) http://www.scotland.gov.uk/Publications/2007/03/16170508/1.

Gilchrist, R., A. Grand, and M. Jess, *Grow Food Nature's Way*, Published by Greenway Consulting (Scotland) CIC, printed by Largs Printing Company, 2011 U.K., 2011.

Harvey, G., *We Want Real Food*, Publ. Constable & Robinson, London, U.K., 2006.

Emsley, J., *Nature's Building Blocks: An A-Z Guide to the Elements*, Publ. Oxford University Press, Oxford, U.K., 2003.

Ferguson, J. and R. Szmidt, *Co-Utilisation of Rockdust, Mineral Fines and Compost*, 2004, http://binnsoilnutrients.com/wp-content/uploads/2011/02/Co-utilization-of-Rockdust-Mineral-Fines-and-Compost.pdf.

Lowenfels, J. and W. Lewis, *Teaming with Microbes*, Publ. Timber Press Linc, Portland, OR, 2010.

Muir, J., *The Eight Wilderness: Discovery Books*, Publ. Diadem Books Ltd., London, U.K. and The Mountaineers Seattle ISBN 0 9 06371 34 1 (UK) ISBN 0 89886 335 X (North America) 1992.

Towers, W., I.C. Grieve, G. Hudson, C.D. Campbell, A. Lilly, D.A. Davidson, J.R. Bacon, S.J. Langan, and D.W. Hopkins, *Scotland's Soil Resources*, Environmental Research Report 2006/01, http://www.scotland.gov.uk/Publications/2006/09/21115639/1. Web only publication.

RECOMMENDED READING

Some of the most advance research into soil remineralization, which I have read, is by Minplus™ working with James Cook University, 2001. See http://www.minplus.com.au/reports/01–173_rirdc_rejuvenating_soils.pdf.

Videos can be found by searching "SEER Rockdust" on You Tube.

22

Stonemeal: Principles, Potential, and Perspectives from Brazil

Suzi Huff Theodoro and Othon Henry Leonardos

CONTENTS

Introduction

This chapter presents the results of research on stonemeal (rock powder) technology that has been developed at the University of Brasilia (UnB), Brazil, for almost two decades. The main objective of this chapter is to present the principles, methodology, results, and regulatory aspects concerning the addition of crushed rocks—which contain a wide variety of nutrients/elements—as a mechanism to achieve a positive change on fertility of weathered or depleted soils. This chapter discusses geochemical, mineralogical, and physical parameters that are important tools for identification of rocks with the most adequate composition for use in soil remineralization or rejuvenation processes and methodological procedures related to the understanding and use of rock dust (also known as agrominerals). Pedological and agricultural production aspects will be presented and discussed as well. The results obtained from the research in Brazil show the potential of stonemeal as an important alternative to ensure sustainable agricultural production, contributing not only to the improvement of both nutritional and environmental security levels in Brazil but also to the reduction of the country's dependence on the import of major chemicals that make up NPK formulations, which have been in the last decade around 70% imported materials. The use of crushed rocks promotes a more appropriate use for materials discarded as by-products of mineral exploration—a fact that, in the Brazilian scenario, is highly significant since those by-products may become inputs for the agricultural sector, enhancing the formation of mineral and agriculture clusters.

Recent data presented by the National Association for the Promotion of Fertilizers (ANDA) in the last Brazilian Fertilizer Congress indicate that there will be large amounts of investment in Brazil in the next 5 years in the agricultural input sector. According to this source, about $18.9 billion will be invested to increase the country's internal fertilizer production capacity. This projection was made from data by the main fertilizer producers. If those projections become facts, Brazil should lessen the current dependence on foreign fertilizer, which has been around 70% over the past 10 years. It is expected that production will increase from 3,425 million tons in 2012 to 9,353 million tons in 2017, while the demand will grow from 12,198 million tons to 14,732 million tons.

Despite these data showing a less dramatic scenario for the fertilizer industry in Brazil, the country's dependence on import of inputs required to make possible the NPK formulations will still be significant.

However, if these projections become true, the problem of high prices remains to be solved since Brazil does not interfere in international market prices. Therefore, the acquisition of such inputs by most farmers—especially by those who make up the group of family farmers (who, according to government data, represent 75% of farm households in the country)—becomes more difficult each year as prices continue to rise worldwide (Manning, 2010).

Stonemeal technology provides a way to solve this problem and make agricultural production more sustainable and economic, since it is based on the assumption that certain kinds of rock can provide quantities and varieties of nutrients to the soil and plants. Remineralization of crushed rocks added to soils allows addition of a large amount of macro- and micronutrients to the soils that have been lost during weathering or anthropogenic processes (Leonardos et al., 1976, 1987, 1991; Leonardos and Theodoro 1999; Theodoro, 2010; Theodoro and Leonardos, 2011). Stonemeal technology can be understood as a mechanism for rejuvenation/remineralization of soils, guaranteed by the addition of multinutrients (agrominerals), which are present in certain types of rocks (Fyfe et al., 2006; Kronberg et al., 1987; Theodoro and Leonardos, 2006; van Straaten, 2010). This input behaves like a bank of nutrients, since it is matched to the quantity demanded by the plants (Theodoro et al., 2010).

Addition of powdered rock facilitates transformation of soil physical characteristics, such as porosity and permeability or permoporosity. This term may be understood as incorporating porosity and permeability—the first referring to the proportion of empty spaces contained in a package of rock (or soil) and the second to the ability of a porous material (rock or soil) to allow percolation of fluids within the system. It is important to remember that porous spaces in the soil are normally filled by fluid or air, which percolates through the system when spaces between pores are rearranged through soil compaction. Since soil is less dense than rocks, percolation of fluids is highly dependent clay particle content in the system. When crushed rocks are added to a sandy soil, the finer particles tend to reorder the arrangement of pores, blocking part of the interconnections. When the soil is mostly made up of clay particles, adding crushed rocks with a larger grain size range helps create larger pores and increases permeability, thereby affecting fluid migration rates.

Addition of crushed rocks with less stable minerals—formed at high pressure and temperature and therefore more easily changeable into secondary materials, such as clay minerals (smectite, vermiculite, and illite groups)—changes soil's permoporosity and increases cationic exchange capacity of macro- and micronutrients. Clays behave like *sponges* that store and release fluids in a gradual way, at a rate compatible with plant assimilation. This tends to favor increased plant resistance to water shortage, especially during dry seasons in tropical environments.

Just as important as aspects regarding nutrition and rearrangement of soil and its fertility is the conservation of natural resources. Another important consideration is the price of inputs. The costs for acquisition of powdered rock are about 60%–80% cheaper than other conventional inputs (Theodoro et al., 2010, 2009). Brazil has immense geodiversity that guarantees the occurrence of various types of rocks (basalts, kamafugites, carbonatites, phonolites, serpentinites, schists, phyllites, marl, and some types of granite and gneisses) with appropriate potential for use in all regions of the country.

Generally, solubility and availability of nutrients in rock powders cause them to be released more slowly than soluble chemical fertilizers. This is compensated by release of a much wider spectrum of macro-, micro-, and trace elements over a much longer period, since these elements are gradually released by rocks by physical breakdown and chemical reaction of minerals with carbon dioxide and organic acids (Formoso, 2006). Acids produced by the roots can directly affect the dissolution of solid phases that contain metals (Hinsinger et al., 2001).

This makes rock powders intelligent or low-soluble—slow-release—fertilizers (Theodoro et al., 2010). This label may be considered by many skeptics as a disadvantage, since slow solubilization of nutrients is, according to their understanding, incompatible with the dynamics of rapid production and may not adequately satisfy all the plants' immediate nutritional requirements. The assumptions disseminated of the Green Revolution—in which the nutrition of each crop (plant) is designed according to a nutritional *diet* based on simplification and homogenization of agroecosystems—often lead agricultural systems to artificiality in the sense of addiction to fast-acting gratification and later on to degradation as essential minor elements are lost from the system and use of pesticides stops natural biological recycling.

Recent studies conducted by Theodoro et al. (2010) and Almeida et al. (2009) show, however, that the process of provision of nutrients can be accelerated, depending on the type of management. The combination of organic and mineral mixtures—organic matter composted with rock, inoculants, and nitrogen-fixing bacteria or green manure—incorporated into the soil along with powdered rock, has shown very promising results, which stimulate the release of nutrients more rapidly into soils. Arcad and Schneider (2006) confirm this effect, showing the attack of bacteria and fungi (*Aspergillus niger*) on apatite crystals—$Ca_5(PO_4)_3(OH,F,Cl)$. The inoculation of bacteria, mycorrhizae, and other microorganisms potentiates solubilization of nutrients at levels that do not jeopardize the environment.

While nutrient release efficiency is essential for management of agroecosystem fertility, strategies to improve production methods should promote both biological nutrient recycling and conservation of natural resources (Petersen and Almeida, 2008). It should be kept in mind, though, that this possibility is not configured as a source of readily available nutrients that aims to replace the high solubility industrial fertilizers.

Materials and Methods

Stonemeal research is multidisciplinary using concepts and methodological procedures of various scientific fields such as geology, geochemistry, agronomy, forestry, agroecology, economics, and ecology. Certainly, some principles and analysis methodologies relate directly to one specific knowledge field, for instance, in the methods of geochemical, related to geology; in fertility analysis, related to agronomy; or even in the measurement of production and costs, a process directly related to economics.

In general, the methodological procedures of analysis, in its various stages, are realized following a pattern: (1) identification of the lithological potential of the region and selection of rocks (secondary data) suitable for use as remineralizers; (2) collection, grinding, and processing of samples, and splitting them for different kinds of analysis; (3) measurement of abrasion pH (shift in acidity of water to which powdered rocks are added); (4) geochemical analysis of rock samples dissolved by strong acids and measured by atomic emission spectrometry; (5) petrographic description of the rocks; (6) characterization of clay or secondary minerals, through x-ray diffractometry (XRD); (7) pedological study of soils where the experiments are done; (8) analysis of soil nutrient fertility using several types of extraction solutions (Mehlich reagent [HCl + 0.05 M H_2SO_4, 0.0125 M, citric acid at 2%, and/or oxalic acid at 5%]) in samples of soils, as well as powdered rocks; (9) determination of macronutrients (P, K, Mg, and Ca) and micronutrients (B, Zn, Cu, Mn, and Fe) immediately available (inductively coupled plasma [ICP] emission spectrometry); (10) analysis of organic matter by calorimetry; (11) analysis of potential acidity and cationic-exchange capacity (CEC) in calcium acetate solution, pH 7.0; (12) experimental plant growth tests in fields or in greenhouse pots, with three or four replicates randomly placed in blocks or pots, with controls; (13) addition of different rock types to soil and incorporation via manual/mechanical processes; (14) selection and planting of agricultural, forestry, or legumes species (seeds or seedlings) in a consortium; (15) in some cases, it is necessary to add coverage—triturated mulch—to the soil to reduce spontaneous weeds or invasive species, reduce irrigation needs, and to conserve soil moisture; (16) management of several phases of crops; (17) foliar tissue analyses; (18) management and systematic control of invasive plants; (19) plant harvesting; (20) separation of bulbs, roots and fruits from the aerial parts or from mulch; (21) analysis of root development (in size and quantity); (22) measurement of productivity (weight, quality of products, and comparison between all treatments applied); (23) preparation of histograms, graphs, and tables to compare productivity and comparison to the average productivity of the region or country; (24) systematic and periodic repetitions of soil fertility analyses to investigate changes in the fertility profile and incorporation of nutrients derived from crushed rock, changes in soil organic matter, and changes in the percolation of fluids and of the acidity profile of soils; and (25) market studies on prices and commercialization of products.

Even when rocks or soils are already well characterized, their chemical, physical, and biological behavior may be different in different soil types, climates, and forms of management. Research on this has been delayed because the Brazilian legal framework still does not provide for use of agrominerals or remineralization products as an agricultural input, although chemical fertilizers are recognized.

Regulatory and Legal Aspects

In Brazil, regulation of the fertilizer sector is supported by three legal instruments: the law of fertilizers (Law No. 6.894/1980); Decree No. 4.954/2004, which regulates this law; and normative instructions, especially IN 35/06, introduced by the Ministry of Agriculture (MAPA). None of these legal instruments have a category that permits the use, marketing, or inspection of powdered rock (agrominerals) since the diversified characteristics of this technology do not fit into any currently existing category of inputs (such as conditioners and fertilizers).

Revision of one or of all of these legal instruments is essential in order to make the use of stonemeal remineralization products viable. Currently, there are some initiatives that seek to change this scenario. A working group coordinated by the Ministry of Mines and Energy (MME) and a series of other public institutions has been discussing and suggesting alternative statements that could change this legal apparatus. Meanwhile, in Congress, two propositions are being proposed that seek to change the Law No. 6.894/1980 or to stimulate the use of powder rock. These are the Law Projects PLS 212/2012 and PL 2727/2011. Considering that the process of making a legislative proposal is, in most cases, quite slow—due to interference from complicated interests by various groups and associations—this change is not expected to happen. Another option would be to change Decree No. 4.954/2004 since this is a prerogative of the executive power.

But despite this subject being considered as strategic for the government, change is being held up, and there are no recent developments to assure that remineralizing will be incorporated into a new text. Finally, another alternative would be to change the Normative Instruction (IN) 35/2006. This would be the easiest and quickest way to modify the current scenario, once powdered rock could be classified under the category of *soil conditioners*, despite some additional safeguards. In this state of lack of regulation, Brazil will remain strongly dependent on imported fertilizers—which are doubtless expensive and which, ultimately, have been causing serious environmental damage such as eutrophication of water bodies and increasing global warming through the greenhouse effect.

Results and Discussion

The weathering processes and mineralogical characteristics of rocks and soils are important for understanding nutrient uptake mechanisms from rock dust to plants. The composition and the alteration stage of the minerals in the rocks (examined by petrographic studies, XRD, geochemical and fertility analyses) gives an indication of the potential and of the rate of liberation of nutrients released from rocks into the soil. Table 22.1 shows the average composition of some kinds of rocks tested in field experiments by researchers from the UnB team. Such indicators provide *clues* to facilitate selection of rocks that are more suitable for different kinds of crops, since they demand an adequate balancing of nutrients.

Knowledge of the chemical composition of rocks along with their mineralogical characterization is not sufficient to guarantee agronomic results. The level of chemical alteration of rocks is also an important parameter that indicates when the presence of secondary materials, important factors for the release of ions, will form into compounds that percolate the soil and be absorbed by plant roots. Petrographic analysis of the rocks in Table 22.1 revealed that they have a wide mineralogical range. The mineral composition of the rocks was initially determined by petrographic analysis: basalts were primarily formed by plagioclase, pyroxene, vitreous material, and zeolites; hydrothermal metabasic rock by pyroxene, feldspar, apatite, and carbonate; kamafugite by olivine, pyroxene, and phlogopite, immersed in a matrix of olivine, diopside, perovskite, richterite, kalsilite, apatite, melilite, and zeolites; and carbonatic mica schist by muscovite, chlorite, dolomite, albite, and quartz. Complementary analyses for identification of weathering and matrix minerals were made by XRD. Both types of basalts showed significant presence of smectite, as well as plagioclase and augite; quartz was a minor constituent of the *fresh* basalt. Kamafugite showed characteristic peaks of smectites plus carbonate and hydroxyapatite. Granite was composed of quartz and alkali feldspar and secondly by biotite, muscovite, and amphibole. In general, the mineralogical composition of these rocks is very close to the averages of these lithotypes. Table 22.2 shows the chemical composition of the main rock-forming minerals (van Straaten, 2010).

TABLE 22.1
Total Chemical Composition of Rocks

	Hydrothermal Metabasic Rock[a]	Fresh Basalt[a]	Kamafugite[a]	Mica Schist[a]	Basalt[a]	Granite[b]
SiO_2	61.8	50.84	34.26	45.07	49.98	72.04
CaO	3.17	8.02	2.15	11.45	8.39	1.82
MgO	1.64	4.43	5.09	7.30	4.65	0.71
Fe_2O	3.78	13.75	19.35	4.47	14.19	1.22
Al_2O_3	12.03	11.75	7.84	10.07	12.20	14.42
MnO_2	0.11	0.22	0.43	0.07	0.24	0.05
Na_2O	1.89	2.26	0.19	0.75	2.16	3.69
K_2O	3.76	1.11	2.80	2.70	1.06	4.12
P_2O_5	2.48	0.27	1.56	0.17	0.30	0.12
TiO_2	0.28	2.59	7.41	0.23	2.78	0.30
Sr	103	200	870	80	200	—
Ni	400	600	480	40	600	—
Cu	600	200	140	ND	180	—
Zn	600	102	180	800	120	
Mo	200	200	200	200	20	—
V	890	220	220	670	420	—
B	90	140	150	80	110	—

Oxids in % and minor elements in ppm.
[a] Data from Theodoro et al. (2010).
[b] Derived from various authors.

TABLE 22.2
Chemical Composition of the Main Minerals Present in Rocks (%)

Mineral	SiO_2	Al_2O_3	Fe_2O_3	CaO	MgO	K_2O	Na_2O	P_2O_5
Quartz—SiO_2	100	—						
Alkali feldspar—(K, Na) $[AlSi_3O_8]$	62–66	18–20		0–3		9–15	9–4	
Plagioclase—$Na[Al\ Si_3O_8]Ca[Al_2Si_2O_8]$	61–40	19–37		0–20		0–4	6–11	
Muscovite—$K_2Al_4\ [Si_6Al_2\ O_{20}](OH,F)_4$	44–46	34–37	0–4	—	0–3	8–11	0–2	
Biotite—$K_2(Mg,Fe^{2+})_{6-4}$ $(Fe^{3+},Al,Ti)_{0-2}[Si_{6-5}Al_{2-3})_{20}]\ (OH,F)_4$	33–36	10–13	3–17	0–2	2–20	6–9		
Amphiboles—$(Mg,Fe^{2+})_7\ [Si_8O_{22}]\ (OH,F)_2$	38–59	0–19	0–22	0–15	2–26	0–2	1–3	
Pyroxene—$(Ca,Mg,Fe)\ Si_2O_6$	45–55	3–10	0–14	16–26	6–20			
Olivines—$(Mg,Fe)_2[SiO_4]$	35–40	—	0–30	—	27–51			
Apatite—$Ca_5(PO_4)_3\ (OH,F,\ Cl)$	—	—		52–55				36–42

Source: Amended from de van Straaten (2010).

The data in Tables 22.1 and 22.2 indicate that the rocks tested are potential sources of the main nutrients (P, K, Ca, and Mg) required for the full growth of plants. Micronutrients (trace elements) occur in much lower proportions in the rocks, and their abundance depends on the predominance of specific types of some minerals.

Undesired geochemical elements in some rocks, such as radioactive or toxic heavy metals, can be regarded harmful to human health as well as to the environment. Further studies must be conducted in order to ensure that the amount of these elements does not exceed the quantity permitted by law; otherwise, rather than bringing benefits to the soil and plants, the technology may facilitate environmental problems (which would affect regional biodiversity) or in extreme cases human health problems.

Organic acids—present in soils due to root exudates (organic acids, sugars, amino acids, phenolic) produced by the roots—act to release agrominerals from soils to plants. Organic acids have the power to stimulate faster release/solubilization of nutrients from minerals, even if considered weak acids. Organic acid content is dependent on the various species of plants/crops present in the soil, on stress conditions, and on the activity of microorganisms that use and produce other organic compounds. Fine-grained material potentiates organic acid action since finely crushed particles have more surface area and react more quickly with organic acids. Larger particles have slower reaction rates, which may be positive from the viewpoint of maintenance of effective fertility (Theodoro et al., 2010; van Straaten, 2010). This interaction between minerals and organic acids in soils make rock powder technology enable more sustainable agriculture, since the solubilization ability of nutrients is a direct function of the amount of organic matter present in soils. Increased microbial activity in soils stimulates higher rates of mineralization and rapid cycling of nutrients, also changing the CEC (Silva et al., 2008). Adsorption of organic constituents can occur in different ways, intensities, and proportions, according to soil characteristics—mineral composition, pH, and ionic strength of the solution (Correa et al., 2008). Besides edaphic factors, characteristics of organic anions (such as the size of the C chain), the arrangement of functional groups, and charge density affect adsorption on soil particles. Another important factor concerns the supply of nitrogen (N) for plants: this nutrient/element is not available in the rocks. In most experiments that showed positive results, an additional source of nitrogen from composted material or green manure was used.

Results

Experiments in Brazil—particularly those taking place at the experimental/demonstration sites of the UnB/CDS stonemeal research group—have achieved significant advances in understanding the incorporation, release, and availability of nutrients (Photo 22.1). The experimental areas implemented by UnB projects have a design that fulfills agronomic criteria and scientific methods, with randomly placed plots or repetition blocks in which performances of the different crops tested (both forest and agricultural species) can be compared, as well as pH changes, alterations in soil fertility and physical characteristics, and the control of invasive species.

Experiments tested rocks that are locally or regionally available: (1) transporting materials over large distances is a limiting factor affecting acquisition costs of these inputs, and (2) the study and use of geological materials available from tailings or by-products provided by beneficiation of ores and materials for geotechnical use favor the formation of local productive arrangements (APLs)—clusters where mineral supply and agricultural use can be integrated. Rock powders (or mix of various rocks) are used to contemplate specific nutrient demands, depending on the region's most commonly used soils and crops. In many cases, the rock powders—agrominerals—are used in combination with organic material or green manure.

PHOTO 22.1 UnB projects and experiment areas; UnB team, students, and farmers.

These experiments were carried out in Bahia (with hydrothermal metabasic rock and phosphate rocks), in Rio Grande do Sul and Paraná (with basalts), in Pará (with phyllites and sediments retained in the reservoir Tucuruí), in Minas Gerais (with kamafugites), and in the Federal District (with kamafugites, metabasic rock, mica schists, and basalts). Generally, in research, 4–5 t crushed rock/ha is used, depending on the fertility profile of the soil and its physical characteristics. This dose of material can be added to the soil in a single application, or it can be split into different application periods during the planting over 4 or 5 years. The performance of agrominerals is compared with other inputs such as NPK in different dosages, with green or organic manure, or with a mix of those along with powdered rock from diverse backgrounds.

The major results show the following: (1) acquisition costs of powdered rock are much smaller than chemical fertilizers; (2) their effects can be extended for up to 4 or 5 consecutive years (due to the slow release of nutrients). Soil fertility levels increase (especially the supply of P, K, Ca, and Mg) until the fourth or fifth year after application. But positive results are still observable for even 10 years after the application of remineralizers; (3) the productivity is equivalent or superior to that of conventional chemical fertilization. In some cases, farmer's net incomes may be up to 30% higher than those obtained by the use of chemical inputs; (4) long cycle crops (such as sugar cane and cassava) and forest species have better yields, since these plants make use of the nutrients for a longer period of time; (5) short cycle crops (maize, rice, beans, and vegetable crops) show similar or slightly higher production compared to chemically fertilized crops; (6) occurrence of roots in plants is enhanced, and roots are more developed than in plants receiving conventional fertilization, probably due to aluminum toxicity reduction and pH correction; (7) moisture content is higher in areas where powdered rock is added to the soil, showing that minerals and/or their alteration products, derived from rocks, retain water/fluids; (8) besides showing a greater amount of green biomass, the plants are more exuberant and have higher tillering (cassava and sugar cane); (9) in some cases, the production cycle of plants was accelerated; (10) association with mulch (cane trash) facilitates the management and control of weeds because it hampers their emergence; (11) association with organic compost and green manure has proven itself crucial for plant development, suggesting that association of organic and mineral fertilization can adequately supply the macro- and micronutrient needs of crops; and (12) no contamination or eutrophication occurs to water resources, because agrominerals have slow and gradual solubility, unlike highly soluble conventional fertilizers.

Several kinds of crops, including cassava, beans, maize, sugar cane, and rice, were tested in a rural settlement (Theodoro, 2010; Theodoro and Leonardos, 2006). Agricultural performance, productivity, and changes in pH and nutrients in soils were compared over 5 years. The performances of kamafugite and NPK (4:14:8) in low-fertility soil (oxisol) were tested and compared. The research also showed that farmers—especially small farmers—not only demand low-impact technologies but also absorb them immediately once they understand that food production and environmental conservation are inseparable components for sustainability (Photo 22.2).

PHOTO 22.2 UnB projects and experiment areas; Cassava produced in a rural settlement, performance comparison between NPK and stonemeal.

The performance of kamafugites was evaluated in agroforestry systems (AFS) consisting in consortiums of agricultural species with both native and exotic trees that made a complex and multistratified system, which aims to resemble the natural ecosystem (Rocha, 2006). To improve soil fertility by the recycling of organic matter in the AFS, the modules were divided into plots with applications of manure and rock dust. The results showed that agricultural crops (maize) and leguminous plants (pork beans and pigeon pea) showed better yields and that the forest species (copaiba, mahogany, and cedar) also showed a faster initial growth. He noted that in plots where the organic and mineral combination was present, there was lower incidence of weeds. The development of healthy and vigorous species ensured the evolution of the system as a whole.

An experiment conducted by Leonardos et al. at Água Limpa farm of the UnB, with species of *Eucalyptus pellita* and *Pinus caribaea*, used three different forms of fertilization over 13 years in oxisols (dust basalt, NPK + basalt powder, and NPK). Crop productivity was determined by diameter and height of trees. The growth in the plot fertilized with NPK was rapid up to the fourth year. After this period, tree growth became less significant when compared with growth observed in plants fertilized only with rock dust, which always grew steadily, though more slowly at the beginning. The durability and permanence of the effects of stonemeal, in terms of assimilation, were effective throughout the years. The mixture of stone powder and NPK showed constantly increasing productivity. The NPK functioned as a kind of *starter* for the plants, and subsequent growth was sustained from nutrients derived from the rock powders.

Basalt dust was tested in agroecological systems with small farmers in southern Paraná and Santa Catarina, Brazil (Almeida et al., 2009). Basalt dust was used with green manure in order to facilitate recovery of soil fertility and enhance improvement of soil physical structure, allowing better root development and providing more access to nutrients and deeper groundwater. The farmers found that the agroecological production was 20% higher on average than conventional methods while having lower production cost and great reduction or elimination of pesticide use and higher resilience. These results generated a great range of social support from the region's farmers still involved with traditional production methods.

Laboratory experiments evaluated the growth of seedlings of two tree species (*Mimosa scabrella* Benth and *Prunus sellowii* Koehne) using powdered basalt and conventional NPK fertilizer (Knapik, 2005). Growth of seedlings of both species tested was higher in plants where the conventional fertilization was made, but foliar analysis showed that there was greater assimilation of nutrients (N, P, K, Ca, Mg, Fe, B, and Mn) in the leaves (especially in *P. sellowii* Koehne) fertilized with basalt dust. The results confirm that the lifetime of the plant in the soil and the availability of microorganisms and acids produced by the roots are important factors for the liberation of nutrients, since many plants can grow on poorly soluble compounds by extracting and incorporating nutrients through their root exudates (Khatounain, 2001; Knapik, 2005). Considering that the research time was very short (5 months), the interaction of the plants with less soluble nutrients from the rocks did not occur significantly, and the only agent that could enhance ion exchange through dissolution, diffusion, and absorption of nutrients was water (through irrigation), which is a poor solvent compared to organic acids in root exudates. In the long run, the water potentiates these effects and, along with soil organic complexes, can form true banks of nutrients, which can be withdrawn or removed by plants, according to their needs (Theodoro, 2010).

The performance of maize and beans was compared between chemical fertilizer (NPK) with tailings from potassium-rich feldspars, pegmatites, kaolinite, quartz, and albite, in both field and greenhouse experiments (Fernandes, 2008). The field experiment was completely lost due to the water scarcity present at the experiment area (Jequitinhonha Valley), but in greenhouse tests, plants fertilized with NPK showed better performance. Foliar analysis showed significant differences among the different treatments, especially assimilation of magnesium (greater in plots treated with by-products of pegmatites) and manganese (in plots with NPK). Both Fernandes and Knapik warned that longer studies should be conducted, since in both cases, the exuberant appearance of aerial parts of plants fertilized with rock powders was evident, but because of the short evaluation period, it was not possible to reach more concrete conclusions.

Comparison of powdered basalt and limestone on change of soil pH showed that basalt, despite not having made a rapid increase in pH, showed a trend toward gradual and long-lasting increase of this parameter, while limestone showed a more stable long-term trend (Duarte et al., 2009).

Evaluating three types of rock (charnockites, gneisses, and steatites) on corn plant growth in successive cultivation in a greenhouse showed that charnockite and gneiss were sources of K and that steatite was a source of Mg and Si, besides positively altering CEC and phosphorus adsorption (Carvalho, 2012). Powdered gneiss promoted plant growth, enhanced absorption of nutrients, and increased desirable chemical changes in the soil. The availability of elements by gneiss, especially macronutrients, was not significantly affected by the management of organic fertilizer. The effectiveness of silicate rock powders appears to increase when the initial levels of nutrients in the soil are low.

Rock waste from coal mining (gray) also showed positive results in South Africa (Harper et al., 2010). The material's effects on soil and productivity of maize and beans were compared under field conditions. Other materials with similar geochemical composition were used (limestone, calmasil—calcium silicate derived from the production of stainless steel—and gypsum) in addition to evaluation of the pH changes. Results showed that productivity in plots of gray, calmasil, limestone, and gypsum was enhanced by approximately 2 t beans/ha and 1–2 t grain maize/ha. The gray rock powder provided an increase of between 0.96 and 1.72 t/ha for bean yields in relation to the yield from traditional methods in these areas, and maize productivity in the second harvest was 5.6–7.8 t/ha. Although the gray presented inferior results for improvement of pH levels among the four tested materials, its effects improving productivity of bean and maize were comparable to the results obtained on plots where agricultural limestone was added to the soil, which may indicate additional supply of P and K.

Positive results for plant growth were obtained in tests made with rocks rich in potassium (Manning, 2010). He cited various authors (Abdel-El and Mouty Greadily, 2008; Badr, 2006; Coroneos et al., 1996; Hinsinger et al., 1996, cited in Manning, 2010; Wang et al., 2000) who had reached similar conclusions.

The UnB team has been researching five kinds of rocks—kamaurgite, carbonatic mica schist, hydrothermal metabasic rock, *fresh* basalt, and weathered basalt—added to soil, mixed with or without manure in different plots (Photo 22.3). To check the results in terms of crop yields, five crops (maize, beans, garlic, carrots, and okra) were tested for 2 years—two harvests each year—in a rotation system. The experimental plots were randomly assigned; each plot had four replicates for each type of rock; and control plots were also installed as well as plots where only manure was added to the soil, a total of 12 different types of treatment in 138 plots of 6 m^2 in a total area of 1402 m^2, including spaces between plots (Figure 22.1).

PHOTO 22.3 UnB projects and experiment areas; rocks and manure mixes.

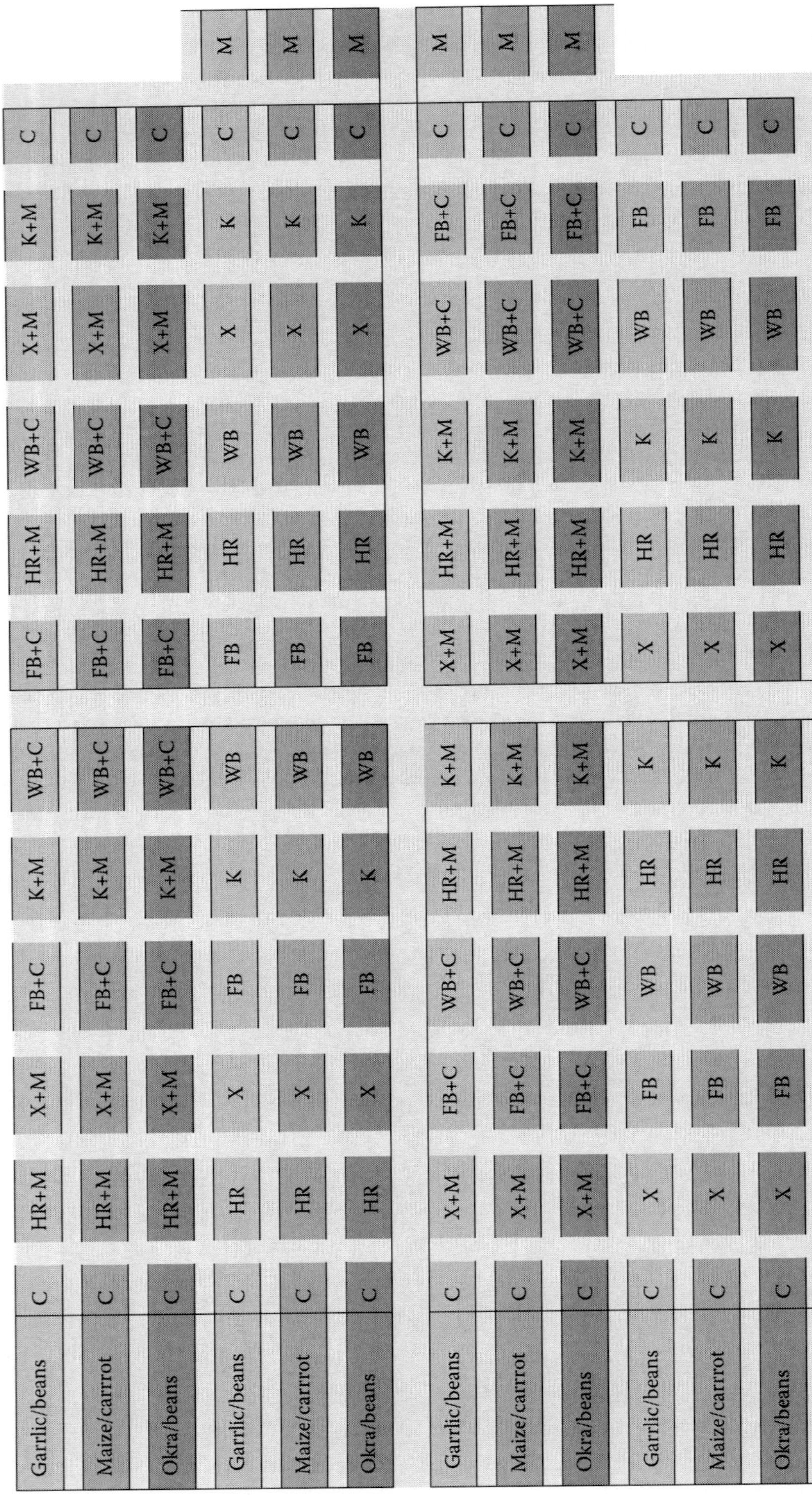

FIGURE 22.1 Design of experimental area. C, Control; X, schist; HR, hydrothermal meta-basic rock; FB, fresh basalt; WB, weathered basalt, K, kamafugite; M, manure.

The rock-to-manure ratio, based on previous research, was 1:2 (5 t/ha of powder rock and 10 t/ha of manure), equivalent to 3 kg of rock and 6 kg of manure in each plot. The incorporation of powdered rocks and of manure (or mixtures of both) was made by broadcasting these materials on the soil surface. In order to reduce spontaneous or invasive species, reduce the need for irrigation, and to conserve soil moisture, a layer of triturated mulch (from grinding *Tithonia diversifolia* and *Pennisetum purpureum*) was laid down covering every plot, which functioned also as a source of additional nitrogen and carbon.

After 1 year, the soil nutrients showed significant increases in fertility, independent of the type of extractor (Mehlich reagent, citric acid, and oxalic acid) used to extract available nutrients. The organic matter content and the pH were also significantly improved. Figure 22.2a through d shows the results. The major nutrients (P, K, Ca, and Mg) increased in all plots, showing that the interaction between rock dust, soil, and plants is rapid, especially if there is organic matter available.

Preliminary results suggest weathering of these rocks was accelerated by manure microbial activity and that they contained primary and secondary mineral phases that were able to remineralize soils that had lost their mineral nutrients by leaching. Soil fertility parameters (pH, CEC, and total organic

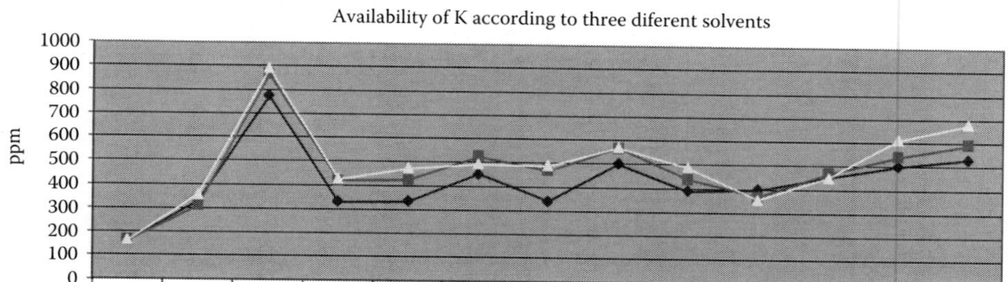

	T	C	M	X	HR	K	WB	FB	X+M	HR+M	K+M	WB+M	FB+M
K (Mehlich l)	164	330	773	330	333	450	340	500	390	397.9	450	500	530
K (Citric acid—2%)	164	310.6	865.1	419.5	422.4	524.2	467.8	565.3	438.4	368.2	467.7	539	591
K (Oxalic acid—5%)	164	354.7	888.6	424.4	471.9	493.2	487.7	567.7	488	354.9	450.3	610.9	679.4

Rocks/treatments

(a) K (Mehlich l) K (Citric acid—2%) K (Oxalic acid—5%)

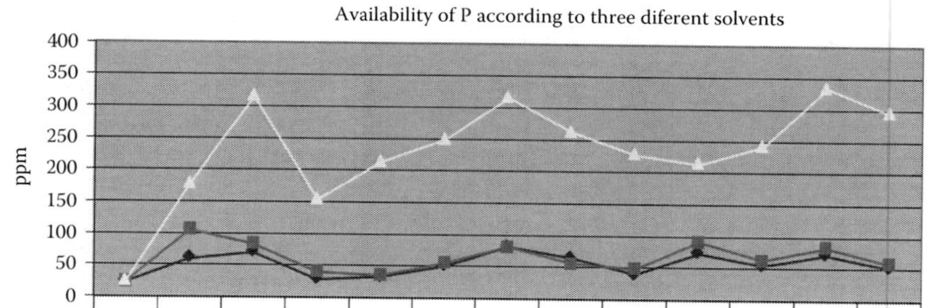

	To	T1	M	X	HR	K	WB	FB	X+M	HR+M	K+M	WB+C	FB+M
P (Mehlich l)	22.8	60	69.9	28.2	32.2	49.8	81	64.7	39	73.5	54.1	72.1	52.2
P (Citric acid—2%)	22.8	105.8	83.7	40	34.7	55.2	80.9	56	48.5	91.3	62.8	84.1	58.2
P (Oxalic acid—5%)	22.8	176.9	314.2	153.5	213	248	315.4	261.3	227.3	213.5	241.2	333.5	293.9

Rock/treatments

(b) P (Mehlich l) P (Citric acid—2%) P (Oxalic acid—5%)

FIGURE 22.2 (a and b) Changes in availability of macronutrients (K, P, Ca, and Mg) measured using three solvents. *(Continued)*

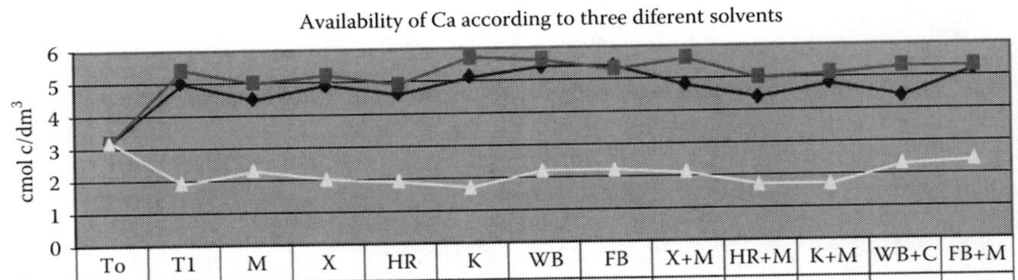

	To	T1	M	X	HR	K	WB	FB	X+M	HR+M	K+M	WB+C	FB+M
Ca (Mehlich l)	3.2	5	4.5	4.9	4.6	5.1	5.4	5.4	4.8	4.4	4.8	4.4	5.2
Ca (Citric acid—2%)	3.2	5.4	5	5.2	4.9	5.7	5.6	5.3	5.6	5	5.1	5.3	5.3
Ca (Oxalic acid—5%)	3.2	1.9	2.3	2	1.9	1.7	2.2	2.2	2.1	1.7	1.7	2.3	2.4

(c)

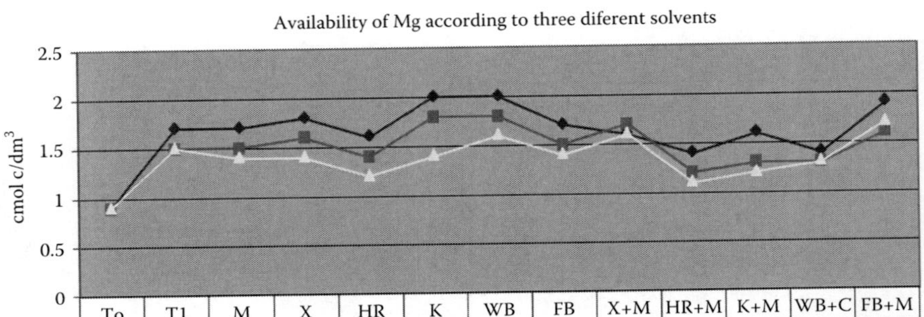

	To	T1	M	X	HR	K	WB	FB	X+M	HR+M	K+M	WB+C	FB+M
Mg (Mehlich l)	0.9	1.7	1.7	1.8	1.6	2	2	1.7	1.6	1.4	1.6	1.4	1.9
Mg (Citric acid—2%)	0.9	1.5	1.5	1.6	1.4	1.8	1.8	1.5	1.7	1.2	1.3	1.3	1.6
Mg (Oxalic acid—5%)	0.9	1.5	1.4	1.4	1.2	1.4	1.6	1.4	1.6	1.1	1.2	1.3	1.7

(d)

FIGURE 22.2 (*Continued*) (c and d) Changes in availability of macronutrients (K, P, Ca, and Mg) measured using three solvents.

carbon) were also positively altered by the use of stonemeal. The only exception was calcium, which decreased when analyzed along with oxalic acid. Availability of soluble phosphorus, measured with the same extractor, increased significantly, probably due to chemical reactions of the oxalic acid, which captures calcium and releases phosphorus when both are in soluble state.

Crop yield results were positive (Figure 22.3), but different rock types (with different nutrient concentrations) had different results on soil fertility and agricultural productivity. All results were positive when compared to control plots (some produced twice as much as the control plot). Carbonatic mica schist showed the best productivity performance for okra and garlic. Kamafugite distinguished itself by higher productivity of maize. *Fresh* basalt and hydrothermal metabasic rock showed the best results for carrots and beans. Weathered basalt showed the lowest productivity for all crops, suggesting that weathering had already removed significant amounts of nutrients. The second season productivity (maize, carrot) was superior compared to the first harvest on these plots. These results indicate that crushed rocks improved production performance in all five types of crops tested, even where there was no addition of manure.

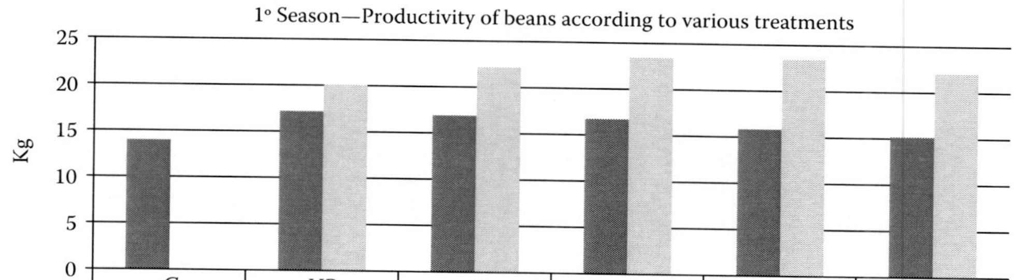

	C	HR	K	X	FB	WB
Rock	13.955	17.185	16.895	16.721	15.845	15.155
Rock + Manure		20.1	22.144	23.455	23.414	22.07

(a)

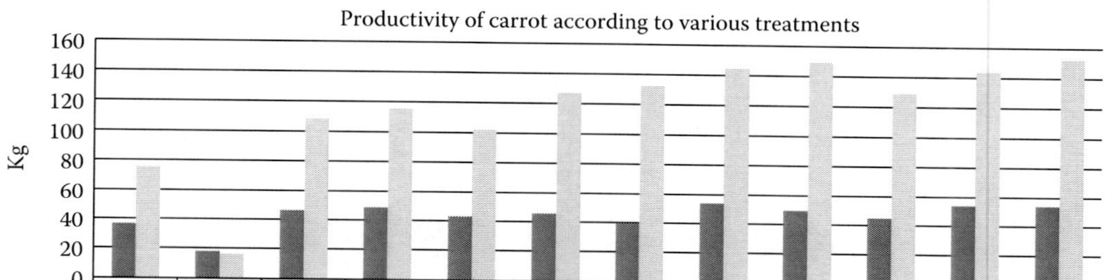

	C	M	X	HR	K	WB	FB	X+ M	HR+ M	K+ M	WB + M	FB + M
1° Season	36.545	17.8	46.32	48.8	43.055	45.63	39.76	54.12	49.445	44.79	54.105	54.09
2° Season	75.345	16	108.47	115.57	102.41	127.07	132.44	144.38	149.34	129.13	143.58	152.29

(b)

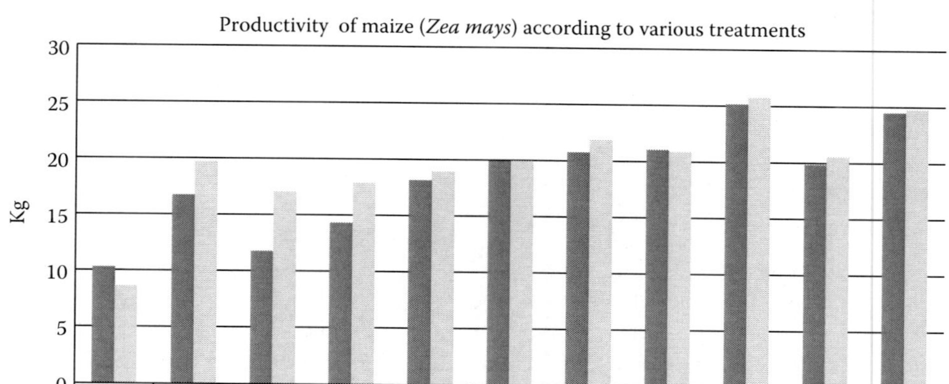

	C	X	HR	K	WB	FB	X+ M	HR + M	K+ M	WB + M	FB + M
1° Season	10.3	16.745	11.765	14.33	18.16	19.93	20.75	21.05	25.155	19.86	24.48
2° Season	8.575	19.73	17.03	17.89	18.98	19.935	21.865	20.885	25.765	20.52	24.77

(c)

FIGURE 22.3 Productivity of five crops, (a) okra, (b) carrot, (c) maize, according to various treatments. *(Continued)*

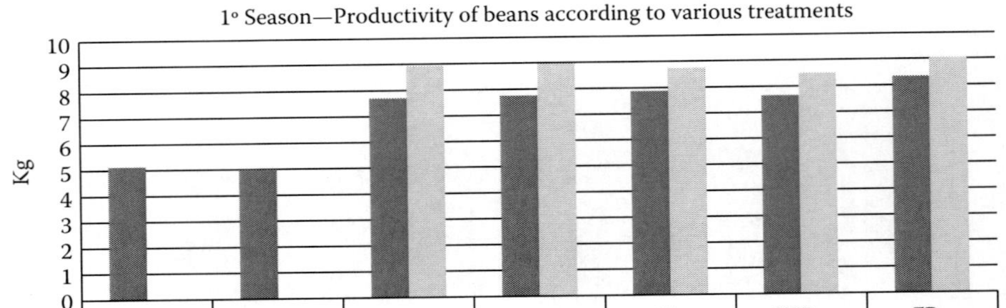

	C	M	X	HR	K	WB	FB
Rock	5.13	5.01	7.715	7.765	7.88	7.66	8.345
Rock + Manure			8.93	8.96	8.75	8.51	9.025

(d)

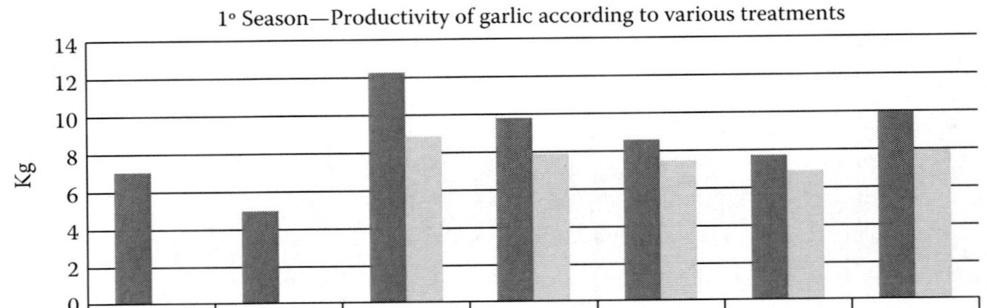

	C	M	X	HR	K	WB	FB
Rock	7.045	4.985	12.26	9.795	8.595	7.705	9.995
Rock + Manure			8.87	7.89	7.45	6.885	7.92

(e)

FIGURE 22.3 (*Continued*) Productivity of five crops, (d) garlic, and (e) beans, according to various treatments.

Conclusion

Stonemeal is a low-cost technology that brings significant production, environmental, and economic results and that opens up the possibility of a new agricultural technology strategy, more suitable for Brazil and many African countries. High costs of conventional fertilizers, foreign dependence for the acquisition of such inputs, and restricted access of small farmers to these products make use of powdered rocks an important option to ensure food security to the population, and producing food more sustainably. Brazilian research results show that remineralization yields similar or even superior production to conventional fertilizers, using widely available local resources in Brazil, costing significantly less (about 70%–80% cheaper) than soluble fertilizers. Stonemeal's long effects in the soil ensure that its benefits can be increased over a period of 5 years.

REFERENCES

Arcand, M.M. and Schneider, K.D. (2006). Plant and microbial-based mechanisms to improve the agronomic effectiveness of phosphate rock: A review. *Journal of the Brazilian Academy of Sciences* 78(4), 1–17.

Carvalho, A.M. (2012). *Rochagem e suas interações no ambiente solo: contribuições para aplicação em agroecossistemas sob manejo agroecológico*. PhD Thesis in the Universidade Federal de Viçosa. 129pp.

Corrêa, M.M, de Andrade, F.V., Mendonça, E.S., Schaefer, C.E.G.R., Pereira, T.T.C., and Almeida, C.C. (2008) Ácidos orgânicos de baixo peso molecular e ácidos húmicos e alterações em algumas propriedades físicas e químicas de latossolos, plintossolo e neossolo quartzarênico. *Revista Brasileira de Solos* 32, 121–131.

Duarte, E.M.G., Carvalho, A.M.X., Simoes, D.F.F., Mello, J.W.V. (2009). Pó de basalto: alternativa agroecológica de baixa emissão de CO2 à correção da acidez do solo. *Revista Brasileira de Agroecologia*, 4, 3994–3998.

Fernandes, M.M. (2008). *Viabilidade Agronômica do Uso do Rejeito de Garimpos do Distrito Pegmatítico de Araçuaí/MG.* PhD Thesis UFMG, 158pp.

Formoso, M.L.L. (2006). Some topics on geochemistry of weathering: A review. *Anais da Academia Brasileira de Ciências* (Brasil), 78(4), 809–820.

Fyfe, W.S, Leonardos, O.H., and Theodoro, S.H. (2006). The use of rocks to improve family agriculture in Brazil. Journal of the Brazilian Academy of Sciences 78(4), 721–730.

Harper, J.R., Fey, M.V., Mbakwe, I., and Awkes, M. (2010). Potential of fly ash as source of alkalinity for amelioration of acid soils of the South African Highveld. In: Martins, E. and Theodoro S.H. (eds.) *Anais do I Congresso Brasileiro de Rochagem.* Brasília-DF, pp. 37–46.

Hinsinger, P., Barros, O.N.F., Benedetti, M.F., Noack, Y., and Callot, G. (2001). Plant-induced weathering of a basaltic rock: Experimental evidence. *Geochimica et Cosmochimica Acta*, 65(1), 137–152.

Gibson, S.A., Thompson, R.N., Leonardos, O.H., Dickin, A.P., and Mitchell, J.G. (1999). The late cretaceous impact of the Tridade Mantle Plume: Evidence from large-volume, mafic, potassic magmatism in SE Brazil. *Journal of Petrology*, 36(1), 189–229.

Khatounian, C.A. (2001). *A reconstrução ecológica da agricultura.* Botucatu: Agroecológica, 349pp.

Knapik, J.G. (2005). Utilização do pó de basalto como alternativa à adubação convencional na produção de mudas de Mimosa scabrella Benth e Prunussellowii Koehne. Master's Thesis. UFPR. 123pp.

Kronberg, B.I., Leonardos, O.H., and Fyfe, W.S. (1987). The chemistry of some Brazilian soils element mobility during intense weathering. *Chemical Geology*, 60, 361–370.

Leonardos, O.H., Fyfe, W.S., and Kronberg, B.I. (1976). Rochagem: O método de aumento da fertilidade em solos lixiviados e arenosos. Anais 29 Congresso. Brasil. Geol., BH, pp. 137–145.

Leonardos, O.H., Fyfe, W.S., and Kronberg, B.I. (1987). The use of ground rocks in laterite systems: An improvement to the use of convetional soluble fertilizers? *Chemical Geology*, 60, 361–370.

Leonardos, O.H., Ulbrich, M.N., and Gaspar, J.C. (1991). The Mata da Corda Vulcanic rocks. In: Leonardos, O.H., Meye, H.O.A., and Gaspar, J.C. (eds.) *Field guidebook, 5th International Kimberlite Conference*, Araxá. Brasília/DF: CPRM Special Publication 3/19, pp. 17–24.

Leonardos, O.H. and Theodoro, S.H. (1999). Fertilizing tropical soils for sustainable development. In: Formoso, M.L.L. and Cerri, C.C. (eds.) *Proceedings of the International Workshop on Tropical Soils*, Rio de Janeiro, Brazil. Anais da ABC, pp. 143–153.

Leonardos, O.H., Ulbrich, M.N., and Gaspar, J.C. (1991). The Mata da Corda Vulcanic rocks. In: Leonardos, O.H., Meye, H.O.A., and Gaspar, J.C. (eds.) *Field Guidebook, Fifth International Kimberlite Conference*, Araxá. Brasília-DF, CPRM Special Publication 3/19, pp. 17–24.

Manning, D. (2010). Stone meal as a source of plant nutrients, especially potash: A mineralogical approach. In: Martins, É. and Theodoro, S.H. Anais do I Congresso Brasileiro de Rochagem. Brasília: Embrapa, pp. 47–54.

Petersen, P. and Almeida, E. de. (setembro de 2008). Revendo o conceito de fertilidade: conversão ecológica do sistema de manejo dos solos na região do Contestado. Agriculturas, 5(3).

Rocha, E.J.P. (2006). Agroflorestas sucessionais no assentamento Fruta D'anta/MG: potenciais e limitações para a transição agroecológica. Dissertação de Mestrado. CDS, UnB, Brasília-DF.

Theodoro, S.H. (2000). A Fertilização da Terra pela Terra: Uma Alternativa de Sustentabilidade para o Pequeno Produtor Rural. PhD thesis, University of Brasilia, Brazil, 231pp.

Theodoro, S.H. and Leonardos, O.H. (2006). Sustainable farming with native rocks: The transition without revolution. *Anais da Acad. Bras. de Ciências* (Rio de Janeiro/RJ), 78(4), 715–720.

Theodoro, S.H. and Leonardos, O.H. (2011). Rochagem: uma questão de soberania nacional. In XIII Congresso brasileiro de geoquímica. Gramado/RS.

Theodoro, S.H., Leonardo, O.H., and Almeida, E. de. (2010). Mecanismos para disponibilização de nutrientes minerais a partir de processos biológicos. Brasília: Embrapa, pp. 173–181.

Theodoro, S.H., Leonardos, O.H., Rego, K.G., and Rocha, E.L. (2009). Stonemeal technology: Changing the poverty scenario in Brazil. In: Van Straaten, P., Oran, N. (eds.) *Proceedings of the Second International Rocks for Crops Conference,* Nairobi and Kisumo, Kenya, pp. 145–170.

van Strateen, P. (2010). Geodiversity, biodiversity and the origin of crops. In: Martins, É. and Theodoro, S.H. (eds.) Anais do I Congresso Brasileiro de Rochagem. Brasília: Embrapa.

23

Chemical Composition of Litter Fall and Inputs of Carbon, Nitrogen, and Mineral Elements in a Secondary Forest of South Cameroon

Damien Henri Odigui Ahanda, Monique Abossolo Angue, and Jean Jacques Braun

CONTENTS

Introduction

Recycling of nutrients via plant litter fall plays an essential role in maintaining ecosystem productivity. The natural baseline of such ecological services is poorly known in many places, but once the forest is cleared and converted to agriculture or forestry, this recycling is greatly reduced and, where the crop is removed and where crop residues and biomass are removed for burning elsewhere, can be completely eliminated. The result is that soil ecosystem nutrient stores are steadily depleted without replenishment, so agricultural and forestry productivity steadily declines and can only be maintained by expensive importation of the lost nutrients.

Forests of the tropical zones annually restitute to the soil a great amount of matter through litter fall (leaves, twigs, fruits, flowers, etc.), rain leaching, root exudates, and dead roots. An important part of this organic debris is quickly mineralized and contributes to increase the soil nutrient content needed for plant growth. The other part slowly forms the thin mull humus (5–15 cm depth) underneath after the humification process.

Paradoxically, despite the importance of litter fall and its critical role in the natural functioning of ferrallitic soil ecosystems, very few studies have been carried out on this subject in Africa (Schwartz, 1991; Reversat, 1993) and in Cameroon (Boyer, 1973). The work carried out in the secondary forest of Nkolbisson since 1997 is a contribution to such studies. The aim of this study is to determine the elementary chemical composition of the litter sampled monthly in the Nkolbisson forest from May 1997 to 1998, to characterize the litter and to quantify the amount of carbon, nitrogen, and total mineral elements restituted to the soil by litter falls.

Material

The region of Nkoabang is situated 13 km SE of Yaoundé. It has an equatorial climate with four seasons. The average annual rainfall is 1420.6 mm, while the annual average temperature is 24.14°C. The vegetation is a mix of deciduous thick forest deeply transformed by man. Among other plant families are found Sterculiaceae, Euphorbiaceae, Caesalpiniaceae, Mimosoideae, Meliaceae, Moraceae, Bombacaceae, Rhamnaceae, Rutaceae, Olacaceae, and Annonaceae. The rocky substratum is made up of metamorphic rocks of the northern-equatorial Pan-African chain (Nzenti and Tchoua, 1998). This substratum is affected by N-S and E-W fractures. The soils are ferrallitic, red, or yellow in well-drained areas. Hydromorphic soils are found in swampy low lands. The relief of this zone is in the range of 600–700 m. It is deeply dissected by a dendritic and highly hierarchized river network. The study site known as Nkolbisson 1 stretches between 3°49′35″ and 3°51′16″ of north latitude and between 11°32′42″ and 11°40′17″ of east longitude. It is situated around 4 km to the SE of Nkoabang. The litter fall sampling site has been marked off at the east flank of the Ekwazong stream, on the midslope of a hillock, at around 725 m altitude.

Methods

Pretreatment of Litter

Litter fall was collected every 2 weeks in 10 traps spread at random in an area of 25 m × 25 m. Litter traps are square wooden boxes of 50 cm × 50 cm carrying a fine mesh (1 mm size). Litter samples are collected and kept in numbered plastic bags. The litter is air dried for 1or 2 weeks. A representative portion of it is crushed (0.2 mm); the powder is kept in marked containers.

Chemical Analysis

C and N contents were determined in the IRAD soil and plant laboratory of Nkolbisson (Yaoundé, Cameroon), litter collected during the first 6 months of the study (May–November 1997). Major and trace elements were determined in the *Laboratoire du Centre de Pédologie Biologique de Nancy* (*France*), in litter over 12 months of study (June 1997 to May 1998).

Carbon was measured using the Walkley–Black method (Walkley and Black, 1934). The principle consists of mineralizing the organic carbon. The oxidation of carbon is done by digestion of 1 g of litter powder with 10 mL of potassium bicarbonate and 20 mL of concentrated sulfuric acid (1.84 density) in an Erlenmeyer flask. Calorimetric measurements were made in a Technicon AutoAnalyzer. Results are given in percent.

For nitrogen determination, 1 g of litter powder and concentrated sulfuric acid and a catalyst (selenium + sodium sulfate) are put in a 1000 mL flask and digested in a Tecator at 370°C for 3 h. The nitrogen is then calorimetrically titrated in a Technicon AutoAnalyzer. Results are given in percent.

The litter powder is mineralized dry: 1 g of powder contained in a ceramic capsule is combusted at 450°C overnight. The ashes are collected and put in a 50 mL Teflon flask. Fifteen milliliters of 30% concentrated hydrochloric acid and 7.5 mL of 40% hydrofluoric acid are added, and the mixture is heated at 120°C for 2 h. The filtered mixture is poured in a Nalgene bottle and adjusted to 1000 mL with distilled water. The elemental measurements are done by atomic absorption, using Sr–HCl range (Rouiller, 1993). Results are given in parts per thousand (‰) for major elements and in ppm for trace elements.

Calculation of Element Input by Litter Fall

This calculation takes into consideration the content of the element:

- For an element content of X%, the annual inputs (A) are calculated by the formula $A = X\% \times 10^{-2} \times P$ (t/ha/year), with P = annual production of litter.
- For an element content of Y‰, the inputs (A) are given by the formula $A = Y‰ \times 10^{-3} \times P$ (t/ha/year).
- For an element content of Z ppm, the inputs (A) are given by the formula $A = Z \text{ ppm} \times 10^{-6} \times P$ (t/ha/year).

Results

The chemical composition of plant residues distinguishes two types of elements, according to their abundance: major elements and trace or oligo elements. The major elements, or macroelements, are the most abundant. Some of them are predominant: C, H, O, and N. Others include both major and minor elements: P, K, S, Mg, Ca, Cl, Na, Si, Al, and Fe.

The trace elements are less abundant. They intervene in plant functioning and represent barely 1% of the weight of vegetal material.

Predominant elements are shown in Figure 23.1. C contents vary between 31.4% and 45.7%, with an average of 38%.

Nitrogen contents vary between 1.6% and 4.1%, with an average of 3%. Monthly C/N ratios range between 7.65 and 21.8, which corresponds to improving litters, easily biodegradable (Duchaufour, 1984).

According to the analytical results (Table 23.1), P, S, Ca, Mg, K, Na, Si, Fe, Mn, and Al represent 61‰ or 6.1% of the weight of litter. Litter is relatively rich in Si, Ca, and Fe, elements whose monthly average contents are greater than 9‰. They represent 57.6% of the total major mineral elements, with 25.5% for Si only. S, K, P, Mg, Al, and Mn represent 42.3% of major elements. Their average contents vary from 2‰ to 7‰. The litters are poor in Na; its contents rarely surpass 0.15‰. Zn, Cd, Pb, Co, Ni, Cr, Cu, Mo, and Ti on average represent 390.4 ppm or 3.6×10^{-2}% of the weight of litter. Zn, Ti, and Cu contents are in order of 20–253 ppm. They represent 87.5% of the entire trace elements. Zn is the most abundant. Cd, Pb, Co, Ni, Cr, and Mo contents are less than 20 ppm and represent 12.17% of trace elements.

The chemical composition of litter of Nkolbisson forest shows variations in the course of different months. Some elements like Mn and Na are relatively constant throughout the year. Others such as Fe, Ca, and Al highly vary between months, independently of the seasons. Some others at least have variations

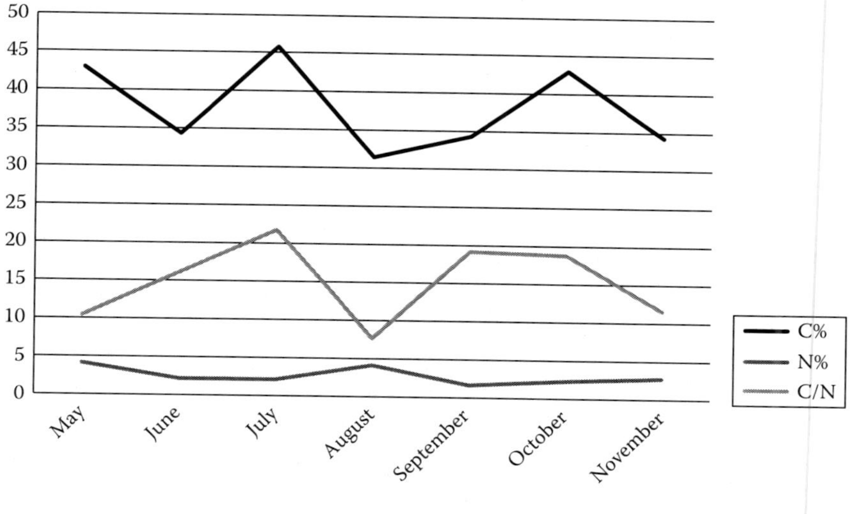

FIGURE 23.1 C and N contents and C/N ratio of litter from May to November 1997.

TABLE 23.1
Average Major Element Contents of Litters and Corresponding Inputs

Elements	C	N	Si	Ca	Fe	S	K	P	Mg	Al	Mn	Na
Average contents (%)	38	2.73	1.55	0.99	0.96	0.71	0.46	0.42	0.41	0.37	0.20	0.02
Inputs kg/ha/year	4085	293	170	110	100	80	50	50	40	40	20	2.15

TABLE 23.2
Average Trace Element Contents of Litters and Corresponding Inputs

Elements	Zn	Ti	Cu	Ni	Co	Pb	Cr	Cd	Mo
Mean values (%)	2.53×10^{-2}	5.9×10^{-3}	1.9×10^{-3}	1.5×10^{-3}	1.1×10^{-3}	1.04×10^{-3}	7.9×10^{-4}	2.0×10^{-4}	1.7×10^{-4}
Inputs kg/ha/year	2.7	0.63	0.21	0.16	0.12	0.11	0.09	0.02	0.02

that are linked to seasons, with Si, K, Mg, P, and S found abundantly in dry season litter. Tables 23.1 and 23.2 show the inputs in different elements from litter fall. The annual inputs of elements to the forest soil by litter fall (Tables 23.1 and 23.2 and Figures 23.2 and 23.3) average 5044.21 kg/ha, with 4085 kg C/ha, 293 kg N/ha, 202.2 kg (Ca, K, Mg, Na), and 460 kg (Si, Fe, S, P, Al, and Mn). The total input of trace elements is 4.0 kg/ha.

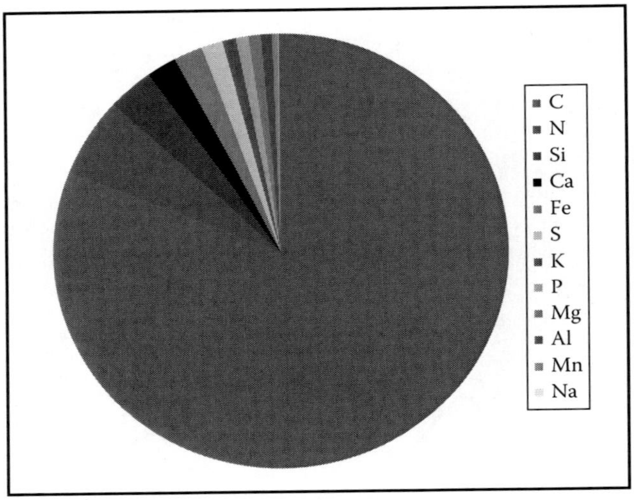

FIGURE 23.2 Average contents.

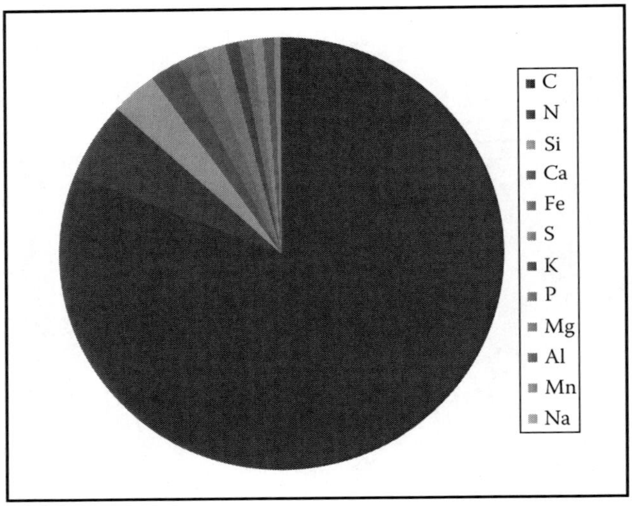

FIGURE 23.3 Inputs (kg/ha/year).

Discussion

The litter of Nkolbisson forest, contrary to those of tropical forests like that of Congo (Schwartz, 1991; Reversat, 1993), Guyana (Leroy, 1991), and Amazonia (Lucas et al., 1993), proves to be poor in C and rich in N, Fe, Ti, and Na. The difference found between different litter chemical compositions might be linked to floristic diversity, to the fraction of litter analyzed, or to the nature of soil.

The forests studied by Leroy and Reversat are monospecific: Leroy (1991) reveals that the litters of *Quala* sp. (Caesalpiniaceae), *Ocotea rubra* (Lauraceae), and *Dicorynia guianensis* (Caesalpiniaceae) have different chemical composition. Reversat (1993) found that *Acacia* litters are richer in nitrogen and lignin than those of *Eucalyptus*, which are rich in carbon and mineral elements. The Nkolbisson forest is species diverse. About 30 different species of 15 families have been identified there. The phenomena of competition and biodegradation are more important here than in monospecific forests.

The chemical composition of litter can vary with the analyzed fraction of the litter. Lucas et al. (1993) reveal the fact that leaves have an elevated content of major and trace mineral elements, while wood is rich in calcium.

Seasonal variation of litter chemical composition was found in the Nsimi secondary forest by Zoetele (Abossolo, to be published). Leaves have higher contents of Si, Fe, Al, and Mg during the rainy season. The twigs and other parts are relatively rich in Ca, Al, and Mg in the dry season, while they are poor in Si. In general, it is noticed that the seasonal variations of litter chemical composition reveal the physiological rhythm of plants (Heller et al., 1990).

Comparing the Nkolbisson's litter to that of the other forests of the humid tropical zones (Luizao and Schubart, 1987; Lucas et al., 1993), element inputs seem to be higher in Nkolbisson. This is due to the high production of litter registered in this forest. Comparing our results to the monospecific litter of cocoa farms (Boyer, 1973), the species-diverse litter of Nkolbisson restitutes more N, P, and Si, while those of cocoa plants produce more Ca (90 kg/ha/year). K and Mg inputs of those different litters are relatively comparable with 10–50 kg/ha/year for K and 25–40 kg/ha/year for Mg. According to their importance in the total litter, leaves can be considered as the principal provider of elements in the soil.

Conclusions

The elementary chemical composition of samples of soil litters collected in the Nkolbisson secondary forest from May 1997 to 1998 has been determined, with the aim of characterizing it and to quantify corresponding annual inputs of carbon, nitrogen, and mineral elements. Results show that there are three main groups of elements in the constitution of litter: predominant elements (C and N), major elements (P, K, S, Mg, Ca, Cl, Na, Si, Al, and Fe), and trace elements (Zn, Cd, Pb, Co, Ni, Cr, Cu, Mo, and Ti).

Carbon contents vary between 31.4% and 45.7%, averaging 38%. Nitrogen contents range between 1.6% and 4.1%, averaging 3%. The corresponding C/N ratios are between 7.65 and 21.8, which are characteristic of improving litters, easily biodegradable.

Si, Ca, and Fe are the most abundant major elements. Their cumulative average content represents 57.6% of the total major elements. Zi, Ti, and Cu are dominant in the trace element group, with average contents varying between 20 and 253 ppm.

This study does not show a clear relation between climatic seasonal variations and the chemical composition of litters. However, some elements, particularly Si, K, Mg, and S, are relatively more abundant in the litter during the dry season compared to their content during rainy seasons. Meanwhile, some other elements like Fe, Ca, and Al show large variations from 1 month to the following, independently of climatic seasons.

Considering the results of chemical analysis, litter falls restitute an average of 4085 kg C/ha/year, 293 kg N/ha/year, 202 kg (Ca, K, Mg, Na)/ha/year, 460 kg (Si, Fe, S, P, Al, Mn)/ha/year, and 4.0 kg of trace elements/ha/year to the soil. The greatest part of these restitutions is attributed to leaves because the total litter is made up of about 61% of leaves.

REFERENCES

Boyer, J. (1973). *Cycle de la matière organique et des éléments minéraux dans une cacaoyère camerounaise.* Café Cacao Thé (Paris) vol. XVII, no. 1, pp. 3–23.

Duchaufour, Ph. (1984). *Abrégé de Pédologie.* Ed. Masson, Paris, France, 216 p.

Heller, R., Esnault, R., and Lance, C. (1990). *Vegetal Physiology Shorter.* 4th edn., Vol. II, Masson Publishers, Paris.

Leroy, C. (1991). *Etudes préliminaires de la biodynamique des humus en forêt guyanaise.* D.E.A de pédologie, Université de Nancy-I, 65p.

Lucas, Y., Luizao, F.J., Chauvel, A., Rouiller, J., and Nahon, D. (April 23, 1993). The relation between biological activity of the rain forest and mineral composition of soils. (Reports) *Science*, 260, 521–523.

Luizao, F.J. and Schubart, H.O.R. (1987). Litter production and decomposition in a terra-firme forest of central Amazonia. Divisão da bioccologia, Instituto Nacional de Pesquisas da Amazõnia, *Experientia*, 43(3), 259–265.

Nzenti, J.-P. and Tchoua, F. (1998). *Signification pétrogénétique des grenatites des migmatites de Yaoundé (chaîne panafricaine Nord-équatoriale du Cameroun).* Géosciences du Cameroun, collect. GEOCAM 1, 379–385.

Reversat, B. (1993). Dynamics of litter and organic matter at the oil-litter interface in fast growing tree plantations on sandy ferrallitic soils (Congo). *Acta Oecologica* 14(2), 179–195.

Rouiller, J. (1993). *Techniques de laboratoire dans l'analyse des sols.* Centre National de la Recherche Scientifique. Centre de Pédologie Biologique. Nancy. 45p.

Schwartz, D. (1991). *Les retombées de litière en tant que source du carbone et de l'azote du sol quantification et périodicité des apports en relation avec les caractéristiques climatiques et édaphiques dans deux parcelles de forêt dense à dimonika (mayombe, congo).* Echanges forêt-atmosphère en milieu tropical humide. DGRST/UMNG - UNESCO/PNUD - CNRS/ORSTOM, pp. 141–158.

Walkley, A. and Black, A. (1934). An examination of the Degtjareff method for determining soil organic matter and a proposed modification of the chromic acid-titration method. *Soil Science*, 37, 29–38.

24

Pyroclastic Rocks as Natural Fertilizer: Case Study of Volcanic Ashes from Tombel Graben (Cameroon Volcanic Line, Central Africa)

David Guimolaire Nkouathio

CONTENTS

Introduction

Food security in Africa, and especially in sub-Saharan Africa, is one of our most crucial challenges, which is why the United Nations has classified it as the first target of Objective 1 of the Millennium Development Goals. The rate of population growth is higher in sub-Saharan Africa than in any other region of the world (FAO, 1998). According to UN evaluations (1997), this population is likely to double in the next 25 years. In order to reduce poverty and hunger, agricultural production will need to increase considerably through sustainable management of soil fertility. Soil fertility is limited by the availability of arable soils, the use of mineral fertilizers, the restoration of soil fertility, water management, and climatic variability. Traditional systems of soil management depend on the use of long periods of fallow land intended to maintain soil fertility, but unfortunately this does not work for Africa due to its high population density. Assessment of available land surface in Africa indicates that there is between 0.28 and 0.52 ha of cultivated land per capita (Fischer and Heilig, 1997). Due to the lack of available land, fallow soils must be put back into agricultural production, and fertility falls, making the land more vulnerable to erosion.

An increase in productivity through the use of fertilizers is hypothetical because of the lack of available resources for farmers. Moreover, the use of chemical fertilizers is strongly limited by the geochemical and mineralogical characteristics of tropical lateritic soils. The fertile soils of the Northern Hemisphere are rich in clay minerals with strong ionic exchange capacity, so they receive the maximum benefit from soluble chemical fertilizers. On the other hand, less fertile tropical soils with very weak ionic exchange capacity are less capable of nutrient uptake. Constant heat stimulates microorganisms that decompose organic matter, so little humus accumulates. In lateritic soils, chemical fertilizers are washed out after a few weeks and the cultivated plants do not receive the necessary nutrients (Schaetzl and Anderson, 2005). The overuse of chemical fertilizers is expensive and detrimental to the environment, infiltrating the water table and causing nitrogen runoff to rivers due to lack of retention by tropical soils.

These factors act to negate Target 3 of Objective 7 of the UN MDG of reducing biodiversity loss, the reason why the UN Conference in Nagoya in 2010 declared 2011–2020 "The Decade of Biodiversity." It is therefore urgent to elaborate techniques of fertilization better adapted to the sub-Saharan pedological and climatic conditions (Haraldsen and Pedersen, 2003; Barral Silva et al., 2005; Kleïr and Thornhill, 2007; Nono et al., 2008). To imitate natural nutrient supply systems, many rocks, notably volcanic, have been tested for use to rematerialize poor soil. An example of the increase in the fertility of soils and therefore some agricultural productivity is given by the use of fine-grained basaltic pyroclastic rocks to restore soil fertility of soils, the subject of this chapter.

Pyroclastic Rocks

Pyroclastic materials are the result of volcanic activity. More than 20 African countries are covered at least in part by volcanic rocks. These volcanoes are the consequence of the activity of a hot spot and/or of the East African Rift. The basaltic scorias are produced by volcanoes of strombolian type. These volcanoes are characterized by alternation of pyroclastic rocks (scoria, lapilli, and ashes) in general of fine grain size (Figure 24.1) and lava flows, forming volcanic cones of variable sizes. The addition of these rocks, especially fine-grained ones, to the soils of these regions allows improvement of soil fertility in areas affected by erosion, bad texture, weak exchange capacity of soils, highly acidic pH, and aluminum and manganese toxicity (Nkouathio et al., 2008b).

A specific case has been achieved with pyroclastic rocks from strombolian cones of the Tombel graben, an important link of the Cameroon volcanic line (CVL). The CVL, which is more than 100 km wide and more than 1200 km long, is located along the Gulf of Guinea in Central Africa. Some of these volcanoes are still active, such as Mount Cameroon (Déruelle et al., 2000; Suh et al., 2003).

Almost all strombolian volcanoes are basaltic. The pyroclastic volcanic rock fragments fall during strombolian eruptions. We classify them according to their grain size (Table 24.1).

In addition, these rocks can be vesicular, resulting from the solidification of lava containing gases, forming scoria.

FIGURE 24.1 A view of erosion on the flanks of a volcanic cone in Tombel graben (Cameroon Volcanic Line).

TABLE 24.1

Granulometric Classification of Ejected Pyroclastic Volcanic Rocks

Blocks (Angular Shape) Volcanic Bomb (Globular Shape)	64 mm
Lapilli	2 mm
Ashes	
Sandy fraction	1/16 mm
Silty fraction	1/256 mm
Clayey fraction	0 mm

Source: Fisher, R.V.B. and Schmincke, A.N., *Pyroclastics Rocks*, Springer Verlag, Berlin, Germany, 1984, 472pp.

Volcanic Cones: An Important Source of Volcanic Ashes

The scoria cones constitute large pyroclastic deposits. They sometimes contain more than 600,000,000 m^3 of pyroclasts (Table 24.2).

The volcanic ashes that have the finest grain size can serve as mineral rock fertilizers. These ashes abound in the cones. In the pyroclastic rocks of Tombel graben, they form more than 10% of all the rocks (Figure 24.2).

These volcanic ashes have basaltic composition. Geochemical analyses of pyroclastic rocks from some cones of the Tombel graben (CVL) (Table 24.3) confirm their basic character (SiO_2 < 49%) with alkali ($K_2O + Na_2O$ = 3.8%–5.1%).

They also possess important reserves of calcium (8%–10%) and magnesium (7%–9%) as shown by their mineralogical composition in the polarized microscope (Nkouathio, 2006) or determined for vitreous groundmass phases by x-ray diffraction spectra (Wandji, 1995). These pyroclastic materials are mainly made up of olivine [$(Mg,Fe)_2SiO_4$], pyroxene (augite ((Ca,Na) (Mg,Fe,Al,Ti) [$(Si,Al)O_3$]), plagioclase [$(Na,Ca)Si_3AlO_8$], titanoferriteoxides (hematite (Fe_2O_3), titanomagnetite (Ti,Fe_3O_4) and ilmenite ($FeTiO_3$) (Figure 24.3).

The majority of these minerals are silica and opaque basic oxide silicates, rich in calcium, magnesium, and ferrous iron. Tests of fertility have therefore been made with these materials to test the contributions of fine-grained pyroclastic materials on soil fertility.

Soils derived from rock alteration carry the chemical imprint of their parent rocks. With progressive weathering of constituent minerals in the rocks, the chemical elements liberated in soil become available for the plants (Leonardos et al., 1987). The change is especially fast when the rocks are of fine grained (volcanic ashes, glacial till, etc.) (Fragstein, 1991).

TABLE 24.2

Volume of Some Strombolian Volcanoes in Tombel Graben

Name of Volcanoes and/or Localization	Latitude	Longitude	Altitude (m)	Height (m): H	Basal Ray (m): R	Crateric Ray (m): r	Volume (m^3): V
Mt Pélé	4°35′16″	9°37′32″	262	122	475	175	43,357,905
Manjocone	4°51′37″	9°50′15″	730	90	400	225	28,333,238
Frontiere cone	4°43′14″	9°41′25″	583	103	375	125	21,909,336
Kopongo	4°36′1″	9°38′25″	143	63	325	75	8,947,648
Loum area	4°43′31″	9°41′43″	619	199	1200	200	358,434,777
Loum area	4°41′42″	9°40′42″	631	251	1500	150	656,459,346
Njombé area	4°48′51″	9°47′31″	260	60	300	—	5,654,866
Manengoteng area	4°48′51″	9°47′31″	540	80	400	—	13,404,128
Mantem area	4°51′12″	9°46′30″	623	103	400	—	17,257,815
Nyassosso area	4°51′5″	9°42′30″	1240	240	1000	—	251,327,412

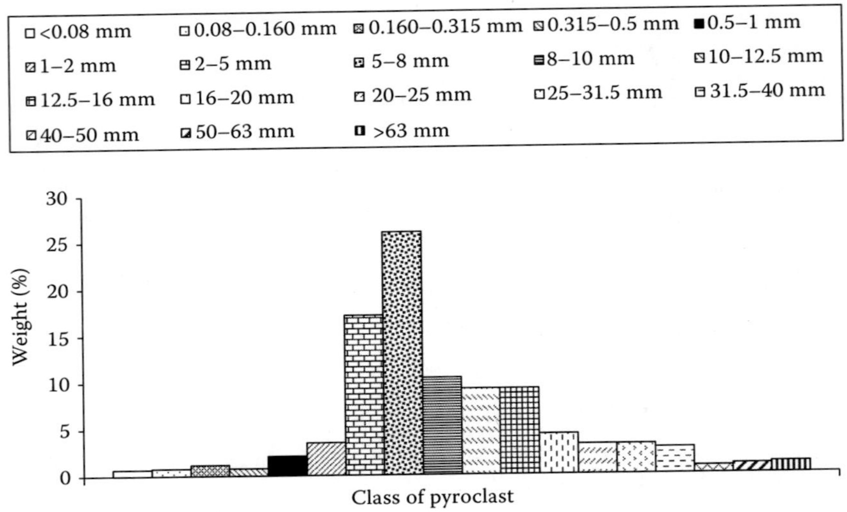

FIGURE 24.2 Histogram of the granulometric analysis of pyroclastic materials from a scoria cone in Tombel graben (CVL).

TABLE 24.3

Chemical Analyses of Some Pyroclastic Rocks of Tombel Graben

Elements	1	2	3	4	5
SiO_2	48.15	47.30	44.80	46.52	45.44
Al_2O_3	15.28	17.20	16.90	15.31	14.87
Fe_2O_3	12.98	12.12	12.47	13.28	13.22
CaO	8.16	8.30	9.75	9.17	9.08
MgO	6.64	7.05	8.57	7.43	7.94
Na_2O	3	4.95	4.63	3.60	3.60
K_2O	0.80	1.66	2.07	1.50	1.50
TiO_2	2.72	0.31	0.35	2.72	2.71
P_2O_5	—	0.31	0.38		
MnO	0.16	0.24	0.24	0.19	0.18
H_2O		—	0.70–0.48	0.31	
Total	97.89	100.14	100.56	99.72	98.54

Source: Nkouathio, D.G. et al., *Bull. Soc. Vaud. Sci. Nat.*, 91, 1, 2008b.

Regions rich in these types of rocks are especially fertile in general, and particularly true in tropical zones (Leonardos et al., 1987). This is the case in the Tombel graben plain of the CVL (Nkouathio et al., 1998, 2002). This graben owes its fertility to the fallout of volcanic ash tephra formed by recent explosive periods (age < 0.5 My; Nkouathio et al., 2008a) of basaltic composition (Nkouathio et al., 1999). This tectonic dynamism is responsible for the formation of many volcanic cones and deposition of volcanic ashes on the plain (Figure 24.1).

Fertilization Tests

To evaluate the benefit of fertilizing with volcanic ash from the Tombel graben, we performed a set of trials. Pot tests containing samples of soils enriched with varying amounts of pyroclasts were tested. Corn plants (*Zea mays*) were observed for growth increases to show growth differences based on the different additions of volcanic rock dust.

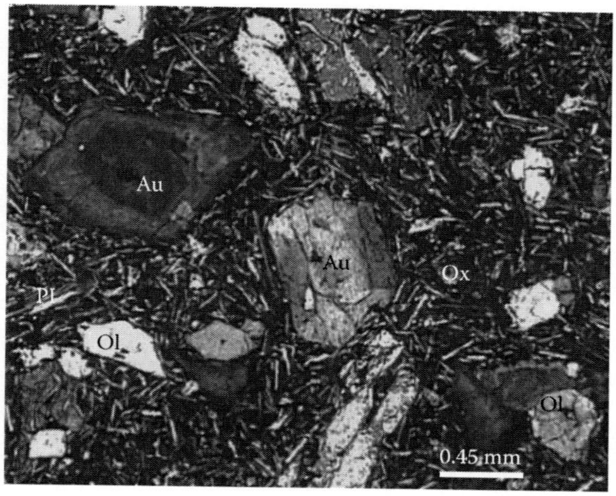

FIGURE 24.3 Microphotography of pyroclast (volcanic bomb) from Tombel graben area (Ol = olivine, Au = augite, Ox = titanoferriteoxides, Pl = plagioclase).

Materials and Methods

Our trials were performed on iron-rich soils from the region of Yaoundé with the addition of volcanic ash of the Tombel graben.

Corn (*Z. mays*) is one of the main crops produced in Cameroon (IFDC/USAID, 1986). Corn productivity is very sensitive to variations in soil fertility, which is why it was selected for this study. The corn used is a variety of the Cameroon Maize Selection type 8704, selected by the Agronomic Research Institute of Nkolbisson in Yaoundé.

The soil selected for testing was sifted (<2 mm). A kilogram of the sifted soil was mixed with either chemical fertilizers or pyroclasts in different fineness and amounts. These pots were exposed to natural conditions on a shelf about 40 cm high, in order to avoid contamination by local soil. The pots received rainwater in addition to being watered. The soils from Yaoundé used in testing were highly damaged from excessive cultivation. They are oxisols (WRB, 1998), that is, ferrallitic-type soils (CPCS, 1967). These soils are rather acidic (*S*; pH < 6, Table 24.4), relatively poor in exchangeable bases ($Ca^{2+} + Mg^{2+} + K^{+} < 3$), and generally very deep (thickness sometimes greater than 10 m). They have good physical structure but are chemically impoverished because they lack nutrients (Onguené Mala, 1993).

pH was measured in a soil/solution of 1/2.5 by measurement with glass electrode and reference electrode.

Exchangeable bases (Ca^{2+}, Mg^{2+}, K^{+}) were extracted in a solution of ammonium acetate (CH_3COONH_4) buffered at pH 7. K^{+} was measured by flame photometry. Ca^{2+} and Mg^{2+} were measured by complexometric titration, using ethylenediaminetetraacetic acid as titrant.

The cationic exchange capacity (CEC) of each soil was measured after extraction of 2 g of soil by 10 mL of solution of 0.2 M $BaCl_2 + NH_4Cl$ and measured by flame absorption spectrometry.
For every test, we experimented on five examples as follows:

1. A soil control (T0) and a soil enriched with classic NPK fertilizers (T2)
2. Four soils with 5%, 10%, 25%, and 50% of natural pyroclastic ash (N1, N2, N3, and N4)
3. Four soils with 5%, 10%, 25%, and 50% of crushed pyroclastic ash (B1, B2, B3, and B4)
4. A test with pure natural pyroclastic ash (100%) (N5)

Four parameters were taken into account by measuring growth and yield. These included the number of leaves per stem, plant height, diameter of the stalk, and weight of dry biomass.

TABLE 24.4

Physical and Chemical Parameters of Pot Soils after the First Experiment

Physical and Chemical Parameters		Reference Values	S	T0	B1	B2	B3	B4	Pi	Pt
pH water		6–7	5.73	5.21	5.14	5.72	6.42	6.38	7.15	6.94
Exchangeable bases (meq/100 g)	Ca^{2+}	3.5–7	1.61	1.61	2.69	2.90	3.60	5	1.56	3.32
	Mg^{2+}	1.5–3	0.91	0.71	1.15	1.38	1.54	1.64	0.51	1.02
	K^+	0.4–0.8	0.15	0.10	0.13	0.15	0.17	0.18	0.12	0.60
Exchangeable acidity (Al^{3+} + H^+) (meq/100 g)			0.50	0.43	0.28	0.15	0.15	0.15	0	0.10
CEC (meq/100 g)		10–15	3.17	2.85	4.25	4.58	5.46	6.97	2.19	5.04
Sum of exchangeable bases (meq/100 g)		10–15	2.67	2.42	3.97	4.43	5.31	6.82	2.19	4.94
Saturation	K %	2–5	4.73	3.5	3.1	3.3	3.1	2.6	5.48	11.9
	Mg %	10–15	28.7	24.9	23.5	30.1	28.2	23.5	23.3	20.2
	Ca %	60–80	50.8	56.5	63.3	63.3	65.9	71.7	71.2	65.2
Base saturation % (S/CEC)		60–80	84.2	84.9	89.9	96.7	97.2	97.8	100	97.9

Reference values according to Bachelor (1951) and Léon (1979).
S, control; Pi, initial pyroclasts; Pt: soaked pyroclasts.

Measurements were made after 2–4 weeks for the first two parameters, then after 1 month for stalk diameter and dry biomass.

The trials were carried out successively three times with the same soil samples to observe their progressive nutrient depletion. After the first test, soils of every treatment were analyzed to compare them to the initial material and evaluate rates of mineralization with respect to the quantity of pyroclastic ash added. Each time, we determined the pH, exchangeable bases (Ca^{2+}, Mg^{2+}, K^+), and exchangeable acids (Al^{3+}, H^+) (Table 24.4).

Results

We observed a significant shift of pH from fairly acidic to very weakly acidic with increased proportion of previously ground pyroclastic ash (Figure 24.4). The exchangeable acidity decreased to stabilize around 0.15 meq/100 g of materials. Above pH 5.5, Al^{3+} becomes insoluble. Reduction of soil aluminum acidity is linked to release of basic cations, notably Ca and Mg.

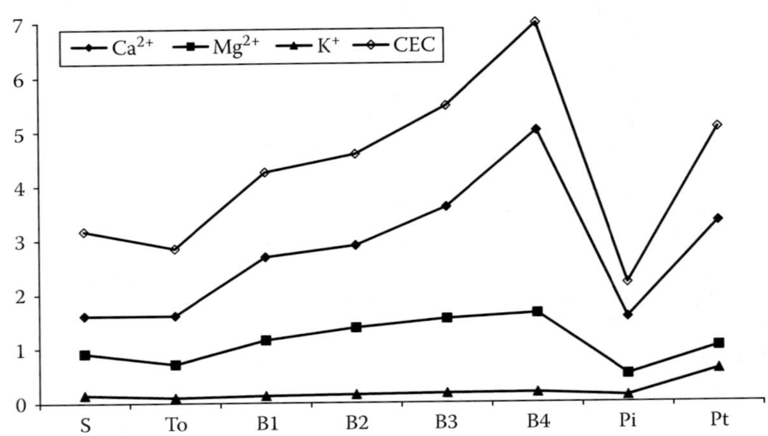

FIGURE 24.4 Evolution of some chemical parameters according to tests.

The data show increased exchangeable bases and CEC with increased pyroclastic materials. The most important concentrations are those of Ca^{2+}, followed by Mg^{2+}, and then K^+ (Figure 24.4). It implies an important release of Mg and Ca by pyroclastic ash. The low rates of K parallel the poor contents of pyroclasts in this element (Table 24.4).

There were only small differences in the number of leaves between the different treatments during the experiment (Figure 24.5a). Nevertheless, it appeared that after 1 month, plants grown in soil–pyroclastic mixtures systematically had an extra leaf. Pure pyroclastic material did not give a better result than that of soils. The efficiency of chemical fertilizers (NPK) was shown by the presence of two extra leaves.

The different treatments had more effect on corn height (Figure 24.5b). Effects of addition of pyroclastic ash became greater throughout the experiment. The best results were found for mixtures of 25% or 50% of pyroclastic material and are insensitive of whether the ash was ground or not. The results in pure pyroclastic ash (poor growth of plants) and the NPK fertilizer (accelerated growth) confirm the previous experiment.

The stalk diameter (Figure 24.5c) and dry biomass (Figure 24.5d) are the parameters most likely to be associated with an increase in yield (Fragstein, 1991). We noted the sensitivity to the addition of pyroclasts. Mixtures of 25% and 50% yielded the best results, especially with ground pyroclastic ash. It should however be noted that the benefits are by far lower than those produced with the NPK manure.

During the second experiment, we observed an increase in yield in the majority of treatments with contributions from pyroclastic ash. This results from progressive liberation of the ions contained in these rocks. On the other hand, the control samples and soils fertilized with NPK fertilizer showed decreasing yields, signifying that their nutritive reserves were progressively exhausted, whereas pyroclastic ash continuously liberated their nutritive elements.

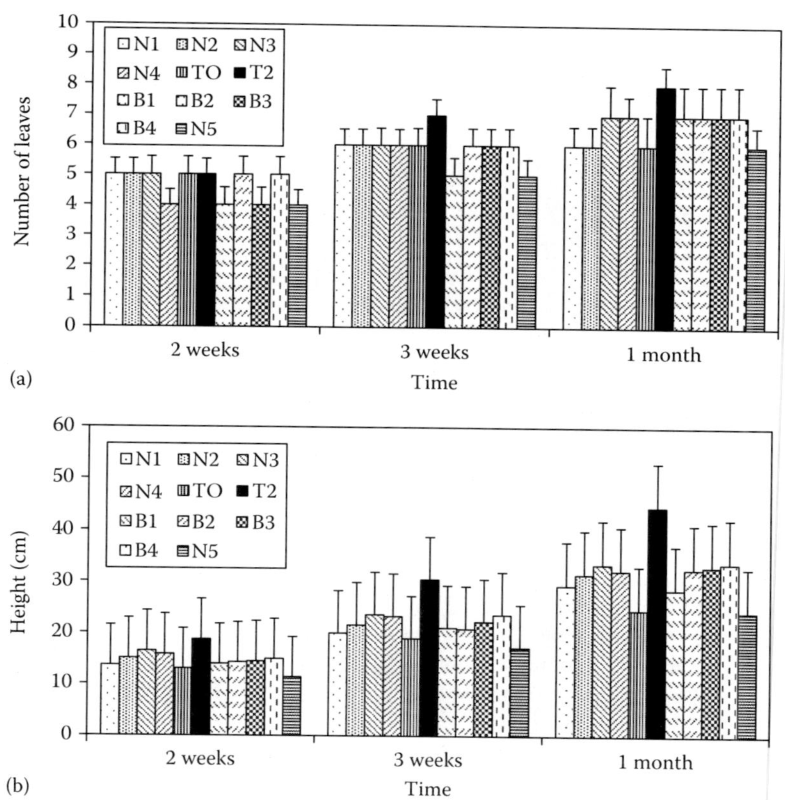

FIGURE 24.5 (a) Effect of the treatment on the number of leaves (after 2 weeks, 3 weeks, 1 month). (b) Effect of the treatments on the height of the plants (after 2 weeks, 3 weeks, 1 month).

(Continued)

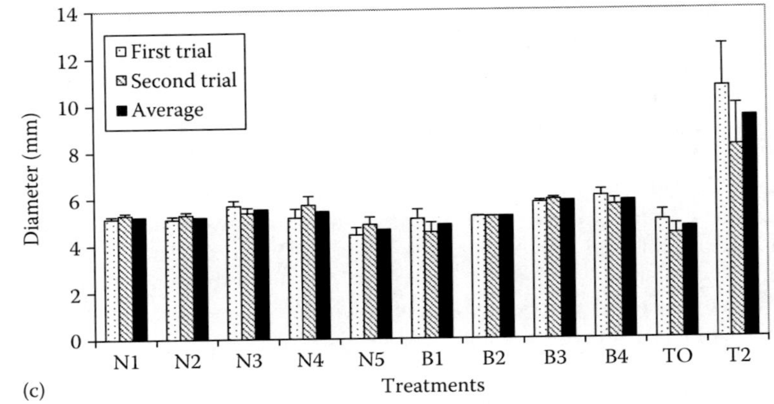

(c)

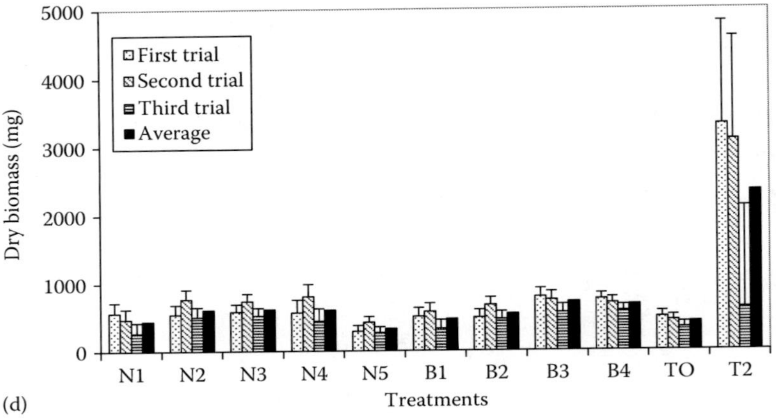

(d)

FIGURE 24.5 (*Continued*) (c) Effect of the treatments on the diameter of the stalk (after 1 month). (d) Effect of the treatments on the dry biomass (after 1 month).

The third experiment showed systematic decrease in yields that must be interpreted with caution, because it could be partly due to abundant rainfall in March and April. These features of the intertropical region may be responsible for important leaching of nutritive elements out of the pots.

The measured parameters of soils analyzed (Table 24.4) after the first test were compared to reference values of Bachelier (1951) and Léon (1979), ranges for which soils are considered to be fertile.

The addition of pyroclastic ash increased pH from 5.2 to 6.4, in the fertile range. This pH increase is due to an increase in basic exchangeable cations, mainly Ca^{2+} and Mg^{2+}. Similarly, the sum of exchangeable bases increased from 2.4 to 6.8 while exchangeable acidity decreased from 0.4 to 0.15. The rate of base saturation also increased. Considering the improvement of their geochemical characteristics, the B3 soils and B4 have a fairly good fertility level.

The depleted control soil is strongly acidic (pH = 5.21), and its CEC is weak.

The pyroclastic ashes are rich in total bases, CaO > 8%, MgO > 6%, K_2O > 1% (Table 24.3). But after being mixed with soil, they are not immediately available for the plants: they constitute only a potential reserve (Ministry of Cooperation and the Development, 1991). In order to replace chemical fertilizers, they must pass through an additional mineralization stage. The different soils analyzed after the first test showed that rate of mineralization of finely ground pyroclastic ash is relatively fast since the concentrations of base elements evolved appreciably only after 1 month of the experiment.

An additional experiment was made to measure mineralization rates. Pyroclastic ash was soaked (Pt) for 9 months in water (to avoid the problems of leaching of the ions by loss of the solution). The comparison of their composition with the initial pyroclastic material (Pi) showed about a doubling of the quantity of exchangeable bases in the solution. This confirmed that solution of the ions from minerals is fast, on the scale of less than 1 year.

These experiments showed that pyroclastic ash can enrich soils and rejuvenate them by the progressive liberation of elements, due to its important reserves of basic elements.

The best results were found with an addition of 25%–50% of pyroclastic material, preferably crushed. Although less efficient than the classic NPK fertilizers, the pyroclastic ashes had the advantage of releasing free nutritive elements in a slow and continuous way. They were exhausted slowly, whereas it is necessary to renew chemical fertilizers frequently. In addition, rock powders do not pollute groundwaters, rivers, lakes, and the coastal zone.

These volcanic rocks, which have minimal exploitation cost (even though it is necessary to take into account expenses of transportation), present an incontestable contribution: without being a perfect surrogate for chemical fertilizers, they can contribute to reduce the needs of the farmers for chemical fertilizers. In particular, they provide valuable sources of calcium and magnesium. Dusting volcanic ash on ferrallitic tropical soils can contribute to reduce the relatively fast impoverishment of these soils due to the rapidity with which the nutrients are leached from mineral weathering.

Conclusion

The addition of pyroclastic ash, especially finely ground, improves soil fertility of regions with poor soils, particularly with problems of erosion, lack of humus, weak CEC, high acidic pH, and aluminum and manganese toxicity. The addition of these volcanic products to soil presents many advantages. The main advantages of using rock fertilizers are as follows:

1. Slow solubility and long-lasting availability of elements needed for plant nutrition
2. The complex geochemical and mineralogical composition of rocks provides nutrients continuously to cultivated plants and provides trace elements often deficient in traditional fertilizers
3. Improvement of ionic exchange capacity of soils through the addition of fine rock clay minerals, better preparing the soils to receive additional nutrients

The use of rock powders as natural agricultural inputs improves physical and chemical characteristics of fertility in ferrallitic soils, most notably for pH, CEC, quantity of exchangeable bases, and rate of saturation. These improvements are due to the release of chemical elements from the pyroclastic ash. Fine basaltic pyroclasts, rich in Ca, Mg, and K, can be recommended as fertilizers for impoverished and acidic soils, and for crops highly dependent on these chemical elements.

REFERENCES

Bachelier G., 1951. Essai d'un tableau d'interprétation des résultats analytiques pour les sols tropicaux. *ORSTOM/IRCAM, IRA*. 68pp.

Barral Silva M.T., Silva Hermo B., García-Rodeja E., and Vázquez Freire N., 2005. Reutilization of granite powder as an amendment and fertilizer for acid soils. *Chemosphere*, 61, 993–1002.

CPCS (Commission de Pédologie et de Cartographie des Sols), 1967. Classification des Sols. *E.N.S.A.*, Paris-Grignon, France.

Déruelle B., Bardintzeff J.-M., Cheminée J.L., Ngounouno I., Lissom J., Nkoumbou C., Etame J. et al., 2000. Eruptions simultanées de basalte alcalin et de hawaiite au mont Cameroun (Mars 28–Avril 17, 1999). *C.R. Acad. Sci. Paris*, 331, 525–531.

FAO, 1998. Soil fertility initiative for sub-saharan Africa. *Proceedings of the SFI/FAO Consultation*. World Soil Resources Report No. 85, Rome, Italy.

Fisher G. and Heilig G., 1997. Population momentum and the demand on land and water resources. In: Greenland D.J., Gregory P.J., and Nye P.H. (eds.), *Land Resources: On the Edge of the Malthusian Precipice*. CAB International, Wallingford, U.K.

Fisher R.V.B. and Schmincke A.N., 1984. *Pyroclastics Rocks*. Springer Verlag, Berlin, Germany, 472pp.

Fragstein V.P., 1991. Silicate rock dusts as natural fertilizers. In: Campe J. (ed.), *Remineralise the Earth, Soil Remineralization Research Packet—Agriculture*. Northampton, MA, pp. 73–76.

Haraldsen T.K. and Pedersen P.A., 2003. Mixtures of crushed rock, forest soils, and sewage sludge used as soils for grassed green areas. *Urban For. Urban Green.*, 2, 41–51.

IFDC/USAID (International Fertilizer Development Center/United State Agency for International Development), 1986. Etude du secteur des engrais au Cameroun. Muscle Shoals, AL.

Kleïr R.A. and Thornhill M., 2007. Production of mechanically activated rock flour fertilizer by high intensive ultrafine grinding. *Miner. Eng.*, 20, 334–341.

Léon J.J., 1979. *Introductory Soil Sciences: A Study Guide & Laboratory Manual*. McMillan Publishing, London, U.K., 264pp.

Leornardos O.H., Fyfe W.S., and Kronberg B.I., 1987. The use of ground rocks in laterite system as improvement to the use of conventional soluble fertilizers. *Chem. Geol.*, 60, 361–370.

Ministry of Cooperation and Development, French Republic, 1991. Mémento de l'agronome, 4e éd. *Collection Techniques rurales en Afrique*, 1635pp.

Nkouathio D.G., 2006. Evolution tectono-magmatique et volcanologique de la Ligne du Cameroun: Comparaison d'un volcanisme de graben (plaine de Tombel) et d'un volcanisme de horst (monts Bambouto). Thèse Doct. Etat, Univ. Yaoundé I, Yaoundé, Cameroun, 231pp.

Nkouathio D.G., Kagou Dongmo A., Bardintzeff J.-M., Wandji P., Bellon H., and Pouclet A., 2008a. Evolution of volcanism in graben and horst along the Cameroon Line: Implication for tectonic evolution and mantle source composition. *Mineral. Petrol.*, 94, 287–303.

Nkouathio D.G., Ménard J.J., Wandji P., and Bardintzeff J.M., 2002. The Tombel graben (West Cameroon): A recent monogenetic volcanic field of the Cameroon Line. *J. Afr. Earth Sci.*, 35–2, 285–300.

Nkouathio D.G., Wandji P., Bardintzeff J.M., Tematio P., Kagou Dongmo A., and Tchoua F.M., 2008b. Utilisation des roches volcaniques pour la reminéralisation des sols ferrallitiques des régions tropicales. Cas des pyroclastites basaltiques du graben de Tombel (Ligne volcanique du Cameroun). *Bull. Soc. Vaud. Sci. Nat.*, 91, 1–14. ISSN 0037-9603.

Nkouathio D.G., Wandji P., Ménard J.J., Bardintzeff J.M., Vicat J.P., Dongmo Kagou A., and Tchoua F.M., 1998. Critères pétrologiques et géochimiques des laves récentes du graben de Tombel (Ouest Cameroun). *Geosciences au Cameroun*, Vicat J.P. et Bilong P. éd., collect. GEOCAM, t1. Press. Univ. Yaoundé I, Yaoundé, Cameroun, pp. 229–234.

Nkouathio D.G., Wandji P., Tchoua F.M., and Kagou Dongmo A., 1999. Revalorisation et intérêt économique des projections des volcans holomagmatiques du graben de Tombel (Ligne du Cameroun). *Géologie et environnements au Cameroun*, Vicat J.P et Bilong P. éd., collect. GEOCAM, t2, Press. Univ. Yaoundé, Yaoundé, Cameroun, pp. 339–345.

Nono A., MvondoZe A., and Towa Foka R., 2008. Some chemical characteristics of selected geological materials in Cameroon (Central Africa) and their potential use for crop production. *Int. J. Biol. Chem. Sci.*, 2(1), 21–32.

Onguené Mala J.B., 1993. Différenciations pédologiques dans la région de Yaoundé (Cameroun): Transformation d'un sol rouge ferrallitique en sol à horizon jaune et relation avec l'évolution du modelé. Thèse Doct. Univ. Paris VI, Paris, France, 357pp.

Schaetzl J.R. and Anderson S., 2005. *Soils: Genesis and Geomorphology*. Cambridge University Press, Cambridge, U.K. 817pp.

Suh C.E., Sparks R.S.J., Fitton J.G., Ayonghe S.N., Annen C., Nana R., and Luckman A., 2003. The 1999 and 2000 eruptions of Mount Cameroon: Eruption behavior and petrochemistry of lava. *Bull. Volc.*, 65, 267–281.

UNO, 1997. *World Population Prospects*. UN Publications, New York.

Wandji P., 1995. Le volcanisme récent de la plaine du Noun (Ouest-Cameroun). Volcanologie, pétrologie, géochimie et pouzzolanicité. Thèse Doct. d'État, Univ. Yaoundé I, Yaoundé, Cameroun, 245pp.

WRB, 1998. World Reference Base for Soil Resources. World Soil Resources, 84, ISSS, ISRIC, FAO, Rome, Italy.

25

Effect of Powders of Basalts, Tuff, Granites, and Pyroclastic Materials on the Yield and Quality of Carrots and Cabbages Grown on Tropical Soils in the Northwest Region of Cameroon

Samuel Tetsopgang, Pierre Kamga, Paul F. Gonang Achoumele, Bonaventure Alemanji, Dieudonne Z. Manjo, and Linda Mazoh

CONTENTS

Introduction

The Northwest Region of Cameroon is covered by soils that developed mostly on basalts. They have been weathered, and the nutrients in the soils have been leached by heavy seasonal rain. This has left soils insufficiently fertile enough to yield the crop productivity needed by the population. To help farmers improve soil productivity and develop sustainable land use, powders from different local rocks mixed with green manure were used to improve the growth productivity of carrots and cabbage and to ameliorate plant nutritional quality. The leaves and stems of the carrots and cabbage treated with rock powder showed a significant improvement over the control soil. Carrot production of 11,666 kg/ha was obtained with basalt and green manure mixture equivalent to the yield treated with tuff powders and green manure. These results were higher than the 8333 kg/ha obtained from the treatment of basalt powder only and 5722 and 4083 kg/ha for the treatments with tuff powder only and the control soil, respectively. While cabbage treated with granite powder yielded 50% increased productivity, soils treated with basalt powders and pyroclastic materials combined yielded also 50% increase versus 0% for the control soil. These field trials demonstrated that the treatment of local soils with rock powders in combination with green manure is an efficient method to improve productivity on these tropical soils.

Tropical soils stretch along the Equator, as do the developing countries with the least productive soil. Most tropical soils are characterized by significant leaching, poor nutrients, and frequently by infertility (Foy, 1984). Soil decomposition is rapid in this geographic zone due to high temperatures, high humidity, and frequent heavy rains. To enhance the fertility of these soils, several amendment techniques and fertilizers have been used to increase crop productivity and therefore improve poor living standards.

However, soil productivity in these tropical zones remains inadequate despite the application of these amendment techniques and fertilizers.

A new method to increase and maintain productivity of tropical soils was then introduced. This method, called soil remineralization, creates these fertile soils by returning minerals to the soil much the same way the Earth does naturally (Hensel, 1982).

To help farmers increase their income and develop sustainable land use practices, rock powders from different locally collected rock types were utilized in this study to improve crop yield and ameliorate the quality of seasonal crops grown in tropical soils of the Northwest Region of Cameroon.

Location of the Study Area

The Northwest Region of Cameroon is located between latitudes 5°33′00″ and 7°03′50″ N and longitudes 09°05′00″ and 11°05′00″ E (Figure 25.1) and is characterized by yearly dry and rainy seasons from November to March and March to October, respectively (Yerima and Ranst, 2005). Average annual

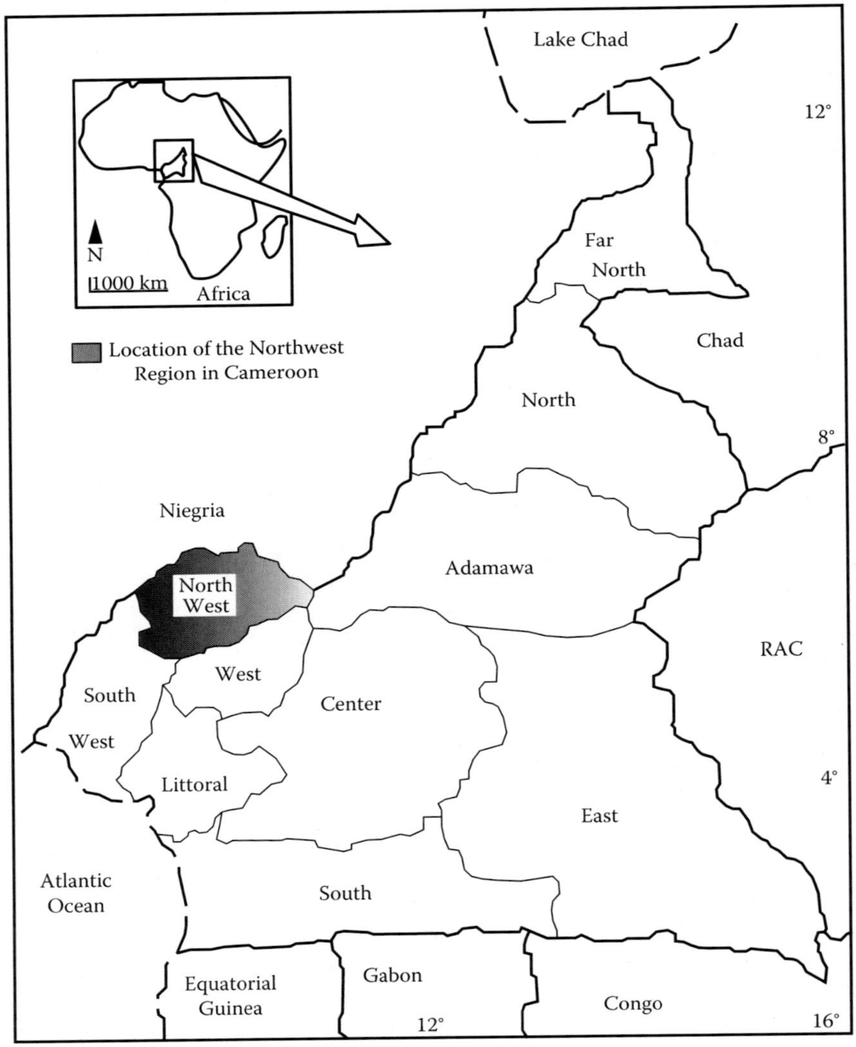

FIGURE 25.1 The location of the Northwest Region Cameroon. Inset map shows Cameroon in Africa.

temperature is 26°C. The region is dominated by savanna vegetation at the altitudes between 1600 and 2200 m. The mountainous landscape is composed mostly of volcanic rock overlying Precambrian igneous rocks (Kamgang et al., 2008). The soils in this zone are classified as oxisols or ferralsols (Yerima and Ranst, 2005). Dark-colored topsoils on basaltic rocks have abundant humus under the cover of savanna vegetation. These soils are highly lateralized and enriched in clays.

Methodology

Fresh samples of different rock types, such as basalt, tuff, granites, and pyroclastic materials exposed in the study area, were collected, with care taken to avoid weathered zones. They were then transported to a factory and crushed into fine grains with a rock mill. The crushed samples were further sieved to remove larger fragments and were again crushed to achieve only the finest powders. These powders were then carried to the farm sites where ridges were formed according to local practices. The area of the farm plot was approximately 46 m^2 made up of four or more ridges according to the number of treatments of the study. Each ridge had the dimensions of approximately 0.8 × 4 m in width and length, respectively.

Rock powders were applied to cabbage in holes of about 2 cm deep, while carrots were treated with a rock powder spray. Already-nursed cabbage seedlings were transplanted into 16 holes of 10 × 5 cm each in diameter and depth, respectively, forming two equal rows separated by 50 cm from each other. *Tithonia diversifolia*, a tropical flower, was used as green manure to supply the plants with nitrogen. The flower was harvested, dried, and milled into powder. The number of treatments and replicates varied according to the different rock types collected in the study area. Width and length of leaves and stems were measured to assess the quality and quantity of the crops grown on the different treatments. The treatments were mixtures between the local soil, rock powders, plants, and manures: plants and local soil alone; plants, rock powder, and local soil; and plants, rock powder, local soil, and green manure.

The average dimensions of the stems and leaves and their quality were assessed during the growth period, as well as the yield obtained after harvesting the crops.

Results

Carrots and cabbage were chosen for this study due to their short periods of growth. The plants were grown on three different sites: Santa, Alatening, and Wum, located within the localities of the Northwest Region of Cameroon. The same methodology was applied to record the quality and yield of the carrots and cabbage.

Carrots

The carrots cultivated in Santa were planted on ridges measuring 0.9 × 2 m in width and length, respectively. This test was carried out for 5 treatments and 3 different replicates for a total of 15 ridges as follows: T_0 = carrot (C) + control soil (CS); T_1 = C + CS + 1 kg of basalt; T_2 = C + CS + 1 kg of tuff; T_3 = C + CS + 1 kg of basalt + 2.5 kg dried leaves of green manure; and T_4 = C + CS + 1 kg of tuff + 2.5 kg green manure. The results for the number and height of stems, as well as the number of leaves, were recorded at the end of the first, second, and third months after seeding (Table 25.1).

Sample sizes of approximately four to five stems were taken randomly during the period of 3 months to measure the average height of stems and the number of leaves for each treatment. Stems with the treatments T_3 and T_4 were the tallest (=24.42 and 22.94 cm) compared to those of T_1 (≈22.33 cm). T_0 and T_2 showed the shortest stems (≈15 cm) during the full period of the test. The number of leaves was also greater in the treatments of T_3 and T_4 (=55 and 54 leaves) than in the others. T_1 showed 49 leaves, with 41 and 43 for T_0 and T_2, respectively. Carrot grown on soil with basalt and tuff

TABLE 25.1

Number of Stems and Leaves and Their Heights during the First, Second, and Third Month of Growth of Carrots on the Five Treatments (T_0, T_1, T_2, T_3, T_4) Representing the Control Soil and Powders of Basalt, Tuff, Basalt + Manure, and Basalt + Tuff + Manure, Respectively, in Santa, Northwest Region, Cameroon

Type of Rock Powders	Period / Parameter	First Month Number of Stems	First Month Height of Stems (cm)	First Month Number of Leaves	Second Month Number of Stems	Second Month Height of Stems (cm)	Second Month Number of Leaves	Third Month Number of Stems	Third Month Height of Stems (cm)	Third Month Number of Leaves
None	T_0	4	2.38	8	4	8.93	24	5	15.42	41
Basalt	T_1	4	3.75	12	6	18.24	46	5	22.33	49
Tuff	T_2	4	2.87	10	5	8.82	27	5	14.88	43
Basalt + manure	T_3	4	4.94	15	6	22.54	65	4	24.42	55
Basalt + tuff + manure	T_4	5	4.37	14	5	22.48	57	4	22.94	54

TABLE 25.2

Yield of Carrots Produced after the Treatments of Soils with Different Powders from Different Rock Types in Addition to *T. diversifolia* in Santa, Northwest Region, Cameroon

Type of Rock Powders	Treatment	Productivity (kg)	Area (m²)	Calculation	Yield (kg/ha)
None	T_0	0.735	1.8	$R_0 = 0.735\ \text{kg} \times 10{,}000/1.8\ \text{m}^2$	4,083
Basalt	T_1	1.500	1.8	$R_1 = 1.500\ \text{kg} \times 10{,}000/1.8\ \text{m}^2$	8,330
Tuff	T_2	1.030	1.8	$R_2 = 1.030\ \text{kg} \times 10{,}000/1.8\ \text{m}^2$	5,722
Basalt + manure	T_3	2.100	1.8	$R_3 = 2.100\ \text{kg} \times 10{,}000/1.8\ \text{m}^2$	11,666
Basalt + tuff + manure	T_4	2.100	1.8	$R_4 = 2.100\ \text{kg} \times 10{,}000/1.8\ \text{m}^2$	11,666

powders treated with green manure showed greatest stem heights and greatest number of leaves. The best yield of 11,666 kg/ha was obtained with basalt powders and green manure treated with tuff, followed by 8,330, 5,722, and 4,083 kg/ha for the treatments of basalt powders, tuff powders, and the control soil, respectively. Table 25.2 presents the different yields of the carrot crops with respect to the treatments added.

The carrots planted in Wum were grown in control soil (T_0), with treatments of basalt powders (T_1), granite powders (T_5), and pyroclastic materials (T_6). Table 25.3 presents the results of the lengths (cm) of 10 stems of selected carrots. The first leaves for all the treatments started to appear after 1 month. They showed the same heights with green color, with those of T_1, T_5, and T_6 fresher than T_0. The growth rate varied after the second month, with the highest stems measuring 10.2, 12.1, 12.5, and 11.2 cm for the treatments T_0, T_1, T_5, and T_6, respectively. The shortest stems measured 4.6, 6.5, 6.9, and 6.5 cm for T_0, T_1, T_5, and T_6, respectively. By the third month, the growth had reached its maximum with the tallest measuring 17.3, 19.0, 19.5, and 18.2 cm for T_0, T_1, T_5, and T_6, respectively. However, the stems of the control soil showed a height of 11.0 cm, less than the 13.3–13.8 cm stem heights observed from those treated with the basalt powder, granite powder, and pyroclastic materials.

The carrots planted in Alatening were grown in control soil (T_0), with treatments of 1.2 kg of green manure in the following combinations: basalt powders + green manure (T_3), granite powders + green manure (T_7), and basalt + granite + green manure (T_8). Table 25.4 presents the lengths of 10 stems of selected carrots. The first leaves were observed after 5 weeks with the same rate of growth in addition to a general greenish color for all treatments. Measured stems had average heights of between 2.7 and 2.9 cm with 4.0 cm as the tallest from the control soil (T_0). During the second month, staining, drying, and shedding of the leaves were noticed, especially from the control soil and some stems from others

TABLE 25.3

Heights (cm) of Selected 10 Stems of Carrots Grown on the Control Soil (T_0) and Soils Treated with Powders of Basalts (T_1), Granite (T_5), and Pyroclastic Materials (T_6) in Wum, Northwest Region, Cameroon

Type of Rock Powders	None			Basalt			Granite			Pyroclastic Materials		
Treatment	T_0			T_1			T_5			T_6		
Period (Month)	First	Second	Third	First	Second	Third	First	Second	Third	First	Second	Third
1	3.3	7.9	9.8	3.2	12.1	15.9	2.7	6.9	13.6	2.0	9.5	10.5
2	2.7	9.3	11.5	2.6	9.1	19.0	2.5	9.3	9.5	1.9	5.5	5.8
3	3.8	6.5	10.3	1.7	6.5	10.7	3.2	10.7	17.8	2.3	7.2	11.5
4	3.1	10.2	14.2	2.5	10.4	17.3	2.2	12.5	7.5	3.4	8.7	9.9
5	3.5	7.6	5.1	3.4	8.3	11.1	3.7	8.2	18.2	2.5	10.9	18.2
6	3.0	8.7	12.3	2.1	6.4	14.7	2.3	12.3	15.5	2.8	9.8	15.7
7	2.9	4.6	8.8	2.4	10.1	6.5	2.6	7.2	7.8	3.2	11.2	13.1
8	3.2	10.0	13.5	3.1	9.7	18.5	3.3	11.5	19.5	3.4	6.7	17.4
9	3.1	9.1	7.6	2.5	7.3	7.5	3.1	9.8	10.3	2.9	10.6	14.3
10	2.1	4.8	17.3	2.9	8.0	16.3	2.3	8.1	13.4	3.6	9.3	16.5
Average height	3.1	7.9	11.0	2.6	8.8	13.8	2.8	9.7	13.3	2.8	8.9	13.3

TABLE 25.4

Lengths (cm) of Selected 10 Stems of Carrots Grown on the Control Soil (T_0) and Soils Treated with Powders of Basalts + Manure (T_3), Granite + Manure (T_7), and Basalt + Granite + Manure (T_8) in Alatening in the Northwest Region, Cameroon

Type of Rock Powders	None			Basalt + Manure			Granite + Manure			Basalt + Granite + Manure		
Treatment	T_0			T_3			T_7			T_8		
Period (Month)	First	Second	Third	First	Second	Third	First	Second	Third	First	Second	Third
1	3.0	10.0	10.7	3.1	10.0	16.5	2.2	7.3	14.6	2.5	11.1	10.4
2	3.1	12.3	12.3	3.2	12.2	17.8	2.3	6.9	12.4	2.6	10.1	13.2
3	2.5	8.9	9.7	1.9	9.0	20.1	2.2	7.3	13.3	2.8	9.3	15.7
4	3.7	9.0	16.1	2.4	8.0	16.2	3.0	7.7	6.9	3.6	9.8	17.8
5	2.1	7.5	14.3	2.1	10.2	17.3	3.5	7.8	10.1	3.0	7.1	19.1
6	2.0	9.2	7.1	3.5	6.7	14.7	3.3	8.0	15.3	2.6	8.2	17.9
7	3.1	6.5	10.3	3.4	9.1	11.1	3.1	6.9	8.3	2.3	9.5	18.1
8	3.1	7.9	12.1	2.1	9.2	19.2	2.0	7.8	9.3	2.3	7.9	11.6
9	4.0	9.0	14.2	2.7	9.7	7.2	3.1	9.0	20.0	3.4	8.6	8.5
10	3.3	9.7	10.2	2.9	7.2	9.1	3.5	6.5	13.4	2.4	9.2	7.9
Average	2.9	9.0	11.7	2.7	9.1	14.9	2.8	7.5	12.4	2.8	9.1	14.0

Manure = *T. diversifolia*.

treated with rock powders. The highest stem was measured from the control soil, with 12.3 cm. The combination of granite powder and green manure showed an average height of 6.5 cm while other treatments had an average height of 9.1 cm. The growth of these carrots reached their greatest by the third month, with the tallest stems measuring 20.1, 20.0, 19.0, and 16.1 cm for T_3, T_7, T_8, and T_0, respectively. After 3 months, the average heights of 14.9 and 14.0 cm were the peak measurements from treatments of basalt powder + green manure and basalt + granite + green manure.

Cabbages

Four beds, representing the different treatments, were planted with cabbage seedlings transplanted from a nursery into holes containing rock powders. The average dimensions of the leaves and stems of the plants were recorded at 2 week intervals over 3 months (Table 25.5). For the first 2 weeks, no significant change in leaf and stem dimensions was observed, as indicated by equal stem lengths and leaf widths for this period. After 4 weeks, the rate of cabbage growth showed a significant change for all the beds, with nearly 7.0, 8.0, 0.5, and 5.0 cm as average lengths and widths of stems and leaves. This change was uniform for all beds in terms of growth rate, while greenness of plants was less for the control bed but uniform for the other treatments. After 6 weeks, results for all treatments were still almost the same with 10.0–15.0 and 8.0–11.5 cm as average length of stems and leaves and 1.0–1.5 and 6.0–11.0 cm as average width of stems and leaves, respectively.

Head formation was also noticed in this period. Cabbage treated with granite powder developed the most heads (eight), cabbage treated with basalt powder and pyroclastic material developed four heads each, and no heads developed for cabbage grown in control soil. The largest and biggest heads developed from cabbage treated with granite powder. After 8 weeks, the cabbage was sprayed with insecticide due to attacks by aphids becoming more prominent on the plants, especially in the control soil more than the other treatments.

Interpretation and Discussions

These field trials were conducted with the treatment of local soils with powders derived from different types of rock and followed the growth of carrot and cabbage crops for a maximum period of 3 months. Stem height/length and width/diameter of leaves were measured.

After the first month of growth, stem heights showed values comparable to those of the control soil in almost all treatments (Figures 25.2 through 25.5). This suggests the nonrelease of nutrients in the ground before the first month of growth. However, this was different with the growth of carrots in Santa. After the first month here, the control soil showed shorter stem with a height of around 2.4 cm, while other treatments had taller stems, about 5 cm height, suggesting faster release of nutrients in these soils (Figure 25.2).

TABLE 25.5

Average Length and Width of Stems and Leaves in Addition to the Number of Fruits and Fruit Diameters of Cabbage Grown on Soil Treated with Powders of Basalt, Granite, and Pyroclastic Materials in Wum, Northwest Region of Cameroon

Type of Rock Powders		**None**				**Basalt**				**Granite**				**Pyroclastic Materials**			
Treatment		T_0				T_1				T_5				T_6			
Period (Month)		**0**	**1**	**1.5**	**2**	**0**	**1**	**1.5**	**2**	**0**	**1**	**1.5**	**2**	**0**	**1**	**1.5**	**2**
Average length (cm)	Stems	5.0	6.7	10.0	14.0	5.0	7.0	12.0	17.0	5.0	7.0	15.0	20.0	5.0	6.8	14.0	18.0
	Leaves	2.5	7.6	8.0	9.5	2.5	8.0	9.5	14.0	2.5	8.0	11.5	18.0	2.5	7.6	9.5	16.0
Average width (cm)	Stems	0.3	0.5	1.0	1.0	0.3	0.5	1.2	1.5	0.3	0.5	1.5	2.0	0.3	0.5	1.2	1.5
	Leaves	2.0	5.0	6.0	7.0	2.0	5.0	7.5	12.0	2.0	5.0	11.0	13.0	2.0	5.0	7.5	14.5
Number of fruits		0				4				8.0				4.0			
Average diameters of fruits (cm)		0				8.3				8.7				8.3			

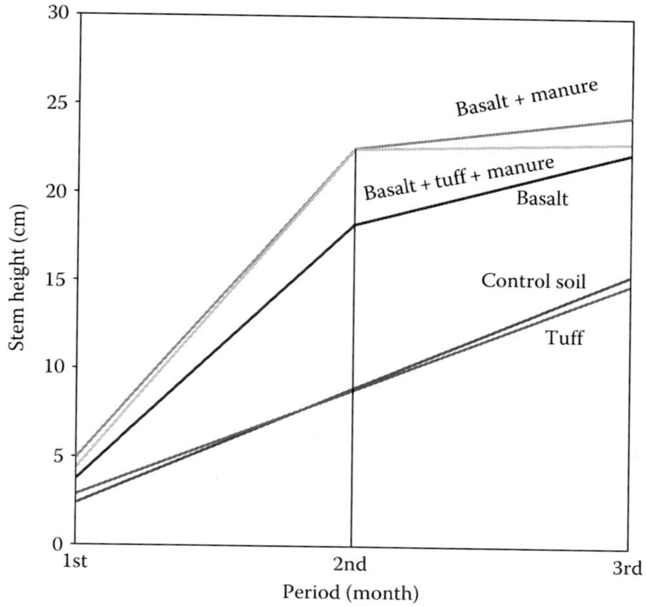

FIGURE 25.2 Stem heights versus the period of growth of carrots on the control soil alone and on the control soil with the mixture of powders of tuff, basalt, basalt + tuff, and/or without the manure in Santa, Northwest Region of Cameroon.

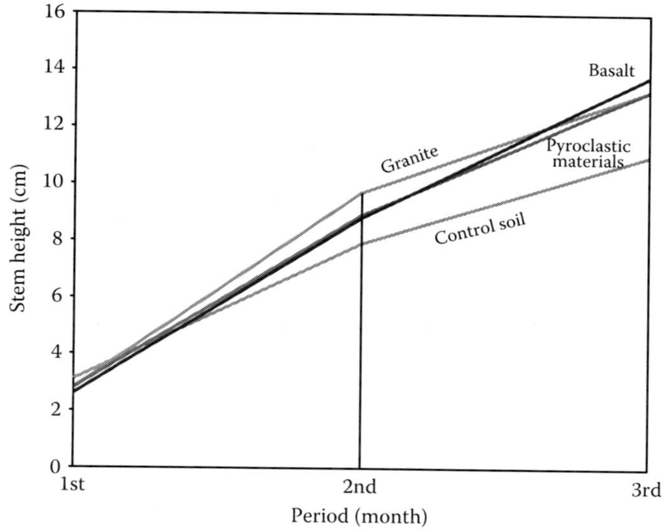

FIGURE 25.3 Stem heights versus the period of growth of carrots on the control soil alone and on the control soil with the mixture of powders of granite, pyroclastic materials, and basalt in Wum, Northwest Region of Cameroon.

After the second month of growth, there was a clear discrepancy between the height of plants in the control soils and the height of plants in the treated soils (Figures 25.2 through 25.5), which suggests that needed nutrients become available to the ground before the 2-month time period. The treatment composed of control soil + basalt + green manure for the carrots in Santa had the tallest plants. This height is comparable to that from basalt powder + tuff + green manure. Medium-sized plants were observed for those treated solely with basalt powder, while the shortest plants were found in the control soil and with the combination of tuff powder and control soil (Figure 25.2).

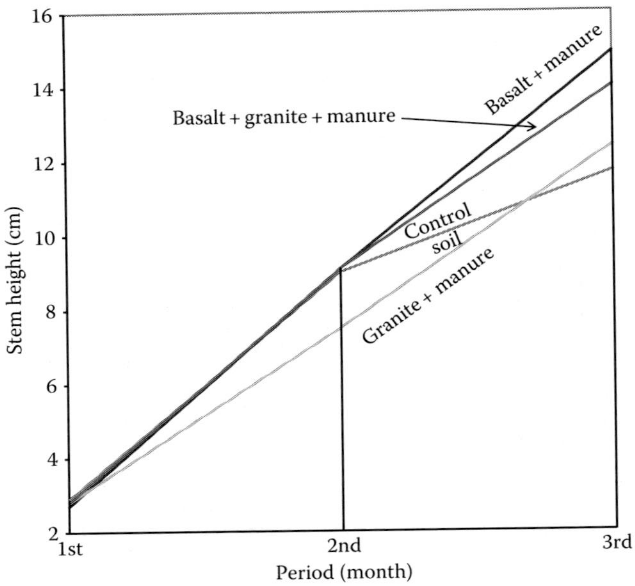

FIGURE 25.4 Stem heights versus the period of growth of carrots on the control soil alone and on the control soil with the mixture of powders of basalt, granite, and manure in Alatening, Northwest Region of Cameroon.

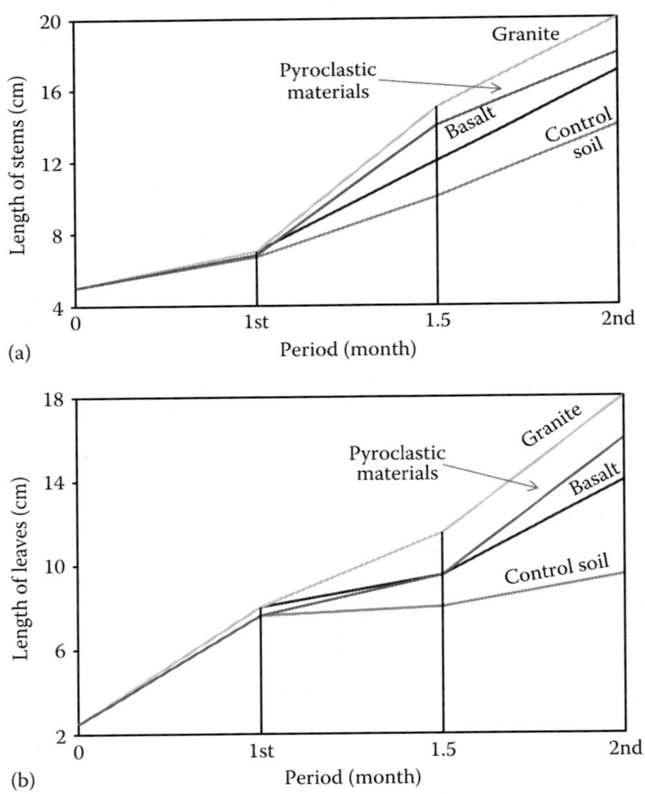

FIGURE 25.5 Length of stems versus period of growth (a) and length of leaves versus period of growth (b) for cabbage grown on the control soil and on the control soil with added powders of basalt, granite, and pyroclastic materials in Wum, Northwest Region of Cameroon.

After the first month, cabbage plants showed comparable stem heights for all treatments, and the discrepancy between heights only appeared after 1 month and 2 weeks (Figure 25.5). This may suggest the low capacity of cabbage to absorb needed nutrients in the soil or the nonrelease of these nutrients before this period. Soils treated with granite powder showed the tallest cabbage heights as well as the largest leaves (Figure 25.5a and b), which suggests that granite powder absorbs and retains the most nutrients for cabbage plants or that the cabbages are better adapted to the more acidic conditions of granite compared to basalt.

The tallest carrot plants were observed after the second month of treatments with basalt powder + green manure and basalt + tuff + green manure (Figure 25.2). These treatments also yielded the highest productivity with about 11,666 kg/ha of carrot (Table 25.2). The shortest plants were observed in the control soil and those treated with tuff powder, yielding the lowest comparable productivities. The treatment with the basalt powder alone had carrot plants taller than those of the control soil and yielded 70% of the highest productivity. Carrot plants treated with rock + manure were also tall, as observed in Figure 25.4, while carrot plants treated with basalt + green manure, basalt + granite + green manure, and even granite + green manure produced taller stems than those of the control soil.

The effect of granite powder was more prominent in the growth of cabbage plants. In fact, the treatments of cabbage with granite powder showed the longest stems and had the largest leaves compared to those treated with basalt powder and pyroclastic materials (Figure 25.5a and b). These treatments of granite powder yielded 50% of the productivity compared to 25%, 25%, and 0% for basalt powder, pyroclastic materials, and control soil, respectively.

Conclusion

Chemical fertilizers are usually recommended to correct soil infertility, but they are too expensive for poor farmers growing in the tropics. Many alternatives have been proposed, among them products from organic waste materials, for example, composts (Hue, 1992; Noble et al., 1996). However, these field trials of local soil treated with rock powders, sourced from basalt, tuff, granite, and pyroclastic materials, in combination with green manure demonstrate the enhanced productivity of local soils, as much as three times greater productivity. These rock powders can therefore be used as an efficient method and affordable alternative to chemical fertilizers to increase productivity and for sustainable management of tropical soils.

ACKNOWLEDGMENTS

The authors thank the farmers of the localities of Santa, Alatening, and Wum in the Northwest Region of Cameroon for giving the plots for the field experimentations. We would also like to thank Dr. Thomas Goreau for the constructive review and Joanna Campe for the constructive comments.

REFERENCES

Foy, C.D., 1984. Physiological effects of hydrogen, aluminum and manganese toxicities in acid soil. In: Adams, F. (ed.), *Soil Acidity and Liming*. American Society of Agronomy, Madison, WI, pp. 57–97.

Hensel, J., 1982. *Bread from Stones: A New and Rational System of Land Fertilization and Physical Regeneration.* Republished by Acres USA, Austin, TX.

Hue, N.V., 1992. Correcting soil acidity of a highly weathered ultisol with chicken manure and sewage sludge. *Communications in Soil Science and Plant Analysis* 23, 241–264.

Kamgang, P., Chazot, G., Njonfang, E., and Tchoua, F., 2008. Geochemistry and geochronology of mafic rocks from Bamenda mountains (Cameroon): Source composition and crustal contamination along the Cameroon volcanic line. *Comptes Rendus Géoscience* 340, 850–857.

Noble, A.D., Zenneck, I., and Randall, P.J., 1996. Leaf litter ash alkalinity and neutralisation of soil acidity. *Plant Soil* 179, 293–302.

Yerima, B.P.K. and Van Ranst, E., 2005. *Introduction to Soil Science: Soils of the Tropics*. Trafford Publishing, Victoria, British Columbia, Canada, 440pp.

26

Rock Fertilizers as an Alternative to Conventional Fertilizers: The Use of Basalt from the Cameroon Volcanic Line for Maize Farming on Ferralitic Soils

Jean Pierre Tchouankoue, Arliane Nicole Tetchou Tchekambou, Monique Abossolo Angue, Christophe Ngansop, and Suzi Huff Theodoro

CONTENTS

Introduction

Cameroon is the breadbasket of Central Africa and the three main agricultural production basins are the Foumbot, Mbo, and Tombel plains, which are covered by recent basaltic rocks. In these areas, the intensive agriculture farming zones (tomatoes, banana, coffee, rubber, palm tree, and papaya) are located in the volcanic plains. Basalts in these areas are likely to be used for fertilization of lateritic soils that occupy most of the country. This can provide small farmers with fertilizer at low cost and provide long-term soil fertility, whose direct benefits include increased agricultural productivity, environmental protection, and saving the country the costs of imported chemical fertilizers.

This chapter characterizes the physicochemical types of basalt that could be optimally used for remineralizing (fertilizer) for lateritic soils. Petrographic and geochemical properties of basalt samples from the Cameroon volcanic line (in the regions of Foumbot and Melong) were studied and a field trial with maize was conducted in the Yaoundé region in order to assess the fertilizing power of each of these basalts.

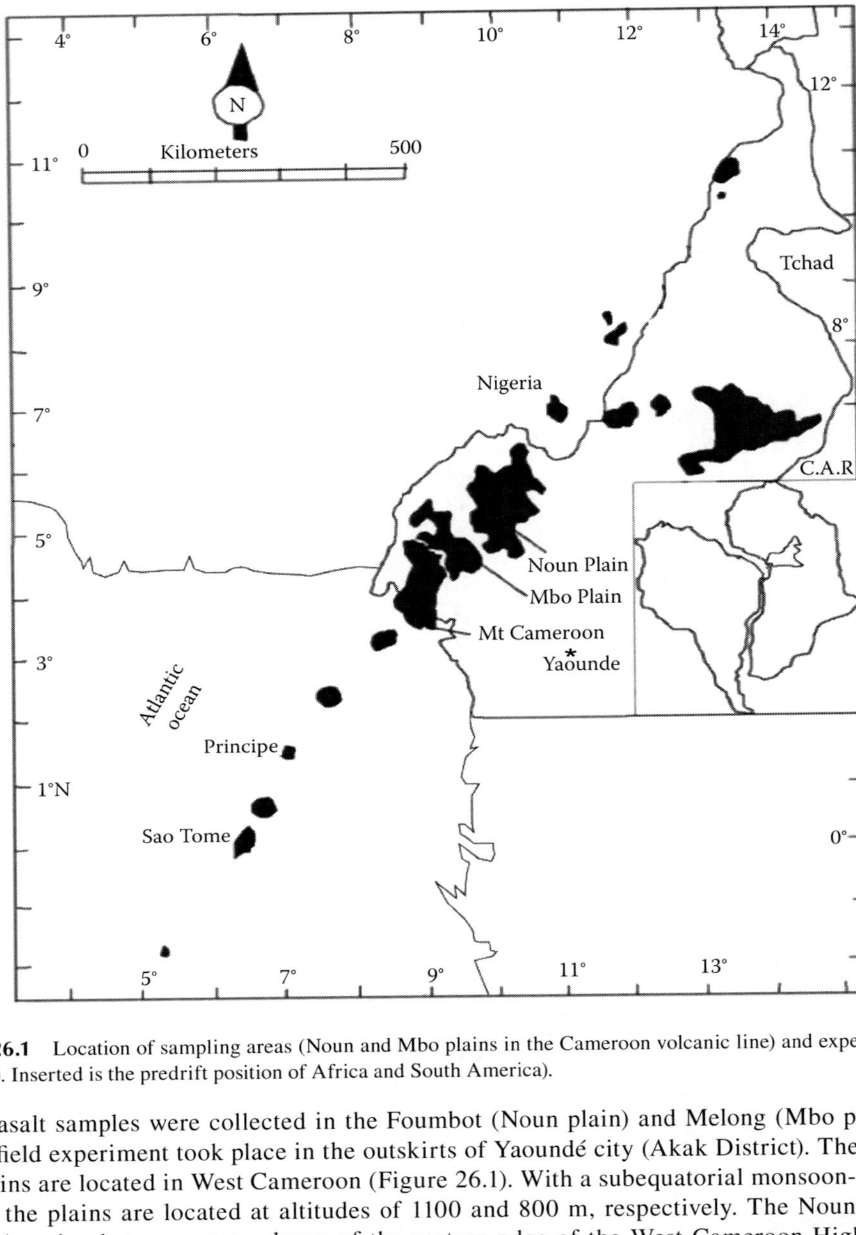

FIGURE 26.1 Location of sampling areas (Noun and Mbo plains in the Cameroon volcanic line) and experimental site (Yaoundé). Inserted is the predrift position of Africa and South America).

The basalt samples were collected in the Foumbot (Noun plain) and Melong (Mbo plain) areas and the field experiment took place in the outskirts of Yaoundé city (Akak District). The Noun and Mbo plains are located in West Cameroon (Figure 26.1). With a subequatorial monsoon-dominated climate, the plains are located at altitudes of 1100 and 800 m, respectively. The Noun plain creates the junction between steep slopes of the eastern edge of the West Cameroon Highlands and the Bamoun plateau. The average 1100 m elevation can be divided into two morphological units: the morphological unit representing more than 1200 m elevation around volcanic Mount Mbepit, which culminates at 1988 m, and the morphological unit of lower than 1200 m corresponds to the Foumbot plain.

Experiments were conducted in the Yaoundé area (Figure 26.1) at a mean altitude of 700 m, characterized by an equatorial climate with four seasons (two dry seasons and two rainy seasons). Meteorological data from 1926 to 2011 show that February is the hottest month (25.73°C) and July the wettest (with a temperature of 23.17°C). The average temperature is 24.45°C.

Geologically, the sampling areas belong to the Cameroon volcanic line, while the area where the experiment took place belongs to the Yaoundé Neoproterozoic unit. The Cameroon line (Figure 26.2) is

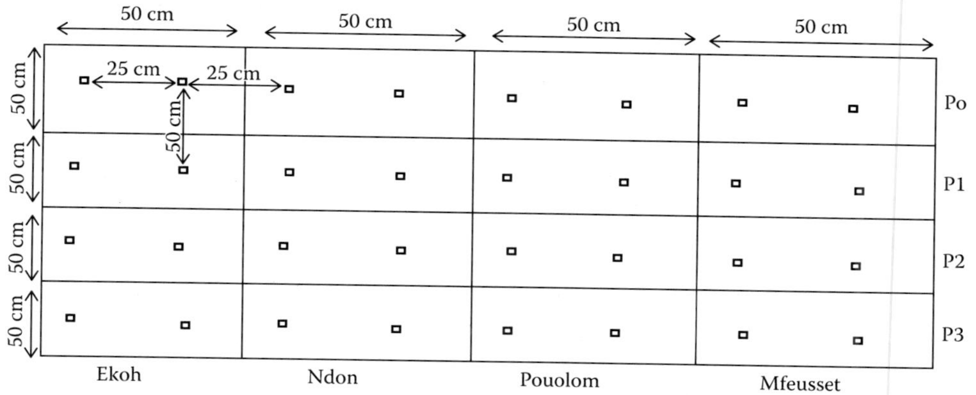

FIGURE 26.2 Disposition of maize plants (small squares) in the experimental area.

a linear megastructure of magmatic, volcanic, and plutonovolcanic with a length of more than 2000 km and a width of about 100 km in certain places, stretching from Pagalu Island in the Gulf of Guinea to Lake Chad (Fitton, 1987; Deruelle et al., 1991, 2007; Lee et al., 1994). The Cameroon Line continues in the ocean as four volcanic islands (Pagalu, São Tomé, Príncipe, and Bioko) and two *guyots* recently discovered (Burke, 2001), all located in the Gulf of Guinea. The continental sector of Cameroon consists of a line of volcanoes.

The Noun plain is located in the southeastern part of the Cameroon Line in West Cameroon (Figure 26.1). *The Cameroon volcanic line* is formed by large quantities of recent (<5 million years) basaltic lavas that essentially lie unconformably on nearly 600 km^2 on top of the granitogneissic basement (Wandji, 1985). Several types of dynamics in volcanic plains are distinguished by Wandji Noun (1985): in the Noun plain, explosive strombolian volcanism allowed the building up of more than 40 volcanic cones of varying importance.

The Mbo plain (Figure 26.1) consists mainly of alkali basalts, hawaiites, hornblende trachytes, and rhyolites younger than 1.5 Ma (Gouhier et al., 1974; Kagou Dongmo, 1998). Explosive activity is strombolian, originally about 70 volcanic cones, or phreatomagmatic, and responsible for the construction of three maars identified in Mount Manengouba (Kagou Dongmo, 2006).

The soils of the Yaoundé region are mainly laterites (Yongue-Fouateu, 1986). They are divided into two groups: red lateritic and yellow ferralitic soils. Red lateritic soils are thicker (more than 15 m), clayey, and acidic (pH < 5.5). The main levels from the bottom to the top are altered and globular clayey upper unit. Their mineralogy consists of kaolinite, hematite, goethite, quartz, and gibbsite (Yongue-Fouateu, 1986; Onguene Mala, 1993). The yellow ferralitic soils are located downslope and are composed of kaolinite, goethite, quartz, and gibbsite (Onguene Mala, 1993). Hydromorphic soils are found in marshy areas. The surface horizon silt has abundant organic matter and deeper horizons are sandy (Onguene Mala, 1993).

Methodology

Geochemical analyses of our samples were performed at the University of Brasilia in Brazil. The technique used was the x-ray fluorescence (XRF). To determine the sizes of different fractions of powdered basalt rock samples previously ground at the mission for local materials promotion (MIPROMALO) in Yaoundé, a sieve analysis was performed in the laboratory of engineering geology and alterology at the Department of Earth Sciences of the University of Yaoundé 1.

The field experiment was conducted with maize (*Zea mays*), Maize Cameroon Selection (CMS) type 8704 variety. The experimental plots at Akak (N 03°59′26″N/011°33′30″E) were 4 m^2 in area. The basalt

samples were applied in varying proportions on these plots. Subsequently, corn kernels were planted at a spacing of 25 cm (Figure 26.2).
Four different experimental plots were distinguished:

P0 = soil + 0 g of basalt powder
P1 = soil + 150 g of basalt powder
P2 = soil + 300 g of basalt powder
P3 = soil + 450 g of basalt powder

Measurements were made of

- The number of leaves per stem
- The length of the leaves
- The weight of ears
- The weight of dry biomass
- Seed weight

At harvest, samples collected were packaged, labeled, and weighed.

Results

Analytical results include the study of basalts in petrographic thin section slides, evaluation of their contents in terms of major and trace elements, and field experimentation of basalt powder and lateritic soil mix for maize farming.

Petrography

In the Noun plain, basalts were sampled in the Mfeusset and Pouolom areas where explosive volcanism produced basalts in the form of ash, lapilli, and volcanic bombs (Figure 26.3a through c). The rock is dark gray-black, ash gray, and vacuolated.

Under the microscope, the rock has a porphyritic texture, as well as microlitic with olivine (25 wt.%), augite (15 wt.%), plagioclase (50 wt.%), and opaque minerals all embedded in a glass representing less than 5% of the rock (Figure 26.4a through d).

In the Mbo plain, samples were collected at Ekoh and Ndon, two volcanic cones with coarse block forms, generally angular, with sizes ranging from centimeters to several meters (Figure 26.3d). The rocks, brownish in color, consist of olivine (5%), brown hornblende (5%), plagioclase (15%), and opaque oxides (5%), defining a dominant microlitic texture (Figure 26.4c and d).

Soils of the experimental site were generally very deep lateritic with thicknesses sometimes exceeding 10 m in the Yaoundé region (Ongene Mala, 1993).

Geochemistry

Geochemistry of basalts was analyzed on four samples (from each of the four cones) at the University of Brasilia in Brazil. As for the analysis of soil developed on gneiss, it is extracted from the literature (International Institute of Tropical Agriculture [IITA], Yaoundé). The results of analyses of basalts are reported in Table 26.1.

Figures 26.5 through 26.7 show variations and main characteristics of the major and trace elements in the studied basalts (reported as weight percent). SiO_2 was analyzed for all samples. Silica varied from 46.13% to 50.35%, being higher in the Mbo plain (50.31%) and lower in the Noun plain (46.13%). Cones in the Mbo plain are richer in K_2O ($1.93\% < K_2O < 1.99\%$) than in the Noun plain ($1.08\% < K_2O < 1.27\%$).

FIGURE 26.3 Outcropping conditions of the basalts. (a–c) Noun plain, (d) Mbo plain. Small size pyroclastic basalts in the Noun plain can be distinguished from dominant large size basalts of the Mbo plain.

Higher aluminum values (15.70%–15.37%) are typical of the Mbo plain, which is consistent with the presence of brown hornblende. MgO is higher in the Noun plain (7.74%) than in the Mbo plain (3.98%) and reflects the abundance of olivine in rocks of the Noun plain. P_2O_5 values are similar for the Mbo (0.76% < P_2O_5 < 0.85%) and Noun plains (0.60% < P_2O_5 < 0.62%). The Noun plain basalts exhibit high levels of calcium (9.68% < CaO < 9.70%) compared to those of the Mbo plain (6.33% < CaO < 7.64%). Amounts of Sr, Cr, Zn, Ni, and V are higher in the Noun plain samples, while Pb and B are higher in the Mbo plain samples.

In the TAS diagram (Figure 26.5), the samples analyzed are all located in the alkaline range. Foumbot lavas have a typical basalt composition, while basalts of the Mbo plain are trachybasalts and trachyandesite. According to Ivrine and Baragar (1971), lavas of our sampling sites are more sodic than potassic (Figure 26.6). The ternary diagram Na_2O–CaO–Al_2O_3 (Figure 26.7) shows that the studied basalts are lower in aluminum in the Noun plain (13.65% < Al_2O_3 < 13.83%) and higher in the Mbo plain (15 37% < Al_2O_3 < 15.70%).

Following the works of Onguene Mala (1993), soils developed in the experimental area are low in exchangeable bases (Ca^{2+} + Mg^{2+} + K + <3), are acidic, and correspond to oxisols.

Particle Size of the Powdered Basalt

Basalt powders were calibrated to separate particles with diameters as follows: >0.4, 0.4–0.315, 0.315–0.2, 0.2–0.08, and <0.08 mm. The following results were obtained: powdered basalt from Ndon (Mbo plain) and Pouolom (Noun plain) have a very fine particle size, 42.11% and 43.20% <0.08 mm;

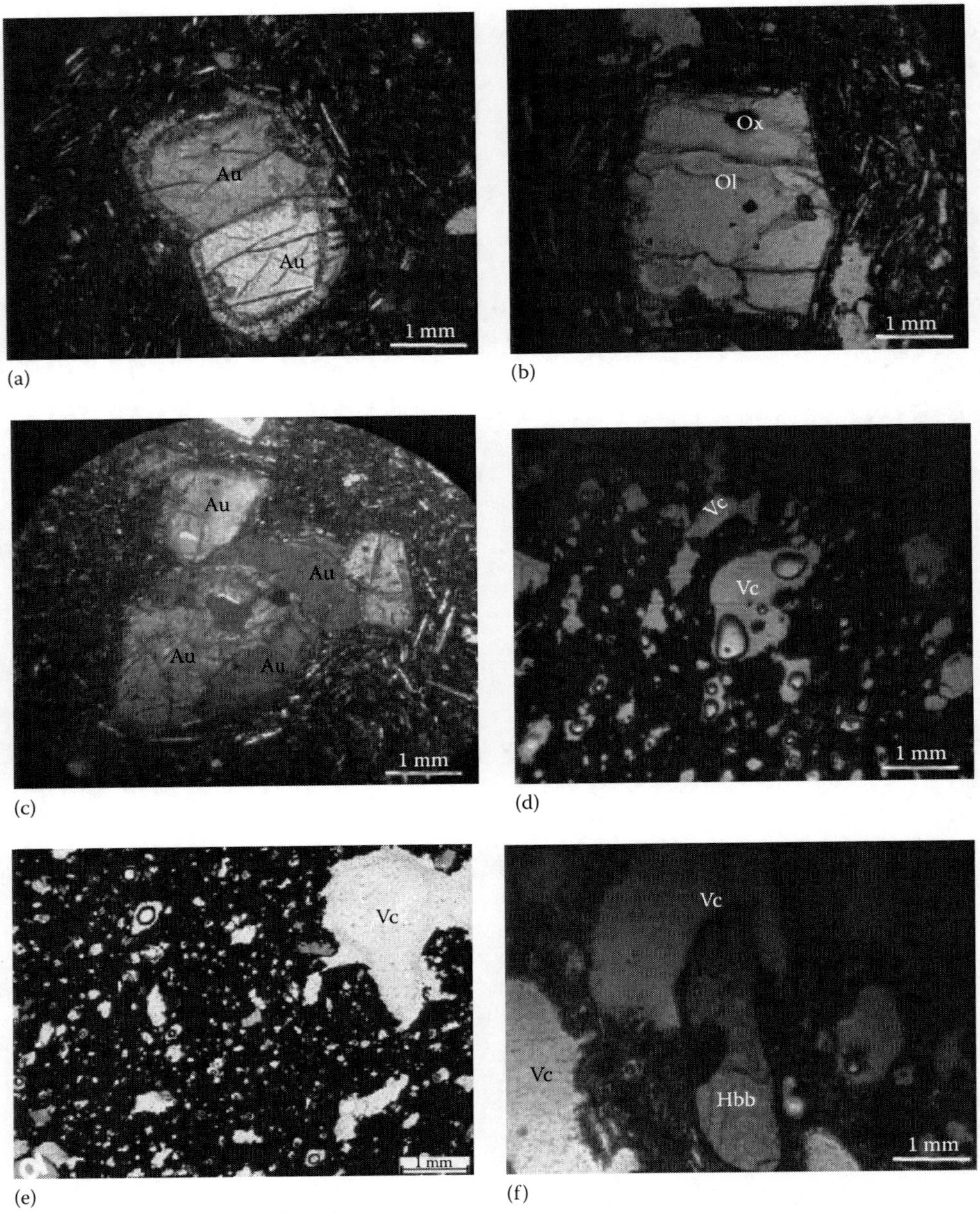

FIGURE 26.4 (a–d) Microlitic porphyritic and vacuolar texture in basalts of the Noun plain and (e and f) dominant microlitic and vacuolar texture for basalts in the Mbo plain. Microliths in (a), (b), and (c) are plagioclases. Au, augite; Ol, olivine; Ox, oxides; Hbb, brown hornblende; Vc, vacuola.

at Mfeusset (Noun plain), basalt powders have a particle size of 29.57% <0.08 mm; and at Ekoh (Mbo plain), 3.04% <0.08 mm.

Description of the Experiment

The experimental plot was cultivated according to traditional cultivation techniques (flat plowing) and the spacing of seedlings of 25 cm. The cultivated area measured 2 m × 2 m and a total of 12 plots of 50 cm × 50 cm each. The application of mineral powder and seedlings (Figure 26.8a) took place on

TABLE 26.1

Major and Trace Elements Analyses of Basalts from the Mbo Plain (Ekoh, Ndon) and Noun Plain (Mfeusset, Pouolom)

	Ekoh Cone	Ndon Cone	Mfeusset Cone	Pouolom Cone
SiO_2	47.75	50.35	46.13	47.06
TiO_2	3.17	2.66	2.83	2.83
Al_2O_3	15.70	15.37	13.65	13.83
FeO^t	12.46	11.59	12.73	12.70
CaO	7.64	6.33	9.70	9.68
MgO	4.20	3.98	7.74	6.84
MnO	0.23	0.21	0.22	0.25
Na_2O	4.04	4.99	3.37	3.37
K_2O	1.93	1.99	1.27	1.08
P_2O_5	0.85	0.76	0.62	0.60
Pb	60	60	60	40
Ba	470	470	470	470
Sr	1000	800	600	730
Cr	20	20	140	80
Ni	60	40	100	100
Cu	20	20	20	20
Zn	140	120	100	120
Mo	20	20	20	20
V	160	130	220	200
B	11	13	11	12

Note: Oxides are in wt.% and trace elements in ppm. FeO^t = total iron as FeO.

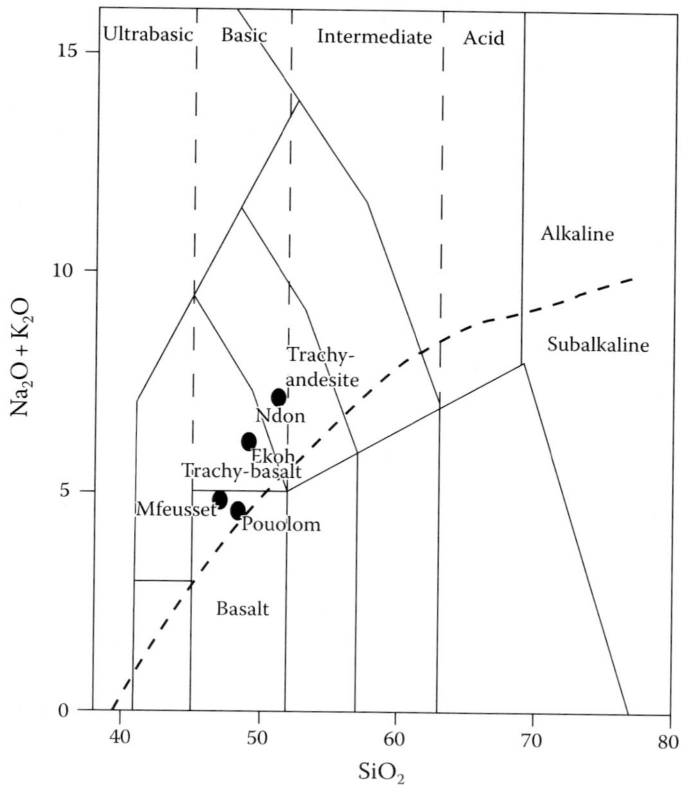

FIGURE 26.5 Diagram TAS. (Data from Le Bas, M.J. et al., *J. Petrol.*, 27, 745–750, 1986.)

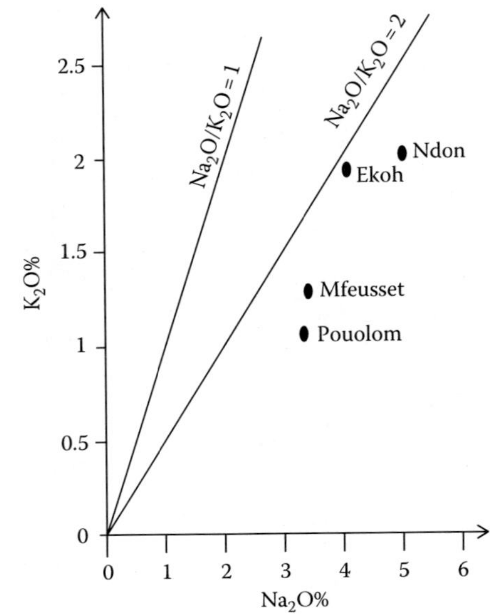

FIGURE 26.6 Na_2O/K_2O diagram of Irvine and Baragar (1971). Studied samples are richer in Na_2O than K_2O.

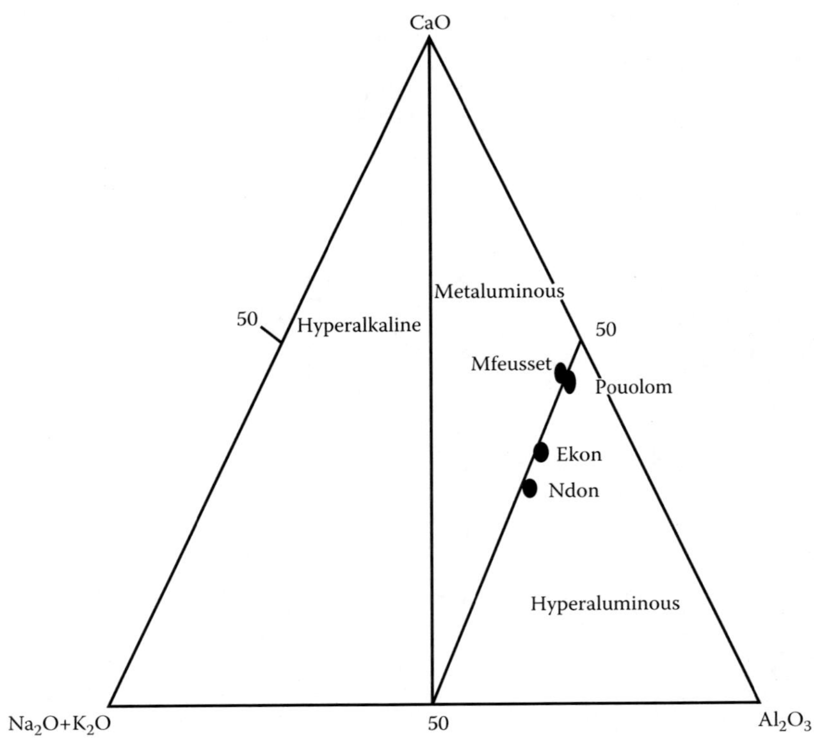

FIGURE 26.7 $CaO + Na_2O + K_2O - Al_2O_3$. The studied rocks fall on the transition between metaluminous and hyperaluminous types.

FIGURE 26.8 Experiment field. (a) Spreading the powder onto plots, (b) seed (seed used from the central part of the ear of corn), (c) weeding field, (d) green leaves of maize having received powdered basalt compared to (e), (e) leaves with purple veins plants in the control plot, and (f) corn in the control plot (left) and maize plot of 300 g of powdered basalt Mfeusset (right).

April 14, 2012. Quantities of powdered rocks used were 450, 300, 150, and 0 g for the control plot (Figure 26.8a). Maize seed chosen for the seed was stripped off its two ends to minimize differences in the germination of seeds (Figure 26.8b). Two grains of corn were subsequently put underground in holes spaced every 25 cm.

Maize growth monitoring was carried out after 1 week, 2 weeks, 1 month, 2 months, and 3½ months (harvest). One week after planting, we saw that the corn stalks in the control plot were less developed than in plots with powdered basalt. The most advanced plots were, respectively, those who received 150, 300, and 450 g of powder. Two weeks after planting, all plots that received the rock powder had leaves that were less dark in color (light yellow) than control plots, and plots with 450 g of rock powder in Foumbot and Melong showed good growth in size and leaves. One month after planting, plots with powdered basalt showed a green to yellowish color, while those in the control plot were colored green. Overall, corn stalks with powdered basalt from the Foumbot plants were much more developed than the rest of the field. Weeding of the field (Figure 26.8c) was followed by thinning by removing less vigorous maize plants. After 2 months, flowering began in corn plots that received the basalt powder, and 5 days later, the control plot began to flower. At this stage, the leaves that were previously greenish-yellow in the corn plots with rock powder were once again green (Figure 26.8d) and for the control plot, leaves showed purple color in the veins (Figure 26.8e). At harvest, 3½ months after planting, crops with powdered rock had significantly larger flowering spikes (Figure 26.8f) than plots with no addition of basalt powder.

At harvest, we measured the size of the leaves (Table 26.2), the weight of dried maize plants (Table 26.3), the weight of ears (Table 26.4), and the weight of maize kernels (Table 26.5). Crops on basalt powder plots showed significantly better results than plots with no basalt powder (Table 26.6), and relative to the efficiency of the basalt powders, samples from Mfeusset (Noun plain) gave the best results (Table 26.7 and Figures 26.8 and 26.9).

TABLE 26.2

Leaf Length (cm) at Harvest

Treatment	150 g	300 g	450 g
Control	76.83	76.83	76.83
Ekoh	93.15	75.35	83.25
Ndon	76.57	67.37	88.02
Pouolom	79.35	69.87	64.60
Mfeusset	92.87	88.47	92.00

TABLE 26.3

Weight of Dry Maize Plants

Treatment	150 g	300 g	450 g
Control	200	200	200
Ekoh	240	165	360
Ndon	180	205	275
Pouolom	295	165	250
Mfeusset	365	250	392.5

TABLE 26.4

Weight of Maize Ears

Treatment	150 g	300 g	450 g
Control	12.5	12.5	12.5
Ekoh	25	—	140
Ndon	—	—	30
Pouolom	70	—	25
Mfeusset	170	125	190

TABLE 26.5

Weight (g) of Maize Grains

Treatment	150 g	300 g	450 g
Control	6.85	6.85	6.85
Ekoh	2.07	—	22.17
Ndon	—	—	9.03
Pouolom	7.03	—	0.73
Mfeusset	62.45	71.38	49.01

TABLE 26.6

Crop Yield (t/ha)

Treatment	0 g	150 g	300 g	450 g
Ekoh	0.274	0.083	0	1.286
Ndon	0.274	0	0	0.361
Pouolom	0.274	0.281	0	0.029
Mfeusset	0.274	2.498	2.855	1.96

TABLE 26.7

Ratio (%) Relative to the Efficiency of Basalt Powder

Treatment	150 g	300 g	450 g
Ekoh	−69	—	369
Ndon	—	—	31
Pouolom	2	—	−89
Mfeusset	811	942	615

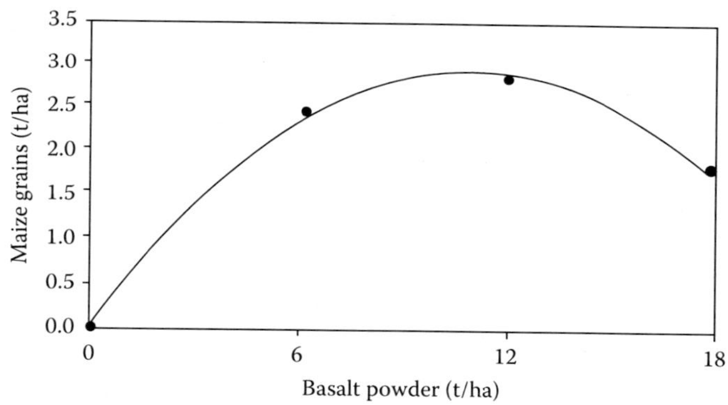

FIGURE 26.9 Maize grain production (t) versus weight/surface of basalt powder for Mfeusset basalts (Noun plain).

Discussion

The results of field trials have shown that addition of powdered basalt to soils developed on gneiss results in higher yields. Based on the overall results, we discuss in the following the influence of pH on yields; the color of leaves throughout the experiment; the relationship between petrography, geochemistry, and fertilization potential; the influence of basalt powder grain size; the toxicity of trace elements; yields; the economic significance of the results; and a comparison with results obtained using similar materials.

Influence of pH

One week after planting, there was a marked difference in the growth of corn plants. Although leaf analysis and pH measurements were not made, a rapid effect due to the uptake of chemical elements from basalt powder was unlikely, and it seemed that in the initial stages of plant growth, changing soil pH may have played the most important role in the growth of maize plants. Nkouathio et al. (2007), in monitoring the growth of corn stalks after addition of powdered basalt, noticed a change in soil pH from 5.2 to 6.4 when the rock powder is added. Moreover, Njofang et al. (2008) in their map of soil pH in Basin Nun noticed a correlation between soil fertility and pH, and regions with pH > 6 always correspond to areas of high natural fertility.

Change of Leaf Color

Two weeks after planting, it was noted that leaves in plots with added basalt powder had a light yellow color, unlike the distinctly green color of the leaves on the control plot. These variations in color may be due to the available nitrogen. Nitrogen in rock powder is low relative to the values in the control plots in relation to plant growth rates, so basalt could have caused in nitrogen deficiency for these more rapidly growing plants. Nitrogen plays an essential role in the initial stages of plant growth (Foth, 1978; Van Straaten, 2007). The purple coloration of corn leaf veins observed 2 months after planting in the control plot is due to phosphorus deficiency (Van Straaten, 2007). This phenomenon was not observed for plots with powdered basalt. Possibly, the increase in soil pH canceled the effects of phosphorus deficiency, or the basalt provided more phosphorus available to the plant after 2 months.

Relationship between Petrography, Geochemistry, and Fertilizing Potential

For proper growth, plants need essential elements like calcium, magnesium, potassium, and phosphorus (Table 26.8). Basaltic rocks of the Noun and Mbo plains are of alkaline sodic types. Chemical analysis shows that these lavas are aluminous and richer in silica in the Mbo plain.

The sample showing the highest oxide contents is Mfeusset in the Noun plain (19.3 wt.%) and for the sum of CaO + MgO (17.44 wt.%). Thus, in the Mfeusset basalts, more than 90% of the total oxide is in the form CaO + MgO, which gives these top basalts the highest fertilizing power, as an increase in pH in acidic soils eliminates the effects of aluminum toxicity (ferralitic soils formed on gneiss).

Calcium and magnesium are present mainly in the plagioclases, olivines, and pyroxenes. The presence of these minerals in abundance in the Mfeusset basalt also reflects the strength of this basalt fertilizer.

Economic Significance of the Results

In Brazil, the standard amount of basalt powder that farmers would use for their crops is 2 t—1 t of basalt powder costs about 85 USD. In our case, for optimum production, we need a quantity of basalt powder equal to about 12 t/ha for a production of nearly 2.8 t of maize. To buy 12 t of powdered basalt would cost a total of 1020 USD, which is equivalent to 510,000 CFA. The cost of 1 kg of maize CMS type 8704 in Cameroon is currently 800 CFA, and selling 2800 kg of maize would produce a profit of 2,240,000 CFA, which means gains for the farmer. And as the effect of basalt powder can extend 3–4 years (Leonardos et al., 1972; Theodoro and Leonardos, 2006), the benefits would increase successively. Because of rapid growth due to shorter growing time, two corn harvests could take place in 1 year (Knapik et al., 2007).

Comparative Results

Nkouathio et al. (2007) performed tests in pots with powdered basalt from Tombel for corn grown in the soil of the region of Yaoundé. Tested over 12 weeks, the results showed that addition of pyroclastics increased pH (5.2–6.4). This addition also increased the amount of exchangeable cations (2.4–6.8) and the exchangeable acidity decreased (0.4–0.15). Although the experiment did not continue through harvest, the results were close to those we obtained after 2 months, indicating that the use of basalt powder as fertilizer has a definite effect on the corn crop yields.

TABLE 26.8

Amounts of Selected Oxides of Studied Basalts

Oxides wt.%	Ekoh	Ndon	Pouolom	Mfeusset
CaO	7.6	6.33	9.68	9.7
MgO	4.2	3.98	6.84	7.74
K_2O	1.9	1.99	1.08	1.27
P_2O_5	0.9	0.76	0.6	0.62
Total	14.6	13.1	18.2	19.3

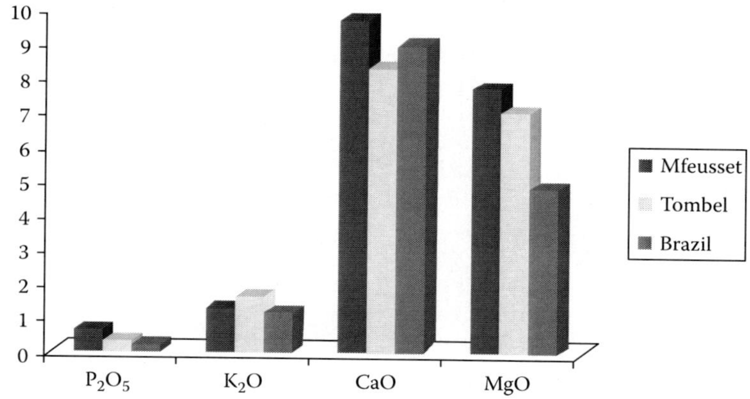

FIGURE 26.10 Comparison of P_2O_5, K_2O, Cao, and MgO contents of basalts from Mfeusset (Foumbot plain) and data from Paraná basalts (Brazil). Mfeusset (this work), Tombel (Nkouathio et al., 2007), Brazil (Knapik et al., 2007).

In southern Brazil, yields of the order of 128%–253% were obtained by adding the rock powder to maize (Silva, 2005; Almeida et al., 2007; Knapik et al., 2007). The source of the rock was Paraná basalts, dated at 125 million years. Figure 26.10 shows that essential element levels of the Brazil basalt are lower than those from Cameroon.

Comparing the levels of essential elements for basalts from Mfeusset (Noun plain), Tombel, and Brazil, the basalts from Mfeusset have the best content of chemical elements essential to the growth of corn.

Conclusion

An evaluation of the fertilization potential of pyroclastic basalts from the Cameroon volcanic line in the Mbo and Noun plains was carried out through petrographic and geochemical study and field tests for maize (*Z. mays*) farming on ferralitic soils formed on gneissic rocks in equatorial south Cameroon (Yaoundé area). Petrographically, the Mbo plain basalts are trachybasalts (Ekoh cone) and trachyandesites (Ndon cone) with similar microlitic and vacuolar texture and constituted plagioclases, olivine, basaltic hornblende, and opaque oxides. In the Noun plain, basalts are of microlitic porphyritic texture (Mfeusset cone) and microlitic texture (Pouolom cone). Vacuoles are a common structure in Noun plain basalts and the mineralogy is mainly plagioclases, augite, olivines, and opaque oxides. Great amounts of glass (30%) exist in all studied basalts, and the petrographic difference in both regions is the restriction of basaltic hornblende and augite to the Mbo and Noun plains, respectively. Mother rock in the farming area is the gneiss with a granolepidoblastic to porphyro-poecilo-granoblastic texture, and its main mineral composition is composed of garnet, kyanite, biotite, orthoclase, and quartz. Geochemically, basaltic rocks are alkaline and sodic; rocks from the Mbo plain are more acidic (SiO_2 = 50.35 wt.%) than those of the Noun plain (SiO_2 = 46.13 wt.%) and SiO_2 content of the gneiss is 61.98 wt.%. Field experiments tested additions of 150, 300, and 450 g of basalt powders to plots of 0.25 m^2 and revealed a noticeably increasing pH (5.2–6.4) in the soil during the early stages of growth of maize plants. Parameters such as leaf length, weight of dry biomass, weight of maize cobs, and weight of corn kernels were recorded at harvest. The results indicate the average output with respect to the cultivated plots: 1.28, 0.36, 0.28, and 2.855 t/ha for Ekoh, Ndon, Pouolom, and Mfeusset basalts, respectively, and 0.274 t/ha for the control plot. From an economic standpoint, the best production (2.498 t/ha) was obtained with an addition of 12 t/ha of basalt powder from Mfeusset cone in the Noun plain.

ACKNOWLEDGMENTS

The authors would like to thank Carlos Augusto Posser da Silveira of Embrapa Clima Temperado (RGS-Brazil) for valuable discussions in the course of this work.

REFERENCES

Almeida, E., Silva, F.J.P., and Ralish, R., 2007. Revitalização dos solos em process de transiçao agroecologica no sul do Brazil. *Agriculturas*, Rio de Janeiro, 4(1): 7–10.

Déruelle, B., Moreau, C., Nkoumbou, C., Kambou, R., Lissom, J., Njonfang, E., Ghogomu, R.T., and Nono, A., 1991. The Cameroon Line: A review. In: Kampunzu, A.B. and R.T. Lubala (eds.), *Magmatism in Extensional Structure Settings: The Phanerozoic African Plate*. Springer, Berlin, Germany, pp. 274–328.

Déruelle, B., Ngounouno, I., and Demaiffe, R., 2007. The Cameroon Hot Line (CHL). A unique example of active alkaline *intraplate structure in both oceanic and con*tinental lithospheres. *Comptes Rendus Géosciences* 339: 589–600.

Fitton, J.G., 1987. The Cameroon line-west Africa: A comparison between oceanic and continental alkaline volcanism. *Geological Society London Special Publications,* 30: 273–291.

Foth, H.D., 1978. *Fundamentals of Soil Sciences*, 6th edn. Wiley, New York, 436pp.

Gouhier J., Nougier D., and Nougier J., 1974. *Contribution à l'étude* volcanologique du Cameroun (Ligne du Cameroun-Adamaoua). *Ann. Fac. Sci. Cameroun*, 17: 3–48.

Irvine, T.N. and Barragar, W.R., 1971. A guide to the chemical classification of common volcanic rocks. *Canadian Journal of Earth Science*, 8: 523–548.

Kagou Dongmo A., 1998. *Etude volcanologique, petrologique et géochimique d'un stratovolcan polygénique de la Ligne du Cameroun*: Thése 3ème cycle, Université Yaoundé I, p. 197.

Le Bas, M.J., Le Maitre, R.W., Streckeisen, A., and Zannetin, B., 1986. A chemical classification of volcanic rocks based on the total Alcali-Silica diagram. *Journal of Petrology*, 27: 745–750.

Lee, D.C., Halliday, A.N., Fitton, J.G., and Poli, G., 1994. Isotopic variation with distance and time in the volcanic islands of the Cameroon Line: Evidence for a mantle plume origin. *Earth and Planetary Science Letters*, 123: 119–138.

Leonardos, O.H., Fyfe, W.S., and Kronberg, B., 1972. Rochagem: método de aumento de fertilidades em solos lixiviados e arenosos. *Congr. Bras. Geol. Ouro Preto. Anais. SBG*. P. 137–145.

Njofang, C., Matschullat, J., Tchouankoue, J.P., and Heilmeier, H., 2008. Soil and plant composition in the Noun river catchment basin, Western Cameroon: Contribution to the development of a biogeochemical baseline. *Environmental Geology* 56(7): 1427–1436.

Nkouathio, D., Wandji, P., Bardintzeff, J.M., Temato, P., kagou Dongmo, A., and Tchoua, F., 2007. Utilisation des roches volcaniques pour la reminéralisation des sols ferralitiques des régions tropicales. Cas des pyroclastites basaltiques du graben de Tombel (Ligne Volcanique du Cameroun). *Bulletin de la Société vaudoise des Sciences Naturelle* 91: 1–14.

Onguene, M., 1993. Différenciation pédologique dans la région de Yaoundé (Cameroun). Transformation d'un sol rouge ferralitique en sol à horizon jaune et relation avec l'évolution du modèle. Thése, University of Paris VI, 253pp.

Silva, F.P. da., 2005. Uso de Pô de basalto em producao de milho com principios agroecologicos. Monografia (Post-graduaçao), Universidade da Uniao da Victoria, 41pp.

Theodoro, S.H. and Leonardos, O.H., 2006. The use of rocks to improve family agriculture in Brazil. *Anais da Academia Brasileira de Ciências* 78: 721–730.

Van Straaten, P., 2007. *Agrogeology: The Use of Rocks for Crops*. Enviroquest Ltd., 352 River Road, Cambridge, Ontario, N3C 2B7 Canada, 440pp.

Wandji, P., 1985. Contribution à l'étude pétrologique, géochimique et géotechnique des projections volcaniques de la région de Foumbot. Thèse Doct. 3ème cycle, Univ. Yaoundé, 158pp. +1 carte.

Yongue-Fouateu, R., 1986. Contribution à l'étude pétrographique de l'alteration et des faciés de cuirassement ferrugineux des gneiss migmatiques de la région de Yaoundé. Thése Doct. 3émecycle, Univ. Ydé, 214pp.

27

*Seawater Concentrate for Abundant Agriculture**

Arthur Zeigler

CONTENTS

Introduction: What is Seawater?

Seawater is water from a sea or ocean. Everything soluble, given enough time, eventually ends up in the sea.

The rivers of the world dump billions of tons of dissolved and particulate minerals into the oceans each year and have for eons. Undersea volcanoes, hot springs, and volcanic eruptions on land have also added billions of tons of elements. More importantly, 90% of all volcanic activities on Earth occur in the oceans. Additionally, seawater is a biological soup of active organic substances.

As a consequence of covering nearly three quarters of Earth's surface, the oceans are the planet's largest interceptors of sunlight, wind, and transducers of solar energy. Transduction is the transfer of energy from one form to another.

One method of solar energy transduction in the oceans is the creation of organic matter through photosynthesis by phytoplankton and photosynthesizing marine bacteria. This may be described as the transmutation of radiant energy into matter.

In addition to all of the elements and complex mineral salts, there are phytoplankton, zooplankton, microbes, and the by-products of their life cycles. Marine phytoplankton comprises hundreds of species of photosynthetic, unicellular organisms with the ability to convert sunlight, warmth, water, and minerals into protein, carbohydrates, vitamins, and amino acids. Each individual plankton is so tiny that without a microscope you would never know that they exist. We are completely unaware of photosynthesizing marine bacteria when we go swimming in the ocean, but they may be the largest single biomass on the planet, exceeding the forests of the world.

In addition to the compounds that are by-products of photosynthesis, the waters of the oceans contain 90 chemical elements in measurable amounts. Every element that occurs in nature is found in the sea. There are also many more minerals in compound form, such as sodium chloride, which have been washed in via rivers or leached out from the vast area of land covered by oceans.

* The following material is excerpted from the book *Seawater Concentrate for Abundant Agriculture* by Arthur Zeigler

Enormous amounts of soluble organic materials are also contributed by the water that flows from the land to the sea. If the reader has ever seen an otherwise clear stream that had a yellowish or brown coloration, they have seen fulvic acid on its way to the sea. Fulvic acid contains over 50,000 different organic substances and has been detected in all seawater. Fulvic acids are one of the largest classes of dissolved organic matter, but they are poorly defined and of unclear origin (Reemtsma et al. 2006).

In total, the seas of the world are estimated to contain as much as 50 quadrillion tons (50,000,000,000,000,000) of dissolved solids. If these dissolved minerals could be removed from the seas and spread on the land as dried solids they would cover all the land surfaces of planet Earth with a layer of minerals 500 ft thick (Swenson 1983).

The principal dissolved solids in seawater are sodium chloride (common salt), calcium carbonate and calcium chloride, potassium sulfate, and the magnesium salts magnesium chloride, magnesium sulfate, and magnesium bromide. Many other elements are present in seawater in trace and ultratrace amounts.

History of Seawater Concentrate Research

Scientific researchers have been examining the effects of seawater, seawater extracts, and seawater concentrates on terrestrial life since the late 1800s, when Dr. René Quinton, the renowned French doctor, biologist, biochemist, and physiologist, observed that seawater had remarkable similarities to blood plasma.

As early as the late 1930s, Dr. Maynard Murray noticed profound effects when seawater was applied to living organisms. Initially, Dr. Murray's experiments used diluted seawater on soils and crops. In an effort to reduce shipping expense, he decided to try working with seawater evaporated to complete dryness. Then only the 3.5% of seawater that is the mineral content would need to be shipped. He located a deposit of impure evaporated sea salt on sun-baked intertidal salt flats in Baja California, Mexico. He called this salt sea solids and used it, during many years of extensive and well-documented research, on many types of crops and soils.

The results were consistently the same: the plants flourished, matured more rapidly, were healthier, were more disease and drought resistant, and produced outstanding taste along with greater yields. In assays testing for nutrients, foods grown with Dr. Murray's sea solids had significantly more minerals, vitamins, and sugars, resulting in a healthier plant and a more nutritious food crop. In addition, he witnessed the same amazing results in all types of livestock and poultry that were offered feed grown in soil enriched with his sea solids. Physiologically, animals were healthier, gained weight more rapidly, and reached maturity sooner (Murray 1976). The main difficulty with Dr. Murray's system was that it required application of 1500 lb/acre of dried sea salt to achieve the desired results.

Using these sea solids was no doubt a significant improvement over shipping bulky seawater inland. However, because of the large sodium chloride content, it was necessary to use large amounts of sea solids in order to obtain enough marine trace minerals to get the desired seawater effect. While this worked in the Midwest and under hydroponic conditions in Florida, one has to wonder if application of such large amounts of sodium chloride would poison the Earth in drier areas such as Southern California or Arizona. Those areas already have a salt buildup problem to the extent that even using manure can be a hazard because of salt content. Each ton of manure contains 50–100 lb of salt. Salinization of soils caused by irrigation can kill most crop plants and has caused the downfall of ancient civilizations in Peru, Mesopotamia, and the American Southwest. Worldwide, up to 110 million acres are affected, and nearly 4 million acres of the *world's* irrigated farmland is lost to excessive salt every year. "Cultivated land is being lost at a similar rate to new land being brought into production" (Umali 1993). These problems are caused by excessive amounts of sodium chloride, the major mineral constituent of seawater. The problem of using seawater concentrate in these areas could be solved if the salt were removed or greatly reduced.

About 12 years ago, I took up the challenge to extract from seawater the bioactive principles without the salt. Proceeding from the work of Dr. Murray and guided by a test for bioactivity recommended by Dr. Philip Callahan, I entered onto a path of research that was to culminate with the creation of the product Sea-Crop. The first goal was to extract the bioactive principles while leaving behind most of the water and sodium chloride. It seemed that this would be the only way a seawater extract would become economical and have a broad-based agricultural application.

Eventually, the research resulted in a bioactive seawater concentrate with reduced salt and loaded with trace minerals. It performed well, compared to Dr. Murray's system, but instead of 1500 lb/acre, only a few gallons per acre of seawater concentrate was needed.

Early experimentation with this concentrate showed that if the extract was allowed to dry out completely, much of the bioactivity was lost. Further testing revealed that if the extract was dried in an oxidizing atmosphere, at a temperature sufficient to burn off all contained carbon, the resulting pure mineral concentrate had little observable bioactivity at the same application rate previously used.

As a result of these experiments, a working hypothesis was developed. It was postulated that active organic substances in seawater, working together with the plentiful trace minerals, were responsible for the observed bioactivity when the extract was used as a plant and soil stimulant. This could be because mineral elements that are organically complexed can be easier for plants to take up.

Further research and several improvements in the process led to discoveries that resulted in Sea-Crop, the seawater concentrate Ambrosia Technology LLC has been making for the last 6 years.

Sea-Crop is a solution to the salt problem as well as to the shipping and application difficulties. In Sea-Crop, 95% of the sodium chloride is excluded, and all of the other minerals are concentrated.

Another advantage that Sea-Crop has over the sea solids is that, because it is a concentrate of only the active principles contained in seawater, much less of it needs to be used. Application rates are typically only 2–4 gal/acre with a few specialty crops requiring up to 6 gal to the acre each year. Even with these low application rates, Sea-Crop is able to perform much the same as Dr. Murray's sea solids did in his field trials or better.

Sea-Crop is certified by the Washington State Department of Agriculture (WSDA) as fully meeting the requirements of the National Organic Program. It has been authorized by the WSDA, the California Department of Food and Agriculture (CDFA), and Oregon Tilth for unlimited use in production of organic food crops.

Benefits of Seawater Concentrate

There are many reasons to use seawater concentrate for agriculture. Among these reasons are increased yield at harvest, improved soil tilth, increased soil microflora, better drought tolerance, increased nutrient density (such as vitamin and mineral content that produces better flavor, keeping qualities and health for the consumer), improved plant health, and overall vigor that makes plants more disease and insect resistant.

While many of these benefits are of primary importance to the gardener, the farmer must first of all look to the bottom line, and that means that his primary interest must be in yield at harvest. Each planting season, he rolls the dice and prays that the promise held in the seed that he plants in the ground can withstand the extremes of drought, flood, wind and hail, chill, and heat so that come harvest time, there is sufficient yield to sustain him or her until the following planting season.

Next for the farmer, after yield per acre, comes the health of the soil. While yield is the short-term yearly focus, in the longer term, if the soil itself is degraded or destroyed, yields cannot be improved or even long maintained.

For the fruit and vegetable grower, nutritional quality is of considerable importance, and improved flavor is certainly a bonus.

For the farmer or rancher who raises forage and feed for use in feeding animals on his own land, nutrient density also becomes an important consideration because animals fed on low-nutrient food must struggle to grow properly and maintain good health.

For the moment, let us set aside all of the other benefits that experience has shown seawater concentrates provide and look only at the one benefit that has historically been of primary importance to the farmer, yield at harvest.

Third-Party Testing Shows Solid Results

The following harvest yield results were obtained during field trials sponsored by third parties. Ambrosia Technology LLC had no involvement in any of them other than to provide the Sea-Crop product.

In 2007, a field trial, sponsored by Cal-Agri Products, was conducted by Dr. Edward McGawley on corn at the University of Louisiana at Baton Rouge with the following results. Two test plots were treated at planting with a single application of Sea-Crop. One was treated with 2 gal/acre and the other with 4 gal/acre. Compared to an untreated control test plot, the treated plots yielded increased corn harvests of 58.15% and 94.49%.

Cal-Agri Products had Dr. McGawley conduct additional field trials at Louisiana State University at Baton Rouge in 2008. In this trial, Sea-Crop was used as a root dip on plug transplants at the dilution rate of 1 part in 50 (2%). The treatment gave the following increases in marketable fruit shown in Figure 27.1.

All of the tests at Louisiana State University (LSU) took place in nematode-infested soil that was inoculated with nematodes prior to or at planting.

Cali-Agri also sponsored testing by Dr. Abdelhaq Hanafi at the University of Morocco at Agadir.

In 2007, eggplant sets were planted in 5 L pots, and eight replicates were then treated with a foliar spray of Sea-Crop that was diluted to a 1% concentration. There was no increase in the amount of fruit, but the fruits of the treated plants were 79.5% larger by weight giving a 79% increase in yield over the control.

In 2008, Dr. Abdelhaq Hanafi also conducted tests for Cal-Agri Products on the following crops:

Runner beans: Sea-Crop was used as a 1% soil drench in the two-leaf stage after emergence and as a 1% foliar spray 3 and 6 weeks after the soil drench. Total Sea-Crop usage was 3 gal/acre and the increase in marketable fruit was 86%. This test demonstrated that Sea-Crop improved plant growth, leaf size, root volume, and root weight while accelerating early production and improving overall yield.

Zucchini: Sea-Crop was evaluated on 120 plants grown hydroponically in coconut husk peat using five different combinations of soil drench and foliar applications as compared with an untreated control. The best performance was achieved when Sea-Crop was applied twice as a 1% foliar application, while the second best was achieved when Sea-Crop was applied twice as a 0.30% soil drench. One foliar application of Sea-Crop at 1% achieved better results than one receiving a soil drench at 0.30% concentration. It was found that Sea-Crop improves leaf size, number of fruits, and total yield, while causing an early flowering and an early production compared to the control.

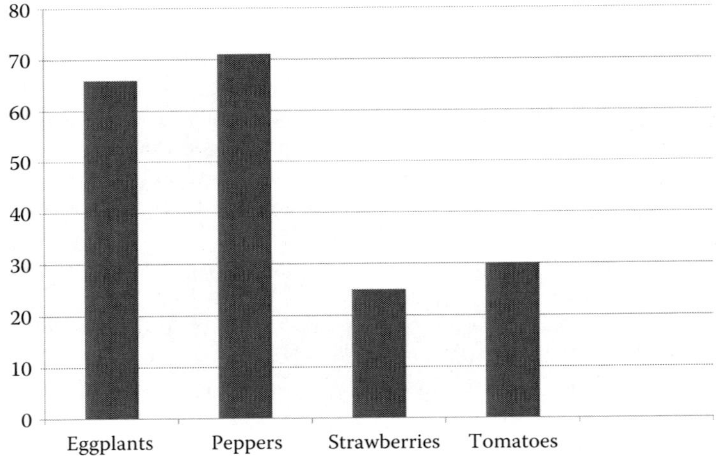

FIGURE 27.1 Increased yield over control. (From Zeigler, A., *Seawater Concentrate for Abundant Agriculture*, 2012.)

Potatoes: In northern Europe, Ambrosia Technology LLC products are distributed by Hak Agro Feed. Starting in 2005, this company tested Sea-Crop on potatoes for 3 consecutive years with excellent results. Barend Hak stated, "We had an increase of 28% three years in a row on a one hectare test plot vis-à-vis the control. The potato grown was Bintje. Twice, 3.75 liter per hectare was treated. First spraying was done when the plants were sprouted and fully above the field. Second spraying was at full blossoming."

In 2010–2011, our international distributor, Collé Agriculture, sponsored testing in Malawi, Africa, by four different entities at six widely separate sites. The objective of the trials was to evaluate the efficacy of Sea-Crop on a variety of locally important crops including maize, peanuts, soybeans, and tobacco. The results follow.

Tobacco: Two tests performed on tobacco by a research institute gave an average yield increase of 38.5% when Sea-Crop was applied at the rate of 4 gal/acre split into three applications, one soil drench and two foliar.

Peanuts: The increased yield obtained by a farmer on a private estate was 42.8% greater than the untreated control. Sea-Crop was used in this trial at the rate of 4 gal/acre. This same farmer also achieved increased yields in soybeans 9.3% and maize (corn) 27%.

Similar results were seen in trials at Bunda Agricultural College and government agricultural research stations of the Ministry of Agriculture and Food Security of the Government of Malawi.

County Line Farm 2012 Winter Trials by Collé Agriculture

In the early part of 2012, Collé Agriculture performed testing at a commercial organic produce farm in Southern California. The testing documented the performance of Sea-Crop under organic growing conditions.

One aim was to determine if yields increased when the only variable was the addition of Sea-Crop. Another goal was to determine if produce grown under commercial conditions and treated with Sea-Crop contained a greater nutritional value.

The crops chosen for testing were baby beets, baby carrots, cauliflower, fennel, spinach, and baby Swiss chard. These crops were grown in fields fertilized with hydrolyzed fish powder, chicken manure, and with both macro and micro mineral soil amendments.

A portion of each crop received treatment with Sea-Crop in addition to the normal fertilization. In all, three applications were made, one soil drench and two foliar applications, totaling 4 gal/acre. The larger, untreated, portion of each crop acted as the control.

As can be seen in Figure 27.2, yields were increased significantly.

For cauliflower and spinach, total yields were used. For the rest, marketable yields, after culling for substandard produce, were used.

The next table of results, Table 27.1, from the same trial shows comparative dry matter content. Dry matter represents everything in the sample other than water. Dry matter includes protein, fiber, fat, minerals, and carbohydrates.

Dry matter increases indicate enhanced flavor, better keeping quality, and greater nutrient density.

Degrees Brix is a measurement of the sugar content of an aqueous solution. During the trial, Brix readings of the crops were taken with an optical device called a Brix refractometer. In addition to sugar content, the optical Brix readings include everything that is soluble in the liquid portion of the plant including vitamins, enzymes, and hormones.

Figure 27.3 shows that the treated crops had elevated Brix levels over the untreated controls. These results are a good confirmation of the dry matter readings in Table 27.1.

As with dry matter content, Brix increases indicate enhanced flavor, better keeping quality, and greater nutrient density.

Next, Table 27.2 shows that significant insect damage reduction was evident in the treated plots.

There was no measurable insect damage in the fennel or spinach crops.

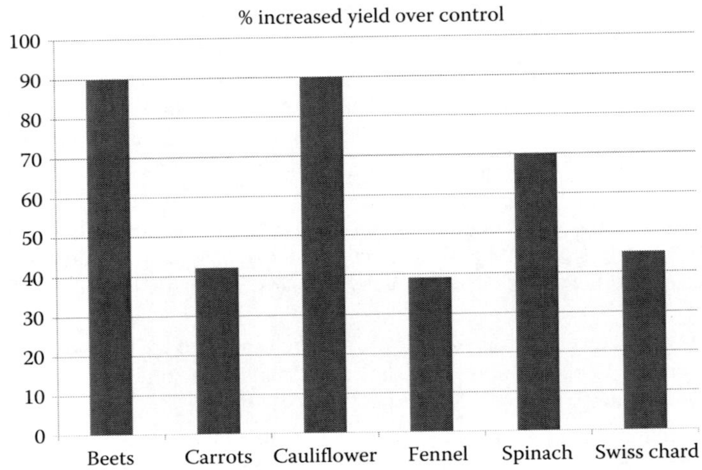

FIGURE 27.2 County Line Farm 2012 Winter Trials. Average for the six species +62.56%. (From Zeigler, A., *Seawater Concentrate for Abundant Agriculture*, 2012.)

TABLE 27.1
County Line Farm 2012 Winter Trials

Dry Matter % Change from Control	
Crop Planted	**Three Applications % Increase**
Beets	+5.97
Carrots	–0.95
Cauliflower	+12.02
Fennel	+2.82
Spinach	+1.42
Swiss chard	+55.21
Average for the six species	+12.72

Source: Zeigler, A., *Seawater Concentrate for Abundant Agriculture*, 2012.

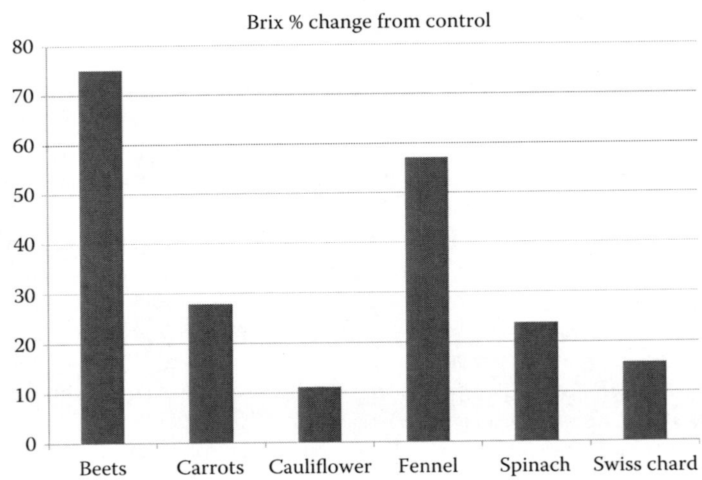

FIGURE 27.3 County Line Farm 2012 Winter Trials. Average for the six species +35.12%. (From Zeigler, A., *Seawater Concentrate for Abundant Agriculture*, 2012.)

TABLE 27.2

County Line Farm 2012 Winter Trials

Insect Damage Reduction % Change from Control		
Crop Planted	**Two Application % Reduction**	**Three Application % Reduction**
Beets	0.0	100.0
Carrots	33.3	33.3
Cauliflower	15.0	45.0
Fennel	—	—
Spinach	—	—
Swiss chard	33.3	33.3
Average for the four species	20.4	52.9

Source: Zeigler, A., *Seawater Concentrate for Abundant Agriculture*, 2012.

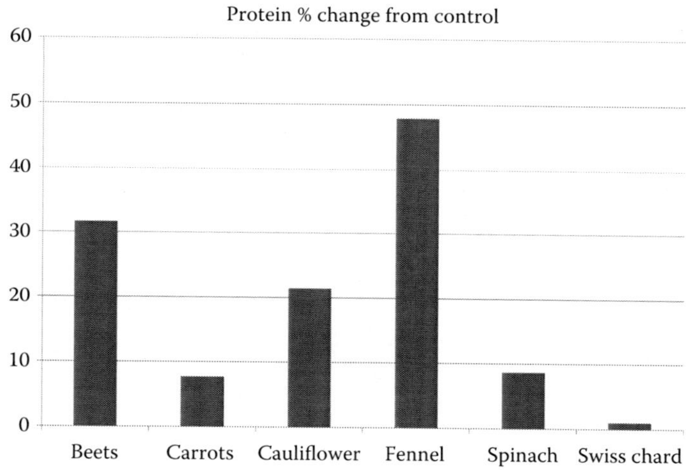

FIGURE 27.4 County Line Farm 2012 Winter Trials. In a world hungry for protein, these are very significant increases. The average protein increase for the six species was 19.73%. (From Zeigler, A., *Seawater Concentrate for Abundant Agriculture*, 2012.)

The results clearly show that in each category—yield, dry matter, Brix, and insect damage reduction—there is a cumulative benefit with additional applications.

Figure 27.4 shows that protein content increases when plants are treated with Sea-Crop.

Figure 27.5 summarizes the County Line Farm testing results.

All of these third-party testing results show that Sea-Crop is able to perform at least as well as Dr. Murray's sea solids did, or better. This illustrates that the bioactive principles contained in seawater can be greatly concentrated for more economical shipment and ease of application.

There is another economic benefit that should be discussed.

In many cases, nitrogen and other inputs may be reduced while maintaining the same or even increased levels of crop production. Depending on the soil and the crop, it may be possible to reduce chemical fertilizer inputs, such as nitrogen, thus providing significant savings and improvement in soil microflora populations.

Over a 2-year period using Sea-Crop, one farmer reduced his nitrogen input by 30% while increasing yields above the state average significantly. A case study of a popcorn farmer in Indiana whose field is managed by John Kempf of Advancing Eco-Agriculture (Table 27.3) illustrates this.

Sea-Crop treatment was started at the beginning of year 2.

During this same time period, yearly potassium input was also reduced by 7.5% and the phosphorus input remained unchanged.

Significantly, in year 3, the crop came in 2 weeks earlier than others in the area that were planted at the same time. This early harvest characteristic is typical with crops where Sea-Crop has been applied.

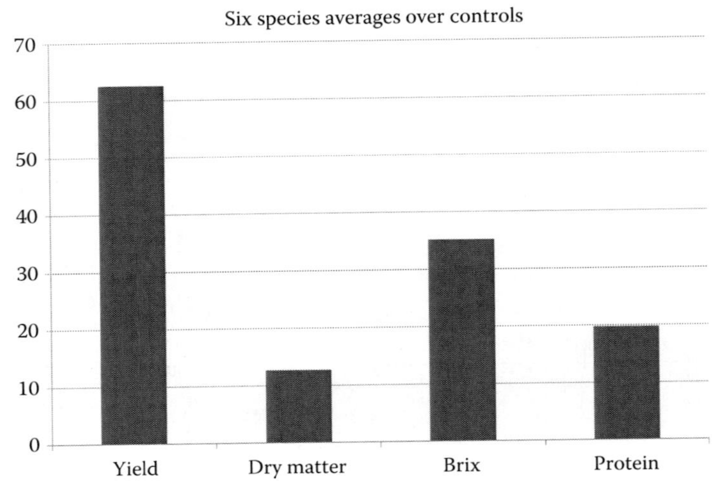

FIGURE 27.5 County Line Farm 2012 Winter Trials. (From Zeigler, A., *Seawater Concentrate for Abundant Agriculture*, 2012.)

TABLE 27.3
Nitrogen Input Reductions

Year	Nitrogen Per Acre (lb)	% Reduction	State Average Yield (lb)	This Farm's Yield (lb)	% Better
1	158	0	1800	2500	39
2	116	26.5	2900	4000	38
3	110.7	30	3000	5000	67

Source: Zeigler, A., *Seawater Concentrate for Abundant Agriculture*, 2012.

Although this farmer was very successful in substantially reducing nitrogen and potassium inputs in a short period of time, Ambrosia Technology LLC recommends not cutting those inputs by more than 20% per year. Farms are each different as to their soils, locations, and weather so the aforementioned results may not be achievable in every case.

In another example, a government agricultural research station in Malawi, Africa, grew corn treated with 4 gal to the acre of Sea-Crop applied as a soil drench 3 weeks after planting and starter fertilizer (92N:21P:0K + 4S). No side dressing was used. The untreated control received both the starter fertilizer and a side dressing of urea so that the control received the full 82 lb of nitrogen per acre that is recommend for corn production in Malawi. The Sea-Crop-treated test received only 20.5 lb of nitrogen but gave a crop at harvest that was 97% of the fully fertilized control.

This amounted to a 75% nitrogen input reduction in a single year with only a slight reduction in yield. Again, it is not recommended for reducing nitrogen by more than 20% in a single year.

County Line Farm 2012 Winter Trials

Photograph 27.1 shows a comparison between carrots planted in the same field on the same day. The carrots on the right received three applications of Sea-Crop, one soil drench at planting followed by two foliar applications (Table 27.4).

Baby Carrot Test Detail:

Seed date—10/19/2011. *Soil drench application date*—10/20/2011, 2 gal/acre Sea-Crop @ 5% strength co-applied with 2 gal/acre molasses @ 5% strength.

Emergence—10/26/2011

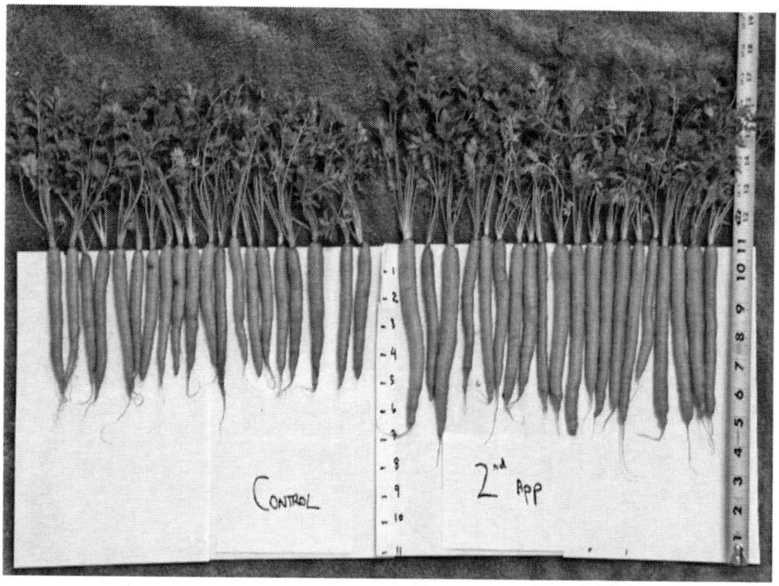

PHOTOGRAPH 27.1 Baby carrots 17 weeks after planting. (From Zeigler, A., *Seawater Concentrate for Abundant Agriculture*, 2012; Photo credit by Julie Brothers.)

TABLE 27.4

Spinach Test% Increase over Control

Category	Two Application Sea-Crop®	Three Application Sea-Crop®
Brix (average of four plots)	+28.6%	+23.8%
Fresh weight (greens portion) 35 representative plants of each category	+43%	+70.1%
Dry matter content increase	+11.8%	+5.9%

Source: Zeigler, A., *Seawater Concentrate for Abundant Agriculture*, 2012.

First foliar app date—11/16/2011, 1 gal/acre Sea-Crop @ 2% strength.

Second foliar app date—12/7/2011, 1 gal/acre Sea-Crop @ 2% strength.

Harvest date—2/13/2012. *Duration of growing cycle*—118 days.

Marketable yield% increase—41.57% (Zeigler 2012). *Brix refractometer reading increase*—30% (Zeigler 2012).

Insect damage% decrease—33% (Zeigler 2012).

Photograph 27.2 compares spinach after the spinach shown on the right received two applications of Sea-Crop. Photograph 27.3 compares the untreated control to treated spinach after it had received three applications of Sea-Crop.

Malawi, Africa 2011

Chitala Research Station is part of Chitedze Agricultural Research Station of the Department of Agricultural Research Services (DARS), which is part of the Ministry of Agriculture and Food Security of the Government of Malawi.

At the Chitala Research Station, as shown in Photograph 27.4, the treated corn on the left has good kernel development all the way to the top of the cob. The kernels are larger, and there are more rows of

PHOTOGRAPH 27.2 Baby spinach harvest results March 2012 @ 57 days with 2 applications. (From Zeigler, A., *Seawater Concentrate for Abundant Agriculture*, 2012; Photo credit by Julie Brothers.)

PHOTOGRAPH 27.3 Baby spinach harvest results March 2012 @ 57 days with 3 applications. (From Zeigler, A., *Seawater Concentrate for Abundant Agriculture*, 2012; Photo credit by Julie Brothers.)

kernels than the normally fertilized control on the right. Sea-Crop enables plants to more nearly achieve their full genetic potential as demonstrated in this example.

Groundnuts (peanuts) are inspected on a farm in Malawi, Africa. In Photograph 27.5, notice the fuller root development and greater number of pods evident on the treated cluster to the left as compared to the untreated cluster on the right. The treated and untreated test plots were side by side in the same field. Each cluster represents one planting station.

In Photograph 27.6, soybeans are inspected on a farm in Malawi, Africa. Notice the fuller root development and greater number of pods evident on the treated cluster to the left as compared to the untreated cluster on the right. The treated and untreated test plots were side by side in the same field. Each cluster represents one planting station.

Morocco

Photograph 27.7 shows how root growth was improved by a Sea-Crop treatment of green beans grown in a commercial greenhouse in Morocco by Dr. Abdelhaq Hanafi.

The left hand holds the root system of an untreated control. The right hand holds a plant that received two treatments of 1% strength Sea-Crop, one soil drench treatment and one foliar application.

From the same green bean study in Morocco at Agadir, Photograph 27.8 shows that there is a dose/benefit relationship when treating with seawater concentrate.

PHOTOGRAPH 27.4 Corn comparison in Malawi, Africa. (From Zeigler, A., *Seawater Concentrate for Abundant Agriculture*, 2012.)

PHOTOGRAPH 27.5 Groundnut comparison in Malawi, Africa. (From Zeigler, A., *Seawater Concentrate for Abundant Agriculture*, 2012.)

The row of plants on the left received two treatments of 1% strength Sea-Crop, one soil drench treatment and one foliar application. The row on the right received only a single soil drench treatment containing the very low concentration of 0.15% Sea-Crop.

Confirming the aforementioned result, a dose/benefit study done on cucumbers by Dr. David Nielson at Davis, California, showed that additional benefit was achieved with each additional increment applied up to 10 gal of Sea-Crop per acre.

Photograph 27.9 shows test results obtained by Cal-Agri in Eugene, Oregon.

PHOTOGRAPH 27.6 Soybean comparison in Malawi, Africa. (From Zeigler, A., *Seawater Concentrate for Abundant Agriculture*, 2012.)

PHOTOGRAPH 27.7 Bean roots compared in Morocco. (From Zeigler, A., *Seawater Concentrate for Abundant Agriculture*, 2012.)

Rosemary (*Salem*) starts were transplanted into 1 gal pots and treated with Sea-Crop at the rate of 3 gal/acre. Cottonseed meal was used on both the untreated control and treated tests as an organic fertilizer. After 30 days, the plants were dug up, the roots washed, and the entire plant desiccated so that dry matter weights could be compared. The percentage gain of the Sea-Crop-treated plants as compared to the controls is listed in the following (Table 27.5).

PHOTOGRAPH 27.8 Green beans compared in Morocco. (From Zeigler, A., *Seawater Concentrate for Abundant Agriculture*, 2012.)

TABLE 27.5
Six Species Averages over Controls

			Dry Matter Increase of Sea-Crop® Treated Plants		
Start Date	**Duration**	**Replications**	**Roots**	**Tops**	**Total**
7-19-07	45 days	6	23.14%	15.29%	17.24%

Source: Zeigler, A., *Seawater Concentrate for Abundant Agriculture*, 2012.

From a personal trial by the agricultural consultant Lawrence Mayhew.

> "The photos (27.10 and 27.11) of the strawberries are an example of what happened. The first photo shows some strawberries approaching the ripening stage."
>
> "The second photo shows the treated strawberries, which does not seem too exciting at first glance unless you realize that not only are the Sea-Crop®-treated berries ripening sooner than the untreated, they are the second crop, while the unripe berries are only yielding their first crop."

These treated strawberries were grown during the Cal-Agri Products field trials at Louisiana State University at Baton Rouge in 2008. In this trial, Sea-Crop was used as a root dip on plug transplants at the dilution rate of 1 part in 50 (2%). That single treatment gave a 25% increase in yield compared to the untreated control (Photograph 27.12).

In addition to the third-party testing already covered, less formal testing was done in other interesting areas of seawater concentrate application.

About 6 years ago, Abe Collins, a Vermont grazier and founder of Collins Grazing, used a Keyline subsoiler plow to aerate and loosen compacted clay soils under pasture in Northern Vermont as part of the Keyline soil formation sequence. The pastures were grazed five times and physically loosened with the subsoiler twice during the growing season, each time sinking the subsoiler shanks a few inches deeper than the apparent bottom of the A-horizon topsoil. Sea-Crop and whole, unpasteurized milk were sprayed on the pastures to stimulate and feed soil organisms and plant life. He sent photographs from trenches on

PHOTOGRAPH 27.9 Rosemary plants compared in Oregon. (From Zeigler, A., *Seawater Concentrate for Abundant Agriculture*, 2012.)

PHOTOGRAPH 27.10 Untreated strawberry plants compared. (From Zeigler, A., *Seawater Concentrate for Abundant Agriculture*, 2012.)

PHOTOGRAPH 27.11 Treated strawberry plants compared. (From Zeigler, A., *Seawater Concentrate for Abundant Agriculture*, 2012.)

PHOTOGRAPH 27.12 Strawberries grown at Cal-Agri Products field trials at LSU. (From Zeigler, A., *Seawater Concentrate for Abundant Agriculture*, 2012.)

the property before and after this treatment, and in the photos, the thickness of the dark topsoil appeared to double in a single growing season. Abe has continued to experiment with different strategies for improving topsoil health and remains impressed by the apparent benefits of Sea-Crop. Monitoring of soil properties is a current focus held by Abe, which he pursues through the establishment of soil carbon baseline and follow-up monitoring plus regenerative agricultural management through the Soil Carbon Challenge and his work with clients. Although the trial described here is anecdotal, it does open a potentially important area of research—carbon accrual in soils by soil microflora stimulation and habitat improvement.

Gary Kline of Black Lake Organics in Olympia, Washington, experimented last year by charging biochar with a 5% strength solution of Sea-Crop and then by growing pepper plants in soil containing the Sea-Crop-charged biochar. These plants were compared to others grown in a variety of special soil blends with different treatments. I did a book signing at his facility last September and had an opportunity to see all of the plants involved in the test. The best looking plants were the ones treated with Sea-Crop and the best of all was the biochar plus Sea-Crop combination. The Sea-Crop-treated plants were the only plants in the test that did not have insect damage. In addition, the plants were larger and healthier looking and bore more peppers. Adding biochar to soil adds important cation exchange capacity. Preloading the biochar with the 89 natural source elements available in seawater concentrate makes perfect sense.

The reader may be wondering at this point if there are any circumstances in which seawater concentrate would fail to give an increased yield, and I must, in all honesty, report that there are.

Soybeans often have 25% or more additional pods when treated with Sea-Crop, and this is very beneficial if there is sufficient rainfall later in the growing season so that the pods fill out properly. However, under dry land farming conditions, if rainfall is lacking, the extra pods may turn out to be a liability rather than a blessing, and there have been instances of slightly reduced harvests under these conditions.

In this chapter, we have seen that seawater concentrate is economically important to the farmer as yield benefits alone, apart from any other benefits that it may have such as increased nutrient density or improved soils.

Nearly a century of experimentation with seawater has shown that it is a wonderful resource that can and must be utilized for more than just making sea salt. Modern methods of making seawater concentrate and years of agricultural trials learning how best to use those concentrates in agriculture have now made possible an inexpensive and effective means for farmers to enrich and protect their soil from depletion and exhaustion.

The future of successful agriculture must, of necessity, involve remineralization of agricultural soils, and the least scarce resource to supply those minerals is the seas of the Earth. Seawater concentrates have the potential to offer relief and a new way of farming to those farmers who want to get off of the industrial/chemical agriculture treadmill.

Farming with seawater concentrate is the modern way to supply complete mineral nutrition to both crop plants and soil biota. Seawater concentrate is a key to a future of agricultural abundance.

The book *Seawater Concentrate for Abundant Agriculture* is the first new book in over three and a half decades written by a primary researcher in the field of seawater-induced fertility. The book is available from the Sea-Crop website http://www.sea-crop.com/book.html.

REFERENCES

Murray, M. *Sea Energy Agriculture*, 1976. Published by Acres, TX, pp. 48–49, 72–73.

Reemtsma T., These, A., Springer, A., and Linscheid, M. Fulvic acids as transition state of organic matter: Indications from high resolution mass spectrometry, August 23, 2006. *Environmental Science and Technology* 40:5839–5845.

Swenson, H. *Why is the Ocean Salty*? 1983. Published by the US Geological Survey Publication, Washington, DC.

Umali, D.L. *Irrigation-Induced Salinity*, 1993. Published by the Food and Agriculture Organization of the United Nations, Roma, Italy.

Zeigler, A. *Seawater Concentrate for Abundant Agriculture*, 2012. Published by Ambrosia Technology LLC.

28

Superior Food Production Using Sea Salt and Plant Extracts

Ioan Hossu

CONTENTS

Increasing Food Production, Declining Nutritional Quality

Increased production of foods for a constantly growing population represents one of the most stringent global crises. As world population constantly grows, the area of fertile land decreases (roads, urban development, industrial sites, etc.), placing increasing stress on agriculture to increase production, and increased use of synthetic fertilizers. However, public awareness and accountability regarding how food is grown has greatly increased. As the world population is better educated, increased awareness regarding the negative effects of chemical fertilizers, herbicides, pesticide, demineralization of soil, loss of biodiversity, mutations, superweeds, and negative impacts on the environment and on the quality and taste of foods lead to a higher demand for natural, organic, biodynamic, and *eco-foods*.

TABLE 28.1

Average Yields of Major Crops in the United States, 1950–1992

Year	Corn (bu/acre)	Wheat (bu/acre)	Soybean (bu/acre)	Alfalfa (ton/acre)
1950	37.6	14.3	21.7	2.1
1964	62.1	26.2	22.8	2.4
1972	96.9	32.7	28.0	2.9
1982	113.2	35.5	31.5	3.4
1992	128.7	41.4	37.0	3.6

Source: Tisdale, S.L. et al., *Soil Fertility and Fertilizers: An Introduction to Nutrient Management*, Prentice Hall, Upper Saddle River, NJ, 1999.

TABLE 28.2

Mineral Content Depletion of Various Food Crops

	Ca		Mg		K		Percent Water	
Produce	**1963**	**1999**	**1963**	**1999**	**1963**	**1999**	**1963**	**1999**
Apples, raw, with skin	7	7	8	5	110	115	84.8	83.9
Beans, snap, green, raw	56	37	32	25	243	209	90.1	90.2
Broccoli, raw	103	48	24	25	382	325	89.1	90.6
Carrots, raw	37	27	23	15	341	323	88.2	87.7
Lettuce, iceberg, raw	20	19	11	9	175	158	95.5	95.8
Oranges, raw	41	40	11	10	200	181	86.0	86.7
Peaches, raw	9	5	10	7	202	197	89.1	87.6
Peas, green, raw	26	25	35	33	316	244	78.0	78.8
Strawberries, raw	21	14	12	10	164	166	89.9	91.5
Tomatoes, red	13	5	14	11	244	222	93.5	93.7

Source: Lyne, J.W. and Barak, P., *Are Depleted Soils Causing a Reduction in the Mineral Content of Food Crops?* University of Wisconsin, Madison, WI, 2000.

While crop yields continue to rise with the development of industrial agriculture (Tisdale et al., 1999, Table 28.1), several studies of historical food composition tables show an apparent decline in food nutrient content over the past 70 years (Lyne and Barak, 2000, Table 28.2).

Nutrition Under Siege (Jack, 1998) examined data published by the USDA—Agricultural Research Service Nutrient Data Laboratory and concluded that the data "show a sharp decline in minerals, vitamins and other nutrients in many foods since the last comprehensive survey published over twenty years ago," which was attributed to "a steady deterioration in soil, air, and water quality." A similar study (Mayer, 1997) compared British data over a 50-year period and noted *significant reductions* in levels of minerals in fruit and vegetables and questioned if modern agriculture could be responsible for the reduction. These declines "may well be a result of the 'mining' of our nation's soil fertility by intensive chemically based agriculture" (Lyne and Barak, 2006).

"All of these minerals are depleted by synthetic fertilizers, pesticides, herbicides, acid rain, and agricultural methods that have ceased to be in dialogue with the earth but only consider quantity, efficiency, and profit. The main effect of ignoring the earth is the depletion of those minerals which animals and people need in order to make catalase—the single most important enzyme needed to prevent incorrect cell duplication, or cancer" (Ferrie, 2006).

"Nutritional research has revealed the critical interrelationship between minerals and optimum human health. There are seventeen known essential minerals and many other trace minerals that are all needed for good health. Minerals interact with each other in many of the body's critical metabolic functions. Minerals may be more vital to physical and mental health than vitamins. The full spectrum of essential minerals is critically important to the maintenance of human health. Because the human body cannot manufacture minerals, deficiencies are common" (Marler and Wallin, 2006).

In the last two decades, many attempts have been made to improve soil quality and the nutritional value of both human and animal foods. In our days, soil remineralization has a high priority for small and medium agricultural communities. In order to address these food quality problems, farmers have to move away from using chemical fertilizers, and consumers have to demand more nutritious naturally grown foods. *Green fertilizers* have gained in popularity.

Formulation of New Fertilizers: IHO-Mineral and IHO-Bio

We started investigating the possibility of producing all-natural fertilizers to address the aforementioned issues. As this project was initiated in Panama, we decided to work with the locally available resources. The main objectives of the project were as follows:

- To use locally available natural resources (sands, soil, salts, plants, and sea water).
- To develop a unique, all-natural liquid fertilizer product. (The product must be less expensive than their chemical counterpart or other organic fertilizers and must be affordable to the farmers.)
- To contribute to the welfare of humankind, by teaching farmers alternative methods of agriculture.
- To conduct research focusing on a wide variety of crops.
- To work with local farmers and involving them in the process from the start of the project.
- To develop products that could be used on farm animals.
- To engage in research on tropical crops—such as Guanabana trees, Borojo trees, and others.

Once the scope of the project was well defined, we started investigating different locally available materials, as well as different methods of extracting the necessary nutrients, to be concentrated into a fertilizer. Part of the work was done in Canada, where different extraction methods were explored. Many product variations were created, and many tests were conducted. Many people participated in testing the different products: family, friends, and others, in different parts of the world: Canada, the United States, Romania, Curacao, the United Kingdom, Sweden, Costa Rica, and other countries.

We tested water from both the Atlantic and the Pacific Oceans as well as from the Black Sea. We conducted tests on different salts from Canadian and Romanian mines and other parts of the world. We even tested salts used in the winter to melt snow, salts which farmers give to their livestock, as well as salts we could find in grocery stores in Canada, Romania, and Panama. We tested different soils, sands, volcanic soils, and sands. We tried to stay away from substances, which could contaminate the environment.

Most of the tests were not successful. Some of the products could not be used as we would have had to use chemicals, which could have contaminated the environment during the manufacturing process. Others were discarded because of the high amount of energy needed to manufacture them, and others could not have been used, marketed, and offered to other people for trials, as they had a very bad odor.

After about 5 years of testing, two different products were selected, IHO-Mineral and IHO-Bio. Both of these products are now registered with the Ministry of Agriculture in Panama, while IHO-Bio is registered in Romania, under the name of IHO-Agro. They are manufactured with well water. Both products contain either sea salt or ocean water extract, other salts, and some organic matter. In addition to the aforementioned ingredients, IHO-Bio has in its composition a well-balanced formula of plant extracts, rich in vitamins, amino acids, and natural plant hormones, as well as some vitamins beneficial to plant health and growth. While IHO-Mineral has a high pH between 9.5 and 10, IHO-Bio has a pH of only 3.8–4.25.

Samples of these products were sent to an accredited testing laboratory in the United States, and proton-induced x-ray emission (PIXE) lab analyses were requested. PIXE is a technique which can be used for nondestructive, simultaneous elemental analysis—sodium through uranium, 72 elements—in solid, liquid, thin film, and aerosol filter samples. Note that the lighter elements such as carbon and nitrogen are

TABLE 28.3
IHO-Agro Mineral Test–PIXE Lab Analysis

Element Name	Energy (keV)	Det. Limit 95% Conf. (ppm)	Concentration Mass
Magnesium	1.25	576.400	0.46%
Sulfur	2.31	31.050	0.40%
Chlorine	2.62	30.970	0.84%
Potassium	3.31	33.920	0.17%
Calcium	3.69	52.600	424.451 ppm
Zinc	8.64	1.523	4.206 ppm
Bromine	11.92	5.009	576.400 ppm

TABLE 28.4
IHO-Agro Bio Test–PIXE Lab Analysis

Element Name	Energy (keV)	Det. Limit 95% Conf. (ppm)	Concentration Mass
Magnesium	1.25	207.000	0.29%
Chlorine	2.62	19.130	1.69%
Potassium	3.31	4.468	113.964 ppm
Calcium	3.69	9.290	170.696 ppm
Iron	6.41	1.193	29.356 ppm
Copper	8.05	0.652	1.947 ppm
Bromine	11.92	1.592	28.146 ppm
Strontium	14.17	4.480	18.600 ppm

not detected with this method. Limits of detection for most elements are in the low ppm range for high organic matrices. Like other spectroscopic techniques used for elemental analysis, PIXE is based upon the physics of the atom, not its chemistry. It involves excitation of the atoms in the sample to produce characteristic x-rays whose intensities may be identified and quantified.

IHO-Mineral PIXE test (Table 28.3) detected seven elements. All these elements are present in very small quantities.

In addition to the aforementioned elements, more than 60 other elements are present; their concentration is measured in parts per billion (ppb). This very small concentration (ultratrace elements) is quite significant in that they provide the plants with trace nutrients which in many cases are deficient in the soils.

IHO-Bio was analyzed in the laboratory in the United States by PIXE (Table 28.4) and in two different Romanian government laboratories. The tests in the Romanian laboratories detected different macronutrients (N, P, and K), in addition to many other complex organic compounds. Just as IHO-Mineral, IHO-Bio contains more than 60 other elements, which can be measured in ppb—ultratrace elements.

However, they are present in different proportions. IHO-Bio is a complex of mineral and organic substance concentrate, where some of the elements can only be measured in ppb.

Product Analysis

During these tests, we observed a number of parameters, such as germination rate and growth, foliage, flowering, yields, leaf size, fruit size, plant size, overall plant health, and many other desirable qualities. In order to determine the best methods of fertilizing, we applied the product directly to the soil, or as a fine mist, for foliar feeding. "Foliar feeding is a technique of feeding plants by applying liquid fertilizer directly to their leaves. It has been known for many years that plants are able to absorb essential elements through their leaves. The absorption takes place through the stomata of the leaves and also through the epidermis. Movement of elements is usually faster through the stomata, but the total absorption may be as great through the epidermis. Plants are also able to absorb nutrients through their bark" (Wikipedia.org- http://en.wikipedia.org/wiki/Foliar_feeding).

"Foliar fertilization can result in rapid nutrient absorption and utilization to correct deficiencies or to merely prevent nutrient shortages during critical periods of growth. However, unlike roots, plant leaves are not adapted to assimilate large amounts of nutrients and meet the bulk of the nutrient requirement" (International Plant Nutrition Institute, 2008).

Most of our tests were done using foliar feeding. In some cases, we chose to soak the seeds prior to sowing, and as the plants emerged, foliar feeding was employed.

Corn, Penonome, Panama: One of the first tests was done in Panama, on a small farm, in Penonome (Figure 28.1). We selected two parcels at a distance of 10 m one from the other. The soil on this particular farm is very poor—about 20 cm of top soil, of very poor quality, and below, a gray, dense soil, which cannot be penetrated by the rain waters, causing surface runoff and erosion. The temperatures in this time of the year were between 25°C and 35°C. On both parcels, the same number of seeds was sown.

The corn was supposed to be harvested in 90–120 days. The control parcel was watered with the same amount of water as the test parcel. The test parcel was watered with a product diluted 1:200. After 4 days, most of the corn on the test parcel already had an average height of about 20 cm, while the control group only had a few seeds that germinated, and the plants were no taller than 5 cm. In the following 2 days, very heavy rain fell. All the control plants drowned, while all the test plants survived. The corn from the test parcel was harvested in 78 days.

Corn, Buena Vista, Panama: The following test was done on a farm in Buena Vista Colon, Panama. The test corn (Figure 28.2) seeds were soaked for 15 min in a solution of IHO-Bio in a dilution of 1:50. They were then sown directly in the ground. Control group (Figure 28.3) was planted directly in the soil, with no special treatment. No other fertilizers were used during the test.

Test corn produced two and three ears, while the control ones only one, which was much smaller than the other ones. There was an increase of 287% in the weight of the dried corn kernels.

We did not measure the total dried plant mass. However, we could observe a substantial increase. This is beneficial in particular to farmers using the corn plants as animal feed. The farm had a very rich black soil, which was never treated with any kind of fertilizers. Test plants were sown in an area that was not

FIGURE 28.1 Corn, Penonome, Panama.

FIGURE 28.2 Test Corn, Buena Vista, Panama.

FIGURE 28.3 Control Corn, Buena Vista, Panama.

properly weeded, but nevertheless, the corn plants were able to grow through the heavy grasses. Test plants were twice as tall as the ones in the control group.

Grape Vine Plants: A small-scale test on 1-year old grape vine was done in Canada. Significant differences in the foliage were observed between the test and the control group. Figure 28.4a shows a leaf sample of the control group, while Figure 28.4b shows a sample leaf of the test group.

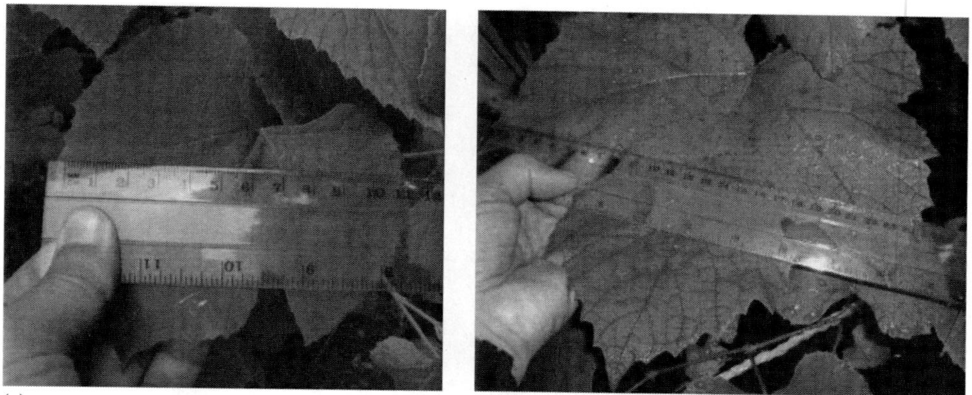

(a) (b)

FIGURE 28.4 (a) Grape leaf control. (b) Grape leaf test.

FIGURE 28.5 Totuma fruits. Control at left, treated at right.

Calabash Tree: A simple trial was conducted on a calabash tree in Panama. This tree yields the *totuma* fruits (Figure 28.5), long used by indigenous people. These fruits have a hard shell, which is used for soup bowls, to drink water, or for artisanal work. Fruits from the same tree were collected 1 year apart. IHO-Mineral 1:200 dilution was applied on the soil, three applications, 3 weeks apart. This test resulted in a 373% fruit volume increase. The ground around the tree was cleared. Early in the morning, IHO-Mineral was mixed with water and poured at the base of the tree.

Sugar Apple Tree (in Panama known as Anon): IHO-Bio, 1:250 dilution was applied to two different trees, of the same age, growing on the same property. The IHO-Bio product was applied three times, 2 weeks apart. The weight increase between the two was 323% (Figure 28.6).

Citrus Trees: Both IHO-Mineral and IHO-Bio were tested on citrus trees such as orange, mandarin, lemon, and lime. Tests were conducted in Panama, Costa Rica, Morocco, and in Canada (in a hot house). The products were diluted 1:200 to 1:230 in all trials. The results reported to us are as follows:

- Young trees came to maturity early and started to bear fruit before they were expected to.
- Trees that had never produced fruit began to bear fruit.
- Fruit production increased dramatically, in some cases, up to 600%.

FIGURE 28.6 Anon fruits. Control at left, treated at right.

Banana Plants: Tests were conducted in Panama City, in a small garden, where they used IHO-Bio, in a dilution of 1:250. The product was applied three times at an interval of 12 days. People conducting the tests reported an increase of more than 35% in banana production.

Noni Trees: IHO-Mineral diluted 1:200 was applied to noni trees on a commercial farm in David, Panama. Within the first month, the noni fruit production increased by about 30%, while the branches grew by a full 12 in.!

Papaya Tests: A test was conducted on a small farm in Buena Vista, Colon, Panama. Papaya seeds were planted in small bags, which were treated with IHO-Bio solution, 1:250. For this test, different varieties of papaya were selected. Both the control and the test group had more than 200 plants each,

FIGURE 28.7 Papaya plant control.

equally divided between the different varieties. The tests started on January 23, 2013. Thirty-six days after seeding, the average plants in the test group (Figure 28.7) were more than 40% taller than the ones in the control group (Figure 28.8).

In addition, the leaves were much larger. For practical reasons, the farmer decided to treat all plants with IHO-Bio solution. The experiment will continue with observations of the size of the papaya fruits once the trees start producing.

Germination Tests, Green Peppers: These tests were done on a small scale in Canada, where 20 seeds were selected for both the control and test group. The soil and environment conditions were the same for both groups (Figure 28.9). In this particular instance, the product used was IHO-Mineral, diluted 1:200. The results were replicated several times by us as well as by other friends from Panama and Costa Rica.

During the next test, we used IHO-Bio, which was diluted 1:250 and examined germination of corn, red pepper, cucumber, squash, pumpkin, zucchini, cucumber, squash, pumpkin, and zucchini. A camera

FIGURE 28.8 Papaya plant test.

FIGURE 28.9 Green pepper germination test. After 8 days, control group at left, test group at right.

took pictures every 1 h. The test group germinated faster than the control group, and after emergence, it was clear the test group was stronger than the control.

Lettuce Test: Canada: We tested different lettuce varieties, using IHO-Bio, diluted 1:200. Most of the seeds germinated in less than 48 h and some in less than 24 h, while the seeds in the control group germinated in 3–5 days. This particular test was repeated several times.

Commercial Trials

Sugar Cane: Grupo CALESA, Panama—http://www.grupocalesa.com/:

In 2011, IHO-Mineral samples were submitted for testing on sugar cane plantations. The tests were run throughout 2011 and 2012. In April 2012, the test results were made available. A summary report of the findings follows:

The aim of this report is to assess the influence on sugarcane production (T cane/ha, applying the IHO-AGRO + + product).

Testing Procedure

The study was set up in plot 121121 (agricultural Class IV [moderate]), whose historical performance is greater than or equal to 70.73 t/ha and <85.5 t/ha in accordance with those reported by Contegazza and Menendes, 2010. The product was applied on sugarcane strain BT7742 first generation. The experimental design was a randomized block with six-treatments (Table 28.5) and replicated five times. Each plot area was $96 m^2$ (6 rows of 10 m long with 1.60 in between rows).

Product application was performed with 16 L Matabi spray pumps, the dose tested was 4 L of IHO-AGRO/ha. The ground application was made in strips near the plant; the following applications were made 15 days apart. The first foliar application was done 50 days after the cut, and the following application was done the same way as on the ground.

Variables evaluated were as follows: stem length in centimeter, stem diameter (in 10 stems per plot for each replication), and the number of stems per meter (were counted all stems from the four center rows of plots). A sample of 1 m from each treatment in the five replications was taken to estimate the weight of the average stem. Finally, cane production in t/ha was estimated, according to the following formula:

$$\text{t cane/ha} = (\text{Average stem weight} \times \text{No of stems/m linear} \times 6250 \text{ (total of 1 ha in linear meters)} \times \% \text{ of seedling population})/2000$$

The harvest took place at 12 months of age in the month of April in 2012, and the test results (t cane/ha) were subjected to analysis of variance of fixed effects. The percentage increase over the control was also determined, as data that have high dispersion in sample plots. In commercial fields where not all factors are controllable, they are in the experimental plots. Therefore, inefficiencies can occur, and the data were only considered for statistical analysis.

TABLE 28.5

Sugarcane IHO-Agro Mineral Application

	Treatments				IHO-Mineral per 96 m²	
Nr. Crt.	IHO-Mineral	Fertilizer[a]	Application	# Applications	Each Application (mL)	Total (mL)
1	0 (test)	—	—	—	0	—
2	0	Standard	—	—	0	—
3	4 L/ha	Standard	Foliar	2	38.4	76.8
4	4 L/ha	Standard	Soil	2	38.4	76.8
5	4 L/ha	Standard	Foliar	3	38.4	115.2
6	4 L/ha	Standard	Soil	3	38.4	115.2

[a] Standard: Recommended fertilizer for the field.

Test Results

The results of analysis of variance showed no significant differences between treatments. It is to be noted that treatments 4, 5, and 6 resulted in an increase over the control of 30.26, 15.7, and 23.9 corresponding to increased applications of IHO-Mineral.

Conclusions

Treatments 4, 5, and 6 result in increases of 30.26%, 15.7%, and 23.9% over the control group, corresponding to applications IHO-Mineral.

Recommendations

1. Evaluate the IHO-Mineral in demonstration plots on different soil types, varieties and strains, as well as different planting materials (cuttings and plantlets).

Sources

Peppers: Cerro Punta, Panama

During the following commercial trial on pepper plants in Panama, IHO-Mineral was mixed with the fertilizers already used on the farm. The control group received the regular fertilizers, while the test group received the regular fertilizers, and in addition IHO-Mineral was applied in a dilution of 1:250. Thirty days into the trials, the test group (Figure 28.10) showed a significant difference in foliage versus the control group (Figure 28.11).

Leaf measurements were taken, and significant differences were observed between the leaves of the control and test groups (Figures 28.12 and 28.13). Some plants produced up to three times more peppers (by number) than the control group (Figure 28.14). The control plant produced three peppers, while test group plants produced up to seven. The fruit wall of the test group (Figure 28.15) was thicker, and the shelf life was longer. In addition, test group plants started maturing earlier, peppers being harvested 1 week earlier than the control peppers.

FIGURE 28.10 Pepper control 30 days into trial.

FIGURE 28.11 Pepper test 30 days into trial.

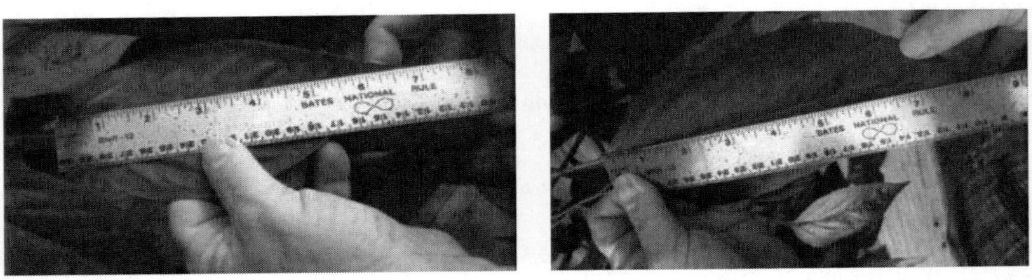

(a) (b)

FIGURE 28.12 Leaf length comparison: (a) Pepper leaf control measurement. (b) Pepper leaf measurement with IHO-AGRO.

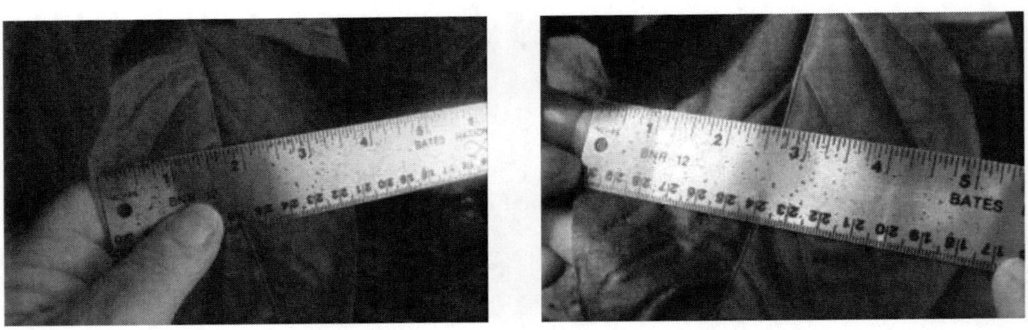

(a) (b)

FIGURE 28.13 Leaf width comparison: (a) Pepper leaf control measurement. (b) Pepper leaf measurement with IHO-AGRO.

Coffee: Boquete, Panama

On a Geisha commercial coffee farm trial in Panama, production with IHO-Mineral more than doubled. In addition, the coffee harvesting in the test group started a week earlier than the control group, and the harvesting period was extended. The control group plants were smaller than their counterpart in the test group. During this trial, no additional fertilizers were applied to the test group. This led us to speculate

FIGURE 28.14 Pepper production on new plants from control group.

FIGURE 28.15 Pepper production from new plants after 30 days after applying IHO-AGRO.

that the coffee plantation had an excess of fertilizers left over from previous years; once IHO-Mineral was applied, the plants could absorb the nutrients needed, left over from the previous fertilization.

Lettuce: Cerro Punta, Panama

Even in the hot Panamanian climate, the lettuce treated with IHO-Mineral was crispier, and the shelf life was extended by 3 days compared to the control group. Production was increased by 43% (measured as the weight of fresh leaves).

TABLE 28.6
Test Results IHO-Mineral on Tomatoes

Harvest #	Control (lbs)	Test (lbs)	Increase (lbs)	Increase (%)
1	92	103	11	111.96
2	393	888	495	225.95
3	1,440	1,936	496	134.44
4	2,611	3,272	661	125.32
5	1,099	1,674	575	152.32
6	1,404	1,966	562	140.03
7	1,169	1,894	725	162.02
8	1,598	2,524	926	157.95
9	1,795	2,712	917	151.09
10	1,006	1,880	874	186.88
Total:	12,607	18,849	6,242	149.51

Tomato: Cerro Punta, Panama

In another commercial trial in Panama, the IHO-Mineral was mixed in with chemical fertilizers. Five thousand tomato plants were selected for controls and 5000 plants for the test group (Table 28.5). The trial was done in Chiriqui in a modern greenhouse. The two groups were observed over a period of several weeks. Special attention was paid at harvest time. Tomatoes from both groups were harvested over a period of 5 weeks, twice per week. The test showed that the plants treated with IHO-Mineral constantly produced more fruits than the control group.

An increase of 149.51% was observed. It was also observed that fruits were in a more *marketable* condition. They were larger and had a firm texture (Table 28.6).

Strawberries: Cerro Punta, Panama

Commercial trials were held in the area of Cerro Punta in Panama. Two separate greenhouses were selected, where the IHO-Mineral was applied, together with other base fertilizers. The production increase was 25.7%, and the fruit BRIX was significantly increased. However, strawberry tests need to be repeated, as the farm did not report the full harvest, nor did they report the full benefits the product had. Pictures taken suggest a much higher increase in production. A typical control group plant can be observed in Figure 28.16, while the fruit size of the control group can be observed in Figure 28.18.

Plants treated with IHO-Mineral can be observed in Figures 28.17 and 28.18. Here we can see an increase in both the number of fruits per plant and the fruit size (Figure 28.19).

FIGURE 28.16 Typical strawberry plant, control group.

FIGURE 28.17 Plants treated with IHO-mineral.

FIGURE 28.18 Strawberries control group.

FIGURE 28.19 Strawberries treated with IHO-mineral.

Rice: Chiriqui, Panama

Rice trials were done on different commercial farms in Panama. IHO-Mineral was used in conjunction with other chemical fertilizers. The product was sprayed using either tractors or from helicopters.

IHO-Mineral was applied three times, first time at sowing time (Figure 28.20), then two more times (Figure 28.21) at 2 weeks interval. The increase in productivity was between 19% and 25%. In one of the trials, a 5 ha parcel was delimited. During the wet season, the control rice was affected by a fungus. In case of rice, different fungi may enter the plant through its walls. This particular test led us to speculate that the 5 ha treated with IHO-Mineral had thicker walls and, therefore, was better protected. However, we cannot draw a definite conclusion without repeating these tests.

Roses: Volcan, Panama

A commercial trial on a rose plantation in Panama yielded more bulbs, livelier colors, larger plants, and larger flowers. In a similar trial in Romania, similar results were reported.

Japanese Roses: Bucharest, Romania

The reports were similar to other roses; however, the comments were: "these are not roses anymore. These are sunflowers!"

Grass/Pasture: Canada, and Panama

Initially, small tests were done on grass that suffered from severe drought conditions. IHO-Bio, in a dilution of 1:250, was used during these tests. Two parcels were delimited. On both of them, the grass was yellow, burned by the strong sun and lack of water. The same amount of water was used for both parcels. The one that was watered with the IHO-Bio solution recuperated in 4 days, twice as fast as the other.

FIGURE 28.20 Spraying field at sowing time.

FIGURE 28.21 Spraying rice field 2 weeks after sowing.

Larger tests were done where farmers sprayed the products on pastures, and as a result, significant benefits were observed—pastures recovered faster after animal grazing.

As the product is diluted in water (1:200–1:250), it is clear that extremely small quantities of micronutrients can produce a significant increase in crop production. Foliar application seems to be more beneficial to plants than direct soil application. We agree with other researchers that nutrients seem to be absorbed more efficiently if applied in a fine mist (foliar application) rather than sprayed on the ground, where most of the nutrients are lost due to ground infiltration or being washed away by rain.

Test Results: National Institute for Research and Development for Soil Science, Agro-Chemistry and Environment Protection, Bucharest, Romania

In order to commercialize a fertilizer in Romania and the EU, one has to conduct a 2 year, back to back study on the proposed product. This was done by entering into a research contract with *The National Institute for Research and Development for Soil Science, Agro-Chemistry and Environment Protection*. In 2011, IHO-Bio was submitted for the first year research contract. The first research work was done on tomatoes, cucumbers, carrots, and spinach. We recommend between 9 and 15 L/ha to be used during a production cycle. During the tests however, the institute used 6 L/ha, during three foliar applications.

Testing methodology consisted in three foliar applications (spraying—fine mist) of IHO-Bio diluted in water, 1:250. Each time 500 L of diluted product was used. In each of the experiments, the results were compared to a control group. All tests were done on black soil, where the agricultural practices were optimal. No base fertilizer was used. The results of the research are presented in Tables 28.6 through 28.9:

1. Tomatoes grown in black soil using three treatments of IHO-Agro, for a total of 6 L/ha, yielded an increase of 11,163 kg fruits/ha (35.4%), corresponding to 1,863.5 kg fruit/L of fertilizer (Table 28.6).
2. Cucumbers grown in black soil using three treatments of IHO-Agro, for a total of 6 L/ha, yielded an increase of 8560 kg fruits/ha (37.2%), corresponding to 1411 kg fruits/L of fertilizer (Table 28.7).
3. Carrots grown in black soil using three treatments of IHO-Agro, for a total of 6 L/ha, yielded an increase of 7506 kg roots/ha (37%), corresponding to 1251 kg roots/L of fertilizer (Table 28.8).
4. Spinach grown in black soil using three treatments of IHO-Agro, for a total of 6 L/ha, yielded an increase of 4312 kg fresh leaves/ha (38.3%), corresponding to 718.7 kg fresh leaves/L of fertilizer (Table 28.9).

As a result of testing, the product was recommended for temporary registration. The 2011 research results are summarized in Table 28.10.

In 2012, corn, tomato, spinach, lettuce, apple trees, and grape vines were tested. The year 2012 was a terrible year for agriculture worldwide. Romania was not spared. Because of the severe drought conditions in some part of the country, some 50% of the crops were lost. The soils the crops were growing on

TABLE 28.7

Tomatoes Test Results, Romania

				Quantity Fertilizers Used (L/ha)			Increase		
Nr.	**Treatment**	**Nr. Applications**	**Concentration Solution (%)**	**One Treatment**	**All Treatments**	**Yield (kg/ha)**	**kg/ha**	**%**	**kg/L Fertilizer**
1	No fertilizer	—	—	—	—	31,524	—	100	—
2	IHO-bio	3	0.4	2	6	42,687	11,163	135.4	1860.5

TABLE 28.8

Cucumbers Test Results, Romania

Nr.	Treatment	Nr. Applications	Concentration Solution (%)	Quantity Fertilizers Used (L/ha)		Yield (kg/ha)	Increase		
				One Treatment	All Treatments		kg/ha	%	kg/L Fertilizer
1	No fertilizer	—	—	—	—	22,728	—	100	—
2	IHO-bio	3	0.4	2	6	31,194	8466	137.2	1411

TABLE 28.9

Carrots Test Results, Romania

Nr.	Treatment	Nr. Applications	Concentration Solution (%)	Quantity Fertilizers Used (L/ha)		Yield (kg/ha)	Increase		
				One Treatment	All Treatments		kg/ha	%	kg/L Fertilizer
1	No fertilizer	—	—	—	—	20,281	—	100	—
2	IHO-bio	3	0.4	2	6	27,787	7506	137	1251

TABLE 28.10

Spinach Test Results, Romania

Nr.	Treatment	Nr. Applications	Concentration Solution (%)	Quantity Fertilizers Used (L/ha)		Yield (kg/ha)	Increase		
				One Treatment	All Treatments		kg/ha	%	kg/L Fertilizer
1	No fertilizer	—	—	—	—	11,241	—	100	—
2	IHO-bio	3	0.4	2	6	15,553	4312	138.3	718.7

in 2012 were less fertile than in 2011. For corn, a total of 2 L of product/ha (in two applications—1 L each) was used; for tomato, lettuce, and spinach, 3 L/ha (in three applications—1 L each); and for grape and apple trees, 6 L/ha (in 3 applications, 2 L each) was used. Despite all these, impressive results were obtained, and the final report recommended the final product registration. The results of 2012 research are presented in Tables 28.11 through 28.17:

1. Apple trees grown in black soil using three foliar treatments of IHO-Agro, for a total of 6 L/ha, yielded an increase of 2866 kg fruits/ha (20.1%), corresponding to 477.6 kg fruit/L of fertilizer (Table 28.11).
2. Corn grown in black soil using two foliar treatments of IHO-Agro, for a total of 2 L/ha, yielded an increase of 350 kg dried kernels/ha (10.8%), corresponding to 175 kg dried kernels/L of fertilizer (Table 28.12).
3. Grape wines grown in loamy chernozem soil using three foliar treatments of IHO-Agro, for a total of 6 L/ha, yielded an increase of 670 kg grapes/ha (12.1%), corresponding to 111.6 kg fruit/L of fertilizer (Table 28.13).
4. Lettuce grown in black soil in a greenhouse, using three foliar treatments of IHO-Agro, for a total of 3 L/ha, yielded an increase of 2554 kg fresh leaves/ha (25.2%), corresponding to 851.3 kg leaves/L of fertilizer (Table 28.14).

TABLE 28.11

Summary Test Results Romania, 2011

Crop	Nr. Applications	Fertilizer Concentration (%)	Fertilizer Used (L/ha) Each Application	Fertilizer Used (L/ha) Total	Yield (kg/ha) Control	Yield (kg/ha) IHO-Agro	Yield Increase kg/ha	Yield Increase %	Yield Increase kg/L Fertilizer
Tomatoes (Adi)	3	0.4	2	6	31,524	42,687	11,163	135.4	1860.5
Cucumbers (Comichon)	3	0.4	2	6	22,728	31,194	8,466	137.2	1411
Carrots (Nantes)	3	0.4	2	6	20,281	27,787	7,506	137	1251
Spinach (Matador)	3	0.4	2	6	11,241	15,553	4,312	138.3	718.7

Source: National Institute for Research and Development for Soil Science, Agro-Chemistry and Environment Protection.

TABLE 28.12

Apple Test Results, Romania

Nr.	Treatment	Nr. Applications	Concentration Solution (%)	Quantity Fertilizers Used (L/ha) One Treatment	Quantity Fertilizers Used (L/ha) All Treatments	Yield (kg/ha)	Increase kg/ha	Increase %	Increase kg/L Fertilizer
1	No fertilizer	—	—	—	—	14,235	—	100	—
2	IHO-bio	3	0.2	2	6	17,101	2866	120.1	477.6

TABLE 28.13

Corn Test Results, Romania

Nr.	Treatment	Nr. Applications	Concentration Solution (%)	Quantity Fertilizers Used (L/ha) One Treatment	Quantity Fertilizers Used (L/ha) All Treatments	Yield (kg/ha)	Increase kg/ha	Increase %	Increase kg/L Fertilizer
1	No fertilizer	—	—	—	—	3260	—	100	—
2	IHO-bio	2	0.2	1	2	3610	350	110.8	175

TABLE 28.14

Grape Test Results, Romania

Nr.	Treatment	Nr. Applications	Concentration Solution (%)	Quantity Fertilizers Used (L/ha) One Treatment	Quantity Fertilizers Used (L/ha) All Treatments	Yield (kg/ha)	Increase kg/ha	Increase %	Increase kg/L Fertilizer
1	No fertilizer	—	—	—	—	5580	—	100	—
2	IHO-bio	3	0.2	2	6	6250	670	112.1	111.6

TABLE 28.15

Lettuce Test Results, Romania

				Quantity Fertilizers Used (L/ha)			Increase		
Nr.	**Treatment**	**Nr. Applications**	**Concentration Solution (%)**	**One Treatment**	**All Treatments**	**Yield (kg/ha)**	**kg/ha**	**%**	**kg/L Fertilizer**
1	No fertilizer	—	—	—	—	10,156	—	100	—
2	IHO-bio	3	0.2	1	3	12,710	2554	125.2	851.3

TABLE 28.16

Spinach Test Results, Romania

				Quantity Fertilizers Used (L/ha)			Increase		
Nr.	**Treatment**	**Nr. Applications**	**Concentration Solution (%)**	**One Treatment**	**All Treatments**	**Yield (kg/ha)**	**kg/ha**	**%**	**kg/L Fertilizer**
1	No fertilizer	—	—	—	—	14,351	—	100	—
2	IHO-bio	3	0.2	1	3	16,680	2509	117.5	836.3

TABLE 28.17

Tomatoes Test Results, Romania

				Quantity Fertilizers Used (L/ha)			Increase		
Nr.	**Treatment**	**Nr. Applications**	**Concentration Solution (%)**	**One Treatment**	**All Treatments**	**Yield (kg/ha)**	**kg/ha**	**%**	**kg/L Fertilizer**
1	No fertilizer	—	—	—	—	21,500	—	100	—
2	IHO-bio	3	0.2	1	3	23,300	1800	108	600

5. Spinach grown in black soil in a greenhouse, using three foliar treatments of IHO-Agro, for a total of 3 L/ha, yielded an increase of 2500 kg fresh leaves/ha (17.5%), corresponding to 836.3 kg leaves/L of fertilizer (Table 28.15).
6. Tomatoes grown in sandy soil in, using three foliar treatments of IHO-Agro, for a total of 3 L/ha, yielded an increase of 1800 kg fruits/ha (8.1%), corresponding to 600.0 kg leaves/L of fertilizer (Table 28.16).

Research Journal of Agriculture Science, 43 (3), 2011, published their results of testing of different foliar fertilizers. The research was done by the same institute, which tested the IHO-Bio product (in Romania registered as IHO-Agro). Although IHO-Bio was not one of the fertilizers studied, it compared favorably to the other fertilizers (Table 28.18).

Benefits IHO-Agro Products versus Agrochemical Fertilizers

We will only give examples on the use of IHO-Agro products on some of the most common crops: corn, wheat, and canola. Typically, fertilizers represent 30%–40% of input costs for many key crops. Fertilizer costs per ha in Ontario are as follows: corn—$305–$316, wheat—$213, and canola $357 (Ontario Ministry of Agriculture, 2012). These costs differ from country to country. Most agrochemical fertilizer prices fluctuate in accordance with the international energy and oil costs.

In contrast, IHO-Agro products are not oil dependent, and the energy used to manufacture these products is minimal. Even in case of natural disasters, should no electricity be available, Industrias

TABLE 28.18

Summary Test Results Romania, 2012

Crop	Nr. Applications	Fertilizer Concentration (%)	Fertilizer Used (L/ha) Each Application	Fertilizer Used (L/ha) Total	Yield (kg/ha) Control	Yield (kg/ha) IHO-Agro	Yield Increase kg/ha	Yield Increase %	Yield Increase kg/L Fertilizer
Corn[b] (Pioneer)	2	0.2	1	2	3,260	3,610	350	110.8	175
Tomato[a] (Rio Grande)	3	0.2	1	3	21,500	23,300	1800	108	600
Lettuce[b] (Silvia—greenhouse)	3	0.2	1	3	10,156	12,710	2554	125.2	851.3
Spinach[b] (Matador)	3	0.2	1	3	14,351	16,860	2509	117.5	836.3
Apple[b] (Idared)	3	0.2	2	6	14,235	17,101	2866	120.1	477.6
Grape[a] (Saugvinon)	3	0.2	2	6	5,580	6,250	670	112.1	111.6

[a] Base fertilizer: N—100, P_2O_2—100, K_2O—100 kg/ha.
[b] Without base fertilizer.

Y Manufacturas Bionaturales SA could continue to manufacture the products, as they are not energy dependent. We do recommend (depending on the crops) 2–4 gal of our product to be used for each Ha. While getting additional benefits from IHO-Agro products, costs of fertilizing will be much lower than using chemical fertilizers. These costs depend on where the products are used, as different import tariffs and taxes will be applied. Even so, the costs will be significantly lower. Farmers will get the assurance that fertilizer costs will not go through the roof every time there will be an increase in the price of oil.

The benefits are as follows:

- Plants more resistant to insects and disease. Farmers may need to use less pesticides and fungicides; therefore, the costs associated with dispersing them are diminished or eliminated.
- Farmers will not incur any additional costs associated with the usage of IHO-Agro products, as they can readily use their existing farm equipment.
- Depending on the extraction method employed in the process, it is to be noted that soils treated with these minerals meet the standards required in the United States, for Organic Certification, as well as the EU Organic Certification requirements—better crop price.
- There is no negative impact on the environment as a result of manufacturing, and there is no environmental pollution associated with the process. Except for water (and perhaps table salt), the process does not generate any types of residue or by-products.
- Plants more robust.
- Nontoxic—IHO-Agro products do not adversely affect the environment.
- More tolerant to weather fluctuations (drought, cold, and excessive rain).
- More transplantation tolerant.
- Larger and better tasting fruits and vegetables.
- Produce lasts longer on the shelf.
- Possible increase in mineral and vitamin content.
- Seeds germinate sooner.
- Plants reach maturity sooner—faster to the market—better crop price.
- Crop yields are increased.

- Better tasting fruits and vegetables.
- Longer producing plants, for example, coffee plants that could start producing earlier, and yield more beans, extending the harvest longer.
- Multiple crops per season, or in greenhouse environments: for example, strawberries, peppers, and lettuce.

Future Work

Both IHO-Mineral and IHO-Bio will be tested on other types of crops and fruit trees, in collaboration with different universities and government research agencies.

We are setting up for tests using a combination of aquaculture, nano-technology, IHO-Mineral, and IHO-Bio products, working toward a self-contained system that will produce enough food, year-round for a family of up to 10 people on a very small area. This will address the growing population and will relieve the stress imposed upon the resources of our planet.

Conclusions

During the more than 5 years of field trials, Industrias Y Manufacturas Bionaturales S.A. has developed two products that help address some of the current food production problems. We have demonstrated that food can be produced without affecting the environment adversely, without need of man-made chemical fertilizers. Crop production increases on an average of 30%–50%, and sometimes more than 300%. Plants can be grown on soils of poor fertility; we tested our products on different crops, and without exceptions, we obtained good results. Our observations are in line with the numerous research papers written about the foliar plant feeding, which has its advantages and disadvantages. Limited tests were done on ornamental shrubs and trees, with promising results. Much work still needs to be done.

REFERENCES

Cortegaza, P. and A. Menéndez. 2010. Classification of agricultural soils Ofelina Ingenio, Volume 1–3. Inedito. 856.

Ferrie, H., 2006. *The Magic of Minerals: The Critical Role of Soil Demineralization in Disease*. Vitality Magazine, Toronto, Ontario, Canada, http://vitalitymagazine.com/article/the-magic-of-minerals/.

International Plant Nutrition Institute, Spring 2008, No. 3—*Better Crop Nutrition Today*.

Jack, A. 1988. Nutrition under siege. *One Peaceful World Journal*, 34:1–8.

Lyne, J.W. and P. Barak. 2000. *Are Depleted Soils Causing a Reduction in the Mineral Content of Food Crops?* University of Wisconsin, Madison, WI, http://www.soils.wisc.edu/~barak/poster_gallery/minneapolis2000a/.

Marler, J.B. and J.R. Wallin. 2006. *Human Health, the Nutritional Quality of Harvested Food and Sustainable Farming Systems*, Nutrition Security Institute, Norcross, GA, http://www.nutritionsecurity.org/PDF/NSI_White%20Paper_Web.pdf.

Mayer, A.-M., September 1997. Historical changes in the mineral content of fruits and vegetables: A cause for concern, In W. Lockeretz, (ed.). *Agriculture Production and Nutrition: Proceedings*. Tufts University, Medford, MA, pp. 69–77, http://www.soils.wisc.edu/~barak/poster_gallery/minneapolis2000a/.

Ontario Ministry of Agriculture, Food and Rural Affairs—*Field Crop Budgets,* http://www.omafra.gov.on.ca/english/busdev/facts/pub60.htm.

Tisdale, S.L., J.L. Havlin, J.D. Beaton, and W.L. Nelson. 1999. *Soil Fertility and Fertilizers: An Introduction to Nutrient Management*, Prentice Hall, Upper Saddle River, NJ.

29

Biochar-Based Amendment Enhances Tomato Transplant Growth and Early Fruiting

Ronald Morse and Jon Nilsson

CONTENTS

Introduction

Field experiments were conducted to evaluate the effects of CharGrow® (CG) on fruiting behavior and marketable fruit yield of *mountain-fresh* tomato. CG is a biochar (BC)-based amendment that contains beneficial soil organisms and microbial foods. In this study, CG served as a bioinoculant for the production of high-quality tomato transplants. Before seeding, CG was thoroughly mixed with potting soil at 2.5% and 5.0% by volume.

In all 4 years, CG-inoculated tomato transplants were greener and approximately 25% taller at field setting than the control (not inoculated). Plant growth in the field was excellent with little disease problems, and multiple fruit harvests were recorded. First-harvest fruit yield (early fruiting) in CG-inoculated tomato plants was increased by 51%, compared to the control.

Increased earliness is attributed to beneficial microorganisms associated with CG that accelerated juvenile plant growth, enabling CG-inoculated tomato plants to flower earlier and more profusely than control plants. Data from this research and other related studies indicate that the incorporated cover crop residues were a key component in facilitating proliferation and expression of beneficial soil microorganisms. In our system, high quantities of cover crop residues were incorporated prior to transplanting tomatoes. This organic food source helped sustain a healthy microbial community and was a source/mechanism to enhance nutrient and water availability for tomato growth. This positive synergy (interaction) between CG inoculation and cover crop–based systems (conservation agriculture, CA—Hobbs, 2007) resulted in both earliness and high total yield of marketable fruit.

Treatments and Research Methods

The CG experiments were conducted over a 4-year period (2007, 2008, 2010, and 2011) in field plots at the Kentland Agriculture Farm, located near Blacksburg, VA. Soil type was similar at all field sites (Hayter loam, having a pH of 6.4, with medium to high levels of phosphorus, potassium, calcium, and magnesium).

Experimental Design and Treatments

Each year, the experimental design was a randomized block, with four replications. Treatments were control (no CG added), CG added to Metromix-360 potting soil at 2.5% and 5.0% by volume, and BC added to the Metromix-360 potting soil at 2.5% and 5.0% (in 2010 and 2011 only).

BC refers to high-carbon materials produced from pyrolysis (heating in the absence of oxygen) of biomass such as wood products and green waste materials. There is considerable interest in using BC as soil amendments because it is a coproduct of renewable energy production from pyrolysis (Graves, 2013).

Field Setting and Cultural Practices

During the first week of June each year, tomato transplants were set by hand in previously established raised beds covered with black plastic mulch. At transplanting, a starter solution consisting of 4-1-1 hydrolyzed fish concentrate (at 2 lb N/acre) was dissolved in water and applied as a liquid drench in all treatments. In all years, no preplant chemical fertilizer was applied. All preplant fertilizer used was organic, derived from soil organic matter (SOM) and organic residues from *in situ* production of legume–grass annual or perennial cover crops seeded in previous months or years. Each year, before laying off raised beds and applying plastic mulch, the cover crops were flail mowed and incorporated using a disk or rototiller (Hobbs, 2007).

Drip irrigation was used to supplement rainfall. Tomato plants were ground grown (not trellised) and sprayed with recommended fungicides and insecticides throughout the growing season. Weeds were managed using recommended herbicides and removed by hand as needed to minimize weed–crop competition. Harvesting was done by hand, beginning in late August and continuing until late September/early October. In these experiments, marketable fruit was defined as unblemished tomato fruit weighing three or more ounces.

Results and Discussion

Data were consistent over the 4-year period of this study and showed that mixing CG in potting soil affected transplant growth and subsequent fruiting habits of field-grown tomato plants. Major results are outlined as follows:

Transplant growth was enhanced: CG-inoculated transplants were greener and approximately 25% taller at field setting than control plants (Figure 29.1). Shortening the time required to produce healthy transplants lowers production costs, without reducing transplant quality or yield potential (Kinet and Peet, 1997).

Early (first harvest) fruit yield was enhanced: Earlier fruiting in CG-inoculated plants was increased by 51%, compared to non-inoculated plants (Table 29.1). Earlier fruiting can be a major economic plus for tomato growers because first-harvest fruit is normally worth more on a per pound or box basis than later harvests. Earlier fruiting has significant economic advantages for tomato producers in regions with short-growing seasons (Grandy et al., 2013).

Total fruit yield was high: Total fruit yield—summation of all harvests—was excellent in both CG-treated and control plots (Table 29.1). Thus, heavy earlier fruiting in CG-inoculated plants did not reduce the total yield. CG-inoculated plants remained vigorous throughout the growing season, resulting in high total yields. This *stay-green* capacity enabled the production of marketable fruit over a prolonged period and would permit extended multiple harvests in areas with long growing seasons (Mattoo and Abdul-Baki, 2006).

Tomato transplants inoculated with fresh BC produced similar results as the control: Thus, BC inoculation had no effect on tomato transplant size at field setting and had no subsequent effect on tomato fruit earliness or total yield (Table 29.2).

FIGURE 29.1 Tomato transplants at field setting—7 weeks from seeding in the greenhouse. From left to right: control (no CG), CG treated (2.5%), and CG treated (5.0%). CG-inoculated transplants were greener and approximately 25% taller than the control when set in the field.

TABLE 29.1

Effects of CG on Early Fruiting (First-Harvest Yield) and Total Tomato Yield, Kentland Agriculture Farm, Blacksburg, VA (4-Year Study)

	First Harvest			Total (All Harvests)	
Year	**Treatment**[a]	**Yield (cwt/ac)**	**Relative Yield (%)**	**Yield (cwt/ac)**	**Relative Yield (%)**
2007	Control (no CG)	97	100	653	100
	CG	137[b]	141	794[b]	122
2008	Control (no CG)	288	100	952	100
	CG	433[b]	150	1017 ns	107
2010	Control (no CG)	183	100	670	100
	CG	282[b]	154	690 ns	103
2011	Control (no CG)	194	100	857	100
	CG	295[b]	152	934 ns	109
Avg.	Control (no CG)	190	100	783	100
(4 year)	CG	287[b]	151	859 ns	110

cwt, hundred weight units (100 lb); relative yield, yield of the control (100), compared to the CG treated; ns, not significant at $p = 0.05$.

[a] Data are average of 2.5% and 5.0% CG.

[b] Statistically significant at $p = 0.05$.

Major Conclusions

Role of BC: In these experiments, tomato transplants were treated by mixing either fresh BC (not charged) or CG (charged or primed BC) with potting soil before seeding tomato transplants in a greenhouse. Compared to the control (not inoculated), CG-inoculated tomato transplants were larger at field setting, and this *accelerated growth effect* persisted throughout the growing season in field-grown plots, resulting in early flowering and fruiting. In contrast,

TABLE 29.2
Effects of BC on Early Fruiting (First-Harvest Yield) and Total Tomato Yield, Kentland Agriculture Farm, Blacksburg, VA (2-Year Study)

	First Harvest			Total (All Harvests)	
Year	Treatment[a]	Yield (cwt/ac)	Relative Yield (%)	Yield (cwt/ac)	Relative Yield (%)
2010	Control (no BC)	183	100	670	100
	BC	217 ns	119	675 ns	101
2011	Control (no BC)	194	100	857	100
	BC	174 ns	90	851 ns	99
Avg.	Control (no BC)	188	100	764	100
(2 year)	BC	195 ns	104 ns	763 ns	100

cwt, hundred weight units (100 lb); relative yield, yield of the control (100), compared to the BC treated; ns, not significant at p = 0.05.
[a] Data are average of 2.5% and 5.0% BC.

BC acting alone (not primed) produced similar effects as the control. Thus, compared to the control, BC inoculation had no effect on tomato transplant size at field setting and had no subsequent effect on tomato fruit earliness or total yield. Based on these experiments, the role of BC is attributed to its function or capacity as a *carrier* to introduce materials such as beneficial microorganisms, growth regulators, and other substances into the rhizosphere of tomato transplants.

Effects of high-biomass cover crop green manure (*using conservation agriculture systems*): In these experiments, most applied fertilizer came from high quantities of cover crop residues (green manure) that were grown and incorporated on site (*in situ*) before setting tomato transplants in early June. In all years and all plots, tomato plants grew well and leaf area duration (*plant health*) persisted until frost (mid October). This prolonged plant health (*stay-green effect*) is attributed to (1) the green-manure soil management system used that provided slow-release nutrients (especially nitrogen) and microbial foodstuff to favor tomato plant growth and (2) a stable food source to enable proliferation and dominance of beneficial microorganisms in the rhizosphere of tomato plants.

Action (mechanism) of CG inoculation: Early tomato fruiting (earliness) was the major crop yield response in CG-inoculated plants. Earliness occurred because the growth rate of tomato plants was *accelerated* through the intervention of beneficial microorganisms and/or plant growth substances generated in the rhizosphere (Bryla and Koide, 1990).

Future Research

Based on this 4-year study, growing CG-inoculated tomato plants in conservation agriculture fields produced both earlier and equal-to-higher total fruit yield, compared to the control plants (not inoculated with CG). We hypothesize that this unique and potentially highly economical synergy effect on fruiting (earliness and total yield) would not occur in conventional low-SOM fields, fertilized only with chemical soil amendments (Grandy et al., 2013). Furthermore, we believe that in low-SOM fields, applying high rates of chemical fertilizers would prevent or minimize the genesis and maintenance of pest-suppressive and nutrient-supplying rhizosphere microorganisms (Maganhotto de Souza Silva and Fay, 2012). This in turn would prevent or minimize (1) early flowering and fruiting of tomato plants and (2) prolonged plant health (stay-green effect) and production of high total fruit yield (Mattoo and Abdul-Baki, 2006).

REFERENCES

Bryla, D.R. and R.T. Koide. 1990. Regulation of reproduction in wild and cultivated tomato (*Lycopersicon esculentum* Mill.) by vesicular-arbuscular mycorrhizal infection. *Oecologia* 84:74–81.

Grandy, A.S., C. Kallenbach, T.D. Loecke, S.S. Snapp, and R.G. Smith. 2013. The biological basis for nitrogen management in agroecosystems, pp. 113–132. In: T.E. Checke, D.C. Coleman and D.H. Wall (eds.), *Microbial Ecology in Sustainable Agroecosystems*. CRC Press, Boca Raton, FL.

Graves, D. 2013. A comparison of methods to apply biochar into temperate soils, pp. 202–260. In: N. Ladygina and F. Rineau (eds.), *Biochar and Soil Biota*. CRC Press, Boca Raton, FL.

Hobbs, P.R. 2007. Conservation agriculture: What it is and why it is important for future sustainable food production. *J. Agric. Sci.* 145:127–137.

Kinet, J.M. and M.M. Peet. 1997. Tomato, pp. 207–258. In: H.C. Wien (ed.), *The Physiology of Vegetable Crops*. CAB International, Wallington, U.K.

Maganhotto de Souza Silva, C.M. and E.F. Fay. 2012. Effect of salinity on soil microorganisms, pp. 177–198. In: Dr. M.C. Hernandez Soriano (ed.), *Soil Health and Land Use Management*. InTech, U.K.

Mattoo, A.K. and A. Abdul-Baki. 2006. Crop genetic response to management: Evidence of root-shoot communications, pp. 221–230. In: U. Uphoff et al. (eds.), *Biological Approaches to Sustainable Soil Systems*. CRC Press, Boca Raton, FL.

30

Organic Restoration Minerals Upgrade Soil

Barry Carter

CONTENTS

Introduction

Natural soil growth is quite slow. Human-caused soil loss is happening more than a hundred times faster than natural soil growth. Climate change is accelerating soil loss due to drought, flooding, and other extreme weather events. Since most people eat food that is grown in soil, hunger is becoming a major factor in our world. Nearly one-seventh of the world's population is hungry or starving.

Many scientists are predicting major social chaos and extinction problems within the next few decades. Starvation is a major driving factor in this. We probably do not have the time to wait until all the sciences are in before we implement new discoveries that might help solve our climate and hunger problems.

Here are some observations, about how soil is formed, which are generally accepted as true:

Mountain minerals erode into the seas. On their way to the seas, these minerals make dirt. Organisms in the dirt convert these minerals into living soil. Plant health depends on these soil minerals. Animal health depends on plant health.

All nutrients are composed of natural minerals. Nearly all of the minerals that plants consume have been solubilized from the soil. Plant and animal health depends on these minerals remaining in (or being returned to) a given area for as long as possible. To maintain optimum soil, plant, and animal health, they must remain in a given area long enough to be replaced by more minerals eroding from the mountains. Or we must artificially replace them in each area.

Minerals that are solubilized move toward the seas more quickly than these same minerals in an insoluble state. Mountain rocks erode quite slowly. This is why we have phrases like *steady as a rock.*

People *harvest* the plant and animal products that depend on these minerals. Then people transport these products away from where they were grown. This depletes essential mineral nutrients from the areas where these products were grown.

The taste of food is quite dependent on the nutrients in that food. Most food does not taste nearly as good as it did 40 years ago.

Let me repeat what I wrote earlier; all soluble minerals get transported, in food, away from the areas where they are essential to the health of the soil, plants, and animal that depend on the solubilized minerals in these areas. Ultimately, they get flushed, with our sewage, into the seas or transported with toxic *waste* to a landfill.

How I Learned about Soil Problems

I am working with hundreds of professional and lay researchers, worldwide, who are exploring the benefits of some newly discovered mineral nutrients that appear to double plant growth and soil carbon sequestration in a couple of years. Most of the reports are anecdotal, but a few have been done by scientific institutions. We call these nutrients the ORMUS minerals.

In 1988, I began working full time as a forest activist. In this work, I became even more aware of the problems that were being created by loss of topsoil, soil compaction, and loss of soil nutrients. In 1995 at the Public Interest Environmental Law Conference at the University of Oregon Law School in Eugene, Oregon, the great environmentalist, David Brower, told us that half of the agricultural topsoil in the United States had been lost in the last 80 years. I first learned about the discovery of some previously unknown mineral nutrients in 1995 when I attended a lecture by an Arizona cotton farmer named David Hudson. Hudson claimed that some of the minerals, in seawater and in the soil, are the precious metal elements in a different state of matter. He claimed that significant amounts of these minerals (up to 5% of the dry matter weight) had also been measured in plant and animal tissues. Metals in an organically complexed chemical form would not be assayed using ordinary soluble chemistry methods. We call these nutrients the ORMUS minerals. Others have confirmed many of David Hudson's claims about the ORMUS elements either by converting the precious metals into the ORMUS state (which is generally quite easy) or by converting these elements in the ORMUS state to metal (which is generally costly and difficult).

Don Nance, an ORMUS provider at oceanalchemy.com, has done conversions in both directions and had ICP mass spectroscopy done on the results. In 2006 and 2009, Don had some assays done on his Golden Tear product, a white precipitate that is made from gold using the peroxide method.

Though the only substances used to make the white precipitate in the peroxide method are gold metal, food-grade hydrogen peroxide, lab-grade hydrochloric acid, and food-grade sodium hydroxide, the resulting precipitate assayed as having only 2.74 parts per billion (ppb) gold in the 2006 assay and 0.21 ppb gold in the 2009 assay. This is a confirmation of David Hudson's claim that the ORMUS elements do not show up using conventional spectroscopic assay methods.

An Australian metallurgist has used the peroxide method to convert 25 metals into their ORMUS state:

Antimony	Gold	Molybdenum	Palladium	Silver
Bismuth	Iridium	Neodymium	Platinum	Tin
Chromium	Iron	Nickel	Rhodium	Titanium
Cobalt	Lead	Niobium	Ruthenium	Tungsten
Copper	Mercury	Osmium	Samarium	Vanadium

He claims that some of the easiest elements to convert from metal to ORMUS are rhodium, iridium, and gold. This might be why they are the more abundant in seawater than calcium.

He also told me that the white powder from each of these elements would dissolve in diethyl alcohol. Alcohol is a natural product of yeast fermentation in soil. One ORMUS producer has added ORMUS in the fermentation process and gotten over twice as much alcohol from the same mash. This suggests that we might want to explore the role of alcohol in moving the ORMUS minerals from the soil into plants.

Don Nance has also done several assays on Dead Sea precipitate and tried converting the ORMUS minerals in it into metal. Here is Don's description of his results from a year-long conversion process converting ORMUS to metal:

> I have taken silver from Masada Dead Sea salt solution which assayed to contain < 0.05 ppb of silver (Ag) and have gotten 279.408 oz/ton from it.
>
> I have taken gold from Masada Dead Sea salt solution which assayed to contain 0.31 ppb of gold (Au) and have gotten 1.4 oz/ton from it.

Don's conversion of these elements from their metallic state to their ORMUS state and from the ORMUS state to their metallic state is some of the best independent evidence we have in support of David Hudson's claims about the abundance of the ORMUS elements in nature.

Though these mineral nutrients are common in nature, they are depleted when we ship food away from where it was grown, and, since this is cutting-edge science, we do not know how important it is to restore these minerals into our soil.

Fortunately, they can be concentrated from commonly available source materials using simple kitchen chemistry. One of these methods (the wet method) consists of taking seawater or whole sea salt dissolved in water and raising its pH to 10.78 using lye (also known as caustic soda or sodium hydroxide) to precipitate out these beneficial minerals. This precipitate contains the ORMUS minerals in a more concentrated form with reduced salt. What this is doing is precipitating out trace metals that are complexed to humic materials.

I learned how to make the white ORMUS precipitate from a 58-year-old miner in north central Washington state on May 18, 1997. I made my first ORMUS precipitate using local water soon afterwards.

I visited this gentleman again in August of 1999, October of 2000, and October of 2001. My first encounter with plant benefits resulting from the application of these sea minerals was during the October 2000 visit when I documented the increased growth of a black walnut tree on his property as a consequence of their use.

The walnuts harvested from the single walnut tree, which was fertilized using seawater precipitate, weighed more than all the walnuts harvested from the six trees on the same property that were not given the seawater precipitate.

Five walnut trees were purchased from the same nursery at the same time. They were all approximately the same age and size. After it was transplanted, one of these five seedlings was given a dispersion of two cups of reduced salt ORMUS from Pacific Ocean water in 5 gal of freshwater. The tree that got the ORMUS is toward the right in the picture as follows and one of the trees that got no ORMUS is on the left (Figure 30.1).

The smaller tree is about 5 ft tall and the larger tree is about 10 ft tall. These trees all grew in a dry interior climate in north central Washington state at around 1000 ft elevation about a mile south of the 49th parallel.

The largest tree received two cups of ORMUS precipitate in the first year. The smallest received none. There have been several additional applications of wash water from the third rinsing of precipitate on the large tree.

Here is an image of four walnuts from these trees (Figure 30.2).

The walnuts on the left are from the trees that did not get the ORMUS, and the ones on the right are from the largest tree that got the most ORMUS. The walnuts on the right are approximately 2½ in. in diameter. The next image is of the unhulled walnuts still on the tree with a man's hand in the image for size comparison (Figure 30.3):

These walnuts are about 7 ft up in the tree, and they look a bit like medium-sized apples.

FIGURE 30.1 Tree with least ORMUS on left and with most ORMUS on right.

FIGURE 30.2 Walnuts with least ORMUS on left and with most ORMUS on right.

FIGURE 30.3 Hand in picture for size comparison.

The ORMUS nuts inside the shells are proportionally larger than their ordinary counterparts. Their taste is comparable. This one tree has consistently yielded the same volume of nuts in their shells as the combined yield of three 15-year-old trees and four other 3-year-old trees. These pictures were taken on October 5, 2000.

A year later, on October 15, 2001, Arthur Zeigler (the owner of sea-crop.com) and I had an opportunity to take more pictures of these trees and the walnuts that grew on them. In Figure 30.4, you can see some of these walnuts on an outdoor picnic table.

Notice how much the large tree has grown compared to the first tree picture. Arthur stood next to the large tree and the small tree to give a better idea of their comparative size (Figure 30.5).

FIGURE 30.4 Hand and penny in picture for size comparison.

FIGURE 30.5 Arthur next to tree without ORMUS on left and next to tree with ORMUS on right.

At the end of the 2001 visit, I was given three walnuts from the small trees and three from the largest tree. I have kept these walnuts.

The picture (Figure 30.6) was taken on June 11, 2008, a few minutes before the three walnuts in each column were weighed. The weight of the three walnuts in each column suggests that the fully dried out walnuts from the tree that was given ORMUS are almost three times as heavy as the three dried walnuts from the tree that was not given ORMUS.

My personal experience with these minerals is that their benefits accumulate over time. I have seen about a fivefold increase in the mass production of my plum tree over a 5-year period.

In 1983 when I moved into my current house, we planted a plum tree seedling in the front yard. This seedling was originally growing near a couple of similar young plum trees in my neighbor's yard directly to the south of my house. All of these trees produced similar plums each year. They were the same size, ripened at the same time, and tasted the same.

31.7 g	91.2 g
Walnuts picked in October 2001.	Picture taken in June 11, 2008.

FIGURE 30.6 Walnuts without ORMUS on left and with ORMUS on right.

In the spring of 2005, I began putting seawater ORMUS precipitate from sea-crop.com on my plum tree. At the end of the first season (on October 7, 2005), I took pictures of two typical plums from my tree. I also picked a couple of plums from the plum trees in my neighbor's yard. These trees are about 50 ft from my tree, but they are the same type of plum tree (Italian prune plums).

The two plums from my neighbor's tree weighed 35 g while the two from my tree weighed 61 g.

On August 16, 2006, I noticed that the plums on my tree seemed to be full size and were starting to darken so I picked a couple of them and a couple from my neighbor's trees for comparison. I realize that it was almost 2 months earlier in the season than when I picked the plums in 2005. This is probably why the two plums from my neighbor's tree only weighed 24.7 g.

I had to check out the plums from my tree because they looked so big. I weighed them and they weighed 129 g. This was more than twice the weight of 2005's plums from my tree and 3.7 times the weight of 2005's plums from my neighbor's tree. It was also 5.2 times the weight of my neighbor's 2006's plums, but they had not reached full size yet while the ones on my tree were much closer to full size.

In 2005, I wanted to have some way to take a picture that would show the size of the plums compared to an object that most people are familiar with, so I took the picture with a couple of golf balls next to the plums. In 2006, I included the golf balls and matched their sizes so that the golf ball in the center overlapped in the 2005 photo (Figure 30.7).

In 2007, a late spring frost nailed all of the plum trees, and only a few plums survived, not enough for a good comparison or pictures.

Plum harvest of 2008 was great, but I did not document it very well. Here is a comparison picture from August 26 (Figure 30.8):

I continued to apply a cup of the Sea-Crop ORMUS to this tree each year after harvest.

In mid June of 2009, I noticed an incredible abundance of mushrooms under my plum tree (Figure 30.9).

I saw this as an indication that the mycorrhizal fungi in the soil were quite active in providing nutrients to the roots of my plum tree. This inspired me to do a more thorough documentation of the results for 2009.

My plum tree started having ripe and edible plums in early September of 2009. These plums were the best I have ever tasted. They were sweet and *smooth*. On September 12, I took some comparison pictures of the plums on my tree and my neighbor's trees that did not get ORMUS.

FIGURE 30.7 Size of the plums compared to golf ball.

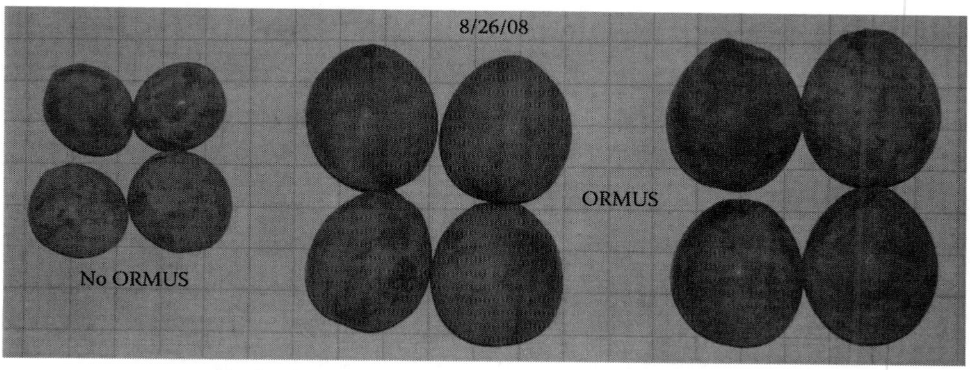

Shady side | Sunny side
37.5 g | 109.3 g | 129.9 g

FIGURE 30.8 Comparison picture.

FIGURE 30.9 Mushrooms under my plum tree.

Without ORMUS
six largest weighed
150.4 g

September 12, 2009
Trees are
50 ft apart.

With ORMUS
six largest weighed
436.8 g

FIGURE 30.10 My neighbor's plums were still 2 weeks away from being ripe.

Parent tree Italian prune
plums 142.2 g

My Italian prune
plums 262.9 g

FIGURE 30.11 Comparison photograph as on September 12, 2010.

Though my plums were ripe, I estimated that my neighbor's plums were still 2 weeks away from being ripe (Figure 30.10).

In 2010, I continued to document the benefits of the ORMUS minerals for my plum tree compared to its parent tree in my neighbor's yard. I took the comparison photograph on September 12, 2010, because I wanted to compare it to the previous photograph from 2009 (Figure 30.11).

You can see that the plums in the previous picture are not as large as the plums in the picture from September 12, 2009.

The following is a table comparing sizes between my plums and my neighbor's plums on September 12, 2009, and September 12, 2010:

Year	Average Weight in Grams of Plums from my Neighbor's Tree	Average Weight in Grams of Plums from My Tree	My Plums Were This Much Larger on September 12 (%)
2009	25.1	72.8	290
2010	15.8	29.2	185

The question arises: why were the plums from both trees so much smaller in 2010 than they were in 2009? I suspect that the springtime weather in 2010 had a great deal to do with the decrease in size.

Baker City, Oregon, is in agricultural Zone 5a. The Italian prune is suitable for USDA Hardiness Zone 5–9. Since Baker City is in the coldest zone that the Italian prune is suitable for, we can assume that there will be years when the bloom times of these trees will be delayed by cold weather into late spring. This weather comparison seems relevant since the last snowfall in 2009 was on March 6, while the last snowfall in 2010 was much later in the spring on May 5.

On September 7, 2010, I took digital pictures of the first 2 ft of my neighbor's plum tree trunk and the first 2 ft of my plum tree trunk. I actually used a yardstick to keep the camera lens in the same distance from each tree so that, when I joined the pictures, the size comparison would be accurate and visible. In Figure 30.12, my neighbor's tree trunk is on the left and mine is on the right.

Then I measured the circumference of both trees at one foot above the ground level. The circumference measurements, in the previous image, translate into 551% more wood in the first 2 ft of my tree compared to the first 2 ft of its parent tree.

As you can see in Figure 30.13, my plums were about seven and a half times as large as my neighbor's plums. Both trees were seriously impacted by the frost and 2 in. of snow we got in mid May of 2011. I suspect that my neighbor's plum tree was affected more significantly than mine.

FIGURE 30.12 My neighbor's Italian prune plum tree on the left is the parent of my tree on the right. My neighbor's tree got no ORMUS while my tree has had ORMUS around the base since 1995. My neighbor's tree is 17.25 in. in circumference while my tree is 40.5 in. in circumference.

FIGURE 30.13 Another comparison picture I took in 2011.

On September 10, 2012, three of my largest plums weighed 120 g and three of my neighbor's largest plums weighed 50 g. I do not have a good picture. Neighbors and others in my town seem to compete to be the first to pick my plums so I do not always get good comparison samples.

Lucky Shamrock

Lately, I have noticed that I only have to water my house plants every 6–8 weeks compared to weekly before I started using the ORMUS minerals on them. I use a pot full of shamrock plants as my gauge. When they start to droop, I water all my house plants. Though my shamrock plants are much taller than they are supposed to be, they have gone from requiring water every week to only requiring it 6–8 weeks since I have been adding the ORMUS minerals. Here is a picture I took just before I watered it on December 31, 2010 (Figures 30.14 through 30.17):

In early 2011, I started putting Garden Shot on my house plants. I put a couple of tablespoons of it in 3 gal of water for watering 24 plants. This product is made with water from a spring near the Snake River in Eastern Idaho. This water is run through a magnetic trap and mixed with worm castings and fish waste.

It appears that the ORMUS minerals can also be *trapped* out of freshwater by spinning the water in a magnetic field. Magnetic traps use the magnetic levitation principle to concentrate the ORMUS in water (Figure 30.18).

This is the ORMUS product that has allowed me to go from watering my shamrock every 2 weeks to only watering it once every 6 weeks. You can find more info on this product on their website at http://www.gardenshot.com/.

After watering my shamrock plant with the Garden Shot product for a few months, I noticed that it was taking much longer before it started drooping. By the end of the summer in 2011, I was only watering it (and all of my other house plants) once every 6 weeks (Figures 30.19 through 30.22).

FIGURE 30.14 Here is a picture I took just before I watered it on December 31, 2010.

FIGURE 30.15 Here is another picture 2 days later.

FIGURE 30.16 And a third picture from January 15.

One morning, I woke up quite early thinking about why these plants seem to thrive on one-sixth of the water. This is what I wrote:

> Nowadays, many people are hungry all of the time. They eat but are not satisfied, so they eat more. But they remain tired and hungry so they have to "sleep it off".
>
> I presume that they are still hungry and tired because they are not getting all of the nutrients they need in the food that they eat.
>
> Plants must be the same. They *eat* nutrients that have been solubilized from the soil. They transpire water into the air as the nutrients are removed for food.

FIGURE 30.17 Shamrock plants in the winter of 1988–1989.

FIGURE 30.18 Cauliflower plant grown with the Garden Shot magnetic trap water.

> If there is insufficient nutrition in the plant food that the roots bring in, more water must be transpired into the air to make room for more solubilized minerals from the soil. Unlike people, plants don't get fat when they eat too much; they just waste water.

I spent a 1½ month giving ORMUS lectures in Australia during October–November of 2008. At one of my lectures, Rob Gourlay, an environmental scientist and founder of the Environmental Research & Information Consortium (ERIC), told us about how he has been combining effective microorganisms

FIGURE 30.19 Here is a picture taken just before I watered it on September 19, 2011.

FIGURE 30.20 Another picture from 3 days after I watered it.

(EM) and ORMUS in some of the ERIC products, and Australian farmers have started asking him, "where is all the water coming from?" after using this product.

Rob speculates that the increased ORMUS in the soil increases carbon sequestration and that the EM organisms play a major role in this process. He also claims that the carbon, in turn, stores more water and nutrients for later release to the plants. You can read more about this on Rob's website at http://eric.com.au/html/papers_soilmap.php.

Rob does not mention on his site that seawater ORMUS is the *secret ingredient* in his soil preparations. More plant and soil results, from the use of these minerals, are linked at http://garden-life.ws/plant-lynx.htm.

FIGURE 30.21 Six weeks later, it started to droop again.

FIGURE 30.22 Here is a picture from the next day after I watered it.

Building Soil Using ORMUS with Rock Dust and Biochar

In February of 2002, I learned about terra preta from a post on a biodynamic forum:

> According to William I. Woods, a soil geographer at Southern Illinois University, ecologist's claims about terrible Amazonian land were based on very little data. In the late 1990's Woods and others began careful measurements in the lower Amazon. They indeed found lots of inhospitable terrain. But they also discovered swaths of "terra preta"— rich, fertile "black earth" that anthropologists increasingly believe was created by human beings.

Since then, I have maintained an interest as more and more people have become involved with building soil using terra preta (now more widely known as biochar). When I learned about biochar, I realized that it may have contributed to the fertility of the soil in areas where I burned leaves and branches in my past gardens.

In March of 2010, I was contacted by a microbiologist named Douglas Beitler who had a website that promoted biochar products. Douglas told me about how he was getting incredible plant growth results using a combination of ORMUS minerals and biochar. Then Douglas started using ORMUS and biochar to convert desert sand in Dubai into productive soil: http://bloomthedesert.com/news/dubai-btd-farm-trails-wvideo.

Douglas and his partner concentrated their own ORMUS from seawater and also used the following ORMUS products:

> SHOT fertilizer from Jan and Loretta Delinger (Idaho USA). We used this everywhere in the outdoor crops. 1 gallon went a long way (www.idaorg.com)
>
> Dan Schreiber, Starseed Gardens (Byron Bay Australia) product called "enormus" www.enormus.org
>
> Tom Waters, Tears of Isis: www.isisnutrients.com

On October 6, 2010, I received a Soil-Age forum email with the following quote:

> The solution to global warming is very simple and cheap, soil sequestration." and "... a 1.6% increase in organic content of soils humans manage would restore atmospheric carbon dioxide levels to normal.

These quotes were from a video and web page by an Australian farmer named Allan Yeomans: http://www.youtube.com/user/AllanYeomans1#p/a/u/0/-o2N6vJDLpE

Allan Yeomans also has a book online at http://www.yeomansplow.com.au/priority-one-contents.htm.

The use of the plow that Allan Yeomans and his father developed is called *keyline plowing*. It actually tills the subsoil more than the topsoil.

Abe Collins, the founder of carbonfarmersofamerica.com, has reported 8 in. of new soil depth in 1 year with the use of keyline plowing combined with concentrated ORMUS minerals. Here is a report that Abe sent me in November of 2009:

> We used milk, Sea-Crop, planned grazing of livestock herds, and keyline subsoil plowing. The combined inputs/management had the effect of converting subsoil to topsoil at about 8 in. per year in 2008. Pretty exciting! Not very scientific information, but compelling enough to warrant some real research (Figures 30.23 through 30.25).

Here is an example of one test showing increased soil carbon resulting from the application of the ORMUS minerals:

Wheat was germinated in soil tests and grown out for 20 days before analysis of the soil. The soil was then sent to an independent soils laboratory for quantitative testing of total bacterial biomass and total fungal biomass.

The control sample received no fertilizer or treatment with sea minerals. This sample was used to establish a baseline. A second test sample received a complete fertilizer and no sea minerals. The third sample received the complete fertilizer and treatment with sea minerals at the rate of 1 gal/acre.

	Total Bacterial Biomass (μg/g)	**Total Fungal Biomass (μg/g)**
Baseline	872	547
Fertilizer only	1183	692
Fertilizer plus Sea-Crop	1282	1075

FIGURE 30.23 Digging the hole.

FIGURE 30.24 Measuring an additional 8 in. of topsoil.

These sea minerals when used with fertilizer increased the total bacterial biomass by 47% over the baseline. Fertilizer alone only increased the total bacterial biomass by 36%.

These sea minerals, when used with fertilizer, increased the total fungal biomass by 97% over the baseline. Fertilizer alone only increased the total fungal biomass by 27%. This is a total increase in fungal plus bacterial biomass of over 66% in only 20 days.

FIGURE 30.25 Grazing and fertilizing.

We should also add the increase in plant root biomass as a result of increased availability of nutrients due to increased fungal and bacterial activity. Several tests have been done with these sea minerals on various plants, but most of these tests were longer than 20 days. For example, the increase in dry root mass of corn was 65.22% after 115 days of growth using these minerals along with N–P–K fertilizer (which was used on the control crop too). The dry matter root mass of purple basil increased by 325.25% in only 30 days!

The previous tests were done by the following supplier of these sea minerals, and I am using the data with their permission: http://www.sea-crop.com/.

Only about 5% of soil organic matter is alive. If we could double this in a year or two, it would also increase the dead organic matter quite quickly. I am convinced that the very low application rate and the open-source methods for concentrating the ORMUS minerals make them the ideal way to sequester more carbon in the soil and that this is the cheapest and easiest way to pull greenhouse gases out of the atmosphere.

Plants mostly *eat* solubilized mineral nutrients from the soil. These solubilized minerals are most likely to be washed to the oceans with soil erosion. Since these minerals are already in a water-soluble state, they typically will remain solubilized in ocean water. Therefore, it should be easy to use seawater to enhance growth in our gardens and local farms. This actually works for a while but, when it is used for a long time, it can make the soil too salty to grow food.

One solution to this problem is to concentrate the minerals from ocean water in a way that also reduces the salt content of the concentrate. Previously, I described a cheap and easy, open-source way to do this (raise the pH of seawater or whole sea salt dissolved in water to 10.78 using sodium hydroxide [lye]). This generates a white precipitate that will settle to the bottom of your container at the rate of about one-fifth of the total starting amount of water, when it is allowed to settle for 8 h.

This ORMUS mineral precipitate is typically applied at the rate of 1–10 gal/acre (10–104 L/h). This is usually applied just once a year for most row crops but will be after each harvest for grains and other crops that get multiple cuttings per year.

I have purchased lye for about $4 per pound. One pound of lye will produce about 10 gal of precipitate from 50 gal of ocean water. This means that anyone with access to ocean water and lye could apply this for a few dollars per acre.

Ocean water contains a mix of the solubilized minerals from every continent. This mix of minerals, according to René Quinton (1866–1925), is approximately the same as the minerals in human blood:

http://www.originalquinton.com/about/history/

As the oceans have become saltier over time, I suspect that land-based plants and animals have adapted to tolerate lower levels of sodium while ocean-based plants and animals have adapted to tolerate higher levels of sodium. This may be the reason why seawater precipitate and products like Garden Shot have greater benefits at lower levels than seawater or sea solids. The application rate for Dr. Maynard Murray's sea solids is mentioned in his book.

While rock dust is a bit cheaper than Sea-Crop, it is more expensive than homemade ORMUS concentrate. Rock dust also costs more to transport and apply than the ORMUS concentrates.

The ORMUS concentrates can be sprayed from an airplane or pumped underground with a keyline plow. I think that the best and cheapest option might be to use a combination of microorganisms, rock dust, and ORMUS precipitate to get to the increased growth quickly and sustain it for as long as possible. Limestone is only a viable option if your soil is too acidic, though it might be worthwhile to concentrate the ORMUS elements from limestone using an acid leach:

The precious metal elements seem to be much more abundant in the ORMUS state. The most common ORMUS element in seawater appears to be ORMUS rhodium. As a metal, rhodium is toxic, but in its ORMUS form, it seems to be quite beneficial to all living things. One ORMUS provider had about 25 years of experience working with the precious metals before he got involved with Ormus. He has done repeated assays on the seawater precipitate he produces and has evidence suggesting that it typically contains from 30% to 40% unassayable ORMUS rhodium, iridium, and gold.

This suggests that the soluble ORMUS minerals in Pacific Ocean water precipitate would be present in the following parts per million portions compared to other common minerals:

Calcium	332.3
Gold	373
Iridium	588
Rhodium	1677
Sodium[a]	2830.8
Magnesium	3215.4
Chloride[a]	4359.4

[a] Present as sodium chloride, reported number adjusted by manufacturer.

I am well aware that all of this needs independent scientific replication before it advances from hypothesis to theory. Even the plant benefits we have documented at http://Garden-Life.ws/plant-lynx.htm need replication and confirmation before the role of the ORMUS elements in plant growth and health is confirmed. We can wait another decade for this confirmation, or we can start using it now to help feed the nearly billion people that are currently starving in the world.

Biochar may be good to sequester the ORMUS minerals but does not assure that they will be in the soil to begin with. I expect that a combination of the ORMUS minerals, paramagnetic rock dust, biochar, fungal inoculations, EMs, and organic/biodynamic methods will be the way we ultimately save humanity and our environment.

31

SEM-EDX Observation of Diatomaceous Earth at Radioactive Paddy Soils in Fukushima, Japan

Kazue Tazaki, Teruaki Takehara, Yasuhito Ishigaki, Hideaki Nakagawa, and Masayuki Okuno

CONTENTS

Introduction

Diatomaceous earth collected from Noto Peninsula, Ishikawa Prefecture, was observed by scanning electron microscopy (SEM) equipped with energy dispersive x-ray analyzer (EDX). These samples had suffered radiation exposure in a radioactive paddy field in Baba, Fukushima Pref., Japan, after the Great East Japan Earthquake, on March 11, 2011. Energy-dispersive x-ray fluorescence (ED-XRF) analysis and Ge semiconductor detector found not only major elements but also radionuclides, such as ^{134}Ce, ^{137}Cs, ^{89}Sr, and ^{90}Sr, in the paddy field in Baba. After 1 year, SEM-EDX quantitative analyses of diatoms and clay minerals (mainly montmorillonite) clearly indicated concentration of radionuclide, such as I, Cs, Ba, Nd, Th, U, Np, and Pu, suggesting absorption of both radionuclide and stable isotope elements from radioactive polluted paddy soils. The leak of radioactivity of ^{137}Cs, part of ^{90}Sr, and the traces of some radionuclide are originated from Fukushima Daiichi Nuclear Power Plant (FDNPP) accident. The diatom–clays–radionuclide interaction in the polluted paddy soils in Fukushima could be effective as the shield for low-level radioactive waste at the geological disposal.

The release of radionuclides into the environment is a subject of intense public concern. Although significant quantities of radionuclide were released as a consequence of nuclear weapon testing in the 1950s and 1960s and via accidental release, for example, from Chernobyl in 1986, the major burden of anthropogenic environmental radioactivity comes from the controlled discharge of process effluents produced by industrial activities allied to the generation of nuclear power (Lioydl and Macaskie, 2000). Nuclear energy and uranium resources may increase in the next few decades because nuclear power generation has relatively low emissions of carbon dioxide. However, if water or air contacts the fuel, the release of massive amounts of radiation could occur (Macfarlane and Miller, 2007; Pusch, 2007).

The threat of environmental pollution from the release and dispersal of uranium and related radionuclide into the biosphere provoked to do research focusing on their geochemical properties and on the interaction of uranium minerals and dissolved species with biological systems. The strong tendency of uranium to form stable aqueous complexes and to precipitate with organic compounds is well known and indicates that U–biotic interactions could play an important role in radionuclide cycling at the earth surface (Haas et al., 1998; Suzuki et al., 2003; Tazaki et al., 2003; Suzuki and Suko, 2006). Uranium is located on the earth surface, which is closely related to microorganisms in soils, groundwater, and organic matter. Filamentous fungi, yeasts, algae, and bacteria were evaluated for uranium tolerance (Lovley, 2000). Spores could survive the very harsh conditions of radiation, heat, pressures, and low water contents and could potentially become active under more favorable conditions (Mulligan, 2007; Mulligan et al., 2007).

Coastal areas in Minamisoma City, Fukushima, Japan, were seriously damaged by the radioactive contamination from FDNPP accident that caused multiple pollutions by the tsunami and radionuclide exposure, after the Great East Japan Earthquake, on March 11 and 12, 2011. FDNPP leaked 17 kinds of radionuclides, such as ^{134}Cs (1.8×10^{16} Bq; half-life time 2.1 years), ^{137}Cs (1.5×10^{16} Bq; half-life time 30.0 years), ^{90}Sr (1.4×10^{14} Bq; half-life time 29.1 years), and ^{95}Zr (1.7×10^{13} Bq; half-life time 64.0 days) to the atmosphere and seawater in Japan (Atomic Energy Safety Agency, 2011). The paddy soils in Fukushima Prefecture have heavily been contaminated by radionuclides, especially by Cs (^{134}Cs, ^{137}Cs) and Sr (^{89}Sr, ^{90}Sr), even though more than 30 km north of the FDNPP.

Few studies and little direct experimental evidence that are available on microbe–clays–radionuclide interaction in the nuclear waste repository were carried out on U, Th, Cs, Sr, Ra, and Rn-bearing bacterial biomineralization of nanometer scale in the context of environmental remediation (REF) (Lioydi and Macaskie, 2000; Lovley, 2000; Pusch, 2007). The objective of this contribution is to introduce the ability of various diatoms associated with clay minerals and microorganisms capable of absorbing both radionuclide and stable isotope from polluted paddy soils in their extreme condition.

In this study, quantitative analysis associated with the elemental content maps by SEM-EDX indicated the possibility of absorption of radioactive pollution in Fukushima paddy soils.

Material and Methods

Physical and Chemical Analyses

Car-borne and hand-borne radiation measurement systems were assembled for easily and rapidly detecting full features of the fission nuclides buried in the ground, soils, and air. An abnormally high radiation was detected at the surface of paddy soils in Minamisoma City in Fukushima Prefecture, which exhibited high levels of radioactive Cs. The measurement of β (γ) ray radiation used Aloka GM survey meter TGS-136 (cpm) and TCS-151(μSv/h) (Tazaki et al., 2011). The radionuclide in the demonstration paddy soils was measured by Ge semiconductor and NAI (TI) Hitachi Aloka medical (CAN-OSP-NAI) and detected Cs (^{134}Cs and ^{137}Cs), but did not detect ^{131}I, on September 16, 2011.

Chemical compositions of the paddy soils and the diatomaceous earth were also determined by EDXRF of JEOL JSX-3201 and XRF JSX-3100RII, using Rh-Kα ray instrument, which operated at an accelerating voltage of 30 kV, under vacuum condition, and FP bulk qualitative and quantitative analyses.

Mineralogical Analysis

X-ray powder diffractometer (XRD) (Rigaku Rinto 1200; Cu-Kα radiation, 40 kV, 30 mA, scan speed of 2°/min) was used for the identification of minerals and clay minerals in diatomaceous earth in Noto Peninsula, Ishikawa Pref. The clay fraction of <2 μm size is oriented on a glass slide.

SEM-EDX Observation

SEM-EDX technique can be used to study the interaction between clays, microorganisms, and radioactive materials in environmental systems, revealing *the role of the geosphere in environmental change.* Electron microscopic techniques in μm- and nm-sized observation of microorganisms show evidence of bioremediation, interactions, cycling, and its role under extreme condition. SEM-EDX is a good tool to observe soils because it can show both micromorphology and the visual elemental composition. Here, we attempted SEM-EDX observations and quantitative analyses of both radionuclide and stable isotope in the demonstration of paddy soils in Baba, Haramachi-ku, Fukushima Prefecture.

SEM-EDX (S-3400N, HITACHI) (HORIBA) was used for morphological observation, quantitative analyses, and elemental content maps, under 15 kV accelerating voltage, current 70–80 μA, analytical time for 1000 s, and area of 10 mm × 10 mm on the carbon double tape with Pt–Pd coating. JEOL transmission electron microscopy (TEM) JEM-2000EX was used for morphological observation inside of nm-sized bacterial cell.

Demonstration Experiments at Paddy Field in Baba

Demonstration experiments at paddy field in Baba, Haramachi-ku, Fukushima, used various kinds of materials that could be expected to shield, absorb, and fix radioactive materials. We set up more than 20 lots using different materials, such as zeolite, fossil shell, and chaff.

Diatomaceous earth sprinkled on the surface of the demonstration paddy soils was collected from Nanao City, Noto Peninsula, in Ishikawa Prefecture. Abundant single-celled algae and various kinds of diatom species associated with clay sediments are found in the diatomaceous earth, showing several layers at outcrop (Figure 31.1a).

Each lot has an area of 2 m × 2 m = 4 m^2 (Figure 31.1b). Lot No. 12 was used for diatomaceous earth collected from Noto Peninsula (Figure 31.1c). The number of 1, 2, and 3 in Figure 31.1c indicates the sampling spots for periodical measurements of radioactivity in lead box, within 1 year (Table 31.2 and Figure 31.3).

FIGURE 31.1 Views of outcrop and demonstration of paddy field. (a) Outcrop of diatomaceous earth at Noto Peninsula, Ishikawa, Japan. (b) Experimental demonstration of paddy field in Baba, Haramachi-ku, Fukushima, Japan, showing 20 lots, which were sprinkled by different materials. (c) Lot No. 12 showed sprinkling diatomaceous earth collected from Noto Peninsula. Numbers 1, 2, and 3 indicated sampling spots for analyses.

The results of radioactive measurements at 11 lots within 2 months using different kinds of materials, 2.5–13 cm in depth, are plotted in Figure 31.2, showing that the most effective reduction in dosage (cpm) was diatomaceous earth (Noto).

The depth profiles of diatomaceous earth in Lot No. 12 are indicated in Figure 31.3, dramatically decreasing the dosage of radioactivity after 1 year. On the other hand, other lots did not clear the decrease of dosage as reported by Tazaki (2012).

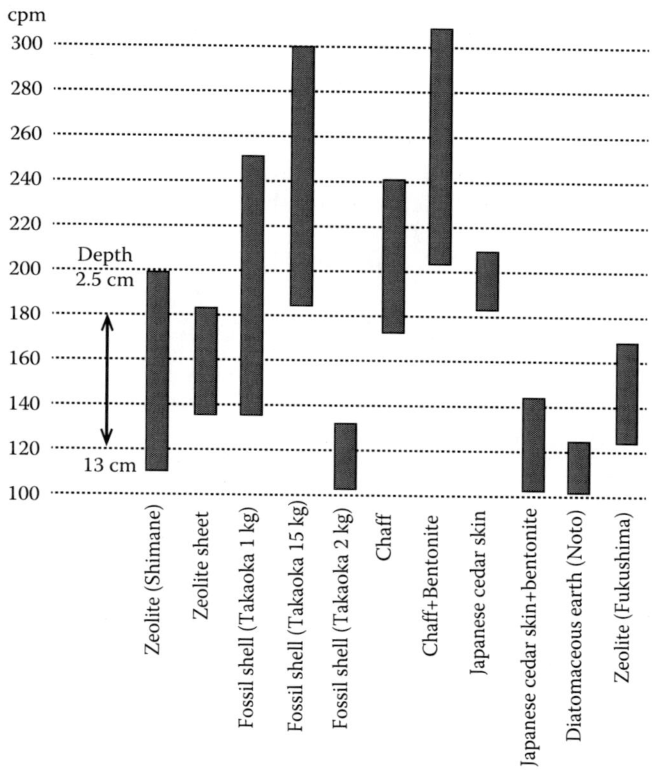

FIGURE 31.2 The decreased dosage of each material that was sprinkled on the surface of paddy soils. The most effective dosage was diatomaceous earth at Noto Peninsula. The dosage was measured in lead box at laboratory.

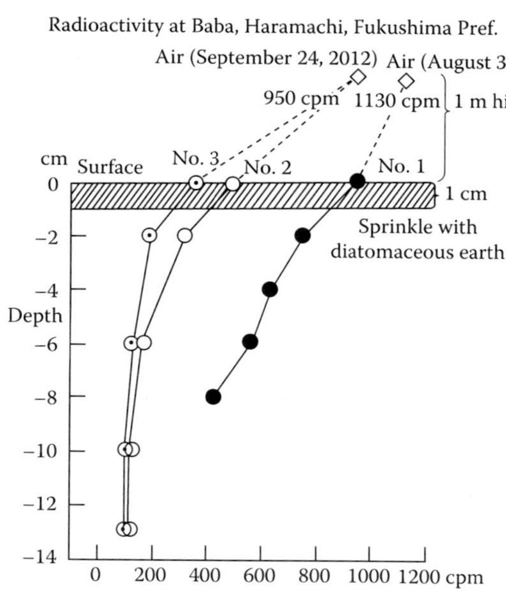

FIGURE 31.3 Comparison of radioactivity at Lot No. 12 sprinkled with diatomaceous earth between August 3, 2011 (●—No. 1) and September 24, 2012 (○—No. 2, ⊙—No. 3). Each core sample with different depths was measured dosages in lead box at laboratory.

Results

Chemical Composition and Radionuclide in Paddy Soils

Elemental composition of paddy soils analyzed by ED-XRF and radionuclides data by Ge semiconductor detector are summarized in Table 31.1. The ED-XRF data of paddy soils at Nagadoro (No. 1) and Baba (No. 3) were radioactive polluted soils, whereas paddy soils in Kanazawa (No. 5) were clean and safe on July 6, 2012. Note that stable isotopes of ^{30}Zn (230–290 mg/kg), ^{37}Rb (240–250 mg/kg), ^{38}Sr (320–440 mg/kg), and ^{40}Zr (490–740 mg/kg) are quite high concentration in each paddy soil.

TABLE 31.1

Elemental Compositions and Radionuclides of Cs-134, Cs-137, Sr-89, and Sr-90 at the Paddy Soils on July 6, 2012, Using ED-XRF and Ge Semiconductor Detector

Element	No. 1 wt(%) Nagadoro	No. 2 wt(%) Nagadoro (Cultivated)	No. 3 wt(%) Baba	No. 4 wt(%) Slide Glass	No. 5 wt(%) Kanazawa City
Na	1.22	1.03	0.567	10.4	1.86
Mg	0.924	0.945	0.974	4.96	1.44
Al	20.0	19.4	24.9	2.74	21.0
Si	52.2	49.2	40.0	63.4	53.1
P	0.294	1.93	0.816	—	0.368
S	0.215	0.276	0.224	0.152	0.121
K	4.44	5.30	3.12	0.698	4.27
Ca	3.96	3.50	2.45	17.1	2.07
Ti	0.948	0.976	1.04	—	0.988
Mn	0.119	0.111	0.138	0.0305	0.157
Fe	6.06	6.04	8.84	0.276	9.07
Sn	—	—	—	0.146	—
Ba	—	—	—	—	0.108
W	1.23	0.442	0.0796	—	—
Cl	—	—	—	—	0.047
Cr	0.044	0.048	0.029	—	0.027
Co	0.052	0.031	—	—	—
Ni	—	—	0.013	—	0.063
Cu	—	—	0.019	—	0.058
Zn	0.023	0.023	0.029	—	0.023
Rb	0.024	0.022	0.025	—	0.024
Sr	0.044	0.038	0.032	0.022	0.057
Zr	0.074	0.072	0.049	0.051	0.031
Pb	—	—	—	—	0.050
Heated 600°C wt(%)	8.2	10.6	16.7	—	5.3
Ge semiconductor detector				**Analytical date**	
Cs-134 (Bq/kg dried)	41,000	47,000	19,000	July 3, 2012	
Cs-134 (Bq/kg dried)	42,000			September 16, 2011	Not detected
Cs-137 (Bq/kg dried)	59,000	68,000	27,000	July 3, 2012	
Cs-137 (Bq/kg dried)	50,000			September 16, 2011	Not detected
Sr-89 (Bq/m^2)	22,000		2,500	June 14, 2011	[a]
Sr-90 (Bq/m^2)	4,800		600	June 14, 2011	[a]

[a] Analyzed by Ministry of Education, Culture, Sports, Science and Technology.

—; not detected.

The ^{134}Cs (41,000–42,000 Bq/kg), ^{137}Cs (50,000–59,000 Bq/kg), ^{89}Sr (22,000 Bq/m^2), and ^{90}Sr (4,800 Bq/m^2) are quite high in Nagadoro (No. 1) in June 2011 and July 2012, whereas no radionuclide was detected in the paddy soils in Kanazawa (No. 5).

The ^{134}Cs (19,000 Bq/kg) and ^{137}Cs (27,000 Bq/kg) were detected on July 3, 2012, whereas the ^{89}Sr (2,500 Bq/m^2) and ^{90}Sr (600 Bq/m^2) were detected on June 14, 2011 in the demonstration paddy soils in Baba (No. 3).

Demonstration Paddy Field in Baba

Diatomaceous earth in Noto Peninsula sprinkled on the surface of the demonstration paddy field in Baba is shown in Figure 31.1. The outcrop of diatomaceous earth in Noto Peninsula contains abundant single-celled algae and various kinds of diatom species associated with clay sediments in several layers (Figure 31.1a). The demonstration paddy field in Baba was set up at No. 20 lot with different materials sprinkled on the surface (Figure 31.1b). The close up photograph of Lot No. 12 with diatomaceous earth is shown in Figure 31.1c.

Results of decrease of radiation of each lot with 0–2.5 cm to 13 cm in depth after 1 month showed various counts between 100 and 300 cpm in lead box, whereas the air radiation dosage at Baba on August 3 was 700 cpm. The diatomaceous earth (Noto) indicated the most effective counts between 100 and 125 cpm, 1–2 cm in thickness (Figure 31.2).

Comparison of depth profiles between the diatomaceous earth samples collected on August 3, 2011 (●; No. 1) and on September 24, 2012 (○; No. 2, ⊙ No. 3) at Lot No. 12 is shown in Figure 31.3. Each core sample was measured; dosages of radioactivity in lead box at laboratory showed dramatically decreased radioactivity within less 10 cm in depth. (○; No. 2, ⊙; No. 3). The rate of decreased counts of radioactivity clearly indicated more than half counts within 1 year, suggesting effect of characteristics of diatomaceous earth. The soils <6 cm in depth are clean as about 100 cpm as control value.

The radionuclides of ^{131}I, ^{134}Cs, and ^{137}Cs in depth profile of paddy soils sprinkling diatomaceous earth after 2.5 month experiment on October 20, 2011, are shown in Table 31.2 (middle). The dosage of ^{131}I was not detected anymore because of short half-life time, whereas ^{134}Cs and ^{137}Cs were dramatically decreased in depth within 0–8 cm.

Radionuclides of ^{134}Cs (8,900 Bq/kg) and ^{137}Cs (11,000 Bq/kg) were detected on the surface of the soils (0–2 cm) without ^{131}I, and ^{134}Cs (3,500 Bq/kg) and ^{137}Cs (4,500 Bq/kg) were detected 2–4 cm in depth without ^{131}I, indicating more than half dosage decreased within 2 cm in depth (Table 31.2). Furthermore, ^{134}Cs (1600 Bq/kg) and ^{137}Cs (2100 Bq/kg) were detected 4–6 cm in depth without ^{131}I, indicating more than half dosage decreased within 4–6 cm in depth. The analytical data indicated that the paddy soils were contaminated by radionuclide up to 8 cm in depth in Baba.

In the lead box, the dosage of paddy soils with depth profile from surface to 13 cm in depth in Baba is shown in Table 31.2 (upper), showing the same phenomena as radionuclide dosages (middle). The effect of sprinkling diatomaceous earth, 1 and 2 cm thick, indicated that the thicker had more effect of decreased dosage (Table 31.2, lowest).

Chemical Point Analysis

The chemical 16-point analyses of few 10–100 μm–sized grains in diatomaceous earth, using electron microscopy, indicated various chemical compositions, as shown in Table 31.3. The analytical data indicated mineral grains associated with clay minerals. Note that point analysis indicated that clay mineral spots are rich in Al and Si (spots of Nos. 10 and 11). The spots of quartz and diatom are rich in Si (spots of Nos. 1, 6, and14). The spots of microorganism are rich in S and K (spots of Nos. 2–5, 15, and 16), respectively.

Chemical Analysis of Diatomaceous Earth

Quantitative analysis of aqueous materials (5 g/250 mL) in Noto diatomaceous earth of 16 samples indicated that the most samples contain high aqueous element of SO_4^{2-}, as shown in Table 31.4.

TABLE 31.2

Radioactivity and Depth Profile of Paddy Soils in Baba, Haramachi-ku, Fukushima Pref., Showing Achievement of Maximum Efficiency within 1–2 cm Thick

Sample	**No. 1**	**No. 1′**	**No. 2**	**No. 3**	
Date	August 8, 2011	September 2, 2011	September 24, 2012	September 24, 2012	
Air (1 m high)	1130 cpm	(2.0 μSv/h)	950 cpm	950 cpm	
Surface 0 cm	950 cpm	850 cpm	500 cpm	380 cpm	Sprinkle of 1 cm depth with diatomaceous earth
Depth—2 cm	750 cpm	750 cpm	330 cpm	200 cpm	
4	630 cpm	640 cpm			
6	570 cpm	580 cpm	190 cpm	140 cpm	
8	430 cpm	420 cpm			
10			140 cpm	130 cpm	
13			140 cpm	130 cpm	
Date	October 20, 2011		Radionuclide		
Sprinkle of diatomaceous earth (1 cm)	Depth	**131-I**	**134-Cs**	**137-Cs**	**(Bq/kg)**
	On the surface 0–2 cm	ND	8900	11,000	
	2–4 cm	ND	3500	4,500	
	4–6 cm	ND	1600	2,100	
	6–8 cm	ND	910	1,200	
	8–10 cm	ND	250	300	
	10–12 cm	ND	210	300	
	12–14 cm	ND	130	150	
	14–16 cm	ND	79	120	
Effect of sprinkling diatomaceous earth					
Sprinkling thickness	cpm	μSv/h	Measured in lead box (September 27, 2011)		
1 cm thick	750	0.75			
2 cm thick	680	0.65			

TABLE 31.3

Chemical Point Analyses of Diatomaceous Earth in Noto, Indicating Different Chemical Compositions of Various Grains, Using Electron Microscopy

No	**O**	**Al**	**Si**	**S**	**K**	**Ti**	**V**	**Cr**	**Fe**	**Ta**	**Total/at%**
1	35.10	2.58	59.96		0.76				1.61		100
2	24.40		24.11	14.13	6.40				30.96		100
3	19.57	1.36	6.58	19.27	9.31		0.13		43.77		100
4	21.09		5.96	19.24	9.59		0.08		44.00		100
5	24.71		23.11		6.26				32.16		100
6	35.71	2.92	58.01		1.56				1.80		100
7	28.39	8.44	47.53	2.17	3.89	0.80	0.08		8.71		100
8	36.22	5.39	49.49	1.42	2.48				5.01		100
9	40.92	8.64	36.74						13.70		100
10	42.14	16.01	34.09					0.15	4.28	3.32	100
11	39.13	16.05	37.42						7.40		100
12	17.24	2.75	10.53	1.95			0.17		67.37		100
13	34.62	6.29	46.82		9.98				2.29		100
14	31.76	4.96	55.28	1.84	1.34		0.19		4.63		100
15	19.81	2.75	12.75	15.73	8.67		0.23		40.05		100
16	21.73	1.25	4.99	18.89	9.41				43.73		100

TABLE 31.4

Quantitative Analyses of Aqueous Materials in Diatomaceous Earth Collected from Noto (5 g/250 mL)

Sample No.	Aqueous	SO_4^{2-}	Fe^{2+}	Fe^{3+}	Ca^{2+}	Na^+	K^+
1	4.04	1.36	0.073	0.011	0.403	0.018	0.004
2	1.94	1.31	0.047	0.009	0.579	0.019	0.005
3	2.61	1.48	0.102	0.006	0.462	0.016	0.015
4	2.04	1.18	0.068	0.005	0.266	0.044	0.038
5	1.92	0.98	0.081	0.007	0.224	0.052	0.029
6	1.64	0.88	0.068	0.004	0.262	0.037	0.041
7	1.48	1.01	0.071	0.003	0.128	0.035	0.058
8	1.26	0.86	0.036	0.003	0.246	0.037	0.084
9	1.71	1.09	0.064	0.005	0.089	0.034	0.088
10	2.58	1.19	0.075	0.003	0.266	0.03	0.074
11	1.86	1.48	0.092	0.005	0.138	0.1	0.058
12	2.11	1.22	0.13	0.005	0.462	0.178	0.078
13	1.83	1.09	0.021	0.028	0.354	0.218	0.053
14	1.96	0.96	0.021	0.005	ND	ND	ND
15	1.82	1.43	0.027	0.006	0	0.385	0.048
16	2.18	1.19	0.008	0.011	0.122	0.445	0.048

Source: After Miyamoto, M., On the basis of quality of Noto-diatomite (diatomaceous earth) and its application, Abstract of *Ishikawa Prefecture Seminar*, pp. 1–3, 2007.

Miyamoto (2007) reported chemistry of Noto diatomaceous earth as follows: LOI (9.5–11.5 mass%), SiO_2 (67–75 mass%), Al_2O_3 (8.0–13 mass%), Fe_2O_3 (3.0–5.0 mass%), TiO_2 (0.35–0.60 mass%), CaO (0.9–1.4 mass%), MgO (0.15–1.5 mass%), K_2O (1.2–1.9 mass%), and Na_2O (0.6–1.0 mass%). Diatom shells, 10–100 μm size, are mainly made of hydrous amorphous silica as the most pure SiO_2 of 94% and H_2O of 6%.

Diatomaceous earth contains not only diatom but also large amount of clays; that is the reason why we use it in this study. Furthermore, the diatomaceous earth indicated relative density of 0.53–0.67 g/cm^3 and strong acidic pH of 3.5–4.5, which have been studied in the application and utilization for materials, such as elimination of radioactive materials, filtration, and absorbing by Noto diatomite after heat treatment (Miyamoto, 2007). Diatomaceous earth (pH of <3) at Suzu City, Noto Peninsula, has been used for soil improvement, compost, fertilizer with oyster shell, desiccant, and combined materials as agriculture application methods (Hasegawa, 2007).

Mineralogical Analysis of Diatomaceous Earth

The mineralogy of diatomaceous earth in Noto, using XRD analysis, contains both minerals and clay minerals (Figure 31.4). Montmorillonite (14.2 and 4.5 Å), gypsum (7.6, 4.27, and 3.07 Å), jarosite (5.1, 3.10, and 2.30 Å), quartz (3.33 Å), cristobalite (4.0 and 2.84 Å), and feldspars (3.23 Å) were identified, respectively. The EG treatment was used for the identification of montmorillonite of 14 Å diffraction, which was shifted to 15 Å after the treatment.

pH and EC of Diatomaceous Earth in Noto

Aqueous examination of diatomaceous earth in Noto (Figure 31.5-1) is compared with rock powder at Dainennji layer (Figure 31.5-2), zeolite at Yonezawa (Figure 31.5-3), and granite at Kawamata town (Figure 31.5-4), indicating characteristic of pH and electrical conductivity (EC), respectively. The diatomaceous earth (Figure 31.5-1) is strongly acidic with a pH of 2 after 174 h experiment and the highest EC of 2500–3500 mS/cm, sharply increasing with experimental time of 174 h, suggesting that montmorillonite and diatom anticipate capability to absorb radionuclide.

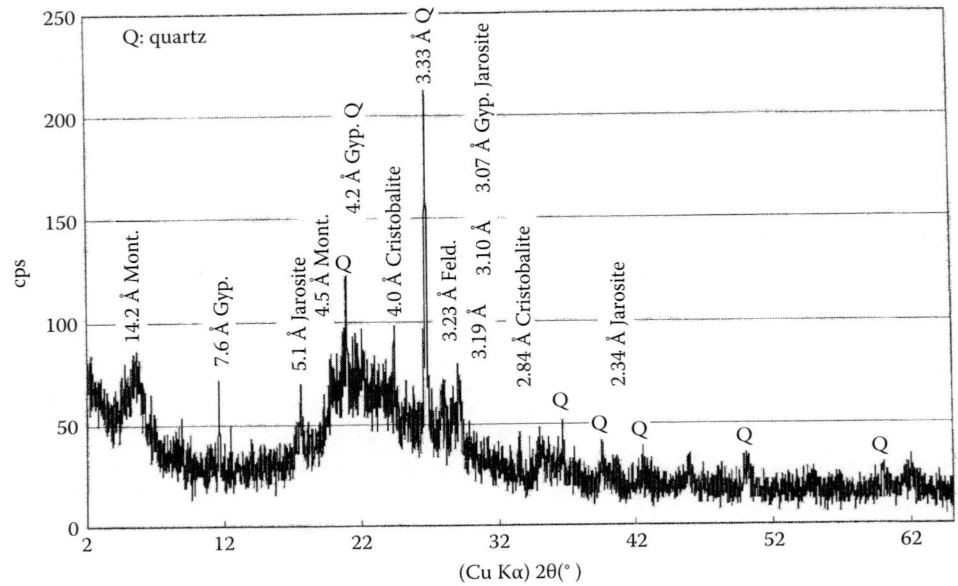

FIGURE 31.4 XRD pattern of diatomaceous earth in Noto identified clay minerals of montmorillonite associated with several kinds of minerals, such as gypsum, jarosite, cristobalite, quartz, and feldspars.

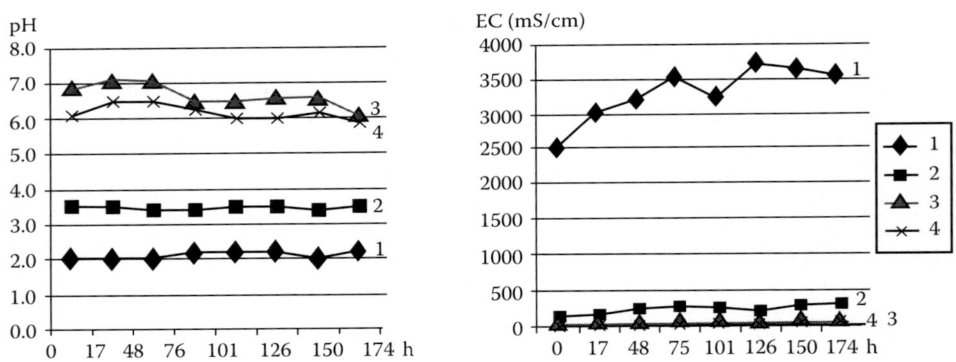

FIGURE 31.5 Comparison of aqueous characteristic examinations between diatomaceous earth (Noto) (1), rock powder (Dainennji layer) (2), zeolite (Yonezawa) (3), and granite (Kawamata town) (4), indicating different pH and EC. The diatomaceous earth is strongly acidic with a pH of 2 and an extremely high EC of 3500 mS/cm for 174 h.

H_2O_2 Treatment of Paddy Soil in Baba and Absorption Test of Diatomaceous Earth in Noto

The paddy soils collected from Baba is a tinge of brown (700 g), showing quite high radioactivity of 1135 cpm, organic content using 3.5% of H_2O_2 solution, and capacity as absorbent using a block of diatomaceous earth in Noto (Figure 31.6). Abundant organic bubbles were found after H_2O_2 treatment, suggesting large amounts of microorganisms and organic materials. The chunk of diatomaceous earth dipped in the muddy water (Figure 31.6, left) absorbed large amount of dosage (Figure 31.6, right), which transferred from the bottom (15 cm) to the surface (0–5 cm). Dosage of 800 cpm at the bottom transferred to upper portion as 970, 380, 230, and 90 cpm for each 5 cm interval after 4 month, whereas 600, 700, 420, 280, and 140 cpm after 10 month. The diatomaceous earth showed high capabilities to adsorb radioactivity.

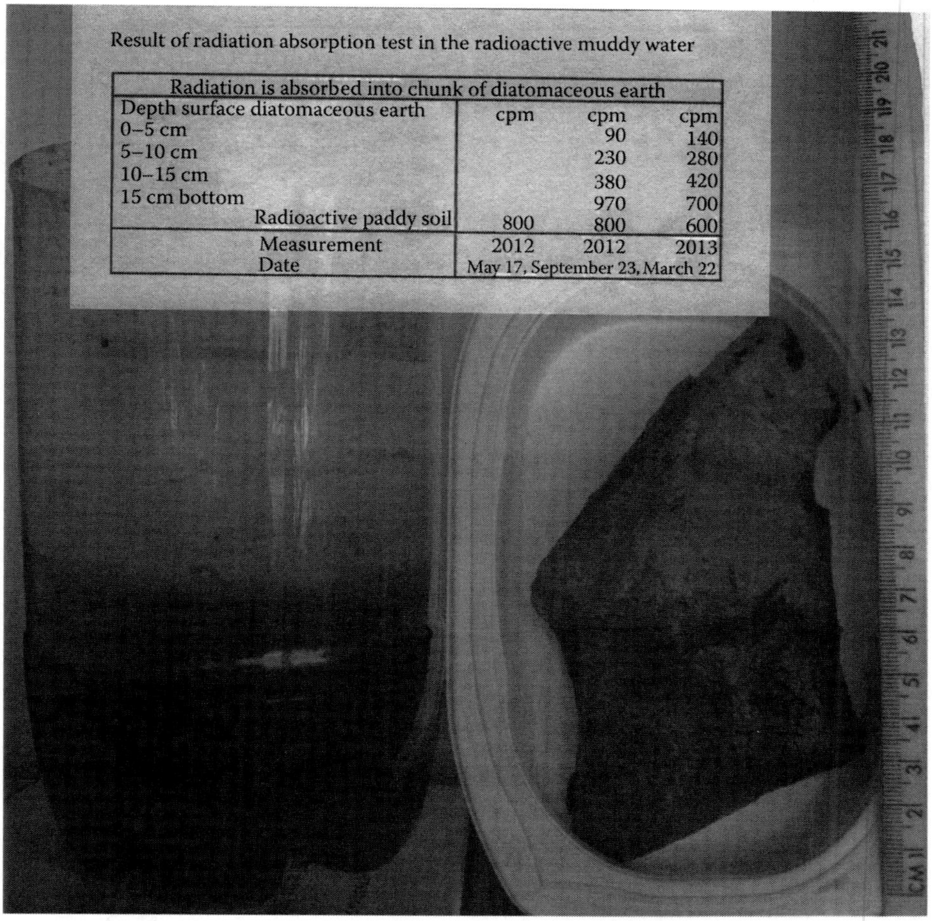

FIGURE 31.6 Photograph of radiation absorption test in the radioactive muddy water collected from Baba (left), indicating dosage of 800 cpm at the first stage on May 17, 2012. After 4 months, the muddy water absorbed to diatomaceous earth and increased dosage to the surface. After 10 months, the radioactive muddy water transferred to top of the chunk, showing 140 cpm (right).

Quantitative analyses of aqueous materials in diatomaceous earth at Noto (5 g/250 mL) showed abundant SO_4^{2-}, and Ca^{2+} was dissolved into the solution (Table 31.4) (after Miyamoto, 2007).

Electron Microscopic Observation of Diatomaceous Earth before Experiment

SEM of diatomaceous earth in Noto showed abundant diatom with clays. Completed shapes of round diatoms, 20 μm in diameter, and elongated diatoms, 40 μm in long, were mixed with small fragments and clayey grains (Figure 31.7a and b). TEM clearly showed that diatoms wear thin films in/around cell (Figure 31.8a through c), whereas bacillus-type bacteria are filled with spherules inside the cell (arrows in Figure 31.8d through f), suggesting biomineralization processes. TEM is able to observe bacteria in nm order, showing not only shape but also materials inside the bacterial cell.

The diatomaceous earth in Noto contained large amount of clays and filamentous microorganisms (Figure 31.9a). The clays showed characteristic crumple shape of montmorillonite associated with gypsum grains (Figure 31.9b). The XRD data confirmed the presence of gypsum (7.6 Å).

The elemental content maps of clay parts around diatoms distributed high concentration of Mg, Al, Si, Sr, Zr, and P, in clean (not radioactively contaminated) samples (Figure 31.9c). Traces of K, Fe, Na, Ca, Ba, S, and Cl were detected in the sample.

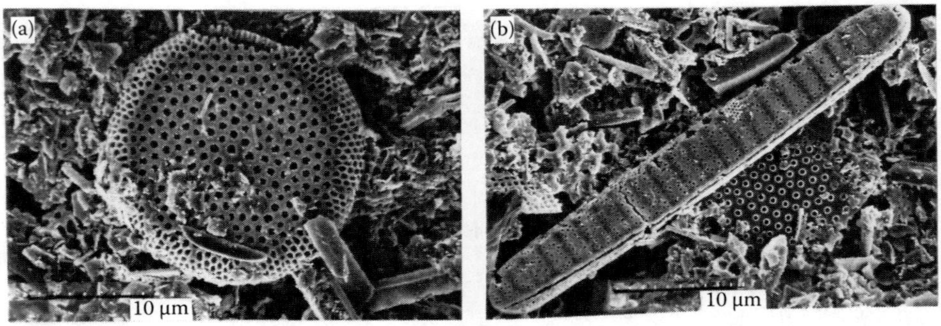

FIGURE 31.7 SEMs of diatomaceous earth in Noto showed various kinds of diatoms associated with clays. (a) marine type, (b) freshwater type.

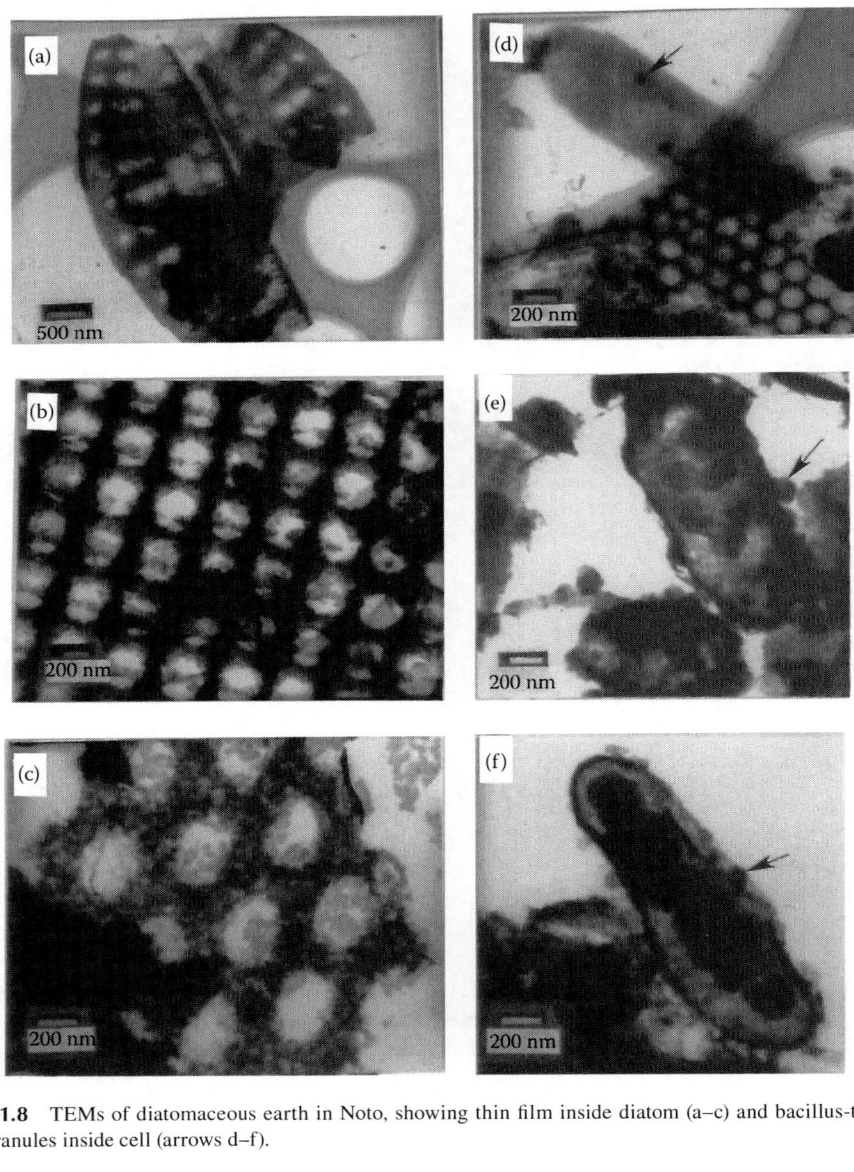

FIGURE 31.8 TEMs of diatomaceous earth in Noto, showing thin film inside diatom (a–c) and bacillus-type bacteria filled up granules inside cell (arrows d–f).

FIGURE 31.9 SEMs of diatomaceous earth show filamentous bacteria in clays (a), montmorillonite and gypsum (b), and diatom (c) in diatomaceous earth in Noto.

SEM-EDX Observation of Diatomaceous Earth in Noto, after One-Year Experimental Demonstration

SEM observation of diatomaceous earth on the surface of contaminated paddy soils in Lot No. 12, after 1-year experimental demonstration, showed clays and diatom (Figure 31.10, the upper right), the spectrum (the bottom right), and quantitative analytical data (left in Figure 31.10). The chemical composition of this area indicated MgK (0.80%), AlK (6.03%), SiK (34.03%), SK (0.75%), ClK (0.90%), KK(1.15%), CaK(0.92%), FeK (2.65%), and ZnL (1.03%), which are main components. Note that trace elements of IL (0.17%), CsL (0.19%), BaL (0.71%), ThM (1.67%), UM (0.08%), and NpM (1.01%) were

	A(%)	B(%)	C(%)	
Mg K	0.80	0.70	1.32	MgO
Al K	6.03	4.81	11.39	Al_2O_3
Si K	34.03	26.07	72.79	SiO_2
P K	0.00	0.00	0.00	P_2O_5
S K	0.75	0.51	1.88	SO_3
Cl K	0.90	0.55	0.00	
K K	1.15	0.63	1.39	K_2O
Ca K	0.92	0.49	1.29	CaO
Mn K	0.00	0.00	0.00	MnO
Fe K	2.65	1.02	3.41	FeO
Zn L	1.03	0.34	1.28	ZnO
Rb L	0.00	0.00	0.00	Rb_2O
Sr L	0.00	0.00	0.00	SrO
Y L	0.00	0.00	0.00	Y_2O_3
Nb L	0.00	0.00	0.00	Nb_2O_5
Pd L	0.00	0.00	0.00	PdO
Ag L	0.00	0.00	0.00	Ag_2O
I L	0.17	0.03	0.00	
Cs L	0.19	0.03	0.20	Cs_2O
Ba L	0.71	0.11	0.80	BaO
Ce L	0.00	0.00	0.00	Ce_2O_3
Nd L	0.00	0.00	0.00	Nd_2O_3
Th M	1.67	0.15	1.90	ThO_2
U M	0.08	0.01	0.10	UO_3
Np M	1.01	0.09	1.18	Np_2O_5
Pu M	0.00	0.00	0.00	Pu_2O_3
O	47.91	64.44		

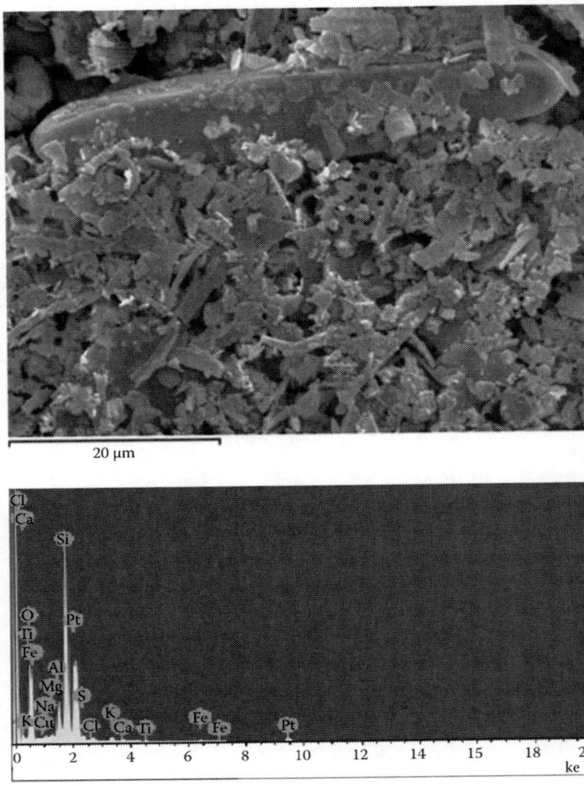

FIGURE 31.10 SEMs showed diatom with clays (the upper right), in radioactive-contaminated Lot No. 12 in Baba. The spectrum and quantitative analyses are shown at the bottom right and left side, respectively. The large amounts of Al and Si and small amounts of K and Fe were found, suggesting diatom and clay components. Note that traces of Cs, Ba, Th, U, and Np were detected. A, mass concentration; B, atomic concentration; C, compound concentration.

detected, even though I, Cs, Np, and Pu are not defined because of no standard for calculation of A (mass%). The elemental content map of this area showed visible high contents of AlKa1 and SiKa1, because of presence of diatom and clays. The Mg, S, Cl, Ca, Fe, Th, and Np are visible in the content map (Figure 31.11).

SEM observation and quantitative analytical data of diatom in Lot No. 12, after one year experiment is showed in Figure 31.12. The chemical composition of this area indicated MgK (0.52%), AlK (5.94%), SiK (35.07%), SK (0.76%), ClK (0.51%), KK (0.89%), CaK (0.46%), MnK (0.02%), FeK (2.43%), and ZnL (0.37%), which are main components. Note that trace elements of IL (0.20%), BaL (0.53%), NdL (0.18%), ThM (1.94%), NpM (1.28%), and PuM (0.39) were detected, even though I, Cs, Np, and Pu are not defined because of no standard for calculation of A (mass%). The elemental content map of this area showed visible high contents of AlKa1 and SiKa1, because of presence of diatom and clays. The Mg, K, Ca, S, I, Th, Nd, and Pu are visible in the content map (Figure 31.13). The small grain at the bottom right concentrated K, Np, and Pu.

The chemical composition of other area indicated MgK (0.49%), AlK (5.74%), SiK (35.78%), SK (0.65%), ClK (0.53%), KK (0.96%), CaK (0.33%), MnK (0.06%), FeK (2.60%), and ZnL (0.36%), which are main components. Note that trace elements of IL (0.07%), BaL (0.64%), NdL (0.07%), ThM (1.76%), UM (0.01%), NpM (1.07%), and PuM (0.02%) were detected, even though I, Np, and Pu are not defined because of no standard for calculation of A (mass%). The elemental content map of this area showed visible high contents of Al Ka1, SiKa1, and ZrLa1, because of presence of diatom and clays (Figures 31.13 and 31.14). The S, Cl, Ca, K, Fe, Ba, Pu, and Np are visible in the content map (Figure 31.11). Note that small grain on the right concentrated Al, K, Pu, and Np (Figure 31.14).

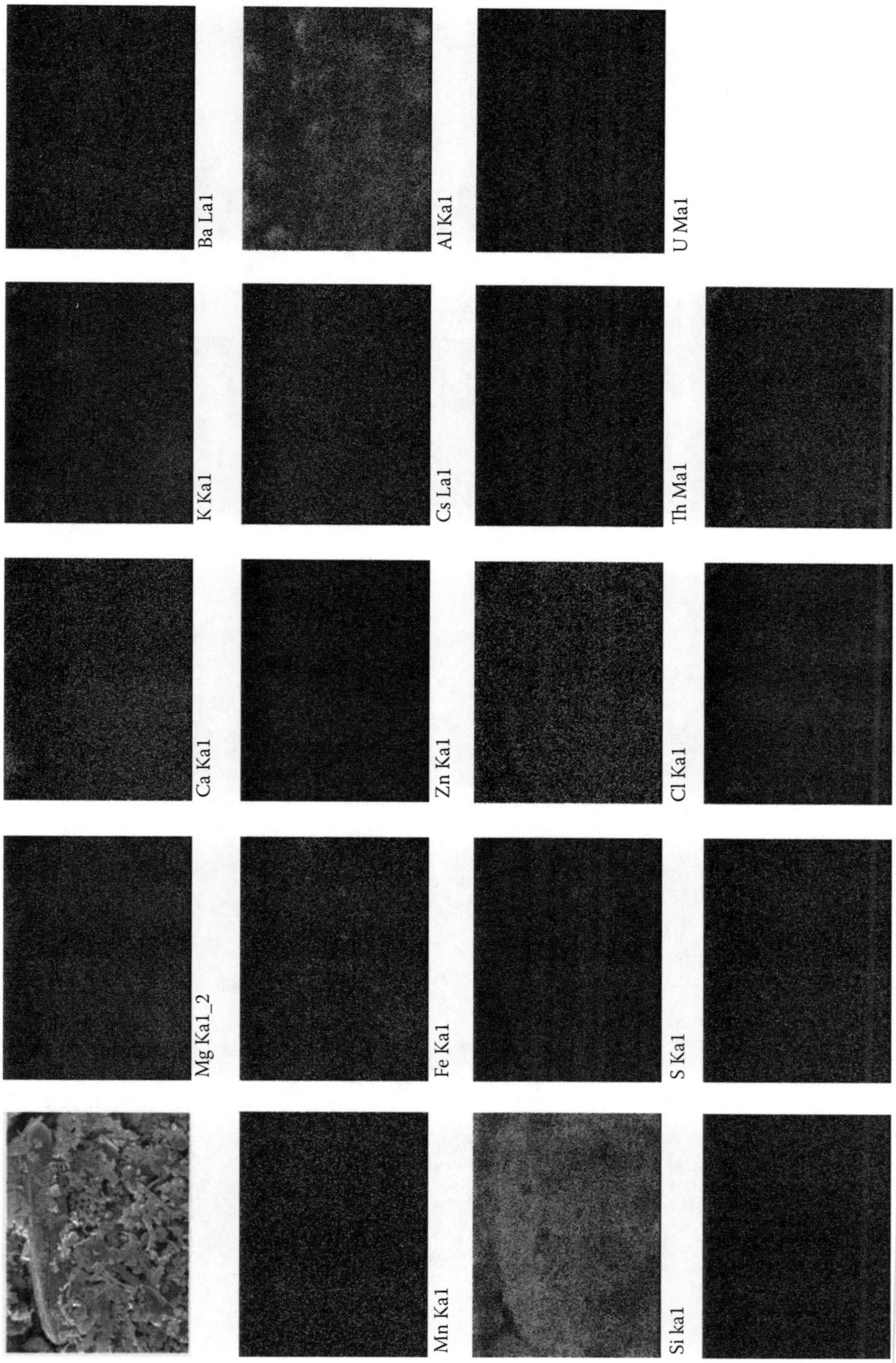

FIGURE 31.11 Elemental content maps showed abundance of Al and Si based on clay and diatom components, associated with trace elements. S, Cl, Ca, K, Fe, Ba, Pu, and Np are visible in the content map.

	A(%)	B(%)	C(%)	
Mg K	0.52	0.46	0.86	MgO
Al K	5.94	4.72	11.22	Al_2O_3
Si K	35.07	26.77	75.03	SiO_2
P K	0.00	0.00	0.00	P_2O_5
S K	0.76	0.51	1.89	SO_3
Cl K	0.51	0.31	0.00	
K K	0.89	0.49	1.08	K_2O
Ca K	0.46	0.25	0.65	CaO
Mn K	0.02	0.01	0.03	MnO
Fe K	2.43	0.93	3.13	FeO
Zn L	0.37	0.12	0.46	ZnO
Rb L	0.00	0.00	0.00	Rb_2O
Sr L	0.00	0.00	0.00	SrO
Y L	0.00	0.00	0.00	Y_2O_3
Nb L	0.00	0.00	0.00	Nb_2O_5
Pd L	0.00	0.00	0.00	PdO
Ag L	0.00	0.00	0.00	Ag_2O
I L	0.20	0.03	0.00	
Cs L	0.00	0.00	0.00	Cs_2O
Ba L	0.53	0.08	0.60	BaO
Ce L	0.00	0.00	0.00	Ce_2O_3
Nd L	0.18	0.03	0.21	Nd_2O_3
Th M	1.94	0.18	2.21	ThO_2
U M	0.01	0.00	0.00	UO_3
Np M	1.28	0.12	1.50	Np_2O_5
Pu M	0.39	0.03	0.42	Pu_2O_3
O	48.49	64.97		

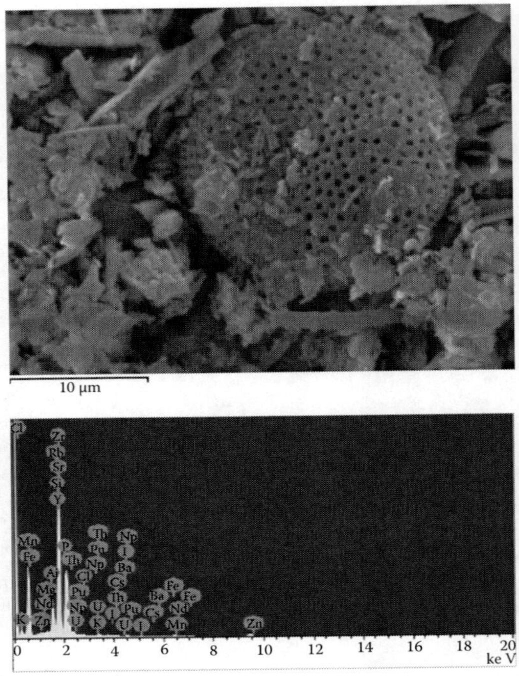

	A(%)	B(%)	C(%)	
Mg K	0.49	0.42	0.80	MgO
Al K	5.74	4.52	10.84	Al_2O_3
Si K	35.78	27.11	76.55	SiO_2
P K	0.00	0.00	0.00	P_2O_5
S K	0.65	0.43	1.62	SO_3
Cl K	0.53	0.32	0.00	
K K	0.96	0.52	1.16	K_2O
Ca K	0.33	0.18	0.47	CaO
Mn K	0.06	0.02	0.08	MnO
Fe K	2.60	0.99	3.53	FeO
Zn L	0.36	0.12	0.45	ZnO
Rb L	0.00	0.00	0.00	Rb_2O
Sr L	0.00	0.00	0.00	SrO
Y L	0.00	0.00	0.00	Y_2O_3
Nb L	0.00	0.00	0.00	Nb_2O_5
Pd L	0.00	0.00	0.00	PdO
Ag L	0.00	0.00	0.00	Ag_2O
I L	0.07	0.01	0.00	
Cs L	0.00	0.00	0.00	Cs_2O
Ba L	0.64	0.10	0.72	BaO
Ce L	0.00	0.00	0.00	Ce_2O_3
Nd L	0.07	0.01	0.09	Nd_2O_3
Th M	1.76	0.16	2.00	ThO_2
U M	0.01	0.00	0.01	UO_3
Np M	1.07	0.10	1.25	Np_2O_5
Pu M	0.02	0.00	0.02	Pu_2O_3
O	48.85	64.98		

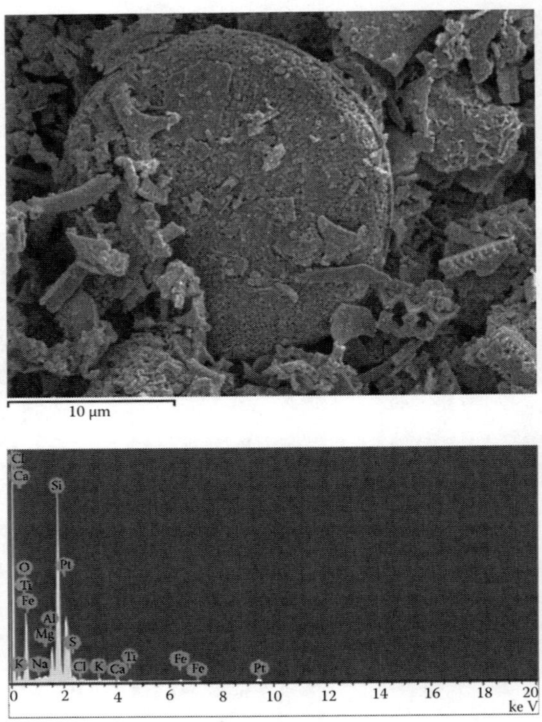

FIGURE 31.12 SEMs showed diatom with clays (the upper right), in radioactive-contaminated Lot No. 12 in Baba. The spectrum and quantitative analyses are shown at the bottom right and left side, respectively. The large amounts of Al and Si and small amounts of S, Cl, K, and Fe were found, suggesting diatom and clay components. Note that traces of I, Ba, Nd, Th, Np, and Pu were detected. A, mass concentration; B, atomic concentration; C, compound concentration.

FIGURE 31.13 Elemental content maps showed abundance of Al and Si based on clay and diatom components, associated with trace elements. Note that small grains showed concentration of K, Ba, Pu, and Np.

FIGURE 31.14 Elemental content maps showed abundance of Al, Si, and Zr based on clay and diatom components, associated with trace elements. The small grain that contains K, Pu, and Np was found whereas only Ba grain can be seen. Note that small grains showed concentration of K, Ba, Pu, and Np.

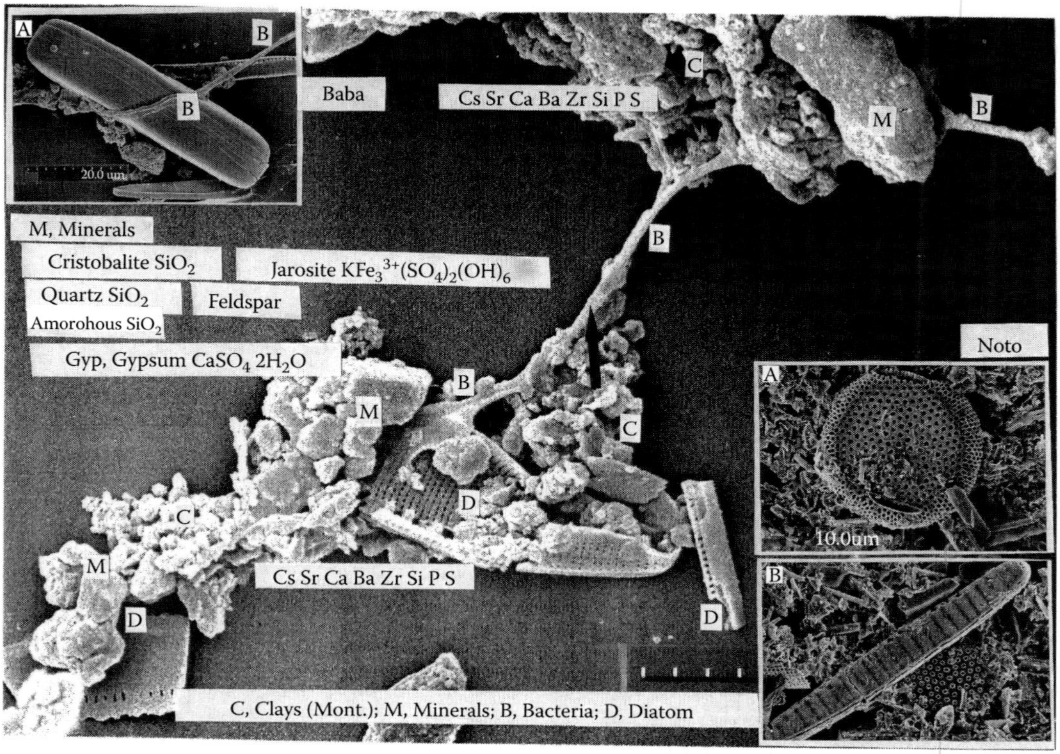

FIGURE 31.15 Clays–microbes–radionuclide interaction processes, displaced with radionuclides in the demonstration paddy soils in Baba.

Schematic SEM-EDX Data of Diatomaceous Earth in Noto, after One-Year Experimental Demonstration

Schematic diagram of elemental content map of diatomaceous earth in Lot No. 12 indicated clays (C), minerals (M), bacteria (B), and diatom (D), as shown in Figure 31.15. The schematic data shown adsorbed both stable isotope and radionuclides, such as Cs and Sr in clays, diatoms, and bacteria.

Discussion

The discovery described here is largely expected to be applied in Fukushima for speeding up the recovery of radioactive polluted soils, which contain clay minerals, diatoms, and microorganisms. SEM-EDX observation showed abundant filamentous and bacillus bacteria, which were covered by clay particles. It is now known that bacteria, eukaryotic algae, fungi, and moss plants absorb Cs^+ (Lovley, 2000; Cygan and Tasaki, 2014). Uptake of Cs^+ is particularly promoted by metabolizing microorganisms. Due to similarity of the K^+ and Cs^+, both ions are taken up by the same metabolism-dependent transport systems. Clays–diatomaceous earth–microbes–radionuclide interaction based on radioactive pollution from FDNPP suggested transportation, concentration, interaction, displacements, and biomineralization processes in the soils discussed in the Figure 31.15.

Clay Minerals: Radionuclide Interactions

Radionuclides of ^{137}Cs and ^{90}Sr are partially adsorbed on the surface of clay minerals and corrosion, and especially ^{137}Cs is highly fixed in the structure of 2:1 typed clay minerals, reacting as the same as K and

NH (Tsumura et al., 1984). The capability of fixing ^{137}Cs is as follows: montmorillonite ≫ > allophone > kaolinite. That is the reason why diatomaceous earth at Noto has higher stability of ^{137}Cs, because of high contents of smectite. The stability of ^{137}Cs and ^{90}Sr depends on coexisting cations in the soils. In the soils at Takada, Niigata Pref., is as follows:

$$Rb^+ = Cs^+ \gg NH_4^+ > K^+ > Ba^{2+} = Sr^{2+} = Na^+ = Ca^{2+} = Mg^{2+} > \text{Cont.}$$

The Cs is not easily transferred, floated, and drifted across the soils by rainfall (2000 mL). But the ^{90}Sr moved and transferred to deeper parts in the depth of 20–30 mm/h (Tsumura et al., 1984). The half-life time of ^{134}Cs (2.1 years), ^{137}Cs (30.0 years), ^{90}Sr (29.1 years), and ^{95}Zr (64.0 days) is different; therefore, radio activities of ^{134}Cs and ^{95}Zr had been reduced by more than half now.

In case of diatomaceous earth at Noto, the contents of both stable isotope and radionuclide of Cs can be fixed on the surface of the paddy soils and decreased the quantity of Cs by time, as confirmed in this study.

Microorganisms–Radionuclides Interaction

Microorganisms can interact with radionuclides via several mechanisms, some of which may be used as the basis of potential bioremediation strategies. Mechanisms of radionuclides–microbe interactions are biological sorption, bioaccumulation, biomineralization, biotransformation, and microbiologically enhanced chemical sorption (Lioydl and Macaskie, 2000; Eby, 2002). Radioactive characteristics of bacterial bioremediation using clays were studied with paddy soil samples collected in Nagadoro, Iitate, Fukushima Prefecture, in Japan. Bacterial cells are entirely encrusted with particles of nm-sized clay minerals (Tazaki, 2012). Based on the TEM data, we can hypothesize that the organic walls were used as nutrients by fossilizing bacteria with clays.

Bacterial leaching methods are commonly used in Canada, Spain, and Brazil mines, such as uranium ore (UO_{2+x}) conditions under pH of <4 and rich in U, Ca, and P, using *Acidithiobacillus* (*Thiobacillus*) *ferrooxidans*. The interactions of U(VI) or Pu(VI) with *Bacillus subtilis*, kaolinite clay, and mixture of *B. subtilis* and kaolinite clay were studied (Yamanaka, 2003). The accumulation of U by the mixture increased with an increase in the amount of *B. subtilis* cells. The treatment of kaolinite with CH_3COOK removed approximately 80% of the associated uranium. TEM-EDS analysis confirmed that most of the U removed from solution was associated with *B. subtilis*. SEM-EDS analysis indicated that most of the Pu in the mixture was associated with *B. subtilis*. These results strongly suggested that U(VI) and Pu(VI) are preferably absorbed to bacterial cells, even though the mechanism of accumulation differs between them (Ohnuki et al., 2007):

$$Ca\,(UO_2)_2(PO_4)_2 + 2Fe_2\,(SO_4)_3 + H_2SO_4 + 4H^+$$

$$\dashrightarrow 2UO_2SO_4 + 4FeSO_4 + 2H_3PO_4 + CaSO_4$$

$$UO_2 + Fe\,(SO_4)_3 \dashrightarrow UO_2SO_4 = 2FeSO_4\,(UO_2SO_4 \text{ is aqueous})$$

$$2H^+ + SO_4^{2-} + Fe^{3+} \dashrightarrow CaCO_3 \text{ add} \dashrightarrow CaSO_4 \cdot 2H_2O \text{ (Gypsum)}$$

$$K^+ \text{ add} \dashrightarrow KF_3\,(SO_4)_2(OH)_6 \text{ (Jarosite)}$$

Diatomaceous earth on the surface of radioactive polluted paddy soils in Baba, associated with Cs, Sr, U, and Pu under acidic pH condition.

The formation of gypsum and jarosite minerals suggested biological sorption, bioaccumulation, biomineralization, biotransformation, and microbiologically enhanced chemical sorption (Figures 31.15 and 31.16).

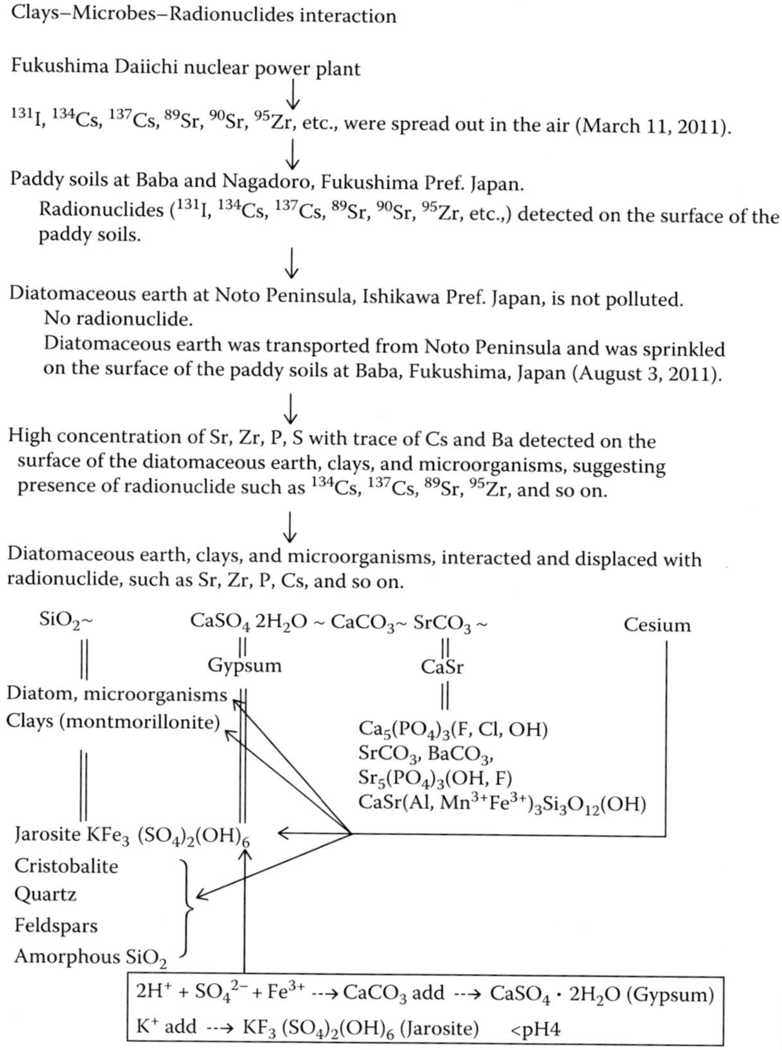

FIGURE 31.16 Clays–microbes–radionuclide interaction processes and description of radioactive-polluted paddy soils in Baba suggested biological sorption, bioaccumulation, biomineralization, biotransformation, and microbiologically enhanced chemical sorption.

Concentration of ^{237}Np

The ^{237}Np in hemp-palm leaves of *Bontenchiku* for fishing gear used by the fifth Fukuryu-Maru (Lucky Dragon) 40 years after *Bravo* had been found and reported by Yamamoto et al. (1996). This gear was exposed to fallout from the second thermonuclear test explosion (Bravo) at Bikini Atoll in the Republic of the Marshall Islands on March 1, 1954. The ^{237}Np content in hemp-palm leaves of *Bontenchiku* sample was determined to be 11.5 ± 0.8 mBq/g, with an activity ratio of ^{237}Np:$^{239,240}Pu$ and an atom ratio of ^{237}Np:^{239}Pu estimated to be $(2.2 \pm 0.2) \times 10^{-3}$ and 0.42 ± 0.04, respectively. The data showed the existence of a chain reaction of ^{238}U and its ratio to be ^{237}Np:^{239}Pu as well as the presence of ^{237}U at the time of fallout from Bravo event in March 1954.

The other radionuclides, such as plutonium isotopes, ^{241}Am and uranium isotopes, ^{137}Cs, and ^{60}Co, were found to be still remaining in this sample. In addition to actinide elements, it is important to note that ^{60}Co could still be detected by nondestructive γ-ray spectrometry, even after an elapse of 40 years from the explosion.

Biomineralization and Bioremediation

Biomineralization and bioremediation of radioactive materials are new applications of traditional mineralogy and clay mineralogy to understand the chemical, physical, and microbiological environmental systems on Earth. The bacteria exposed to radionuclides may become resistant to or even capable of chemically transforming and detoxifying radionuclides. The bacteria produce extracellular polymers around the cells, which form capsules and slime layers, defending them from radiation. It is possible that radioactive biofilms and microbial mats are capable of immobilization of radioactive materials and can be used to counteract the disastrous effects of radionuclides polluted water and soils.

Bioaccumulation of uranium by lower plants on the weathered granite and microorganisms in the soils and biomineralization of radioactive sulfide minerals have been shown in strongly acidic Tamagawa Hot Springs, radiochemical aspect, and biomineralization of microbial mats in radioactive hot springs in Japan (Tazaki et al., 2003, 2007; Tazaki and Watanabe, 2004; Yamamoto et al., 2005; Tazaki, 2009).

Radioactive Waste

Microbial characteristic of bentonite-based buffer materials employed in a number of large-scale experiments was studied in order to address the potential for microbial activity and its possible consequences in a Canadian deep geological repository. Results showed that cultural populations of heterotrophic aerobes, anaerobes, and sulfate-reducing bacteria were present at all locations examined (Stroes-Gascoyne, 2010). The organic matter and microorganisms may be resistant to microbial degradation, surrounding buffer to a low level of radiation. The radiation resistance of the natural microbial population in the buffer and of a radiation-resistant organism added to the buffer was studied in order to establish if radiation would affect the microbial activity in the buffer.

In this study, the discovery of diatom–clays–radionuclide ecosystem described here, using SEM-EDX in the radioactive polluted paddy soils in Fukushima could be effective as the shield for low-level radioactive waste at the geological disposal. While the benefits of local and indigenous knowledge in reducing disaster risk are increasingly being acknowledged, a challenge remains: to integrate such knowledge constructively with scientific knowledge and policy.

Acidic Condition of Diatomaceous Earth

The release of U, Th, Ra, Rn, and As to the environments is clearly related to the physical stability of the tailings disposal facility, including engineered covers, and the chemical stability of mineral phase controlling elemental concentrations in solution. Fe oxyhydroxides are key phases that control radionuclide and hazardous element concentrations in solution. In low-pH waters, desorption of radionuclides and hazardous elements occur, increasing their concentration, as shown in Figure 31.5. Seepage of such acid water may affect the surrounding environment, including soils, surface water, and groundwater that are direct threats to nearby ecosystems. Recently, effort has focused on better containment of radioactivity by means of managed drainage and retention ponds. Uraninite particles precipitated on an indigenous bacterium grown in amended groundwater from the Tuba City mill tailings site, Arizona, United States (Abdelouttall, 1998).

Conclusions

Coastal areas in Minamisoma City, Fukushima, Japan, were seriously damaged by the radioactive contamination from FDNPP that caused multiple pollutions by both tsunami and radionuclide exposure, after the Great East Japan Earthquake, on March 11 and 12, 2011. The diatomaceous earth at Noto Peninsula has high properties to adsorb both radionuclide (I, Cs, Ba, Nd, Th, U, Np, and Pu) and stable isotope elements confirmed by ED-XRF, Ge semiconductor detector, and SEM-EDX quantitative analysis associated with elemental content maps, based on radiation measurements for 1 year at demonstration of paddy field in Baba, Fukushima. The diatom–clays–radionuclide interaction in the radioactive polluted paddy soils could be effective as the shield for low-level radioactive waste at the geological disposal. This type of remediation is safe, inexpensive, simple and easy, and sustainable and uses local natural materials.

ACKNOWLEDGMENTS

The authors wish to thank the Syoken Co., K. Kiji, S. Yamamoto, Dr. H. Takahashi, K. Imoto, K. Gonjou, and Kelly Brandon for their kind support on this research.

REFERENCES

Abdelouttall, E. (1998) Chemical reactions of uranium in ground water at a mill tailing sites. *Journal of Contaminant Hydrology*, **34**, 343–361.

Atomic Energy Safety Agent. (2011) Results of emitted radionuclide examination on June 6, 2011. http://www.meti.go.jp/press/2011/06/20110606008/20110606008.html.

Cygan, R.T. and Tazaki, K. (2014) Interactions of kaolin minerals in the environment. *Elements*, **10**, 195–200.

Haas, J.R., Bailey, E.H., and Purvis, O.W. (1998) Bioaccumulation of metals by lichens: Uptake of aqueous uranium by *Peltigera membranacea* as a function of time and pH. *American Mineralogist*, **83**, 1494–1502.

Lioydi, J.R. and Macaskie, L.E. (2000) Bioremediation of radionuclide-contaminated wastewaters. In: Lovley, D.R., ed., *Environmental Microbe-Metal Interactions*. pp. 277–327. ASM Press, Washington, D.C..

Lovley, D.R. (2000) *Environmental Microbe-Metal Interactions*. ASM Press, Washington, DC, 400pp.

Macfarlane, A.M. and Miller, M. (2007) Nuclear energy and uranium resources. *Elements*, **3**, 185–192.

Miyamoto, M. (2007) On the basis of quality of Noto-diatomite (diatomaceous earth) and its application. Ishikawa Prefecture. Report in Japanese, pp. 1–3.

Mulligan, C.N. (2007) *Environmental Biotreatment: Technologies for Air, Water, Soil, and Wastes*. Tokai University Press, Tokyo, Japan, 403pp.

Mulligan, C.N., Yong, R.N., and Fukue, M. (November 26–28, 2007) Some effects of microbial activity on the evolution of clay-based buffer properties in underground repositories. In: *Workshop on Long-Term Performance of Smectitic Clays Embedding Canisters with High Radioactive Waste*. Lund, Sweden, pp. 298–305.

Ohnuki, T., Ozaki, T., Nankai, T., Kozaki, N., Sakamoto, F., Suzuki, Y., and Francis, A.J. (2007) Association of actinides with microorganisms and clay. *Proceedings of the Symposium*, Lund, Sweden. pp. 266–273.

Pusch, R. (2007) Workshop on long-term performance of smectitic clays embedding canisters with highly radioactive waste. *Proceedings of the Symposium*, Lund, Sweden. 355pp.

Strocs-Gascoyne, S. (2010) Microbial occurrence in bentonitic-based buffer, backfill and scaling materials from large-scale experiments at AECL's underground research laboratory. *Applied Clay Science*, **47**, 36–42.

Suzuki, Y., Kelly, S.D., Kemmer, K.M., and Banfield, J.F. (2003) Microbial populations stimulated for hexavalent uranium reduction in uranium mine sediment. *Applied and Environmental Microbiology*, **69**, 1337–1346.

Suzuki, Y. and Suko, T. (2006) Geomicrobiological factors that control uranium mobility in environment: Update on recent advances in the bioremediation of uranium-contaminated sites. *Journal of Mineralogical and Petrological Sciences*, **101**, 299–307.

Tazaki, K. (2009) Observation of microbial mats in radioactive hot springs. *Science Reports of Kanazawa University*, **53**, 25–37.

Tazaki, K. (2012) Possibility method of reduced radionuclide in paddy soils using microorganisms. In: Saitou, K., ed., *Countermeasure Against Radioactive Pollution after the Great East Japan Earthquake*. KK NTS, Tokyo, Japan, pp. 120–136.

Tazaki, K., Ishiguro, T., and Saji, I. (2003) Bioaccumulation of uranium by lower plants on the weathered granite and microorganisms in the soils. In: Li, S., Shen, J., and Xu, H., eds., *Mineralogy and Geochemistry: Resources, Environment and Life*. Geological Publishing House, China pp. 37–58.

Tazaki, K., Shirotori, T., and Charn, S.K. (2007) Radiochemical aspect and bio-mineralization at hot springs in Japan. In: *Workshop on Long-Term Performance of Smectite Clays Embedding Canisters with Highly Radioactive Waste*, Geo-development International AB in Co-Operation with Swedish Nuclear Fuel and Waste Management Co. (SKB). Lund, Sweden and Posiva Oy, Finland, pp. 274–285.

Tazaki, K., Takahashi, M., Suzuki, S., and Imoto, K. (2011) Measurement of atmospheric beta-ray after the Great East Japan Earthquake, Fukushima, Japan. *Earth Science*, **65**, 175–177.

Tazaki, K. and Watanabe, H. (2004) Biomineralization of radioactive sulfide minerals in strong acidic Tamagawa Hot Springs. *Science Reports of Kanazawa University*, **49**, 1–24.

Tsumura, A., Komamura, M., and Kobayashi, H. (1984) Behavior of radioactive Sr and Cs between soils and soils-botanical systems. *Agro-Technical Report Series B.*, **36**, 57–113.

Yamamoto, M., Ishiguro, T., Tazaki, K., Komura, K., and Ueno, K. (1996) ^{237}Np in hemp-palm leaves of *Bontenchiku* for fishing gear used by the fifth Fukuryu-maru, 40 years after "Bravo". *Health Physics Society*, **70**, 744–748.

Yamamoto, M., Tomita, J., Sakaguchi, A., Shirotori, T., Shiraki, K., and Tazaki, K. (2005) Radiochemical aspect in Hokutolite collected recently from Peito Hot Spring, Taiwan. In: *Proceeding of 100 Years Anniversary Symposium of Hokutolite in Taiwan*, pp. 1–6.

Yamanaka, T. (2003) *Introduction to Microbiology Concern with Environment*. KK Kodansha, 129pp.

32

Healing the Earth by Healing the Waters: Recycling Waste Nutrients Using Biochar and Limestone

Kirk D.S. Jones

CONTENTS

Introduction

Exploitation and mismanagement of Earth's natural resources has gone too far. The human race has evolved into such a large, dominant network; increasingly aware of our effects on Earth's ecosystems, we realize that we must collectively manage our finite resources for the benefit of all life. We will have no choice but to mimic and design with ecology, ensure species diversity, support abundant polycultural food systems, optimize global nutrient recycling, and transform pollution while creating natural storage systems for carbon. Using our infinite talent and imagination, we can guide the Earth's healing evolution. When the Earth is well respected and taken care of through proper management, she is capable of easily providing for everyone, if it is the collective will of the people to care for her and her diverse forms of life, thereby caring for ourselves and each other.

This chapter offers one solution for a myriad of problems. It presents original research on developing a nutrient and carbon sequestering agricultural runoff filtration media that, when applied to land, can enhance crop yields and support long-term soil and water health. This media can also be used in wastewater treatment, aquatic habitat restoration, subsurface remediation, and other innovative nutrient cycling systems. The result of this research is to advocate the use of biochar and biogenic limestone as part of a total biogeochemical cycling plan in order to consciously steward the Earth in perpetuity.

Need for Action

Water bodies around the world are suffering from excess nutrients, while farmers struggle to keep nutrients in their soil. To tackle the concomitant problems of global climate change, eutrophication, rising background toxicities, ozone depletion, land use change, species diversity loss, and opportunist species gains, a universal global bioethics is a necessary foundation on which to develop global natural resource management plans. Unified global environmental policy is necessary. Richard Grantham, striving for global proactive geotherapy, advocated the creation of a *World Ecological and Demographic Council* to help shift humanity toward long-term planning for biospheric survival (Grantham 1992a,b). Global action taken by the Montreal Protocol to begin to reverse stratospheric ozone depletion is a proof that humanity can come together and implement effective legislation to safeguard this planet (Mäder et al. 2010). We must utilize this cooperative potential to solve the larger problem of global warming.

Limits

Atmospheric carbon dioxide (CO_2) has reached 400 ppm (Scripps Inst. Oceanography/UC San Diego 2013), while global climate scientists have repeatedly warned that we need to return to below 350 ppm to avoid ecological climate catastrophe (Hansen et al. 2008; Webb III and Bartlein 1992). Above 350 (as we have been experiencing since 1988) means more droughts, fires, floods, storms, rising sea levels, shrinking glaciers, ocean acidification, soil erosion, and overall general stress on human and other natural ecosystems. As far back as our climate records stretch, CO_2 and temperature closely correspond, while CO_2 has not been above 400 ppm in *millions* of years (Hansen et al. 2008; Zhang et al. 2013). The recent increase is mostly due to the burning of fossil fuels, but a significant fraction of greenhouse gases are also coming from deforestation and agriculture (Rogner et al. 2007).

Conventional agriculture is reaching its limits. Globally, groundwater depletion in dry lands has more than doubled from 1960 to 2000 (Wada et al. 2010). Global population is projected to reach 9.3 billion by 2050 while agricultural soils are being lost 10–40 times faster than they are being created (Pimentel and Burgess 2013). Practices such as clearing native vegetation, tilling soil, overgrazing, and applying agrochemical pesticides and fertilizers are causing soil to dry out, erode, compact, lose beneficial microbial life, oxidize soil organic matter, and release greenhouse gases (Kibblewhite, Ritz, and Swift 2008; Lal 2003; Oldeman 1992). When soils lose their ability to retain water, runoff increases. This exacerbates soil and bank erosion while creating water pollution with suspended solids, nutrients, and pesticides. Soil loss and degradation are heightened with tropical temperatures and heavy rainfall.

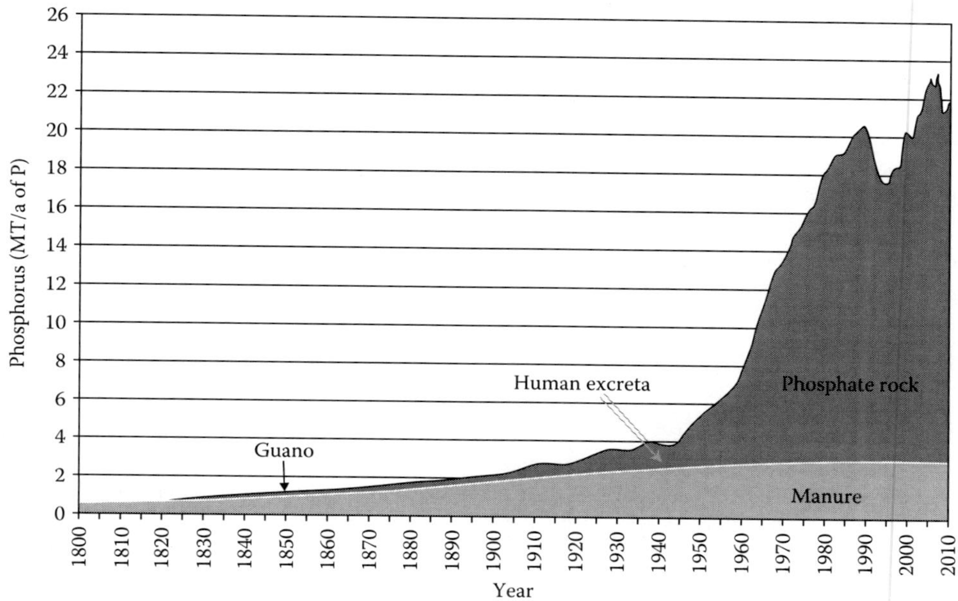

FIGURE 32.1 The exponential rise in phosphate rock fertilizer production mirrored the repurposing of large-scale ammonia production from munitions to agriculture post-WWII due to the need to maintain N–P–K fertilizer ratios. (From Cordell, D. et al., *Glob. Environ. Change*, 19(2), 292, 2009.)

Organic matter and organic nutrients are removed from the agricultural system with every harvest and are not returned in sufficient amounts to be sustainable. These crops are fed to humans and other animals, yet the waste products (excrement, food scraps, and other residues), rich in the very life-sustaining elements that agriculture needs, are largely lost to landfills, pristine surface waters, and coveted drinking water supplies (through erosion and leaching). In landfills, this wasted organic matter rots into global warming gases that are occasionally (under)utilized for electricity generation. In rivers, lakes, and oceans, these nutrients cause opportunist algal overgrowth and toxic blooms, resulting in species loss, dead zones, fish kills, coral death, and danger to human and other life.

Another result of the problem of treating food scraps and human and animal urines and manures as wastes is that one very essential element, phosphorus (P), is reaching its own global limits (Figure 32.1). Most of the world's agriculture depends on the mining of economically available phosphate rock and subsequent processing into highly bioavailable forms of P. The estimated 50–100 years of *fossil* phosphate rock left on Earth is concentrated in a handful of countries, and as the price goes up due to dwindling supplies, the quality is going down with less P and more toxic heavy metals. Worse still, every ton of phosphoric acid produced in this fashion creates 5 tons of radioactive phosphogypsum that is typically stockpiled (Cordell et al. 2009).

What We Can Do

Carbon

Numerous options exist for global carbon management. Typical climate change mitigation plans include cutting energy consumption, phasing out fossil fuels, phasing in the renewable energy infrastructure, and sequestering CO_2. Long-term sequestration is facilitated through slowing deforestation while energizing sustainable forest use, reforestation, and afforestation efforts. Short-term sequestration payback can be realized through carbon farming methods such as no-till, deep-banded biochar application, intensive rotational grazing (with keyline plowing), cover cropping with green manures, intercropping, and forest gardening, while instituting principles of conservation agriculture, agroecology, agroforestry, and permaculture.

Improved agriculture and forestry techniques could reduce CO_2 by about 50 ppm by 2150 (Hansen et al. 2008). This would have to be accompanied by replacing coal, petroleum, and natural gas with renewable energy sources to stop fossil carbon emissions.

Currently, we can replace the fossil fuel economy by rebuilding a fuel infrastructure utilizing a mixture of biomass, geothermal, passive and active solar, wind, hydro, and tidal. Estimates of the potential contribution of biomass to the current U.S. energy supply vary from 21% to 60% depending on land and crop assumptions and whether including agricultural residues, forestry residues, and municipal solid wastes (Kreith 2011). The Royal Society states that CO_2 can be reduced by 50–150 ppm through the development of biomass energy with carbon sequestration and reduced by 10–50 ppm through the use of biochar (Royal Society (Great Britain) 2009).

Nutrients

Similar conservation strategies must be implemented with respect to nitrogen and phosphorus. Organic and inorganic sources must be recycled and not landfilled or lost to aquatic ecosystems. When fertilizers are applied to agricultural lands, these nutrients are either taken up by plants, locked in the soil, or are lost from the soil by runoff, leaching, or volatilization. Runoff losses are mitigated through fertilizer application timing, fertilizer application method, tillage, and the presence of vegetated buffers. Leaching of nutrients can be slowed by applying biochar and calcium carbonate to soils, alone or combined with fertilizers or manure. Biochar, like activated carbon (AC), can adsorb many organic and inorganic compounds, while data in this chapter show that calcium carbonate adsorbs and prevents the leaching of phosphate and possibly ammonium. Volatile release of nitrogen compounds is also mitigated by biochar. Ammonia from the breakdown of organic matter can be directly adsorbed, and biochar has been shown to reduce the release of the potent greenhouse gas nitrous oxide (N_2O) by an average of 55% in 15 different soils (Cayuela et al. 2013).

Phosphate availability is dependent on soil pH and dominant cations. If the soil is acidic, phosphate will be very tightly bound to iron or aluminum (Figure 32.2). If the soil is alkaline, phosphate is typically bound to calcium, but not as tightly. Phosphate is most available at a pH of 6.2–7.2. In order to increase the availability of phosphate in acid soils, a liming agent such as calcium hydroxide, biochar, or calcium carbonate can be added. Phosphate availability is also mediated by plant roots and the presence of soil microorganisms. Plant roots and soil microbes can solubilize sparingly soluble aluminum, iron, and calcium phosphates and mineralize phosphorus from organic substances through the excretion of organic acids and enzymes (Njira 2013). Nearly all plants take up phosphate through symbiotic exchange with mycorrhizal fungi. The mycorrhizae solubilize bound phosphorus and nitrogen in exchange for carbohydrates from the plant (Hause and Fester 2004).

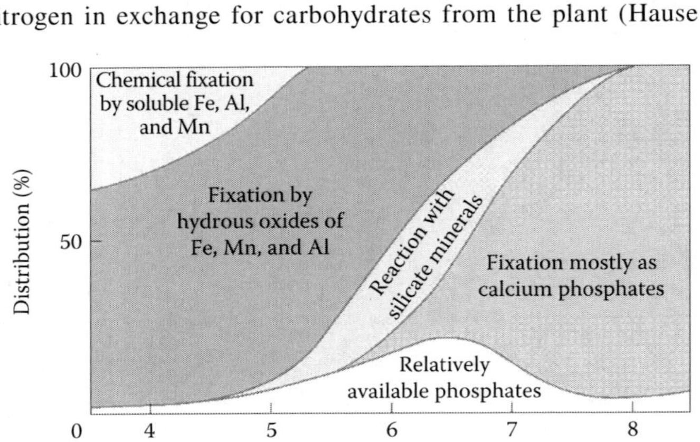

FIGURE 32.2 Solubility of phosphorus in the soil as a function of pH. Phosphorus availability is limited by iron and aluminum at acidic pH and calcium at alkaline pH. Phosphorus is most available at a pH near 7. (From Schlesinger, W.H., *Biogeochemistry: An Analysis of Global Change*, 3rd edn, Elsevier/Academic Press, New York, 2012.)

Chemical fertilizers inhibit the growth of mycorrhizal networks, while addition of biochar has been shown to benefit mycorrhizae (Warnock et al. 2007).

Nutrients taken up by plants are found in harvested crops or in crop residues. Crop residues can be left in place, tilled into the soil, composted, or processed into biofuels and/or biochar. The harvested portion enters the food and feed system, where, ideally, the food scraps are returned to compost and the human and animal excreta and remains are also returned to agriculture, in order to maintain sustainable nutrient cycles.

Human and domestic animal excreta along with food and agricultural wastes contain more than enough phosphorus to provide all of the agricultural food requirements for humanity, if they are recycled to arable land and not lost to landfills and surface waters (Cordell et al. 2009). Full of bioavailable macronutrients, urine contains most of the N and P that animals excrete. A Swedish study on human urine shows a typical N–P–K of 14:2:4 (Höglund 2001), while another study of five developing countries shows that urine contains 88% of excreted N and 69% of excreted phosphorus (Jönsson 2004). As the population grows, agricultural lands are shrinking; therefore, more nutrients can be applied per acre every year, increasing the potential for agricultural productivity. Due to the implications for global food and water security, the World Health Organization has produced the manual, *International Guidelines for the Safe Use of Wastewater, Excreta and Greywater*. The manual provides an "integrated preventative management framework for maximizing the public health benefits of wastewater, excreta, and greywater use in agriculture and aquaculture" (World Health Organization 2006). The benefits to society include requiring less water use, less water treatment, less wastewater treatment, less energy use, less need for fertilizer imports, and more sustainable food production with year-round irrigation closer to the population.

Closing the human nutrient cycle is necessary. Whereas economically extractable phosphate rock (apatite) is a limited resource, human and animal excreta are renewable and readily available in our communities in a form that would take little processing to reuse (Figure 32.3). With proper composting, nearly all pathogens and most pharmaceuticals and pesticides are safely removed at low cost (Diaz et al. 1996; Dolliver et al. 2008; VanMiert and Jimmerson 2014). Biochar can be added to this system to absorb urine, adsorb ammonia and other odors, add stable carbon, and speed the composting process by providing habitat for beneficial microorganisms. Biochar can also adsorb recalcitrant pharmaceuticals and pesticides so as to prevent leaching (Jones et al. 2011; Yao et al. 2012).

Optimization of regional nutrient cycling systems should be encouraged. As has been done in the past, cities will offer centralized pickup locations for agricultural fertilizer transport networks. For example, extensive canals (Figure 32.4) were utilized for the transport of human excrement on barges from cities

FIGURE 32.3 Closing the human nutrient cycle. (From Jenkins, J.C., *The Humanure Handbook: A Guide to Composting Human Manure*, Joseph Jenkins, Inc., White River Junction, VT [Distributed by] Chelsea Green Pub., 2005.)

FIGURE 32.4 Canals in Shanghai utilized for nutrient transport from cities to farms by barge for thousands of years. (From King, F.H., *Farmers of Forty Centuries, Or, Permanent Agriculture in China, Korea and Japan*, Mrs. F. H. King, Madison, WI, 1911, http://archive.org/stream/cu31924073872685#page/n215/mode/2up.)

to farms throughout China for thousands of years until the twentieth century (King 1911). This efficient system functioned with no fossil carbon usage yet cycled a majority of nutrients back to agriculture. Biochar incorporation (to inhibit nutrient and vapor loss) and composting (to reduce pathogens) may be desired modern innovations to this ancient system.

When designing systems for the reuse of excreta, separation is key. Urine is a sterile liquid and perfect for fertigation with minimal processing. Researchers around the world have found that agricultural yields are similar when fertilizing with urine versus chemical fertilizers (Heinonen-Tanski et al. 2010; Malisie et al. 2007). In Vermont, the Rich Earth Institute has noted hay yields of two to six times that of unfertilized plots (Nace and Noe-Hays 2012, 2013). On the other hand, feces are high in organic matter and need processing to remove pathogens. Once processed, this makes an ideal soil amendment. However, current industrialized wastewater treatment includes mixing the wet fraction and dry fraction into drinking water and sending to a series of chemically intensive, stirred and aerated bacterial bioreactors which are expensive to operate (hence many in developing countries don't) and generate massive amounts of carbon dioxide. These facilities often take water from factories, commercial businesses, and runoff, so that the precipitated sludge is contaminated and finds less than optimal reuse. Research is starting to be done on the use of composting toilets in urban environments (Anand and Apul 2014). The cost of replacing the current system will be offset by the rising value of the fertilizer and the reduction in water use.

These nutrients have real (monetary) value. Just like deposits are put on milk bottles, when buying food, we are essentially putting a deposit on those nutrients and therefore should receive compensation from farmers and fertilizer producers as an incentive to recycle. Public restrooms in Tamil Nadu, India, previously charged per use, but now with the ecological sanitation (EcoSan) community compost toilets, they are paying users for their contributions instead! The EcoSan Research Programme is working toward the Millennium Development Goal of providing sanitation for the 2.5 billion who go without sanitary facilities everyday (SCOPE 2009).

But recycling human and animal waste cannot be optional like deciding whether to throw a newspaper into the garbage can or recycling bin. This is the future of life. Phosphorus is critical in genetic material (DNA), protein synthesis (RNA), energy metabolism (ATP), teeth and bones (hydroxyapatite), and cell membranes (phospholipids). It is required for life and is a critical component in soil fertility; therefore, it must be mandatory to recycle it. The more we let pass into our water bodies (and landfills), the more difficult and costly it will be to retrieve.

Biochar for Carbon and Nutrient Capture

Biochar is a high internal surface area charcoal produced for use as a carbon sequestering soil amendment that improves soil fertility by retaining water and nutrients, increasing cation exchange capacity, building soil carbon, improving soil architecture and drainage, while providing habitat for soil microbial ecologies. Biochar is produced as a bioenergy coproduct during biomass pyrolysis. The pyrolysis process heats biomass in the absence of oxygen to yield char, bio-oil, and noncondensable gases. Numerous system designs exist to maximize one or two of the three.

Bio-oil is the cheapest liquid fuel derived from biomass today (Kreith 2011). It can be used straight as fuel oil for electricity generation, upgraded to gasoline, or used for specialty chemicals. Pyrolysis gas can be used to heat the operation or be catalytically steam reformed to syngas to synthesize alcohols or Fischer–Tropsch liquid fuels. On a farm scale, waste heat can be used to heat a greenhouse, barn, or other infrastructure; while on a municipal or regional scale, the waste heat can be utilized in a combined heat and power system or to heat other industrial processes.

To ensure the environmental sustainability of biochar, feedstocks must be sustainably sourced, transported, and processed. Using existing biomass waste materials is ideal, and growing new fuel crops on marginal lands is possible. Growth of fuel crops must always be greater than or equal to harvest, without degrading existing ecosystems or competing with human food production. Potential feedstocks include crop residues (stover, bagasse, nut shells, and fruit pits), wood chips from yard and forestry wastes, livestock manures, and native perennial grasses. One study shows how shifting Appalachia from an environmentally and socially destructive coal economy to an ecologically restorative biochar economy can remediate mountaintop removal coal mines and replace them with a successionally designed agro-eco-park model (Todd et al. 2010). This type of interaction of combined ecological and economic restoration through the afforestation of human-degraded landscapes will be more and more necessary in the future and should perhaps become commonplace.

The recent interest in biochar has been sparked by research into the anthropogenic and fertile Amazonian dark earths, *terra preta*, which was found to hold onto nutrients and water and lock carbon in typically highly leached tropical soils for hundreds and thousands of years. It is now being looked at as a viable carbon sequestration tool due to its recalcitrance to decomposition. Longevity estimates for biochars made under different pyrolysis conditions in various soils around the world range from less than 300 to over 10,000 years (Lehmann et al. 2009).

Amazonian *terra preta* today contains more carbon, nitrogen, calcium, potassium, and phosphorus, with higher cation exchange capacities and pH values than surrounding soils. This affinity for cations, along with high surface area, porosity, and presence of both polar and nonpolar surface sites, contributes to biochar's effects on nutrient absorption. One study found biochar with and without manure adsorbed phosphate and ammonium but had little effect on nitrate (Major et al. 2009). Another study on biochar, manure, and Midwestern soils showed significant abilities of biochar to retain nutrients during a 45-week leaching trial, despite having significant nutrient concentrations itself (Laird et al. 2010). Chemical and microbiological activity contributed to significant reductions in the leaching of nitrate, total dissolved phosphate, ammonium, potassium, magnesium, calcium, and organic carbon. In a rooftop garden stormwater study, biochar blended with a standard green roof grow mix reduced leaching of nitrate (79%), phosphate (43%), and organic carbon (72%) (Beck et al. 2011).

Survey of Potential Phosphate Adsorption Media

Impetus for Original Research

Excessive nutrient loading and coliform bacteria pollution are leading causes of impaired water quality in Vermont, the United States, and around the world. Nutrient management strategies have been effective in reducing point source pollution; however, significant amounts of nutrients remain unaccounted for, and adverse effects from nutrient pollution are widespread. Even with management strategies in place, coliform bacteria may be released during every storm, affecting water uses downstream. This study explores several potential water treatment media for the purpose of finding more sustainable alternatives to remove dissolved phosphate and ammonium ions and coliform bacteria from solution.

Phosphorus is the macronutrient that typically limits (or co-limits) biological productivity in freshwater and many coastal environments. In nature, phosphorus is typically found within the phosphate molecule (PO_4). In aquatic environments, phosphate can be part of an organic compound, an inorganic compound, or by itself as a soluble ion (PO_4^{-3}). Phosphorus needs to be in this soluble reactive phosphorus (SRP) or orthophosphate form in order to be taken up by plants or algae. In terms of treatment, organic and inorganic particulates can settle out relatively easily, but once phosphate is dissolved it needs to be adsorbed, chemically precipitated, or biologically taken up.

Bench Experiments

These 10 experiments examined the effectiveness of different natural and man-made adsorption substrates for the removal of SRP in a high-flow (agricultural runoff) environment. The experiments were lab based with potential application at a Vermont dairy farm on Lake Champlain:

1. Elemental analysis of potential media
2. Media effects on effluent pH
3. Column trial SRP reductions
4. SRP breakthrough curves
5. Time dependence
6. Performance of biochar mixes
7. Normalized flow
8. Ammonium breakthrough curve
9. Coliform reduction tests
10. SRP adsorption isotherms

Elemental Analysis of Potential Media

Phosphate's highly negative charge (PO_4^{-3}) typically repels other negative charges and attracts positive charges. Biochar is generally considered to have a negative surface charge corresponding to its high cation exchange capacity. An ideal compound to adsorb SRP would be rich in cations and poor in soluble phosphorus. Calcium, magnesium, aluminum, and iron salts are usually used to precipitate out phosphates in municipal wastewater treatment operations. An elemental analysis was performed on two biochars, six potential biochar feedstocks, and one volcanic pumice. The results are seen in Table 32.1.

The post-field-use, farm-created biochar (*biochar-used*, without caked-on silt) had been in an agricultural seasonal runoff ditch for months and thus may have been rinsed off many of its ions (Ca, P, Al, and Fe). The not-yet-used, farm-created biochar (*biochar-new*) may have had a higher concentration of some ions (Ca, Al, and Fe) due to its higher relative ash content post-pyrolysis. The Guatemalan pumice was high in K, Mg, Na, Al, and Fe, while the willow, buckthorn, and fir were high in Ca, P, and K. The fir sample was also relatively high in Mg, while reed was high in Na, Ca, P, and K, and bamboo was high in P and K.

TABLE 32.1

Elemental Analysis of Various Biochars, Biomass, and Pumice Showing Concentrations of Calcium (Ca), Phosphorus (P), Potassium (K), Magnesium (Mg), Sodium (Na), Aluminum (Al), and Iron (Fe) (Units: mg/kg)

Sample ID	Ca	P	K	Mg	Na	Al	Fe
Pumice	668	59	2703	1541	2,417	2269	4116
Cedar	777	39	262	111	25	25	37
Willow	6733	904	1700	526	25	77	96
Buckthorn	6269	986	2396	473	25	50	87
Fir	5734	1048	3930	1028	183	25	71
Reed	4260	1111	4737	306	15,510	25	36
Bamboo	150	1244	9824	326	25	25	21
Biochar-new	7818	134	537	394	25	146	210
Biochar-used	2061	35	634	422	25	47	61

Despite its high P concentration, buckthorn was used as a biochar feedstock in this study because it was a readily available nuisance species at the farm.

pH Effects

For experiment 2, 11 substrates were chosen and placed in 11 vertical flow columns (diameter = 3 cm, height = 50 cm). Then 40 mL of 1.0 mg/L PO_4-P (phosphate-phosphorus) solution was poured three times, creating three successive pulse flows. This apparatus and the pulse flow method were used in steps 2 through 9. The phosphorus solution had an average pH of 6.33, and most substrates increased the pH (Figure 32.5). Listed from low to high pH, these media were lab-grade iron filings, lab-created buckthorn biochar (BBC), Guatemalan pumice, farm-created biochar (BC-new), farm-created used biochar (BC-silty) that had been previously used in a runoff filter and had caked-on silt, lab-grade bauxite, heat-activated bituminous coal-based AC, lab-created partially oxidized hardwood pellet biochar, Atlantic fine crushed coral sand, electric arc furnace (EAF) steel slag, and wood-burning power plant fly ash.

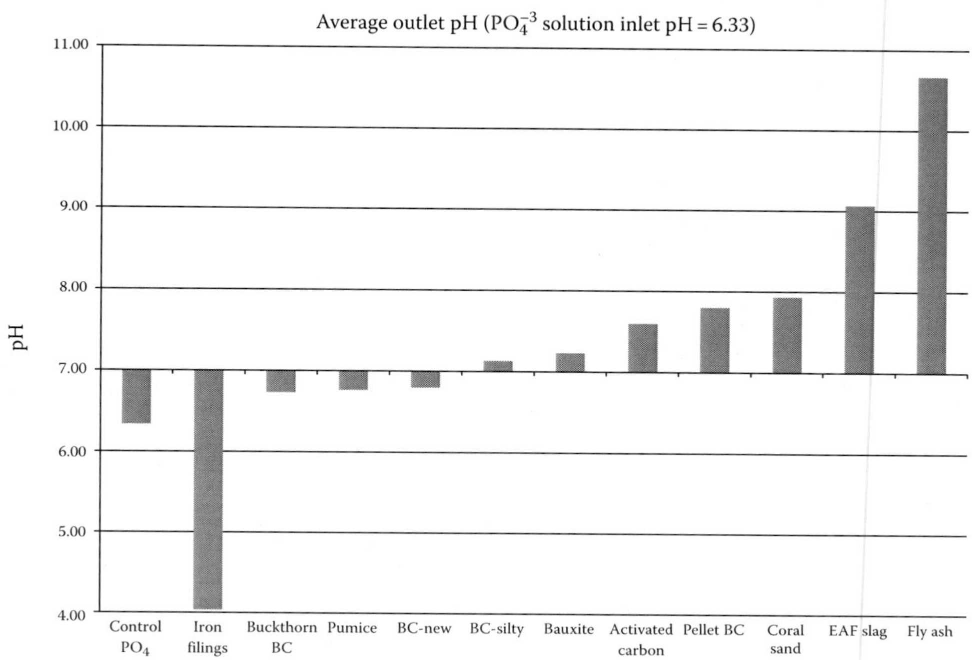

FIGURE 32.5 The average outlet pH of the 11 initial substrates tested.

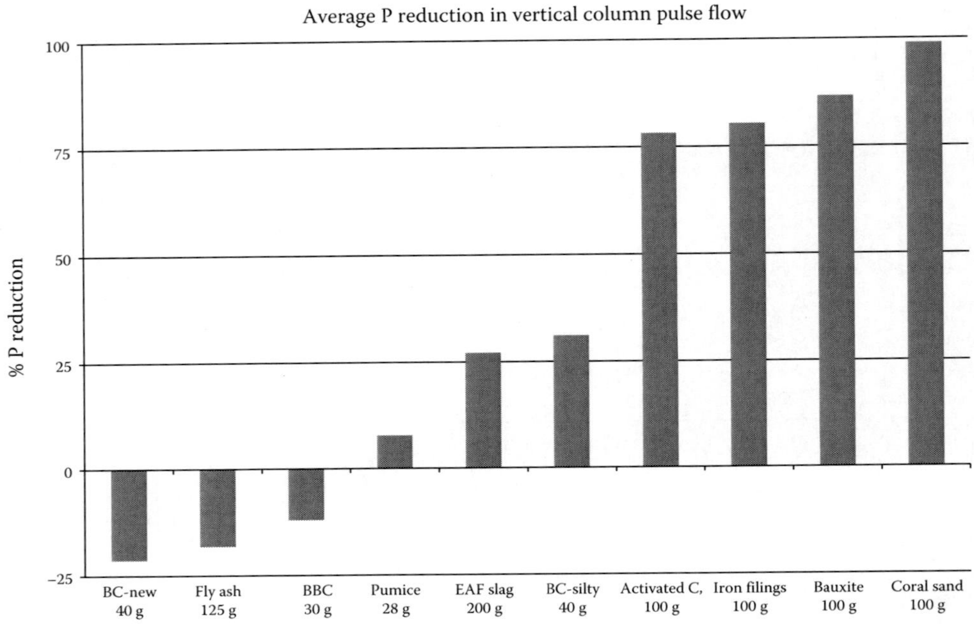

FIGURE 32.6 Column trial average SRP reductions by substrate.

SRP Reduction

The initial testing of filter materials for phosphorus adsorption (shown in Figure 32.6) showed the highest average percent removal by fine coral sand (99%), followed by bauxite (87%), iron filings (80%), and activated charcoal (78%). Each of the types of biochar tested, with the exception of BC-silty, showed no decrease in effluent phosphorus concentration and instead showed slight increases. The partially oxidized wood pellet biochar tripled the P concentration and was removed from further study. The pumice showed a slight decrease in P concentration (8%), whereas EAF slag and used silty biochar removed 27% and 31%, respectfully.

Although pumice showed a high metal content in the elemental analysis, it showed little pH change and captured little phosphate. The substrates that increased pH the most were not necessarily the best at removing SRP. Fly ash saw an increase in SRP during one trial and reductions during the following two trials. Although the iron filings reduced the pH by more than two orders of magnitude, they were successful at removing SRP. The *new* biochars increased pH slightly while increasing SRP slightly. The used silty biochar probably removed SRP because of cations in the caked-on silt.

When scaling up filtration designs, it is necessary to consider what mass of substrate will remove what mass of phosphorus. According to Figure 32.7, most of the substrates stay in the same order, with fine coral sand being the most effective by weight; however, with the exception of the phosphorus-donating biochars and fly ash, EAF slag traded with ultralight pumice stone as the least effective fast SRP adsorbing media per mass. Interestingly, the silty biochar was quite effective at SRP reduction per mass.

SRP Breakthrough Curves

Also when scaling up, it is necessary to consider how long the substrate will be an effective adsorption medium. One method of testing this is to make a breakthrough curve (Figure 32.8). Breakthrough curves plot the effluent concentration versus either time or, in this case, volume treated. In the diagram, V_b (or breakthrough volume) is when the pollutant starts to be seen in the effluent. V_x is the volume at which the pollutant is not treated at all and the filtration media are considered exhausted. In practice, one would let the filter go until the discharge concentration is nearing the discharge limit. For example, one of the point source discharge limits for Lake Champlain is 0.8 mg PO_4-P/L. If agricultural *nonpoint* pollution streams were managed as point sources and the filter influent was at 1 mg/L, most of the substrates

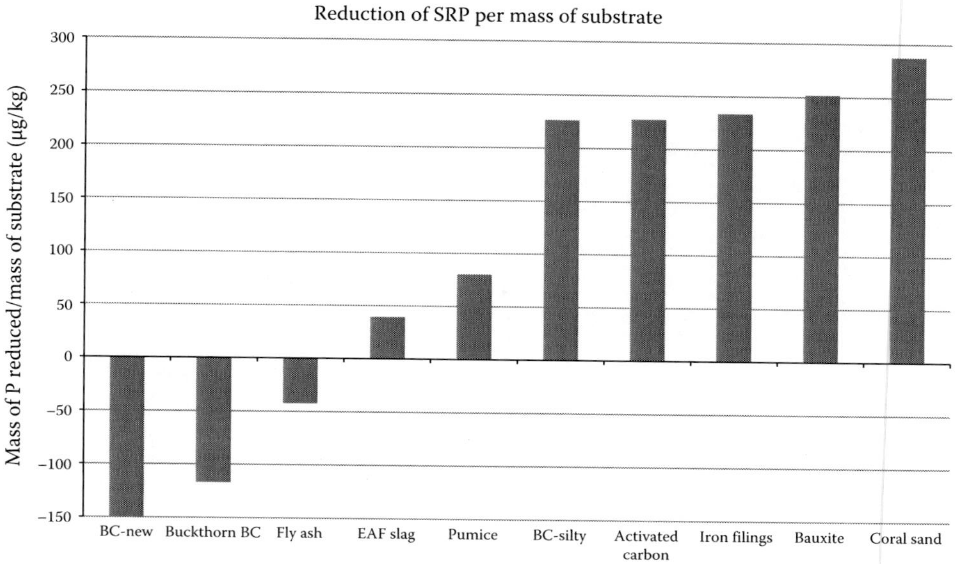

FIGURE 32.7 The reduction of SRP relative to total substrate mass in vertical column pulse flow.

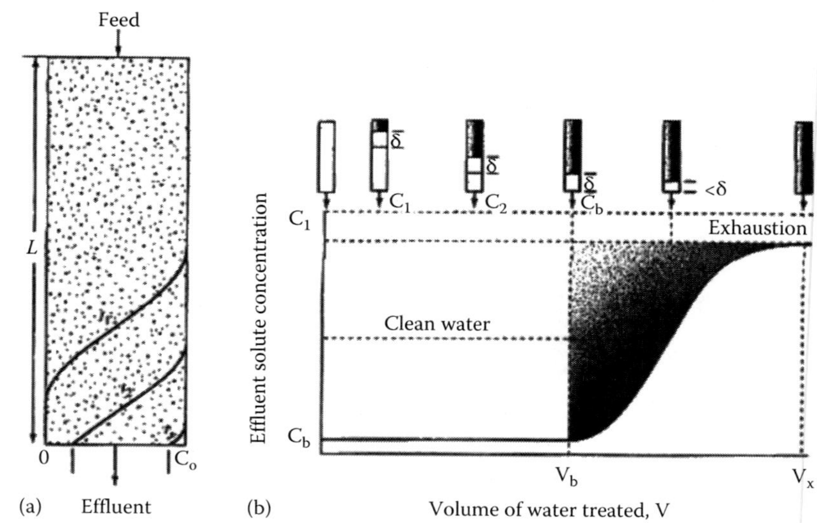

FIGURE 32.8 (a) This depicts an adsorption column with solute concentration profiles (active treatment zones) at time t_1, t_2, and t_3. At time t_1, the area above the curve is considered exhausted while the bed below is clean. At time t_2, the first amount of solute is detected in the effluent and at t_3 the entire bed is almost exhausted. (b) The breakthrough curve depicts the effluent solute concentration from initial start up to breakthrough (when the solute begins to be detected in the effluent) to exhaustion (when treatment has ceased and the bed is ready to be recharged). The movement of the active zone (δ) is depicted in the top portion of the panel. (From Sincero, A.P. and Sincero, G.A., *Physical-Chemical Treatment of Water and Wastewater*, CRC Press, Boca Raton, FL, 2003.)

tested would bring these streams into compliance (Figure 32.9). With new regulations set at 0.2 mg/L discharge, only coral sand and coral sand mixes would meet the criteria.

Figure 32.9 shows that even after soaking BBC for 2 h, the effluent SRP levels were still increasing. Soaking may release some phosphate, but without rinsing, that phosphate can re-adsorb. The bauxite and steel slag slowly lost effectiveness, while fine coral sand went from removing 97% to 99%. Coral sand improved the SRP reduction capabilities of BBC enough to make a nice curve, while AC acts similarly.

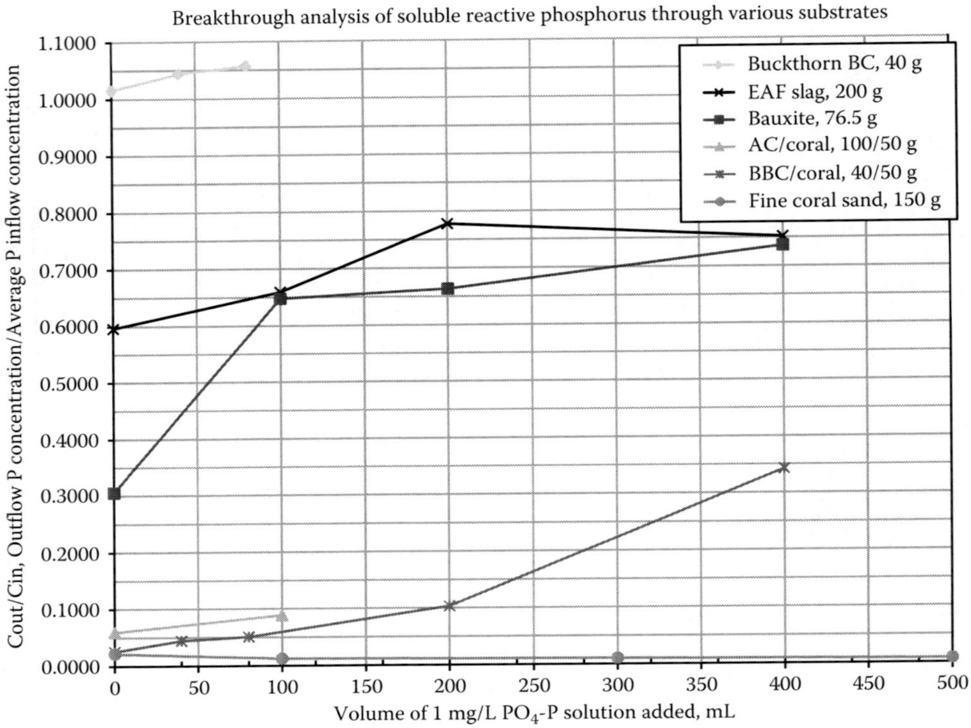

FIGURE 32.9 SRP breakthrough curves for four substrates and two carbon/coral sand mixes.

Time Dependence

Knowing the retention time and total solute reduction gives the rate of adsorption and helps determine the size of the filter. The average P reduction for each substrate was divided by the retention time to yield the rate of P reduction (Figure 32.10). BBC had a negative phosphorus reduction and cannot be displayed on this log plot. Bauxite and steel slag had the best phosphorus reductions per second because of their large particle size and high bulk porosity. The mixture of AC and coral sand had the longest retention time due to damming from small, mixed particle sizes. The fine coral sand by itself was poorly sorted (with a bimodal particle distribution of 0.1 and 0.6 mm) and had a slow flow rate. Well-sorted media allow higher flow, which is useful in a high-flow seasonal runoff environment or an engineered filtration system after a sedimentation basin, tile drain, or constructed wetland.

Biochar Mixtures

Because of the variable adsorption by mass and the benefits of using biochar as a filter and soil amendment, we decided to test equal masses of EAF slag, bauxite, and coral sand with equal masses of biochar (Figure 32.11). While the coral sand mix varied considerably from a 56% to a 77% reduction, the bauxite (50% and 51% reductions) and the slag (13% and 14% reductions) showed little variation.

Normalized Flow

Because of the slow rate of flow through the fine coral sand (typical particle diameter <1 mm) and high amount of SRP retention, coarse coral sand (~5 mm) and limestone granules (~10 mm) were tested. The coarse Florida crushed coral sand showed great improvement in time and rate of SRP retention over the fine coral sand (Figure 32.12). Florida coral sand provided the most SRP reduction, while Shelburne limestone (having less internal porosity and therefore less reactive surface area) was used up quickly.

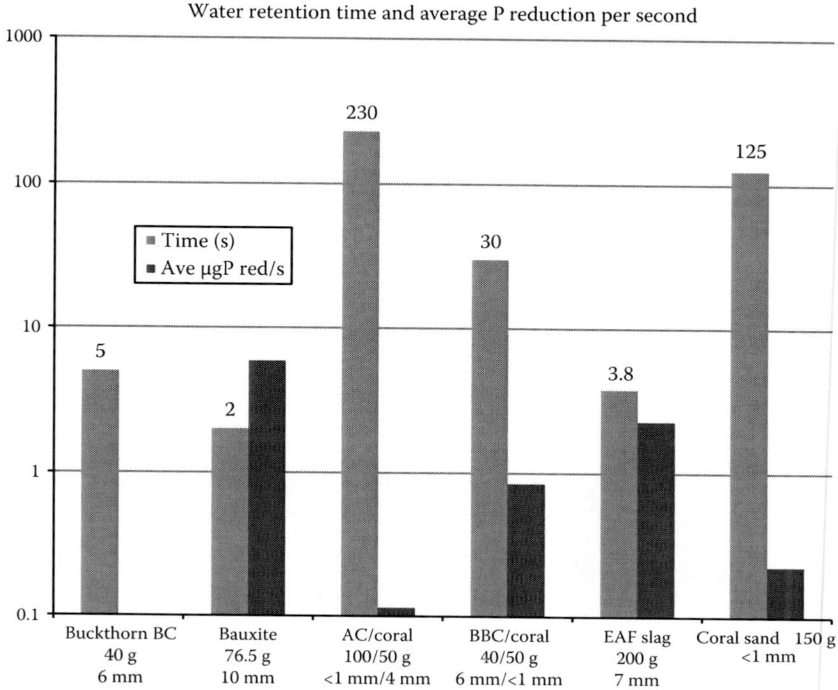

FIGURE 32.10 The amount of time for water to pass through each column (light gray) and the corresponding rate of SRP reduction (dark gray).

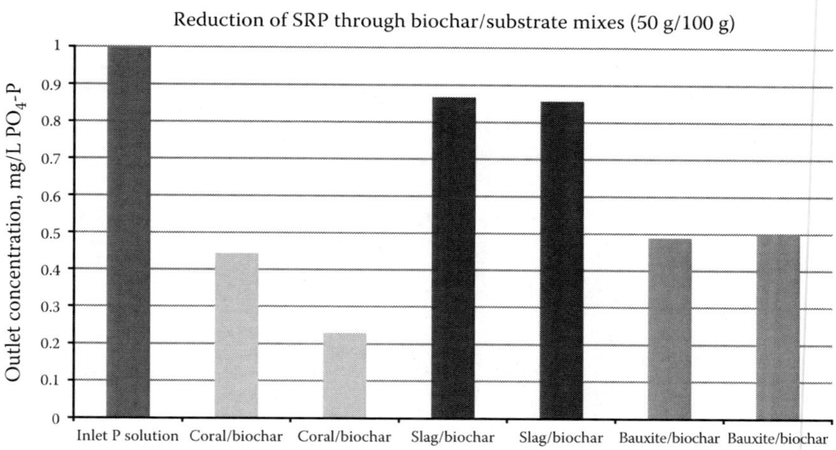

FIGURE 32.11 SRP reduction trials of equal-mass biochar mixes.

The Roanoke slag pieces were quite large (20 mm) and thus had lower surface-to-mass ratios and higher flow rates than the Taiwan slag (7 mm) and the other media. The coarse coral sand was mostly composed of seashells with intricate surface structures and rough-surfaced coral stones; thus, it probably had a higher effective surface area than the other media.

Ammonium Adsorption Breakthrough Curve

In the previous tests, the new biochars did not significantly adsorb orthophosphate (SRP). The next test was to see if the biochar could adsorb ammonium to still be useful in improving *nonpoint* source

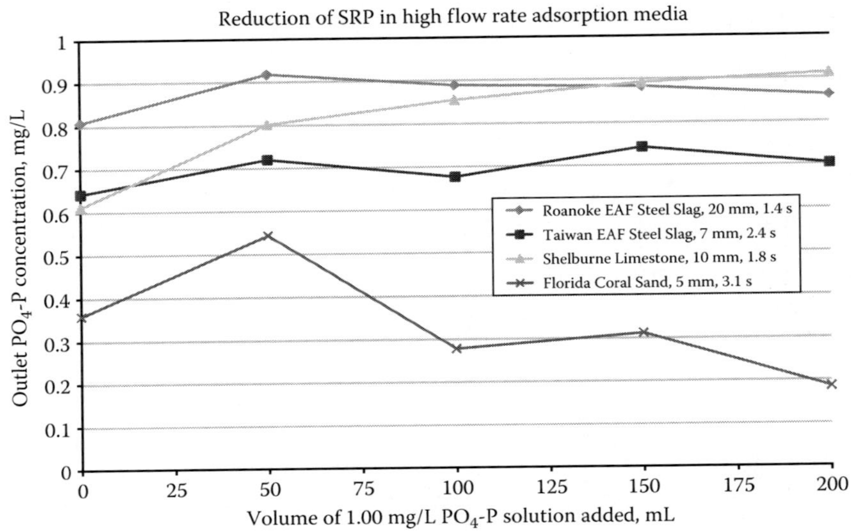

FIGURE 32.12 Outlet SRP concentrations of high-flow, coarse media.

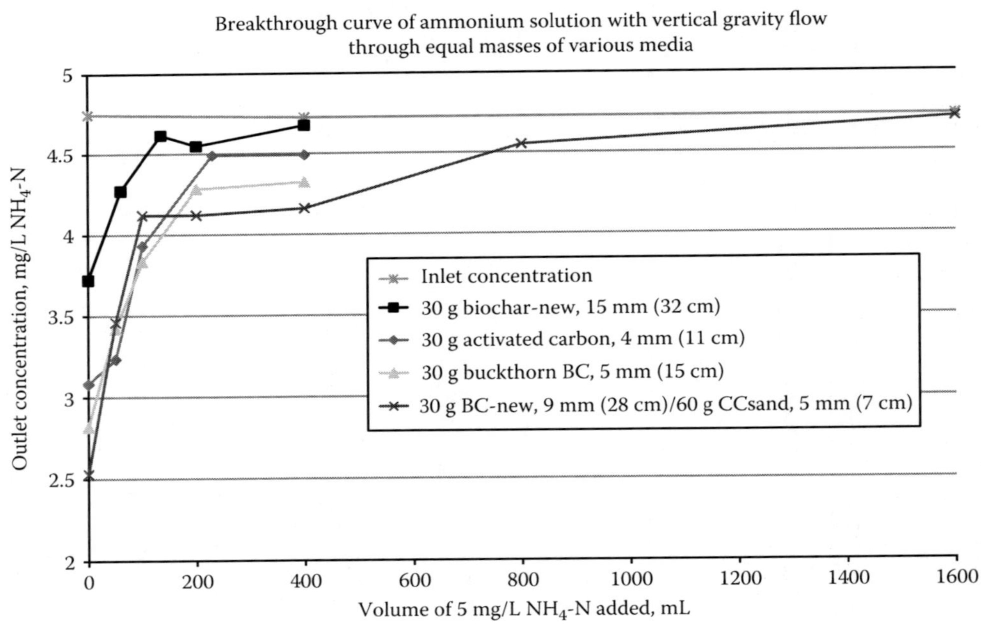

FIGURE 32.13 Ammonium breakthrough curves measured with pulse gravity flow through 32 vertical centimeters of 15 mm farm-made biochar, 11 cm of 4 mm AC, 15 cm of 5 mm BBC, and 28 cm of 9 mm farm biochar with 7 cm of 5 mm coarse coral sand.

agricultural runoff water quality. To test this hypothesis, additional fast-flow, one-pass column trials were performed with 5 mg NH_4^+-N/L (ammonium-nitrogen) solution. The ammonium solution breakthrough curves for various media are shown in Figure 32.13.

All of the media adsorbed ammonium. None of the media adsorbed more than 50% of the inlet concentration, and all reached near exhaustion after 400 mL. Smaller particle sizes seem to be more effective. This is perhaps true because of larger reactive surface areas and slower flow rates, allowing enough contact time for adsorption to occur. It follows then that when the solution was poured too fast at

50 mL in the biochar/coarse coral sand column, the outlet concentration was higher because there was less time to react. When the solution was poured at the normal rate again, the concentration leveled out and then tapered up to the inlet concentration. The results are inconclusive of coral sand reactivity with ammonium, but biochar and AC both seem to adsorb ammonium in significant amounts.

Coliform Filtration Tests

The reduction of total coliform and *Escherichia coli* was measured in fast gravity flow column tests with water from an agricultural stream (North Branch Potash Brook, Chittenden County, Vermont). All of the media seemed to reduce total coliform levels significantly (Figure 32.14) yet reduced *E. coli* levels only slightly (Figure 32.15). The variation across media may be influenced by particle diameter. The 10 mm diameter biochar has less active surface area per bulk volume and would tend to impose less friction on the flowing media and possess less adherent surface area for microbial capture. The columns in this study were only 20 cm tall, so when combined with the high flow rate, there was little chance for complete coliform removal.

SRP Adsorption Isotherms

Twenty-four hour SRP adsorption isotherms were performed using a standard batch equilibrium technique (Nair et al. 1984) in order to determine the steady-state adsorption capacities of four materials.

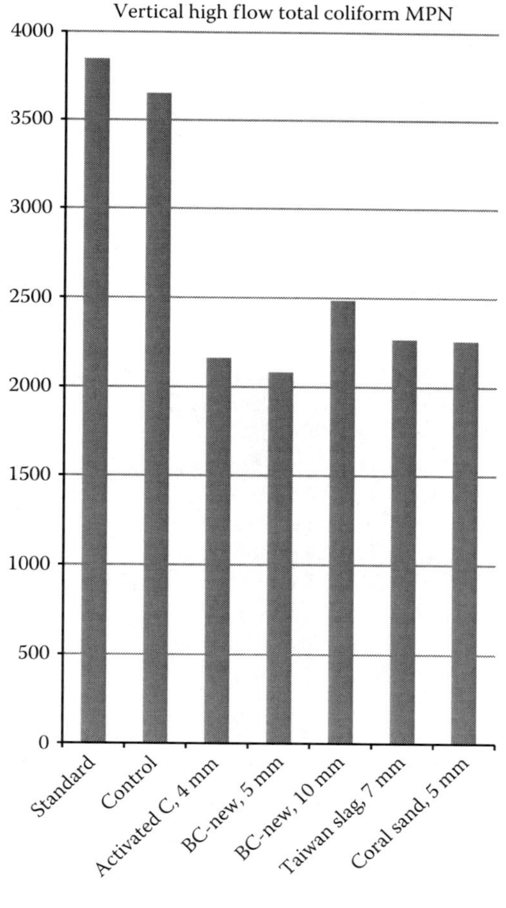

FIGURE 32.14 Total coliform reduction test through columns of various substrates.

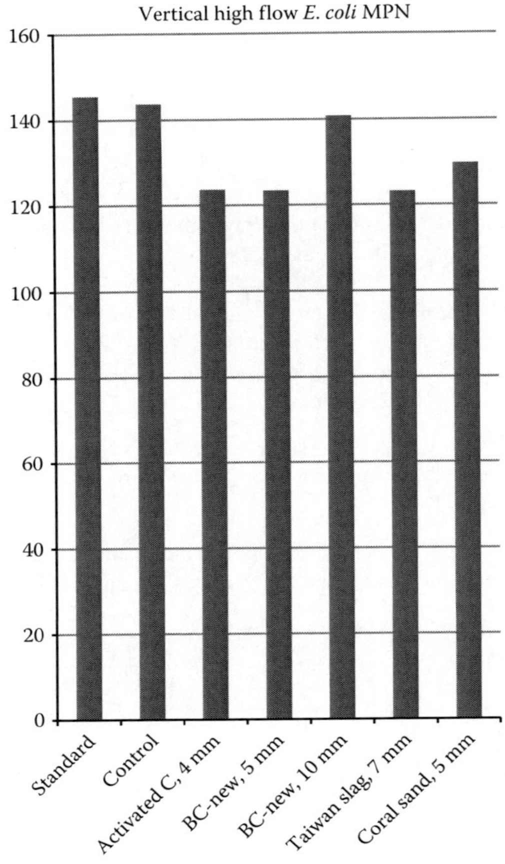

FIGURE 32.15 *Escherichia coli* reduction tests through columns of various substrates.

Fine and coarse coral sands had very similar and high average P adsorption at all initial concentrations, from that of a typical lake (0.01 mg P/L) to a typical manure pit (1000 mg P/L). The highest measured adsorption values for fine and coarse coral sand were 22.7 and 22.3 mg PO_4-P/g substrate, respectively, at 1000 mg PO_4-P/L initial solution concentration. The highest values for AC and soaked BBC were 10.0 and 0.66 mg/g, respectively, also at 1000 mg/L. The isotherm curves for AC and BBC had already leveled off at 1000 mg/L input concentration and were extrapolated to have adsorption maxima around 11.9 and 0.64 mg/g, respectively. The isotherm curves for fine and coarse coral sands, however, were still going up beyond 1000 mg PO_4-P/L, as their extrapolated maxima both exceeded 100 mg/g.

The coral sands showed 91.2% and 89.7% SRP reductions at high concentration (1000 mg/L) as well as 82.7% and 81.0% reductions at low concentration (10 μg/L), whereas AC showed a 95.5% reduction at medium concentration (1 mg/L). BBC successfully adsorbed phosphate from solutions of 1 mg/L and above, with a maximum reduction of 30.5% at 10 mg/L.

Discussion

Biochar and Phosphorus

Although the biochar had high calcium and low phosphorus contents, all column tests showed *new* biochar as a phosphate source. This is simply due to the significant difference in volatilization temperatures between carbon and phosphorus during pyrolysis. A large portion of plant phosphorus is bound to organic molecules with ester and pyrophosphate bonds. Organic carbon begins to volatilize around 100°C when heated, whereas phosphorus will begin to volatilize around 700°C (DeLuca et al. 2009). Many chars are produced below 700°C and therefore have high concentrations of soluble phosphate. Different feedstocks will also show higher amounts of soluble nutrients depending on species and pyrolysis conditions. Nitrogen compounds begin to volatilize around 200°C and sulfur compounds above 375°C. When used in these systems, it is necessary to rinse off SRP (and other remaining soluble nutrients) and use the rinsed water for fertigation. The isotherm experiments showed that presoaked biochar immersed and shaken in phosphate solution for 24 h had net adsorption at 1 mg PO_4/L and above. This result suggests that the presoaked biochar still had a surface phosphate concentration greater than 0.1 mg/L.

Activated Carbon Adsorption

ACs have been engineered to optimize reactive surface area in order to maximize adsorption capacity. In this study, AC (from heat-activated bituminous coal) performed well because of its small particle size (3–4 mm) and very high reactive surface area. Its generally nonpolar structure adsorbed a surprising amount of ions. Activating biochar may give it more of the properties of AC. Higher adsorption properties can be achieved through heating at higher temperature, heating longer at lower temperatures, or through chemical or steam activation post-pyrolysis. AC is not known for removing highly charged anions. Arsenate, chemically similar to phosphate, is a major pollutant of concern in drinking water. Some researchers modify the surface of ACs by impregnating them with iron oxide in order to adsorb arsenate (Chen et al. 2007). Other researchers have begun to prepare specialized chars from waste materials. Biochar produced from anaerobically digested sugar beet tailings has been shown to remove phosphate, possibly due to magnesium oxide crystals in their surface structure (Yao et al. 2011).

Bauxite and Iron Oxide

Iron and aluminum-containing compounds did well at phosphorus adsorption in this study. Experiment 3 showed bauxite (aluminum ore) removing 87% of phosphate, iron filings removing 80%, and coral sand 99%. However, the iron filings in experiment 2 lowered effluent pH to 4. Such acidity is out of the comfort range of most aquatic life forms and makes treatment more challenging. Iron and aluminum adsorb phosphate best at low pH; however, the phosphate becomes tightly bound and is not easily plant available. Aluminum is not an essential plant nutrient and in fact can be toxic to plants in acid soils. Iron is also more likely than aluminum or calcium to release phosphorus, when it undergoes redox changes to anaerobic conditions (Ballantine and Tanner 2010). Geology tends to combine phosphate with calcium, and major phosphate ore deposits are calcium minerals (of marine origin). At moderate soil pH, calcium phosphate compounds are fairly available to plants, and availability can be moderated by organic matter content and soil biota. When considering the goal of cycling nutrients in a predictable and ecologically sensitive manner, natural calcium compounds seem the most logical to combine with phosphorus.

Marine Limestone

Coral sand, like limestone and other shell sands, primarily consists of calcium carbonate and to a lesser extent magnesium carbonate. The coral sands tested here contain aragonite and high magnesium calcite. Coral and shell sands have high phosphorus sorption capacity due to their high calcium and magnesium content (Liu et al. 2012; Roseth 2000; Vohla et al. 2011) and may limit biological productivity in shallow tropical and subtropical coastal systems due to their regulation of soluble phosphorus (McGlathery et al. 1994; Millero et al. 2001). Carbonate marine sediments can be biogenic or a result of oolitic precipitates. Biogenic sediments (mollusk shells, coral skeletons, and foraminiferal tests) have higher internal surface areas, while oolitic sediments are small and smooth with high bulk external surface area. Deposits of both are quite vast, found on land and in the deep and shallow oceans (above

FIGURE 32.16 Scanning electron microscope image of foraminiferal ooze (calcium carbonate tests) from the Gulf of Mexico (Greco 2014).

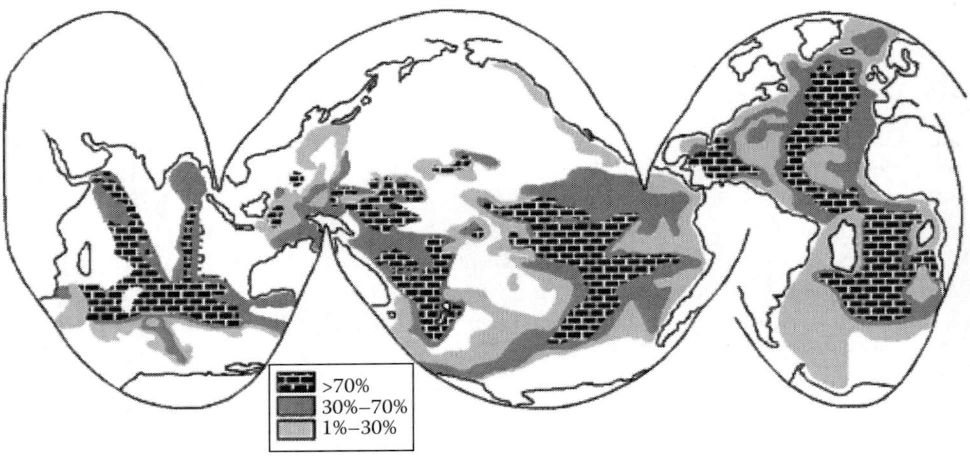

FIGURE 32.17 Distribution of calcium carbonate (by weight %) in ocean surface sediments (Dittert et al. 2005).

the carbonate compensation depth). As seen in Figures 32.16 and 32.17, high-surface-area biogenic calcium carbonate deposits cover 50% of the ocean bottoms (Hallock and Boersma 2005). Oolitic aragonite sand drifts precipitate in submillimeter particle size in warm carbonate-rich waters such as at the Bahama Banks.

Aragonite is a readily soluble biogenic calcium carbonate with a crystal structure that rapidly adsorbs phosphate and efficiently buffers pH. Land-derived limestone works for phosphate removal, but not as well as the higher-surface-area biogenic or oolitic deposits. Depending on locality, crushed limestone can be used for phosphorus capture if readily available locally. Otherwise, marine carbonate sands are used around the world as an inexpensive, sustainably harvested liming media. Figures 32.18 and 32.19 show photos of the fine and coarse coral sands used in this study, with a U.S. penny for scale.

FIGURE 32.18 Closeup photo of fine coral sands used in this study.

FIGURE 32.19 Closeup photo of coarse coral sands used in this study.

Adsorption Capacities

Phosphate adsorption media have been reviewed extensively for use in agricultural and wastewater treatment applications, especially for constructed wetlands (Ballantine and Tanner 2010; Johansson 2006; Vohla et al. 2011). Vohla et al. reviewed 64 papers and categorized media for phosphorus removal in wastewater treatment wetlands as natural (N), man-made (M), or industrial by-products (I). Adsorption capacities recorded in the study are as follows: 0.04–420 g P/kg for fly ash/slag (I), 0.5–335 g P/kg for shell sand (N), 119.6 g P/kg for heated opoka (N/M), 0.4–4.8 g P/kg for apatite (N), 0.05–4.5 g P/kg for Filtralite (M), 0.02–4 g P/kg for sand (N), 2.15 g P/kg for zeolite (N), 0.25–0.682 g P/kg for limestone (N), 0.16–0.61 g P/kg for bauxite (N), and 0.001–0.005 g P/kg for soils (N) (Vohla et al. 2011). The industrial by-products show great promise in phosphorus adsorption; however, use and reuse may be limited due to heavy metal content, extreme pH, and high densities. The results of the review show wide variability depending on the source materials and study conditions, but they indicate that the research presented in this chapter falls in the indicated ranges, that biogenic calcium carbonate is very effective, and that biochar may be comparable to other well-studied adsorption media when used to treat high concentration effluents.

Phosphate typically binds to calcium carbonate through chemisorption (McGlathery et al. 1994) and can bind to active carbons by dispersive adhesion or ion exchange. Apatite (calcium phosphate) crystals are thought to lower the activation energy for precipitating phosphorus (Vohla et al. 2011). Calcite crystals may provide nucleation surfaces for further apatite crystal growth. Millero et al. (2001) suggest that calcium and magnesium cations on calcium carbonate surfaces further promote phosphate adsorption from solution forming surface phosphorus-rich layers on the carbonate crystals. This new apatite growth can accrete in layers similar to oolitic precipitates. Phosphate has also been shown to adsorb to calcium carbonate and slowly replace carbonate, changing the existing chemical structure (McGlathery et al. 1994).

Engineered Systems

Nutrient adsorption media can be used in a variety of systems in natural and man-made environments. Municipal systems such as traditional wastewater treatment plants can utilize adsorption bed columns. Agricultural systems for treating wastewater, tile drainage, and stormwater typically utilize engineered wetlands. In all engineered systems, the maximum longevity of the filter is realized when there is maximum pretreatment of the influent. Minimum pretreatment would include settling out particulate P and other solids in a sedimentation or retention basin, typically followed by direct plant uptake of soluble P in an emergent or floating plant wetland. This can then be followed by a cell with replaceable media cartridges to adsorb P. When phosphate adsorption media are used as the main substrate of a constructed wetland, they typically reach their sorption capacity before the lifetime of the wetland is realized, often in only 2–5 years. Drizo et al. (1999) and Ballantine and Tanner (2010) recommend using a dedicated replaceable phosphorus adsorption bed as the last cell of a treatment wetland train. To prevent clogging and overland flow, media particle size must be uniform and large enough to maximize hydraulic conductivity while offering high specific surface area (Drizo et al. 1999). The high surface area of coral sands and biochar can improve biological nitrification and denitrification, removing excess nitrogen from effluents.

These phosphorus sorption media can be integral parts of other engineered systems. Use of fresh media in innovative river and lake restoration systems, such as floating islands (Zhao et al. 2012) and eco-restorers (Todd et al. 2003), can improve already advanced levels of treatment. In addition, when plants are harvested from this system, they can be pyrolyzed and used as a high-nutrient, beneficial soil amendment (Masto et al. 2013). Biochar and limestone can be used in wetlands to protect streams from metals, toxins, and nutrients in road runoff. On agricultural lands, a large amount of nutrients are lost to subsurface groundwater flow. Fresh media can be placed in a deep ditch or incorporated into soils around the perimeter of fields to inhibit losses to surface waters (Steinhilber and Weismiller 2000). This is best combined with a vegetated riparian buffer with mixed species (for root and nutrient uptake variety) and possibly agroforestry management (so land *lost* to the buffer stays in economically productive uses) (Hoffmann et al. 2009; Schultz et al. 2004).

Soil Amendment

Once the media have been used to cleanse waters of excess nutrients, biochar and limestone can be considered concentrated, slow-release fertilizers. The fertilized media can be used directly in land-based agriculture to improve soil architecture, moisture retention, pH, cation and anion exchange capacities, nutrient profiles, and beneficial microbial activity. The media can also be utilized as hydroponic growth media for urban/vertical farms and greenhouses, or the nutrients can be chemically extracted and concentrated (Brogowski and Renman 2004). Depending on the waste stream treated, enriched media may need to be composted or further processed before soil application or other beneficial reuse.

Brogowski and Renman state that adsorbing P onto natural materials is more environmentally friendly and more effective for releasing P to plants in agricultural production compared to artificial chemicals. Depending on how nutrients adhere to the substrate, they may have higher availability for plant root uptake. Nutrients attached by physical adsorption can become plant available when the dissolved nutrient in the soil pore water is lower than on the particle surface. Nutrients attached through ion exchange become available as the media come into contact with other anions with higher selectivity for the specific substrate. Nutrients attached by chemisorption may become available through dissolution by weak organic acids or displacement by a conjugate base. One study addressing biogenic carbonate sediment P availability to seagrasses in coastal Bermuda found that 2% of total sediment P was loosely adsorbed SRP, 15%–20% was strongly adsorbed SRP, and the remainder was mostly in the form of authigenic calcium fluorapatite minerals formed by *in situ* precipitation, which are soluble in weak organic acid. According to the study, calcium fluorapatite is the primary sink of P in carbonate sands. The fluorapatite P pool was found to dissolve in the rhizosphere and was capable of supplying decades of P availability (Jensen et al. 1998). This extended P availability is supplied by a sediment P adsorption level of about 0.3–0.6 g P/kg media, derived from the data in the study. When adsorption capacities are maximized through the use of an engineered system, long-term available P storage ability of soils can be maintained.

In order to increase soil fertility and regenerative capabilities, microbe/plant synergies in the root zone need proper pH and organic matter content (Gershuny 1999). It is essential to have strong microbial communities including nitrogen-fixing and phosphate-solubilizing bacteria and mycorrhizal and nonmycorrhizal phosphate-solubilizing and phosphate-transporting fungi (Njira 2013; Richardson and Simpson 2011). These microbes scavenge inorganic and organic phosphates through organic acid production and other pH lowering mechanisms. In fact, it has been shown that bacteria (such as nitrogen-fixing rhizobia) can solubilize phosphate that is then taken up by external mycorrhizal mycelia and transferred to plants (Ahiabor and Hirata 2003; Eckardt 2005; Rodríguez and Fraga 1999). Soil management practices that increase microbial activity such as crop rotation, intercropping, application of organics (crop residues or compost), and biofertilizers (like compost tea) should be encouraged.

Raising pH to free nutrients bound up in acid soils is accomplished through liming (typically by spreading calcium carbonate). Forty percent of global arable lands are considered acidic (Atwell et al. 1999), and more are becoming acidic due to high anthropogenic air pollution. Coral sand and limestone are already ubiquitous as soil liming agents around the world. Careful application is necessary because overliming can speed oxidation and decomposition of soil organic matter and humus, whereas proper liming releases nutrients and stimulates bacteria and plant growth, to create more organic material. Using the P-enriched media (applying it to the root zone where it is needed) would be beneficial over the highly leachable acid superphosphates that require heavy liming and therefore further exacerbate humus decomposition (Gilman 2011). Biochar can be produced to have an acid, neutral, or alkaline pH (depending largely on pyrolysis temperature), but it is typically found to have a pH around 8.

Before biochar is applied to soil, it must have nutrients added to it. If it is incorporated in soil without added nutrients, biochar competes with plants for available nutrients. If nutrient dense urine is to be used in agriculture, it may be beneficial to add it to biochar first. When biochar absorbs urine, it adsorbs odors and volatile nitrogen and will reduce leaching of valuable nutrients, while urine would eliminate loss of airborne biochar dust. A similar effect would be achieved when blending with liquid manures, compost tea, or other nutrient-rich fluids. Biochar and limestone can be mixed with solid manures or composts and then top-dressed in trenches, topsoil mixed, or deep banded. The deeper applications will be more efficient in bringing phosphorus to the root zone of the plants and hence aiding plant availability.

Kibblewhite, Ritz, and Swift defined soil health as the capacity of soil to provide ecosystem services while maximizing agricultural production and described the importance of soil health for maintaining four major functions: carbon transformations, nutrient cycles, soil structure maintenance, and the regulation of pests and disease (Kibblewhite et al. 2008). Integrating composted, nutrient-saturated biochar and limestone should enhance biological processes and abiotic interactions leading to the maintenance of the four major functions while maximizing agricultural yields.

Conclusion

> All the human and animal manure which the world wastes, restored to the land instead of being cast into the water, would suffice to nourish the world.
>
> **Les Misérables (Hugo 1862)**

In order to preserve and restore global freshwater and coastal marine ecosystems, *wasted* nutrients must be sequestered as close to the source as possible and returned to agriculture, improving soil fertility while reversing climate change. With atmospheric carbon dioxide crossing 400 ppm, peak oil upon us, and peak phosphorus looming on the horizon, global soils need to support the ever-growing demands of feeding the human population, providing renewable energy, medicine, building and weaving materials, preserving biodiversity, and sequestering carbon. Recycling nutrients with biochar and biogenic limestone can restore ecosystems while supporting natural economies and regenerative agriculture.

Many substrates can capture phosphorus and other nutrients. Biochar and limestone make the most sense from sustainability and reuse standpoints. Biochar is a renewable energy coproduct that can be derived from waste biomass. It is similar to AC and thus has many water treatment capabilities, including nutrient adsorption. Biogenic calcium carbonate's high surface area adsorbs phosphate efficiently and holds onto it until biologically or chemically released. Together or separately, these two substrates, once saturated, can be used as a slow release, nutrient dense soil amendment that improves water retention, ion exchange capacity, microbial food webs, overall soil health, crop production, and provision of ecosystem services. Utilizing this technology will create jobs in the nutrient cycling industry, reduce the need to mine already depleted phosphate stores, and begin to clear the waters of the planet, while sustainably feeding our population and reversing climate change.

The importance of nutrient cycling today is unparalleled. As humanity is expanding, we are consuming resources at alarming rates. Our island, the Earth, is teaching us the lessons of responsible management. We must heed these lessons and take conscious charge of recycling our material flows for the highest benefit and least entropic loss. Entropy is best mitigated by capturing *wastes* at the source, where they are most concentrated. Urine and manure reuse is ideal to maintain the inherent evolutionary connection between plants and animals. Once these nutrients make it to surface and subsurface waters, capture and reuse through vegetation uptake and media adsorption will be the best management practices. Vegetation in constructed wetlands or floating island-type systems uptake significant quantities of nutrients and can be harvested and returned to land-based agriculture. Aquatic vegetation is not dependable for uptake in all seasons or situations, so adsorption media become a necessary assurance against nutrient loss year-round and are best built into systems to create redundancy.

Needed Research

In order to scale up this technology and recycle nutrients more completely throughout the biosphere, more studies are needed on how to economically recycle the *riches* of the cities back to agricultural production while reducing pollution and consumption of drinking water resources. Replicable engineered systems must be designed and funding put in place for the construction and integration of nutrient reclamation and reuse technologies, such as large-scale excreta recycling to agriculture and aquaculture, agricultural nonpoint source nutrient control, and aquatic remediation. A comparison should be made on the benefits of various regionally available calcium-rich media (crushed limestone, oyster shells, oolites, coral sand, and foram tests) and the effectiveness of different crystal structures and surface areas. Another comparison should be made on the benefits of various biochars (varying feedstock as well as pyrolysis conditions) on nutrient adsorption and soil application. Research should be performed in determining best methodologies for reusing the nutrient-saturated media, including optimal soil pH, method of application, rate of application, the presence of necessary microbes and root associations, and types of crops, for optimal long-term P availability.

REFERENCES

Ahiabor, B. D. and H. Hirata. 2003. Associative influence of soluble phosphate, rock phosphate, and arbuscular mycorrhizal fungus on plant growth and phosphorus uptake of three tropical legumes. *West African Journal of Applied Ecology* 4: 75–90.

Anand, C. K. and D. S. Apul. 2014. Composting toilets as a sustainable alternative to urban sanitation: A review. *Waste Management* 34 (2): 329–343. doi:10.1016/j.wasman.2013.10.006.

Atwell, B. J., P. E. Kriedemann, and C. G. N. Turnbull, eds. 1999. *Plants in Action: Adaptation in Nature, Performance in Cultivation*. Melbourne, Victoria, Australia: Macmillan Education Australia. http://plantsinaction.science.uq.edu.au/edition1/?q = content/16-5-1-soil-acidi-cation.

Ballantine, D. J. and C. C. Tanner. 2010. Substrate and filter materials to enhance phosphorus removal in constructed wetlands treating diffuse farm runoff: A review. *New Zealand Journal of Agricultural Research* 53 (1): 71–95. doi:10.1080/00288231003685843.

Beck, D. A., G. R. Johnson, and G. A. Spolek. 2011. Amending greenroof soil with biochar to affect runoff water quantity and quality. *Environmental Pollution* 159 (8–9): 2111–2118. doi:10.1016/j.envpol.2011.01.022.

Brogowski, Z. and G. Renman. 2004. Characterization of opoka as a basis for its use in wastewater treatment. *Polish Journal of Environmental Studies* 13 (1): 15–20.

Cayuela, M. L., M. A. Sánchez-Monedero, A. Roig, K. Hanley, A. Enders, and J. Lehmann. 2013. Biochar and denitrification in soils: When, how much and why does biochar reduce N2O emissions? *Scientific Reports* 3 (April). http://dx.doi.org/10.1038/srep01732.

Chen, W., R. Parette, J. Zou, F. S. Cannon, and B. A. Dempsey. 2007. Arsenic removal by iron-modified activated carbon. *Water Research* 41 (9): 1851–1858. doi:10.1016/j.watres.2007.01.052.

Cordell, D., J.-O. Drangert, and S. White. 2009. The story of phosphorus: Global food security and food for thought. *Global Environmental Change* 19 (2): 292–305. doi:10.1016/j.gloenvcha.2008.10.009.

DeLuca, T. H., M. D. MacKenzie, and M. J. Gundale. 2009. Biochar effects on soil nutrient transformations. In *Biochar for Environmental Management: Science and Technology*, J. Lehmann and S. Joseph, eds., pp. 251–266. Oxford, U.K.: Earthscan.

Diaz, L. F., G. M. Savage, and C. G. Golueke. 1996. Stabilization of hazardous wastes through biotreatment. In *Science of Composting*, M. Bertoldi, P. Sequi, B. Lemmes, and T. Papi, eds., pp. 849–862. Glasgow, Scotland: Chapman & Hall.

Dolliver, H., S. Gupta, and S. Noll. 2008. Antibiotic degradation during manure composting. *Journal of Environment Quality* 37 (3): 1245. doi:10.2134/jeq2007.0399.

Drizo, A., C. A. Frost, J. Grace, and K. A. Smith. 1999. Physico-chemical screening of phosphate-removing substrates for use in constructed wetland systems. *Water Research* 33 (17): 3595–3602.

Eckardt, N. A. 2005. Insights into plant cellular mechanisms: Of phosphate transporters and arbuscular mycorrhizal infection. *The Plant Cell Online* 17 (12): 3213–3216. doi:10.1105/tpc.105.039297.

Gershuny, G. 1999. *The Soul of Soil: A Soil-Building Guide for Master Gardeners and Farmers*, 4th edn. White River Junction, VT: Chelsea Green Pub. Co.

Gilman, S. 2011. *Organic Soil-Fertility and Weed Management*, Northeast Organic Farming Association, ed. Updated and rev. Organic Principles and Practices Handbook Series. White River Junction, VT: Chelsea Green Pub.

Grantham, R. 1992a. Geotherapy as evolutionary choice. *Global Environmental Change* 2 (1): 2–4.

Grantham, R. 1992b. Geotherapy and global bioethics: Towards acceptable survival. *Global Environmental Change* 2 (4): 258–261.

Hallock, P. and A. Boersma. 2005. Deep sea sediments and paleoceanography: Biogenic sediments. University of Puerto Rico—Mayaguez, Mayaguez, PR. http://geology.uprm.edu/Morelock/dpseabiogenic.htm.

Hansen, J., M. Sato, P. Kharecha, D. Beerling, R. Berner, V. Masson-Delmotte, M. Pagani, M. Raymo, D. L. Royer, and J. C. Zachos. 2008. Target atmospheric CO_2: Where should humanity aim? *The Open Atmospheric Science Journal* 2 (1): 217–231. doi:10.2174/1874282300802010217.

Hause, B. and T. Fester. 2004. Molecular and cell biology of arbuscular mycorrhizal symbiosis. *Planta* 221 (2): 184–196. doi:10.1007/s00425-004-1436-x.

Heinonen-Tanski, H., S.K. Pradhan, and P. Karinen. 2010. Sustainable sanitation—a cost-effective tool to improve plant yields and the environment. *Sustainability* 2 (1): 341–353. doi:10.3390/su2010341.

Hoffmann, C. C., C. Kjaergaard, J. Uusi-Kämppä, H. C. B. Hansen, and B. Kronvang. 2009. Phosphorus retention in riparian buffers: Review of their efficiency. *Journal of Environment Quality* 38 (5): 1942. doi:10.2134/jeq2008.0087.

Höglund, C. 2001. *Evaluation of Microbial Health Risks Associated with the Reuse of Source-Separated Human Urine*. Stockholm, Sweden: Tekniska Högsk.

Hugo, V. 1862. *Les Miserables*. Vol. V. 5 vols. Brussels, Belgium: A. Lacroix, Verboeckhoven & Cie. http://www.readprint.com/chapter-5810/Les-Miserables-Victor-Hugo.

Jenkins, J. C. 2005. *The Humanure Handbook: A Guide to Composting Human Manure*. White River Junction, VT: Joseph Jenkins, Inc.; [Distributed by] Chelsea Green Pub.

Jensen, H. S., K. J. McGlathery, R. Marino, and R. W. Howarth. 1998. Forms and availability of sediment phosphorus in carbonate sand of Bermuda seagrass beds. *Limnology and Oceanography* 43 (5): 799–810.

Johansson, W. L. 2006. Substrates for phosphorus removal—Potential benefits for on-site wastewater treatment? *Water Research* 40 (1): 23–36. doi:10.1016/j.watres.2005.11.006.

Jones, D. L., G. Edwards-Jones, and D. V. Murphy. 2011. Biochar mediated alterations in herbicide breakdown and leaching in soil. *Knowledge Gaps in Soil C and N Interactions* 43 (4): 804–813. doi:10.1016/j.soilbio.2010.12.015.

Jönsson, H. 2004. *Guidelines on the Use of Urine and Faeces in Crop Production*. Stockholm, Sweden: Stockholm Environment Institute.

Keese, S. 2013. VPR: Brattleboro Group researches innovative fertilizer application. *Vermont Public Radio*. March 8. http://www.vpr.net/news_detail/97732/brattleboro-group-researches-innovative-fertilizer/.

Kibblewhite, M. G., K. Ritz, and M. J. Swift. 2008. Soil health in agricultural systems. *Philosophical Transactions: Biological Sciences* 363 (1492): 685–701. doi:10.2307/20208462.

King, F. H. 1911. *Farmers of Forty Centuries, Or, Permanent Agriculture in China, Korea and Japan*. Madison, WI: Mrs. F. H. King. http://archive.org/stream/cu31924073872685#page/n215/mode/2up.

Kreith, F. 2011. *Principles of Sustainable Energy*. Mechanical Engineering Series, p. 46. Boca Raton, FL: CRC Press.

Lal, R. 2003. Soil erosion and the global carbon budget. *Environment International* 29 (4): 437–450. doi:10.1016/S0160-4120(02)00192-7.

Laird, D., P. Fleming, B. Wang, R. Horton, and D. Karlen. 2010. Biochar impact on nutrient leaching from a midwestern agricultural soil. *Geoderma* 158 (3–4): 436–442. doi:10.1016/j.geoderma.2010.05.012.

Lehmann, J., C. Czimczik, D. Laird, and S. Sohi. 2009. Stability of biochar in the soil. In *Biochar for Environmental Management: Science and Technology*, J. Lehmann and S. Joseph, eds., pp. 183–205. Oxford, U.K.: Earthscan.

Liu, Q., L. Guo, Y. Zhou, Y. Dai, L. Feng, J. Zhou, J. Zhao, J. Liu, and G. Qian. 2012. Phosphate adsorption on biogenetic calcium carbonate minerals: Effect of a crystalline phase. *Desalination and Water Treatment* 47 (1–3): 78–85. doi:10.1080/19443994.2012.696798.

Mäder, J. A., J. Staehelin, T. Peter, D. Brunner, H. E. Rieder, and W. A. Stahel. 2010. Evidence for the effectiveness of the Montreal protocol to protect the ozone layer. *Atmospheric Chemistry and Physics* 10 (24): 12161–12171. doi:10.5194/acp-10-12161-2010.

Malisie, A. F., M. Prihandrijanti, and R. Otterpohl. 2007. The potential of nutrient reuse from a source-separated domestic wastewater system in Indonesia—case study: Ecological sanitation pilot plant in Surabaya. *Water Science and Technology* 56 (5): 141–148. doi:10.2166/wst.2007.566.

Major, J., C. Steiner, A. Downie, and J. Lehmann. 2009. Biochar effects on nutrient leaching. In *Biochar for Environmental Management: Science and Technology*, J. Lehmann and S. Joseph, eds., pp. 272–287. Oxford, U.K.: Earthscan.

Masto, R. E., S. Kumar, T. K. Rout, P. Sarkar, J. George, and L. C. Ram. 2013. Biochar from water hyacinth (*Eichornia crassipes*) and its impact on soil biological activity. *CATENA* 111 (December): 64–71. doi:10.1016/j.catena.2013.06.025.

McGlathery, K. J., R. Marino, and R. W. Howarth. 1994. Variable rates of phosphate uptake by shallow marine carbonate sediments: Mechanisms and ecological significance. *Biogeochemistry* 25 (2): 127–146. doi:10.2307/1469169.

Millero, F., F. Huang, X. Zhu, X. Liu, and J.-Z. Zhang. 2001. Adsorption and desorption of phosphate on calcite and aragonite in seawater. *Aquatic Geochemistry* 7: 33–56.

Nace, K. and A. Noe-Hays. 2012. Urine nutrient reclamation project–2012 experiment, Rich Earth Institute: Fertilizer from urine. http://richearthinstitute.org/?page_id=27.

Nace, K. and A. Noe-Hays. 2013. *Urine Source Separation and Reuse: A Community Scale Pilot Project.* In Tech, Waterbury, VT.

Nair, P. S., T. J. Logan, A. N. Sharpley, L. E. Sommers, M. A. Tabatabai, and T. L. Yuan. 1984. Interlaboratory comparison of a standardized phosphorus adsorption procedure. *Journal of Environmental Quality* 13 (4): 591–595.

Njira, K. O. W. 2013. Microbial contributions in alleviating decline in soil fertility. *British Microbiology Research Journal* 3 (4): 724–742.

Pimentel, D. and M. Burgess. 2013. Soil erosion threatens food production. *Agriculture* 3 (3): 443–463. doi:10.3390/agriculture3030443.

Richardson, A. E. and R. J. Simpson. 2011. Soil microorganisms mediating phosphorus availability update on microbial phosphorus. *Plant Physiology* 156 (3): 989–996. doi:10.1104/pp.111.175448.

Rodŕiguez, H. and R. Fraga. 1999. Phosphate solubilizing bacteria and their role in plant growth promotion. *Biotechnology Advances* 17 (4): 319–339.

Rogner, H.-H., D. Zhou, R. Bradley, P. Crabbé, O. Edenhofer, B. (Australia) Hare, L. Kuijpers, and M. Yamaguchi. 2007. Introduction. In *Climate Change 2007: Mitigation. Contribution of Working Group III to the Fourth Assessment Report of the Intergovernmental Panel on Climate Change*, B. Metz, O.R. Davidson, P.R. Bosch, R. Dave, and L.A. Meyer, eds., pp. 97–111. Cambridge, United Kingdom and New York, NY: Cambridge University Press.

Roseth, R. 2000. Shell sand: A new filter medium for constructed wetlands and wastewater treatment. *Journal of Environmental Science and Health, Part A* 35 (8): 1335–1355. doi:10.1080/10934520009377039.

Royal Society (Great Britain). 2009. *Geoengineering the Climate: Science, Governance and Uncertainty.* London, U.K.: Royal Society.

Schultz, R. C., T. M. Isenhart, W. W. Simpkins, and J. P. Colletti. 2004. Riparian forest buffers in agroecosystems—Lessons learned from the bear creek watershed, central Iowa, USA. *Agroforestry Systems* 61–62 (1–3): 35–50. doi:10.1023/B:AGFO.0000028988.67721.4d.

SCOPE. 2009. ESR news: Toilet users get paid money. EcoSanRes, Stockholm Environment Institute. http://www.ecosanres.org/pongalgift.htm.

Scripps Inst. Oceanography/UC San Diego. 2013. The keeling curve. Accessed December 18. http://keelingcurve.ucsd.edu/.

Silva, G. 2012. The peaks and valleys of phosphorus fixation: Understanding the peaks and valleys of phosphorus fixation in the soil, as affected by the soil pH, is one of the critical steps to manage this essential plant nutrient. Michigan State University Extension. http://msue.anr.msu.edu/news/the_peaks_and_valleys_of_phosphorus_fixation.

Sincero, A. P. and G. A. Sincero. 2003. *Physical-Chemical Treatment of Water and Wastewater.* Boca Raton, FL: CRC Press.

Steinhilber, P. and R. Weismiller. 2000. On-farm management options for controlling phosphorus inputs to the bay. In *Agriculture and Phosphorus Management: The Chesapeake Bay*, A. N. Sharpley, ed. Boca Raton, FL: Lewis Publishers.

Todd, J., E. J. G. Brown, and E. Wells. 2003. Ecological design applied. *Ecological Engineering* 20 (5): 421–440. doi:10.1016/j.ecoleng.2003.08.004.

Todd, J., S. Doshi, and A. McInnis. 2010. Beyond coal: A resilient new economy for Appalachia. *Solutions* 1 (4): 45–52.

VanMiert, J. and J. Jimmerson. 2014. Compost fundamentals: Composter's needs—Pesticides and herbicides. Washington State University: Whatcom County Extension. Accessed January 15. http://whatcom.wsu.edu/ag/compost/fundamentals/consideration_pesticides.htm.

Vohla, C., M. Kõiv, H. J. Bavor, F. Chazarenc, and Ü. Mander. 2011. Filter materials for phosphorus removal from wastewater in treatment wetlands—A review. *Ecological Engineering* 37 (1): 70–89. doi:10.1016/j.ecoleng.2009.08.003.

Wada, Y., L. P. H. van Beek, C. M. van Kempen, et al. 2010. Global depletion of groundwater resources: Global groundwater depletion. *Geophysical Research Letters* 37 (20): doi:10.1029/2010GL044571.

Warnock, D. D., J. Lehmann, T. W. Kuyper, and M. C. Rillig. 2007. Mycorrhizal responses to biochar in soil – Concepts and mechanisms. *Plant and Soil* 300 (1–2): 9–20. doi:10.1007/s11104-007-9391-5.

Webb III, T. and P. J. Bartlein. 1992. Global changes during the last 3 million years: Climatic controls and biotic responses. *Annual Review of Ecology and Systematics* 23 (January): 141–173. doi:10.2307/2097285.

World Health Organization. 2006. *Guidelines for the Safe Use of Wastewater, Excreta and Greywater.* Geneva, Switzerland: World Health Organization.

Yao, Y., B. Gao, H. Chen, L. Jiang, M. Inyang, A. R. Zimmerman, X. Cao, L. Yang, Y. Xue, and H. Li. 2012. Adsorption of sulfamethoxazole on biochar and its impact on reclaimed water irrigation. *Journal of Hazardous Materials* 209–210 (0): 408–413. doi:10.1016/j.jhazmat.2012.01.046.

Yao, Y., B. Gao, M. Inyang, A. R. Zimmerman, X. Cao, P. Pullammanappallil, and L. Yang. 2011. Biochar derived from anaerobically digested sugar beet tailings: Characterization and phosphate removal potential. *Bioresource Technology* 102 (10): 6273–6278. doi:10.1016/j.biortech.2011.03.006.

Zhang, Y. G., M. Pagani, Z. Liu, S. M. Bohaty, and R. DeConto. 2013. A 40-million-year history of atmospheric CO_2. *Philosophical Transactions of the Royal Society A: Mathematical, Physical and Engineering Sciences* 371 (2001): 20130096. doi:10.1098/rsta.2013.0096.

Zhao, F., S. Xi, X. Yang, W. Yang, J. Li, B. Gu, and Z. He. 2012. Purifying eutrophic river waters with integrated floating island systems. *Ecological Engineering* 40 (March): 53–60. doi:10.1016/j.ecoleng.2011.12.012.

33

Australian Journey toward Commercially Viable Carbon Farming

Cindy Eiritz

CONTENTS

> The single largest opportunity for CO_2 emissions reduction in Australia is through bio-sequestration, and the replenishment of soil carbon in particular.
>
> **Greg Hunt, Minister for the Environment, 2014**

Introduction

Bound by the Kyoto Protocol, Australia is one of 41 developed countries that have individual emission reduction targets defined by specific accounting rules. Such countries can meet their obligations by reducing emissions or by increasing their emission allowance through activities that reduce atmospheric carbon. In this context, Australia is almost unique in being a *developed country* (defined under Annex 1) that is a net exporter of energy, specifically coal. Under Articles 3.3 and 3.4 of the Kyoto Protocol, land use, land use change, and forestry (LULUCF) were offered to countries as options to fulfill their commitments. Throughout decades of climate change negotiations, Australia has looked to this provision as a way of meeting its international obligations.

The environment had been a political issue to voters in Australia long before the first Intergovernmental Panel on Climate Change (IPCC) report in 1990. The Carbon Farming Initiative (CFI) has evolved as a pragmatic way to restore the environment, improve the economic viability of the agricultural sector, and meet international obligations while minimizing the impact of emission reduction on other sectors of the economy. The journey has been long, but there is optimism in the agricultural sector and supporting industries about the potential of carbon farming.

This chapter provides a brief history of the politics and practice of carbon farming in Australia and the ways in which various governments and land managers have tried to meet these international obligations.

Requirements under the Kyoto Protocol

Scope of Accounting

Under the Kyoto Protocol, there is a general requirement that when accounting for emissions by sources, and removals by sinks under Articles 3.3 and 3.4, soil carbon must be accounted for in addition to accounting for the carbon stored in the vegetation itself (United Nations Framework Convention on Climate Change, 1998). Article 3.3 refers to forestry activities, and Article 3.4 refers to additional voluntary activities in land use. The IPCC Good Practice Guidance for LULUCF requires that approaches for reporting must be adequate, consistent, complete, and transparent (IPCC, 2003).

Many countries have struggled with databases, inventories, and sampling methods to meet these requirements, finding the difficulties with the *scope of accounting* and *cost of measuring* deterrent to voluntary activities (Zanchi, 2002).

The complexity of measuring and monitoring diverse ecosystems and land use change led to these considerations being excluded from early policy development. At the 1997 Kyoto negotiations (United Nations Framework Convention on Climate Change [UNFCCC]), Australian negotiators were able to secure the *Australia Clause* (Talberg, 2009). This allowed countries with net emissions from land use change, compared to the 1990 baseline, to use any reduction in land clearing to offset emissions from burning fossil fuels. Consequently, Australia's overall average emissions barely increased from 1990 to 2012 because increasing domestic emissions were offset by a 79% decrease in deforestation resulting from land clearing reforms. Australia's commitment was met, through *Australia Clause* offsets, even though energy emissions increased by 60% and transport emissions increased by 45% (Australian National Greenhouse Accounts, 2012).

Deforestation, particularly in developing countries, was identified as accounting for approximately 18% of global greenhouse gas emissions. Australia adopted the position that reducing emissions from deforestation and forest degradation (REDD+) "should be designed to maximise the benefits that can flow to local, Indigenous and forest-dependent communities" (Commonwealth Department of Environment, 2014). Australia has invested $273 million in the International Forest Carbon Initiative (Commonwealth Department of Environment, 2014). The Abbott government is preparing to implement tree planting programs as part of meeting emission reduction targets. Clearly, Australia's interest in REDD+ is not only for the benefit of developing countries.

Initially, the accounting rules required developed countries to account for emissions due to natural disturbance events, such as fire, drought, and extreme weather, as well as human activities. The Australian government's position was that having to account for these was unreasonable because such phenomena were out of land managers' control and could result in significant anomalies (Prime Minister's Science, Engineering and Innovation Council, 1999). In December 2012, the Australian government successfully negotiated to ensure that for the first Kyoto reporting period (2009–2012), emissions from natural phenomena would not be included in Australia's calculated emissions with respect to the target (Macintosh, 2012).

Capacity to Measure

Soil carbon has long been thought too difficult and too expensive to measure because of its enormous variability in time and space, depending on plant and microbial processes at microscopic levels. Australian governments, universities, and private enterprise have invested heavily in efforts to find an accurate and affordable method of measurement to support Australia's objective of meeting international obligations through land use.

The Department of Climate Change (Commonwealth), the Commonwealth Scientific and Industrial Research Organisation (CSIRO), and the Australian National University jointly developed the National Carbon Accounting System (NCAS), which then formed the basis for the Global Carbon Monitoring System. This was the result of a 10-year, three-phase, $35 million program developing technologies to transform large archives of data from the Landsat satellites to provide a way of accounting for all land-based emissions (CSIRO, 2011).

Greenhouse gas sources from the land constitute around 30% of Australia's emission profile from activities such as land clearing, cropping, grazing, and forestry. The Soil Carbon Research Program, the largest and most extensive soil sampling and analysis program undertaken in Australia, measured stocks of soil carbon with more than 20,000 samples taken from a wide range of soil types and farming operations across more than 4000 different locations in selected farming regions (Department of Environment, 2014).

A new and inexpensive analytical method has been developed to provide information to the NCAS model. Using midinfrared (MIR) spectroscopy, the CSIRO team (2014) has been able to generate a spectral signature of any soil, similar to a fingerprint. When this *fingerprint* is combined with previous measurements of carbon fractions across a range of soil types and analyzed using a complex mathematical process, the amount of carbon and its allocation to carbon fractions can be predicted easily for additional soils (CSIRO, 2014).

Carbon Link, Resource Consulting Services, Minnamurra Pastoral Co., CSIRO, and Peabody Energy have undertaken the world's first, large-scale, accurate carbon mapping project in Mudgee (NSW), trialing gamma detector and electromagnetic sensor equipment to measure soil carbon as a more cost-effective and accurate practice of mapping soil carbon than taking soil samples alone. Using information from the mapping project, a soil sampling design was submitted for public consultation and became an approved methodology in July 2014 (CarbonLink, 2014).

There is an expectation that the NASA satellite technology coming on line in 2015 will lower the cost of measuring soil carbon even further. The satellite technology has the capacity to monitor increments of stable soil carbon on 1 ha pixels to a depth of 3 m. Verification will be achieved by ground truthing the veracity of algorithms specific to Australia's different soil and landscape type including via calibrations with existing data from soil samples and wet chemistry. According to Walter Jehne, retired soil scientist and director of Healthy Soils Australia,

> That is why it is far more accurate, cheaper and useful not to sample, but measure the stable soil carbon over the whole area of each comparative paddock. This can then give you an accurate measure of how management systems are changing these values over time, which is what offsets need to be based on. Because we have satellite data going back to the 1970s we can also confirm changes and offsets retrospectively. We are measuring the exact same total area at each time so all sampling errors over space and time cancel out giving you much better data far more cost effectively. (Personal communication, February 24, 2014)

Carbon Farming Initiative

Australia's CFI is the first national offset scheme in the world to include carbon credits derived broadly from natural resource management. The CFI is a voluntary scheme (under Article 3.4) that provides economic rewards to farmers and landholders who take steps to reduce greenhouse gas emissions (Commonwealth Government Department of Environment, 2014).

The CFI emerged from a long history of scientific research, innovative farm practices, widespread public concern about the environment, and a very strong carbon farming lobby.

Under the CFI, land managers can earn carbon credits from activities such as reducing livestock emissions, increasing efficiency of fertilizer use, enhancing carbon in agricultural soil, and storing carbon through revegetation and reforestation. Carbon farming activities that reduce greenhouse gas emissions are referred to as abatement activities. They reduce emissions in one of two ways: by storing carbon in soil or plants (sequestration projects) or by reducing emissions of carbon and other harmful greenhouse gases (emission reduction or avoidance projects). Importantly, land managers who help reduce Australia's greenhouse gas emissions through approved CFI activities receive a benefit—a tradeable carbon credit (Department of Environment, 2014).

The Carbon Credits (CFI) Bill 2011 was one of a package of three related bills to establish a voluntary carbon offset scheme with the purpose of creating incentives for carbon abatement or avoidance projects in land use sectors. The bill had three stated objectives:

- To implement certain commitments Australia has under the UNFCCC and the Kyoto Protocol
- To create incentives for carbon abatement projects
- To increase carbon abatement while still protecting Australia's natural environment and enhancing resilience to climate change impacts (p. 4)

The concluding comments are provided in the following:

> The CFI is a carrot rather than a stick approach to atmospheric carbon reduction. It creates the incentive to take positive action by rewarding the agricultural and forestry sectors with carbon credits. As such it creates a supply of credits and a framework for trading them. However, the scheme, on its own, does nothing to stimulate demand. It merely assumes that demand for carbon credits exists through the voluntary market and the limited overlap with the Kyoto Protocol system... As landholders gradually develop a better understanding of the mechanism, the scheme may grow in importance and deliver the confidence stakeholders require for further investment. This process may be expedited by the establishment of an economy-wide carbon price, in which case the CFI may be become an valuable instrument for reducing Australia's total greenhouse gas emissions... the CFI is essential for engaging the land sector in greenhouse gas abatement and helping Australia meet its Kyoto commitment. To leave this sector with no means of participating in the growing international carbon market would be irresponsible.

Background

Early Scientific, Conservation, Regeneration, and Environmental Activity

The earliest report about Australian soils was published in 1845 (Lee, 1998), and what became the CSIRO Division of Soils was established in 1927 (Lee, 1998). A growing awareness in both the scientific and farming communities of the need to improve land management practices saw Australian farmers receive state government grants or loans for farm-scale soil conservation with free extension advice as early as 1945.

Some weed and pest control initiatives as well as tree-growing programs also operated, and these were followed by river catchment–based programs, which emerged in the 1960s (Youl et al., 2006).

The environmental movement in Australia began as a conservation movement when the Royal National Park in Sydney was proclaimed in 1879 following the lead of the United States in establishing Yellowstone National Park. Large-scale killing of koalas following World War I led to a public outcry and the movement became mainstream.

Bush regeneration started in Australia in the 1960s with the development of the *Bradley Method* of weed control to restore remnant native vegetation (Radi, 1993).

The Landcare movement was started by farmers working to control soil erosion and salinity through revegetation projects in the 1960s in Victoria. The movement gained momentum, spreading to other states and eventually going national. An alliance between the National Farmers Federation and the Australian Conservation Council in 1989 was able to secure bipartisan support and substantial government funding for *a decade of Landcare* (Commonwealth Government Department of Agriculture, 2008).

Many other conservation organizations emerged including the Wilderness Society (1976), Conservation Volunteers Australia (1982), Greening Australia (1983), and the Bush Heritage Trust (1991). This personal, *hands-on, hip pocket* involvement in environmental issues by ordinary people forced environmental issues onto the political agenda.

In 1999, a significant piece of legislation impacting land use, the Environment Protection and Biodiversity Conservation Act 1999 (EPBC Act), was passed. This legislation, still in force, enables the Australian federal government to join with the states and territories in providing a truly national scheme of environment and heritage protection and biodiversity conservation:

> The increasing awareness of and concern for environmental issues in Australia has found expression in a broad range of community led activities. They have also led to legislation, regulation and expenditure by governments, at national, state and local level, to protect the environment. (Department of Environment, 2002)

Early Farming Movements

The organic movement, which began in Europe, had begun to be adopted by some farmers in Australia during the 1960s and 1970s in response to consumer demand. *No-dig gardening* was promoted in Australia by Esther Deans in the 1970s, and *Permaculture One* was published by Bill Mollison and David Holmgren in 1978. The Organic Federation of Australia was established as a peak body in 1998.

Hydrology and Natural Sequence Farming

Water for Every Farm was published by Percy Yeomans in 1973, detailing work he had done developing *key line design* since the 1940s. Experimentation with natural sequence farming began in the 1970s and has become part of regenerative land management practices in the last 20 years (Andrews, 2014).

Restoration of the natural chain of ponds and weirs, typical of Australian landscapes prior to agriculture. (From Soils for Life, 2012.)

At Jillamatong (NSW), eroding headwalls were stabilized with natural weirs using natural sequence farming techniques, and contour channels were built to spread water naturally back onto the flood plains. The regeneration has restored a healthy semipermanent stream and natural flows replenish subsurface soil moisture, sustaining yearlong perennial pasture. At times of drought where lack of feed and water saw neighbors unable to run any stock for up to 11 months, cattle on Jillamatong were still being fattened (Soils for Life, 2012). For further information on this case study, please see http://www.soilsforlife.org.au/downloads/7%20Jillamatong.pdf.

Conservation Farming

Conservation farming, a global movement that started in Australia in the 1970s, became formalized when Conservation Farmers Incorporated was registered in 1979 (Conservation Farmers, Inc. 2014). Practices encompassed by this approach include no-tillage, farm forestry, minimum and reduced tillage, trap cropping for insect control, cover and green manure cropping, alley cropping, contour farming and strip cropping, organic and biodynamic farming, stubble mulching, integrated pest management, and crop and pasture rotation (O'Gara, 1998).

Presently in Australia, conservation farming is conducted on about 8.5 million hectares, and over 4000 farmers are represented by state and national organizations. Broadacre cropping contributed over 50% of the gross value of Australian agriculture production in 2009–2010. Across Australia, 76% of cropped land is now managed with no cultivation, apart from when seeds are sown (Barson et al., 2012). Achieving this level of adoption took 15 years.

Pasture Cropping

At Winona, NSW, Colin Seis and Daryl Cluff developed the innovative *pasture cropping* land management technique. The original woodland had been transformed to crops sown every 3–5 years, ploughing and working the soil up to five times. Colin began a transition to perennial pastures and away from inputs. Pasture cropping combines cropping and grazing into one symbiotic management method. The paddock is prepared for cropping by short-term time-controlled grazing of a large mob of sheep (100–150 a hectare). The sheep graze and trample the perennial pasture down to a height of 100 mm. The added nutrients from the manure and urine create a litter and mulch layer. Crops are then directly drilled into the perennial pasture. Sheep are reintroduced for a short period after harvest to graze the remaining native pasture. The soil, vegetation, and animals are now healthy and productivity has increased significantly, primarily because costs have been substantially reduced. *Pasture cropping* is becoming so popular that there is a ground swell of over 1500 farmers using the technique in Australia and it is starting to be adopted in other countries (Soils for Life, 2012). For further information on this case study, please see http://www.soilsforlife.org.au/downloads/12%20Winona.pdf.

Harvesting showing the perennial grass under the sown crop. (From Soils for Life, 2012.)

Cell Grazing

The concept of cell grazing was first introduced in Australia with an address by Dr. Stan Parsons to the Northern Territory Cattleman's Association. There was a strong opposition from scientists and advisors at the time, but innovative graziers began putting the system in place in the early 1990s. The principles have been steadily evolving through practice, and considerable progress has been made in Australia since 1994 (McCosker, 2000).

The Soils for Life case study of Dukes Plain describes the success they have had with cell grazing. The property was divided into 97 paddocks, 20–40 ha each, grouped into three cells to manage the various mobs of cattle. Grazing in each paddock started being constantly adjusted through comprehensive monitoring and measuring. Now rainfall and feed availability govern stocking rate, applied at 26 stock days per hectare per 100 mm of rainfall. The rotation is generally 2–3 days of grazing followed by 60 days of recovery. Production increased within 2 years and the investment in infrastructure paid off in 3 years. After 9 years, productivity has increased 30%, and soil organic matter is now 400% higher than the district average of 1%. About 10% of properties are now using cell grazing (Soils for Life, 2012). For further information on this case study, please see http://www.soilsforlife.org.au/downloads/1%20Dukes%20Plain.pdf.

Cell grazing. Cattle have become very docile; they wait patiently to be moved to the next paddock. (From Soils for Life, 2012.)

Carbon Farming Lobby

Over the past decade, a number of strong organizations have emerged to inform and advocate on behalf of land managers. The combined strength of these organizations has been instrumental in gaining bipartisan support for carbon farming. Australian farmers have actively campaigned for their right to sell soil carbon since 2005 (Carbon Farmers of Australia, 2014).

The Carbon Farmers of Australia organization conducted the first study tour of the US carbon industry in 2006 and secured the first order for Australian soil carbon from the Chicago Climate Exchange. The first sales of credits followed in March 2007. The world's first Carbon Farming Conference was held at Mudgee that year with the first formal training program on soil carbon launched the following year. They have since convened nine gatherings of scientists and land managers to promote a spirit of collaboration and knowledge exchange. The first *Carbon Farming Handbook* was published by CFA in 2009, and their advocacy helped to secure $26 million in funds for CSIRO research into soil carbon for trade. International recognition came with the organization's director's appointment to FAO rangelands and conservation farming advocacy groups (Carbon Farmers of Australia, 2014).

Other influential organizations include Soils for Life, a project of Outcomes Australia, working closely with Healthy Soils Australia and the Soil Carbon Initiative at the US Studies Centre (University of Sydney). Research is focused on how changes in agricultural practices and policy can lead to increased levels of carbon in soils. Outcomes Australia is headed by former Governor General Michael Jeffries, who has recently been reappointed Soils advocate by the government. Robert Hill, who heads the US Studies Centre, campaigned for the *Australia Clause* and signed the Kyoto Protocol on behalf of Australia as Minister for Environment in 1998 (Blair, 2006). This is not an exhaustive list of organizations and individuals involved in promoting carbon farming, but provides some indication of the strength of the lobby.

Political Journey

It would be a mistake to imagine that Australia decided that carbon farming would be a good idea and proceeded to implement the necessary legislation. The political journey that led to the CFI becoming law has been long and tortuous.

The first Senate inquiry into climate was initiated in 1988 with the first Australian greenhouse gas emission reduction proposal (for a 20% reduction in 1988 emission levels by 2005) submitted to the Cabinet in 1989 (Talberg et al., 2013). By 2000, another Senate inquiry released a report that was highly critical of the lack of action to that date and made 104 recommendations (Talberg et al., 2013).

Coalition Government under John Howard

The coalition (liberal and national parties) came into government in 1996. Although Australia signed the Kyoto Protocol in April 1998, Prime Minister John Howard steadfastly refused to ratify it throughout his term of office. He said that "the reason it is not in Australia's interests to ratify the Kyoto protocol is that, because the arrangements currently exclude—and are likely under present settings to continue to exclude—both developing countries and the United States, for us to ratify the protocol would cost us jobs and damage our industry. That is why the Australian government will continue to oppose ratification. One of the things that makes Australia almost unique in this context is that as a developed country we are a major net exporter of energy" (Howard, 2002).

In April 1998, Australia had become the first country to establish a government agency dedicated to reducing greenhouse gas emissions, the Australian Greenhouse Office (AGO). The *Measures for a Better Environment* package was announced as part of the coalition's 2000–2001 budget. More than half of this funding was to be dedicated to the Greenhouse Gas Abatement Program to "to reduce Australia's net greenhouse gas emissions by supporting practical and effective activities that will deliver substantial greenhouse gas emission reductions or substantial carbon sink enhancement" (Greenhouse Gas Abatement Program, 2011).

In January 2003, the state of New South Wales implemented the Greenhouse Gas Reduction Scheme, the world's first mandatory emission trading scheme that used a baseline and credit system rather than a cap-and-trade system (Government of New South Wales, 2013).

An independent review of the AGO released 4 years after its inception noted criticism of the pace of program delivery but considered the pace was acceptable "under what has been an ambitious timeframe for a recently formed organisation" (Department of Environment, 2002).

The report also recommended the involvement of the Department of Foreign Affairs and Trade in international greenhouse issues. This concern with international relations was demonstrated in August 2002, when a new climate change strategy was announced jointly by the Minister for Environment and the Minister for Foreign Affairs ahead of the World Summit on Sustainable Development in Johannesburg (Australian Parliament, 2002).

In 2003, there were press reports that an emissions trading scheme (ETS) plan was presented to the Cabinet by the Federal Treasury, the Industry Department, and the Environment Department with support from six other government portfolios. The reports claimed that following a meeting with the Industry Department, the prime minister (Howard) rejected the plan (The Age, November 15, 2006).

Russia ratified the Kyoto Protocol at the end of 2004, bringing it into force. In response to a written question, the Minister for Environment said, "Australia will meet its Kyoto greenhouse emissions target. Australia has not ratified the Protocol to underline that while it has some positive elements, it is flawed as a basis for an effective global response to climate change. Global greenhouse gas emissions are on track to increase by around 40% on 1990 levels by the first commitment period of the Kyoto Protocol, and the Protocol will change this growth by only 1%. Despite this, there is no pathway for action by all major emitters" (Australian Parliament, 2005).

Two months before the 2007 election, the National Greenhouse and Energy Reporting Bill became law, which put in place a process to account for emissions. With an election looming, John Howard promised to establish a national emission trading scheme and to set up a climate change fund to receive revenue from the ETS and reinvest the funds into climate change initiatives. At the same time, the Australian Financial Review reported that Howard rejected a proposal taken to the Cabinet by the Minister for Environment Malcolm Turnbull to ratify the Kyoto agreement (Garnaut, 2008b).

Labor Government under Kevin Rudd

In December 2007, the Rudd Labor Government was voted in on a mandate of action on climate change and Australia ratified the Kyoto Protocol immediately.

The interim report of the Garnaut Review was released in February 2008. In the absence of support from the Howard government, this review, by eminent economist Prof. Ross Garnaut, had been set up by the states and territories. The final Garnaut report was released on September 30, 2008, with Chapter 22 devoted to rural land use. An extract is provided as follows:

> Rural Australia faces pressures for structural change from both climate change and its mitigation. Effective mitigation would greatly improve the prospects for Australian agriculture, at a time when international demand growth in the Platinum Age is expanding opportunities. Choices for landowners will include production of conventional commodities, soil carbon, bioenergy, second-generation biofuels, wood or carbon plantations, and conservation forests. There is considerable potential for biosequestration in rural Australia. The realisation of this potential requires comprehensive emissions accounting. The realisation of a substantial part of the biosequestration potential of rural Australia would greatly reduce the costs of mitigation in Australia. It would favourably transform the economic prospects of large parts of remote rural Australia. Full utilisation of bio sequestration could play a significant role in the global mitigation effort. This is an area where Australia has much to contribute to the international system. (Garnaut, 2008a)

The first Australian ETS legislation, the Carbon Pollution Reduction Scheme Bill (2009), was introduced into the Parliament and was rejected by the (coalition dominated) Senate. It was reintroduced into the House of Representatives in October. Media reports indicated that the Leader of the Opposition, Malcolm Turnbull, was negotiating amendments to the legislation with the government. In a dramatic leadership spill, Turnbull lost the leadership to Tony Abbott by one vote, and the following day, the legislation was again rejected by the Senate, a week before the 15th UNFCCC in Copenhagen (Talberg et al., 2013).

At the end of April 2010, Prime Minister Kevin Rudd announced that the Carbon Pollution Reduction Scheme would be postponed until the end of the Kyoto commitment period in 2012. Journalist Paul Kelly described the move as "one of the most spectacular backdowns by a prime minister in decades" and declared that Rudd was "a prime minister without the courage to champion the policy that defined him" (Australian Broadcasting Corporation, 2010).

Two months later, Rudd was replaced as prime minister by Julia Gillard following an internal party challenge. In the 2010 election campaign, the coalition called the Carbon Pollution Reduction Scheme "a great big tax on everything." Their proposed alternative was an *emission reduction fund*. Businesses that reduced emissions would be able to sell their CO_2 abatement to the government and those who increased emissions would incur a penalty. The *direct action* policy referred to soil carbon as "a once in a century replenishment of our soils" (Liberal Party of Australia, 2010).

The 2010 election was focused on leadership rather than issues. Women wanted to support the first female prime minister, but people who had been supporters of Kevin Rudd were upset by his demise. Tony Abbott's long career in politics meant most voters could remember something they did not like about him. No one won the 2010 election. The result was a hung parliament.

Labor Government under Julia Gillard with Greens and Independent Support

With the support of the Greens and two independents, the Labor party, under Julia Gillard, was able to form a government. From a position of power, the Greens negotiated a written agreement with Labor, which included the statement, "Australia must tackle climate change and that reducing carbon pollution by 2020 will require a price on carbon agreement" (The Australian Greens and the Australian Labor Party, 2010).

As part of the agreement between the Greens and Labor, a Multi-Party Climate Change Committee was formed to examine ways to introduce a carbon price. Agreement was reached for a fixed carbon price for unlimited permits starting in 2012, followed by a transition to a market-based cap-and-trade scheme in 2015. The Climate Commission was launched in February 2011 as an independent body to provide advice and information on climate change (Talberg et al., 2013).

The Clean Energy Act 2011, a suite of 18 bills providing a framework for an emission trading scheme, was passed in November 2011. Under this legislation, organizations emitting over 25,000 tonnes of carbon a year became *liable entities* and were required to purchase permits. A price on carbon came into effect on the 1st of July with an unlimited number of carbon units made available at a fixed price of $23. Without a cap, the scheme functioned similarly to a carbon tax. This mechanism was immediately compromised by excessive exemptions. The option for *liable entities* to meet their obligations through buying overseas Clean Development Mechanisms meant they could simply pay for poor practices rather than change and that there was no incentive to buy local carbon offsets. The Australian ETS was then linked with the European Union ETS in August (Talberg et al., 2013) in a deal that removed Australia's price floor, making local carbon offsets even less attractive.

The CFI Bill became law in September 2011.

As prime minister, Julia Gillard survived leadership challenges in February 2012 and March 2013, but with an election announced for September and poor standing in opinion polls, a leadership spill in June resulted in Kevin Rudd regaining the position of prime minister. In the election that followed, the Australian Labor Party lost 17 seats and Tony Abbott led the coalition into government.

Coalition Government under Tony Abbott

Eleven days after the election, hours after Abbott was sworn in as prime minister, the new government ordered the Clean Energy Finance Corporation to cease investments, announced plans to abolish the Climate Change Authority, abolished the Climate Commission, and began drafting legislation to repeal the Clean Energy Act 2011 and the associated legislation that Labor had worked for 4 years to pass. The absence of support in the Senate meant that it was July 2014 before the first of the repeal bills was passed, amid more dramatic scenes.

The Abbott government intends to retain the Clean Energy Regulator (2014), established in April 2012, as the government body responsible for administering legislation to reduce carbon emissions and increase the use of clean energy. The Clean Energy Regulator administers the National Greenhouse and Energy Reporting Scheme (established in 2007 as a single, national framework for corporations to report on greenhouse gas emissions, energy use, and energy production).

The CFI is being retained but is being amended, together with associated legislation, to replace the carbon pricing mechanism with the Emissions Reduction Fund (Parliament of Australia, 2014).

The Emissions Reduction Fund is the *centerpiece* of the Australian government's direct action plan. The purpose of the fund is to purchase emission reductions from approved projects undertaken by business, community organizations, local councils, and other members of the community. Eligible projects include those that upgrade commercial buildings, improve energy efficiency of industrial facilities and houses, reduce electricity generator emissions, capture landfill gas, reduce waste coal mine gas, implement reforestation and revegetation of marginal lands, improve Australia's agricultural soils, upgrade vehicles and improve transport logistics, and manage fires in savanna grasslands (Australian Government Department of Environment, 2014). Public consultation on the ERF has been completed and legislation is being drafted.

Under the proposed arrangements, the government will continue to support voluntary carbon markets to ensure that there is a supply of credible carbon units available for businesses and other organizations who wish to offset their emissions. It will eventually be possible for overseas buyers to own Australian carbon credit units but not for at least 3 years because the government is concerned about meeting present targets. Future arrangements will be part of a 2015 review.

Once the Emissions Reduction Fund is operational, it is envisaged that the CFI will be *folded* into the ERF so there will be just one program with a view to streamline the existing administrative arrangements, which have been onerous. In their response to the proposed Emissions Reduction Fund green paper in November 2013, Carbon Farmers of Australia observed

> With the best of intentions, the Carbon Farming Initiative has been over-engineered in the name of integrity and robustness, with every change adding more complexity. The continual stream of regulatory 'improvements' creates uncertainty, expense and delay. The pursuit of the perfect became the enemy of the good. (Kiely, 2013)

Simplified reporting and audit arrangements to be established under the Emissions Reduction Fund will apply automatically to CFI projects. The government intends to continue working closely with the land sector to develop methodologies. Priority areas for development include increasing soil carbon, reducing livestock emissions, and expanding opportunities for environmental and carbon sink plantings and reforestation and avoided deforestation:

> The land sector makes a significant contribution to Australia's emissions and has significant potential to generate emissions reductions to 2020. Many land sector emissions reduction activities deliver valuable co-benefits. For example, reforestation can reduce erosion and improve water quality, address salinity and provide habitat for native species. Reducing livestock emissions can increase meat production. Increasing soil carbon can improve soil health, water retention and plant growth. Many land projects also provide important community benefits. For example, savanna fire projects provide employment for Indigenous communities as well as reducing fire risks to people and property. (p. 58, Emissions Reduction White Paper 2014)

Current Scientific Research and Debate

A barrier to advancing soil carbon sequestration trading in Australia is the significant differences of scientific opinion on the amount of carbon that well-managed soil can sequester. Big agribusiness- and government-funded soil scientists tend to focus on high-input, chemical-based farming, and very little research funding has been available to study emerging innovative practices. While the physical and chemical contents of different soils are well understood, the significance of soil microbiology is only now beginning to be recognized.

The $26 million Soil Carbon Research Project created a database of thousands of soil samples, but the data were dominated by samples of soil that had been the subject of traditional farming practices; consequently, the mean measured soil carbon improvement was found to be only 0.1–0.3 tonnes per hectare per annum (Soil Carbon Research Progra, 2012).

In a personal communication with the Carbon Coalition Against Global Warming, a CSIRO researcher disclosed that the project did not initially test soil samples that had been subject to innovative regenerative land management practices such as grazing management and pasture cropping (Carbon Farmers of Australia, 2011). Nevertheless, an interesting trend was identified in their latest studies. The researcher described to Carbon Farmers of Australia the *long tail* of the distribution data—a cohort of high-performance soil carbon managers, with sequestration figures up to 200% greater than previously assessed averages. Two distinct subpopulations appear to be represented in the data: conventional farmers, in the hump of the curve, and innovators in the tail (Carbon Farmers of Australia, 2012).

When best practice, rather than standard practice, is assessed in detail and when the *outliers* of the data bell curves are considered as being indicative of a *new norm* (rather than statistical anomalies), the evidence suggests a completely different and new paradigm. Innovative farmers question if CSIRO's sample set, which simply confirms the low baseline of status quo practices, was consciously or subconsciously biased by conservative researchers. Many believe that the project's results in no way reflect the high carbon sequestration rates possible with regenerative land management. It is a concern that these data are now used by those who claim that soil carbon farming cannot deliver significant greenhouse gas abatement (Carbon Farmers of Australia, 2012).

High Hopes

In spite of all the obstacles, the Clean Energy Regulator has now approved over 150 CFI projects, and more than 6.5 million Australian carbon credit units have been issued to these projects (Australian Government Clean Energy Regulator, 2014, Register of Offset Projects).

Although there was an initial pessimism about demand for carbon credits through the voluntary market, "an ever-increasing number of businesses - large and small - in Australia are actively participating in the voluntary carbon credits market, either through a genuine desire to take their environmental responsibilities seriously, or by recognising the obvious reputation, branding and general PR benefits such activity brings with it" (Brand Carbon, 2014).

A strong indication of future possibilities is the purchase of the Dukes Plain cell grazing property by Origin/Shell. The company will have ownership of the perpetual income from the carbon being sequestered through good pasture management and the regrowth of Brigalow forest, without even considering income from the cattle operation. In order to manage potential risk, corporations are already factoring in the cost of offsets, which may be required in order to sell fossil fuel products in the future. As there is no legally binding requirement to offset at present, Origin/Shell will be free to sell the verified offsets from Dukes Plain on the market. The purchase also positions the company to benefit from environmental credibility for its foresight in acquiring such an asset. It is astonishing that there is any doubt about the viability of carbon farming when large corporations are already positioning themselves to exploit the emerging opportunities (ABC, 2014).

Conclusion

A century of soil science has provided a wealth of information. Where once CSIRO Division of Soils dominated research and Department of Primary Industries extension agents disseminated information to people on the land, there is now a collaborative effort between land managers and advisors. CSIRO is still a preeminent research organization but universities are eager to enhance their reputations by making a contribution. Australian farmers and graziers were early adopters of information technology, and a great deal of research is now being done on farms by well-informed managers, themselves. Information circulates freely, and there are many instances of individual experiments undertaken in pursuit of better financial returns outperforming academic field trials.

Businesses are working directly with producers to develop new products, and independent consultants and trainers work directly with land managers to develop solutions. All of these enterprises, as well as the brokers gearing up to handle carbon credits, are now well positioned to take their knowledge and products into international markets.

A recent article in The Australian summarized this position:

> The question, then, is not can it work, but when? Australia leads on being able to measure and monitor increases in soil carbon, and to account for the abatement against national greenhouse accounts. With an international soil policy community searching for solutions, Australia will have much to offer, not only in its direct action for climate change, but also in how the world can secure its soil and achieve sustainability. (Koch, 2013)

ACKNOWLEDGMENTS

The author wishes to acknowledge the work of Johanna Kieboom who kindly ghostwrote the first drafts of this chapter.

REFERENCES

Andrews, P. 2014. Natural sequence farming. http://www.nsfarming.com, accessed on August 15, 2014.

Australian Broadcasting Commission. 2014. Energy giant shell considers sale of Australian petrol stations, fuel terminals. http://www.abc.net.au/news/2014-01-31/shell-looking-to-sell-service-stations/5231278, accessed on August 15, 2014.

Australian Broadcasting Corporation. 2010. http://www.abc.net.au/lateline/content/2010/s2887625.htm, Lateline April 30, 2010.

Australian Government. 2014. Emissions reduction fund white paper. http://www.environment.gov.au/system/files/resources/1f98a924-5946-404c-9510-d440304280f1/files/emissions-reduction-fund-white-paper_0.pdf, accessed on August 15, 2014.

Australian Government Clean Energy Regulator. 2014. Register of offsets projects. http://www.cleanenergyregulator.gov.au/Carbon-Farming-Initiative/Register-of-Offsets-Projects/Pages/default.aspx, accessed on August 15, 2104.

Australian Greens and The Australian Labor Party. 2010. 'The 'Parties' Agreement, September 1, 2010. http://www.aph.gov.au/About_Parliament/Parliamentary_Departments/Parliamentary_Library/pubs/rp/rp1314/ClimateChangeTimeline, accessed on August 15, 2014.

Australian Parliament. 2002. Global greenhouse challenge: The way ahead for Australia. http://parlinfo.aph.gov.au/parlInfo/search/display/display.w3p;query=Id%3A%22media%2Fpressrel%2FHZ676%22, accessed on August 15, 2014.

Australian Parliament. 2005. Hansard, June 15, 2005, p. 250. http://parlinfo.aph.gov.au/parlInfo/search/display/display.w3p;query=Id%3A%22chamber%2Fhansardr%2F2005-06-15%2F0232%22, accessed on August 15, 2014.

Barson, M., Mewett, J., and Paplinska, J. 2012. Land management practice trends in Australia's broad acre cropping industries: Caring for our Country sustainable practices fact sheet 3. Department of Agriculture, Fisheries and Forestry, ACT. http://nrmonline.nrm.gov.au/catalog/mql:2866, accessed on April 4, 2014.

Blair, R. 2006. Renegotiating the Kyoto Treaty. McKean and Park Lawyers. http://www.google.com.au/url?sa=t&rct=j&q=&esrc=s&source=web&cd=1&ved=0CB4QFjAA&url=http%3A%2F%2Fwww.mckeanpark.com.au%2Farticles%2FRenegotiatingtheKyotoTreaty.pdf&ei=MBjuU8f_CJChogST4YLoDQ&usg=AFQjCNGngsBsa2xiDQauMgbPeYwVQVhWJA&bvm=bv.73231344,d.cGU.

Brand Carbon. 2014. Credits: First things first—What exactly is a carbon credit? http://brandcarbon.com.au/credits.php, accessed on August 15, 2014.

Carbon Farmers of Australia. 2011. Carbon coalition against global warming, November 16, 2011. http://www.carboncoalition.com.au/carbon-farmers-of-australia-blog/carbon-coalition-against-global-warming, accessed on June 15, 2014.

Carbon Farmers of Australia. 2012. Look out for the outliers—They prove it! http://www.carbonfarmersofaustralia.com.au/carbon-farmers-of-australia-blog/look_out_for_outliers_they_prove_it, accessed on August 15, 2015.

Carbon Farmers of Australia. 2013. The 100 year rule: Deal killer. http://100yearrule.blogspot.com.au/%20%20%20%20%20%20%20http://www.carbonfarmersofaustralia.com.au/carbon-farmers-of-australia-blog/tag/100_Years_Rule/, accessed on June 15, 2014.

Carbon Farmers of Australia. 2014. http://www.carboncoalition.com.au/, accessed on June 17, 2014.

CarbonLink. 2014. Carbon baselining. http://www.carbonlink.com.au/carbon-baselining, accessed on August 15, 2014.

Commonwealth Budget. 2000–2001. The commonwealth's environmental expenditure. Appendix 6—Greenhouse and Atmosphere. http://www.budget.gov.au/2000-01/minst/html/enviro-10.htm, accessed on June 16, 2014.

Commonwealth Government Clean Energy Regulator. 2014. How does it work? http://www.cleanenergyregulator.gov.au/Carbon-Farming-Initiative/How-does-it-work/Pages/default.aspx, accessed on August 15, 2014.

Commonwealth Government Department of Agriculture. 2008. Evaluation report on a decade of landcare plan: National overview. http://www.daff.gov.au/natural-resources/landcare/publications/eval-decade-plan, accessed on August 15, 2014.

Commonwealth Government Department of the Environment. 2014. International Forest Carbon Initiative. http://www.climatechange.gov.au/international/actions/australias-contribution/reducing-emissions-deforestation-and-forest-degredation/international-forest-carbon-initiative.

Conservation Farmers, Inc. 2014. http://www.cfi.org.au/.

CSIRO. 2014. Australia's national carbon accounting system leads the world. http://www.csiro.au/Outcomes/Climate/NCASpartnership.aspx, accessed on August 15, 2014.

Department of Environment. 2002. Australian actions to combat desertification and land degradation. National Report by Australia on Measures Taken to Support Implementation of the United Nations Convention to Combat Desertification, Commonwealth Intergovernmental Working Group for the UNCCD, April 2002. http://www.environment.gov.au/resource/australian-actions-combat-desertification-and-land-degradation.

Department of Environment. 2014. Reducing emissions from deforestation and forest degradation in developing countries. http://www.climatechange.gov.au/international/actions/australias-contribution/reducing-emissions-deforestation-and-forest-degredation, accessed on August 15, 2014.

Garnaut. 2008a. Garnaut climate change review. Interim report to the Commonwealth, State and Territory Governments of Australia.

Garnaut 2008b. Garnaut climate change review. Final Report, September 30, 2008, Chapter 22, Transforming rural land use, key points page 531. http://www.rossgarnaut.com.au/ClimateChange.html.

Government of New South Wales. 2013. Strengths, weaknesses and lessons learned. Final Report, New South Wales Greenhouse Gas Reduction Scheme, July 2013, Independent Pricing and Regulatory Tribunal of New South Wales. http://www.ipart.nsw.gov.au/Home/Industries/Greenhouse_Gas_Reduction_Scheme/NSW_Greenhouse_Gas_Reduction_Scheme_-_Strengths_weaknesses_and_lessons_learned_-_Final_Report_-_July_2013.

Greenhouse Gas Abatement Program. 2011. The commonwealth budget, 2000–2001, Chapter 2. http://www.budget.gov.au/2000-01/minst/html/enviro-02.htm.

Howard, J. 2002. Questions without notice—Environment: Kyoto protocol. House Hansard, Parliament of Australia, June 5, 2002. http://parlinfo.aph.gov.au/parlInfo/search/display/display.w3p;query=Id%3A%22chamber%2Fhansardr%2F2002-06-05%2F0030%22.

Intergovernmental Panel on Climate Change. 2003. Good practice guidance for land use, land use change and forestry. http://www.ipcc-nggip.iges.or.jp/public/gpglulucf/gpglulucf_files/0_Task1_Cover/Cover_TOC.pdf, accessed on August 15, 2014.

Kiely, M. 2013. Carbon coalition against global warming. http://carboncoalitionoz.blogspot.com.au/2013/11/getting-back-down-to-erf.html, accessed on August 15, 2014.

Koch, A. The Australian, 2013. Answer to carbon conundrum is beneath our feet. http://www.theaustralian.com.au/national-affairs/opinion/answer-to-carbon-conundrum-is-beneath-our-feet/story-e6frgd0x-1226772530058?nk=7c001f564af3caf29a4c001e26098903, accessed on August 15, 2014.

Lee, K. 1998. A history of the CSIRO division of soils: 1927–1997 CSIRO land and water. Adelaide Technical Report 43/98, November 1998. http://www.google.com.au/url?sa=t&rct=j&q=&esrc=s&source=web&cd=1&ved=0CCAQFjAA&url=http%3A%2F%2Fwww.clw.csiro.au%2Fpublications%2Ftechnical98%2Ftr43-98.pdf&ei=XO3tU4WlKojl4QT0sIG4Bg&usg=AFQjCNETuqUW_S5UgKvqQgkvYVh-HTP10A&bvm=bv.73231344,d.bGE.

Liberal Party of Australia. 2010. Point 10: Reducing carbon emissions. http://www.liberal.org.au/our-plan/environment, accessed June 15, 2014.

Macintosh, A. 2012. LULUCF offsets and Australia's 2020 abatement task. Australian National University Centre for Climate Law and Policy. http://law.anu.edu.au/sites/all/files/coast/lulucfoffsets_17_feb_2011.pdf, accessed on August 15, 2014.

McCosker, T. 2000. Cell grazing—The first 10 years in Australia. *Tropical Grasslands*, 34, 207–218. http://www.google.com.au/url?sa=t&rct=j&q=&esrc=s&source=web&cd=1&ved=0CB4QFjAA&url=http%3A%2F%2Fwww.tropicalgrasslands.asn.au%2FTropical%2520Grasslands%2520Journal%2520archive%2FPDFs%2FVol_34_2000%2FVol_34_03-04_00_pp207_218.pdf&ei=bxHuU4qMC4yJogTW7oCQBw&usg=AFQjCNE1_rN4nGtjN2YmKcmGg3fKjZ8Tbg&bvm=bv.73231344,d.cGU.

Nature Conservation Council of New South Wales. 2014. http://www.nature.org.au/.

O'Gara, F. 1998. *Striking the Balance Conservation Farming and Grazing Systems for the Semi-Arid Tropics of the Northern Terrritory*. Department of Primary Industries and Fisheries, Darwin, Northern Territory, p. 17. http://www.google.com.au/url?sa=t&rct=j&q=&esrc=s&source=web&cd=2&ved=0CCMQFjAB&url=http%3A%2F%2Fwww.nt.gov.au%2Fd%2FContent%2FFile%2Fp%2FTech_Bull%2FIP11.pdf&ei=aQXuU93BD8u48gW3lYLQDQ&usg=AFQjCNHiRtIafx3UIoasLfJiqcl_DJsqyg&bvm=bv.73231344,d.dGc.

Parliament of Australia. 2014. Carbon farming initiative bill amended. http://www.aph.gov.au/Parliamentary_Business/Bills_Legislation/Bills_Search_Results/Result?bId=r5280, accessed on August 15, 2014.

Prime Minister's Science, Engineering and Innovation Council. 1999. From defence to attack: Australia's response to the greenhouse effect. Agenda Item 2, June 25, 1999.

Radi, H. 1993. Bradley, Eileen Burton (1911–1976). *Australian Dictionary of Biography*, 13 (MUP). http://adb.anu.edu.au/biography/bradley-eileen-burton-9566.

Soils for Life. 2012. Case studies of Jillamatong, Winona, and Dukes Plain. http://www.soilsforlife.org.au/case-studies.html.

Talberg, A. 2009. The Kyoto Protocol Accounting Rules. Parliament of Australia. http://www.aph.gov.au/About_Parliament/Parliamentary_Departments/Parliamentary_Library/pubs/BN/~/link.aspx?_id=C5066CCDFBF2437BAFEB744911ADEA67&_z=z, accessed on August 15, 2014.

Talberg, A., Loynes, K., and Hui, S.2013. Australian climate change policy: A chronology. Parliamentary Library, Canberra, Australian Capital Territory, Australia.

United Nations Framework Convention on Climate Change. 1998. http://unfccc.int/essential_background/convention/background/items/1349.php.

Youl, R. et al. 2006. Landcare in Australia founded on local action. SILC and Rob Youl Consulting Pty Ltd., Wallington, Victoria, Canada. http://www.google.com.au/url?sa=t&rct=j&q=&esrc=s&source=web&cd=1&ved=0CB4QFjAA&url=http%3A%2F%2Fwww.daff.gov.au%2F__data%2Fassets%2Fpdf_file%2F0017%2F29141%2Flandcare_in_australiaJune08.pdf&ei=EO_tU5XaG4Pd8AWY44DwAg&usg=AFQjCNGTByamvCVdsYCuAZnNaR7PYONcfw&bvm=bv.73231344,d.dGc.

Zanchi, G. 2002. Assessing forest carbon stock and temporal changes: Methods and regulations. Atti del 42.mo corso di Cultura in Ecologia. http://intra.tesaf.unipd.it/sanvito/dati/Atti_42CultEcol_2006.pdf, accessed on August 15, 2014.

WEBSITE LINKS

Healthy Soils Australia—http://www.healthysoils.com.au/index.html.
Outcomes Australia—http://www.outcomesaustralia.com.au/.
Soil Carbon Initiative—http://ussc.edu.au/events-special/soil-carbon-initiative.
Soils for Life—http://www.soilsforlife.org.au/.

34

Conclusions: Regreening Earth and Growing Our Way out of Global Crisis

Thomas J. Goreau, Ronal W. Larson, and Joanna Campe

CONTENTS

We must cultivate our gardens

Voltaire, Candide, 1759

Basic Conclusions

Soils and wetlands have around two to four times more carbon than the atmosphere or biomass. They are the only place where CO_2 can be stored quickly enough to avert the worst effects of global climate change, while providing many benefits for humanity.

Soils and wetlands are *still* not identified as carbon sinks in the United Nations Framework Convention on Climate Change (UNFCCC)! So, soil carbon projects are not eligible for carbon credits under international agreements.

CO_2 cannot be stabilized at safe levels without large-scale restoration of ecosystems and their ability to regulate global temperature, rainfall, sea level, food, renewable energy, water supplies, soil fertility, and marine productivity.

The methods to greatly increase soil carbon and restore ecosystems have been developed in every single habitat. But large-scale restoration of degraded habitats is *still* not yet on the climate change agenda. Action cannot wait any more!

Immediate Action Needed to Reverse Global Climate Change

The slow-motion crisis we have unleashed through the use of fossil fuels is far more serious for our entire planet's long-term future than the political, economic, and religious events that obsess shortsighted policy makers and funding agencies. They are flying away from solving problems too big for them to grasp, and if they remain asleep at the wheel, a disastrous crash is inevitable.

No habitat is more complex and variable in its physical, chemical, or biological properties than soil, and sound management requires an integrated approach that optimizes all these properties simultaneously (Pauli, 1967; Brady and Weil, 2007). The papers in this book show that there are a wide variety of methods capable of removing carbon from the atmosphere and storing it in the soil that (1) have the capacity to store enough carbon to reduce atmospheric CO_2 to safe levels, (2) can act fast enough to prevent the worst damages of runaway climate change, but (3) must be implemented very quickly on a global scale to be effective, and (4) require a mechanism of carbon accounting to ensure that the polluter pays and that the amounts are really stored.

Our hope is for the world community to act immediately in order to minimize the costs and damage that will result from past and ongoing maladapted practices. Only a truly global solution can work. Our best chance is to grow our way out of the mess we have made by growing more plants to restore soil fertility and biodiversity, regreening the wastelands we have created, and replenishing the natural integrated planetary-scale biogeochemical processes we have eviscerated.

We are confident this can be done in decades if we are serious and start now.

Fundamental Recommendations

At the Global Level

1. It is imperative for the United Nations Framework Convention on Climate Change (UNFCCC) to officially recognize that soils are a carbon sink and implement complete global strategies to greatly increase soil carbon with the goal of CO_2 stabilization at safe levels. A united global effort is essential to avert a certain slow-motion catastrophe in order to create a better future for our descendants and all the other organisms that share our planet. The net biogeochemical metabolism of soils determines the future of our climate, food, and water and affects everyone who breathes air, drinks water, and eats food.
2. Complete accounting of all greenhouse gas sources and sinks is needed as a tool for verification and correction, to replace the partial and incomplete accounting now used, which is so flawed that it cannot serve its purpose.
3. A global legal mechanism is needed for all those who produce CO_2 to directly pay those who remove it, to ensure that those who have turned green land brown, and turned our common atmosphere and ocean resources into polluted sewers, pay to reverse the effects of what they have destroyed or stolen from future generations.

At the Local Level

1. A holistic approach that combines all of these methods together, in all soil types and ecosystems, is essential, as no one component by itself supplies all of the plant's needs. These methods work everywhere, but each mixture of biochar, rock dusts, leaf sprays, compost, and biological inoculation will have somewhat different results, depending on local geology, geography, hydrology, soil, climate, and the biology of the soil and of the crops.
2. It must be understood that biochar properties are highly variable, and to be most effective it must be fully charged with all the elements the plants need and the microbes to pass them on to the roots.
3. Worldwide research efforts are urgently needed to optimize these methods for the widest range of conditions so that the most effective combinations can be placed in the farmers' hands.
4. We cannot count on governments to address issues they simply cannot grasp, what is needed is a real grass-roots, bottom-up, locally managed restoration mass underground movement of informed people, a groundswell of popular opinion driven by a green ethos, to regreen our planet and recarbonize our soil.

Call for Action by the United Nations Framework Convention on Climate Change

We therefore call on all parties to UNFCCC to

1. Identify soil and wetlands as carbon sinks that are eligible for carbon credits
2. Mandate international funding agencies to urgently elevate spending on projects to increase carbon storage in soils, wetlands, and biomass
3. Mandate complete accounting of *all* greenhouse gas sources and sinks

REFERENCES

Brady, N. C. and R. Weil, 2007, *The Nature and Properties of Soil*, 14th edn., Prentice Hall, Upper Saddle River, NJ.

Pauli, F. W., 1967, *Soil Fertility: A Biodynamic Approach*, Hilger, London, U.K.

Voltaire, J. M., 1759, *Candide, ou l'optimisme*, Sirene, Paris, France.

Index

D